# Ancient Orogens and Modern Analogues

## IUGS/GSL publishing agreement

This volume is published under an agreement between the International Union of Geological Sciences and the Geological Society of London and arises from IGCP 453 project entitled 'Ancient orogens and modern analogues'.

GSL is the publisher of choice for books related to IUGS activities, and the IUGS receives a royalty for all books published under this agreement.

Books published under this agreement are subject to the Society's standard rigorous proposal and manuscript review procedures.

It is recommended that reference to all or part of this book should be made in one of the following ways:

MURPHY, J. B., KEPPIE, J. D. & HYNES, A. J. (eds) 2009. *Ancient Orogens and Modern Analogues*. Geological Society, London, Special Publications, **327**.

DOSTAL, J., KEPPIE, J. D. & FERRI, F. 2009. Extrusion of high-pressure Cache Creek rocks into the Triassic Stikinia-Quesnellia arc of the Canadian Cordillera: implications for terrane analysis of ancient orogens and paleogeography. *In*: MURPHY, J. B., KEPPIE, J. D. & HYNES, A. J. (eds) 2009. *Ancient Orogens and Modern Analogues*. Geological Society, London, Special Publications, **327**, 71–87.

GEOLOGICAL SOCIETY SPECIAL PUBLICATION NO. 327

# Ancient Orogens and Modern Analogues

EDITED BY

## J. B. MURPHY
St Francis Xavier University, Canada

## J. D. KEPPIE
Universidad Nacional Autonoma de Mexico, Mexico

and

## A. J. HYNES
McGill University, Canada

2009
Published by
The Geological Society
London

# THE GEOLOGICAL SOCIETY

The Geological Society of London (GSL) was founded in 1807. It is the oldest national geological society in the world and the largest in Europe. It was incorporated under Royal Charter in 1825 and is Registered Charity 210161.

The Society is the UK national learned and professional society for geology with a worldwide Fellowship (FGS) of over 9000. The Society has the power to confer Chartered status on suitably qualified Fellows, and about 2000 of the Fellowship carry the title (CGeol). Chartered Geologists may also obtain the equivalent European title, European Geologist (EurGeol). One fifth of the Society's fellowship resides outside the UK. To find out more about the Society, log on to www.geolsoc.org.uk.

**The Geological Society Publishing House** (Bath, UK) produces the Society's international journals and books, and acts as European distributor for selected publications of the American Association of Petroleum Geologists (AAPG), the Indonesian Petroleum Association (IPA), the Geological Society of America (GSA), the Society for Sedimentary Geology (SEPM) and the Geologists' Association (GA). Joint marketing agreements ensure that GSL Fellows may purchase these societies' publications at a discount. The Society's online bookshop (accessible from www.geolsoc.org.uk) offers secure book purchasing with your credit or debit card.

To find out about joining the Society and benefiting from substantial discounts on publications of GSL and other societies worldwide, consult www.geolsoc.org.uk, or contact the Fellowship Department at: The Geological Society, Burlington House, Piccadilly, London W1J 0BG: Tel. +44 (0)20 7434 9944; Fax +44 (0)20 7439 8975; E-mail: enquiries@geolsoc.org.uk.

For information about the Society's meetings, consult *Events* on www.geolsoc.org.uk. To find out more about the Society's Corporate Affiliates Scheme, write to enquiries@geolsoc.org.uk.

Published by The Geological Society from:
The Geological Society Publishing House, Unit 7, Brassmill Enterprise Centre, Brassmill Lane, Bath BA1 3JN, UK

(*Orders*: Tel. +44 (0)1225 445046, Fax +44 (0)1225 442836)
Online bookshop: www.geolsoc.org.uk/bookshop

**British Library Cataloguing in Publication Data**

A catalogue record for this book is available from the British Library.
ISBN 978-1-86239-289-2

Typeset by Techset Composition Ltd, Salisbury, UK
Printed by MPG Books Ltd, Bodmin, UK


**Distributors**

*North America*
For trade and institutional orders:
The Geological Society, c/o AIDC, 82 Winter Sport Lane, Williston, VT 05495, USA
*Orders*: Tel. +1 800-972-9892
   Fax +1 802-864-7626
   E-mail: gsl.orders@aidcvt.com

For individual and corporate orders:
AAPG Bookstore, PO Box 979, Tulsa, OK 74101-0979, USA
*Orders*: Tel. +1 918-584-2555
   Fax +1 918-560-2652
   E-mail: bookstore@aapg.org
   Website: http://bookstore.aapg.org

*India*
Affiliated East-West Press Private Ltd, Marketing Division, G-1/16 Ansari Road, Darya Ganj, New Delhi 110 002, India
*Orders*: Tel. +91 11 2327-9113/2326-4180
   Fax +91 11 2326-0538
   E-mail: affiliat@vsnl.com

# Contents

## Proterozoic orogens

# Ancient orogens and modern analogues: an introduction

J. BRENDAN MURPHY[1]*, J. DUNCAN KEPPIE[2] & ANDREW J. HYNES[3]

[1]*Department of Earth Sciences, St. Francis Xavier University, Antigonish,
Nova Scotia, Canada, B2G 2W5*

[2]*Universidad Naçional Autónoma de México, México, D.F. 04510*

[3]*Department of Earth and Planetary Sciences, McGill University,
Montreal, PQ, Canada, H3A 2A7*

*\*Corresponding author (e-mail: bmurphy@stfx.ca)*

**Abstract:** Plate-tectonics principles have been routinely applied to the study of Phanerozoic orogenic belts and, more controversially, to Precambrian orogens as far back as the Early Archaean.
Recent advances in a variety of fields have vastly improved our understanding of ancient orogenic
belts, so that realistic modern analogues can be entertained. This volume presents up-to-date syntheses of some classic modern and ancient orogenic belts as well as examples of some of the processes responsible for their evolution.

Plate-tectonics principles provide a unifying
conceptual framework for understanding the evolution of modern oceanic lithosphere and Cenozoic
orogens (Dewey & Bird 1970). Since Tuzo Wilson
(1966) first proposed that the evolution of the
Appalachian–Caledonide orogen of eastern North
America and western Europe records the birth and
death of an earlier ocean (now known as Iapetus),
plate-tectonics principles have been routinely
applied to the study of Phanerozoic orogenic belts
(e.g. van Staal *et al.* 1998), and, more controversially, to Precambrian orogens as far back as the
Early Archaean (Condie & Kröner 2008; Wyman
*et al.* 2008).

For pre-Jurassic orogenic belts, no seafloor-
spreading record is available. Furthermore, as we
go further back in time, faunal and palaeomagnetic
data become more sparse and difficult to apply.
Tectonic models are therefore less well constrained
with increasing age. However, recent advances in
geochronological and palaeomagnetic techniques,
and conceptual advances in other fields such as
geodynamics, have vastly improved the available
tools for the study of ancient orogenic belts, so that
realistic modern analogies can now be drawn with
some confidence.

This volume is an outgrowth of IGCP Project
453 (1999–2004) entitled 'Uniformitarianism revisited: a comparison between modern and ancient
orogens'. The main goal of the project was to
enhance our understanding of the causes and effects
of modern and ancient mountain belts, and how
these relationships have varied with time. Today,
most geoscientists apply these principles back to
the Early Proterozoic or Late Archaean. However,
the state of preservation of these orogens is such

that resolution of orogenic complexities has not
matched that of modern orogens. Detailed models
that include the effects of, for example, delamination, transform faults, subduction of oceanic and
aseismic ridges, overriding of plumes and subduction erosion have rarely been considered in
ancient orogens, although they have a demonstrably
profound effect on the styles of Cenozoic orogens.
On the other hand, studies of deeply eroded
ancient orogens provide insights into the hidden
roots of modern orogens, providing a complementary perspective.

## Modern orogens

### Subduction tectonics

This volume treats examples of subduction-related
tectonics from modern orogens, including the role
of structure in determining the chemical composition of volcanism, a description of flat-slab subduction, and the potential relationship between
ridge-trench collision and the formation of fold-
and-thrust belts.

The Taupo Volcanic Zone (TVZ) of northern
New Zealand is a classic modern example of an
ensialic island arc that is undergoing extension.
The TVZ results from oblique subduction of the
Pacific plate beneath New Zealand's North Island.
Magmatism is attributed to the rise of hot mantle
beneath thinned continental crust (c. 16 km, e.g.
Harrison & White 2006) that generates crustal melts
as well as melts from the decompressed mantle
(Cole *et al.* 1998, 2005). In the northern and south-
ern TVZ, the magmatism is predominantly andesi-
tic, whereas in the central zone it is predominantly

*From*: MURPHY, J. B., KEPPIE, J. D. & HYNES, A. J. (eds) *Ancient Orogens and Modern Analogues.*
Geological Society, London, Special Publications, **327**, 1–8.
DOI: 10.1144/SP327.1   0305-8719/09/$15.00 © The Geological Society of London 2009.

rhyolitic. Spinks *et al.* (2005) show that the regions with the greatest extension in the TVZ produce calderas up to 50 km wide that are dominated by felsic magmatism, whereas areas with the greatest transtension produce andesitic stratovolcanoes. The strain rate along local faults influences magma composition and eruption style (Paterson & Tobsich 1992; Petford *et al.* 2000). In this volume, **Cole & Spinks** provide a comprehensive overview of the structural and volcanological features of TVZ, and include recent high precision radiometric data. Using these data, the authors draw attention to the close correspondence between volcanism and structure in TVZ, pointing out that vents are commonly aligned along local faults and that many calderas have boundaries with a rectangular geometry that reflect local fault patterns. They describe a model in which basalt rises into the lower crust causing partial melting of the lower crust. These crustal melts pond at the mid-crustal brittle-ductile transition zone, and rise into the upper crust exploiting structural weaknesses.

The influence of subduction-zone geometry on the tectonic style of orogenesis at convergent margins has long been recognized. Since Barazangi & Isacks (1976, 1979) documented flat slab segments along the Andes continental margin, the causes and effects of flat-slab subduction have been a matter of intense debate. Recent analysis of modern flat-slab subduction zones has drawn attention to their spatial and temporal correlation with subducting oceanic plateaus (e.g. Gutscher *et al.* 2000; Ramos & McNulty 2002). The modern Andean margin has several flat-slab segments, up to 500 km wide, with each correlated to subduction of anomalously buoyant oceanic crust, represented by oceanic plateaus (Pilger 1981; Gutscher *et al.* 2000). In this volume, **Ramos & Folguera** identify and characterize modern flat slab segments, incipient flat slab segments, and ancient flat slab segments, including an Early Permian example. They propose that most of the Andes have experienced a stage of flat subduction at some stage in their evolution. They also identify the characteristics of slab steepening, which, beneath thick crust, results in delamination and bimodal magmatism, but beneath thin crust results in extensional tectonism and voluminous mafic volcanism.

Fold-and-thrust belts are a common manifestation of deformation in the overriding plate above subduction zones (e.g. Price & Mountjoy 1970; Allmendinger *et al.* 1997). **Mandujano-Velazquez & Keppie** point out a disconnect between models proposed to explain ancient fold-and-thrust belts (which are commonly attributed to collisions between continents or terranes), and Mesozoic–Cenozoic belts which are typically interpreted to reflect changes in factors such as Benioff-zone geometry, convergence rate, and the degree of coupling between the subducting and overriding plates (e.g. McQuarrie 2002). The authors present structural and stratigraphic sections across the Middle Miocene Chiapas fold-and-thrust belt that are constrained by seismic and well data, and they propose that the belt was formed as a result of ridge-trench collision, thereby providing a potential modern analogue to be considered in ancient belts.

## Accretionary tectonics

The Cordillera of North America are considered a classic example of an orogen dominated by terrane accretion and has resulted in the growth of North America by an average of about 500 km since the end of the Palaeozoic (e.g. Coney *et al.* 1980). Obducted vestiges of oceanic lithosphere are commonly interpreted to reflect suture zones between adjacent terranes. This interpretation is valid if the subducted oceanic lithosphere is exhumed up the subduction channel (e.g. Federico *et al.* 2007). In this volume, however, **Dostal *et al.*** propose that an entire terrane (the Cache Creek terrane) in the Canadian Cordillera could represent subduction-zone rocks that were extruded into the upper plate. The Cache Creek terrane is a high-pressure blueschist-eclogitic terrane (Johnston & Borel 2007) that occurs between the Quesnellia and Stikinia terranes. Middle–Upper Triassic arc magmatic sequences in Quesnellia and Stikinia (Takla Group) have very similar geochemical and isotopic characteristics that are typical of primitive arcs with limited continental crust involvement. Extrusion of the HP rocks into the upper plate indicates that rather than two terranes (Stikinia and Quesnellia), the Takla Group represents one arc split at the site of extrusion of the Cache Creek rocks. Dostal *et al.* point out that existing models (post-Middle Jurassic duplication of the arc by strike-slip faulting, oroclinal or synformal folding) are inconsistent with both palaeomagnetic and faunal data. Building on the numerical models of Stöckhert & Gerya (2005) and Gerya & Stöckhert (2006), they propose an alternative model in which oblique eastward subduction and high-pressure metamorphism of the Cache Creek accretionary prism and forearc was followed by extrusion into the upper plate arc and exhumation by the Middle Jurassic. In this scenario, the high-pressure rocks occur between similar units and do not separate terranes or define an oceanic suture.

## Collisional tectonics

Three contributions from the Mediterranean region provide examples of the complexity and heterogeneity of continent-continent collisional zones. **Stampfli & Hochard** provide detailed plate reconstructions (rather than traditional continental drift reconstructions) of the Mediterranean region from late Triassic to Miocene. This time period includes the opening of the Neotethys and North Atlantic oceans, followed by the Pyrenean and Alpine orogenies. The reconstructions describe the evolution of several distinct tectonic domains and the relationships between them, and imply differential transport of thousands of kilometres for the plates and the terranes within them. Other interesting implications of the reconstructions are that: (i) the Neotethys and North Atlantic oceans were never directly connected; (ii) a series of basins opened and closed, due to subduction zone roll-back and back-arc extension; and (iii) the Alpine orogen is an example of the juxtaposition of exotic tectonic elements.

The Calabrian orocline is one of the most distinctive structural elements in the Mediterranean region and reflects the bending of the Apennine and Sicilian mountain chains. Although its status as an orocline is well accepted (e.g. Tapponnier 1977; Eldredge *et al.* 1985), the mechanisms responsible for its development remain enigmatic. Oroclines, as defined by Carey (1955, 1958), are arcuate orogenic belts that were formerly more linear than they are today. They are commonly best established where palaeomagnetic data show a change of declination around the oroclinal bend (Eldredge *et al.* 1985). The Calabrian orocline development previously has been viewed as a thin-skinned, allochthonous structure (Eldredge *et al.* 1985), that may have involved subduction zone roll-back (Kastens *et al.* 1988). **Johnston & Mazzoli** present a geometric model for development of the orocline that also accounts for the coeval development of the Tyrrhenian Sea, a tract of oceanic lithosphere that began to form about 7 to 9 million years ago. They propose that the orocline formed by eastward buckling of an originally north-south continental ribbon that involved the crust and some of the lithospheric mantle and was mechanically facilitated by the presence of a subduction zone along the length of its eastern margin.

According to **Dilek & Sandvol**, the tectonic evolution of the Eastern Mediterranean in the late Mesozoic–Cenozoic is analogous to the development of the Himalayan–Indonesian orogen. According to their model, the tectonic evolution of the Eastern Mediterranean was governed by collisions of Gondwana-derived continental terranes with Eurasia as well as by intra-oceanic subduction and closure of the intervening Neotethyan oceanic basins. Jurassic–Cretaceous ophiolites, developed in the proto-arcs and forearcs of these subduction zones, were emplaced during collision of the passive margin with the trench, and partial subduction of these margins yielded high-pressure metamorphic assemblages. Paleocene–Eocene collisions

resulted in crustal thickening and plateau formation, and was followed by slab break-off which resulted in extension. The mid-Miocene collision of Arabia with Eurasia at 13 Ma resulted in crustal shortening and regionally extensive strike-slip fault systems that accommodated the tectonic escape of the Anatolian plate. The authors compare this setting with that of the Tibetan plateau.

## Ancient orogens

### Palaeozoic orogens

In the absence of an ocean floor record, palaeocontinental reconstructions for the Palaeozoic Era are not as precise as those available for the Mesozoic and Cenozoic. Nevertheless, since the landmark papers of Wilson (1966) and Dewey (1969), plate-tectonic principles have been routinely applied to Palaeozoic orogenic belts. Over the past twenty years, a combination of faunal, palaeomagnetic, lithostratigraphic, geochronological and petrological data have provided first-order constraints on palaeocontinental reconstructions, and although some important issues remain, a broad consensus has emerged (e.g. McKerrow & Scotese 1990; Scotese 1997; Cocks & Fortey 1990; van Staal *et al.* 1998; Cocks & Torsvik 2002; Stampfli & Borel 2002). Tectonic history of the Palaeozoic Era is dominated by the breakup of the supercontinent Pannotia (Powell 1995; Dalziel 1997; Cawood *et al.* 2001) between 600–550 Ma, followed by the assembly and amalgamation of Pangaea by the Late Carboniferous–Permian. These reconstructions provide the backdrop for models for the evolution of the Palaeozoic orogens presented herein.

**Puchkov** reviews recent advances in the understanding of the Uralian orogen, which, he points out, has many of the characteristics expected from the Wilson Cycle. Its evolution begins with the development and destruction of the Palaeouralian ocean which involved collisions between Baltica, Siberia and Kazakhstania. Puchkov stresses that the Uralian orogen also displays some very distinctive features that preclude a simple geodynamic connection with the Late Palaeozoic Variscide orogenic belts of western Europe as has been previously proposed (e.g. International Committee of Tectonic Maps 1982). Late Cambrian–Early Ordovician continental rifting was followed by Middle Ordovician passive margin development, Late Ordovician subduction, arc–continent collision in the Late Devonian–Early Carboniferous, and continent-continent collision beginning in the mid-Carboniferous. Renewed and relatively recent uplift is thought to be a far-field effect of the collision between India and Asia. The Uralian orogen also has some rare and enigmatic characteristics, such as the presence of a cold isostatically equilibrated root, the abundance of oceanic mafic complexes, the limited crustal shortening in the southern Urals, and the Silurian platinum-rich belt hosted by subduction-related layered plutons.

The papers by **Hofmann** *et al.* and **Pereira** *et al.* provide a detailed analysis of the late-stage tectonic evolution in two classical areas of the Late Palaeozoic Variscan orogen (Bohemian Massif and Iberia, respectively). Global reconstructions show that the Variscan orogeny was due to the destruction of the Rheic Ocean, and was a key event in the formation of Pangaea. Hofmann *et al.* present new geochronological data that constrain the timing of the final pulse of the Variscan orogeny in the northern (Saxo-Thuringian Zone) of the Bohemian Massif, which they relate to closure of the Rheic Ocean and terminal collision between Gondwana and Laurussia. These data, together with field relationships, allow an understanding of the relationships between late-stage strike-slip motion, plutonism and lateral extrusion, as well as the formation of post-orogenic basins.

The evolution of the Variscan orogen in SW Iberia is widely recognized as a world-class example of a transpressional orogen, and it is one of the few places in which an oceanic suture, reflecting the collision between Gondwana (Iberian massif) and Laurentia (South Portuguese zone) is documented (Quesada 1990; Quesada & Dallmeyer 1994). Pereira *et al.* focus on the post-collisional (Visean) extensional phases of the origin, most specifically the evolution in the footwall of several metamorphic core complexes, which occur within a 250 km long belt of metamorphic rocks in the Ossa-Morena zone along the southwestern flank of the Iberian Massif. They interpret new U–Pb SHRIMP data from zircon rims and monazites to reflect flushing of hydrothermal fluids along the detachment zones that bound each of these core complexes. While accepting the popular view that the syn-collisional style of deformation is compatible with regional sinistral transpression (Quesada & Dallmeyer 1994), they also point out that post-collisional sinistral shear could provide the local regions of extension where the core complexes developed. This extensional event is manifest at different crustal levels and is also held responsible for coeval basin formation and magmatism. Taken together, the data presented by Pereira *et al.* provide important evidence for post-collisional intra-orogenic shear along the Gondwanan margin in the terminal phases of closure of the Rheic Ocean.

In recent years, there has been an increased understanding of the relationship between Palaeozoic orogenesis in central America and coeval events in the Appalachian–Caledonide orogen. The pioneering work of Ortega-Gutiérrez (1975) showed

that the evolution of the Middle American terranes of southern Mexico records the creation and destruction of a Palaeozoic ocean. Palaeozoic metasedimentary rocks occur in a number of these terranes, including the Acatlán Complex of the Mixteca terrane, which is the largest inlier of Palaeozoic rocks in Mexico and underlies an area almost the size of Belgium. Broad tectonostratigraphic similarities to parts of the Appalachian-Ouachita orogen and close proximity to a *c.* 1 Ga basement (Oaxacan Complex) have led to models in which the Acatlán Complex was interpreted in terms of Laurentia–Gondwana (Amazonia) collision (e.g. Yañez *et al.* 1991). Ortega-Gutiérrez *et al.* (1999) proposed that the complex represents a vestige of the Iapetus suture formed during a Late Ordovician–Early Silurian collisional orogeny between Laurentia and Oaxaquia, a crustal block of Gondwanan affinity. More recently, Talavera-Mendoza *et al.* (2005) proposed that the complex records suturing of both peri-Laurentian and peri-Gondwanan arcs to Laurentia during closure of both the Iapetus and Rheic oceans. In this volume, **Nance** *et al.* focus on the provenance of the Palaeozoic metasedimentary rocks in the Middle America terranes by compiling and reviewing available detrital zircon data. These data confirm the Gondwanan affinity of the metasedimentary rocks, but indicate an origin on the southern flank of the Rheic Ocean, rather than Iapetus. In this scenario, these rocks are interpreted to have been transferred to Laurentia during and following the amalgamation of Pangaea.

Ever since the papers of Williams (1964, 1979), Wilson (1966) and Dewey (1969) were published, the Appalachian orogen has been a type area for the application of modern plate tectonic concepts to Palaeozoic orogens. **van Staal** *et al.* present an overview paper that synthesizes recent interpretations of various parts of the northern Appalachians. The paper, which combines field and petrological studies with comprehensive and precise geochronology, shows that prior to the amalgamation of Pangaea, the development of the Appalachian orogen was typical of an accretionary orogen, analogous to the Mesozoic–Cenozoic evolution of the western Pacific Ocean. The authors discuss four episodes of orogenesis, each related to the accretion of a microcontinent, one peri-Laurentian (Dashwoods) during the Ordovician Taconic orogeny, and three peri-Gondwanan terranes (Ganderia, Avalonia and Meguma) which docked at various times to the eastern margin of Laurentia between the Late Ordovician and Devonian. Sequential accretionary tectonics led to pulses of deformation and metamorphism.

Despite their obvious importance to the assembly of Pangaea, studies of the mafic complexes, interpreted by most authors as ophiolites, formed during the lifespan of the Rheic Ocean, have lagged behind the comprehensive studies of ophiolites within the Iapetan realm. In this volume, there are two papers that provide such syntheses. Sanchez-Martinez *et al.* provide a detailed analysis of the range of ophiolites that occur in allochthonous complexes in NW Iberia (Galicia). Murphy *et al.* provide an overview of mafic complexes in the Rheic Ocean realm, from Mexico to Bohemia.

The allochthonous complexes of NW Iberia within the Variscan suture zone provide one of the best-preserved expressions of the collision between Laurussia and Gondwana (Arenas *et al.* 1986, 2007). **Sanchez-Martinez *et al.*** show that the ophiolites have a wide range in ages, tectonic setting and subsequent tectonothermal evolution, precluding the simple interpretation that they reflect only closure of the Rheic Ocean during Pangaean assembly. This study provides an important example of the complexity of collisional suture zones. The oldest ophiolite is Mesoproterozoic in age, has supra-subduction zone affinities, and is thought to have been generated near the West African craton before the *c.* 1.1 Ga amalgamation of the supercontinent Rodinia. An Early Cambrian ophiolite was probably generated in an arc setting, but a Late Cambrian ophiolite has MORB affinities and may have formed in the Iapetus-Tornquist ocean. Other Late Cambrian-Early Ordovician ophiolites bodies have arc affinities, and together with coeval metamorphism, are thought to reflect the opening of the Rheic Ocean as a back-arc basin. A Devonian ophiolite was formed during contraction of the Rheic Ocean, shortly preceding collision between Laurussia and Gondwana.

**Murphy *et al.*** point out that mafic complexes widely interpreted to represent vestiges of the Rheic Ocean are widespread, from the Acatlán Complex in Mexico to the Bohemian Massif in eastern Europe. Most of these complexes are either Late Cambrian-Early Ordovician or Late Palaeozoic in age. With the exception of those in NW Iberia, these Late Cambrian-Early Ordovician mafic complexes are not ophiolites. Their geochemical and Sm-Nd isotopic signatures indicate that they are rift-related continental tholeiites, derived from an enriched *c.* 1.0 Ga subcontinental lithospheric mantle, and are associated with crustally-derived felsic volcanic rocks. They are interpreted to reflect magmatism along the Gondwanan margin during the formation of the Rheic Ocean, and to have remained along that margin as Avalonia and other peri-Gondwanan terranes drifted northward. The Late Palaeozoic mafic complexes (Devonian and Carboniferous), however, do preserve many of the characteristics of ophiolites. They are characterized by derivation from an anomalous ultra-depleted mantle, and may reflect narrow tracts of oceanic crust that originated along the Laurussian margin, but were thrust over Gondwana during Variscan orogenesis. The Carboniferous ophiolites may have formed in a strike-slip regime within relict ocean basins during closure of the Rheic Ocean.

## Proterozoic orogens

Over the past 20 years, the existence of a supercontinent, Rodinia, and the corresponding global ocean, Mirovoi (McMenamin & McMenamin 1990), between *c.* 1.1–0.75 Ga has gained acceptance (see Li *et al.* 2008 for a review). However, the configuration of this supercontinent is highly controversial. Hoffman (1991) attributed global-scale orogenesis between 1.3–1.0 Ga to the amalgamation of Rodinia, and the distribution of Late Neoproterozoic passive margins around Laurentia to its *c.* 0.75 Ga breakup. Since that time, several authors have modified Hoffman's original reconstruction, but its basic premise remains intact. In contrast to most reconstructions that are based on geological data, **Evans** produces a reconstruction of Rodinia that is built primarily upon the available palaeomagnetic database, although geological relations are included as supporting evidence. The Evans reconstruction is a radical departure from previous versions of Rodinia and is sure to generate much discussion. In this reconstruction, Rodinia was a long-lived supercontinent that contained all major Precambrian continents (see e.g. Pisarevsky *et al.* 2003). Perhaps the most controversial features are: (i) the lack of a collider for the Grenville orogen of eastern North America; (ii) the inverted position of Australia and Antarctica relative to Baltica; and (iii) the positioning of Amazonia, West Africa, and Rio de la Plata cratons close to western (rather than eastern) Laurentia.

The Grenville orogen is widely held to be a classic Proterozoic example of a continent-continent collision (e.g. Ludden & Hynes 2000 and references therein; Li *et al.* 2008). **Rivers** provides a thorough summary of the thermal evolution of the Grenville orogen exposed in Canada, which, on the basis of global reconstructions, he interprets to have developed in late Mesoproterozoic to early Neoproterozoic time by the collision between Laurentia and Amazonia during the assembly of Rodinia (Li *et al.* 2008; compare Evans, this volume). He points out that its characteristics such as width (>600 km), abundance of high grade metamorphic rocks, and duration (≥100 My) is typical of large hot long-duration orogens as defined by Beaumont *et al.* (2006). Such orogens develop in orthogonal collisional settings, and in ideal cases, the amount of new crust added (by thrusting) is balanced by crust removed by erosion at orogenic fronts where a mid-crustal channel of hot, low-viscosity rocks becomes exposed. Rivers adopts this 'channel-flow' model for the Grenville and also identifies features consistent with imbalances between crust added and removed.

The paper by **Occhipinti & Reddy** provides a link between the Palaeoproterozoic Capricorn Orogen of Western Australia and its reactivation in the Neoproterozoic, as documented by precise $^{40}Ar/^{39}Ar$ mica ages. The Capricorn Orogen comprises rocks deformed and metamorphosed during several Palaeoproterozoic orogenic events ranging from 2.0–1.6 Ga (e.g. Occhipinti *et al.* 2004). Together with the Archaean Pilbara and Yilgarn cratonic blocks, the Palaeoproterozoic Capricorn Orogen form the West Australian Craton (Myers 1993). This craton was reworked during the *c.* 1.0 Ga oblique collision of Western Australia with another continent (Kalahari or Greater India). The authors point out that although the evolution of such reworked cratonic margins can be complex, the timing of events affecting those margins may be recorded by low-temperature events in the cratonic interior. These low-temperature 'events' are difficult to detect by structural or textural analysis, but precise $^{40}Ar/^{39}Ar$ mineral ages can be utilized to document the timing of reactivation and fluid flow events. The authors further propose that the reactivation is a far-field effect of the continental collisions and, more generally, that the results of this study demonstrate that the evolution of orogens can be constrained by thermochronological studies in adjacent cratons.

The Capricorn orogen is one of several Paleoproterozoic collisional belts that formed during the 2.0–1.8 Ga amalgamation of supercontinent Nuna (also known as Columbia, Bleeker 2003; Rogers & Santosh 2002). **Corrigan *et al.*** provide an up-to-date synthesis of the accretionary and collisional evolution of the largest of these belts, the Trans-Hudson Orogen of the Canadian Shield. As noted by the authors, these orogenic events occurred during a fundamental interval in the evolution of the Earth's systems, with important changes in the ocean and atmosphere (e.g. Anbar & Knoll 2002), biosphere (e.g. Konhauser *et al.* 2002) as well as the growth of cratons into large stable continents (e.g. Hoffman 1988). The Trans-Hudson orogen orogen is widely recognized as the oldest example of a relatively complete Wilson-Cycle (e.g. Ansdell 2005; Corrigan *et al.* 2005; St-Onge *et al.* 2006), from *c.* 2.45–1.92 Ga passive margin assemblages deposited along Archean cratonic margins, to formation and accretion of *c.* 2.00–1.88 Ga oceanic and 1.88–1.83 Ga continental arc systems, followed by 1.83–1.80 terminal collision. The paper by Corrigan *et al.* is an example of the complexity and scale of Palaeoproterozoic orogenic processes, and the type of geodynamic operative during this era.

This special publication is an outcome of IGCP Project 453 of the International Geoscience Programme of the International Union of Geological Sciences and UNESCO. We acknowledge the Natural Sciences and Engineering Research Council, Canada (JBM, AH), the St. Francis Xavier University Council for Research (JBM) and Papiit and CONACyT grants (JDK). We thank all those who organized meetings, field trips and special volumes in connection with this project. We also thank to Margarete Patzak (UNESCO) and Angharad Hills (Geological Society of London) for their help and support. We are especially grateful to the following reviewers who generously gave of their time and expertise: Valerio Acocella, Kevin Ansdell, Andrea Arnani, Sandra Barr, Pete Betts, Myron (Pat) Bickford, Gilles Borel, Peter Cawood, David Corrigan, Brian Cousens, Allen Dennis, Richard Ernst, Mary Ford, John Gamble, Laurent Godin, Gabi Gutierrez-Alonso, Stephen Johnston, Fred McDowell, Brendan McNulty, Damian Nance, Francisco Pereira, Sergei Pisarevsky, Russell Pysklywec, Cecilio Quesada, Victor Ramos, Paul Robinson, Paul Ryan, Rob Strachan, Ben van der Pluijm, Rob Van Der Voo and several anonymous reviewers.

# References

ALLMENDINGER, R. W., JORDAN, T. E., KAY, S. M. & ISACKS, B. L. 1997. The evolution of the Altiplano–Puna Plateau of the Central Andes. *Annual Review of Earth and Planetary Science*, **25**, 139–174.

ANBAR, A. D. & KNOLL, A. H. 2002. Proterozoic ocean chemistry and evolution: A bioinorganic bridge? *Science*, **297**, 1137–1142.

ANSDELL, K. M. 2005. Tectonic evolution of the Manitoba-Saskatchewan segment of the Paleoproterozoic Trans-Hudson Orogen, Canada. *Canadian Journal of Earth Sciences*, **42**, 741–759.

ARENAS, R., GIL IBARGUCHI, J. I. *ET AL.* 1986. Tectonoestratigraphic units in the complexes with mafic and related rocks of the NW of the Iberian Massif. *Hercynica*, **II**, 87–110.

ARENAS, R., MARTÍNEZ CATALÁN, J. R. *ET AL.* 2007. Paleozoic ophiolites in the Variscan suture of Galicia (northwest Spain): distribution, characteristics and meaning. *In*: HATCHER, R. D., JR., CARLSON, M. P., MCBRIDE, J. H. & MARTÍNEZ CATALÁN, J. R. (eds) *4-D Framework of Continental Crust*. Geological Society of America Memoirs, **200**, 425–444.

BARAZANGI, M. & ISACKS, B. 1976. Spatial distribution of earthquakes and subduction of the Nazca plate beneath South America. *Geology*, **4**, 686–692.

BARAZANGI, M. & ISACKS, B. 1979. Subduction of the Nazca plate beneath Peru evidence from spatial distribution of earthquakes. *Geophysical Journal of the Royal Astronomic Society*, **57**, 537–555.

BEAUMONT, C., NGUYEN, M. H., JAMIESON, R. A. & ELLIS, S. 2006. Crustal flow modes in large hot orogens. *In*: LAW, R. D., SEARLE, M. P. & GODIN, L. (eds) *Channel Flow, Ductile Extrusion and Exhumation in Continental Collision Zones*. Geological Society, London, Special Publications, **268**, 91–145.

BLEEKER, W. 2003. The late Archean record: a puzzle in ca. 35 pieces. *Lithos*, **71**, 99–134.

CAREY, S. W. 1955. The orocline concept in geotectonics. *Proceedings of the Royal Society of Tasmania*, **89**, 255–258.

CAREY, S. W. 1958. The tectonic approach to continental drift. *In*: CAREY, S. W. (ed.) *Continental Drift – A Symposium*. University Tasmania Press, Hobart, Tasmania, 178–355.

CAWOOD, P. A., McCAUSLAND, P. J. A. & DUNNING, G. R. 2001. Opening Iapetus: constraints from the Laurentian margin of Newfoundland. *Geological Society of America Bulletin*, **113**, 443–453.

COCKS, L. R. M. & FORTEY, R. A. 1990. Biogeography of Ordovician and Silurian faunas. *In*: McKERROW, W. S. & SCOTESE, C. R. (eds) *Paleozoic Paleogeography and Biogeography*. Geological Society, London, Memoirs, **12**, 97–104.

COCKS, L. R. M. & TORSVIK, T. H. 2002. Earth geography from 500 to 400 million years ago: a faunal and palaeomagnetic review. *Journal of the Geological Society, London*, **159**, 631–644.

COLE, J. W., BROWN, S. J. A., BURT, R. M., BERESFORD, S. W. & WILSON, C. J. N. 1998. Lithic types in ignimbrites as a guide to the evolution of a caldera complex, Taupo volcanic centre, New Zealand. *Journal of Volcanology and Geothermal Research*, **80**, 217–237.

COLE, J. W., MILNER, D. M. & SPINKS, K. D. 2005. Calderas and caldera structures: a review. *Earth Science Reviews*, **69**, 1–26.

CONDIE, K. C. & KRÖNER, A. 2008. When did plate tectonics begin? Evidence from the geologic record. *In*: CONDIE, K. C. & PEASE, V. (eds) *When Did Plate Tectonics Begin on Planet Earth?* Geological Society of America, Special Papers, **440**, 281–294.

CONEY, P. J., JONES, D. L. & MONGER, J. W. H. 1980. Cordilleran suspect terranes. *Nature*, **288**, 329–333.

CORRIGAN, D., HAJNAL, Z., NÉMETH, B. & LUCAS, S. B. 2005. Tectonic framework of a Paleoproterozoic arc-continent to continent-continent collisional zone, Trans-Hudson Orogen, from geological and seismic reflection studies. *Canadian Journal of Earth Sciences*, **42**, 421–434.

DALZIEL, I. W. D. 1997. Overview: Neoproterozoic-Paleozoic geography and tectonics: review, hypotheses and environmental speculations. *Geological Society of America Bulletin*, **109**, 16–42.

DEWEY, J. F. 1969. Evolution of the Appalachian-Caledonian orogen. *Nature*, **222**, 124–129.

DEWEY, J. F. & BIRD, J. M. 1970. Mountain belts and the new global tectonics. *Journal of Geophysical Research*, **75**, 2625–2647.

ELDREDGE, S., BACHTADSE, V. & VAN DER VOO, R. 1985. Paleomagnetism and the orocline hypothesis. *Tectonophysics*, **119**, 153–179.

FEDERICO, L., CRISPINI, L., SCAMBELLURI, M. & CAPONNI, G. 2007. Ophiolite mélange zone records exhumation in a fossil subduction channel. *Geology*, **35**, 499–502.

GERYA, T. & STÖCKHERT, B. 2006. Two-dimensional numerical modeling of tectonic and metamorphic histories of active continental margins. *International Journal of Earth Sciences*, **95**, 250–274.

GUTSCHER, M. A., SPAKMAN, W., BIJWAARD, H. & ENGDAHL, E. R. 2000. Geodynamic of flat subduction: seismicity and tomographic constraints from the Andean margin. *Tectonics*, **19**, 814–833.

HARRISON, A. & WHITE, R. S. 2006. Lithospheric structure of an active backarc basin: The Taupo Volcanic Zone, New Zealand. *Geophysical Journal International*, **167**, 968–990.

HOFFMAN, P. F. 1988. United plates of America, the birth of a craton: Early Palaeozoic assembly and growth of Laurentia. *Annual Review of Earth and Planetary Science*, **16**, 543–603.

HOFFMAN, P. F. 1991. Did the breakout of Laurentia turn Gondwanaland inside-out? *Science*, **252**, 1409–1412.

INTERNATIONAL COMMITTEE OF TECTONIC MAPS. 1982. *Tectonics of Europe and Adjacent Areas. Variscides, epi-Paleozoic Platforms and Alpides*. Explanatory note to the International tectonic map of Europe and adjacent areas, scale 1:2 500 000. 'Nauka' Publishing House.

JOHNSTON, S. T. & BOREL, G. D. 2007. The odyssey of the Cache Creek terrane, Canadian Cordillera: implications for accretionary orogens, tectonic setting of Panthalassa, the Pacific superswell, and break-up of Pangea. *Earth and Planetary Science Letters*, **253**, 415–428.

KASTENS, K., MASCLE, J. *ET AL.* 1988. ODP Leg 107 in the Tyrrhenian Sea: insights into passive margin and back-arc basin evolution. *Geological Society of America Bulletin*, **100**, 1140–1156.

KONHAUSER, K. O., HAMADE, T., RAISWELL, R., MORRIS, R. C., FERRIS, F. G., SOUTHAM, G. & CANFIELD, D. E. 2002. Could bacteria have formed the Precambrian banded iron formations? *Geology*, **30**, 1079–1082.

LI, Z.-X., BOGDANOVA, S. V. *ET AL.* 2008. Assembly, configuration, and break-up history of Rodinia: a synthesis. *Precambrian Research*, **160**, 179–210.

LUDDEN, J. & HYNES, A. J. (eds) 2000. The Abitibi-Grenville transect. *Canadian Journal of Earth Sciences*, **37**, 115–516.

McMENAMIN, M. A. S. & McMENAMIN, D. L. S. 1990. *The Emergence of Animals: The Cambrian Breakthrough*. Columbia University Press, New York.

McKERROW, W. S. & SCOTESE, C. R. (eds) 1990. *Palaeozoic Palaeogeography and Biogeography*. Geological Society, London, Memoirs, **12**, 1–435.

McQUARRIE, N. 2002. The kinematic history of the central Andean fold-thrust belt, Bolivia: implications for building a high plateau. *Geological Society of America Bulletin*, **114**, 950–963.

MYERS, J. S. 1993. Precambrian history of the West Australian craton and adjacent orogens. *Annual Review of Earth and Planetary Sciences*, **21**, 453–485.

OCCHIPINTI, S. A., SHEPPARD, S., PASSCHIER, C., TYLER, I. M. & NELSON, D. R. 2004. Palaeoproterozoic crustal accretion and collision in the southern Capricorn Orogen: the Glenburgh Orogeny. *Precambrian Research*, **128**, 237–255.

ORTEGA-GUTIÉRREZ, F. 1975. *The pre-Mesozoic Geology of the Acatlán Area, South Mexico*. PhD thesis, University of Leeds, UK.

ORTEGA-GUTIÉRREZ, F., ELÍAS-HERRERA, M., REYES-SALAS, M., MACIAS-ROMO, C. & LÓPEZ, R. 1999. Late Ordovician-Early Silurian continental collision orogeny in southern Mexico and its bearing on

Gondwana-Laurentia connections. *Geology*, **27**, 719–722.

PATERSON, S. R. & TOBSICH, O. T. 1992. Rates of processes in magmatic arcs: implications for the timing and nature of pluton emplacement and wall rock deformation. *Journal of Structural Geology*, **14**, 291–300.

PETFORD, N., CRUDEN, A. R., MCCAFFREY, K. J. W. & VIGNERESSE, J. L. 2000. Granite magma formation, transport and emplacement in the Earth's crust. *Nature*, **408**, 669–673.

PILGER, R. H. 1981. Plate reconstruction, aseismic ridges, and low angle subduction beneath the Andes. *Geological Society of America Bulletin*, **92**, 448–456.

PISAREVSKY, S. A., WINGATE, M. T. D., POWELL, C. MCA., JOHNSON, S. & EVANS, D. A. D. 2003. Models of Rodinia assembly and fragmentation. *In*: YOSHIDA, M., WINDLEY, B. & DASGUPTA, S. (eds) *Proterozoic East Gondwana: Supercontinent Assembly and Breakup*. Geological Society, London, Special Publications, **206**, 35–55.

POWELL, C. MCA. 1995. Are Neoproterozoic glacial deposits preserved on the margins of Laurentia related to the fragmentation of two supercontinents? Comment. *Geology*, **23**, 1053–1054.

PRICE, R. A. & MOUNTJOY, E. W. 1970. Geologic structure of the Canadian Rocky Mountains between Bow and Athabasca rivers-a progress report. *Geological Association of Canada Special Paper*, **6**, 1–166.

QUESADA, C. 1990. Precambrian terranes in the Iberian Variscan foldbelt. *In*: STRACHAN, R. A. & TAYLOR, G. K. (eds) *Avalonian and Cadomian Geology of the North Atlantic*. Blackie and Son, Glasgow, 109–133.

QUESADA, C. & DALLMEYER, R. D. 1994. Tectonothermal evolution of the Badajoz–Córdoba shear zone (SW Iberia): characteristics and $^{40}$Ar/$^{39}$Ar mineral age constraints. *Tectonophysics*, **231**, 195–213.

RAMOS, V. A. & MCNULTY, B. (eds) 2002. Flat-slab subduction in the Andes. *Journal of South American Earth Sciences*, **15**, 1–155.

ROGERS, J. J. W. & SANTOSH, M. 2002. Configuration of Columbia, a Mesoproterozoic supercontinent. *Gondwana Research*, **5**, 5–22.

SCOTESE, C. R. 1997. *Continental Drift*. 7th ed., PALEOMAP Project, Arlington, Texas.

SPINKS, K. D., ACOCELLA, V., COLE, J. W. & BASSETT, K. N. 2005. Structural control of volcanism and caldera development in the transtensional Taupo Volcanic Zone, New Zealand. *Journal of Volcanology and Geothermal Research*, **144**, 7–22.

STAMPFLI, G. M. & BOREL, G. D. 2002. A plate tectonic model for the Paleozoic and Mesozoic constrained by dynamic plate boundaries and restored synthetic oceanic isochrones. *Earth and Planetary Science Letters*, **196**, 17–33.

STÖCKHERT, B. & GERYA, T. V. 2005. Pre-collisional high pressure metamorphism and nappe tectonics at active continental margins: a numerical simulation. *Terra Nova*, **17**, 102–110.

ST-ONGE, M. R., SEARLE, M. P. & WODICKA, N. 2006. Trans-Hudson Orogen of North America and Himalaya-Karakoram-Tibetan Orogen of Asia: structural and thermal characteristics of the lower and upper plates. *Tectonics*, **25**, TC4006, doi: 10.1029/2005TC001907.

TALAVERA-MENDOZA, O., RUIZ, J., GEHRELS, G., MEZA-FIGUEROA, D., VEGA-GRANILLO, R. & CAMPA-URANGA, M. F. 2005. U–Pb geochronology of the Acatlán Complex, and implications for the Paleozoic paleogeographic and the tectonic evolution of southern Mexico. *Earth and Planetary Science Letters*, **235**, 682–699.

TAPPONNIER, P. 1977. Evolution tectonique du système alpin en Méditerranée: poinçonnement et écrasement rigide-plastique. *Bulletin de la Société Géologique de France*, **7**, 437–460.

VAN STAAL, C. R., DEWEY, J. F., MAC NIOCAILL, C. & MCKERROW, W. S. 1998. The Cambrian-Silurian tectonic evolution of the Northern Appalachians and British Caledonides: history of a complex, west and southwest Pacific-type segment of Iapetus. *In*: BLUDELL, D. J. & SCOTT, A. C. (eds) *Lyell: The Past is the Key to the Present*. Geological Society, London, Special Publications, **143**, 199–242.

WILLIAMS, H. 1964. The Appalachians in Newfoundland – a two sided symmetrical system. *American Journal of Science*, **262**, 1137–1158.

WILLIAMS, H. 1979. Appalachian Orogen in Canada. *Canadian Journal of Earth Sciences*, **16**, 792–807.

WILSON, J. T. 1966. Did the Atlantic Ocean close and then re-open? *Nature*, **211**, 676–681.

WYMAN, D. A., O'NEILL, C. & AYER, J. A. 2008. Evidence for modern-style subduction to 3.1 Ga: a plateau-adakite-gold (diamond) association. *In*: CONDIE, K. C. & PEASE, V. (eds) *When Did Plate Tectonics Begin on Planet Earth?* Geological Society of America Special Paper, **440**, 129–148.

YAÑEZ, P., RUIZ, J., PATCHETT, P. J., ORTEGA-GUTIÉRREZ, F. & GEHRELS, G. 1991. Isotopic studies of the Acatlán Complex, southern Mexico: implications for Paleozoic North American tectonics. *Geological Society of America Bulletin*, **103**, 817–828.

# Caldera volcanism and rift structure in the Taupo Volcanic Zone, New Zealand

J. W. COLE[1]* & K. D. SPINKS[2]

[1]*Department of Geological Sciences, University of Canterbury, Private Bag 4800, Christchurch 8140, New Zealand*

[2]*Mighty River Power Ltd, PO Box 445, Hamilton, New Zealand*

*\*Corresponding author (e-mail: jim.cole@canterbury.ac.nz)*

**Abstract:** The Taupo Volcanic Zone (TVZ) is an active continental volcanic arc/back-arc basin in central North Island, New Zealand. It is the youngest area of volcanic activity that extends southwards from the Coromandel Volcanic Zone (CVZ), where andesitic volcanism began *c*. 18 Ma and rhyolitic volcanism *c*. 10 Ma. It is an extensional basin (average *c*. 8 mm a$^{-1}$) with numerous, predominantly normal (dip $>60°$) faults within the Taupo Rift, but with some strike-slip component. TVZ can be divided into three parts. In the north (Whakatane Graben – Bay of Plenty) and south (Tongariro volcanic centre) volcanism is predominantly andesitic, while in the central part it is predominantly rhyolitic. This central area comprises eight caldera centres; the oldest of which (Mangakino caldera; 1.62–0.91 Ma) may be transitional between CVZ and TVZ. Kapenga caldera (*c*. 700 ka) is completely buried by younger volcanics, but is probably a composite structure with most recent subsidence related to volcano-tectonic processes. Of the remaining five caldera centres, Rotorua, Ohakuri and Reporoa are all simple, sub-circular structures which collapsed *c*. 240 ka, and are each associated with one ignimbrirte outflow sheet (Mamaku, Ohakuri and Kaingaroa, respectively). Okataina and Taupo are caldera complexes with multiple ignimbrite eruptions and phases of collapse. The three simple calderas are extra-rift, occurring outside the main fault zone in the centre of the Taupo Rift system, while the two caldera complexes are both intra-rift. There is a close relationship between volcanism and structure in TVZ, and many of the structural caldera boundaries have rectangular geometry reflecting the fault pattern. Intrusion of high-alumina basalts as dykes, parallel to the fault trend, may have had a strong influence in causing rhyolitic eruptions in central TVZ.

The Taupo Volcanic Zone (TVZ) is the currently active zone of calc-alkaline volcanism and intra-arc rifting (Fig. 1), resulting from oblique subduction (*c*. 42 mm a$^{-1}$; De Mets *et al*. 1990) of the Pacific plate beneath the North Island, New Zealand (Fig. 2), and is the most active rhyolitic system on Earth (Houghton *et al*. 1995). TVZ is a zone of active rifting within continental lithosphere that is extending at an average rate of *c*. 8 mm a$^{-1}$ (Darby & Meertens 1995; Darby *et al*. 2000) in a NW–SE direction.

Wilson *et al*. (1995) consider the earliest volcanic activity was andesitic and began *c*. 2 Ma ago, with rhyolitic volcanism commencing *c*. 1.6 Ma from the central part of TVZ, accompanied by rifting from *c*. 1.1 Ma. The exact timing of initiation of TVZ remains uncertain because much of the early geological history is concealed by eruptives from the largest (Whakamaru) eruption at *c*. 0.34 Ma. Also, the relationship to the Coromandel Volcanic Zone (CVZ) in both space and time is unclear (Briggs *et al*. 2005). Andesitic volcanism in CVZ began *c*. 18 Ma and rhyolitic volcanism *c*. 10 Ma

(Adams *et al*. 1994). Volcanism in CVZ was largely NNW-oriented and becomes progressively younger southwards (Fig. 3; Skinner 1986; Briggs *et al*. 2005), culminating with eruptions in the Tauranga and Kaimai centres (Fig. 3) between 2.9 and 1.9 Ma (Briggs *et al*. 2005). The earliest rhyolitic centre considered part of TVZ is Mangakino (Fig. 3), with ages ranging from 1.62–0.91 Ma, but this centre is on the western side of TVZ, and may itself be transitional between CVZ and TVZ.

Many geophysical papers (e.g. Stratford & Stern 2006) refer to the Central Volcanic Region (CVR). This is a wedge-shaped area comprising low-density, low-velocity volcanics that occupy the eastern side of Coromandel, and TVZ (Fig. 3), as defined by gravity and seismicity (Cole *et al*. 1995), and hence corresponds to the period of rhyolitic volcanism in the last 10 Ma. Studies of volcanic rocks in CVZ and TVZ suggest there is no obvious time break between volcanism in the two zones (Fig. 3). If there is a complete transition, it may be better to delineate the TVZ purely on visible structure, and to restrict it to volcanism between clearly

*From*: MURPHY, J. B., KEPPIE, J. D. & HYNES, A. J. (eds) *Ancient Orogens and Modern Analogues.*
Geological Society, London, Special Publications, **327**, 9–29.
DOI: 10.1144/SP327.2   0305-8719/09/$15.00 © The Geological Society of London 2009.

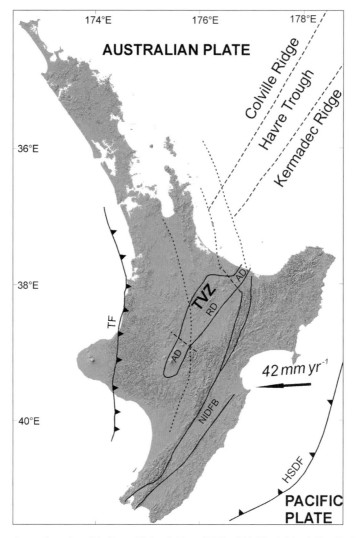

**Fig. 1.** Location and tectonic setting of the Taupo Volcanic Zone (TVZ) within North Island, New Zealand. Velocity of the Pacific plate relative to the Australian plate after De Mets *et al.* (1990). AD, andesite-dominated northern and southern part of TVZ; RD, rhyolite-dominated central part of TVZ; TF, Taranaki Fault, a long-lived structure last active in the Miocene as a thrust fault (e.g. Knox 1982); NIDFB, North Island Dextral Fault Belt, a zone of dominantly strike-slip faulting related to the rotation of the eastern North Island; HSDF, Hikurangi Subduction Deformation Front, which marks the edge of subduction of the Pacific Plate. After Reyners *et al.* (2006).

defined NNE-trending faults, as indicated in Figure 3. This would equate with the 'Modern TVZ' of Wilson *et al.* (1995).

This review considers current views on the 3D structure of TVZ and discusses the location and nature of caldera volcanism in central TVZ.

## Subsurface structure of TVZ

Our understanding of what is beneath TVZ has largely come from geophysical studies, including

gravity (e.g. Rogan 1982; Stern 1986; Bibby *et al.* 1995), electrical modelling (e.g. Bibby *et al.* 2000), earthquake tomography (e.g. Reyners *et al.* 2006; Sherburn *et al.* 2003, 2006), explosion seismology (e.g. Harrison & White 2004; Stratford & Stern 2004, 2006) and magnetotelluric studies (Ogawa *et al.* 1999).

The primary reason for TVZ existence is the Wadati–Benioff zone, which descends westwards from the Hikurangi Trough (Fig. 2) to a depth of at least 300 km, and lies 80–100 km beneath TVZ

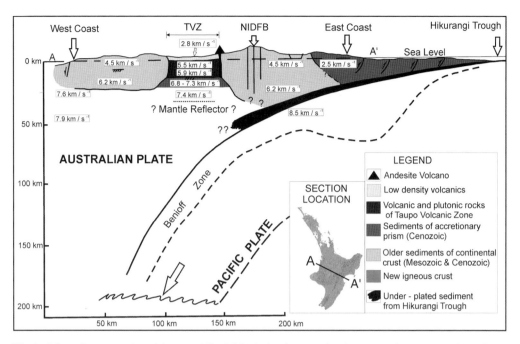

**Fig. 2.** Schematic cross-section of the central North Island, showing crustal and upper mantle structure as determined by seismology. From Cole (1984) with addition of new data from Stratford & Stern (2004).

(Reyners *et al.* 2006). This subduction zone has now been well defined by a dense seismograph deployment (code named CNIPSE, for Central North Island Passive Seismic Experiment) which was carried out in 2001 (Reyners *et al.* 2006) concurrently with a number of other active and passive seismic experiments in the same region (Henrys *et al.* 2003). It has been imaged as a high $Vp$ ($>8.5$ km s$^{-1}$), low $Vp/Vs$ feature, the latter probably reflecting dehydration of the downgoing slab (Reyners *et al.* 2006). In contrast, the mantle wedge is imaged as a low $Vp$, high $Vp/Vs$ feature, with the lowest $Vp$ (7.4 km s$^{-1}$) and highest $Vp/Vs$ ratio (1.87) at about 65 km depth, possibly representing rising plumes of magma or hot upwelling, decompressing mantle (Reyners *et al.* 2006). Material with $Vp > 8.0$ km s$^{-1}$ directly above the dipping plane is interpreted as more viscous material sinking into the wedge, which is largely insulating the slab from the high-temperature part of the wedge.

Explosion seismology also suggests a significant mantle reflector at *c.* 35 km beneath TVZ (Fig. 2) within the mantle wedge (Stratford & Stern 2004), but Harrison & White (2004) believe this simply represents the base of the under-plated basaltic crust. This may reflect a pool of magma within the mantle wedge. These authors also suggest the upper part of the mantle is heterogeneous, with *c.* 1–2% partial melt, and that the transition to the lower crust is broad without a sharp reflection

Moho (see also Price *et al.* 2005). Instead there is a distinct layer between *c.* 16 and 20 km depth, which they interpret as a zone of mafic underplating, at the top of which is a very strong reflector. Rocks above 16 km are interpreted to be a mix of intruded igneous rock, volcanics and greywacke.

Structure within the top 10 km is largely determined from gravity data, earthquake tomography and electrical modelling. Gravity data indicates a broad area of low gravity (*c.* −350 μN/kg; Bibby *et al.* 1995; Stagpoole & Bibby 1999), with smaller areas of even lower gravity generally corresponding to the centres of rhyolitic volcanism (Rogan 1982). These areas are also characterized by low-$Vp$ indicating low-density, volcaniclastic materials that have filled areas of caldera collapse (Sherburn *et al.* 2003). There is a seismogenic zone at *c.* 6.5 km that may represent the maximum depth that brittle fracture may occur (Bibby *et al.* 1995; Bryan *et al.* 1999), and below this is probably quartzo-feldspathic continental crust.

Additional information about the upper 2–5 km has come from extensive geothermal drilling (e.g. Wood *et al.* 2001) and from studies of lithic blocks incorporated in ignimbrites erupted during caldera-forming events (e.g. Brown *et al.* 1998*a*; Cole *et al.* 1998).

No indication has yet been found of molten or semi-molten magma bodies near the surface, although some plutons (e.g. Ngatamariki) are

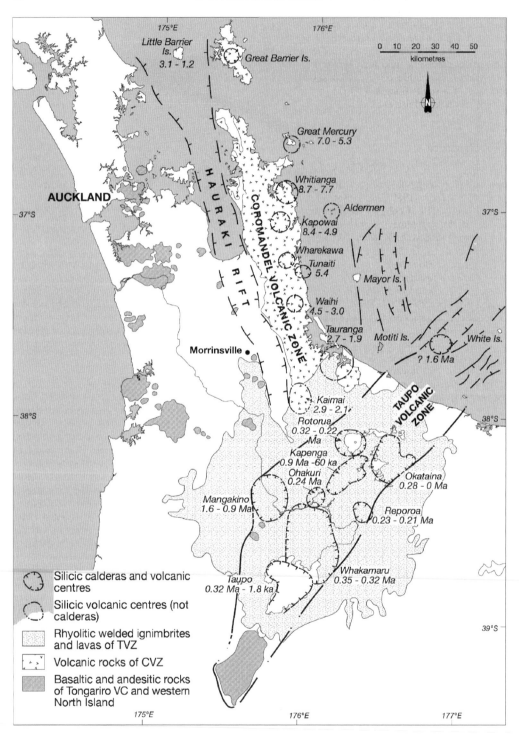

**Fig. 3.** Map of northern and central North Island showing the Coromandel and Taupo volcanic zones, and the location of the main rhyolitic calderas since 10 Ma. Modified from Briggs *et al.* (2005), with permission of *New Zealand Journal of Geology and Geophysics*.

characterized by a high V$p$ anomaly (Sherburn et al. 2003). A rapid increase in conductivity occurs at about 10 km depth, about 2 km below the base of the seismogenic zone, and well above the base of the quartzo-feldspathic crust, which suggests the presence of a small fraction (<4%) of connected melt at this level (Heise et al. 2007).

## TVZ structure

Surface faulting in TVZ has largely been identified from aerial photograph interpretation (e.g. Healy et al. 1964; Nairn 1989, 2002), satellite imagery (e.g. Oliver 1978; Cochrane & Tianfeng 1983; Spinks 2005) supplemented by field observation and interpretation (e.g. Rowland & Sibson 2001; Acocella et al. 2003; Spinks et al. 2005; Villamor & Berryman 2006a, b; Nicol et al. 2006), information from subsurface drilling (e.g. Grindley & Browne 1976; Wood et al. 2001) and seismic data such as that from the $M_L$ 6.3 Edgecumbe earthquake of 2 March 1987 (Anderson & Webb 1989; Nairn & Beanland 1989).

Faults (Fig. 4a) range from major topographic scarps with substantial displacements, such as the Paeroa Fault, to lineaments inferred to be the morphologic expression of faults, but because of the very high production rate of rhyolitic volcanism (0.28 m$^3$s$^{-1}$; Wilson et al. 1995; Wilson 1996), only the most recent are observed. The structure of TVZ is thus defined by a system of NNE-trending faults that extend from Ruapehu in the south to the Bay of Plenty coast with a 15–20 km wide axial rift zone, referred to as the Ruamoko Rift Zone by Rowland & Sibson 2001 or more commonly the Taupo Fault Belt (Grindley 1960) or Taupo Rift (Villamor & Berryman 2006b; Acocella et al. 2003; Spinks et al. 2005). The axial part of the rift is roughly symmetric with a central graben axis but is partitioned along strike into a number of structural domains or rift segments that are often coincident with major volcanic centres, for example Okataina and Taupo caldera complexes (Spinks et al. 2005).

Dips of the faults within this zone appear to be steep (>60°), and the presence of en echelon fault traces (Cole 1990) displaced streams (Villamor & Berryman 2001), and the pitch of slip vectors (Acocella et al. 2003) indicate some dextral strike-slip component. Acocella et al. (2003) indicate that $\beta$-values (angle between the direction perpendicular to the trend of the rift segment and extension direction) vary from 0° to 25° (Fig. 4b). They suggest that dextral shear is not confined to individual structures but appears to be distributed across TVZ.

To the east of TVZ is a zone of primarily dextral strike-slip faults (Figs 1 & 5a) referred to as the North Island Shear Belt (NISB) by Cole (1990) and the North Island Dextral Fault Belt (NIDFB) by Cashman et al. (1992), Beanland (1995) and Beanland et al. (1998). These faults are probably closely related to rotation on the eastern side of the North Island, but most strike-slip displacement has been in the last 1–2 Ma, and the faults have not contributed to major displacements within the North Island (Nicol et al. 2006).

The inter-relationship between this belt and TVZ is unclear. Some of the faults of the NIDFB terminate on the eastern side of TVZ, but there is no direct evidence that they continue beneath it.

## Relationship between faulting and volcanism

TVZ can be divided into three broad areas (Fig. 1): (a) A wide (<50 km) central segment (from Okataina to Taupo inclusive) dominated by rhyolitic volcanism including several calderas and caldera complexes; (b) lateral segments (c. 20 km wide) in the north (Whakatane graben – Bay of Plenty); and (c) south (Tongariro Volcanic Centre) dominated by andesitic volcanism.

### Whakatane graben – Bay of Plenty

This area is dominated by the andesite–dacite volcanoes of Edgecumbe, the partially buried Manawahe, Motuhora (Whale Island) and White Island, although volcanic activity continues north for at least 150 km to the Whakatane seamount (Wright 1992; Gamble et al. 1993). Faults at the surface are hard to see onshore because of the high sedimentation rate in the graben, but in 1987 the Edgecumbe earthquake created fault traces which were slightly oblique to that of the graben (Beanland et al. 1990; Rowland & Sibson 2001), and faulting is clear from geothermal drilling logs (Wood et al. 2001). Interpretation of the faults by Nairn & Beanland (1989) suggest their subsurface portions were block faults dipping at 45 ± 10° on the SE side of the graben (significantly shallower than dips indicated in the central part of TVZ), with steeper antithetic faults on the NW side.

The area offshore is extending 13 ± 6 mm a$^{-1}$, with up to 4.6 ± 2.1 mm a$^{-1}$ of dextral motion, parallel to the rift axis, and with the most intense faulting on the eastern side (Lamarche et al. 2006). Like its onshore equivalent, this is a zone of subsidence (Wright 1990) and thick sediment accumulation (Wright 1993). Beyond the continental–oceanic crust boundary (at 36°45'S) the belt merges with the volcanic arc of the southern Kermadec ridge (Parsons & Wright 1996).

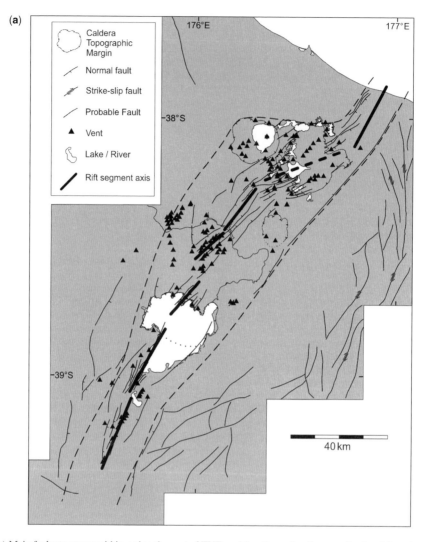

**Fig. 4.** (a) Main fault structures within and to the east of TVZ, and locations of main vents for rhyolitic and basaltic volcanism. (b) Rift segments of TVZ showing their extensional directions and the amount of shear ($\beta$) in each segment. After Spinks *et al.* (2005).

## Tongariro Volcanic Centre

Volcanism here is dominated by andesite massifs of Tongariro and Ruapehu, with older centres to the north and west and associated satellite vents. Recent fault traces have a strong NNE trend with a rift axis of 027° (Rowland & Sibson 2001), and aligned vents have a similar orientation (Cole 1990). Earlier volcanic cones (e.g. Kakaramea and Pihanga) have a NW trend, and there is some evidence that the earliest vents on Tongariro had similar trends (Cole 1978). This may reflect the importance of cross fractures early in the history

of Tongariro Volcanic Centre. Alternatively, in these early vents we may be seeing relics of the Coromandel structural grain.

Villamor & Berryman (2006a) note that fault initiation in the Tongariro Volcanic Centre is younger than to the north, indicating continued southward propagation with time. Extension within the central area, referred to as the Mt. Ruapehu Graben by Villamor & Berryman (2006b), is estimated at $2.3 \pm 1.2$ mm a$^{-1}$. The graben abruptly terminates at its southern end with three sets of cross-cutting faults (Villamor & Berryman 2006b).

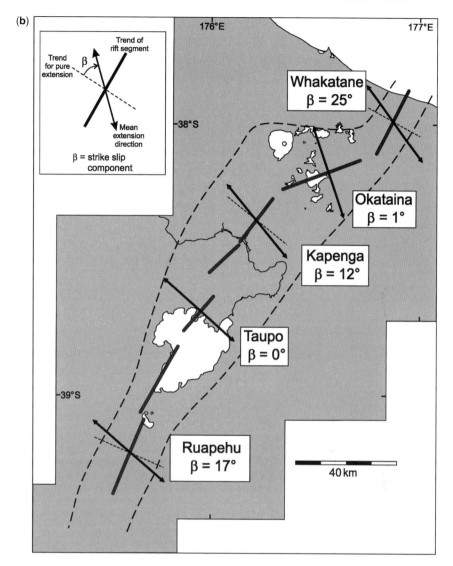

**Fig. 4.** (*Continued*).

## Central TVZ

The central TVZ is structurally complex with a number of variable oriented and offset rift segments (Spinks *et al.* 2005). At the surface it is overwhelmingly dominated by rhyolitic volcanism, manifest as a number of volcanic centres comprising clustered rhyolite lava domes, explosive vents and large calderas surrounded and filled by lavas and the deposits of explosive eruptions, including large ignimbrites. Small amounts of high-alumina basalt (Cole 1972; Houghton *et al.* 1987) are also associated with these rhyolite centres. These were erupted along lineaments parallel to the local rift axis. Near Taupo, for example, a series of basalt cones and phreatomagmatic deposits collectively form the K Trig basalts (Cole 1972; Brown *et al.* 1994). Vents for these basalts trend at 023°, parallel to the faults on the north side of Lake Taupo (Cole 1984). Further north, at Tarawera volcano, there was an eruption on 10 June 1886 when basalt was erupted from a series of en echelon dykes trending between 073° and 080°, within a fissure trending 057° across the mountain parallel to the local fault trend (Nairn & Cole 1981). Most of the individual dykes are <2 m wide, and appear to have filled dilational fractures without detectable horizontal or vertical shear in the fissure walls.

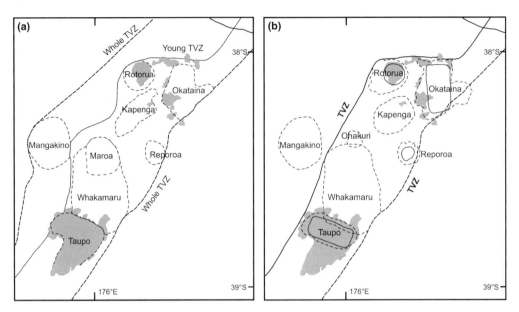

**Fig. 5.** Distribution of calderas within TVZ. (**a**) Calderas shown by Wilson *et al.* (1995). (**b**) Distribution of calderas within this paper. Dashed lines are maximum extent of calderas; solid lines are possible structural boundaries. After Spinks *et al.* 2005.

Vents for rhyolite domes are also commonly aligned parallel to the local faults, as at Maroa (Leonard 2003) and in the Okataina centre (Nairn 2002; Spinks *et al.* 2005).

## Caldera volcanoes of central TVZ

Wilson *et al.* (1995) identified eight calderas in TVZ (Fig. 5a), expressed at the surface by clustering of known or inferred vent locations and/or at depth by geophysically defined basement depressions. As such, the surface expression of calderas at the surface in TVZ varies dramatically, and the morphology and structure of only four (Taupo, Okataina, Rotorua, Reporoa) is understood with any level of precision (Spinks *et al.* 2005); other collapse structures, while recognized, are obscured by younger pyroclastic units or largely destroyed by subsequent activity.

The oldest recognized caldera collapse structure is the Mangakino Caldera (1.6–0.91 Ma), which is attributed as the source of at least five major ignimbrites and airfall deposits (Briggs *et al.* 1993) but is little more than a shallow basin at the surface and is defined largely on gravity data (Rogan 1982; Wilson *et al.* 1984). It is outside TVZ, as defined in this paper, and will not be discussed further.

Recent work within the earlier proposed Maroa Caldera does not indicate any separate caldera-forming event or caldera structure associated with emplacement of the Maroa Dome Complex

(Leonard 2003). This is therefore only discussed in the context of the Whakamaru Caldera.

Our interpretation of the distribution of calderas in TVZ is shown in Figure 5b.

### Kapenga 'Caldera'

The 'Kapenga' Caldera was postulated entirely on geophysical evidence (Rogan 1982; Wilson *et al.* 1984), and while several ignimbrites erupted between 0.89 and 0.71 Ma (Houghton *et al.* 1995), are attributed to a now buried structure, no more recent deposits are unequivocally related to it. Gravley *et al.* (2007) suggest that much of Kapenga's current surface morphology is related to collateral (non-eruptive) subsidence associated with the 240 ka rhyolitic eruptions from Rotorua and Ohakuri (described below), which reactivated faulting associated with the older structure.

### Whakamaru Caldera and Maroa Dome Complex

Whakamaru Caldera (Fig. 6) formed during the largest caldera-forming rhyolitic ignimbrite eruption in TVZ history at *c.* 0.34 Ma, but has no clear topographic margin. It was proposed by Wilson *et al.* (1986) on the basis of the thickness and distribution of the Whakamaru-group ignimbrites, exposed east and west of central TVZ (Fig. 6), which constitute the largest eruptive episode in TVZ history, with an exposed volume of at least

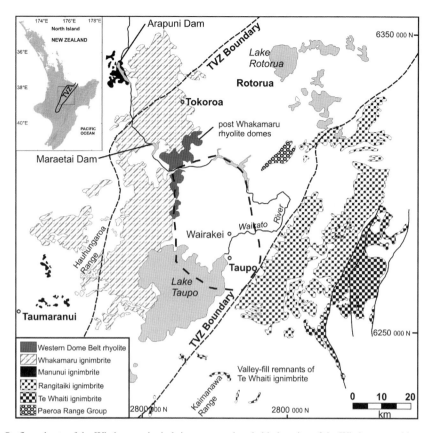

**Fig. 6.** Outflow sheets of the Whakamaru ignimbrite group and probable location of the Whakamaru caldera complex (dashed line). Redrawn from Brown *et al.* (1998*b*).

1000 km$^3$. Several episodes of collapse occurred at Whakamaru Caldera, with dome emplacement between collapse events (Brown *et al.* 1998*b*). Other caldera boundaries are defined on drillhole thickness of Whakamaru ignimbrite, which is particularly thick inside the structure from observations of geothermal drilling at Wairakei, Mokai and Rotokawa (Brown 1994).

The proposed Whakamaru Caldera is defined to the west by the Western Dome Belt (WDB), a 32 km long curvilinear chain of simple and compound silicic domes inferred from field evidence to post-date the Whakamaru ignimbrite eruptions (Wilson *et al.* 1986; Brown *et al.* 1998*b*), a result of post-collapse volcanism localized along the western caldera margin. Eruption of the domes has clearly been controlled by a north–south-trending fault system, which has subsequently ruptured to displace the domes. Domes immediately east of the faults are probably younger features. Given that the domes and faults are aligned along a lineation oblique to the regional trend, if the lavas of the WDB are post-collapse features (Fig. 6), then

faulting likely relates to continued movement along the margins of the Whakamaru Caldera.

Maroa Dome Complex (location shown in Fig. 5a) is an accumulation of simple and composite silicic lava domes (Leonard 2003). Domes are strongly aligned along NNE trends and extensively faulted along the same lineation, indicating the same regional structure is responsible for controlling vent locations and subsequent deformation. Vent and fault lineations are sub-parallel with faulting in the Kapenga graben to the NE.

## Rotorua Caldera

Rotorua Caldera is perhaps the most conspicuous caldera in TVZ, accentuated the local topography and by its large, sub-circular caldera lake, Lake Rotorua (Fig. 7). The caldera is located 15–20 km NW of the junction between Okataina and Kapenga rift axes, and formed during and immediately following the eruption of the $0.24 \pm 11$ Ma Mamaku Ignimbrite (Milner *et al.* 2002; date from Leonard 2003), with a minimum eruption volume

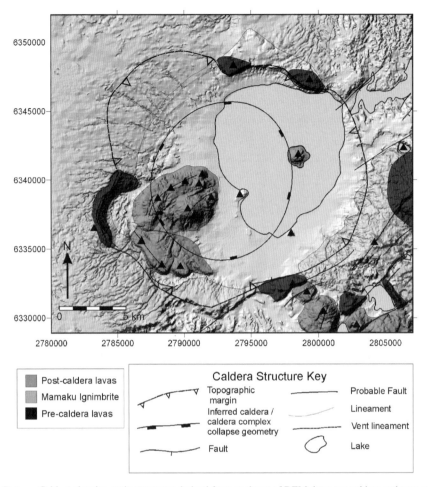

**Fig. 7.** Rotorua Caldera showing main structures derived from analyses of DEM data; pre-caldera and post-caldera lavas and distribution of Mamaku ignimbrite outflow sheet. Grid coordinates shown in the New Zealand map grid (NZMG).

(including intra-caldera ignimbrite) of 145 km³ DRE (Milner *et al.* 2003). Some authors have proposed earlier events at Rotorua Caldera to account for older ignimbrites in the area (Wood 1992), or that Rotorua is not a caldera at all (Hunt 1992), but a detailed study by Milner *et al.* (2002, 2003) strongly indicates Rotorua is a single-event caldera, and the source of the Mamaku Ignimbrite.

A number of pre-caldera rhyolite domes are exposed in the vicinity of Rotorua Caldera (Fig. 7). Milner *et al.* (2002) showed that those on the rim of the caldera are geochemically distinct from each other and from the Mamaku magma system. The post-caldera Ngongotaha and Puke-hangi rhyolite dome complexes, and lavas exposed at Kawaha Point are geochemically similar to the Mamaku Ignimbrite system and may reflect a final

eruptive phase from the Mamaku magma system (Milner *et al.* 2002). Smaller rhyolite domes are geochemically distinct and thought to be much younger.

Stratigraphic evidence outlined by Milner *et al.* (2002) indicates that caldera collapse occurred throughout the eruption and emplacement of the Mamaku Ignimbrite during a single eruptive episode. Milner *et al.* (2002) describe asymmetric caldera collapse that was deepest in the SW of the caldera, with a component of downsag expressed in the overlying Mamaku Ignimbrite. Mamaku Ignimbrite geochemistry indicates the eruption of a single, compositionally zoned magma reservoir, represented by three petrogenetically related pumice types. An andesitic juvenile component in upper parts of the Mamaku Ignimbrite is thought to reflect a discrete magma injected into the residual silicic chamber and tapped during later phases of the eruption during advanced stages of caldera collapse (Milner *et al.* 2003).

Rotorua is a simple sub-circular caldera with dimensions of *c*. 20 × 16 km. The caldera floor is dominated by the *c*. 9 km diameter caldera lake and the youthful morphologies of the post-caldera rhyolite dome complexes. The caldera is characterized by a North–South elongate negative residual gravity anomaly to the west and SW of Lake Rotorua, including Rotorua city and the post-caldera rhyolite dome of Ngongotaha (Rogan 1982; Hunt 1992; Davy & Caldwell 1998). The surrounding gravity contours are not specifically concentric to the caldera margin, defining an asymmetric rise in basement towards the NE and NW caldera margins. The basement gradient is shallowest towards the east and steepest around the southwestern margin. In the NW of the caldera several low scarps sub-parallel to the caldera margin are located within the area defined by Milner *et al.* (2002) as having deformed by downsag into the caldera during collapse.

## Ohakuri Caldera

The Ohakuri Caldera is a newly recognized structure in the Whangapoa basin, 25 km SSW of Rotorua (Figs 5b & 8), which is considered the source of the *c*. 240 ka Ohakuri pyroclastic deposits (Gravley *et al.* 2006, 2007). It overlies the northern margin of the older Whakamaru Caldera and lies adjacent to the axis of the modern Taupo Fault Belt. It is difficult to define either the topographic or structural caldera margins precisely because of subsequent deposition of volcaniclastic sediments, but its presence is inferred from the distribution of air fall deposits, distribution of and transport directions within the Ohakuri Ignimbrite, and from geophysical data.

Stratigraphic evidence presented by Gravley *et al.* (2007) suggests that the Mamaku Ignimbrite and the Ohakuri pyroclastic deposits were erupted in rapid succession, from the Rotorua and Ohakuri Calderas respectively. They suggest that the most feasible way of achieving this linkage is through volcano-tectonic processes, in which contemporaneous faulting triggered almost concurrent events 25 km apart.

## Reporoa Caldera

Reporoa Caldera (Nairn *et al.* 1994) is located at the northern end of the Taupo-Reporoa depression, *c*. 15 km east of the Kapenga segment axis (Fig. 9), and is the source of the 0.23 ± 0.01 Ma

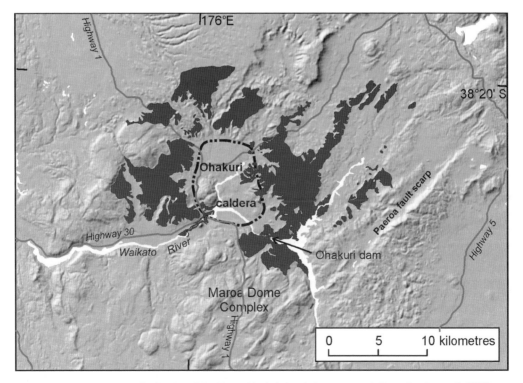

**Fig. 8.** Ohakuri Caldera and distribution of the Ohakuri ignimbrite (darker ornament). From Gravley *et al.* (2007).

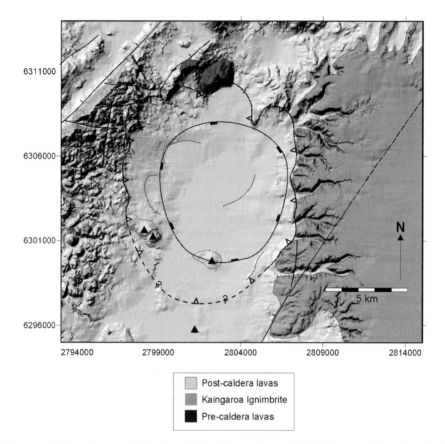

**Fig. 9.** Reporoa Caldera showing main structures derived from analyses of DEM data; pre-caldera and post-caldera lavas and distribution of Kaingaroa ignimbrite outflow sheet. Symbols and grid as in Figure 7.

(Houghton *et al.* 1995) Kaingaroa Ignimbrite, with a total eruptive volume of *c.* 100 km³ (Nairn *et al.* 1994; Beresford & Cole 2000). Kaingaroa Ignimbrite extends radially for 20–30 km beyond the caldera, mostly to the east of Reporoa Caldera where it caps the Kaingaroa Plateau (Fig. 9).

Pre-caldera volcanism in the Reporoa area comprises rhyolite lavas unrelated to the Kaingaroa magma system or the formation of Reporoa Caldera (Beresford & Cole 2000) and older ignimbrites from caldera sources to the west (Wilson *et al.* 1986; Ritchie 1996; Beresford *et al.* 2000). Minor (<2 km³) post-caldera rhyolite domes are geochemically and isotopically distinct from the Kaingaroa magma system. Lithic componentry data for the Kaingaroa Ignimbrite presented by Beresford & Cole (2000) suggest multiple stages in the eruption event: (1) an initial single vent phase; (2) a multiple vent or ring fracture phase on the eastern side with asymmetric caldera collapse leading to eastward-directed pyroclastic flows; and (3) piston collapse accompanied by radially directed pyroclastic flows.

The caldera has a small but distinctive negative gravity anomaly (Nairn *et al.* 1994; Stagpoole 1994; Stagpoole & Bibby 1999) and a low V*p* anomaly consistent with low density, low V*p* caldera fill (Sherburn *et al.* 2003). The gravity anomaly corresponds well with the topographic expression of the caldera, with a gentle and largely open western margin and a steep eastern margin, consistent with asymmetric collapse (Beresford & Cole 2000). Nairn *et al.* (1994) interpret post-caldera rhyolite domes and a buried dome complex to have erupted along fractures related to the caldera rim and a supposed inner caldera ring fault.

Reporoa is a morphologically simple sub-circular caldera (Fig. 9), with approximate dimensions of 11 × 13 km and well-preserved 250 m high collapse scarps along the northern boundary. The north–south long axis of the caldera (eccentricity E = 0.81) is oblique to the regional trend of faults to the west; in the east a NE-trending fault scarp merges with the north–south trending eastern caldera margin. The flat-floored caldera has a well-defined topographic margin in the north and east, but is open to the west and south. The caldera margin is neither dissected by younger faults nor does it truncate older structures. Minor lineaments within the caldera may record modern subsidence or reflect the lacustrine history of the basin.

## Okataina Caldera Complex

Okataina Caldera Complex (OCC) is a complex of overlapping and nested collapse structures, largely filled by the products of post-caldera rhyolite volcanism (Fig. 10). The composite structure is the result of two main collapse events associated with the $0.28 \pm 0.01$ Ma Matahina Ignimbrite (Bailey & Carr 1994; date from Houghton *et al.* 1995)

and the 65 ka Rotoiti eruption (Nairn 1981, 1989; date from Houghton *et al.* 1995), and modified by substantial intra-caldera rhyolite volcanism (e.g. Jurado-Chichay & Walker 2000; Nairn 1989, 2002). Older (>300 ka) collapse events are likely, but potential deposits and precise collapse margins are obscured and/or overprinted by subsequent activity. Magmatic volume estimates for the Matahina and Rotoiti events (including intra-caldera estimates) are $150 \text{ km}^3$ (Bailey & Carr 1994) and $120 \text{ km}^3$ (Froggatt & Lowe 1990), respectively; other eruptives from within and adjacent to OCC account for at least $150 \text{ km}^3$ (e.g. Nairn 1989; Froggatt & Lowe 1990; Bellamy 1991; Jurado-Chichay & Walker 2000). The Matahina and Rotoiti ignimbrites extend predominantly to the east and north, and appear related to

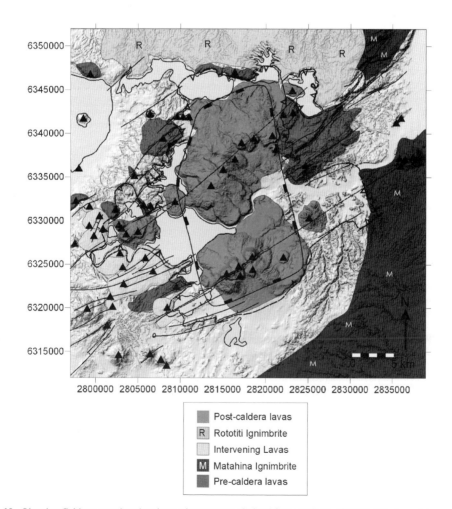

Fig. 10. Okataina Caldera complex showing main structures derived from analyses of DEM data, pre-caldera and post-caldera lavas and distribution of Matahina and Rotoiti ignimbrite outflow sheets. Symbols and grid as in Figure 7.

overlapping but distinct sources in the southern and northern parts of the caldera complex respectively (Nairn 1989).

A number of rhyolite dome lavas scattered around the rim of the caldera complex record volcanism in the Okataina area predating the first collapse event (Nairn 1989, 2002). No precise dates exist for these lavas, but Bowyer (2001) showed that they are chemically distinct and relate to discrete magma batches. Geochemical variation is small however, and no significant variability exists between pre-caldera lavas adjacent to the Okataina and Rotorua calderas; this would imply that these lavas are not related specifically to a Rotorua or Okataina 'volcanic centre'. Only one pre-caldera magma batch has a similar chemistry with the Matahina magma system (e.g. Nairn 1981; Bowyer 2001).

Lavas and associated pyroclastics erupted between the major caldera-forming events are predominantly exposed to the SW of the caldera where it intersects the Kapenga axial rift segment and relate to multiple magma batches (Bellamy 1991; Bowyer 2001). Following the caldera collapse event associated with the eruption of the Rotoiti Pyroclastics, a major phase of explosive volcanism ensued from sources within the caldera complex prior to the development of the two large rhyolite lava massifs that currently fill the caldera. Two main magma types were erupted during the explosive phase, generating fourteen eruptive episodes (Smith et al. 2002). During more recent effusive activity, multiple magmas were often involved with a single eruptive episode. The two documented caldera-forming events at Okataina are spatially overlapping but are significantly separated temporally (>200 ka), reflecting geochemically distinct magma systems (Burt et al. 1998). Volcanism in the Okataina area from the earliest to the most recent eruptives, including the caldera forming events, therefore records the eruption of multiple discrete magma chambers rather than the progressive tapping of a single large chamber.

A distinct large negative residual gravity anomaly (Rogan 1982; Davy & Caldwell 1998) and low $Vp$ anomaly (Sherburn et al. 2003) define a north–south elongated depression consistent with the mapped Okataina Caldera margin and filled with a large volume of low $Vp$, low density, volaniclastic sediment. A clear low $Vp$ anomaly at 4 km effectively corresponds to a minimum depth extent of the collapse structure (Sherburn et al. 2003). Gravity data presented by Nairn (2002) indicates a north–south elongate negative gravity anomaly roughly concentric to the topographic margin, and centred beneath Tarawera Volcanic Complex and the southern part of Haroharo Volcanic Complex (Fig. 10). The contours open on the west side of the structure but gravity highs separate the structure from basement highs to the east and Rotorua Caldera to the west. Topographic embayments in the caldera complex margin are also peripheral to the main basement depression indicated by gravity and $Vp$ data.

The topographic margin at Okataina (Nairn 2002) is variably manifest as scalloped slump scars in pre-caldera rhyolite domes and ignimbrites, eroded caldera walls, rectilinear fault scarps coincident with regional faulting (Fig. 10), and in the SW by the steep margins of post-caldera constructional rhyolite domes. As such, the composite topographic margin defines a depression considerably modified from the original multiple collapse structure. Distinct embayments occur on each side of OCC where it is intersected by regional faulting of the axial rift within the Okataina transfer zone. These are contiguous with two intra-caldera dome complexes forming two overlapping linear vent zones, which transect the caldera complex as the lateral continuation of the adjacent rift segment axes. The boundaries of individual collapse events are complex and largely overprinted by subsequent volcanism and tectonism, but caldera reconstructions suggest the major collapses are centred on the axes of the intersecting rift segments. Davy & Bibby (2005) indicate from a seismic reflection survey that Mamaku Ignimbrite dips <6°E across the western side of Lake Tarawera, and then becomes cut off, suggesting a caldera structural boundary at this point. Lakes at Okataina exhibit a moat pattern where they have ponded between the topographic rim of the caldera and caldera-filling post-caldera constructional volcanism. Lakes filling earlier manifestations of the structure may have been much larger.

## Taupo Caldera Complex

Taupo Caldera Complex (Fig. 11) has been frequently active in the past c. 65 ka (Wilson et al. 1986; Houghton et al. 1995), while its poorly constrained early eruptive history indicates activity over c. 300 ka (Wilson et al. 1986; Cole et al. 1998). The caldera-forming Oruanui eruption at 26.5 ka (calibrated; Wilson 1993) generated a c. 430 km³ fall deposit, a 320 km³ bulk volume non-welded density current deposits (mostly ignimbrite) and c. 420 km³ of caldera-fill material erupted, equivalent to c. 530 km³ of magma (Wilson 2001). The Oruanui event is thus largely responsible for the modern caldera morphology. Wilson (1993) has identified 28 separate eruptions since the Oruanui eruption; the most recent and largest of these, the caldera modifying 35 km³ Taupo Ignimbrite eruption, occurred about 1800 years ago from vents near the Horomatangi Reefs in the eastern

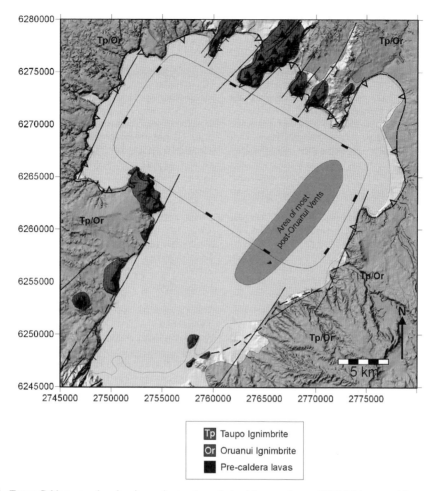

**Fig. 11.** Taupo Caldera complex showing main structures derived from analyses of DEM data, pre-caldera and post-caldera lavas and distribution of Oruanui and Taupo ignimbrite outflow sheets. Symbols and grid as in Figure 7.

part of the lake (Wilson & Walker 1985; Smith & Houghton 1995).

The early history (>65 ka) of volcanism in the vicinity of modern Lake Taupo is represented mainly by domes and limited pyroclastics scattered around the lake (Sutton *et al.* 1995). Two ignimbrites exposed on the margin of the caldera are commonly attributed to a Taupo source (Sutton *et al.* 1995; Cole *et al.* 1998), although their limited extent means their relationship to the current caldera complex is ambiguous. Pre-caldera rhyolite lavas form a series of headlands along the northern caldera margin and to the SW of the caldera, while the large caldera-filling domes and flows characteristic of OCC are noticeably absent. Vents for the post-Oruanui explosive eruptions are inferred by isopach data to be concentrated along a NE-trending lineation (Fig. 11) in the eastern part of modern Lake Taupo (Wilson 1993). Construction of a

dome complex in this area during the post-Oruanui phase is suggested by lithic componentry data for the Taupo Ignimbrite (Cole *et al.* 1998) with its likely destruction during the Taupo eruption. Lithic componentry analysis of both the Oruanui and Taupo ignimbrites identifies different lithic suites, interpreted by Cole *et al.* (1998) as reflecting dissimilar sub-caldera geology beneath mutually exclusive collapse structures.

Petrological studies (e.g. Sutton *et al.* 1995, 2000; Charlier *et al.* 2005) show a complex magmatic system, involving the stepwise appearance of compositionally distinct magma batches with short crustal-residence times. Eruptives prior to the Oruanui eruption form distinct compositional and spatial groups, while a large isotopically homogeneous magma body was generated prior to the Oruanui caldera-forming eruption (Sutton *et al.* 1995). Some of the pre-Oruanui domes exposed on the northern caldera margin, and widespread tephras erupted between 65 ka and the Oruanui eruption, are the same composition as the Oruanui magma, and thus record the coalescence of a large magma chamber. Distinct magmas were erupted during the same period from different areas around Taupo, suggesting that magma batches may have been erupted during the Oruanui eruption. Eruptives of the post-Oruanui sequence form four temporally grouped magma types that are compositionally distinct from the Oruanui magma. The youngest magma, associated with the Taupo eruption, represents the largest homogeneous magma accumulation in the post-Oruanui sequence (Sutton *et al.* 1995).

A large trapezoidal-shaped negative Bouguer gravity anomaly, the most intense negative gravity anomaly in TVZ (Davy & Caldwell 1998), is documented over the northern part of Lake Taupo. The gravity anomaly is consistent with a caldera collapse structure elongate NW–SE, perpendicular to the axial rift zone in this segment. The gravity data do not facilitate identification of individual collapse structures, and Davy & Caldwell (1998) consider the structures are nested, with the Taupo eruption producing additional subsidence in the NE part of the modern lake. Geophysical data also demonstrate differential subsidence towards the caldera in the southern part of the lake, and a NW–SE-trending structural boundary marking the southern caldera margin (Davy & Caldwell 1998). Seismic reflection, gravity and magnetic data

presented by Davy & Caldwell (1998) all suggest that a line between Karangahape Cliffs and Motutere Point marks a major structural boundary perpendicular to the trend of TVZ.

The southern part of Lake Taupo occupies a NE-trending fault-bounded depression intersecting the southern caldera margin; the western margin of this structure is a continuation of fault systems dissecting the Tongariro Volcanic Centre to the south. A prominent structural feature is the divergence in fault trend (*c.* 20°) to the south and north of the caldera complex. This bend effectively forms the boundary between the Tongariro and Taupo South Rift segments. Vents for pre-caldera lavas to the south and north of the caldera complex lie along NE-trending lineaments; the vents for post-Oruanui eruptions mostly occur along the eastern edge of the caldera complex (Wilson 1993).

## Discussion

The characteristics of TVZ calderas in the last 300 ka (Taupo, Okataina, Rotorua, Ohakuri and Reporoa) clearly indicate the complex eruptive and structural history of Okataina and Taupo compared to Rotorua, Ohakuri and Reporoa. These data show that these caldera structures can be divided into two groups: (1) extra-rift calderas (Reporoa and Rotorua) are simple, relatively small, sub-circular, monogenetic structures; and (2) intra-rift caldera complexes (Okataina and Taupo) are large, multiple

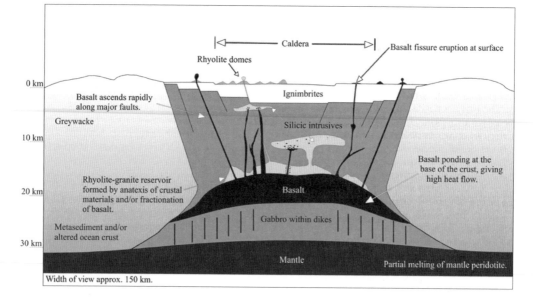

**Fig. 12.** Schematic cross-section of TVZ indicating likely features within the crust and upper mantle. After Hiess *et al.* (2007).

collapse structures, with rectangular geometries, clear coupling to regional structure, and broadly homogeneous magmas. This distinction demonstrates the role of active regional tectonics in influencing the location, structure and development of caldera systems within a rift zone. Ohakuri is a simple structure, but is intra-rift.

With an increasing number of high precision radiometric dates from ignimbrites and rhyolite domes (e.g. Houghton *et al.* 1995; Leonard 2003), it has become evident that caldera-forming events are not uniformly spread in time and space across TVZ. The earliest caldera-forming events occurred in the most central part of TVZ and the currently active areas are to the north and south, at Okataina and Taupo. Equally, as shown by the above authors, there are periods of more intense caldera-forming activity (e.g. 0.89–0.71 Ma; 0.34–0.32 Ma; 0.28–0.24 Ma). This may well relate to variability in rate of extension. It must be remembered however that rhyolite dome/tephra eruptions occur throughout the history of TVZ, and in any one area there may be activity over a considerable period of time (as at Kapenga). The caldera-forming event is therefore just one episode in the activity, so using the terms 'pre'- and 'post'-caldera volcanism needs to be used with caution.

Magmatism and geological structure are closely related in the TVZ, as shown by the common alignment of vents along local faults. This is true of many rhyolite domes and of high-alumina basalt fissure eruptions (Cole 1972; Nairn & Cole 1981; Houghton *et al.* 1987). Spinks *et al.* (2005) also show that there is a correlation between areas of maximum extension, particularly Okataina and Taupo, and eruptive volume. A combination of these observations suggests that basalt rises into the lower crust, providing high heat flow. This is likely to cause partial melting of parts of the lower crust, composed of metasediments and igneous rocks (Fig. 12), which rise and pond at the ductile/brittle transition zone in the mid-crust. At this point they become strongly influenced by the many faults in central TVZ, and so erupt at the surface along linear fault zones. The greatest volume of magma will occur where the greatest amount of extension takes place, and this in turn will create the greatest volume of eruptives, and the largest and most complex calderas. The shapes of these calderas will also be strongly influenced by the local faults.

Early drafts of the paper have been read by H. Bibby and D. Gravley, and their comments are much appreciated. The paper was reviewed by J. Gamble and V. Acocella, and their helpful and constructive suggestions have considerably improved the paper. Assistance with figure preparation by S. Hampton is also acknowledged.

## References

ACOCELLA, V., SPINKS, K. D., COLE, J. W. & NICOL, A. 2003. Oblique back arc rifting of Taupo Volcanic Zone, New Zealand. *Tectonics*, **22**, 1045, doi: 0.1029/2002TC001447.

ADAMS, C. J., GRAHAM, I. J., SEWARD, D. & SKINNER, D. N. B. 1994. Geochronological and geochemical evolution of late Cenozoic volcanism in the Coromandel Peninsula, New Zealand. *New Zealand Journal of Geology and Geophysics*, **37**, 359–379.

ANDERSON, H. & WEBB, T. 1989. The rupture process of the 1987 Edgecumbe Earthquake, New Zealand. *New Zealand Journal of Geology and Geophysics*, **32**, 43–52.

BAILEY, R. A. & CARR, R. G. 1994. Physical geology and eruptive history of the Matahina Ignimbrite, Taupo Volcanic Zone, North Island, New Zealand. *New Zealand Journal of Geology and Geophysics*, **37**, 319–344.

BEANLAND, S. 1995. *The North Island Dextral Fault Belt, Hikurangi subduction margin, New Zealand.* PhD thesis, Victoria University of Wellington, New Zealand.

BEANLAND, S., BLICK, G. H. & DARBY, D. J. 1990. Normal faulting in a back-arc: geological and geodetic characteristics of the 1987 Edgecumbe earthquake, New Zealand. *Journal of Geophysical Research*, **95**, 4693–4707.

BEANLAND, S., MELHUISH, A., NICOL, A. & RAVENS, J. 1998. Structure and deformational history of the inner forearc region, Hikurangi subduction margin, New Zealand. *New Zealand Journal of Geology and Geophysics*, **41**, 325–342.

BELLAMY, S. 1991. *Some studies of the Te Wairoa ignimbrites and the associated volcanic geology of the SW Okataina volcanic centre, Taupo Volcanic Zone.* MSc thesis, University of Waikato, Hamilton, New Zealand.

BERESFORD, S. W. & COLE, J. W. 2000. Kaingaroa Ignimbrite, Taupo Volcanic Zone, New Zealand: evidence for asymmetric caldera subsidence of the Reporoa Caldera. *New Zealand Journal of Geology and Geophysics*, **43**, 471–481.

BERESFORD, S. W., COLE, J. W. & WEAVER, S. D. 2000. Weak chemical and mineralogical zonation in the Kaingaroa Ignimbrite, Taupo Volcanic Zone, New Zealand. *New Zealand Journal of Geology and Geophysics*, **43**, 639–680.

BIBBY, H. M., CALDWELL, T. G., DAVEY, F. J. & WEBB, T. H. 1995. Geophysical evidence of the structure of the Taupo Volcanic Zone and its hydrothermal circulation. *Journal of Volcanology and Geothermal Research*, **68**, 29–58.

BIBBY, H. M., RISK, G. F., CALDWELL, T. G., OGAWA, Y., TAKAKURA, S. & UCHIDA, T. 2000. Deep electrical structure beneath the Taupo Volcanic Zone and the source of geothermal heat. *Proceedings of the 22nd Geothermal Workshop*, 81–86.

BOWYER, D. A. 2001. *Petrologic, geochemical, and isotopic evolution of rhyolite lavas from the Okataina, Rotorua, and Kapenga volcanic centres, Taupo Volcanic Zone, New Zealand.* PhD thesis, University of Waikato, Hamilton, New Zealand.

BRIGGS, R. M., GIFFORD, M. G., MOYLE, A. R., TAYLOR, S. R., NORMAN, M. D., HOUGHTON, B. F. & WILSON, C. J. N. 1993. Geochemical zoning and eruptive mixing in ignimbrites from Mangakino volcano, Taupo Volcanic Zone, New Zealand. *Journal of Volcanology and Geothermal Research*, **56**, 175–203.

BRIGGS, R. M., HOUGHTON, B. F., McWILLIAMS, M. & WILSON, C. J. N. 2005. ⁴⁰Ar/³⁹Ar ages of silicic rocks in the Tauranga-Kaimai area, New Zealand: dating the transition between volcanism in the Coromandel Arc and the Taupo Volcanic Zone. *New Zealand Journal of Geology and Geophysics*, **48**, 459–469.

BROWN, S. J. A. 1994. *Geology and geochemistry of the Wkahamaru-group ignimbrites, and associated rhyolite domes, Taupo Volcanic Zone, New Zealand*. PhD thesis, University of Canterbury, Christchurch, New Zealand.

BROWN, S. J. A., SMITH, R. T. & COLE, J. W. 1994. Compositional and textural characteristics of the strombolian and surtseyan K-Trig basalts, Taupo Volcanic Centre, New Zealand: implications for eruption dynamics. *New Zealand Journal of Geology and Geophysics*, **37**, 113–126.

BROWN, S. J. A., BURT, R. M., COLE, J. W., KRIPPNER, S. J. P., PRICE, R. C. & CARTWRIGHT, I. 1998a. Plutonic lithics in ignimbrites of the Taupo Volcanic Zone, New Zealand; sources and conditions of crystallisation. *Chemical Geology*, **148**, 21–41.

BROWN, S. J. A., COLE, J. W., WILSON, C. J. N. & WOODEN, J. 1998b. Geochemistry of the Whakamaru group ignimbrites, Taupo Volcanic Zone, New Zealand; evidence for non-sequential tapping of a large silicic magma system. *Journal of Volcanology and Geothermal Research*, **84**, 1–37.

BRYAN, C. J., SHERBURN, S., BIBBY, H. M., BANNISTER, S. C. & HURST, A. W. 1999. Shallow seismicity of the central Taupo Volcanic Zone, New Zealand: its distribution and nature. *New Zealand Journal of Geology and Geophysics*, **42**, 533–542.

BURT, R. M., BROWN, S. J. A., COLE, J. W., SHELLEY, D. & WAIGHT, T. E. 1998. Glass-bearing plutonic fragments from ignimbrites of the Okataina volcanic complex, Taupo Volcanic Zone, New Zealand; remnants of a partially molten intrusion associated with preceding eruptions. *Journal of Volcanology and Geothermal Research*, **84**, 209–337.

CASHMAN, S. M., KELSEY, H. M., ERDMAN, C. F., CUTTEN, H. N. C. & BERRYMAN, K. 1992. Strain partitioning between structural domains in the forearc of the Hikurangi subduction zone. *Tectonics*, **11**, 242–257.

CHARLIER, B. L. A., WILSON, C. J. N., LOWENSTERN, J. B., BLAKE, S., VAN CALSTEREN, P. W. & DAVIDSON, J. P. 2005. Magma generation at a large hyperactive silicic volcano (Taupo, New Zealand) revealed by U–Th and U–Pb systematics in zircons. *Journal of Petrology*, **46**, 3–32.

COCHRANE, G. R. & TIANFENG, W. 1983. Interpretation of structural characteristics of the Taupo Volcanic Zone, New Zealand, from Landsat imagery. *International Journal of Remote Sensing*, **4**, 111–128.

COLE, J. W. 1972. High-alumina basalts of Taupo Volcanic Zone, New Zealand. *Lithos*, **6**, 53–64.

COLE, J. W. 1978. Andesites of the Tongariro Volcanic Centre, New Zealand. *Journal of Volcanology and Geothermal Research*, **3**, 121–153.

COLE, J. W. 1984. Taupo-Rotorua Depression – an ensialic marginal basin of North Island, New Zealand. *In*: KOKELAAR, P. & HOWELLS, M. F. (eds) *Marginal Basin Geology: volcanic and associated sedimentary and tectonic processes in modern and ancient marginal basins*. Geological Society, London, Special Publications, **16**, 109–120.

COLE, J. W. 1990. Structural control and origin of volcanism in the Taupo volcanic zone, New Zealand. *Bulletin of Volcanology*, **52**, 445–459.

COLE, J. W., DARBY, D. J. & STERN, T. A. 1995. Taupo Volcanic Zone and Central Volcanic Region: backarc Structures of North Island, New Zealand. *In*: TAYLOR, B. (ed.) *Backarc Basins: Tectonics and Magmatism*. Plenum Press, New York, 1–27.

COLE, J. W., BROWN, S. J. W., BURT, R. M., BERESFORD, S. W. & WILSON, C. J. N. 1998. Lithic types in ignimbrites as a guide to the evolution of a caldera complex, Taupo volcanic centre, New Zealand. *Journal of Volcanology and Geothermal Research*, **80**, 217–237.

DARBY, D. J. & MEERTENS, C. M. 1995. Terrestrial and GPS measurements of deformation across the Taupo back-arc and Hikurangi fore-arc regions in New Zealand *Journal of Geophysical Research*, **100**, 8221–8232.

DARBY, D. J., HODGKINSON, K. M. & BLICK, G. H. 2000. Geodetic measurement of deformation in the Taupo Volcanic Zone, New Zealand: the north Taupo network revisited. *New Zealand Journal of Geology and Geophysics*, **43**, 157–170.

DAVY, B. W. & BIBBY, H. 2005. Seismic reflection imaging of the Haroharo Caldera boundary beneath Lake Tarawera, Okataina Volcanic Centre, New Zealand. *New Zealand Journal of Geology and Geophysics*, **48**, 153–166.

DAVY, B. W. & CALDWELL, T. G. 1998. Gravity, magnetic and seismic surveys of the caldera complex, Lake Taupo, North Island, New Zealand. *Journal of Volcanology and Geothermal Research*, **81**, 69–89.

DE METS, C., GORDON, R. G., ARGUS, D. F. & STEIN, S. 1990. Current plate motions. *Geophysical Journal International*, **101**, 425–478.

FROGGATT, P. C. & LOWE, D. J. 1990. A review of late quaternary silicic and some other tephra formations from New Zealand; their stratigraphy, nomenclature, distribution, volume, and age. *New Zealand Journal of Geology and Geophysics*, **33**, 89–109.

GAMBLE, J. A., WRIGHT, I. C. & BAKER, J. A. 1993. Seafloor geology and petrology in the oceanic to continental transition zone of the Kermadec-Havre-Taupo Volcanic Zone Arc System, New Zealand. *New Zealand Journal of Geology and Geophysics*, **36**, 417–435.

GRAVLEY, D. M., WILSON, C. J. N., ROSENBERG, M. D. & LEONARD, G. S. 2006. The nature and age of Ohakuri Formation and Ohakuri Group rocks in surface exposures and geothermal drillhole sequences in the central Taupo Volcanic Zone, New Zealand. *New Zealand Journal of Geology and Geophysics*, **49**, 305–308.

GRAVLEY, D. M., WILSON, C. J. N., LEONARD, C. S. & COLE, J. W. 2007. Double Trouble: paired ignimbrite eruptions and collateral subsidence in the Taupo Volcanic Zone, New Zealand. *Geological Society of America Bulletin*, **119**, 18–30.

GRINDLEY, G. W. 1960. *Sheet 8 – Taupo. Geological map of New Zealand 1:250:000*. Department of Scientific and Industrial Research, Wellington.

GRINDLEY, G. W. & BROWNE, P. R. L. 1976. Structural and hydrological factors controlling the permeabilities of some hot-water systems. *Proceedings of the 2nd UN Symposium on Development and Use of Geothermal Resources*, 377–386.

HARRISON, A. J. & WHITE, R. S. 2004. Crustal structure of the Taupo Volcanic Zone, New Zealand: stretching and igneous intrusion. *Geophysical Research Letters*, **31**, 1–4. doi: 10.1029/2004GL019885.

HEALY, J., SCHOFIELD, J. C. & THOMPSON, B. N. 1964. *Sheet 5 – Rotorua. Geological map of New Zealand 1:250,000*. Department of Scientific and Industrial Research, Wellington.

HEISE, W., BIBBY, H. M., CALDWELL, T. G., BANNISTER, S. C., OGAWA, Y., TAKAKURA, S. & UCHIDA, T. 2007. Melt distribution beneath a young continental rift: the Taupo Volcanic Zone, New Zealand. *Geophysical Research Letters*, **34**; L14313. doi: 10.1029/2007GL029629.

HENRYS, S. H., REYNERS, M. & BIBBY, H. 2003. Exploring the plate boundary structure of the North Island, New Zealand. *EOS, Transaction of the American Geophysical Union*, **84**, 289–295.

HIESS, J., COLE, J. W. & SPINKS, K. D. 2007. High-alumina basalts of the Taupo Volcanic Zone, New Zealand: influence of the crust and crustal structure. *New Zealand Journal of Geology and Geophysics*, **50**, 327–342.

HOUGHTON, B. F., WILSON, C. J. N., LLOYD, E. F., GAMBLE, J. A. & KOKELAAR, B. P. 1987. A catalogue of basaltic deposits within Taupo Volcanic Zone. *New Zealand Geological Survey Record*, **18**, 95–101.

HOUGHTON, B. F., WILSON, C. J. N., MCWILLIAMS, M. O., LANPHERE, M. A., WEAVER, S. D., BRIGGS, R. M. & PRINGLE, M. S. 1995. Chronology and dynamics of a large silicic magmatic system, Central Taupo Volcanic Zone, New Zealand. *Geology*, **23**, 13–16.

HUNT, T. M. 1992. Gravity studies in the Rotorua area, New Zealand. *Geothermics*, **21**, 65–74.

JURADO-CHICHAY, Z. & WALKER, G. P. L. 2000. Stratigraphy and dispersal of the Mangaone Subgroup pyroclastic deposits, Okataina Volcanic Centre, New Zealand. *Journal of Volcanology and Geothermal Research*, **104**, 319–383.

KNOX, G. J. 1982. Taranaki Basin, structural style and tectonic setting. *New Zealand Journal of Geology and Geophysics*, **25**, 125–140.

LAMARCHE, G., BARNES, P. & BULL, J. M. 2006. Faulting and extension rate over the last 20,000 years in the offshore Whakatane Graben, New Zealand continental shelf. *Tectonics*, **25**, TC4005, doi: 10.1029/2005TC001886.

LEONARD, G. S. 2003. *The evolution of Maroa Volcanic Centre, Taupo Volcanic Zone, New Zealand.*

PhD thesis, University of Canterbury, Christchurch, New Zealand.

MILNER, D. M., COLE, J. W. & WOOD, C. P. 2002. Asymmetric, multiple-block collapse at Rotorua Caldera, Taupo Volcanic Zone, New Zealand. *Bulletin of Volcanology*, **64**, 134–149.

MILNER, D. M., COLE, J. W. & WOOD, C. P. 2003. Mamaku Ignimbrite: a caldera-forming ignimbrite erupted from a compositionally zoned magma chamber in Taupo Volcanic Zone, New Zealand. *Journal of Volcanology and Geothermal Research*, **122**, 243–264.

NAIRN, I. A. 1981. *Some studies of the geology, volcanic history and geothermal resources of the Okataina volcanic centre, Taupo Volcanic Zone, New Zealand*. PhD thesis, Victoria University, Wellington, New Zealand.

NAIRN, I. A. 1989. *Sheet V16AC – Mount Tarawera, geological map of New Zealand, 1:50,000*. Department of Scientific and Industrial Research, Wellington, New Zealand.

NAIRN, I. A. 2002. *Geology of the Okataina Volcanic Centre*. Institute of Geological and Nuclear Sciences Geological Map, **25**, 156p + map.

NAIRN, I. A. & BEANLAND, S. 1989. Geological setting of the 1987 Edgecumbe earthquake, New Zealand. *New Zealand Journal of Geology and Geophysics*, **32**, 1–13.

NAIRN, I. A. & COLE, J. W. 1981. Basalt dykes in the 1886 Tarawera Rift. *New Zealand Journal of Geology and Geophysics*, **24**, 585–592.

NAIRN, I. A., WOOD, C. P. & BAILEY, R. A. 1994. The Reporoa Caldera, Taupo Volcanic Zone: source of the Kaingaroa Ignimbrites. *Bulletin of Volcanology*, **56**, 529–537.

NICOL, A., MAZENGARB, C., CHANIER, F., RAIT, G., URUSKI, C. & WALLACE, L. 2006. Large (>50 km) post Oligocene strike-slip displacements in the North Island; Fact or Fiction. Abstract, Geosciences '06 – Our planet, our future. *Geological Society of New Zealand Miscellaneous Publication*, **122A**, 59.

OGAWA, Y., BIBBY, H. M. *ET AL*. 1999. Wide-band magnetotelluric measurements across the Taupo Volcanic Zone, New Zealand – preliminary results. *Geophysical Research Letters*, **26**, 3673–3676.

OLIVER, P. J. 1978. Seismotectonic, structural, volcanological and geomorphological study of New Zealand. *In*: ELLIS, P. J. *ET AL*. (eds) 'Landsat II over New Zealand, Monitoring our Resources from Space'. Department of Scientific and Industrial Research Bulletin, **221**, 298.

PARSONS, L. M. & WRIGHT, I. C. 1996. The Lau-Havre-Taupo back-arc basin: a southward propagating, multi-stage evolution from rifting to spreading. *Tectonophysics*, **263**, 1–22.

PRICE, R. C., GAMBLE, J. A., SMITH, I. E. M., EGGINS, S. M. & WRIGHT, I. C. 2005. An integrated model for the temporal evolution of andesites and rhyolites in New Zealand's North Island. *Journal of Volcanology and Geothermal Research*, **140**, 1–24.

REYNERS, M., EBERHART-PHILLIPS, D., STUART, G. & NISHIMURA, Y. 2006. Imaging subduction from the trench to 300 km depth beneath the central North

Island, New Zealand, with V$p$ and V$p$/V$s$. *Geophysical Journal International*, **165**, 565–583.

RITCHIE, A. 1996. *Volcanic geology of the Waiotapu and related ignimbrites, Taupo Volcanic Zone*. MSc thesis, University of Canterbury, Christchurch, New Zealand.

ROGAN, M. 1982. A geophysical study of the Taupo Volcanic Zone, New Zealand. *Journal of Geophysical Research*, **87**, 4073–4088.

ROWLAND, J. & SIBSON, R. H. 2001. Extensional fault kinematics within the Taupo Volcanic Zone, New Zealand: soft-linked segmentation of a continental rift system. *New Zealand Journal of Geology and Geophysics*, **44**, 271–283.

SHERBURN, S., BANNISTER, S. & BIBBY, H. 2003. Seismic velocity structure of the central Taupo Volcanic Zone, New Zealand, from local earthquake tomography. *Journal of Volcanology and Geothermal Research*, **122**, 69–88.

SHERBURN, S., REYNERS, M., EBERHART-PHILLIPS, D. & BANNISTER, S. 2006. Seismic velocity structure of the Taupo Volcanic Zone. Abstract, Geosciences '06 – Our planet, our future. *Geological Society of New Zealand Miscellaneous Publication*, **122A**, 76.

SKINNER, D. N. B. 1986. Neogene volcanism of the Hauraki Volcanic region. *In:* SMITH, I. E. M. (ed.) *Late Cenozoic volcanism in New Zealand*. Royal Society of New Zealand Bulletin, **23**, 21–47.

SMITH, R. T. & HOUGHTON, B. F. 1995. Vent migration and changing eruptive style during the 1800 ka Taupo eruption: new evidence from the Hatepe and Rotongaio phreatoplinian ashes. *Bulletin of Volcanology*, **57**, 432–439.

SMITH, V. C., SHANE, P. & SMITH, I. E. M. 2002. Tephrostratigraphy and geochemical fingerprinting of the Mangaone Subgroup tephra beds, Okataina Volcanic Centre, New Zealand. *New Zealand Journal of Geology and Geophysics*, **45**, 207–219.

SPINKS, K. D. 2005. *Rift Architecture and Caldera Volcanism in the Taupo Volcanic Zone, New Zealand*. PhD thesis, University of Canterbury, Christchurch, New Zealand.

SPINKS, K. D., ACOCELLA, V., COLE, J. W. & BASSETT, K. N. 2005. Structural control on volcanism and caldera development in the transtensional Taupo Volcanic Zone, New Zealand. *Journal of Volcanology and Geothermal Research*, **144**, 7–22.

STAGPOOLE, V. M. 1994. Interpretation of refraction seismic and gravity data across the eastern margin of the Taupo Volcanic Zone. *Geothermics*, **23**, 501–510.

STAGPOOLE, V. M. & BIBBY, H. M. 1999. *Residual gravity anomaly map of the Taupo Volcanic Zone, New Zealand, 1:250 000, version 1.0*. Institute of Geological & Nuclear Sciences Geophysical Map, **13**. Institute of Geological & Nuclear Sciences Limited, Lower Hutt, New Zealand.

STERN, T. A. 1986. Geophysical studies of the upper crust within the Central Volcanic region. *In:* SMITH, I. E. M. (ed.) *Late Cenozoic volcanism in New Zealand*. Bulletin of the Royal Society of New Zealand, **23**, 92–111.

STRATFORD, W. R. & STERN, T. A. 2004. Strong seismic reflections and melts in the mantle of a continental back-arc basin. *Geophysical Research Letters*, **31**, L06622.

STRATFORD, W. R. & STERN, T. A. 2006. Crust and upper mantle structure of a continental backarc: central North Island, New Zealand. *Geophysical Journal International*, **166**, 469–484.

SUTTON, A. N., BLAKE, S. & WILSON, C. J. N. 1995. An outline geochemistry of rhyolite eruptives from Taupo volcanic centre, New Zealand. *Journal of Volcanology and Geothermal Research*, **68**, 153–175.

SUTTON, A. N., BLAKE, S., WILSON, C. J. N. & CHARLIER, B. L. A. 2000. Late Quaternary evolution of a hyperactive rhyolite magmatic system: taupo volcanic centre, New Zealand. *Journal of the Geological Society, London*, **157**, 537–552.

VILLAMOR, P. & BERRYMAN, K. 2001. A late Quaternary extension rate in the Taupo Volcanic Zone, New Zealand, derived from fault slip rate. *New Zealand Journal of Geology and Geophysics*, **44**, 243–269.

VILLAMOR, P. & BERRYMAN, K. 2006a. Late Quaternary geometry and kinematics of faults at the southern termination of the Taupo Volcanic Zone, New Zealand. *New Zealand Journal of Geology and Geophysics*, **49**, 1–21.

VILLAMOR, P. & BERRYMAN, K. 2006b. Evolution of the southern termination of the Taupo Rift, New Zealand. *New Zealand Journal of Geology and Geophysics*, **49**, 23–37.

WILSON, C. J. N. 1993. Stratigraphy, chronology, styles and dynamics of late Quaternary eruptions from Taupo volcano, New Zealand. *Philosophical Transactions of the Royal Society of London A*, **343**, 205–306.

WILSON, C. J. N. 1996. Taupo's atypical arc. *Nature*, **379**, 27–28.

WILSON, C. J. N. 2001. The 26.5 ka Oruanui eruption, New Zealand: an introduction and overview. *Journal of Volcanology and Geothermal Research*, **112**, 133–174.

WILSON, C. J. N. & WALKER, G. P. L. 1985. The Taupo eruption, New Zealand. *Philosophical Transactions of the Royal Society of London, Series A*, **314**, 199–228.

WILSON, C. J. N., ROGAN, M. A., SMITH, I. E. M., NORTHEY, D. J., NAIRN, I. E. & HOUGHTON, B. F. 1984. Caldera volcanoes of the Taupo Volcanic Zone, New Zealand. *Journal of Geophysical Research*, **89**, 8463–8484.

WILSON, C. J. N, HOUGHTON, B. F. & LLOYD, E. F. 1986. Volcanic history and evolution of the Maroa-- Taupo area, Central North Island. *In:* SMITH, I. E. M. (ed.) *Late Cenozoic volcanism in New Zealand*. Royal Society of New Zealand Bulletin, **23**, 194–223.

WILSON, C. J. N., HOUGHTON, B. F., MCWILLIAMS, M. O., LANPHERE, M. A., WEAVER, S. D. & BRIGGS, R. M. 1995. Volcanic and structural evolution of Taupo Volcanic Zone, New Zealand: a review. *Journal of Volcanology and Geothermal Research*, **68**, 1–28.

WOOD, C. P. 1992. Geology of the Rotorua Geothermal System. *Geothermics*, **21**, 25–41.

WOOD, C. P., BRAITHWAITE, R. L. & ROSENBERG, M. D. 2001. Basement structure, lithology and permeability at Kawerau and Ohaaki geothermal fields, New Zealand. *Geothermics*, **30**, 461–481.

WRIGHT, I. C. 1990. Late Quaternary faulting of the offshore Whakatane Graben, Taupo Volcanic Zone,

New Zealand. *New Zealand Journal of Geology and Geophysics*, **33**, 245–256.

WRIGHT, I. C. 1992. Shallow structure and active tectonism of an offshore continental back-arc spreading system: the Taupo Volcanic Zone, New Zealand. *Marine Geology*, **103**, 287–309.

WRIGHT, I. C. 1993. Southern Havre Trough-Bay of Plenty (New Zealand): structure and seismic stratigraphy of an active back-arc basin complex. Chapter 11. *In*: BALANCE, P. (ed.) *South Pacific Sedimentary Basins of the World 2*. Elsevier, Amsterdam, 195–211.

# Andean flat-slab subduction through time

VICTOR A. RAMOS & ANDRÉS FOLGUERA*

*Laboratorio de Tectónica Andina, Universidad de Buenos Aires – CONICET*
*Corresponding author (e-mail: andes@gl.fcen.uba.ar)*

**Abstract:** The analysis of magmatic distribution, basin formation, tectonic evolution and structural styles of different segments of the Andes shows that most of the Andes have experienced a stage of flat subduction. Evidence is presented here for a wide range of regions throughout the Andes, including the three present flat-slab segments (Pampean, Peruvian, Bucaramanga), three incipient flat-slab segments ('Carnegie', Guañacos, 'Tehuantepec'), three older and no longer active Cenozoic flat-slab segments (Altiplano, Puna, Payenia), and an inferred Palaeozoic flat-slab segment (Early Permian 'San Rafael'). Based on the present characteristics of the Pampean flat slab, combined with the Peruvian and Bucaramanga segments, a pattern of geological processes can be attributed to slab shallowing and steepening. This pattern permits recognition of other older Cenozoic subhorizontal subduction zones throughout the Andes. Based on crustal thickness, two different settings of slab steepening are proposed. Slab steepening under thick crust leads to delamination, basaltic underplating, lower crustal melting, extension and widespread rhyolitic volcanism, as seen in the caldera formation and huge ignimbritic fields of the Altiplano and Puna segments. On the other hand, when steepening affects thin crust, extension and extensive within-plate basaltic flows reach the surface, forming large volcanic provinces, such as Payenia in the southern Andes. This last case has very limited crustal melt along the axial part of the Andean roots, which shows incipient delamination. Based on these cases, a Palaeozoic flat slab is proposed with its subsequent steepening and widespread rhyolitic volcanism. The geological evolution of the Andes indicates that shallowing and steepening of the subduction zone are thus frequent processes which can be recognized throughout the entire system.

## Introduction

The pioneer work of Barazangi & Isacks (1976, 1979) described the first two well documented segments along the Andes without late Cenozoic arc magmatism and adscribed them to flat-slab subduction (Fig. 1). This cold subduction was associated with a subhorizontal Benioff zone identified in the retroarc area that was characterized by large and frequent intracrustal earthquakes driven by important basement shortening. As a result, important foreland basement uplifts took place in late Cenozoic times giving rise to the present Sierras Pampeanas (González Bonorino 1950; Jordan *et al.* 1983*a*, *b*). Another detailed seismotectonic study in the northern Andes recognized a flat-slab segment in the northern Colombian Andes with similar characteristics (Pennington 1981).

Multidisciplinary research performed during the last two decades, mainly based on seismological and geological data on the continents, and oceanographic studies in the adjacent areas, depict the present setting of these three segments, where shallowing of the Benioff zone was closely related to collision of aseismic ridges (Pilger 1981, 1984). However, it was only recently that geological evidence was obtained along the Andes showed steepening of past subhorizontal subduction.

The objective of the present study is to characterize geological processes linked to shallowing and steepening of the subduction zones and their geological consequences. We aim to characterize these parameters along the Andes in order to be able to identify palaeo flat slab segments during the Phanerozoic. Based on these premises, three palaeo flat slabs were identified in Cenozoic times. Even further, it is speculated that a late Palaeozoic flat slab could have developed in the Central Andes. These new data enhance the importance of flat-slab subduction through time, and indicate that it is not an anomalous feature of the present-day margin, but has been an important feature of the geological record and its frequency is higher than expected.

## Present flat-slab subduction segments

Seismological data clearly show that there are three distinct segments with horizontal subduction along the Andean margin: the Bucaramanga, Peruvian and Pampean segments (Gütscher *et al.* 2000; Ramos 1999*a*). There is also a striking transition to a subhorizontal subduction in the Ecuadorian Andes (Gütscher *et al.* 1999*a*) that will be described to show the initial geological processes linked to the beginning of shallowing. These segments will be

*From*: MURPHY, J. B., KEPPIE, J. D. & HYNES, A. J. (eds) *Ancient Orogens and Modern Analogues.*
Geological Society, London, Special Publications, **327**, 31–54.
DOI: 10.1144/SP327.3   0305-8719/09/$15.00 © The Geological Society of London 2009.

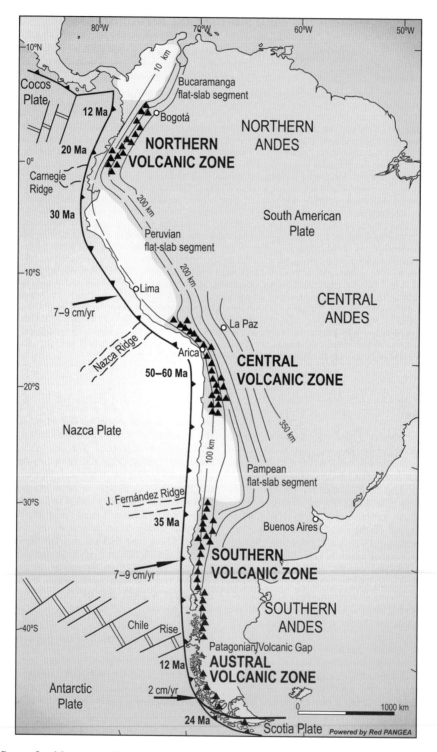

**Fig. 1.** Present flat-slab segments along the Andes (modified from Barazangi & Isacks 1976; Pennington 1981; Ramos 1999*a*; Gütscher *et al.* 2000).

described from south to north, in order to move from better known segments to less known settings.

## Pampean flat-slab segment

This segment was one of the first where systematic data were collected to reconstruct the tectonic history associated with flat-slab subduction in the Andes (Isacks *et al.* 1982; Jordan *et al.* 1983*a*, *b*). The segment is located between 27° and 33°30′S latitude along the Pampean foreland. The highest segment of the Main Andes coincides with the central part of the Pampean flat slab, where mountain peaks, such as the Aconcagua (6967 m a.s.l.),

the Mercedario (6850 m) and the La Ramada (6400 m) among others, correspond to tectonically uplifted areas with Miocene to Late Palaeozoic rocks above 6000 m (Ramos *et al.* 1996*a*). The description of the geological evidence will encompass the magmatic, sedimentological and structural history (Fig. 2), later linked to the oceanic features associated with the shallowing.

*Magmatic evidence.* The recognition of volcanic gaps in the Quaternary volcanic arc of the Andes emphasized the presence of cold subduction that coincides with the flat-slab segments (Barazanghi & Isacks 1976). Subsequent studies were able to

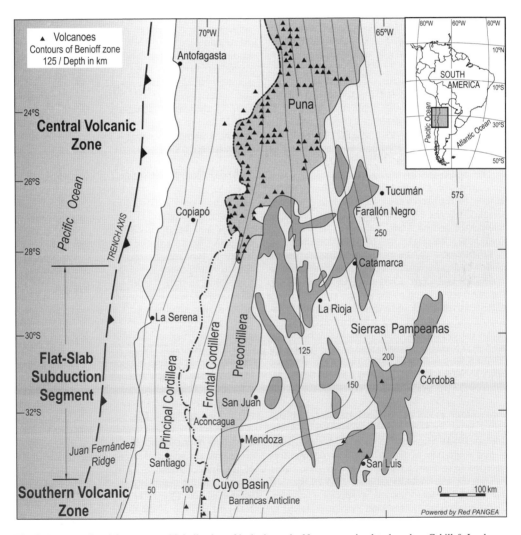

**Fig. 2.** Pampean flat-slab segment with indication of isobaths to the Nazca oceanic plate based on Cahill & Isacks (1992) (compare with the Benioff geometry proposed by Pardo *et al.* 2002 and Alvarado *et al.* 2005*a*, *b*); main basement uplifts of Sierras Pampeanas (Jordan *et al.* 1989), and location of the Precordillera fold and thrust belt (Ramos *et al.* 2002).

recognize that the gap had existed since Late Miocene times (Jordan *et al.* 1983*b*). Detailed petrographic studies performed in the late Cenozoic arc show that the geochemical signature changes in the main arc through time (Kay *et al.* 1987), and that the arc expanded towards the foreland region (Kay & Gordillo 1994). The geochemistry shows that the La/Yb ratios increased in Early to Late Miocene arc rocks, at the same time that crustal stacking thickened the crust (Kay *et al.* 1991; Kay & Mpodozis 2002; Litvak *et al.* 2007).

Geochronological data show that the main andesitic arc was active from 22–8.6 Ma (Fig. 3), although volumes of erupted magmas were drastically reduced throughout this time (Ramos *et al.* 1996*a*). A minor late rhyolitic eruption of Vacas Heladas Ignimbrites at 7.67 Ma was the last activity in the area (Ramos *et al.* 1989). Subsequent hydrothermal mineralization was widespread along the segment in El Indio, Valle del Cura and Maricunga mineral districts (Mpodozis *et al.* 1995; Kay & Mpodozis 2001). The latest activity east of the previous main arc was the eruption of the Cerro de Vidrio rhyolitic dome dated at 2.0 ± 0.2 Ma (Ar–Ar in glass) by Bissig *et al.* (2002) in Valle del Cura. Both rhyolitic episodes are interpreted as minor melts of the crust.

The expansion of the arc magmatism is first associated with a second dehydration front. At the latitude of the Aconcagua for example, the main Middle Miocene arc was characterized by large volumes of andesites and dacites in the Principal Cordillera, whereas in the Precordillera at *c.* 130 km east of the main arc, small volcanic centres and subvolcanic bodies were emplaced in Paramillos and Cerro Colorado (Kay *et al.* 1991). The main arc, as well as the second volcanic front, shifted eastward. The shifting and subsequent cessation of the magmatic arc simultaneously moved from west to east, and from north to south, ending at 5 ± 0.5 Ma west of Sierra de Aconquija (27°20′S lat.), 4.7 ± 0.3 Ma at the Pocho volcanic field (31°30′S lat.), and 1.9 ± 0.2 Ma in Sierra del Morro at 33°10′S lat. (Ramos *et al.* 2002).

*Sedimentary evolution.* Several retroarc foreland basins were formed along the flat-slab segment (Jordan 1984). Besides the general Andean trend of east migration of the synorogenic depocentres recorded from Late Cretaceous to Neogene times through the entire Andes (Ramos 1999*b*), the flat-slab segment superimposed a special character. East migration of the foreland system is linked to fragmentation of the foreland basement (Jordan *et al.* 1989). Detailed magnetostratigraphic studies show that subsidence rates were exceptional during the broken foreland stage (Reynolds *et al.* 1990). Locally, some depocentres recorded more

than 10 000 m of continental fluvial deposits, such as in the Neogene depocentre of Sierra de Los Colorados at 29°S lat. (Ramos 1999*b*).

The beginning of the broken foreland stage coincided with the eastward advance of the shallowing of the subducted slab beneath the retroarc area. Sedimentological studies show that the Early Miocene foreland basin was cannibalized during the Miocene, with the largest subsidence rates experienced during the Middle Miocene inception of the Pampean flat-slab at these latitudes (Fig. 4).

Some basin remnants in the western interior areas between the Frontal Cordillera and the Precordillera, such as the Iglesia Valley basin, were reactivated as piggy-back basins by out-of-sequence thrusts (Beer *et al.* 1990; Zapata & Allmendinger 1996). There is also a great variation in the timing of deformation when the sedimentary record is compared from north to south. Synorogenic deposition gets younger to the east and to the south (Vergés *et al.* 2001), when comparing time of deposition along the Río San Juan and Jachal further north. The same trend is regionally observed along the entire segment (Jordan *et al.* 2001; Ramos 1999*b*).

*Tectonic history.* The timing of shortening in the Principal and Frontal cordilleras and the Precordillera show some striking relations when analyzed in conjunction with: (1) the shortening rates of the fold and thrust belts; (2) the propagation of the orogenic front; (3) the subsidence rate of the adjacent foreland basin; and (4) the uplift of Sierras Pampeanas (Fig. 5). The shortening of this fold-and-thrust belt was concentrated in a thin-skinned belt within the Principal Cordillera prior to the shallowing. This period recorded a shortening rate of 5.5–5.75 mm/a, and a slow propagation rate of 2.5 mm/a of the thrust or orogenic mountain front. The propagation rate increased to 13.3 mm/a soon after the beginning of shallowing, while the shortening was reduced to 3.6 mm/a. This change from thin to thick skinned shortening is also reflected in the subsidence rate of the foreland basin (Fig. 5).

This data – when compared with the tectonic evolution of the adjacent oceanic region – show close time and space relationships between collision of the Juan Fernández aseismic ridge against the margin and the beginning of the shallowing of the subducted slab (Yañez *et al.* 2001). The south and eastward shifting of the magmatic arc, the time of deformation and basin evolution accompany the migration of the Juan Fernández ridge along and beneath the upper plate, as clearly demonstrated by Pilger (1984), Gütscher *et al.* (2000) and Kay & Mpodozis (2002). The most active neotectonic

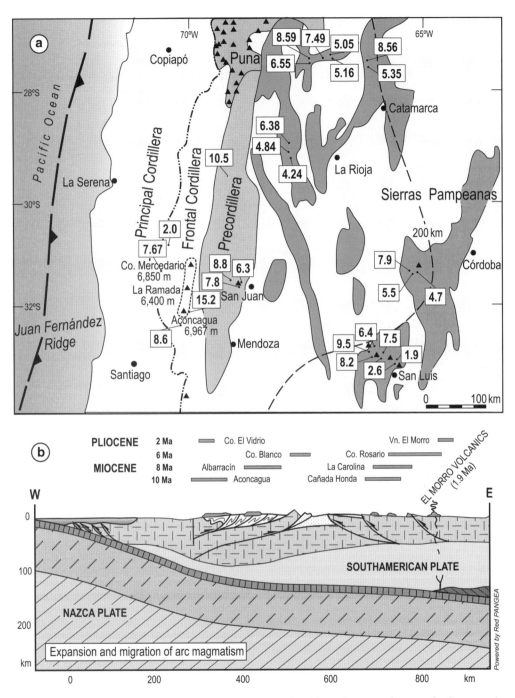

**Fig. 3.** Evolution of arc magmatism through time in the Pampean flat-slab: (**a**) Representative ages after Ramos *et al.* (2002) with indication of the isobath of 200 km depth corresponding to the oceanic slab; (**b**) Cross-section at crustal scale showing the expansion and migration of the main volcanic centers during the shallowing of the oceanic slab. Main elevations in the High Andes not related to the Quaternary volcanoes are also indicated.

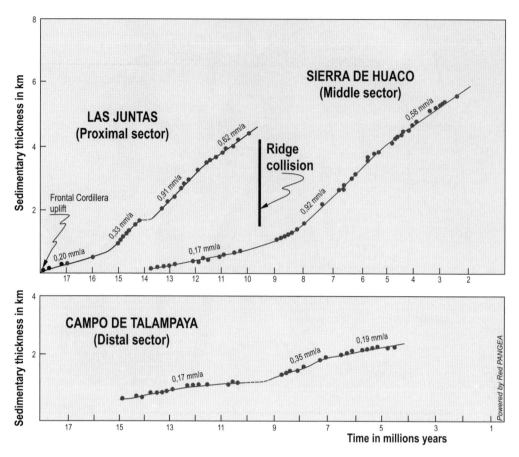

**Fig. 4.** Subsidence rates in the proximal, intermediate and distal areas of the Bermejo broken foreland basin, with indication of the beginning of flat-slab subduction at these latitudes (modified from Ramos 1999b). Seismostratigraphic data after Reynolds et al. (1990).

area corresponds to the Pie de Palo uplift, an area of high intracrustal seismicity (Kadinsky-Cade et al. 1985; Regnier et al. 1992) and a western Sierras Pampeanas block where an average uplift rate of 1.0 mm/a during the last 3 Ma has been observed (Ramos et al. 2002; Siame et al. 2006a). Pie de Palo is just above the track of the Juan Fernández ridge, as indicated by the coincidence between high density of earthquake epicentres and the projection of the oceanic feature (Kirby et al. 1996), and is located where the ridge is presently shallowing the subducting slab.

*Peruvian flat-slab segment*

This segment is encompassed between the Gulf of Guayaquil at 5°S and Arequipa at 14°S latitudes. It has been described by Barazanghi & Isacks (1976, 1979) based on global data of the ISC catalogue, and with more precision using local networks

by Dorbath et al. (1986, 1991). This survey demonstrated that the subduction zone starts under the trench with a 30° dip until approximately 100 km depth (Fig. 6), where it becomes horizontal beneath the Eastern Cordillera and the Subandean zone (Dorbath et al. 1991). Pilger (1984) showed the kinematics between the Nazca Ridge collision and the shallowing of the central Peru segment. This region was examined again by Gütscher et al. (1999b), who challenged the previous proposal and instead of the collision of an aseismic ridge proposed that the large Peruvian flat-slab segment was the result of the Nazca Ridge and the Inca Plateau subduction. Precise timing of the Nazca Ridge collision, and constraints in the length of the ridge, support that collision started at c. 11.2 Ma at about 11°S, moving later to the present position, as depicted by Hampel (2002). The segment north of this latitude requires a collision of a plateau or other oceanic feature.

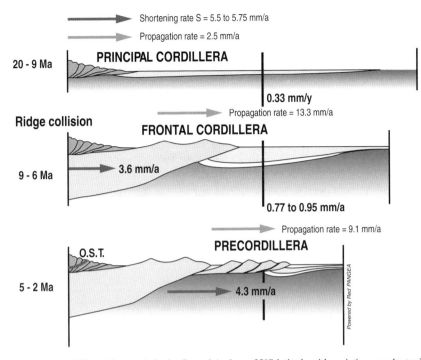

**Fig. 5.** The Aconcagua fold and thrust belt in the Central Andes at 32°S latitude with variations on shortening and propagation rates through time (after Ramos *et al.* 1996*b* and Hilley *et al.* 2004) and the subsidence rates in the foreland basin after Irigoyen *et al.* (2002).

The Peruvian flat-slab segment shares many common features with the Pampean flat slab. The second highest part of the Andes coincides with the Cordillera Blanca, with mountains such as Huascarán (6778 m a.s.l.), which is only 110 m lower than the Aconcagua massif, and other Late Miocene granitic peaks over 6000 m. The Cordillera Blanca is in the central part of an important basement high, which includes the Marañón Massif further to the east in the Eastern Cordillera. The Cordillera de Marañón is a basement uplift that exposed middle crustal rocks very similar in composition and metamorphic degree to the Sierras Pampeanas. The Peruvian segment also coincides with an area of no-arc volcanism, at least since latest Miocene times. Radiometric ages document several Cenozoic pulses of eastward magmatic migration (Aleman 2006). The cessation (*c.* 12 Ma) of magmatism in the northern part of the flat slab correlates with the complete subduction of the Inca Plateau and the arrival of the Nazca Ridge. As in the Pampean flat slab, the cessation of the main magmatic activity in the volcanic arc is followed by the emplacement of minor crustal melts of acidic composition. An example are the granites of Cordillera Blanca where McNulty *et al.* (1998) and Giovanni *et al.* (2006) reported U–Pb

zircon ages as young as 6 Ma. The magmatic lull following Nazca Ridge subduction began at the end of the Miocene. Most of the emplacement of the Cordillera Blanca Batholith and coeval ignimbrites took place during the southern sweep of the Nazca Ridge (Aleman 2006).

Neotectonics in the forearc where the Nazca Ridge intersects the trench are described by Macharé *et al.* (1986). Further support includes active tectonics and uplift in the foreland region in the Fitzcarrald arch in the Subandean region, where the aseismic ridge is being presently subducted. Evidence consists of a radial drainage network and deformation of Pliocene–Recent fluvial deposits on both sides of this structural high (Espurt *et al.* 2007). Both forearc and foreland geology, together with the distribution of late Cenozoic arc volcanoes, highlight the relationships between aseismic ridge subduction, active uplift and cessation of magmatism.

*Bucaramanga segment*

The early proposal of Pennington (1981), based on limited seismological data, showed a shallow subduction zone beneath northern Colombia. This fact has been confirmed by the seismological studies of

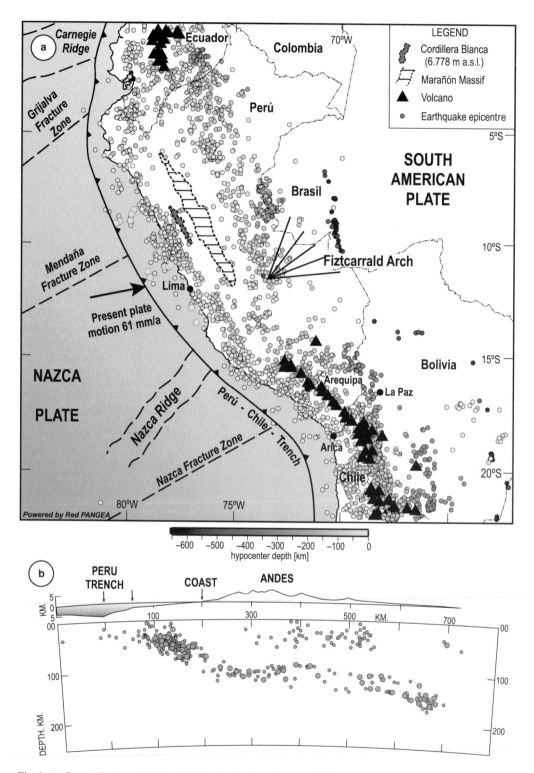

**Fig. 6.** (a) General features of the Peruvian flat slab based on Hampel (2002) and Aleman (2006). See the coincidence between the projection of the Nazca Ridge into the foreland and the uplift of the Fitzcarrald arch and associated alluvial fan (Espurt *et al.* 2007). (b) Geometry of Benioff–Wadatti zone beneath central Peru at 14–12°S latitude (based on Dorbarth *et al.* 1991).

Gütscher *et al.* (2000) in the Northern Andes north of 5°N, and the analysis made by Corredor (2003), who shows the shallow subduction produced by the recent subduction of the Caribbean plate beneath the Northern Andes. The dense concentration of intracrustal earthquakes of the Bucaramanga nest (Fig. 7) is associated with basement deformation and uplift of the Eastern Cordillera, characteristic of flat-slab subduction. However, an alternative hypothesis was advanced by Taboada *et al.* (2000), where most of this intraplate deformation at Bucaramanga was explained as a palaeo-Benioff zone associated with an old but still active subduction between the Panamá

microplate and South America, after the middle Miocene collision of the Chocó block (12–13 Ma, Duque Caro 1990).

The cessation of the late Cenozoic magmatic arc north of Cerro Bravo and Nevados de Ruiz (Mendez Fajury 1989), as well as the intense widespread neotectonic intracrustal activity, is better explained by the flat slab model of Gütscher *et al.* (2000). Their regional seismic tomography depicts a cold mantle and lower crust in this segment.

The latest volcanic activity is exposed near Boyacá in the retroarc region in the northern part of the Eastern Cordillera. The Tunja and Paipa volcanoes, among others, are associated with

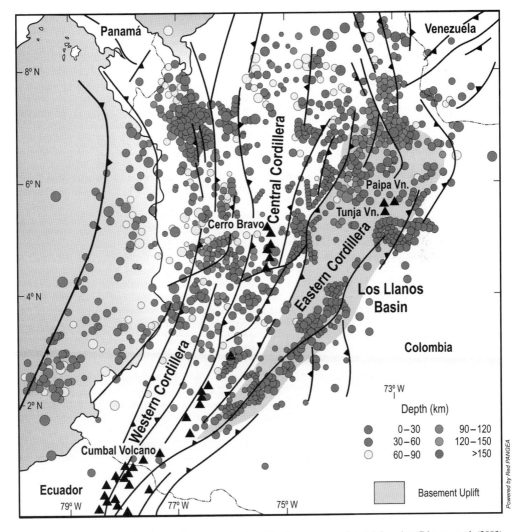

**Fig. 7.** Seismic activity and main morphostructural units of the Bucaramanga flat slab based on Dimate *et al.* (2003). Volcanic arc based on Mendez Fajury (1989) and retroarc volcanoes based on Cepeda (2004).

pyroclastic flows that range in age from 2.5–2.1 Ma for the oldest eruption, to pyroclastic flows younger than 1.5 Ma (Cepeda *et al.* 2004). This dominant Pliocene–Quaternary explosive volcanic activity has a high-K rhyolitic composition that resembles the last magmatic activity, characterized by the rhyolitic dome described by Bissig *et al.* (2002) in the Pampean flat slab. The rhyolitic composition of both areas, the residual thermal fields, and the mechanism of emplacement are very similar in both regions (Cepeda *et al.* 2004). This volcanic activity is better explained by the shallowing of the subducted Bucaramanga Pacific slab than by the inception of a new volcanic arc, as the result of the subduction zone that is being developed from the Caribbean margin of Colombia and Venezuela (Audemard & Audemard 2001).

Therefore, the flat slab hypothesis explains the active present uplift of the northern segment of the Eastern Cordillera, the cessation of arc magmatism, the neotectonic features associated with the tectonic inversion of previous rifts (Sarmiento-Rojas *et al.* 2006), the large intracrustal seismicity (Dimate *et al.* 2003) and the complex latest Cenozoic structure of the Pie-de-monte Llanero (Martínez 2006).

## Incipient flat-slab subduction segments

One of the best lines of evidence of early-stage shallowing is documented inland of the collision of the Carnegie aseismic ridge (Gütscher 1999a). The volcanic arc of Colombia is composed of a line of individual volcanoes from Cerro Bravo at 5°N to Cumbal at 2°30'N latitude (Fig. 7). South of the border with Ecuador, it changes to a complex volcanic arc system, which is expanded towards the foreland. Active volcanoes are emplaced on the Western Cordillera, the Inter-Andean Valley, the Eastern Cordillera (or Cordillera Real) and in the Subandean zone across 120 km from the volcanic arc front.

Individual volcanoes, such as the Cayambé and Quimsacocha volcanoes, show a trend from old calc-alkalic volcanic rocks to a more recent new edifice with a typical adakitic signature (Beate *et al.* 2001; Samaniego *et al.* 2002). Although the origin of this adakitic signal was early ascribed to slab melting, this has been questioned with their formation being attributed to melting of thickened continental crust or forearc subduction erosion (Ramos 2004). Both processes, crustal thickening and forearc crustal erosion, are consistent with flat subduction (Kay & Mpodozis 2002).

The variation in the dip of the subducted slab has been addressed by the change in petrological characteristics, such as the depth of generation and degree of partial melting in the asthenospheric

wedge (Bourdon *et al.* 2003), and in the expansion of the volcanic arc that coincides with the projection of the Carnegie ridge, an aseismic oceanic ridge that is now obliquely colliding against the margin (Gütscher *et al.* 1999a). The forearc crust is over-thickened only in the segment where the Carnegie ridge (Fig. 1) is colliding against the margin, as demonstrated by wide-angle seismic data recently collected offshore (Gailler *et al.* 2007). This collision is also related to the abnormal present uplift of the Cordillera Real and the Subandean block that controls the Pastaza alluvial megafan (Bés de Berc *et al.* 2005). Uplift rates during the Pleistocene of 1.37–1.4 cm/a, associated with an exhumation of the late Cenozoic alluvial plain of 500 m, are closely linked to the Carnegie ridge collision (Christophoul *et al.* 2002; Baby *et al.* 2004). Important intracrustal seismic activity is related to the basement structure of the Cutucu high. This uplift may correlate with the fission track data for the Cordillera Real that shows more than 9 km uplift in late Cenozoic times (Spikings *et al.* 2001).

Another segment with incipient evidence of shallowing is the Guañacos segment, located between 36° and 38°30'S latitudes. It is characterized by strong neotectonic and intracrustal activity in both: i) the forearc region at the Nahuel Buta Cordillera and offshore Cretaceous–Paleogene Arauco Basin (36°30'–37°30'S; Melnick *et al.* 2006a) and ii) the western retroarc zone at the Guañacos fold and thrust belt (36°–38°S; Folguera *et al.* 2004a). The two sectors correspond to ancient deformed belts that have been suddenly reactivated in Late Pliocene to Quaternary times. The offshore Arauco Basin, which was previously uplifted in the Late Cretaceous, as indicated by fission track ages (Glodny *et al.* 2007), has been shortened since 3.6 Ma at a rate of 0.8 mm a$^{-1}$ as an eastward vergent fold and thrust belt. On the other hand, recent neotectonics characterized the Guañacos fold and thrust belt, which was a Palaeogene basin inverted during Late Miocene times. The Pleistocene magmatic arc has migrated about 30 km to the east in this segment regarding the Pliocene volcanic front. Petrological studies performed in the Cenozoic arc at these latitudes show crustal thickening and subduction erosion, both processes consistent with shallowing of the subduction zone (Kay *et al.* 2005). Gravimetric studies show that the 36°–38°30'S segment is characterized by a long wavelength residual gravimetric anomaly that can only be explained (see density model in Alasonati Tašárová 2007; Hackney *et al.* 2006) by the shallowing by 10° of the subduction angle of the Nazca subducted plate. Therefore, the anomalous concentration of crustal earthquakes linked to unusual neotectonic activity in a 200 km wide subducted segment, may indicate incipient shallowing at the transition between the Central and Patagonian Andes since Late Pliocene times.

Another segment with an incipient flat slab is the Transmexican volcanic belt in central Mexico, which is related to the collision of the Tehuantepec aseismic ridge, although a different mechanism for uplift has been proposed (Ferrari 2006). A detailed analysis of this segment is outside the scope of this paper.

## Past flat-slab subduction segments

There is a strong correlation between the segment with current arc volcanism in the Central Andes (see central volcanic zone in Fig. 1) and the area of past flat-slab subduction extending from southern Peru to northern Argentina (Fig. 8). A summary of the geological processes involved in the changes from normal to flat, and from flat to normal subduction, will be discussed updating the proposal of James & Sacks (1999) (also see Sebrier et al. 1988).

### Altiplano flat-slab segment of Southern Peru

A period of flat-slab subduction was recorded in southern Peru and northern Bolivia, between 14° and 20°S latitudes (James & Sacks 1999). The evidence was similar to the previous described segments: (1) rapid cessation of the magmatic arc between 45 and 35 Ma; (2) widespread deformation and crustal thickening in the Eastern Cordillera; (3) the tectonothermal Zongo San Gabán effect that pervasively resets the Ar–Ar ages along 450 km, overprinting Permian and Triassic metamorphic rocks with a cryptic 38 Ma age; and (4) no igneous rocks of this age are known in this segment. This effect was interpreted as the result of heat advection by fluids at 38 Ma that predated the activity of the sub-Andean fold and thrust belt (Farrar et al. 1988). These processes were explained by a shallowing of the subducted slab that became subhorizontal at about c. 35 Ma and lasted until c. 25 Ma.

The steepening of the subduction zone was evidenced by widespread bimodal volcanism where rhyolites and basalts cover a wide area. As a result, great volumes of rhyolites up to 530 $km^3$ were spread on the present Altiplano and western slope of Eastern Cordillera between 26 and 22 Ma (Sandeman et al. 1995). During flat subduction the overlying lithosphere is hydrated by dewatering of the flat slab (James & Sacks 1999). Consequent steepening and expansion of the mantle wedge controlled the flow of hot asthenosphere and melting of the hydrated lithosphere beneath the Altiplano and Eastern cordilleras. Volcanic arc retreat is reflected by the shifting to the trench of the Tacaza arc between 29 and 15 Ma, the Upper Barroso arc

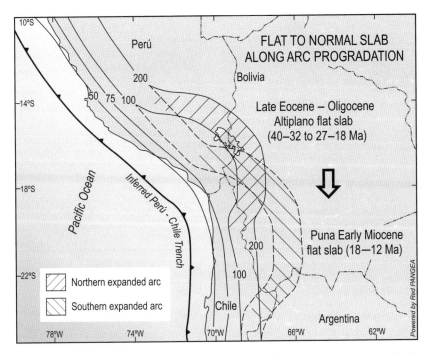

**Fig. 8.** Segments that recorded flat subduction in Late Eocene to Early Miocene times that correspond to the present Central Volcanic Zone (based on James & Sacks 1999 and Kay et al. 1999).

(10–6 Ma) and Lower Barroso in the last 3 Ma to meet the present frontal arc during the Pleistocene.

The main points of these processes are the weakening of the lithosphere during steepening of the subduction, delamination of the lithosphere and part of the lower crust (Kay & Kay 1993), and the collapse of the crust to form the Subandean fold and thrust belt. For further details, see James & Sacks (1999) and Kay *et al.* (1999).

### Puna flat-slab segment of southern Bolivia– northern Argentina

This trend of shallowing progressed to the south, where another period of flat subduction was recognized between 20° and 24°S (Kay *et al.* 1999). The shallowing took place between 18 and 12 Ma, as recognized by the cessation of the magmatism, crustal shortening and deformation of the southern Altiplano and northern Puna. Precise timing of the deformation established by Allmendinger *et al.* (1997), Baby *et al.* (1995) and Oncken *et al.* (2006), together with the palaeogeography of the

foreland basin (De Celles & Horton 2003) enabled the onset of the deformation in the Subandean region to be constrained to after 10 Ma.

Again, the same processes indicate that strong deformation in the axial part of the Puna and Eastern cordilleras were related to shallowing of the subduction zone, while steepening produced important hot asthenospheric flow, which in contact with the hydrated lithosphere (Oncken *et al.* 2006), led to important crustal and lithospheric delamination. As a result, huge rhyolitic calderas and ignimbritic fields are associated with thermal uplift and the consequent horizontal collapse and weakening of the crust with the deformation of the Subandean belt (Isacks 1988; Kay *et al.* 1999; Beck & Zandt 2002; Garzione *et al.* 2006).

### Payenia segment

Arc related rocks were emplaced more than 550 km away from the trench during Late Miocene times, from 34°30′ to 37°45′S (Fig. 9), suggesting shallow subduction processes at that time (Kay

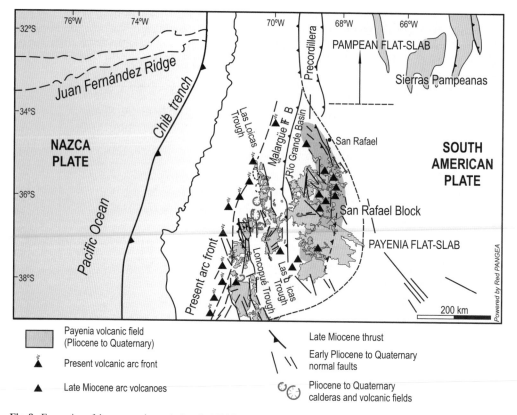

**Fig. 9.** Expansion of the magmatic arc during the Middle to Late Miocene showing the location of exhumed andesitic to dacitic arc rocks on the San Rafael block. Subsequent extensional structures, within plate basaltic flows, and huge rhyolitic calderas and ignimbritic flows along the main Andes suggest steepening of the subducted slab.

2001; Kay *et al.* 2006*a*, *b*). Intermediate positions of the arc are located on the eastern slope of the Andes near the drainage divide area (Nullo *et al.* 2002) to the east of the Late Oligocene arc, emplaced mainly on the western Andean slope. Easternmost centres were emplaced over the San Rafael block, a basement block that cannibalized the distal section of the Rio Grande foreland basin. The uplift of this block was associated with the foreland migration of the Malargüe fold and thrust belt to the east (Kozlowski *et al.* 1993; Manceda & Figueroa 1995). The San Rafael block was exhumed in Late Miocene times (Dessanti 1956; González Díaz 1964; Polanski 1964; Yrigoyen 1993, 1994). The Middle Miocene age assigned to the synorogenic sequences at the San Rafael block (Soria 1984; Marshall *et al.* 1986) points to a Late Miocene exhumation that coincides with the age of the

dacites and andesites emplaced in the San Rafael block between 13 and 4 Ma.

In addition, the main phase of deformation in the eastern section of the Malargüe fold and thrust belt at these latitudes has been constrained to 13–10 Ma (Giambiagi *et al.* 2008), which indicates a genetic relationship between the initial phase of arc expansion, uplift of the main Andes, sedimentation in the adjacent foreland basin, and the breaking of the foreland area.

During latest Miocene–Early Pliocene times, this compressional crustal stage changed to an extensional regime with the development of extensional troughs across the area that had previously recorded arc expansion (Fig. 10) (Bermúdez *et al.* 1993; Melnick *et al.* 2006*b*; Folguera *et al.* 2008). Arc dynamics were characterized during this period by fast retreat to the present position on the western

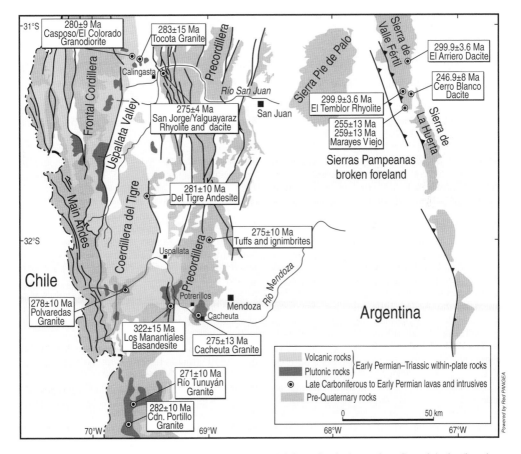

**Fig. 10.** Distribution of Upper Palaeozoic magmatic rocks and deformation in the southern Central Andes (based on Caminos 1979; Ramos *et al.* 1988; Varela *et al.* 1993; Mpodozis & Ramos 1989; Mpodozis & Kay 1990). Maximum expansion of arc volcanic rocks in the Early Permian was followed by subsequent extensional regime associated with the Choiyoi volcanic province.

flank of the Andes. Extensional deformation is associated at depth with crustal attenuation as well as anomalous sublithospheric heating inferred by teleseismic and tomographic analysis (Gilbert *et al.* 2006; Yuan *et al.* 2006). Gravimetric studies show high positive residual anomalies with areas submitted to extension, inferring an area of continuous asthenospheric upwelling in coincidence with the area of previous arc expansion (Folguera *et al.* 2007*a*). This extensional setting hosted rhyolitic associations derived from crustal melts at the highest collapsed sector of the Andes in the Las Loicas trough (Fig. 9) (Hildreth *et al.* 1984, 1991, 1999), whereas in foreland sectors it was associated with poorly differentiated mantle derived products (González Díaz 1972; Rossello *et al.* 2002; Kay *et al.* 2006*b*).

These two contrasting stages of deformation and arc dynamics, which occurred during the last 15 Ma between 34°30′ and 37°45′S, point to a scenario in which progressive shallow subduction from 15–5 Ma was followed by sudden slab steepening during the last 4 Ma, associated with the partial collapse of the orogen at these latitudes.

## Palaeozoic flat-slab subduction segment?

Palaeozoic deformations exhumed by Andean events through the Southern Central Andes have been connected to collisional episodes (Ramos *et al.* 1984; Ramos 2004). Early Permian deformations of the San Rafael tectonic phase have also been related to collision of an unidentified X terrane (Mpodozis & Kay 1990). This deformation exhibits some peculiarities in the foreland sedimentation and is associated with a phase of orogenic collapse that led Mpodozis & Kay (1990) to propose structural instabilities after orogenic development. The analysis of the late Palaeozoic orogenies in other areas of Gondwana led Cawood & Buchan (2007) to argue that deformation is not always related to a collisional event. Furthermore, in this segment of the Andes, little attention has been paid to coeval arc dynamics, which constitutes a direct indicator of Benioff zone variations through time. The Early Permian San Rafael tectonic phase is associated with unique processes that resemble more those of Andean tectonics than those occurred in Palaeozoic times at these latitudes (Ramos & Folguera 2007): (1) arc related volcanic assemblages cover diachronically Early Permian compressive deformation features; (2) Early Permian arc abnormally expanded to extend through the entire region and probably its front shifted to the east; (3) extensional processes followed the main phase of orogenic building and intraplate rhyolitic sequences were erupted through the area of previous

arc expansion; and (4) resetting of remanent magnetization in the area suggests abnormal lithospheric heating that preceeds eruption of intraplate melts. These facts point to a flat subduction cycle in Early Permian times, followed by slab steepening and consequent orogenic collapse in the Late Permian to Early Triassic, as proposed by Martínez *et al.* (2006).

*Sedimentary evolution.* Late Carboniferous–Early Permian 7000–8000 m thick marine to non-marine sequences are hosted along the eastern slope of the Principal Cordillera of Mendoza and San Juan (Fig. 10). Those are locally covering a Late Proterozoic basement indicating an important erosional hiatus prior to their deposition. The broad area uplifted in the Main Andes was the source of these sequences, which are characterized by coarsening-up cycles. This episode of mountain building, known as the San Rafael orogenic phase (280–270 Ma: Azcuy & Caminos 1987; Llambías *et al.* 1993; Cortés & Kleiman 1999), ended in the Lower Permian with an important angular unconformity.

From west to east these sequences were gathered in the Loma de los Morteritos and El Plata formations, with palynomorphs indicative of a Late Carboniferous to Early Permian age. These units, located on the eastern slope of the Frontal Cordillera, formed the maximum depocentre of the Late Palaeozoic in the region (Polanski 1958; Caminos 1965; Folguera *et al.* 2004*b*). To the east, Late Palaeozoic thicknesses fall in the western Precordillera region (Fig. 10), where several coarsening-up tectonostratigraphic units do not reach 500 m. These sequences, as determined by invertebrate and palynomorph associations (Ottone 1987), are coeval with the magmatic rocks and the structural deformation of the region.

This main depocentre of several thousand metres flanked the Early Permian belt of deformation, and pinch out to the platform area. The foreland basin started with shore sediments over which deltaic bodies and turbiditic lobes prograded, ending with braided fluvial systems (Heredia *et al.* 2002). Moreover, the dominance of westward palaeocurrents and lithoclasts of crystalline basement indicate that the basement may have been exhumed east of the Early Permian orogenic front, potentially as an incipient Sierras Pampeanas system, similar to the present setting of the Pampean flat slab (Fig. 2).

Lower Permian mesosiliceous lavas are part of the basal section of the Choiyoi Group. The upper part of this unit accumulated either in the Frontal Cordilleran or Precordilleran areas in a contrasting tectonic regime when compared to the basal member. As revealed by the structural style of the Andean fold and thrust belt at these latitudes, the main basement thrusts are the result of tectonic inversion of extensional faults that controlled the main depocentres of the Choiyoi Group (Cristallini & Ramos 2000; Rodríguez Fernández *et al.* 1997).

*Magmatic evidence.* Several studies have pointed out that Lower Permian calc-alkaline series, gathered with different names in the southern Precordilleran region, have unconformably covered the San Rafael unconformity in the Frontal Cordillera (Coira & Koukharsky 1976; Vilas & Valencio 1982; Cortés 1985; Kay *et al.* 1989; Rapalini & Vilas 1991; Sato & Llambías 1993; Sotarello *et al.* 2005). In addition, other isolated minor volcanic bodies with similar chemical patterns and Early Permian age have been found to the east up to 250 km away from their westernmost position (Fig. 10), on the Precordillera and Sierras Pampeanas domains (Rubinstein & Koukharsky 1995; Castro de Machuca *et al.* 2007). The magmatic arc was located mainly westward of the Frontal Cordillera during the Carboniferous (Hervé *et al.* 1987), which implies a strong eastward shifting and expansion from the Late Carboniferous to the Early Permian (Rodrígez Blanco 2004). Early Permian sequences are in turn separated by an erosional hiatus from an extensive intraplate rhyolitic association of the Choiyoi Group of Late Permian to Early Triassic age (Rapalini & Vilas 1991). On geochemical grounds, the plutonic and volcanic rocks of the Choiyoi Group define a large within plate volcanic province (Kay *et al.* 1989; Mpodozis & Ramos 1989) that covers important sectors of the Main Andes and Precordillera regions (Fig. 10). The area of Early Permian arc expansion coincides with a phase of extensional collapse with peak igneous activity around 260–240 Ma at these latitudes (Martínez 2004).

*Tectonic history.* A wide volcanic arc, in excess of 200 km, developed in Late Carboniferous–Early Permian times and has been exhumed along the Pampean flat slab zone. The volcanic sequences are interfingered in the west with a 7000–8000 thick turbiditic to deltaic succession whose easternmost section is preserved at the eastern Frontal Cordillera (Fig. 10). Towards the east, the volcanic rocks were emplaced over folded and thrust sequences deformed during the San Rafael orogenic phase. The sedimentary depocentre, characterized by the stacking of coarsening-up cycles, was affected by the Early Permian deformation. This basin was formed during the arc expansion stage, with its subsidence controlled by orogenic loading. It experienced rapid thinning towards the east in the present eastern Precordillera.

Regional analysis of the Late Carboniferous to Early Permian tectonics shows some striking facts. There are major crustal anisotropies east of the Main Andes that correspond to sutures formed as a result of Late Proterozoic to Early Palaeozoic terrane amalgamation (Ramos 1988). These sutures were reactivated with important strike-slip displacements in the Late Palaeozoic. The dominant right lateral displacements were caused by the oblique convergence of the subducting Pacific (Panthalassa) oceanic plate (Rapalini & Vilas 1991), which originated several deep transtensional depocentres (Fernández Seveso *et al.* 1993; Fernández Seveso & Tankard 1995). The depocentres are associated with alkaline eruptions typical of extensional intraplate settings (Koukharsky *et al.* 2001; Ramos *et al.* 2002), found in the Paganzo Basin. Fernández Seveso *et al.* (1993) discuss the relation between Early Permian compressive thrusting in the western Andean sector and transtension at the eastern foreland area. They propose that the origin of the extension could have been related to breaking up of the foreland basement due to crustal downwarping as found in modern analogues. This transtension in the Paganzo Basin would be a passive response in the foreland area to orogenic loading of the San Rafael thrust wedge. An alternative hypothesis would be to consider a high partitioned subduction system where displacements perpendicular to the trench would have been absorbed in the San Rafael fold and thrust belt; lateral displacements imposed by oblique convergence between plates would have been concentrated and localized in ancient lithospheric boundaries (Rapalini & Vilas 1991; Fernández Seveso *et al.* 1993). In this context, high oblique convergence and strong coupling associated with shallow subduction would be the condition for the development of a high strain partitioned subduction regime during Late Carboniferous–Early Permian times.

Arc expansion, stacking of the western sector of the fold and thrust belt during San Rafael tectonic phase, formation of foreland basins, and transtensional to transpressional reactivation of Proterozoic-early Palaeozoic sutures in the foreland area, ended in the tectonic wedge collapse. As a result, a multitude of rift systems were filled by the Choiyoi Group.

Rotation of half grabens produced erosional unconformities that separate Early Permian volcanics from the rest of the late Palaeozoic sequence. This zone of orogenic collapse coincides with the area of arc expansion and San Rafael orogenic compressional deformations, suggesting a common mechanism. Therefore, slab steepening and consequent asthenospheric injection in the broadened asthenospheric wedge, after shallow subduction, are the mechanisms proposed for the origin of the anomalously voluminous rhyolitic magmas of the Choiyoi Group and its extensional tectonic control. As a result, delamination of the lower crust took place after thickening and eclogitization during the San Rafael compressive phase. Sublithospheric heating due to slab steepening explains the massive crustal melting, as the lower crust was directly in contact with the rising asthenospheric flux (Martínez *et al.* 2006).

## Normal to flat-slab transition

Several examples of different ages and several distinct segments of the Andes show that the transit from normal subduction to flat-slab subduction is associated with a series of events:

*Migration of the volcanic front and expansion of the arc magmatism.* It is important to note that migration of the arc is indicated by the location of the largest volume of magmatic rocks; although magmatism in the previous setting may last for several million years, but with insignificant volumes. Such migration involves a decreasing volume of magmatic rocks that parallel the decline in dehydration in the subducted slab. This migration can be correlated to crustal weakening of the foreland and subsequent faulting. Geochemical signature of these magmas changes with the distance to the trench as well as the depth of generation (Kay & Mpodozis 2002). Final products may be as far as 600 km from the trench, as in the Bucaramanga segment (Jaramillo & Rojas 2003; Cepeda *et al.* 2004), and up to 750 km in the Pampean flat slab (Kay & Gordillo 1994).

*Uplift of the Main Andes.* Tectonic uplift is well documented in the Peruvian segment and in the Pampean flat slab, where the Cordillera Blanca and the High Cordillera of Mendoza and San Juan encompass the highest sectors of the Andes with the Huascarán (6778 m) and the Aconcagua (6967 m) mountains. The main difference between these two segments is that the Peruvian one registers some extensional collapse of the Cordillera Blanca (Siame *et al.* 2006*b*), while the Aconcagua shows no evidence of extension (Ramos *et al.* 1996*b*). This could imply that extension is more related to slab buoyancy from ridge subduction of the pre-thickened continental crust, as proposed by McNulty & Farber (2002), than to orogenic collapse in the sense of Dewey (1988).

*Broken foreland.* Although the Sierras Pampeanas is one of the most typical features of the Pampean flat slab (Jordan *et al.* 1983*a*, *b*), most other areas have recorded basement uplifts. The Peruvian segment is characterized by the Marañón Massif (3400 m a.s.l.), a basement uplift of the Eastern Cordillera produced in Late Miocene times almost along the suture between an allochthonous terrane

and the Gondwana margin. The larger area and elevations up to 5250 m reached by the Sierras Pampeanas in the Sierra de Aconquija could be related to the more segmented nature of the basement with several sutures and ophiolitic belts reactivated first as extensional faults during the opening of the South Atlantic, and later, as a thrust during the shallowing of the oceanic slab (Ramos *et al.* 2002). Other segments such as the Bucaramanga are related to the reactivation and uplift of the Eastern Cordillera of Colombia by tectonic inversion of extensional faults, partially coinciding with sutures (Cortés *et al.* 2006; Ramos & Moreno 2006). Even in a small segment as the Payenia flat-slab, the uplift of the San Rafael Block coincided with the maximum expansion of the arc. There is a close relationship between arc migration, thermal weakening of the crust and basement uplift (James & Sacks 1999; Ramos *et al.* 2002) during the shallowing of the oceanic slab. Some pervasive tectonothermal effects, such as the Zongo–San Gabán (Farrar *et al.* 1988) and the San Rafael effect (Rapalini & Astini 2005), are associated with this stage.

*Basin subsidence.* The increase in subsidence has a clear relationship with the approximation of the thrust front, as shown in several Subandean basins (Irigoyen *et al.* 2002; Jordan 1995). However, the subsidence achieves a critical collapse when the basement is broken and maximum thicknesses are obtained. This is seen in the Pampean flat slab, where more than 10 000 m of sediments in the synorogenic deposits of the Bermejo foreland basin have been reported by Ramos *et al.* (2002). There are incomplete records in other segments, but De Celles & Horton (2003) described several thousand metres in the Oligocene and Early Miocene of the Altiplano. The Payenia segment nicely depicts the migration and cannibalization of the previous basins until the broken foreland stage is reached.

## Flat to normal slab transition

On the other hand, the processes related to the transition from flat-slab to normal subduction are less well known, but have interesting characteristics:

*Rhyolitic flare-up.* One of the first results of steepening of the subducted oceanic slab is the presence of large crustal melts that are suddenly erupted over the flat-slab area in thick continental crust (Kay *et al.* 1999). Recent studies demonstrate that these large lower crustal melts are associated with lithospheric removal, sinking of the eclogitized lower crust, and crustal delamination, as earlier proposed by Kay & Kay (1993).

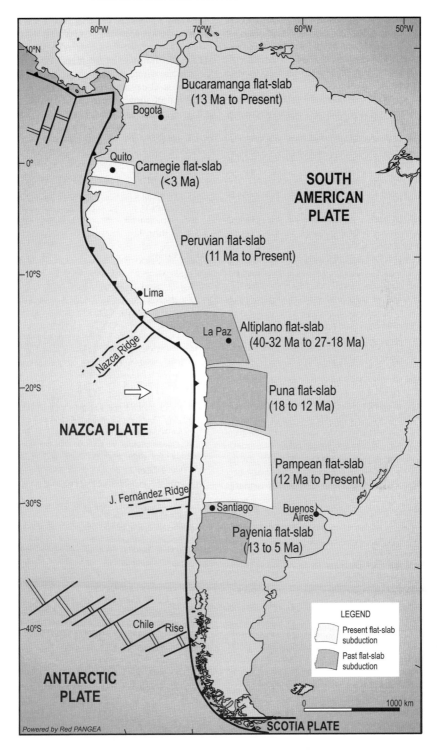

**Fig. 11.** Segments that have experienced shallowing of the subduction zone during Cenozoic times along the Andes. Note the almost continuous outline of flat-slabs.

*Thermal uplift.* This effect is a direct consequence of the lithospheric removal (Isacks 1988), although it has only been well documented in the Altiplano–Puna segment (Whitman *et al.* 1996; Allmendinger *et al.* 1997). Different geophysical tools have been used to confirm this evidence (see review in Oncken *et al.* 2006). Evidence of thermal uplift has not been documented in other segments. Reduced uplift in a thermal weakened area has been recently proposed in the Payenia segment with reduced geophysical datasets by Folguera *et al.* (2007*b*).

*Extensional regime.* The onset of the steepening of the subducted slab in some areas is associated with the vertical collapse by extension of the previous contracted structures. This is seen in the Payenia segment, where the pre-Miocene peneplain, uplifted in the Late Miocene, is segmented by normal faults (Ramos & Folguera 2005). Although the Puna has evidence of Pliocene extensional faulting that has been interpreted in different ways (Allmendinger *et al.* 1997), it is important here to note that extension occurs immediately after the thermal uplift of the area.

*Intense deformation shifted to the foreland.* The best example of migration of deformation that post-dates thermal uplift and some extension in the axial area, is the formation of the southern Subandean fold and thrust belt. The spatial and temporal relationships are clearly seen in southern Bolivia (Beck & Zandt 2002). In some other segments, this relationship is not evident, although in the Payenia flat slab, the Guañacos fold and thrust belt was developed after the emplacement of calderas and rhyolitic domes as well as the San Rafael block. This belt along the axis of the Andean Cordillera has evidence of neotectonic activity (Folguera *et al.* 2004*a*).

*Widespread mafic within plate floods.* The segment with thin crust, even after gentle shortening, shows an important basaltic flood linked to the inception of the steepening. These basaltic floods, in the Payenia segment, indicate a mantle-derived poorly-evolved magma of mafic composition and within plate signature (Kay *et al.* 2006*a*, *b*). Acidic rocks of Pliocene to Quaternary age in this area are scarce and are mainly small crustal melts as in Cerro Peceño, in the San Rafael Block.

In conclusion, it is interesting to show that when the present and past Cenozoic segments that had experienced flat-slab subduction are posted along the Andes (Fig. 11), an almost continuous belt of flat-slabs is outlined. The area that does not show evidence is Patagonia, although some studies postulate that the northern Patagonian massif between 40° and 43°S has experienced some shallowing during late Paleogene times (de Ignacio *et al.* 2001). There is no obvious trend or wave of shallowing, except among the Altiplano, Puna and Pampean segments, where there is some defined younging to the south. The other segments, at the present level of knowledge, show a random inception.

Funding for this research was provided by grants ANPCYT PICT 14144, CONICET PIP 5965 and UBACyT × 160. The authors are grateful to C. Mpodozis (Sipetrol, Chile) and S. M. Kay (Cornell University, USA) for many years of fruitful discussions on these topics, as well as to the researchers of Laboratorio de Tectónica Andina (University of Buenos Aires). The critical reviews of B. McNulty and P. Cawood are greatly appreciated.

# References

ALASONATI TAŠÁROVÁ, Z. 2007. Towards understanding the lithospheric structure of the southern Chilean subduction zone (36°S–42°S) and its role in the gravity field. *Geophysical Journal International*, doi: 10.1111/j.1365-246X.2007.03466.

ALEMAN, A. M. 2006. The Peruvian flat-slab. *Backbone of the Americas, Asociación Geológica Argentina – Geological Society of America Symposium, Abstract with Programs*, p. 17, Mendoza.

ALLMENDINGER, R. W., JORDAN, T. E., KAY, S. M. & ISACKS, B. L. 1997. The evolution of the Altiplano-Puna Plateau of the Central Andes. *Annual Reviews Earth Planetary Sciences*, **25**, 139–174.

ALVARADO, P., CASTRO DE MACHUCA, B. & BECK, S. 2005*a*. Comparative seismic and petrographic crustal study between the Western and Eastern Sierras Pampeanas region (31°S). *Revista de la Asociación Geológica Argentina*, **60**, 787–796.

ALVARADO, P., BECK, S., ZANDT, G., ARAUJO, M. & TRIEP, E. 2005*b*. Crustal deformation in the south-central Andes backarc terranes as viewed from regional broad-band seismic waveform modeling. *Geophysical Journal International*, **163**, 580–598.

AUDEMARD, F. E. & AUDEMARD, F. A. 2001. Structure of the Mérida Andes, Venezuela: relation with the South America–Caribbean geodynamic interaction. *Tectonophysics*, **345**, 299–327.

AZCUY, C. & CAMINOS, R. 1987. Diastrofismo. *In*: ARCHANGELSKY, S. (ed.) *El sistema Carbonífero en la República Argentina*. Academia Nacional de Ciencias, Córdoba, 239–251.

BABY, P., MORETTI, I. *ET AL.* 1995. Petroleum system of the northern and central Bolivian Subandean zone. *In*: TANKARD, A. J., SUÁREZ SORUCO, R. & WELSINK, H. J. (eds) *Petroleum basins of South America*. American Association of Petroleum Geologists, Memoir, **62**, 445–458.

BABY, P., RIVADENEIRA, M. & BARRAGÁN, R. 2004. La Cuenca de Oriente: Geología y Petróleo. *Travaux de l'Institut Francais d'Etudes Andines*, **144**, 1–195.

BARAZANGI, M. & ISACKS, B. 1976. Spatial distribution of earthquakes and subduction of the Nazca plate beneath South America. *Geology*, **4**, 686–692.

BARAZANGI, M. & ISACKS, B. 1979. Subduction of the Nazca plate beneath Peru evidence from spatial distribution of earthquakes. *Geophysical Journal of Royal Astronomic Society*, **57**, 537–555.

BEATE, B., MONZIER, M., SPIKINGS, R., COTTEN, J., SILVA, J., BOURDON, E. & EISSEN, J.-P. 2001. Mio-Pliocene adakite generation related to flat subduction in southern Ecuador: the Quimsacocha volcanic center. *Earth and Planetary Science Letters*, **192**, 561–570.

BECK, S. L. & ZANDT, G. 2002. The nature of orogenic crust in the central Andes. *Journal of Geophysical Research*, **107**, doi: 10.1029/2000JB000124.

BÈS DE BERC, S., SOULA, J. C., BABY, P., SOURIS, M., CHRISTOPHOUL, F. & ROSERO, J. 2005. Geomorphic evidence of active deformation and uplift in a modern continental wedge-top-foredeep transition: example of the eastern Ecuadorian Andes. *Tectonophysics*, **399**, 351–380.

BEER, J. A., ALLMENDINGER, R. W., FIGUEROA, D. E. & JORDAN, T. 1990. Seismic stratigraphy of a Neogene Piggyback basin, Argentina. *American Association Petroleum Geologist, Bulletin*, **74**, 1183–1202.

BERMÚDEZ, A., DELPINO, D., FREY, F. & SAAL, A. 1993. Los basaltos de retroarco extraandinos. *In*: RAMOS, V. A. (ed.) *Geología y Recursos Naturales de Mendoza*. XII° Congreso Geológico Argentino y II° Congreso de Exploración de Hidrocarburos, Relatorio, 161–172, Buenos Aires.

BISSIG, T., CLARK, A. H. & LEE, J. K. W. 2002. Cerro de Vidrio rhyolitic dome: evidence for Late Pliocene volcanism in the central Andean flat-slab region, Lama-Veladero district, 29°20'S, San Juan Province, Argentina. *Journal of South American Earth Sciences*, **15**, 571–576.

BOURDON, E., EISSEN, J.-P., GUTSCHER, M.-A., MONZIER, M., HALL, M. L. & COTTEN, J. 2003. Magmatic response to early aseismic ridge subduction: the Ecuadorian margin case (South America). *Earth and Planetary Science Letters*, **205**, 123–138.

CAHILL, T. & ISACKS, B. L. 1992. Seismicity and the shape of the subducted Nazca plate. *Journal of Geophysical Research*, **97**, 17503–17529.

CAMINOS, R. 1965. Geología de la vertiente oriental del Cordón del Plata, Cordillera Frontal de Mendoza. *Revista de la Asociación Geológica Argentina*, **20**, 351–392.

CAMINOS, R. 1979. Cordillera frontal. *In*: TURNER, J. C. M. (ed.) *Segundo Simposio de Geología Regional Argentina*. Academia Nacional de Ciencias, Córdoba, **1**, 397–453.

CASTRO DE MACHUCA, B., CONTE-GRAND, A., MEISSL, E., PONTORIERO, S., SUMAY, C. & MORATA, D. 2007. Manifestaciones del magmatismo neopaleozoico (Carbonífero superior-Pérmico) en la sierra de la Huerta, sierras Pampeanas occidentales, Provincia de San Juan. Caracterización petrológica de los pórfidos Marayes Viejo y El Arriero. *Revista de la Asociación Geológica Argentina*, **62**, 447–459.

CAWOOD, P. A. & BUCHAN, C. 2007. Linking accretionary orogenesis with supercontinent assembly. *Earth-Science Reviews*, **82**, 217–256.

CHRISTOPHOUL, F., BABY, P. & DÁVILA, C. 2002. Stratigraphic responses to a major tectonic event in a foreland basin: the Ecuadorian Oriente Basin from Eocene to Oligocene times. *Tectonophysics*, **345**, 281–298.

CEPEDA, N. 2004. *Estratigrafía de las vulcanitas asociadas al Volcán de Paipa, municipios de Paipa y Tuta, Departamento de Boyacá, Colombia*. PhD thesis, Universidad Nacional de Colombia, (unpublished), Bogotá.

CEPEDA, H., PARDO, N. & JARAMILLO, J. M. 2004. The Paipa Volcano, S.A. International Association of Volcanology and Chemistry of the Earth Interior. *IAVCEI International meeting 2004*, Poster Session, Pucón.

COIRA, B. & KOUKHARSKY, M. 1976. Efusividad tardío-hercínica en el borde oriental de la cordillera Frontal, zona del arroyo del Tigre, provincia de Mendoza, República Argentina. *Congreso Geológico Chileno, Abstracts*, **1**, 105–123.

CORREDOR, F. 2003. Seismic strain rates and distributed continental deformation in the northern Andes and three-dimensional seismotectonics of northwestern South America. *Tectonophysics*, **372**, 147–166.

CORTÉS, J. 1985. Vulcanitas y sedimentitas lacustres en la base del Grupo Choiyoi al sur de l estancia de Tambillos, Provincia de Mendoza. República Argentina. *IV Congreso Geológico Chileno, Abstracts*, **1**, 89–108.

CORTÉS, J. & KLEIMAN, L. 1999. La orogenia Sanrafaélica en los Andes de Mendoza. *14° Congreso Geológio Argentino, Abstracts*, **1**, 31.

CORTÉS, M., COLLETTA, B. & ANGELIER, J. 2006. Structure and tectonics of the central segment of the Eastern Cordillera of Colombia. *Journal of South American Earth Sciences*, **21**, 437–465.

CRISTALLINI, E. & RAMOS, V. A. 2000. Thick-skinned and thin-skinned thrusting in the La Ramada fold and thrust belt: crustal evolution of the High Andes of San Juan, Argentina (32°S). *Tectonophysics*, **317**, 205–235.

DESSANTI, R. 1956. Hoja Cerro Diamante 1 sheet 1: 250,000. Provincia de Mendoza. *Servicio Nacional Minero Geológico, Boletín*, **85**, 1–79.

DE CELLES, P. G. & HORTON, B. 2003. Early to middle Tertiary foreland basin development and the history of Andean crustal shortening in Bolivia. *Geological Society of America, Bulletin*, **115**, 58–77.

DE IGNACIO, C., LÓPEZ, I., OYARZUN, R. & MÁRQUEZ, A. 2001. The northern Patagonia Somuncura plateau basalts: a product of slab-induced, shallow asthenospheric upwelling? *Terra Nova*, **13**, 117–121.

DEWEY, J. F. 1988. Extensional collapse of orogens. *Tectonics*, **7**, 1123–1139.

DIMATE, C., RIVERA, L. *ET AL.* 2003. The 19 January 1995 Tauramena (Colombia) earthquake: geometry and stress regime. *Tectonophysics*, **363**, 159–180.

DORBATH, C., DORBATH, L., CISTERNAS, A., DEVERCHERE, J., DIAMENT, M., OCOLA, L. & MORALES, M. 1986. On crustal seismicity of the Amazonian foothill of the central Peruvian Andes. *Geophysical Research Letters*, **13**, 1023–1026.

DORBATH, L., DORBATH, C., JIMENEZ, E. & RIVERA, L. 1991. Seismicity and tectonic deformation in the Eastern Cordillera and the sub-Andean zone of Central Perú. *Journal of South American Earth Sciences*, **4**, 13–24.

DUQUE CARO, H. 1990. The Chocó block in the north-western corner of South America: structural, tectonos-traigraphic and paleogeographic implications. *Journal of South American Earth Sciences*, **3**, 71–84.

ESPURT, N., BABY, P. *ET AL.* 2007. How does the Nazca Ridge subduction influence the modern Amazonian foreland basin? *Geology*, **35**, 515–518.

FARRAR, E., CLARK, A. H., KONTAK, D. J. & ARCHIBAL, D. A. 1988. Zongo-San Gabán zone: eocene foreland boundary of the Central Andean orogen, northwest Bolivia and southeast Peru. *Geology*, **16**, 55–58.

FERNÁNDEZ SEVESO, F. & TANKARD, A. 1995. Tectonics and stratigraphy of the Late Paleozoic Paganzo basin of Western Argentina and its regional implications. *In*: TANKARD, A. J., SUÁREZ SORUCO, R. & WELSINK, H. J. (eds) *Petroleum Basins of South America*. American Association of Petroleum Geologists, Memoir, **62**, 285–301.

FERNÁNDEZ SEVESO, F., PÉREZ, M., BRISSON, I. & ALVAREZ, L. 1993. Sequence stratigraphy and tectonic analysis of the Paganzo basin, Western Argentina. *12° International Congress on the Carboniferous-Permian systems (Buenos Aires), Comptes Rendus*, **2**, 223–260.

FERRARI, L. 2006. Laramide and Neogene shallow subduction in Mexico: constraints and contrasts. *Backbone of the Americas, Asociación Geológica Argentina – Geological Society of America Symposium, Abstract with Programs*, p. 17, Mendoza.

FOLGUERA, A., RAMOS, V. A., HERMANNS, R. L. & NARANJO, J. 2004*a*. Neotectonics in the foothills of the southernmost central Andes (37°–38°S): evidence of strike-slip displacement along the Antiñir-Copahue fault zone. *Tectonics*, **23**, TC5008, doi: 10.1029/2003TC001533.

FOLGUERA, A., ETCHEVERRÍA, M. *ET AL.* 2004*b*. Hoja Geológica Potrerillos 3369–15. *Servicio Geológico Minero Argentino, Boletín*, **301**, 1–142.

FOLGUERA, A. & INTROCASO, A. 2007*a*. Crustal attenu-ation in the Southern Andean retroarc determined from gravimetric studies (38°–39°30′S): The Lonco-Luán astenospheric anomaly. *Tectonophysics*, **439**, 129–147, doi: 10.1016/j.tecto.2007.04.001.

FOLGUERA, A., ALASONATI TAŠÁROVÁ, S., HESE, F., HACKNEY, R., GOTZE, H., SCHMIDT, S. & RAMOS, V. A. 2007*b*. Crustal attenuation along the Andean retroarc zone 36°–39°S from gravimetric análisis. *20° Colloquium on Latinoamerican Earth Sciences, Abstracts*, 38–41, Kiel.

FOLGUERA, A., BOTTESI, G., ZAPATA, T. & RAMOS, V. A. 2008. Crustal collapse in the Andean back-arc since 2 Ma: Tromen volcanic plateau, Southern Central Andes (36°40′–37°30′S). *Tectonophysics (Special Issue on Andean Geodynamics)*, **459**(1–4), 140–160, doi: 10.1016/j.tecto.2007.12.013.

GAILLER, A., CHARVIS, P. & FLUEH, E. R. 2007. Seg-mentation of the Nazca and South American plates along the Ecuador subduction zone from wide angle seismic profiles. *Earth and Planetary Science Letters*, **260**, 444–464.

GARZIONE, C. N., MOLNAR, P., LIBARKIN, J. C. & MACFADDEN, B. J. 2006. Rapid late Miocene rise of the Bolivian Altiplano: evidence for removal of

mantle lithosphere. *Earth and Planetary Science Letters*, **241**, 543–556.

GIAMBIAGI, L., BECHIS, F., GARCÍA, V. & CLARK, A. 2008. Temporal and spatial relationships of thick- and thin-skinned deformation: a case study from the Malargüe fold and thrust belt, Southern Central Andes. *Tectonophysics, Special Issue ISAG*, **459**(1–4), 123–139.

GILBERT, H., BECK, S. & ZANDT, G. 2006. Lithospheric and upper mantle structure of central Chile and Argentina. *Geophysical Journal International*, **165**, 383, doi: 10.1111/j.1365-246X.2006.02867.x.

GLODNY, J., GRÄFE, K., ECHTLER, H. & ROSENAU, M. 2007. Mesozoic to Quaternary continental margin dynamics in South-Central Chile (36–42°S): the apatite and zircon fission track perspective. *Inter-national Journal of Earth Sciences (Geologische Rundschau)*, doi: 10.1007/s00531-007-0203-1.

GONZÁLEZ BONORINO, F. 1950. Algunos problemas geológicos de las Sierras Pampeanas. *Revista de la Asociación Geológica Argentina*, **5**, 81–110.

GONZÁLEZ DÍAZ, E. F. 1964. Rasgos geológicos y evolu-ción geomorfológica de la Hoja 27 d (San Rafael) y zona occidental vecina (Provincia de Mendoza). *Revista de la Asociación Geológica Argentina*, **19**, 151–188.

GONZÁLEZ DÍAZ, E. F. 1972. Descripción geológica de la Hoja 30d Payún-Matrú, Provincia de Mendoza. *Direc-ción Nacional de Geología y Minería, Boletín*, **130**, 1–88.

GÜTSCHER, R., MALAVIEILLE, J., LALLEMEND, S. & COLLOT, J. Y. 1999*a*. Tectonic segmentation of the North Andean margin: impact of the Carnegie Ridge Collision. *Earth and Planetary Science Letters*, **168**, 255–270.

GÜTSCHER, M. A., OLIVET, J. L., ASLANIAN, D., EISSEN, J. P. & MAURY, R. 1999*b*. The 'lost Inca Plateau': cause of flat subduction beneath Peru? *Earth and Planetary Science Letters*, **171**, 335–341.

GÜTSCHER, M. A., SPAKMAN, W., BIJWAARD, H. & ENGDAHL, E. R. 2000. Geodynamic of flat subduction: seismicity and tomographic constraints from the Andean margin. *Tectonics*, **19**, 814–833.

HACKNEY, R., ECHTLER, H. *ET AL.* 2006. The segmented overriding plate and coupling at the South-Central Chilean margin (36°–42°S). *In*: ONCKEN, O., CHONG, G. *ET AL.* (eds) *The Andes – Active Subduc-tion Orogeny*. Frontiers in Earth Sciences Series, **1**, Springer, Berlin.

HAMPEL, A. 2002. The migration history of the Nazca ridge along the Peruvian active margin: a re-evaluation. *Earth and Planetary Science Letters*, **203**, 665–679.

HEREDIA, N., RODRÍGUEZ FERNÁNDEZ, L. R., GALLAS-TEGUI, G., BUSQUETS, P. & COLOMBO, F. 2002. Geological setting of the Argentine Frontal Cor-dillera in the flat-slab segment (30°00′–31°30′S latitude). *Journal of South American Earth Sciences*, **15**, 79–99.

HERVÉ, F., GODOY, E., PARADA, M., RAMOS, V. A., RAPELA, C., MPODOZIS, C. & DAVIDSON, J. 1987. A general view on the Chilean Argentinian Andes, with emphasis on their early history. *In*: MONGER, J. & FRANCHETEAU, J. (eds) *Circum-Pacific orogenic belts and evolution of the Pacific Basin*. American Geophysical Union, Geoynamics Series, **18**, 97–113.

HILDRETH, W., GRUNDER, A. & DRAKE, R. 1984. The Loma Seca Tuff and the Calabozos Caldera: a major ash-flow and caldera complex in the southern Andes of Central Chile. *Geological Society of America, Bulletin*, **95**, 45–54.

HILDRETH, W., DRAKE, R., GODOY, E. & MUNIZAGA, F. 1991. Bobadilla caldera and 1.1 Ma ignimbrite at Laguna del Maule, Southern Chile. *VI° Congreso Geológico Chileno Actas*, 62–63.

HILDRETH, W., FIERSTEIN, J., GODOY, E., DRAKE, R. & SINGER, B. 1999. The Puelche volcanic field: extensive Pleistocene rhyolite lava flows in the Andes of Central Chile. *Revista Geológica de Chile*, **26**, 275–309.

HILLEY, G. E., STRECKER, M. R. & RAMOS, V. A. 2004. Growth and erosion of fold-and-thrust belts with application to the Aconcagua fold-and-thrust belt, Argentina. *Journal of Geophysical Research* **109**, B01410, doi: 10.1029/2002JB002282.

IRIGOYEN, M. V., BUCHAN, K. L., VILLENEUVE, M. E. & BROWN, R. L. 2002. Cronología y significado tectónico de los estratos sinorogénicos neógenos aflorantes en la región de Cacheuta-Tupungato, provincia de Mendoza. *Revista de la Asociación Geológica Argentina*, **57**, 3–18.

ISACKS, B. 1988. Uplift of the Central Andean plateau and bending of the Bolivian orocline. *Journal Geophysical Research*, **93**, 3211–3231.

ISACKS, B., JORDAN, T., ALLMENDINGER, R. & RAMOS, V. A. 1982. La segmentación tectónica de los Andes Centrales y su relación con la placa de Nazca subductada. *V° Congreso Latinoamericano de Geología, Actas*, **3**, 587–606.

JAMES, D. E. & SACKS, S. 1999. Cenozoic formation of the Central Andes: a geophysical perspective. *In*: SKINNER, B. *ET AL.* (eds) *Geology and Mineral Deposits of Central Andes*. Society of Economic Geology, Special Publication, **7**, 1–25.

JARAMILLO, J. M. & ROJAS, P. 2003. Depósitos de cenizas volcánicas en la Cordillera Oriental. *9° Congreso Colombiano de Geología, Resúmenes*, p. 150, Medellín.

JORDAN, T. E. 1984. Cuencas, volcanismo y acortamientos cenozoicos, Argentina, Bolivia y Chile, 20°28' lat. S. *9° Congreso Geológico Argentino, Actas*, **2**, 7–24.

JORDAN, T. E. 1995. Retroarc foreland and related basins. *In*: SPERA, C. & INGERSOLL, R. V. (eds) *Tectonics of Sedimentary Basins*. Blackwell Scientific, Cambridge, MA, 331–362.

JORDAN, T. E., ISACKS, B., RAMOS, V. A. & ALLMENDINGER, R. W. 1983*a*. Mountain building in the Central Andes. *Episodes*, **1983**(3), 20–26.

JORDAN, T. E., ISACKS, B. L., ALLMENDINGER, R. W., BREWER, J. A., RAMOS, V. A. & ANDO, C. J. 1983*b*. Andean tectonics related to geometry of subducted Nazca plate. *Geological Society of America, Bulletin*, **94**, 341–361.

JORDAN, T., ZEITLER, P., RAMOS, V. A. & GLEADOW, A. J. W. 1989. Thermochronometric data on the development of the basement peneplain in the Sierras

Pampeanas, Argentina. *Journal of South American Earth Sciences*, **2**, 207–222.

JORDAN, T. E., SCHULENEGGER, F. & CARDOZO, N. 2001. Unsteady and spatially variable evolution of the Neogene Andean Bermejo Foreland basin, Argentina. *Journal of South Americn Earth Sciences*, **14**, 775–798.

KADINSKY-CADE, K., REILINGER, R. & ISACKS, B. 1985. Surface deformation associated with the November 23, 1977, Caucete, Argentina, earthquake sequence. *Journal of Geophysical Research*, **90**, 12 691–12 700.

KAY, R. W. & KAY, S. M. 1993. Delamination and delamination magmatism. *Tectonophysics*, **219**, 177–189.

KAY, S. 2001. *Tertiary to recent magmatism and tectonics of the Neuquén basin between 36°05' and 38°S latitude*. Repsol-YPF Buenos Aires. Unpublished Report, 77 p.

KAY, S. M. & GORDILLO, C. E. 1994. Pocho volcanic rocks and the melting of depleted continental lithosphere above a shallowly dipping subduction zone in the Central Andes. *Contributions to Mineralogy and Petrology*, **117**, 25–44.

KAY, S. M. & MPODOZIS, C. 2001. Central Andean ore deposits linked to evolving shallow subduction systems and thickening crust. *GSA Today*, **11**(3), 4–9.

KAY, S. M. & MPODOZIS, C. 2002. Magmatism as a probe to the Neogene shallowing of the Nazca plate beneath the modern Chilean flat-slab. *Journal of South American Earth Sciences*, **15**, 39–57.

KAY, S. M., MAKSAEV, V., MOSCOSO, R., MPODOZIS, C. & NASI, C. 1987. Probing the evolving Andean lithosphere: Mid-late Tertiary magmatism in Chile (29°–30°30'S) over the modern zone of subhorizontal subduction. *Journal Geophysical Research*, **92**(B7), 6173–6189.

KAY, S., RAMOS, V. A., MPODOZIS, C. & SRUOGA, P. 1989. Late Paleozoic to Jurassic silicic magmatism at the Gondwana margin: analogy to the Middle Proterozoic in North America? *Geology*, **17**, 324–328.

KAY, S. M., MPODOZIS, C., RAMOS, V. A. & MUNIZAGA, F. 1991. Magma source variations for mid to late Tertiary volcanic rocks erupted over a shallowing subduction zone and through a thickening crust in the Main Andean Cordillera (28–33°S). *In*: HARMON, R. S. & RAPELA, C. (eds) *Andean Magmatism and its Tectonic Setting*. Geological Society of America, Special Paper, **265**, 113–137.

KAY, S. M., MPODOZIS, C. & COIRA, B. 1999. Neogene magmatism, tectonism, and mineral deposits of the Central Andes (22°S to 33°S). *In*: SKINNER, B. *ET AL.* (eds) *Geology and Mineral Deposits of Central Andes*. Society of Economic Geology, Special Publication, **7**, 27–59.

KAY, S. M., GODOY, E. & KURTZ, A. 2005. Episodic arc migration, crustal thickening, subduction erosion, and magmatism in the south-central Andes. *Bulletin of the Geological Society of America*, **117**, 67–88.

KAY, S. M., MANCILLA, O. & COPELAND, P. 2006*a*. Evolution of the Backarc Chachahuén volcanic complex at 37°S latitude over a transient Miocene shallow subduction zone under the Neuquén Basin. *In*: KAY, S. M. & RAMOS, V. A. (eds) *Evolution of an Andean Margin: A Tectonic and Magmatic View* *from the Andes to the Neuquén Basin (35°–39°S lat.).* Geological Society of America, Special Paper, **407**, 215–246.

KAY, S. M., BURNS, M. & COPELAND, P. 2006*b*. Upper Cretaceous to Holocene Magmatism over the Neuquén basin: evidence for transient shallowing of the subduction zone under the Neuquén Andes (36°S to 38°S latitude). *In*: KAY, S. M. & RAMOS, V. A. (eds) *Evolution of an Andean Margin: A Tectonic and Magmatic View from the Andes to the Neuquén Basin (35°–39°S lat.).* Geological Society of America, Special Paper, **407**, 19–60.

KIRBY, S., ENGDAHL, E. R. & DENLINGER, R. 1996. Intermediate-depth intraslab earthquakes and arc volcanism as physical expressions of crustal and uppermost mantle metamorphism in subducting slabs. *In*: BEBOUT, G. E., SCHOLL, D. W., KIRBY, S. H. & PLATT, J. P. (eds) *Subduction Top to Bottom.* Geophysical Monograph Series, **96**, 195–214.

KOUKHARSKY, M., TASSINARI, C., BRODTKORB, M. & LEAL, P. 2001. Basaltos del Neopaleozoico-Triásico temprano? en las sierras del norte de Córdoba y de Ambargasta, Sierras Pampeanas Orientales: Petrograía y edades K/Ar. *Revista de la Asociación Geológica Argentina*, **56**, 400–403.

KOZLOWSKI, E., MANCEDA, R. & RAMOS, V. A. 1993. Estructura. *In*: RAMOS, V. A. (ed.) *Geología y Recursos Naturales de Mendoza.* XII° Congreso Geológico Argentino y I° Congreso de Exploración de Hidrocarburos, Relatorio, 235–256, Buenos Aires.

LITVAK, V. D., POMA, S. & MAHLBURG KAY, S. 2007. Paleogene and Neogene magmatism in the Valle del Cura region: new perspective on the evolution of the Pampean flat-slab, San Juan province, Argentina. *Journal of South American Earth Sciences*, **24**, 117–137.

LLAMBÍAS, E., KLEIMAN, L. & SALVARREDI, J. 1993. El magmatismo gondwánico. *In*: RAMOS, V. A. (ed.) *Geología y Recursos Naturales de Mendoza.* XII° Congreso Geológico Argentino y II° Congreso de Exploración de Hidrocarburos, Relatorio, 53–64, Buenos Aires.

MACHARÉ, J., SEBRIER, M., HUAMÁN, D. & MERCER, J. L. 1986. Tectónica cenozoica de la margen peruana. *Boletín de la Sociedad Geológica del Perú*, **76**, 45–77.

MANCEDA, R. & FIGUEROA, D. 1995. Inversion of the Mesozoic Neuquén rift in the Malargüe fold-thrust belt, Mendoza, Argentina. *In*: TANKARD, A. J., SUÁREZ, R. & WELSINK, H. J. (eds) *Petroleum Basins of South America.* American Association of Petroleum Geologists, Memoir, **62**, 369–382.

MARSHALL, L., DRAKE, R. & CURTISS, G. 1986. $^{40}K$–$^{40}Ar$ calibration of Late Miocene-Pliocene Mammal-bearing Hayquerías and Tunuyán Formations, Mendoza Province, Argentina. *Journal of Paleontology*, **60**, 448–457.

MARTÍNEZ, A. 2004. Secuencias volcánicas permo-triásicas de los cordones del Portillo y del Plata, Cordillera Frontal, Mendoza: su interpretación tectónica. PhD thesis, Universidad de Buenos Aires.

MARTÍNEZ, A., RODRÍGUEZ BLANCO, L. & RAMOS, V. A. 2006. Permo-Triassic magmatism of the Choiyoi Group in the cordillera Frontal of Mendoza, Argentina: geological variations associated with

changes in Paleo-Benioff zone. *Backbone of the Americas, Asociación Geológica Argentina – Geological Society of America Symposium, Abstract with Programs.* Mendoza, 1–77.

MARTINEZ, J. A. 2006. Structural evolution of the Llanos foothills, Eastern Cordillera, Colombia. *Journal of South American Earth Sciences*, **21**, 510–520.

MCNULTY, B. A. & FARBER, D. 2002. Active detachment faulting above the Peruvian flat-slab. *Geology*, **30**, 567–570.

MCNULTY, B. A., FARBER, D. L., WALLACE, G. S., LOPEZ, R. & PALACIOS, O. 1998. Role of plate kinematics and plate-slip-vector partitioning in continental magmatic arcs; evidence from the Cordillera Blanca, Peru. *Geology*, **26**, 827–830.

MELNICK, D., BOOKHAGEN, B., ECHTLER, H. & STREICKER, M. 2006*a*. Coastal deformation and great subduction earthquakes, Isla Santa María, Chile (37°S). *Geological Society of America, Bulletin*, **118**, 1463–1480.

MELNICK, D., CHARLET, F., ECHTLER, H. & BATIST, M. 2006*b*. Incipient axial collapse of the main cordillera and strain partitioning gradient between the central and Patagonian Andes, Lago Laja, Chile. *Tectonics*, **25**, TC 5004, doi: 10.1029/2005TC001918: 1–22.

MÉNDEZ FAJURY, R. A. 1989. Catálogo de volcanes activos de Colombia. *Boletín Geológico Ingeominas*, **30**(3), 1–75.

MPODOZIS, C. & KAY, S. 1990. Provincias magmáticas ácidas y evolución tectónica del Gondwana: Andes Chilenos (28°–31°S). *Revista Geológica de Chile*, **17**, 153–180.

MPODOZIS, C. & RAMOS, V. A. 1989. The Andes of Chile and Argentina. *In*: ERICKSEN, G. E., CAÑAS PINOCHET, M. T. & REINEMUD, J. A. (eds) *Geology of the Andes and its relation to Hydrocarbon and Mineral Resources.* Circumpacific Council for Energy and Mineral Resources, Earth Sciences Series, **11**, 59–90.

MPODOZIS, C., CORNEJO, P., KAY, S. M. & TITTLER, A. 1995. La franja de Maricunga: síntesis de la evolución del frente volcánico Oligoceno–Mioceno de la zona sur de los Andes Centrales. *Revista Geológica de Chile*, **22**, 273–313.

NULLO, F., STEPHENS, G., OTAMENDI, J. & BALDAUF, P. 2002. El volcanismo del Terciario superior del sur de Mendoza. *Revista de la Asociación Geológica Argentina*, **57**, 119–132.

ONCKEN, O., CHONG, G. *ET AL.* (eds) 2006. *The Andes – Active Subduction Orogeny.* Frontiers in Earth Sciences Series, Springer, Berlin.

OTTONE, E. G. 1987. *Estudios Bioestratigráficos y Paleoambientales de la Formación Santa Máxima, Paleozoico Superior, provincia de Mendoza, República Argentina.* PhD thesis, Universidad de Buenos Aires.

PARDO, M., COMTE, D. & MONFRET, T. 2002. Seismotectonic and stress distribution in the central Chile subduction zone. *Journal of South American Earth Sciences*, **15**, 11–22.

PENNINGTON, W. D. 1981. Subduction of the Eastern Panama Basin and seismotectonics of Northwest South America. *Journal of Geophysical Research*, **86**, 10753–10770.

PILGER, R. H. 1981. Plate reconstructions, aseismic ridges, and low angle subduction beneath the Andes. *Geological Society of America, Bulletin*, **92**, 448–456.

PILGER, R. H. 1984. Cenozoic plate kinematics, subduction and magmatism: South American Andes. *Journal of the Geological Society, London*, **141**, 793–802.

POLANSKI, J. 1958. El bloque varíscico de la Cordillera Frontal de Mendoza. *Revista de la Asociación Geológica Argentina*, **12**, 165–196.

POLANSKI, J. 1964. Descripción geológica de la Hoja 26c, La Tosca (Prov. de Mendoza). *Dirección Nacional de Geología y Minería, Boletín*, **101**, 1–86.

RAMOS, V. A. 1988. Tectonics of the Late Proterozoic – Early Paleozoic: a collisional history of Southern South America. *Episodes*, **11**(3), 168–174.

RAMOS, V. A. 1999a. Plate tectonic setting of the Andean Cordillera. *Episodes*, **22**(3), 183–190.

RAMOS, V. A. 1999b. Los depósitos terciarios sinorogénicos de la región andina. *In*: CAMINOS, R. (ed.) *Geología Argentina*. Instituto de Geología y Recursos Minerales, Anales, **29**, 651–682.

RAMOS, V. A. 2004. Cuyania, an exotic block to Gondwana: review of a historical success and the present problems. *Gondwana Research*, **7**, 1009–1026.

RAMOS, V. A. & FOLGUERA, A. 2005. Tectonic evolution of the Andes of Neuquén: constraints derived from the magmatic arc and foreland deformation. *In*: VEIGA, G. ET AL. (eds) *The Neuquén Basin: A case study in sequence stratigraphy and basin dynamics*. Geological Society, London, Special Publications, **252**, 15–35.

RAMOS, V. A. & FOLGUERA, A. 2007. Evidence of flat and steep subduction on the Andes: geological processes and tectonic inferences. *20° Colloquium on Latinoamerican Earth Sciences, Abstracts*, 17–18, Kiel.

RAMOS, V. A. & MORENO, M. 2006. Tectonic evolution of the Colombian Andes, Editorial. *Journal of South American Earth Sciences*, **21**, 319–321.

RAMOS, V. A., JORDAN, T. E., ALLMENDINGER, R. W., KAY, S. M., CORTÉS, J. M. & PALMA, M. A. 1984. Chilenia: Un terreno alóctono en la evolución paleozoica de los Andes Centrales. *9° Congreso Geológico Argentino, Actas*, **2**, 84–106.

RAMOS, V., MUNIZAGA, F. & MARÍN, G. 1988. Las riolitas neopaleozoicas de la sierra de la Huerta (provincia de San Juan): evidencias de una metalogénesis aurífera gondwánica en Sierras Pampeanas. *3° Congreso Nacional de Geología Económica, Actas*, **1**, 149–159.

RAMOS, V. A., KAY, S. M., PAGE, R. N. & MUNIZAGA, F. 1989. La Ignimbrita Vacas Heladas y el cese del volcanismo en el Valle del Cura, provincia de San Juan. *Revista de la Asociación Geológica Argentina*, **44**, 336–352.

RAMOS, V. A., AGUIRRE URRETA, M. B. ET AL. 1996a. Geología de la Región del Aconcagua, Provincias de San Juan y Mendoza. *Dirección Nacional del Servicio Geológico, Anales*, **24**, 1–510.

RAMOS, V. A., CEGARRA, M. & CRISTALLINI, E. 1996b. Cenozoic tectonics of the High Andes of west-central Argentina, (30°–36°S latitude). *Tectonophysics*, **259**, 185–200.

RAMOS, V. A., CRISTALLINI, E. & PÉREZ, D. J. 2002. The Pampean flat-slab of the Central Andes. *Journal of South American Earth Sciences*, **15**, 59–78.

RAPALINI, A. E. & ASTINI, R. A. 2005. La remagnetización sanrafaélica de la Precordillera en el Pérmico: nuevas evidencias. *Revista de la Asociación Geológica Argentina*, **60**, 290–300.

RAPALINI, A. & VILAS, J. 1991. Tectonic rotations in the Late Paleozoic continental margin of southern South America determined and dated by paleomagnetism. *Geophysical Journal International*, **107**, 333–351.

REGNIER, M., CHATELAIN, J. L., SMALLEY, R., MING CHIU, J., ISACKS, B. L. & PUEBLA, N. 1992. Seismotectonic of the Sierra Pie de Palo, a basement block uplift in the Andean foreland, Argentina. *Bulletin of Seismological Society of America*, **82**, 2549–2571.

REYNOLDS, J. H., JORDAN, T. E., JOHNSON, N. M., DAMANTI, J. F. & TABBUTT, K. D. 1990. Neogene deformation on the flat-subduction segment of the Argentine-Chilean Andes: magneteostratigraphic constraints from Las Juntas, La Rioja Province, Argentina. *Geological Society of America, Bulletin*, **102**, 1607–1622.

RODRÍGUEZ BLANCO, L. 2004. *Geología, estructura y volcanismo permotriásico entre los los Cerros Puntudo y Colorado, San Juan*. Trabajo Final de Licenciatura, Universidad de Buenos Aires, unpublished, Buenos Aires.

RODRÍGUEZ FERNÁNDEZ, L., HEREDIA, N., ESPINA, R. & CEGARRA, M. 1997. Estratigrafía y estructura de los Andes Centrales Argentinos entre los 30° y 31° de latitud sur. *Acta Geológica Hispánica*, **32**, 51–75.

ROSSELLO, E., COBBOLD, P., DIRAISON, M. & ARNAUD, N. 2002. Auca Mahuida (Neuquén Basin, Argentina): a Quaternary shield volcano on a hydrocarbon-producing substrate. *5° International Symposium on Andean Geodynamics, Extended Abstracts*, 549–552.

RUBINSTEIN, N. & KOUKHARSKY, M. 1995. Edades K/Ar del volcanismo neopaleozoico en la Precordillera noroccidental sanjuanina (30°S; 69°03'W). *Revista de la Asociación Geológica Argentina*, **50**, 270–272.

SAMANIEGO, P., MARTIN, H., ROBIN, C. & MONZIER, M. 2002. Transition from calc-alkalic to adakitic magmatism at Cayambe volcano, Ecuador: insights into slab melts andmantle wedge interactions. *Geology*, **30**, 967–990.

SANDEMAN, H. A., CLARK, A. H. & FARRAR, E. 1995. An integrated tectono-magmatic model for the evolution of the Southern Peruvian Andes (13°–20°S) since 55 Ma. *International Geology Review*, **37**, 1039–1073.

SARMIENTO-ROJAS, L. F., VAN WESS, J. D. & CLOETINGH, S. 2006. Mesozoic transtensional basin history of the Eastern Cordillera, Colombian Andes: inferences from tectonic models. *Journal of South American Earth Sciences*, **21**, 383–411.

SATO, A. & LLAMBÍAS, E. 1993. El Grupo Choiyoi, provincia de San Juan: equivalente efusivo del Batolito de Colangüil. *12° Congreso Geológico Argentino y 2° Congreso de Exploración de Hidrocarburos, Actas*, **4**, 156–165.

SÉBRIER, M., MERCIER, J., MACHARÉ, D., BONNOT, J. & CABRERA Y, J. L. BLANC. 1988. The state of stress in an overriding plate situated above a flat-slab: the Andes of central Peru. *Tectonics*, **7**, 895–928.

SIAME, L. L., BELLIER, O. & SEBRIER, M. 2006a. Active tectonics in the Argentine Precordillera and Western Sierras Pampeanas. *Revista de la Asociación Geológica Argentina*, **61**, 604–619.

SIAME, L. L., SÉBRIER, M., BELLIER, O. & BOURLES, D. 2006b. Can cosmic ray exposure dating reveal the normal faulting activity of the Cordillera Blanca Fault, Peru? *Revista de la Asociación Geológica Argentina*, **61**, 536–544.

SORIA, M. F. 1984. Vertebrados fósiles y edad de la Formación Aisol, provincia de Mendoza. *Revista de la Asociación Geológica Argentina*, **38**, 299–306.

SOTARELLO, G., BELVIDERI, I., DE MACHUCA, B., FERNÁNDEZ, G., MACHUCA, E. & MCGUINTY, W. 2005. Avances en el conocimiento y exploración del depósito epitermal de oro y plata 'Casposo', Departamento de Calingasta, Provincia de San Juan. *8° Congreso Argentino de Geología Económica, Actas*, 299–305.

SPIKINGS, R. A., WINKLER, W. & SEWARD, D. 2001. Along-strike variations in the thermal and tectonic response of the continental Ecuadorian Andes to the collision with heterogeneous oceanic crust. *Earth and Planetary Science Letters*, **186**, 57–73.

TABOADA, A., RIVERA, L. A. *ET AL.* 2000. Geodynamics of the northern Andes: subductions and intracontinental deformation (Colombia). *Tectonics*, **19**, 787–813.

VARELA, R., CINGOLANI, C., DALLA SALDA, L., ARAGÓN, E. & TEIXEIRA, W. 1993. Las monzodioritas y monzogabros de Cacheuta, Mendoza: edad, petrografía e implicancias tectónicas. *12° Congreso Geológico Argentino y 2° Congreso de Exploración de Hidrocarburos, Actas*, **4**, 75–80.

VERGÉS, J., RAMOS, E., SEWARD, D., BUSQUETS, P. & COLOMBO, F. 2001. Miocene sedimentary and tectonic evolution of the Andean Precordillera at 31°S, Argentina. *Journal of South American Earth Sciences*, **14**, 735–750.

VILAS, J. & VALENCIO, D. 1982. Implicancias geodinámicas de los resultados paleomagnéticos de formaciones asignadas al Paleozoico tardío-Mesozoico temprano del centro-oeste Argentino. *5° Congreso Latinoamericano de Geología, Actas*, **3**, 743–758.

WHITMAN, D., ISACKS, B. & KAY, S. M. 1996. Lithospheric structure and along-strike segmentation of the central Andean plateau. *Tectonophysics*, **259**, 29–40.

YAÑEZ, G., RANERO, G. R., VON HUENE, R. & DIAZ, J. 2001. Magnetic anomaly interpretation across a segment of the southern Central Andes (32–34°S): implications on the role of the Juan Fernández ridge in the tectonic evolution of the margin during the Upper Tertiary. *Journal of Geophysical Research*, **106**, 6325–6345.

YRIGOYEN, M. 1993. Los depósitos sinorogénicos terciarios. *In*: RAMOS, V. A. (ed.) *Geología y Recursos Naturales de Mendoza*. XII° Congreso Geológico Argentino y II° Congreso de Exploración de Hidrocarburos, Relatorio, 123–148, Buenos Aires.

YRIGOYEN, M. 1994. Revisión estratigráfica del Neógeno de las Huayquerías de Mendoza septentrional, Argentina. *Ameghiniana*, **31**, 125–138.

YUAN, X., ASCH, G. *ET AL.* 2006. Deep seismic images of the Southern Andes. *In*: KAY, S. M. & RAMOS, V. A. (eds) *Late Cretaceous to Recent Magmatism and Tectonism of the Southern Andean Margin at the Latitude of the Neuquen Basin (36–39°S)*. Geological Society of America, Special Paper, **407**, 61–72.

ZAPATA, T. R. & ALLMENDINGER, R. W. 1996. The thrust front zone of the Precordillera thrust belt, Argentina: a thick-skinned triangle zone. *American Association of Petroleum Geologists, Bulletin*, **80**, 359–381.

# Middle Miocene Chiapas fold and thrust belt of Mexico: a result of collision of the Tehuantepec Transform/Ridge with the Middle America Trench

J. J. MANDUJANO-VELAZQUEZ[1] & J. DUNCAN KEPPIE[2]*

[1]*Subdirección de Exploración y Producción, Instituto Mexicano del Petróleo. Avenida Lázaro Cárdenas Norte 152, San Bartolo Atepehuacán, Delegación Gustavo A. Madero, C. P. 07730, México D. F.*

[2]*Departamento de Geología Regional, Instituto de Geología, Universidad Nacional Autónoma de México, C. P. 04510, México D. F.*

*\*Corresponding author (e-mail: keppie@glinx.com)*

**Abstract:** The Middle Miocene, thin-skinned, Chiapas fold-and-thrust belt (Gulf of Mexico–southeastern Mexico–Belize) consists of WNW-trending folds and thrusts, and East–West sinistral transcurrent faults resulting from N60°E shortening. Balanced cross-sections indicate that shortening varies from 48% (SW) to *c.* 8% (NE) with a total shortening of 106 km, and that thrusts merge into a basal décollement in the Callovian salt horizon. The Middle Miocene age of the deformation is synchronous with collision of the Tehuantepec Transform/Ridge with the Middle America Trench off Chiapas. The presently exposed Tehuantepec Transform/Ridge varies from a transform fault across which the age of the oceanic crust changes producing a step (down to the east) to a ridge resulting from compression following a change in plate motion and a series of seamounts. On the other hand, the earthquake data show that the part of the Tehuantepec Transform/Ridge subducted during the past 5 Ma is a step with no accompanying ridge. Whereas collision of a ridge segment with the trench is inferred to be responsible for the 13–11 Ma deformation in the upper plate, its termination at 11 Ma suggests an along-strike transition to a step. Collision of the Tehuantepec Transform/Ridge also appears to have terminated arc magmatism along the Pacific coast of Chiapas. The similarity between the petroleum-producing, Cantarell structure in the Sonda de Campeche and the buried foldbelt in the Sierra de Chiapas suggests there is considerable further hydrocarbon potential.

## Introduction

The Middle Miocene, Chiapas–Tabasco–Campeche fold-and-thrust belt (Chiapas foldbelt hereafter) in Mexico (Fig. 1) is remarkable because it formed during a restricted time interval (*c.* 2.5 Ma), is of restricted size (300 × 600 km) and is host to some of the largest petroleum resources in the world. Its origin is enigmatic. Origins inferred for fold-and-thrust belts elsewhere in the world vary depending on their age. The older ones are related to collisions, such as continent–continent, arc–continent and terrane–continent collisions. On the other hand, younger fold-and-thrust belts bordering subducting oceanic plate containing little or no continental lithosphere have been attributed to other factors, such as a shallowing of the dip of the Benioff zone, an increased convergence rate and absolute plate motion, a change in the convergence vector, overriding of an active plume or seamount chain, an increase in friction between the overriding and subducting plates, and the thickness of the upper continental plate (Sobolev & Babeyko 2005). Recognition of these genetic factors in ancient orogens requires a better understanding of modern analogs, such as the Chiapas foldbelt, which forms the topic of this paper.

The genesis of the Chiapas foldbelt has been linked to the inferred eastward relative movement of the Chortis block (mainly Honduras) from a position off southwestern Mexico during the last 45 Ma (Ross & Scotese 1988; Pindell *et al.* 1988, 2006; Schaaf *et al.* 1995). The contrast in duration of these events (*c.* 2.5 versus 45 Ma) poses a problem for this model. Alternatively, Keppie & Moran-Zenteno (2005) proposed that the fold-and-thrust belt resulted from collision of the Tehuantepec Ridge with the Middle America Trench. To resolve the origin of the Chiapas foldbelt, we present an analysis of seismic reflections data from two NE–SW composite sections across the Sonda de Campeche and the Sierra de Chiapas (Fig. 1) that document the thin-skinned nature of the deformation, whereas well data indicate its restricted duration. This information is then considered in a plate tectonic context. The final integration of all the information suggests that collision of the Tehuantepec Ridge with the trench may have induced short-lived deformation of the overriding plate.

*From*: MURPHY, J. B., KEPPIE, J. D. & HYNES, A. J. (eds) *Ancient Orogens and Modern Analogues.* Geological Society, London, Special Publications, **327**, 55–69.
DOI: 10.1144/SP327.4   0305-8719/09/$15.00 © The Geological Society of London 2009.

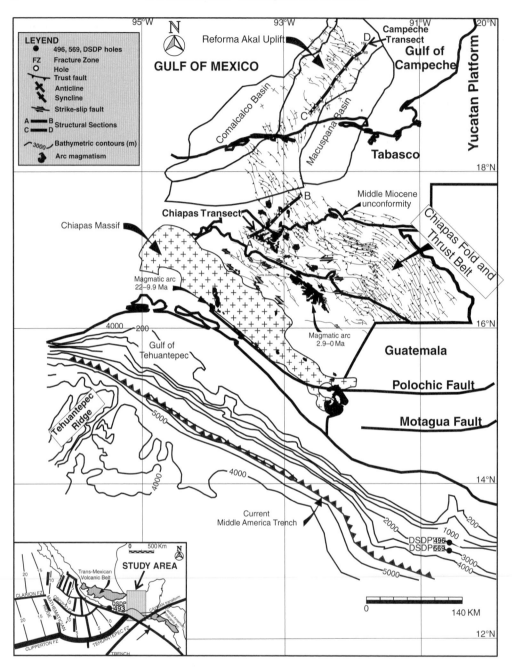

**Fig. 1.** Map of eastern Mexico showing the locations of the Chiapas foldbelt, the reflection seismic sections, the Tehuantepec Transform/Ridge, the Chiapas massif and various other tectonic and geographical elements.

## Geological setting

The NW-trending Chiapas foldbelt lies close to the southern margin of the North American Plate and is approximately parallel to the Middle America

Trench that marks the boundary with the northward subducting Cocos Plate (Fig. 1). The Chiapas foldbelt is bound to the south by the Motagua Fault Zone that forms the northern border of the Chortis block, to the SW by the Permo-Triassic Chiapas

Massif and the Gulf of Tehuantepec, to the west by the Miocene–Recent Trans-Mexican Volcanic Belt, and to the north and east it is unconformably overlain by Late Neogene sediments. Where it is exposed in southern Mexico, Guatemala and Belize, the foldbelt consists of a series Z-shaped, NW- to E-trending folds and thrusts that verge both to the NE and to the SW associated with common E–W striking, sinistral transcurrent faults (de la Rosa *et al.* 1989) (Fig. 1). The foldbelt has been traced northwestwards beneath the Late Neogene rocks into the Sonda de Campeche and the well studied NE-trending Reforma–Akal uplift, where the Cantarell structure hosts major petroleum resources. The Reforma–Akal uplift is bounded by the Comalcalco and Macuspana basins that formed during Neogene extensional growth faulting associated with salt tectonics (Ricoy 1989; Galloway *et al.* 1991). In southeastern Mexico, constrains on the age of the fold is defined by an unconformity between folded Latest Cretaceous to Early Middle Miocene (12–15 ± 1 Ma: K–Ar ages on biotite: Ferrusquía 1996) and Upper Miocene–Quaternary sediments and volcanic rocks (*c.* 11–0 Ma) (Fig. 2b). Synchronous deformation in the Sonda de Campeche is similarly bracketed between folded Jurassic–Middle Miocene (Serravallian: 13.65–11.61 Ma: Gradstein *et al.* 2004) rocks and unconformably overlying Late Miocene (Tortonian: 11.61–7.25 Ma; Gradstein *et al.* 2004)–Recent sediments (Mitra *et al.* 2005; Mandujano-Velazquez & Keppie 2006) (Fig. 2c). In contrast to the deformation recorded in the Chiapas foldbelt, contemporaneous Late Cretaceous and Tertiary rocks in the Gulf of Tehuantepec are not folded; however there are two unconformities: between the Eocene and Early Miocene, and between the Early and Late Miocene (Sanchez-Barreda 1981; Keppie & Morán-Zenteno 2005) (Fig. 2a).

The rocks deformed in the Chiapas foldbelt range in age from Early Jurassic (possibly Triassic) to Middle Miocene. The oldest rocks consist of alluvial fan and fluvial red beds deposited in grabens and half grabens bordered by normal faults formed during opening of the Gulf of Mexico. Red beds overlain by Middle–Upper Jurassic evaporites (including salt) pass upwards into Upper Jurassic and Middle Cretaceous platformal carbonates, and shales and sandstone of Upper Cretaceous–Palaeogene age. In the Gulf of Tehuantepec, conglomerates with tonalite, dacite and basalt fragments occur. The Upper Jurassic rocks formed the source of most of the hydrocarbons (Mitra *et al.* 2005). The Cretaceous–Palaeocene boundary is marked by carbonate breccias derived from the Chixhulub impact crater (Grajales *et al.* 2000), which forms the main reservoir for hydrocarbons in the Sonda de Campeche. Palaeocene shales

form the regional topseal for the reservoir rocks. The contact beneath the Middle Miocene is generally an unconformity that locally rests upon sediments as old as Eocene. This unconformity marks a period of deformation that produced the Chiapas foldbelt. The overlying Middle Miocene to Holocene sediments consist of shale, sandstone, limestone and conglomerate deposited in environments that range from basin through shelf, alluvial, fluvial to lacustrine.

Development of the Chiapas foldbelt at *c.* 13–11 Ma coincides with a beginning of a hiatus in arc magmatism defined by Damon & Montesinos (1978: Cenozoic Igneous in Fig. 1). Between *c.* 17.8 and 12.4 Ma, arc magmatism in the Chiapas massif and eastern Oaxaca State formed a western continuation of the Central American Volcanic Arc (Damon & Montesinos 1978; Keppie *et al.* 2009). However, between *c.* 10 and 2.8 Ma, a hiatus in arc magmatism produced a magmatic arc gap between the eastern end of the Trans-Mexican Volcanic Belt and the western end of the Central American Volcanic Arc in Guatemala. Sporadic alkalic–adakitic arc magmatism in the gap was reinitiated at 2.8 Ma within the Chiapas foldbelt in the NW-trending Chiapanecan volcanic arc, which lies *c.* 200 km north of the present Pacific coast (Mora *et al.* 2007) (Fig. 1).

Offshore in the Cocos Plate, the Tehuantepec Transform Fault juxtaposes contrasting ages of the ocean floor due to a combination of mid-ocean ridge offset and different rates of spreading north and south of the transform (Manea *et al.* 2005). It is also coincident with a topographic ridge (hence its alternate name, the Tehuantepec Ridge) resulting from a combination of: (i) compression produced by a change in the relative plate motions at 13 Ma (Manea *et al.* 2005); and (ii) the presence of a seamount chain (Keppie & Morán-Zenteno 2005). As its character varies along its length, we hereafter use the term, Tehuantepec Transform/Ridge. However, the inferred location of the subducted Tehuantepec Transform/Ridge beneath Mexico over the past 15 Ma varies between authors. Keppie & Morán-Zenteno (2005) infer that, following a Middle Miocene plate reorganization, the pre-13 Ma old segment of the Tehuantepec Transform/Ridge rotated counterclockwise about a pole in the mouth of the Gulf of California from an ENE–North trend. Rotation would have resulted in a westward migration of its intersection with the Middle America Trench. Manea & Manea (2006) projected the Tehuantepec Ridge northeastwards beneath Mexico, omitting the sharp bend in the transform that formed during the Middle Miocene plate reorganization: this results in an easterly migration of the Tehuantepec/Middle America Trench intersection in the last 13 Ma. Earthquake data indicate

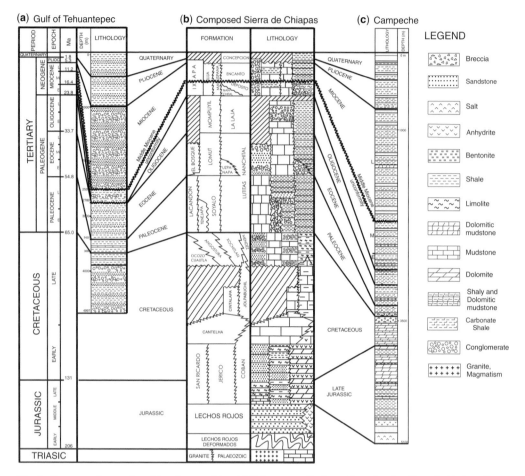

**Fig. 2.** Representative stratigraphic columns of the: (**a**) Tehuantepec Platform (from Sanchez-Barreda 1981, well Salina Cruz-1); (**b**) Sierra de Chiapas; (**c**) Sonda de Campeche. Abbreviations: E, Early; M, Middle; L, Late.

that instead of a ridge, a N–S step in the Benioff Zone extends northwards beneath Mexico from the exposed Tehuantepec Transform/Ridge at the Middle America Trench (Bravo *et al.* 2004). The dip of the Benioff zone changes from west to east across the step from *c.* 11°N to *c.* 40°N (Pardo & Suárez 1995; Rebollar *et al.* 1999*a*). Using the convergence rate of 67 km/Ma (Pindell *et al.* 1988), the step-like geometry extends back to at least 5 Ma. These observations are consistent with the location of the Tehuantepec Transform/Ridge inferred by Keppie & Morán-Zenteno (2005); however, it highlights its changing nature along strike.

## Interpretation of reflection seismic data

Stratigraphic and structural interpretation was supported by lithological and chronostratigraphical data from seven wells in the Sierra de Chiapas and ten in the Sonda de Campeche (Figs 3 & 5). Two composite NE-SW PEMEX reflection seismic sections, roughly perpendicular to the trend of the folds, were compiled: Section A–B is 97.75 km long across the Sierra de Chiapas, and the other (C–D) is 172 km in length across the Sonda de Campeche (Figs 4 & 6).

### Sierra de Chiapas

In the Sierra de Chiapas, the deepest wells (#11 & 17) reached Kimmeridgian and Tithonian (Upper Jurassic) carbonates. The Cretaceous is typically dolomite in the Early Cretaceous passing upwards into mixed carbonate and dolomite in the Middle Cretaceous, and marl and breccia in the Upper Cretaceous (up to 2142 m) (Fig. 3). The Tertiary is characterized by clastic rocks (shale, sandstone

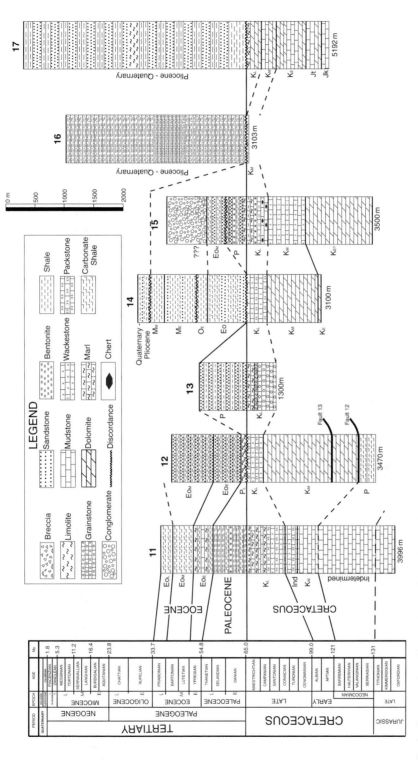

**Fig. 3.** Chronostratigraphic correlation of wells in the Sierra de Chiapas (Fig. 4). Abbreviations: E, Early; M, Middle; L, Late; Jk, Kimmeridgian; Jt, Tithonian; K, Cretaceous; P, Palaeocene; Eo, Eocene; O, Oligocene; M, Miocene; Ind, undifferentiated.

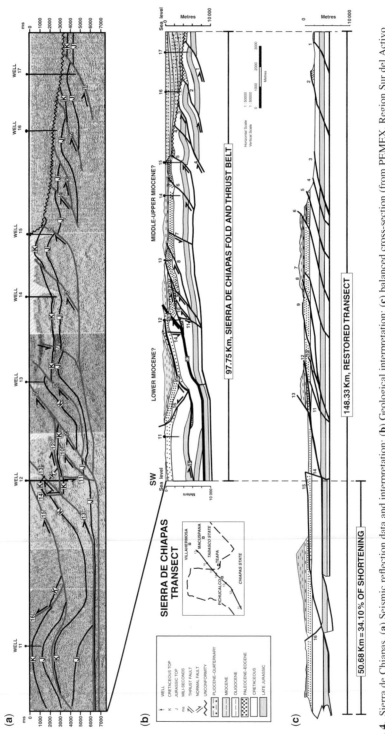

**Fig. 4.** Sierra de Chiapas. (**a**) Seismic reflection data and interpretation; (**b**) Geological interpretation; (**c**) balanced cross-section (from PEMEX, Region Sur del Activo Regional de Exploración, Coordinación de Plays Establecidos: Medina *et al.* 1997).

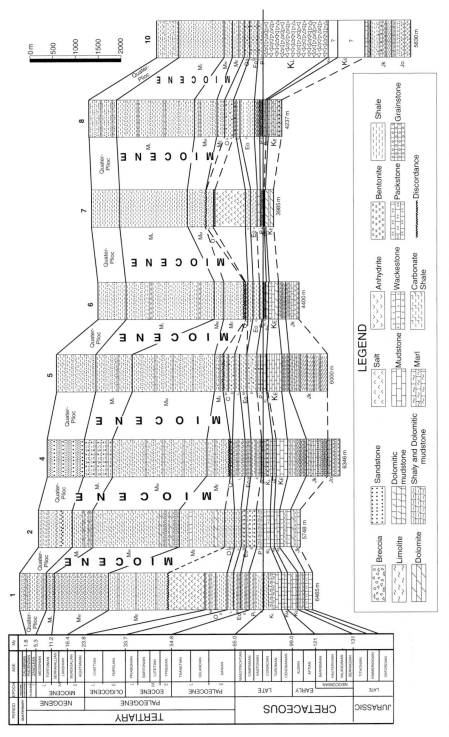

**Fig. 5.** Chronostratigraphic correlation of wells in the Sonda de Campeche section (Fig. 6). Abbreviations: E, Early; M, Middle; L, Late; Jo, Oxfordian; Jk, Kimmeridgian; Jt, Tithonian; K, Cretaceous; Eo, Eocene; O, Oligocene; P, Palaeocene; M, Miocene; and Quater–Plioc, Quaternary–Pliocene.

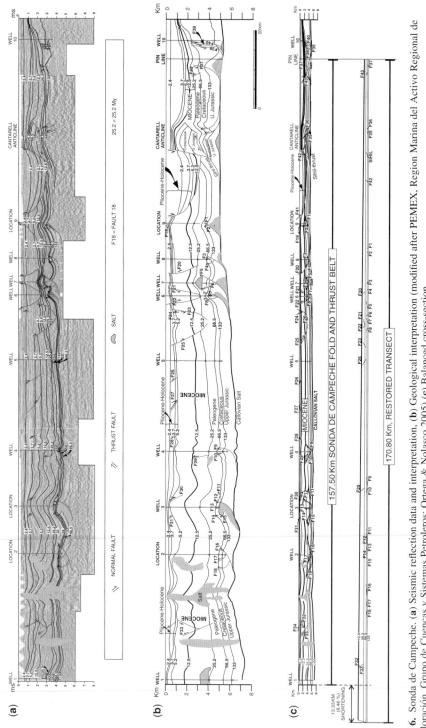

**Fig. 6.** Sonda de Campeche. (**a**) Seismic reflection data and interpretation. (**b**) Geological interpretation (modified after PEMEX, Region Marina del Activo Regional de Exploración, Grupo de Cuencas y Sistemas Petroleros: Ortega & Nolasco 2005) (**c**) Balanced cross-section.

and conglomerate) with minor limonite and bentonite horizons. Unconformities are present at various levels: at the base of the Palaeocene (well #12), at the base of the Eocene (wells #14 and 15), at the base of the Early Miocene (well #14) and at the base of the Pliocene/Quaternary (wells #14, 16 and 17). In places, the Eocene rests directly upon the Cretaceous (well #14). The unconformities lead to considerable variation in thickness: Palaeocene 0–800 m, Eocene up to 1175 m, Oligocene 0–185 m, Early Miocene 550 m, Middle Miocene >250 m and Pliocene–Quaternary >3000 m.

The Chiapas fold-and-thrust belt is characterized by NE verging folds and thrusts with trends, varying from East–West through NW–SE to NNW–SSE. Sinistral transcurrent faults strike East–West and, rarely, NW–SE (Fig. 1). The reflection seismic section (A–B) reveals numerous folds and associated thrusts. Thrust duplexes are floored by a décollement zone beneath the Upper Jurassic, which is inferred to be located in the Middle-Upper Jurassic salt horizon (Fig. 4). Towards the southwestern end of the section, the thrusts appear to have been folded during back-thrusting. Two normal faults downthrown to the south are present in the middle of the section (faults 7 and 11). The northern half of the section shows that all of these structures are unconformably overlain by Miocene–Present sediments that overstep and overlap towards the south.

A balanced section was produced using standard techniques (Marshak & Woodward 1988) (Fig. 4c), and reveals that the present 97.75 km section was originally 148.33 km in length, that is a shortening of 50.68 km (=34.1%). The shortening increases from 6% in the NE to 48% in the SW.

## Sonda de Campeche

In the Sonda de Campeche, the deepest wells reached Callovian–Oxfordian salt, anhydrite, sandstone, mudstone and shale (Fig. 5). Salt is visible in the seismic reflection data along the base of the section, and as diapers that, in the southwestern part, penetrate up through the entire stratigraphic sequence. The salt diapirs have associated anticlines. Salt also occurs as lenses in the Palaeogene and Miocene (Fig. 6). The Kimmeridgian is represented by clastic rocks overlain by carbonate rocks, anhydrite and dolomite, which continues upwards into the 150–400 m-thick Tithonian sediments. The Cretaceous consists of 315–751 m-thick carbonates that pass upwards into shaly carbonate and marl rocks; however, towards the northeastern end of the section the total thickness of the Cretaceous increases to 1420 m where thick meteorite impact breccias occur. The 395–1294 m-thick Palaeogene, 475 m-thick Eocene and 2085 m-thick

Early Miocene consist of shale, carbonate shales, marl, sandstone and bentonite with rare carbonate. Middle Miocene sandstone and shale unconformably overlying the Early Miocene range from 1302 m in the SW to 175 m in the NE. In contrast, the Late Miocene lithologies are thickest in the NE, tapering to 175 m in the SW. The Pliocene–Quaternary sediments are 750 m thick.

The structures observed in the reflection seismic section along line C–D may be divided into two types: (i) folds and thrusts in rocks older than Middle Miocene; and (ii) listric normal faults that cut part or all of the stratigraphic section (Fig. 6). Most of the anticlines have salt cores and are associated with ramps in underlying NE-vergent thrusts that root into a décollement in the salt horizon. In the hanging wall of many of these NE-vergent thrusts are small SW-vergent thrusts. The largest and best documented of the NE-vergent thrusts is the Sihil Thrust, which ramps upwards to the east and places Upper Jurassic rocks upon Early Miocene sediments in the Cantarell anticline (Mitra et al. 2005). This Sihil thrust is truncated by the Middle Miocene unconformity. Two SW-vergent thrusts occur below the Sihil Thrust on the northern limb of the Cantarell anticline.

Listric normal faults are common in the upper part of the section, where some delimit graben and half graben located over older synclines (Fig. 6). Whereas some of these faults are truncated by Holocene sediments, others extend to the surface indicating continuing activity. The variations in the thickness of the Middle Miocene–Present units is especially marked in the half graben on the south side of the Cantarell anticline and shows that the bounding faults were active throughout their deposition. Listric normal faults at the northeastern end of the section cut only Mesozoic and Palaeogene rocks and are interpreted as being associated with the boundary between the Macuspana Basin and the Yucatan Platform.

In order to calculate the amount of shortening, a balanced section was produced in a two-stage process using standard procedures (Marshak & Woodward 1988). First, extension during listric normal faulting was removed, and then the thrusts were moved back to connect displaced units and unfolding the strata (Fig. 6c). In this way we determined that the present distance of 157.5 km between C and the Pin Line near D was originally 170.8 km, that is 13.3 km (= 7.8%) of shortening. The magnitude of shortening is similar to that at the northeastern end of the Chiapas section (A–B) and indicates that the Chiapas foldbelt continues 200 km to the NE beneath the Pliocene–Quaternary cover.

The similarity in the structure of the Chiapas foldbelt both onland in the Sierra de Chiapas, and offshore in the Sonda de Campeche, suggests that there should be a similar potential for hydrocarbons. However, in the exposed part of the belt, the absence of an adequate topseal and apparent lack of maturation of organic matter may account for the present lack of known reserves. On the other hand, eastward projection of the foldbelt from the Sonda de Campeche suggests considerable unexplored potential beneath the Middle Miocene unconformity.

## Evidence for subduction erosion

Subduction erosion along the Mexican and Guatemalan Pacific margin appears to have occurred over the past 20 Ma (Clift & Vannucchi 2004), and so may be a factor in development of the Chiapas foldbelt: this is investigated now. Estimates of subduction erosion are based on the subsidence history of the forearc using fossils as depth indicators; subsidence has been related to landward migration of the trench as a result of frontal subduction erosion (see Clift & Vannucchi 2004 for the methodology, which we follow herein). Depths of deposition has been documented by the Deep Sea Drilling Project (DSDP leg #66, holes #489 and #493 at c. 99°W) and off Guatemala (DSDP leg 84, hole #569 at 90.84°W). Clift and Vannucchi's (2004) calculations give average rates of landward

retreat of the trench over the last c. 23 Ma of 1 km/Ma off Mexico and 0.9 km/Ma off Guatemala. Within error, this is similar to the average rate calculated from the data recorded in DSDP leg 67, hole #496 at 90.8°W (Fig. 7); however, detailed data from this hole indicate that the subsidence rate was variable and yields a faster landward migration rate of c. 3 km/Ma between 18 and 11 Ma. This faster rate of trench migration mirrors the faster rate at c. 22–13 Ma calculated for hole #493 (Clift & Vannucchi 2004). These legs are located west (leg #66 Acapulco, México) and east (leg 84, San José, Guatemala) of the Gulf of Tehuantepec, respectively, and it assumed that similar subduction rates occurred south of the Gulf of Tehuantepec. These results suggest that the period of subduction erosion predated development of the Chiapas foldbelt, and thus subduction erosion is probably not a factor in its genesis.

## Tectonic interpretation

The orientation of the finite strain ellipse in the Chiapas foldbelt may be deduced from the orientation of the faults and folds. Assuming little or no rotation of the E-W sinistral faults, constrain the shortening direction to N60°E (Fig. 1). This direction is c. 22° clockwise of the N38°E relative plate

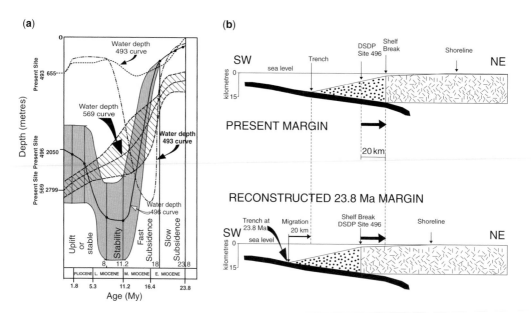

**Fig. 7.** Subsidence Evolution: (**a**) Mexican margin, DSDP 493, [(----------) Clift & Vannucchi 2004], (— ·· — ·· — ·· — ·· —) this paper); Guatemala margin: DSDP 569 (Clift & Vannucchi 2004), DSDP 496 (this paper); (**b**) Subduction-Erosion of the Guatemala Margin. Abbreviations: E, Early; M, Middle; L, Late. N. B. The palaeobatymetric curves calculated in this work (Mexican margin 493 and Guatemala margin 569) were constructed based on the benthic foraminifers reported in these wells, which were drilled by the Deep Sea Drilling Project.

motion between the Cocos and North American plates at 10–20 Ma (Pindell *et al.* 1988; Engebretson *et al.* 1985) (Fig. 1). The difference is explicable in terms of the general difference between strain axes in finite strain versus infinitesimal strain (approximately parallel to stress directions)

associated with slightly oblique convergence (Jiang *et al.* 2001). The total shortening across the Chiapas foldbelt (106 km) may be calculated by adding the 51.7 km of shortening along A–B, the 13.3 km of shortening along C–D, the *c.* 30 km of shortening between A and the Chiapas massif

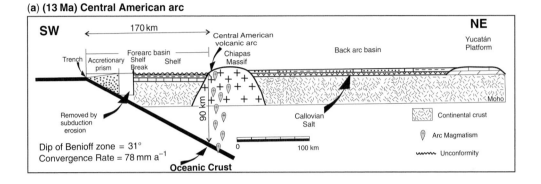

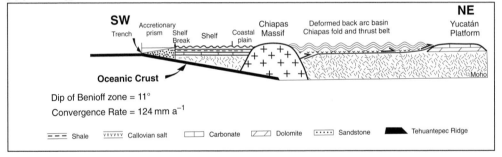

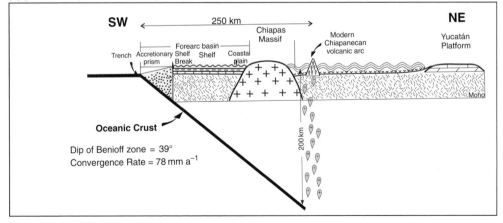

**Fig. 8.** Tectonic evolution of Chiapas (**a**) before, (**b**) during and (**c**) after development of the Chiapas and Campeche fold-and-thrust.

assuming 50% shortening (similar to the 48% short-ening at the southwestern end of A–B) and 11 km between B and C assuming a 7.8% shortening (as in C–D). Seismic data indicates no shortening across the Gulf of Tehuantepec (Sánchez-Barreda 1981) and none has been reported across the Chiapas massif. This is analogous to that observed in California where thrusts east of the Sierra Nevada batholith merge into a basal décollement that passes beneath a more rigid block under the Sierra Nevada batholith and the Great Valley (Fay *et al.* 2006). In order to restore the position of the Middle America Trench prior to 18 Ma, the 20 km loss to subduction erosion between 18 and *c.* 11 Ma must be added to the 106 km of shortening accommodated by the Chiapas fold and thrust belt. Thus the Middle America Trench at 18 Ma lay 126 km SSW of its present location off Chiapas. Assuming the present arc-trench width of 150 km is applicable to the Early Miocene arc (17.8–12.4 Ma) and 90 km for the slab depth beneath the

magmatic arc, the dip of the Benioff zone before 13 Ma may be calculated as 31° (Fig. 8a).

The dip of the Benioff zone during development of the Chiapas foldbelt cannot be estimated directly. However, its present dip west of the Tehuantepec Ridge is *c.* 11° (Pardo & Súarez 1995; Bravo *et al.* 2004). We assume a dip of 11° was present in the Middle Miocene (Fig. 8b).

The present *c.* 39° dip of the Benioff zone beneath Chiapas has been derived by Bravo *et al.* (2004) using earthquake data (Fig. 8c). The slab beneath the Central American Volcanic Arc in Guatemala to the south dips at 45° (Bevis & Isacks 1984; Burbach *et al.* 1984) and decreases to 11° west of the Tehuantepec Transform/Ridge (Pardo & Suárez 1995). These changes along strike to the south increase in slab dip can be interpreted as resulting from the progressive westward migration of the Tehuantepec Transform/Ridge, which would have resulted in a gradual decrease in slab dip of the Benioff zone before passage of the ridge followed

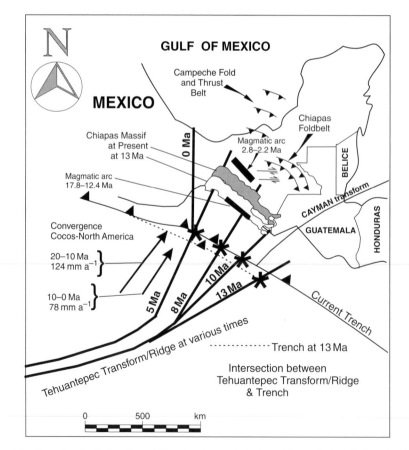

**Fig. 9.** Reconstructions showing the locations of the volcanic arcs, the Chiapas Massif, the Middle America Trench and the Tehuantepec Transform/Ridge (modified from Keppie & Morán-Zenteno 2005) at 13–0 Ma.

by an increase in slab dip after the ridge had passed through (Fig. 9). In our model, we apply these dips to the last 10 Ma.

The Middle Miocene age of the Chiapas foldbelt is synchronous with: (a) the 12.5–11 Ma, Middle Miocene plate reorganization during which the Pacific-Cocos pole of rotation moved *c.* 600 km southwards along the Pacific coast resulting in a decrease in convergence rate between the Cocos and North American plates (Fig. 1) (Mammerickx & Klitgord 1982); (b) collision of the Tehuantepec Transform/Ridge with the Middle America Trench off Chiapas near the Guatemala-Mexico border (Keppie & Morán-Zenteno 2005); (c) the start of an hiatus in arc magmatism in Chiapas; and (d) the end of a period of increased subduction erosion. Subduction erosion affected the whole Mexican, Guatemalan, Nicaraguan and Costa Rican margins suggesting that there is no causal link with either the Chiapas foldbelt or collision of the Tehuantepec Transform/Ridge. The decrease in the convergence rate during the Middle Miocene would not appear to be a factor in producing the Chiapas foldbelt.

The difference in age of the subducted oceanic plate across the Tehuantepec Transform/Ridge is *c.* 10 Ma at the Middle America Trench and produces a step down to the east in the sea-floor elevation of *c.* 1100 m (Manea *et al.* 2005). A similar N–S step in the top of the subducted Cocos Plate, documented from earthquake data, shows an increase from *c.* 25 km beneath the coast of the Gulf of Tehuantepec to *c.* 100 km at 18°N (Bravo *et al.* 2004). This indicates that the offset in ages across the Tehuantepec Transform/Ridge results in an increase in the dip of the Benioff zone from 11° on the western side to *c.* 30° on the eastern side. Slab dip of the Cocos plate increases eastwards to 45° at the Guatemalan border. Earthquake data indicates that the dip of the Benioff zone gradually increases westwards beneath the Trans-Mexican Volcanic Belt (Pardo & Suárez 1995). The geometry of the Benioff zone across the subducted Tehuantepec Transform/Ridge with its step down to the east is similar to that of central South America where a step down to the south occurs across the subducted Juan Fernandez Ridge (Ramos & Aleman 2000). A decrease in the Benioff zone dip is accompanied by an increase in shortening across the Andes that produced a fold-and-thrust belt (Ramos *et al.* 2002), thereby providing an analogy for the Chiapas foldbelt.

Assuming that a similar geometry in the Benioff zone across the Tehuantepec Transform/Ridge has existed since 15 Ma, the reconstructions of Keppie & Morán-Zenteno (2005) indicate that the intersection between Tehuantepec Transform/Ridge and

the Middle America Trench migrated westwards along the Chortis block and Chiapas margins from *c.* 15–12 Ma to 12–0 Ma (Fig. 9), respectively. Our plate reconstructions imply that the intersection would have migrated west through time. If collision of the Tehuantepec ridge with the trench was responsible for formation of the Chiapas fold-and-thrust belt, then thrust belt development should also be diachronous. Although our data are restricted to the western part of the foldbelt, such diachronism is not apparent because the Chiapas foldbelt does not appear to continue into the Chortis block (Donnelly *et al.* 1990) and deformation appears to have been of short duration (13–11 Ma). The reason for initiation and termination of the deformation may perhaps be related to the variation in the geometry of the Tehuantepec Transform/Ridge. Thus, when a ridge was being subducted it induced deformation in the upper plate. However, when the transform was only a step with no accompanying ridge, no deformation occurred in the upper plate: this certainly applies to the period 5 Ma – Present because the earthquake data indicate that the geometry of the subducted Tehuantepec Transform/Ridge is a step with no accompanying ridge (Bravo *et al.* 2004).

In conclusion, the origin of the Chiapas foldbelt as a result of subduction of a transform fault where it is accompanied by a compressional ridge and sea-mount chain indicates the importance of the topography of the subducting plate in the genesis of modern and ancient fold-and-thrust belts. This may find application to short-lived episodes of deformation in the long pre-collisional history of orogenic belts.

We are grateful to A. O. Pérez, A. A. López, J. H. García of PEMEX and A. A. P. Luna of Instituto Mexicano del Petróleo for allowing us to use the seismic data Chiapas and the Sonda de Campeche. We thank M. E. V. Meneses for a constructive review of the structural model and R. Strachan and S. T. Johnston for comments on the manuscript that helped in revision of the paper.

# References

BEVIS, M. & ISACKS, B. L. 1984. Hypocentral trend surface analysis: probing the geometry of Benioff zones. *Journal of Geophysical Research*, **89**, 6153–6170.

BRAVO, H., REBOLLAR, C. J., URIBE, A. & JIMÉNEZ, O. 2004. Geometry and state of stress of the Wadati-Benioff zone in the Gulf of Tehuantepec, Mexico. *Journal of Geophysical Research*, **109**, doi: 10.1029/2003JB002854, 1–14.

BURBACH, G. B., FROHLICH, C., PENNINGTON, W. D. & MATUMOTO, T. 1984. Seismicity and Tectonics of the Subducted Cocos Plate. *Journal of Geophysical Research*, **89**, 7719–7735.

CLIFT, P. D. & VANNUCCHI, P. 2004. Controls on tectonic accretion versus erosion in subduction zones: implications for the origin and recycling of the continental crust. *Review of Geophysics*, **42**, 1–31.

DAMON, P. E. & MONTESINOS, E. 1978. Late Cenozoic volcanism and metallogenesis over an active Benioff zone in Chiapas, Mexico. *Arizona Geological Society Digest*, **11**, 155–168.

DE LA ROSA, J. L., EBOLI, M. A., DÁVILA, S. M., DENGO, G., YAMAZAKI, M. F. & BALINAS, G. R. 1989. *Geología del Estado de Chiapas: Subdirección de Construcción y departamento de Geología de la Comisión Federal de Electricidad, superintendencia de estudio Zona Sureste*. Tuxtla Gutiérrez, Chiapas, México, 1–89.

DONNELLY, T. W., HORNE, G. S., FINCH, R. C. & LÓPEZ-RAMOS, E. 1990. Northern Central America; the Maya and Chortis blocks. *In*: DENGO, G. & CASE, J. E. (eds) *The Caribbean region: the geology of North America*. Geological Society of America, 37–76.

ENGEBRETSON, A. C., COX, A. & GORDON, R. G. 1985. Relative motions between oceanic and continental plates in the Pacific Basin. *Geology Society of America, Special Paper*, **206**, 1–59.

FAY, N., HUMPHREYS, E. D. & COBLENTZ, D. D. 2006. The Pacific-North America plate boundary in the western United States: dynamics of the Sierra Nevada block and implications for western US tectonics. *Geological Society of America*. Abstracts with programs Speciality Meeting No. 2, p. 109. Paper No. 14–8.

FERRUSQUÍA, V. 1996. Contribución al conocimiento geológico de Chiapas, el área Ixtapa-Soyaló. Universidad Nacional Autónoma de México. *Boletín del Instituto de Geología*, **109**, 1–130.

GALLOWAY, W. E., BEBOUT, D. G., FISHER, W. L., DUNLAP, J. B., JR, CABRERA, C. R., LUGO, R. J. E. & SCOTT, T. M. 1991. Cenozoic, *In*: SALVADOR, A. (ed.) *The Gulf of Mexico Basin*. The Geology of North America, **J**, 245–324.

GRADSTEIN, F. M., OGG, J. G., SMITH, A. G., BLEEKER, W. & LOURENS, L. J. 2004. A New Geologic Time Scale, with special reference to Precambrian and Neogene. *Episodes*, **27**, 83–100.

GRAJALES, N. J. M., CEDILLO, P. E., ROSALES, D. C. & MORÁN, C. D. 2000. Chicxulub impact: the origin of reservoir and seal facies in the southeastern Mexico oil fields. *Geology*, **28**, 307–310.

JIANG, D., LIN, S. & WILLIAMS, P. F. 2001. Deformation path in high strain zones, with reference to slip partitioning in transpressional plate-boundary regimes. *Journal of Structural Geology*, **23**, 991–1005.

KEPPIE, J. D. & MORÁN-ZENTENO, D. J. 2005. Tectonic implications of alternative Cenozoic reconstructions for southern Mexico and the Chortis Block. *International Geology Review*, **47**, 473–491.

KEPPIE, J. D., MORÁN-ZENTENO, D. J., MARTINY, B. & GONZÁLEZ-TORRES, E. 2009. Synchronous 29–19 Ma arc hiatus, exhumation and subduction of forearc in southwestern Mexico. *In*: JAMES, K. H., LORENTE, M. A. & PINDELL, J. L. (eds) *The Origin and Evolution of the Caribbean Plate*. Geological Society, London, Special Publications, **328**, 169–179.

MAMMERICKX, J. & KLITGORD, K. D. 1982. Northern East Pacific Rise: evolution from 25 m.y. to the present. *Journal of Geophysical Research*, **87**, 6751–6759.

MANDUJANO-VELAZQUEZ, J. J. & KEPPIE, D. J. 2006. Cylindrical and conical fold geometries in the *Cantarell* structure, Southern Gulf of Mexico: implications for hydrocarbon exploration. *Journal of Petroleum Geology*, **29**, 215–226.

MANEA, M., MANEA, V. C., FERRARI, L., COSTOGLODOV, V. & BANDY, W. L. 2005. Tectonic evolution of the Tehuantepec Ridge. *Earth and Planetary Science Letters*, **238**, 64–77.

MANEA, V. C. & MANEA, M. 2006. Origin of the modern Chiapanecan Volcanic arc in southern México inferred from thermal models. *Geological Society of America, Special Paper*, **412**, 27–38.

MARSHAK, S. & WOODWARD, N. 1988. Introduction to cross-section balancing. *In*: MARSHAK, S. & MITRA, G. (eds) *Basic Methods of Structural Geology*. Prentice Hall, Englewood Cliffs, 303–332.

MEDINA, F. U., HAM, W. J. M., NAMSON, J. & ALCÁNTARA, G. J. R. 1997. Proyecto Simojovel. Petróleos Mexicanos (PEMEX), Región Sur del Activo Regional de Exploración, Coordinación de Plays Establecidos, (unpublished report).

MITRA, S., CORREA, F. G., HERNÁNDEZ, G. J. & MURILLO, A. A. 2005. Three-dimensional structural model of the Cantarell and Sihil structures, Campeche Bay, Mexico. *American Association of Petroleum Geology Bulletin*, **89**, 1–26.

MORA, J. C., JAIMES, V. M. C., GARDUÑO, M. V. H., LAYER, P. W., POMPA, M. V. & GODINEZ, M. L. 2007. Geology and geochemistry characteristics of the Chiapanecan Volcanic Arc (central area), Chiapas Mexico. *Journal of Volcanology and Geothermal Research*, doi: 10.1016/j.jvolgeores.2006.12.009.

ORTEGA, V. & NOLASCO, J. 2005. *Modelo geológico de la Sonda de Campeche*. Petróleos Mexicanos (PEMEX), Región Marina del Activo Regional de Exploración, Grupo de Cuencas y Sistemas Petroleros, unpublishing report.

PARDO, M. & SUÁREZ, G. 1995. Shape of the subducted Rivera and Cocos plates in southern Mexico: seismic and tectonic implications. *Journal of Geophysical Research*, **100**, 12 357–12 373.

PINDELL, J. L., CANDE, S. C. W., PITMAN, W. C., ROWLEY, D. B., DEWEY, J. F., LEBRECQUE, J. & HAXBY, W. 1988. A plate-kinematic framework for models of Caribbean evolution. *Tectonophysics*, **155**, 121–138.

PINDELL, J. L., LORCAN, K., WALTER, M., KLAUS, S. & GRENVILLE, D. 2006. Foundations of Gulf of Mexico and Caribbean evolution: eight controversies resolved. *Geologica Acta*, **4**, 303–341.

RAMOS, V. A. & ALEMAN, A. 2000. Tectonic evolution of the Andes. *In*: CORDANI, U. G., MILANI, E., THOMAZ, F. A. & CAMPOS, D. A. (eds) *Tectonic evolution of South America*. 31st International Geological Congress, Rio de Janeiro, Brazil. Geological Society of America, Colorado, 635–685.

RAMOS, V. A., CRISTALLINI, E. O. & PÉREZ, D. J. 2002. The Pampean flat-slab of the central Andes. *Journal of South American Earth Sciences*, **15**, 59–78.

REBOLLAR, C. J., ESPÍNDOLA, V. H., URIBE, A., MENDOZA, A. & PÉREZ, V. A. 1999a. Distribution of stresses and geometry of the Wadati–Benioff zone under Chiapas, Mexico. *Geophysical International*, **38**, 95–106.

RICOY, U. J. 1989. *Tertiary Terrigenous Depositional Systems of the Mexican Isthmus Basins*. PhD thesis, University of Texas, Austin, Texas.

ROSS, M. I. & SCOTESE, C. R. 1988. A hierarchical tectonic model of the Gulf of Mexico and Caribbean regions. *Tectonophysics*, **135**, 139–168.

SÁNCHEZ-BARREDA, L. A. 1981. *Geologic evolution of the continental margin of the Gulf of Tehuantepec in southern Mexico*. PhD thesis, University of Texas, Austin, Texas.

SCHAAF, P., MORAN, Z. D., HERNÁNDEZ, B. M., SOLÍS, P. G., TOLSON, G. & KÖHLER, H. 1995. Paleogene continental margin truncation in southwestern Mexico: geochronological evidence. *Tectonics*, **14**, 1339–1350.

SOBOLEV, S. V. & BABEYKO, A. Y. 2005. What drives orogeny in the Andes? *Geology*, **33**, doi: 10.1130/G21557AR.1, 617–620.

# Extrusion of high-pressure Cache Creek rocks into the Triassic Stikinia–Quesnellia arc of the Canadian Cordillera: implications for terrane analysis of ancient orogens and palaeogeography

JAROSLAV DOSTAL[1]*, J. DUNCAN KEPPIE[2] & FILIPPO FERRI[3]

[1]*Department of Geology, Saint Mary's University, Halifax, Nova Scotia B3H 3C3, Canada*

[2]*Departamento de Geología Regional, Instituto de Geologia, Universidad Nacional Autonoma de Mexico, 04510 Mexico, D.F., Mexico*

[3]*British Columbia Ministry of Energy and Mines, 1810 Blanchard Street, Victoria, British Columbia V8W 9N3, Canada*

*\*Corresponding author (e-mail: jdostal@smu.ca)*

**Abstract:** The volcanic Triassic Takla Group constitutes a significant part of Stikinia and Quesnellia, two major terranes in the Canadian Cordillera that are separated by high-pressure rocks of the Cache Creek terrane containing Asian fauna. The geochemical and isotopic characteristics of the Takla Group in Quesnellia and Stikinia are similar, that is, tholeiitic basalts characterized by low abundances of strongly incompatible trace elements, negative Nb anomalies, +6 to +8 $\varepsilon_{Nd}$ values, the low initial Sr isotopic ratios, and relatively horizontal chondrite-normalized heavy REE patterns, all features typical of relatively primitive arcs with little or no continental crust involvement. These similarities have led to several geometric models: post-Middle Jurassic duplication by strike-slip faulting, and oroclinal or synformal folding. However, they are all inconsistent with either palaeomagnetic or faunal data, and the presence of a Triassic overstep sequence, which indicates amalgamation *c.* 50 ma before emplacement of the youngest oceanic rocks of the Cache Creek terrane. An alternative model is proposed: oblique eastward subduction of the Cache Creek accretionary prism and fore-arc producing high-pressure metamorphism, followed by extrusion into the arc and exhumation by the Middle Jurassic. This model implies that these high-pressure rocks, rather than marking an oceanic suture between disparate arc terranes, support a para-autochthonous origin.

The Canadian Cordillera is composed of various tectono-stratigraphic terranes that were accreted to the western margin of Laurentia (North America) before the early Cenozoic (Colpron *et al.* 2006, 2007). The geology of the northern Cordillera has recently been synthesized by Colpron *et al.* (2006, 2007), who divided it into five first-order tectonic entities (Fig. 1): (1) ancestral North America (Laurentia), which includes the western craton margin and overlying miogeocline and fringing para-autochthonous terranes; (2) allochthonous pericratonic terranes (Intermontane terranes, which includes the Yukon-Tanana terrane, presumed basement to the Quesnel and Stikine arc terranes addressed in this paper) that are separated from Laurentia by the discontinuous, Upper Devonian-Permian, Slide Mountain back-arc terrane; (3) Insular and Northern Alaska terranes, which evolved in the Arctic during the Palaeozoic; (4) oceanic and accretionary complexes (including the oceanic Cache Creek terrane that presently lies between Quesnellia and Stikinia, two major terranes of the Canadian Cordillera); and (5) Late accreted arc and accretionary terranes that form the western and southern fringe of the Cordillera.

There is currently considerable debate about the palaeogeography of the northern Cordillera between 100–55 Ma (e.g. Cowan *et al.* 1997; Colpron *et al.* 2006). Some palaeomagnetic data suggest that a large region of the Cordilleran hinterland (Baja, British Columbia) made a $2100 \pm 700$ km southward excursion from its present latitude relative to Laurentia during the Late Cretaceous, returning northward during the Palaeocene (Enkin 2006). On the other hand, other such data suggest that the Yukon-Tanana terrane has been para-autochthonous since 215 Ma (Late Triassic: Norian) and that, in the interval 100–55 Ma, the Intermontane terranes rotated $35 \pm 14°$ clockwise as they moved $915 \pm 775$ km northwards relative to Laurentia, followed by a further $16 \pm 6°$ clockwise rotation in the Cenozoic (Symons *et al.* 2005). However, about half of the northward relative motion may be accounted for by the 430 km dextral displacement along the Tintina Fault (Gabrielse *et al.* 2006), with an additional 130–150 km on the Fraser-Pinchi Fault system (Coleman & Parrish 1991; Enkin *et al.* 2006). The latter data are consistent with other Upper Triassic palaeomagnetic data that indicate little or no relative motion between

*From:* MURPHY, J. B., KEPPIE, J. D. & HYNES, A. J. (eds) *Ancient Orogens and Modern Analogues.*
Geological Society, London, Special Publications, **327**, 71–87.
DOI: 10.1144/SP327.5   0305-8719/09/$15.00 © The Geological Society of London 2009.

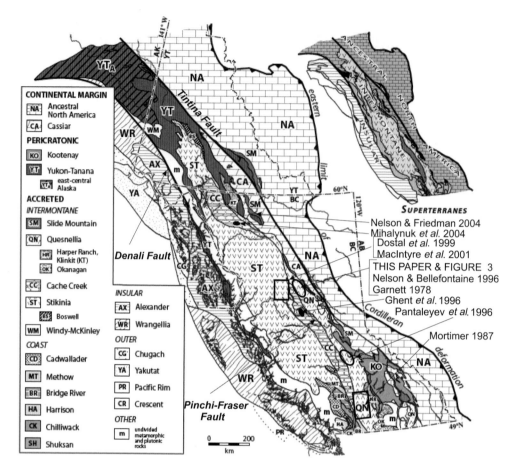

**Fig. 1.** Simplified terrane map of the Canadian Cordillera (after Colpron et al. 2007), showing the location of previous geochemical studies and Figure 3. Inset shows terrane grouping and tectonic realms.

the Intermontane terranes and Laurentia (Irving & Wynne 1992). These data are also consistent with the recognition of a Triassic overstep sequence that ties the Intermontane terranes to North America. Thus, provenance of Late–Middle Devonian detrital zircons in Lower Triassic (Smithian) rocks of the Selwyn basin (overlying the Palaeozoic miogeocline) can only be found in the Yukon-Tanana terrane that underlies the Quesnel and Stikine terranes (Beranek & Mortensen 2006). These conclusions are inconsistent with the suggestion that Stikinia, Quesnellia and Cache Creek are exotic, accreting to one another in western Panthalassa at 230 Ma, and then traveling c. 10 000 km across Panthalassa to accrete with western North America by c. 150 Ma (e.g. Johnston & Borel 2007, and references therein).

In this paper we present geochemical data for Middle–Upper Triassic volcanic arc rocks of the Takla Group in an inboard terrane (Quesnellia)

and compare these with similar data from Takla Group rocks located in a more outboard terrane (Stikinia). All of these rocks were originally mapped as one unit, the Takla Group, based on lithological and age similarities (Church 1975; Monger 1977; Monger & Church 1977), even though they are separated by the Mississippian–Lower Jurassic high-pressure (HP) rocks of the Cache Creek terrane and a major dextral fault (Pinchi-Fraser Fault: Fig. 1). These observations have led to a wide variety of tectonic models, including the development of one arc that was subsequently either offset by strike-slip motions (e.g. Wernicke & Klepacki 1988; Beck 1991, 1992; Irving et al. 1996), bent into an orocline (Nelson & Mihalynuk 1993; Mihalynuk et al. 1994; Nelson et al. 2006), or synformal folding of the Cache Creek klippe that was thrust over the arc during the Middle Jurassic (Samson et al. 1991; Gehrels et al. 1991).

The Takla Group volumetrically forms significant portions of both Quesnellia and Stikinia (Fig. 1). Correlative rocks of the Takla Group to the south include the Nicola Group in Quesnellia (Mortimer 1987) and to the north the Stuhini Group of Stikinia (Mortimer 1986). Whereas igneous geochemical data are available at several localities in Quesnellia (Fig. 1: Mortimer 1987; Pantaleyev *et al.* 1996; Nelson & Bellefontaine 1996), geochemical data is only available in central Stikinia (Dostal *et al.* 1999; MacIntyre *et al.* 2001). Thus, the purpose of this paper is to present more extensive geochemical data on the volcanic rocks of the Quesnel Takla Group from the Germansen Landing-Manson Creek and Aiken Lake area in north–central British Columbia and compare them with Stikine Takla volcanic rocks from the McConnel Creek map area, just to the west (Fig. 1).

Geochemical data presented in this paper indicate that the Takla Group volcanic rocks in both Quesnellia and Stikinia were erupted in an almost identical volcanic arc setting and represent one arc terrane rather than two. Geochemical variations indicate a west-facing arc above an east-dipping subduction zone. Accretion of the Intermontane terranes to western Laurentia by late Lower Triassic, combined with palaeomagnetic and faunal province data suggesting little or no lateral offset relative to Laurentia, pose difficult problems for several models. We propose an alternative model in which the Cache Creek rocks were subducted and extruded into the arc during the Upper Triassic–Lower Jurassic.

# Geological setting

## Quesnellia and Stikinia

Quesnellia is separated from autochthonous-para-autochthonous Laurentian margin rocks by the discontinuous, ophiolitic Slide Mountain terrane (Fig. 1). Furthermore, Quesnellia and Stikinia are separated by a belt of HP ophiolitic Cache Creek rocks that cuts diagonally across them in southern British Columbia (Fig. 1). Thus, although Quesnellia and Stikinia are typically *c.* 300 km wide, the width of Quesnellia tapers out in northern British Columbia as the width of Stikinia tapers to zero kilometres in southern British Columbia. The upper Palaeozoic–Lower Mesozoic geological record of Quesnellia and Stikinia consists of (Fig. 2): (a) Mississippian–Permian, volcano-sedimentary, arc-related and pericratonic rocks (Stikine assemblage, Asitka Group, Harper Ranch Group, Lay Range Assemblage: Hart 1997; Colpron *et al.* 2006; Roots *et al.* 2006; Beatty *et al.* 2006); and (b) Middle–Upper Triassic Takla, Stuhini, and Nicola groups unconformably overlain by Lower–Middle Jurassic Hazelton Group (Mortimer 1987; Monger *et al.*

1991; Ferri & Melville 1994; Pantaleyev *et al.* 1996; MacIntyre *et al.* 2001; MacIntyre 2006; Beatty *et al.* 2006; Breitsprecher *et al.* 2007). Both sequences consist of arc-related, mafic-intermediate pyroclastic rocks, massive flows and epiclastic rocks, and lie on or laterally interfinger with argillites, limestones and minor volcanic-derived epiclastic rocks containing fauna with Laurentian affinities (Carter *et al.* 1992; Stanley & Senowbari-Daryn 1999). The Permo-Carboniferous rocks of Quesnellia have been correlated with similar rocks (Finlayson and Klinkit units) of the Yukon–Tanana terrane (Simard *et al.* 2003), which unconformably overlie Neoproterozoic-Lower Devonian distal Laurentian margin rocks of the Snowcap assemblage (Colpron *et al.* 2006, 2007). Similarly, the Stikine assemblage is inferred to overlie the Yukon–Tanana terrane (Monger & Struik 2006).

## Slide Mountain terrane

Uppermost Devonian–Permian rocks of the Slide Mountain terrane (Fig. 2; ultramafic rocks, oceanic basalts, radiolarian cherts, argillites and carbonates: Sylvester Allochthon, Nina Creek Group, Slide Mountain Group, Fennell Group: Ferri 1997; Nelson *et al.* 2006; Colpron *et al.* 2006, 2007) reflect cold water conditions in the Permian (Orchard 2006). These rocks are inferred to have formed in a back-arc basin behind the Permo-Carboniferous arc in Quesnellia-Stikinia-Yukon-Tanana that collapsed in Late Permian–Early Triassic times. Overthrusting of Stikinia over the Slide Mountain terrane is recorded by clasts of Upper Permian Stikine arc rocks and eclogites in Upper Permian conglomerates deposited on the Slide Mountain units (Murphy *et al.* 2006). A lower Norian sandy limestone, which is part of a sequence lying unconformably upon Slide Mountain, contains Carboniferous detrital zircons inferred to have been derived from the Intermontane terranes (Beranek & Mortensen 2006).

## Cache Creek terrane

In southern British Columbia, the Cache Creek terrane cuts diagonally across Stikinia-Quesnellia (Fig. 1). It is composed of a series of thrust sheets cut by strike-slip faults that are made up of several assemblages consisting of Pennsylvanian–Early Jurassic (Pliensbachian) slates, siltstones, greywackes, cherts and carbonates (Fig. 2) (Struik *et al.* 2001) containing Tethyan fauna (Cordey *et al.* 1987; Stanley 1994; Orchard *et al.* 2001). These are accompanied by Permian and Upper Triassic oceanic island basalts (OIB) with some transitional to N–type mid-oceanic ridge basalts (N–MORB), Upper Permian primitive arc lavas

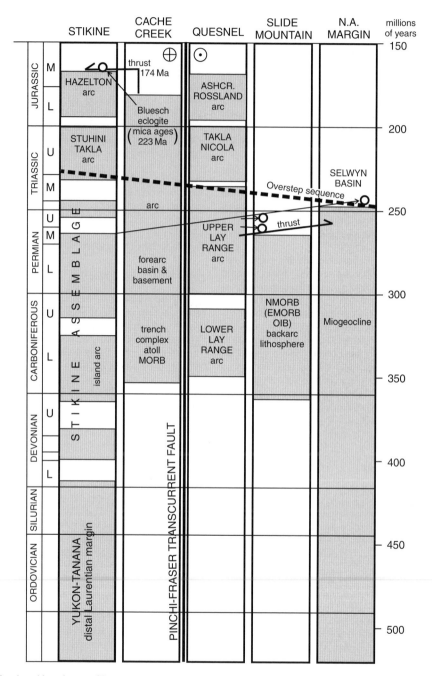

**Fig. 2.** Stratigraphic columns of Intermontane terranes and North American para-autochthonous and autochthonous margin (sources given in text). O = pebbles; ⊕ and ⊙ indicate transcurrent motion on the Pinchi-Fraser fault, away and towards reader, respectively. Heavy dashed line indicates the base of the Triassic overstep sequence.

(Sitlika assemblage) and undated, supra-subduction zone, ultramafic rocks (Schiarizza & Massey 2000; Tardy *et al.* 2001). Some of these rocks underwent HP blueschist and eclogite facies metamorphism (200–300 °C at 8–10 kb = 28–35 km, and 450–565 °C at 12–13 kb = 42–46 km, respectively: Ghent *et al.* 1993, 1996) with white micas yielding Upper Triassic ages (221 ± 2–224 ± 2 Ma: Ghent *et al.* 1996). Slightly older white mica ages have been recorded in the Bridge River blueschists that occur along the southern margin of Stikinia (Archibald *et al.* 1991). Similar rocks occur in northern British Columbia (English & Johnston 2005), where unroofing of some of the HP rocks in the earliest Jurassic is tightly constrained. Here fossiliferous Pliensbachian–Toarcian cherts (*c.* 196–176 Ma) underwent blueschist metamorphism (at 200 °C and 5 kb = 17.5 km depth) that yielded white mica ages of 174 ± 2 Ma, which was the source for HP eclogite and rare blueschist detritus that was deposited in the adjacent Whitehorse Trough before *c.* 171 Ma in front of westward vergent, Lower-Middle Jurassic thrusts (Mihalynuk *et al.* 2004; English & Johnston 2005). The eastern margin of the Cache Creek terrane is generally defined by steeply dipping faults with *c.* 140 km Tertiary dextral displacements, which are generally inferred to have been superimposed on easterly vergent thrusts (Price & Monger 2003). Restoration of 140 km dextral offset places the Bridge River terrane along-strike of the Cache Creek terrane.

All of these terranes, Stikinia, Cache Creek, Quesnellia, Slide Mountain and para-autochthonous–autochthonous terranes, are unconformably overlain by an upper Lower–Upper Triassic overstep sequence that provides a depositional link between them (Fig. 2) (Beranek & Mortensen 2006).

## Local Geology

In north–central British Columbia, Upper Triassic volcanic and sedimentary rocks in both Quesnellia and Stikinia terranes have been assigned to the Takla Group and are juxtaposed along the dextral Pinchi Fault, which passes southwards into the Cache Creek terrane (Fig. 1).

### Takla Group in Quesnellia

The Takla Group in Quesnellia (QTG) in the study area (Fig. 3; Germansen Landing-Manson Creek and Aiken Lake areas) was described by Ferri & Melville (1994), who estimated the total thickness of the group to be at least 5 km. The QTG overlies and is in fault contact with the late Palaeozoic (Mississippian to Permian) Lay Range Assemblage (Ferri 1997), which contains internally imbricated fault slices composed mainly of phyllite, chlorite

schist and metavolcanic rocks (Figs 2 & 3). Like the Stikine Takla Group, the QTG is separated from the overlying Lower to Middle Jurassic volcano-sedimentary sequences by a regional unconformity (Monger *et al.* 1991; Lang *et al.* 1995; Nelson & Bellefontaine 1996). Ferri & Melville (1994) divided the QTG into two units, the Slate Creek Succession and the Plughat Mountain Succession. The lower Slate Creek Succession is composed mainly of fine-grained clastic sedimentary rocks containing conodont-bearing limestones that yielded a Middle (Ladinian) to Late (Carnian) Triassic age. The upper part of the unit interfingers with the contemporaneous and dominant Plughat Mountain Succession, which consists of mafic volcanic and volcaniclastic rocks with minor amount of limestone and shale. Comparisons with similar volcanic rocks to the south (Bailey 1988; Panteleyev & Hancock 1989) led Ferri & Melville (1994) to conclude that the Plughat Mountain Succession is of Norian age (Late Triassic). This age was later confirmed by fossil collections (Ferri *et al.* 2001*a*, *b*) and also by palaeontological data for QTG (Stanley & Nelson 1996; Stanley & Senowari-Daryan 1999) collected south of the study area.

### Takla Group in Stikinia

The Takla Group in Stikinia (STG) is situated on the western side of the Pinchi Fault. It unconformably overlies the Lower Permian Asitka Group (Fig. 3), which is made up of cherts, limestones and felsic volcanic rocks, and is in fault contact with the Cache Creek terrane (Fig. 1). Like QTG, STG is composed of basaltic lava flows, volcaniclastic rocks and argillites deposited in subaerial and submarine environments (Dostal *et al.* 1999). Fossil collections from the STG indicate an Upper Triassic age in the late Carnian to early Norian stages (Monger & Church 1977; Monger *et al.* 1991). Monger & Church (1977) divided the STG into three formations. The lowermost unit (Dewar Formation) contains mainly tuffaceous siltstones and argillites. The overlying Savage Mountain Formation, up to 3000 m thick, is made up of pillowed and massive basalt, breccia and pyroclastic rocks. The uppermost unit of the STG is the Moosevale Formation, which is comprised of up to 1800 m thick of marine and nonmarine, mafic to intermediate, volcaniclastic rocks with a few lava flows. Dostal *et al.* (1999) documented that the lava flows from the Savage Mountain and Moosevale formations have overlapping compositions.

## Petrography

Volcaniclastic rocks volumetrically dominate the Pughat Mountain Succession, and most interbedded

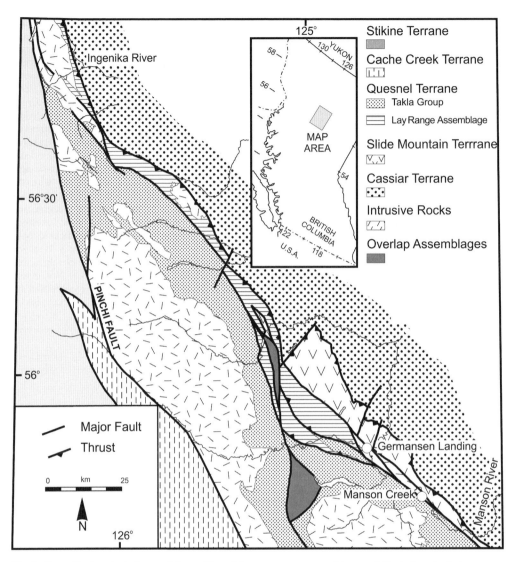

**Fig. 3.** Generalized geological map of Quesnellia and the westernmost part of Ancestral North America between Manson Creek and the Ingenika River (modified after Ferri & Melville 1994 and Ferri *et al.* 2001*a*, *b*). The Pinchi Fault marks the terrane boundary of Stikinia (west) and Quesnellia (east). QTG. samples are from the area between Germansen Landing and 126°W. STG samples (see Dostal *et al.* 1999) are from the area of Stikinia west of the Pinchi Fault. The inset is a map of British Columbia showing the location of the map area.

lava flows of the QTG are distinctly porphyritic basaltic rocks with clinopyroxene and plagioclase phenocrysts. The phenocrysts are commonly 2–5 mm in size, but may reach 1 cm in length. In addition, amphibole and rare olivine are also present in some rocks. The groundmass is composed of clinopyroxene, plagioclase, altered glass and minor opaques, usually replaced by secondary minerals. Accessory opaque minerals (Fe–Ti oxides) are present both as microphenocrysts and in the

groundmass. Some samples exhibit a pilotaxitic texture characterized by subparallel orientation of the plagioclase microcrystals (Ferri & Melville 1994).

Plagioclase and clinopyroxene phenocrysts are typically euhedral to subhedral and the latter often display a glomeroporphyric texture. Pyroxene is augite with compositional range of $Wo_{41-46}$ $En_{40-46} Fs_{10-15}$ with $Al_2O_3$ ranging from 1.4– 4.3 wt.% and with low $TiO_2$ (<1.0 wt.%). The composition of QTG clinopyroxene is comparable to that from the STG basaltic rocks (Dostal *et al.* 1999). Relatively rare, light-brown amphiboles occur as elongate euhedral to subhedral phenocrysts (2–4 mm in length) in plagioclase-phyric lavas with minor clinopyroxene (<10%). Clasts in the volcaniclastic rocks are mostly >1 cm in size, and have the same petrographic and mineralogical characteristics as the lava flows.

All the lava flows were affected by very low-grade metamorphism: prehnite, pumpellyite and calcite are the most common index minerals in QTG rocks. Prehnite and pumpellyite are abundant in both the groundmass and amygdules, whereas calcite frequently fills veinlets and amygdules. Other secondary minerals include chlorite, which is abundant in the groundmass, and sericite (or saussuritization) replacing plagioclase.

The petrographic characteristics of the mafic rocks of the STG are similar to those of QTG. They both contain distinct coarse, platy plagioclase and (or) clinopyroxene phenocrysts, although the STG rocks from the type area lack amphibole. The STG rocks were metamorphosed to zeolite and prehnite–pumpelleyite facies (Monger 1977; Dostal *et al.* 1999).

## Analytical methods and alteration

Samples were selected from a suite of specimens collected during the detailed mapping of the Germansen Landing–Manson Creek and Aiken Lake areas (Fig. 3) by the British Columbia Geological Survey, in particular by Ferri & Melville (1994) and Ferri *et al.* (2001*a*, *b*). In addition, for comparative purposes, we have augmented data from our previous study (Dostal *et al.* 1999) by additional analyses of some samples of the Stikine Takla Group. The STG samples are from the McConnel Creek map sheet just west of the QTG study area (Figs 1 & 3).

The major and partial trace element (Cr, Ni, Sc, V, Rb, Ba and Sr) analyses of the whole rock samples were done by X-ray fluorescence spectrometer (Table 1). The additional trace element analyses for rare earth elements (REE), Th, Hf, Zr, Nb and Y were performed by inductively coupled plasma-mass spectrometry at the Memorial University of Newfoundland (Table 1). The method and quality of the data were described by Longerich *et al.* (1990).

Eight whole-rock QTG basalts were selected for Nd and Sr isotopic analyses (Table 2). Samarium and Nd abundances and Nd and Sr isotope ratios were determined by isotope dilution mass spectrometry in the laboratory of the Department of Earth Sciences at the Memorial University of

Newfoundland. The concentrations of Rb and Sr were obtained by X-ray fluorescence. A description of the analytical technique was given by Kerr *et al.* (1995). Measured $^{143}Nd/^{144}Nd$ values were normalized to a natural $^{146}Nd/^{144}Nd$ ratio of 0.7219. Replicate analyses of La Jolla standard yielded $^{143}Nd/^{144}Nd = 0.511849 \pm 9$. Replicate runs for the NBS 987 standard gave $^{87}Sr/^{86}Sr = 0.710250 \pm 11$. Initial Nd and Sr isotopic ratios and $\varepsilon_{Nd}$ values were calculated assuming an age of 220 Ma for the Takla Group (Monger & Church 1977).

Like for the STG rocks (Dostal *et al.* 1999), low-grade metamorphism of the QTG did not significantly modify the whole-rock chemical composition. Evaluation of petrogenesis and tectonic setting of the mafic rocks is based mainly on trace elements (e.g. high-field-strength elements [HFSE] and REE), which are considered to be relatively 'immobile' in hydrothermal fluids (Winchester & Floyd 1976).

## Geochemistry

The QTG volcanic rocks in the the Germansen Landing–Manson Creek and Aiken Lake areas are mainly basaltic rocks with $SiO_2$ <56 wt.% (LOI-free), although most fall within a narrow $SiO_2$ range of 48–53 wt.% (Figs 4 & 5). Their Mg# values (Mg/Mg + $Fe^T$) are variable, ranging between 0.75 and 0.35. The low contents of transition elements, such as Ni and Cr in many samples, relative to primitive mantle melts (BVSP 1981), suggest that the rocks underwent extensive fractionation. All the samples have >400 ppm Ni, a lower limit for melts in equilibrium with mantle lherzolite. Ni and Cr show positive correlations with Mg#, indicating the early crystallization of olivine and clinopyroxene. The QTG rocks display a tholeiitic trend on the $SiO_2$ versus $FeO^T/MgO$ plot (Fig. 5). Their low abundances of $TiO_2$ (0.6–1 wt.%), HFSE and other incompatible trace elements (Table 1) are also typical of island arc tholeiitic suites (Pearce & Peate 1995).

Using trace element abundances, the QTG volcanic rocks can be separated into two groups, which both occur throughout the studied area. The first group displays nearly horizontal chondrite-normalized REE patterns (Fig. 6) with $(La/Yb)_n$ *c.* 1–2 and $(La/Sm)_n$ *c.* 0.75–1.25, accompanied by low La (<6 ppm), Zr (<60 ppm) and $SiO_2$ (<50.5 wt.%). The rocks of the second group have light REE-enriched patterns (Fig. 6), with $(La/Yb)_n$ *c.* 3–5 and $(La/Sm)_n$ *c.* 1.5–2.5. These rocks also have higher abundances of La (7–17 ppm), Zr (60–130 ppm) and $SiO_2$ (>51.5 wt.%), and higher Zr/Y ratios. However,

**Table 1.** *Chemical composition of Quesnel Takla Group volcanic rocks*

| Sample | (2–6) | (5–4) | (19–14) | (15–7) | (23–18) | (11–10) | (14–3) | (38–1) | (38–5) | (38–12) | (1–2) | (10–9) | (7–9) | (23–7) |
|---|---|---|---|---|---|---|---|---|---|---|---|---|---|---|
| Group | 1 | 1 | 1 | 1 | 1 | 1 | 2 | 2 | 2 | 2 | 2 | 2 | 2 | 2 |
| $SiO_2$ (%) | 44.71 | 48.96 | 48.87 | 46.78 | 48.37 | 46.84 | 50.2 | 49.69 | 50.57 | 48.97 | 52.03 | 50.82 | 49.9 | 50.62 |
| $TiO_2$ | 0.82 | 0.77 | 0.84 | 0.56 | 0.87 | 0.89 | 0.92 | 0.7 | 0.82 | 0.89 | 0.74 | 0.63 | 0.91 | 0.64 |
| $Al_2O_3$ | 14.03 | 13.92 | 14.53 | 12.24 | 16.54 | 14.08 | 13.44 | 16.82 | 14.55 | 12.92 | 16.76 | 11.38 | 15.16 | 16.33 |
| $Fe_2O_3^T$ | 8.76 | 11.29 | 10.53 | 9.01 | 9.85 | 11.1 | 9.32 | 9.01 | 10.06 | 9.56 | 8.85 | 10.94 | 10.8 | 8.83 |
| MnO | 0.18 | 0.16 | 0.19 | 0.18 | 0.16 | 0.18 | 0.16 | 0.16 | 0.17 | 0.14 | 0.14 | 0.21 | 0.19 | 0.21 |
| MgO | 7.37 | 6.41 | 7.58 | 13.76 | 6.72 | 8.89 | 5.13 | 2.85 | 7.23 | 3.26 | 3.53 | 8.57 | 6.12 | 3.86 |
| CaO | 12.81 | 11.25 | 10.55 | 9.68 | 8.95 | 12 | 5.98 | 4.89 | 7.96 | 9.31 | 7.16 | 11.72 | 7.55 | 6.33 |
| $Na_2O$ | 2.82 | 2.47 | 3.05 | 0.94 | 3.36 | 2.43 | 3.16 | 4.14 | 2.55 | 2.23 | 2.45 | 1.03 | 3.49 | 3.32 |
| $K_2O$ | 0.43 | 1.97 | 0.87 | 1.34 | 1.56 | 0.48 | 2.81 | 4.7 | 2.53 | 2.1 | 1.54 | 2.74 | 1.97 | 4.76 |
| $P_2O_5$ | 0.15 | 0.24 | 0.19 | 0.19 | 0.17 | 0.15 | 0.31 | 0.75 | 0.32 | 0.27 | 0.33 | 0.36 | 0.58 | 0.71 |
| LOI | 7.06 | 1.69 | 2.22 | 5.08 | 2.67 | 2.68 | 7.7 | 5.34 | 2.31 | 9.45 | 5.82 | 1.28 | 3.02 | 3.96 |
| Total | 99.14 | 99.13 | 99.42 | 99.76 | 99.22 | 99.72 | 99.13 | 99.05 | 99.07 | 99.1 | 99.35 | 99.68 | 99.69 | 99.57 |
| Mg# | 0.62 | 0.53 | 0.59 | 0.75 | 0.57 | 0.61 | 0.52 | 0.39 | 0.59 | 0.4 | 0.44 | 0.46 | 0.39 | 0.36 |
| Cr (ppm) | 260 | 168 | 229 | 801 | 90 | 324 | 130 | 49 | 222 | 98 | 39 | 380 | 120 | 62 |
| Ni | 104 | 48 | 83 | 141 | 78 | 97 | 61 | 28 | 86 | 28 | 31 | 89 | 42 | 27 |
| Sc | 41 | 39 | 40 | 29 | 41 | 63 | 30 | 14 | 32 | 26 | 22 | 36 | 32 | 12 |
| V | 274 | 329 | 317 | 199 | 302 | 338 | 281 | 308 | 271 | 240 | 141 | 260 | 282 | 217 |
| Rb | 11 | 73 | 20 | 21 | 14 | 10 | 51 | 158 | 58 | 43 | 18 | 41 | 54 | 90 |
| Ba | 188 | 520 | 392 | 410 | 796 | 213 | 937 | 794 | 643 | 671 | 664 | 1063 | 338 | 755 |
| Sr | 240 | 482 | 284 | 652 | 338 | 305 | 374 | 843 | 565 | 318 | 1338 | 581 | 399 | 879 |
| Nb | 3.42 | 1.26 | 1.75 | 0.89 | 1.2 | 1.43 | 7.6 | 2.83 | 6.76 | 8 | 7.33 | 7 | 6.5 | 1.9 |
| Hf | 1.64 | 1.19 | 1.04 | 1.05 | 1.41 | 1.23 | 2.32 | 1.8 | 2.56 | 2.64 | 2.87 | 0.87 | 1.85 | 1.34 |
| Zr | 52 | 44 | 44 | 49 | 51 | 50 | 93 | 56 | 90 | 86 | 121 | 59 | 110 | 56 |
| Y | 16 | 19 | 19 | 17 | 20 | 20 | 20 | 16 | 21 | 19 | 26 | 12 | 14 | 9 |
| Th | 0.78 | 0.93 | 0.62 | 0.54 | 1.34 | 0.88 | 2.49 | 1.64 | 2.54 | 1.95 | 4.78 | 1.34 | 2.32 | 1.22 |
| La | 4.96 | 5.5 | 3.6 | 2.54 | 3.7 | 4.19 | 13.07 | 11.47 | 13.32 | 10.94 | 15.56 | 6.51 | 14.15 | 9.09 |
| Ce | 11.68 | 12.6 | 9.55 | 6.64 | 9.26 | 10.4 | 26.48 | 21.84 | 26.57 | 24.68 | 32.42 | 14.14 | 28.44 | 17.68 |
| Pr | 1.71 | 1.85 | 1.49 | 1.01 | 1.4 | 1.57 | 3.23 | 2.79 | 3.26 | 3.19 | 3.96 | 1.96 | 3.48 | 2.25 |
| Nd | 8.49 | 9.14 | 7.9 | 5.48 | 7.43 | 7.49 | 13.87 | 12.27 | 14.12 | 14.16 | 15.29 | 9.01 | 14.52 | 9.7 |
| Sm | 2.44 | 2.75 | 2.4 | 1.83 | 2.29 | 2.28 | 3.39 | 3.01 | 3.43 | 3.44 | 3.62 | 2.45 | 3.45 | 2.3 |
| Eu | 0.91 | 0.9 | 0.79 | 0.59 | 0.74 | 0.83 | 1.06 | 1.02 | 1.08 | 1.08 | 1.1 | 0.77 | 1.08 | 0.8 |
| Gd | 3.31 | 3.87 | 3.3 | 2.18 | 3.06 | 2.91 | 4.4 | 3.48 | 4.23 | 4.08 | 3.9 | 3.16 | 4 | 2.81 |
| Tb | 0.49 | 0.51 | 0.5 | 0.34 | 0.47 | 0.44 | 0.6 | 0.49 | 0.57 | 0.55 | 0.56 | 0.4 | 0.56 | 0.38 |
| Dy | 3.41 | 3.26 | 3.15 | 2.49 | 3.25 | 2.87 | 3.49 | 2.98 | 3.52 | 3.32 | 3.83 | 2.41 | 3.61 | 2.32 |
| Ho | 0.69 | 0.68 | 0.64 | 0.46 | 0.63 | 0.56 | 0.75 | 0.57 | 0.74 | 0.69 | 0.78 | 0.51 | 0.69 | 0.47 |
| Er | 2.05 | 1.95 | 1.89 | 1.4 | 1.92 | 1.67 | 2.06 | 1.72 | 2.13 | 1.94 | 2.32 | 1.35 | 2.02 | 1.39 |
| Tm | 0.3 | 0.29 | 0.25 | 0.2 | 0.27 | 0.23 | 0.32 | 0.25 | 0.34 | 0.27 | 0.32 | 0.21 | 0.27 | 0.19 |
| Yb | 1.98 | 1.85 | 1.7 | 1.32 | 1.78 | 1.55 | 2.09 | 1.58 | 2.14 | 1.73 | 2.17 | 1.29 | 1.83 | 1.29 |
| Lu | 0.28 | 0.25 | 0.24 | 0.19 | 0.26 | 0.21 | 0.33 | 0.24 | 0.31 | 0.25 | 0.34 | 0.18 | 0.27 | 0.19 |

*Notes:* $Fe_2O_3^T$ = total Fe as $Fe_2O_3$; Mg# = Mg/(Mg + $Fe^T$); $Fe^T$ = total Fe.

**Table 2.** *Nd and Sr isotopic composition of basaltic rocks of the Quesnel Takla Group*

| Sample No. | Group | Nd(ppm) | Sm(ppm) | $^{143}Nd/^{144}Ndm$ | $^{147}Sm/^{144}Nd$ | $\varepsilon_{Nd}$ | $^{143}Nd/^{144}Ndi$ | Rb(ppm) | Sr(ppm) | $^{87}Sr/^{86}Srm$ | $^{87}Sr/^{86}Sri$ |
|---|---|---|---|---|---|---|---|---|---|---|---|
| 19–14 | 1 | 8.08 | 2.42 | 0.512978 (10) | 0.1812 | 7.07 | 0.512717 | 20 | 284 | 0.704290 (12) | 0.703653 |
| 23–18 | 1 | 7.45 | 2.31 | 0.512951 (12) | 0.1479 | 7.48 | 0.512738 | 14 | 338 | 0.704191 (10) | 0.703816 |
| 11–10 | 1 | 7.52 | 2.32 | 0.512942 (11) | 0.1478 | 7.31 | 0.512729 | 10 | 305 | 0.703952 (11) | 0.703655 |
| 10–9 | 2 | 8.99 | 2.38 | 0.512898 (11) | 0.1604 | 6.09 | 0.512667 | 41 | 581 | 0.704152 (15) | 0.703513 |
| 23–7 | 2 | 10.91 | 2.66 | 0.512964 (14) | 0.1473 | 7.76 | 0.512752 | 90 | 879 | 0.704618 (10) | 0.703691 |
| 7–9 | 2 | 14.27 | 3.41 | 0.512946 (12) | 0.1445 | 7.47 | 0.512738 | 54 | 399 | 0.704923 (9) | 0.703698 |
| 14–3 | 2 | 13.91 | 3.4 | 0.512931 (10) | 0.1462 | 7.14 | 0.51272 | 51 | 374 | 0.705824 (10) | 0.703622 |
| 38–5 | 2 | 14.2 | 3.44 | 0.512978 (17) | 0.1465 | 8.04 | 0.512767 | 58 | 565 | 0.704367 (10) | 0.703438 |

*Note:* $^{143}Nd/^{144}Ndm$ and $^{87}Sr/^{86}Srm$ are measured Nd and Sr isotopic ratios, respectively; $^{143}Nd/^{144}Ndi$ and $^{87}Sr/^{86}Sri$ are initial Nd and Sr isotopic ratios, respectively, and $\varepsilon_{Nd}$ is the fractional difference between the $^{143}Nd/^{144}Nd$ of rock and the bulk earth at the time of crystallization. $\varepsilon_{Nd}$, $^{143}Nd/^{144}Ndi$ and $^{87}Sr/^{86}Sri$ assume an age of 220 Ma for QTG rocks. Values of $\varepsilon_{Nd}$ were calculated using modern $^{143}Nd/^{144}NdCHUR = 0.512638$ and $^{147}Sm/^{144}NdCHUR = 0.1967$. Concentrations of Nd and Sm were determined by isotope dilution, and those of Rb and Sr by X-ray fluorescence. Precision of concentrations of Nd and Sm is $\pm 1\%$.

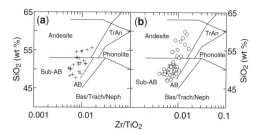

**Fig. 4.** $Zr/TiO_2$ versus $SiO_2$ (wt.%) diagrams of Winchester & Floyd (1977) for the volcanic rocks from Quesnel (**a**) and Stikine (**b**) Takla Group. Sub-AB, subalkaline basalt; AB, alkaline basalt; TrAn, trachyandesite; Bas, basanite; Trach, trachyte; Neph, nephelinite.

both groups have comparable abundances of major and many trace elements including $TiO_2$ as well as Th/La, Ba/La, Tb/Lu (Fig. 7) and Nb/La ratios. Similar values of the Th/La ratios, a sensitive indicator of upper crustal contamination, in the two groups suggest that the differences are not due to crustal contamination. Both subgroups display negative Nb anomalies on mantle-normalized trace element diagrams (Fig. 8). These patterns are characteristic of rocks emplaced in a relatively primitive arc environment (Pearce & Peate 1995). Their slight enrichment in large-ion-lithophile elements (LILE), accompanied by depletion of Nb relative to light REE, require derivation from a mantle source modified through subduction processes that led to an enrichment of LILE. The nearly horizontal heavy REE patterns of the QTG rocks

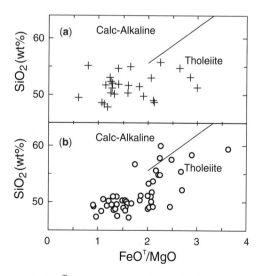

**Fig. 5.** $FeO^T/MgO$ versus $SiO_2$ (wt.%) for the volcanic rocks of the Quesnel (**a**) and Stikine (**b**) Takla Group. The line separating the calc-alkaline and tholeiitic fields is after Miyashiro (1974).

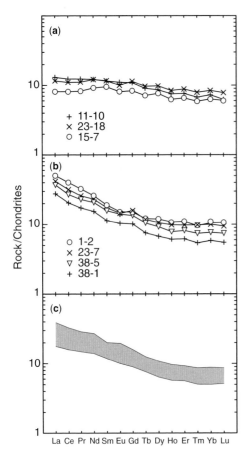

**Fig. 6.** Chondrite-normalized REE patterns of the basaltic rocks of the Takla Group. Normalizing values after Sun & McDonough (1989). (**a**) QTG basalt group 1; (**b**) QTG basalt group 2; (**c**) range of STG basalts (after Dostal *et al.* 1999).

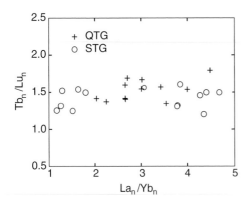

**Fig. 7.** Variations of $La_n/Yb_n$ versus $Tb_n/Lu_n$ (n-chondrite-normalized) in the QTG (+) and STG (o) basalts. Composition of STG rocks is from Dostal *et al.* (1999).

argue against a significant role for garnet in their genesis and suggest mantle melting in the stability field of spinel peridotite at a depth of less than 60 km (White *et al.* 1992). The differences between the two QTG groups cannot be readily explained by either fractional crystallization, variable degrees of partial melting or crustal contamination, but are consistent with a heterogeneous mantle source.

The QTG basalts have positive initial $\varepsilon_{Nd}$ values (+6 to +8) and the low initial Sr isotopic ratios (0.7034–0.7038; Fig. 9; Table 2) with the two QTG groups having overlapping isotopic characteristics. Differences in trace element distributions between the two basalt groups with similar isotope ratios suggest that the subduction-related trace element enrichment of a mantle source occurred

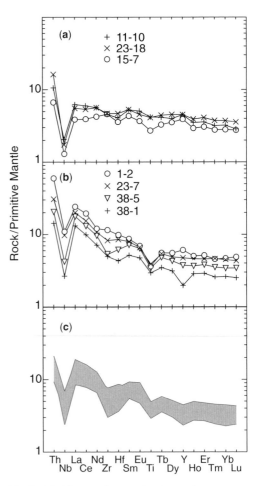

**Fig. 8.** Primitive mantle-normalized trace element patters of the Takla Group rocks. Normalizing values after Sun & McDonough (1989). (**a**) QTG basalt group 1; (**b**) QTG basalt group 2; (**c**) range of STG basalts (after Dostal *et al.* 1999).

shortly prior to melting so that the metasomatic process had not produced isotopic enrichment. Similar high positive $\varepsilon_{Nd}$ values are present in some modern island arc tholeiites, such as those from the Mariana arc in the Pacific, where they are usually attributed to subduction of juvenile material (basalts with little sediment) with no involvement of old continental crust (Pearce & Peate 1995).

## Comparison of STG and QTG

The Takla Group in Stikinia and Quesnellia were originally mapped as a single unit (Lord 1948; Monger 1977). These two units show many stratigraphic similarities (Monger 1977; Monger & Church 1977; Ferri & Melville 1994; Dostal *et al.* 1999). They are of the same age and both are composed of mafic volcanic flows, volcanogenic sandstones and argillites. Both Takla units show an evolution from the lower marine assemblages to upper subaerial sections, and contain similar fossil associations (Orchard 2006). Volcanic assemblages are characterized by clinopyroxene- and feldspar-phyric basalts with clinopyroxene typically forming large phenocrysts.

In addition to petrographic and stratigraphic commonalities, QTG and STG rocks have similar chemical (major, trace and REE elements and Nd and Sr isotopes) compositions (Figs 4–8). Both suites are dominated by basaltic rocks with geochemical characteristics of island arc tholeiites. Emplacement of both QTG and STG in an island arc environment is supported by the dominance of basaltic over andesitic rocks and the association of volcanic rocks with shallow marine to subaerial sediments (Dostal *et al.* 1999). The compositional similarities of QTG and STG basaltic rocks with some modern island arc tholeiites suggest a similar origin; that is, by melting of an upper mantle that was modified by subduction processes. Both the STG and QTG have overlapping Nd and Sr isotopic data, high positive $\varepsilon_{Nd}$, low initial Sr ratios (Fig. 9), and their flat hREE patterns (Fig. 6), all of which are consistent with an island arc setting and derivation from a similar spinel peridotite mantle source. Dostal *et al.* (1999) inferred that the Takla Group can be correlated with the other Upper Triassic volcanic units in both Stikinia and Quesnellia, in particular with the Stuhini Group of Stikinia and the Nicola Group in Quesnellia (Fig. 2) (Mortimer 1986, 1987). All these rocks are of similar age and exhibit lithological and chemical similarities.

## Discussion and Conclusions

The geochemical data presented here support correlations between the Takla Group in Quesnellia and Stikinia and suggest that they represent a single island arc. This arc must have been in a peri-Laurentian location given the presence of the Triassic overstep sequence extending from the Yukon-Tanana terrane to western margin of Laurentia (Beranek & Mortensen 2007).

### Arc polarity

In southernmost British Columbia, the Quesnellian segment has a width of *c.* 270 km and is bounded on its western side by HP rocks of the Cache Creek and Bridge River terranes: the Stikinian segment is absent (Fig. 1). Based upon variations in the geochemistry of volcanic rocks in the Nicola Group (low-K calc-alkaline in the west changing to shoshonitic in the east), Mortimer (1987) inferred an east-dipping Benioff zone beneath the Quesnellian segment. This is consistent with the inference that the Bridge River rocks represent an accretionary prism (Mortimer 1986; Garver & Scott 1995) separated from the Nicola arc by the Cadwallader–Methow terrane (fore-arc volcanic and sedimentary rocks containing some HP rocks: Gabrielse & Yorath 1992; Garver & Scott 1995; Monger & Sturik 2006).

Similar shoshonitic, Upper Triassic rocks may be traced northwards along the eastern side of the Quesnellian arc segment throughout British Columbia (Garnett 1978; Pantaleyev *et al.* 1996; Nelson & Bellefontaine 1996; Nelson & Friedman 2004). In central British Columbia, the Upper Triassic volcanic rocks of the Takla Group in the Stikinian segment are less alkaline than those to the east (Nelson & Bellefontaine 1996; Dostal *et al.* 1999; MacIntyre *et al.* 2001; Nelson & Friedman 2004). Furthermore, the present width of the arc is similar to that in the southernmost Quesnellian segment. However, here the Quesnellian and Stikinian arc, segments are generally separated by the Cache Creek HP rocks (Fig. 1).

### Evaluation of existing models

Existing hypotheses proposed to explain the present distribution of Upper Triassic arc rocks separated by the HP accretionary complex of the Cache Creek terrane (strike-slip duplication and oroclinal bending) both require relative latitudinal displacements between the two arc segments. The strike-slip duplication implies >1500 km northward motion of the Stikinian segment relative to the Quesnellian segment (Wernicke & Klepacki 1988; Beck 1992). On the other hand, the oroclinal model requires that the Stikinian segment lay *c.* 1500 km north of the Quesnellian segment in the Upper Triassic and was bent into an orocline during the Lower-Middle Jurassic with final closure by 187–174 Ma

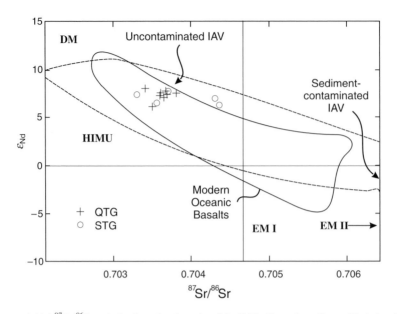

**Fig. 9.** $\varepsilon_{Nd}$ versus initial $^{87}Sr/^{86}Sr$ ratio for the volcanic rocks of the Takla Group from Quesnellia (+) and Stikinia (o: after Dostal *et al.* 1999) showing also for comparison the composition of modern oceanic basalts (Zindler & Hart 1986), island-arc volcanic rocks [IAV; dashed line; sediment-contaminated (e.g., Sunda arc) and uncontaminated (e.g., Aleutians, the Marianas) after Samson *et al.* 1989]. Position of mantle end-member components (DM, HIMU, EM I and EM II) is after Zindler & Hart (1986).

(Mihalynuk *et al.* 1994). However, palaeomagnetic data from volcanic rocks of the Upper Triassic Stuhini and Lower Jurassic Hazelton groups of the Stikinian segment show zero to minimal southerly offsets relative to Laurentia: c. 300–700 km and 0–500 km, respectively (Irving & Wynne 1992). Similarly, Upper Triassic Nicola Group rocks yielded a palaeolatitude consistent with its present location relative to Laurentia (Irving & Wynne 1992). This implies negligible offset between the Stikinian and Quesnellian segments since Triassic times, a result that is consistent with recent Lower Jurassic palaeomagnetic data from either side of the Fraser–Pinchi Fault that limits dextral movements to <3° (Enkin *et al.* 2006). Thus, these data are inconsistent with both the strike-slip and oroclinal models.

*Extrusion model*

Although an intra-arc rift interpretation for the Cache Creek terrane may satisfy the palaeomagnetic constraints, it cannot account for the prolonged subduction indicated by the range of HP rocks (>50 Ma) and Carboniferous–Jurassic Asian fauna that implies a wide ocean, possibly as wide as 3500 km using a reasonable translation rate of 7 cm a$^{-1}$ (Gordon 1998). The geometrical enigma can be resolved if one considers that the Cache

Creek HP rocks were extruded *into* the upper plate rather than thrust over it (Fig. 10). Most extrusion models for the emplacement of HP rocks involve subduction of one continental block beneath another followed by collision, slab breakoff and extrusion back up the subduction channel (e.g. Ernst *et al.* 1997; Burov *et al.* 2001; Hynes 2002; Liou *et al.* 2004): in this case the HP rocks mark an oceanic suture between separate terranes. However, HP rocks extruded into the upper plate above an active subduction zone have recently been discovered (Keppie *et al.* 2008), and have been recorded in numerical models (Stöckhert & Gerya 2005; Gerya & Stöckhert 2006). In these cases, the HP rocks neither separate disparate terranes nor mark an oceanic suture; instead, they occur between similar units. Protoliths of such HP rocks may include a wide range of different environments, including beheaded seamounts (subducting plate) and forearc-arc material removed from the upper plate by subduction erosion. Such protoliths have been recorded in the Cache Creek HP rocks (Tardy *et al.* 2001; Struik *et al.* 2001; English & Johnston 2005). We propose that the Cache Creek HP rocks represent trench complex and fore-arc material removed from the leading edge of the upper plate by subduction erosion that was taken down to depths up to 50 km before being extruded into the arc in the upper plate (Fig. 10). Whereas

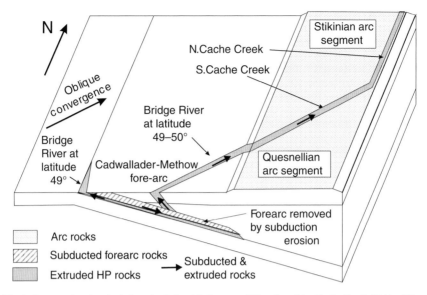

**Fig. 10.** Block diagram showing the inferred relationship between HP rocks of the Cache Creek/Bridge River terrane that have been extruded into the volcanic arc rocks of the Stikinia-Quesnellia during the Late Triassic and Early-Middle Jurassic. The HP rocks are inferred to represent parts of the subducting slab (e.g. decapitated seamounts and associated reef sediments), and parts of the fore-arc that were subducted to depths of 17–46 km before being extruded into the upper plate: the cross-cutting relationship between the HP rocks and the Stikine-Quesnel arc is inferred to have resulted from oblique convergence between the Palaeo-Pacific and North American plates.

subduction erosion is likely in a convergent arc, extrusion into the upper plate may be aided by extension of the arc. Such extrusion may be been further facilitated by trench rollback.

Preservation of blueschists requires relatively rapid exhumation, and the numerical models of Gerya & Stöckhert (2006) show that a complete cycle from the surface through 50–100 km depth and back to near the surface can occur in c. 20 Ma. Geochronological data for the southern Cache Creek rocks suggest some of the HP metamorphism took place at c. 223 Ma, although some rocks (Sowchea succession: greenschist-lower amphibolite facies) containing Pliensbachian–Toarcian radiolarian (c. 183 ± 2 Ma) were also involved in the process (Struik et al. 2001), indicating post-183 Ma metamorphism. In the northern Cache Creek, rocks with similar-aged radiolarian underwent blueschist metamorphism dated at 174 ± 2 Ma before being unroofed and eroded into rocks at least 171 Ma old (Mihalynuk et al. 2004). Thus the HP metamorphism appears to have lasted at least c. 50 Ma. These data are consistent with the continuity of synchronous arc magmatism (with a minor hiatus in the latest Triassic-earliest Jurassic in Stikinia and Quesnellia) throughout the Late Triassic and Early Jurassic. Given that arc magmatism generally requires a depth of >90 km for the Benioff zone (Tatsumi & Eggins 1995), the maximum depth of c. 17 km

reached by some of the Cache Creek rocks within the Stikine-Quesnel arc is puzzling because one might expect extruded HP rocks to have also come from >90 km. However, an analogous situation occurs along the southern Caribbean plate boundary in northeastern Venezuela, where oblique subduction has been related to extension of the arc and oblique–lateral extrusion of HP rocks (Lallemant & Guth 1990). Such a model is consistent with the obliquity of the Cache Creek HP rocks relative to the Stikine–Quesnel arc (Fig. 1), and may indicate dextral plate convergence during the Late Triassic and Early Jurassic. Such a model is also consistent with the south to north transition from a normal sequence of HP trench complex-forearc–arc in southernmost British Columbia and northernmost USA through an abnormal forearc–HP–forearc between 49–50°N to arc–HP–arc in the rest of British Columbia. Thus, a model involving oblique extrusion of the HP Cache Creek rocks into the Stikine–Quesnel arc appears to overcome most of the apparently contradictory data and explains the enigmatic relationships reported for the Intermontane Belt of British Columbia. A test for this model involves the nature of the faults bounding the HP rocks: thrust below and listric normal above. Although a thrust lower boundary is evident in existing mapping, the upper boundary is obscured by subsequent, Cenozoic vertical faulting; however, further examination of this boundary may reveal the nature of the Middle Jurassic boundary.

HP rocks in ancient orogens are generally inferred to mark an oceanic suture between terranes that originated on opposite sides of an ocean basin. However, the proposal that the HP Cache Creek rocks were extruded into the overriding plate implies Quesnellia and Stikinia are one terrane, not two. It also leads to simpler palaeogeographic reconstructions, in which subduction occurred along an active ocean–continent boundary, not involving collision. Such a model has implications for other ancient orogens, for it is now imperative to distinguish between HP rocks extruded up along the subduction channel (Ernst *et al.* 1997) from those extruded into the overriding plate (Keppie *et al.* 2008; this paper). Criteria include distinct versus similar 'terranes', and a precollisional versus collisional origin for HP belts.

The study was supported by Natural Sciences and Engineering Research Council of Canada grant to Dostal and by the British Columbia Geological Survey Branch. We are grateful to R. Corney for cheerful technical assistance and to V. Owen, B. Murphy, B. Cousens and R. Strachan for their constructive reviews.

# References

ARCHIBALD, D. A., SCHIARIZZA, P. & GARVER, J. I. 1991. $^{40}Ar/^{39}Ar$ Evidence for the Age of Igneous and Metamorphic Events in the Bridge River and Shulaps Complexes, Southwestern British Columbia. British Columbia Ministry of Energy, Mines and Petroleum Resources, Geological Fieldwork 1990, **1991-1**, 75–83.

BAILEY, D. G. 1988. Geology of the Central Quesnel Belt, Hydraulic, South-central British Columbia (93A/12). British Columbia Ministry of Energy, Mines and Petroleum Resources, Geological Fieldwork 1987, **1988-1**, 147–153.

BEATTY, T. W., ORCHARD, M. J. & MUSTARD, P. S. 2006. Geology and tectonic history of the Quesnel terrane in the area of Kamloops, British Columbia. In: COLPRON, M. & NELSON, J. L. (eds) Paleozoic Evolution and Metallogeny of Pericratonic Terranes at the Ancient Pacific Margin of North America, Canadian and Alaskan Cordillera. Geological Association of Canada Special Paper, **45**, 483–504.

BECK, M. E., JR. 1991. Case for northward transport of Baja and coastal southern California: paleomagnetic data, analysis, and alternatives. Geology, **19**, 506–509.

BECK, M. E., JR. 1992. Tectonic significance of paleomagnetic results for the western conterminous United States. In: BURCHFIEL, B. C., LIPMAN, P. W. & ZOBACK, M. L. (eds). The Cordilleran Orogen: Conterminous US. (The Geology of North America). Geological Society of America, **G-3**, 683–697.

BERANEK, L. P. & MORTENSEN, J. K. 2006. Triassic overlap assemblages in the northern Cordillera: Preliminary results from the type section of the Jones Lake Formation, Yukon and Northwest territories (NTS 105I/13). In: EMOND, D. S., BRADSHAW, G. D., LEWIS, L. L. & WESTON, L. H. (eds) Yukon Geological Survey, 79–91.

BERANEK, L. P. & MORTENSEN, J. K. 2007. Investigating a Triassic overlap assemblage in Yukon: On-going field studies and preliminary detrital zircon age data. In: EMOND, D. S., LEWIS, L. L. & WESTON, L. H. (eds) Yukon Exploration and Geology 2006. Yukon Geological Survey, 83–92.

BREITSPRECHER, K., SCOATES, J. S., ANDERSON, R. G. & WEIS, D. 2007. Geochemistry of Mesozoic intrusions, Quesnel and Stikine Terranes (NTS 082; 092; 093), South-central British Columbia: Preliminary Characterization of Sampled Suites. British Columbia Ministry of Energy, Mines and Petroleum Resources, Geological Fieldwork 2006, **2007-1**, 247–257.

BUROV, E., JOLIVET, L., LE POURHIET, L. & POLIAKOV, A. 2001. A thermomechanical model of exhumation of high pressure (HP) and ultra-high pressure (UHP) metamorphic rocks in Alpine-type collision belts. Tectonophysics, **342**, 113–136.

BVSP 1981. Basaltic Volcanism on the Terrestrial Planets. Pergamon Press, New York, 1–1286.

CARTER, E. S., ORCHARD, M. J., ROSS, C. A., ROSS, J. R. P., SMITH, P. L. & TIPPER, H. W. 1992. Paleontological signatures of terranes. In: GABRIELSE, H. & YORATH, C. J. (eds) Geology of the Cordilleran Orogen in Canada. (The Geology of North America). Geological Society of America. **G-2**, 28–38.

CHURCH, B. N. 1975. Geology of the Sustut Area. British Columbia Department of Mines and Petroleum Resources, Geology, Exploration and Mining in British Columbia, **1974**, 305–309.

COLEMAN, M. E. & PARRISH, R. R. 1991. Eocene dextral strike-slip and extensional faulting in the Bridge River Terrane, southwest British Columbia. Tectonics, **10**, 1222–1238.

COLPRON, M., NELSON, J. L. & MURPHY, D. C. 2006. A tectonostratigraphic framework for the pericratonic terranes of the northern Canadian Cordillera. In: COLPRON, M. & NELSON, J. L. (eds) Paleozoic Evolution and Metallogeny of Pericratonic Terranes at the Ancient Pacific Margin of North America, Canadian and Alaskan Cordillera. Geological Association of Canada Special Paper, **45**, 1–23.

COLPRON, M., NELSON, J. L. & MURPHY, D. C. 2007. Northern Cordilleran terranes and their interactions through time. GSA Today, **17**, 4–10.

CORDEY, F., MORTIMER, N., DEWEVER, P. & MONGER, J. W. H. 1987. Significance of Jurassic radiolarians from the Cache Creek terrane, British Columbia. Geology, **15**, 1151–1154.

COWAN, D. S., BRANDON, M. T. & GARVER, J. I. 1997. Geological tests of hypotheses for large coastwise displacements; a critique illustrated by the Baja British Columbia controversy. American Journal of Science, **297**, 117–173.

DOSTAL, J., GALE, V. & CHURCH, B. N. 1999. Upper Triassic Takla Group volcanic rocks, Stikinine Terrane, north-central British Columbia: geochemistry, petrogenesis and tectonic implications. Canadian Journal of Earth Sciences, **36**, 1483–1494.

ENGLISH, J. M. & JOHNSTON, S. T. 2005. Collisional orogenesis in the northern Canadian Cordillera:

implications for Cordilleran crustal structure, ophiolite emplacement, continental growth, and the terrane hypothesis. *Earth and Planetary Science Letters*, **232**, 333–344.

ENKIN, R. J. 2006. Paleomagnetism and the case for Baja British Columbia. *In*: HAGGART, J. W., MONGER, J. W. H. & ENKIN, R. J. (eds) *Paleogeography of the North American Cordillera: Evidence For and Against Large-scale Displacements*. Geological Association of Canada Special Paper, **46**, 233–253.

ENKIN, R. J., MAHONEY, J. B. & BAKER, J. 2006. Paleomagnetic signature of the Silverquick/Powell Creek succession, south-central British Columbia: reaffirmation of Late Cretaceous large-scale terrane translation. *In*: HAGGART, J. W., MONGER, J. W. H. & ENKIN, R. J. (eds) *Paleogeography of the North American Cordillera: Evidence For and Against Large-scale Displacements*. Geological Association of Canada Special Paper, **46**, 201–220.

ERNST, W. G., MARAYAMA, S. & WALLIS, S. 1997. Buoyancy-driven, rapid exhumation of ultrahigh-pressure metamorphosed continental crust. *Proceedings of the National Academy of Science USA*, **94**, 9532–9537.

FERRI, F. 1997. Nina Creek Group and Lay Range Assemblage, north-central British Columbia: remnants of late Paleozoic oceanic and arc terranes. *Canadian Journal of Earth Sciences*, **3**, 854–874.

FERRI, F. & MELVILLE, D. M. 1994. *Bedrock Geology of the Germansen Landing – Manson Creek Area, British Columbia*. British Columbia Ministry of Energy, Mines and Petroleum Resources, Bulletin, **91**, 147.

FERRI, F., DUDKA, S., REES, C., MELDRUM, D. & WILLSON, M. 2001*a*. *Geology of the Uslika Lake Area, North-Central British Columbia*. British Columbia Ministry of Energy, Mines and Petroleum Resources, Geoscience Map 2001–4.

FERRI, F., DUDKA, S., REES, C. & MELDRUM, D. 2001*b*. *Geology of the Aiken Lake area, North-Central British Columbia*. British Columbia Ministry of Energy, Mines and Petroleum Resources, Geoscience Map 2001–10.

GABRIELSE, H. & YORATH, C. J. (eds) 1992. *Geology of the Cordilleran Orogen in Canada*. Geology of Canada No. 4, Geological Survey of Canada, Ottawa, 1–844.

GABRIELSE, H., MURPHY, D. C. & MORTENSEN, J. K. 2006. Cretaceous and Cenozoic dextral orogen-parallel displacements, magmatism and paleogeography, north-central Canadian Cordillera. *In*: HAGGART, J. W., MONGER, J. W. H. & ENKIN, R. J. (eds) *Paleogeography of the North American Cordillera: Evidence For and Against Large-scale Displacements*. Geological Association of Canada Special Paper, **46**, 255–276.

GARNETT, J. A. 1978. *Geology and Mineral Occurrences of the Southern Hogem Batholith*. British Columbia Ministry of Mines and Petroleum Resources, Bulletin, **70**, 1–75.

GARVER, J. I. & SCOTT, T. J. 1995. Trace elements in shale as indicators of crustal provenance and terrane accretion in the southern Cordillera. *Geological Society of America Bulletin*, **107**, 440–453.

GEHRELS, G. E., MCCLELLAND, W. C., SAMSON, S. D. & PATCHETT, P. J. 1991. U–Pb geochronology of detrital zircons from a continental margin assemblage in the northern Coast Mountains, southeastern Alaska. *Canadian Journal of Earth Sciences*, **28**, 1285–1300.

GERYA, T. & STÖCKHERT, B. 2006. Two-dimensional numerical modeling of tectonic and metamorphic histories of active continental margins. *International Journal of Earth Sciences*, **95**, 250–274.

GHENT, E. D., STOUT, M. Z. & ERDMER, P. 1993. Pressure-temperature evolution of lawsonite-bearing eclogites, Pinchi Lake, British Columbia. *Journal of Metamorphic Geology*, **11**, 279–290.

GHENT, E. D., ERDMER, P., ARCHIBALD, A. & STOUT, M. 1996. Pressure-temperature and tectonic evolution of Triassic lawsonite-aragonit blueschists from Pinchi Lake, British Columbia. *Canadian Journal of Earth Sciences*, **33**, 800–810.

GORDON, R. G. 1998. The plate tectonic approximation: plate nonrigidity, diffuse plate boundaries, and global reconstructions. *Annual Reviews in Earth and Planetary Sciences*, **26**, 615–642.

HART, C. J. R. 1997. *A Transect Across Northern Stikinia: Geology of the Northern Whitehorse Map-area, Southern Yukon Territory (105D/13-16)*. Exploration and Geological Services Division, Yukon Region, Indian and Northern Affairs Canada, Bulletin, **8**, 1–122.

HYNES, A. 2002. Encouraging the extrusion of deep-crustal rocks in collisional zones. *Mineralogical Magazine*, **66**, 5–24.

IRVING, E. & WYNNE, P. J. 1992. Paleomagnetism: review and tectonic implications. *In*: GABRIELSE, H. & YORATH, C. J. (eds) *Geology of the Cordilleran Orogen in Canada. (The Geology of North America)*. Geological Society of America, **G-4**, 61–86.

IRVING, E., WYNNE, P. J., THORKELSON, D. J. & SCHIARIZZA, P. 1996. Large (1000 to 4000 km) northward movements of tectonic domains in the northern Cordillera, 83 to 45 Ma. *Journal of Geophysical Research*, **101**, 17 901–17 916.

JOHNSTON, S. T. & BOREL, G. D. 2007. The odyssey of the Cache Creek terrane, Canadian Cordillera: implications for accretionary orogens, tectonic setting of Panthalassa, the Pacific superswell, and break-up of Pangea. *Earth and Planetary Science Letters*, **253**, 415–428.

KEPPIE, J. D., DOSTAL, J., MURPHY, J. B. & NANCE, R. D. 2008. Synthesis and tectonic interpretation of the westernmost Paleozoic Variscan orogen in southern Mexico: from rifted Rheic margin to active Pacific margin. *Tectonophysics*, http://dx.doi.org/10.1016/j.tecto.2008.01.012.

KERR, A., JENNER, G. A. & FRYER, B. J. 1995. Sm-Nd isotopic geochemistry of Precambrian to Paleozoic granitoid suites and the deep-crustal structure of the southeast margin of the Newfoundland Appalachians. *Canadian Journal of Earth Sciences*, **32**, 224–245.

LALLEMANT, H. G. A. & GUTH, L. R. 1990. Role of extensional tectonics in exhumation of eclogites and blueschists in an oblique subduction setting: northeastern Venezuela. *Geology*, **18**, 950–953.

LANG, J. R., LUECK, B., MORTENSEN, J. K., RUSSEL, J. K., STANLEY, C. R. & THOMPSON, J. F. H. 1995. Triassic-Jurassic silica-undersaturated and silica-saturated alkalic intrusions in the Cordillera of British Columbia. *Geology*, **23**, 451–454.

LIOU, J. G., TSUJIMORI, T., ZHANG, R. Y., KATAYAMA, I. & MARUYAMA, S. 2004. Global UHP metamorphism and continental subduction/collision: the Himalayan model. *International Geology Review*, **46**, 1–17.

LONGERICH, H. P., JENNER, G. A., FRYER, B. J. & JACKSON, S. E. 1990. Inductively coupled plasma-mass spectrometric analysis of geological samples: a critical evaluation based on case studies. *Chemical Geology*, **83**, 105–118.

LORD, C. S. 1948. *McConnell Creek Map-area, Cassiar District, British Columbia*. Geological Survey of Canada, Memoir, **251**, 1–72.

MACINTYRE, D. G. 2006. *Geology and Mineral Deposits of the Skeena Arch, West-Central British Columbia: A Geoscience BC Digital Data Compilation Project*. British Columbia Ministry of Energy, Mines and Petroleum Resources, Geological Fieldwork 2005, **2006-1**, 303–312.

MACINTYRE, D. G., VILLENEUVE, M. E. & SCHIARIZZA, P. 2001. Timing and tectonic setting of Stikine Terrane magmatism, Babine-Takla lakes area, central British Columbia. *Canadian Journal of Earth Sciences*, **38**, 579–601.

MIHALYNUK, M. G., NELSON, J. & DIAKOW, L. J. 1994. Cache Creek terrane entrapment: Oroclinal paradox within the Canadian Cordillera. *Tectonics*, **13**, 575–595.

MIHALYNUK, M. G., ERDMER, P., GHENT, E. D., CORDEY, F., ARCHIBALD, D. A., FRIEDMAN, R. M. & JOHANNSON, G. G. 2004. *Coherent French Range Blueschist: Subduction to Exhumation in <2.5 m.y.?* Geological Society of America Bulletin, **116**, 910–922.

MIYASHIRO, A. 1974. Volcanic rock series in island arcs and active continental margins. *American Journal of Sciences*, **274**, 321–357.

MONGER, J. W. H. 1977. *The Triassic Takla Group in McConnell Creek Map Area, North-Central British Columbia*. Geological Survey of Canada, Paper, **76–29**, 145.

MONGER, J. W. H. & CHURCH, B. N. 1977. Revised stratigraphy of the Takla Group, north-central British Columbia. *Canadian Journal of Earth Sciences*, **14**, 318–326.

MONGER, J. W. H. & STRUIK, L. C. 2006. Chilliwack terrane: a slice of Stikinia? A tale of terrane transfer. In: COLRON, M. & NELSON, J. L. (eds) *Paleozoic Evolution and Metallogeny of Pericratonic Terranes at the Ancient Pacific Margin of North America, Canadian and Alaskan Cordillera*. Geological Association of Canada Special Paper, **45**, 351–368.

MONGER, J. W. H., WHEELER, J. O. *ET AL.* 1991. Cordilleran terranes. In: GABRIELSE, H. & YORATH, C. J. (eds) *Geology of the Cordilleran Orogen in Canada. (The Geology of North America)*. Geological Society of America, **G-2**, 281–327.

MORTIMER, N. 1986. Late Triassic arc-related potassic igneous rocks in the North American Cordillera. *Geology*, **14**, 1035–1038.

MORTIMER, N. 1987. The Nicola Group: Late Triassic and Early Jurassic subduction-related volcanism in British Columbia. *Canadian Journal of Earth Sciences*, **24**, 2521–2536.

MURPHY, D. C., MORTENSEN, J. K., PIERCEY, S. J., ORCHARD, M. J. & GEHRELS, G. E. 2006. Mid-Paleozoic to early Mesozoic tectonostratigraphic evolution of Yukon-Tanana and Slide Mountain terranes and affiliated overlap assemblages, Finlayson Lake massive sulphide district, southeastern Yukon. In: COLPRON, M. & NELSON, J. L. (eds) *Paleozoic Evolution and Metallogeny of Pericratonic Terranes at the Ancient Pacific Margin of North America, Canadian and Alaskan Cordillera*. Geological Association of Canada Special Paper, **45**, 75–105.

NELSON, J. L. & BELLEFONTAINE, K. A. 1996. The geology and mineral deposits of north-central Quesnellia; Tezzeron Lake to Discovery Creek, Central British Columbia. *British Columbia Geological Survey Branch Bulletin*, **99**, 112.

NELSON, J. L. & FRIEDMAN, R. 2004. Superimposed Quesnel (late Paleozoic-Jurassic and Yukon-Tanana (Devonian-Mississippian) arc assemblages, Cassiar Mountains, northern British Columbia: field, U–Pb, and igneous petrochemical evidence. *Canadian Journal of Earth Sciences*, **41**, 1201–1235.

NELSON, J. L. & MIHALYNUK, M. 1993. Cache Creek ocean: closure or enclosure. *Geology*, **21**, 173–176.

NELSON, J. L., COLPRON, M., PIERCEY, S. J., DUSEL-BACON, C., MURPHY, D. C. & ROOTS, C. F. 2006. Paleozoic tectonic evolution and metallogenic evolution of the pericratonic terranes in Yukon, northern British Columbia and eastern Alaska. In: COLPRON, M. & NELSON, J. L. (eds) *Paleozoic Evolution and Metallogeny of Pericratonic Terranes at the Ancient Pacific Margin of North America, Canadian and Alaskan Cordillera*. Geological Association of Canada Special Paper, **45**, 323–360.

ORCHARD, M. J. 2006. Late Paleozoic and Triassic conodont faunas of Yukon and northern British Columbia and implications for the evolution of the Yukon-Tanada terrane. In: COLPRON, M. & NELSON, J. L. (eds) *Paleozoic Evolution and Metallogeny of Pericratonic Terranes at the Ancient Pacific Margin of North America, Canadian and Alaskan Cordillera*. Geological Association of Canada Special Paper, **45**, 229–260.

ORCHARD, M. J., STRUIK, L. C., RUI, L., BAMBER, B., MAMET, B., SANO, H. & TAYLOR, H. 2001. Paleontological and biogeographical constraints on the Carboniferous to Jurassic Cache Creek terrane in central British Columbia. *Canadian Journal of Earth Sciences*, **38**, 551–578.

PANTELEYEV, A. & HANCOCK, K. D. 1989. *Quesnel Mineral Belt: Summary of the Geology of the Beaver Creek-Horselly Map Area*. British Columbia Geological Survey Branch, Geological Fieldwork 1988, **1989-1**, 159–166.

PANTELEYEV, A., BAILEY, D. G., BLOODGOOD, M. A. & HANCOCK, K. D. 1996. *Geology and Mineral Deposits of the Quesnel River-Horsefly Map Area, Central Quesnel Trough*. British Columbia Ministry of Employment and Investment, Bulletin, **97**, 156.

PEARCE, J. A. & PEATE, D. W. 1995. Tectonic implications of the composition of volcanic arc magmas. *Annual Review of Earth Sciences*, **23**, 251–285.

PRICE, R. & MONGER, J. 2003. *A transect of the southern Canadian Cordillera from Calgary to Vancouver.* 2003 Annual Meeting of GAC/MAC Calgary, fieldguide, 16.

ROOTS, C. F., NELSON, J. L., SIMARD, R.-L. & HARMS, T. A. 2006. Continental fragments, mid-Paleozoic arcs and overlapping late Paleozoic arc and Triassic sedimentary strata in the Yukon-Tanana terrane of northern British Columbia and southern Yukon. *In*: COLPRON, M. & NELSON, J. L. (eds) *Paleozoic Evolution and Metallogeny of Pericratonic Terranes at the Ancient Pacific Margin of North America, Canadian and Alaskan Cordillera.* Geological Association of Canada Special Paper, **45**, 153–177.

SAMSON, S. D., MCCLELLAND, W. C., PATCHETT, P. J., GEHRELS, G. E. & ANDERSON, R. G. 1989. Evidence from neodymium isotopes for mantle contributions to Phanerozoic crustal genesis in the Canadian Cordillera. *Nature*, **337**, 705–709.

SAMSON, S. D., PATCHETT, P. J., MCCLELLAND, W. C. & GEHRELS, G. E. 1991. Nd isotopic characterization of metamorphic rocks in the Coast Mountains, Alaskan and Canadian Cordillera: ancient crust bounded by juvenile terranes. *Tectonics*, **10**, 770–780.

SCHIARIZZA, P. & MASSEY, N. 2000. Volcanic facies and geochemistry of the Sitlika assemblage, central British Columbia: part of a primitive intra-oceanic arc in western Cache Creek terrane. *Geological Society of America Abstracts with Programs, Cordilleran Section*, **32**(16), A66.

SIMARD, R.-L., DOSTAL, J. & ROOTS, C. F. 2003. Development of late Paleozoic volcanic arcs in the Canadian Cordillera: an example from the Klinkit Group, northern British Columbia and southern Yukon. *Canadian Journal of Earth Sciences*, **40**, 907–924.

STANLEY, G. D. J. 1994. Late Paleozoic and early Mesozoic reef-building organizations and paleogeography: the Tethyan-North American connection. *Courier Forschungsinstitut Seckenberg*, **172**, 69–75.

STANLEY, G. D., JR. & NELSON, J. L. 1996. *New Investigations on Eaglenest Mountain, Northern Quesnel Terrane: An Upper Triassic Reef Facies in the Takla Group, Central British Columbia (93N/IE).* British Columbia, Ministry of Energy, Mines and Petroleum Resources, Geological Survey Branch, Geological Fieldwork 1995, **1996-1**, 127–135.

STANLEY, G. D., JR. & SENOWBARI-DARYN, B. 1999. Upper Triassic reef fauna from Quesnel terrane, Central British Columbia, Canada. *Journal of Paleontology*, **73**, 787–802.

STÖCKHERT, B. & GERYA, T. V. 2005. Pre-collisional high pressure metamorphism and nappe tectonics at active continental margins: a numerical simulation. *Terra Nova*, **17**, 102–110.

STRUIK, L. C., SCHIARIZZA, P. ET AL. 2001. Imbricate architecture of the upper Paleozoic to Jurassic oceanic Cache Creek Terrane, central British Columbia. *Canadian Journal of Earth Sciences*, **38**, 495–514.

SUN, S. S. & MCDONOUGH, W. F. 1989. Chemical and isotopic systematics of oceanic basalts: implications for mantle composition and processes. *In*: SAUNDERS, A. D. & NORRY, M. J. (eds) *Magmatism in the Ocean Basins.* Geological Society, London, Special Publications, **42**, 313–345.

SYMONS, D. T. A., HARRIS, M. J., MCCAUSLAND, P. J. A., BLACKBURN, W. H. & HART, C. J. R. 2005. Mesozoic-Cenozoic paleomagnetism of the Intermontane and Yukon-Tanana terranes, Canadian Cordillera. *Canadian Journal of Earth Sciences*, **42**, 1163–1185.

TARDY, M., LAPIERRE, H., STRUIK, L. C., BOSCH, D. & BRUNET, P. 2001. The influence of mantle plume in the genesis of the Cache Creek oceanic igneous rocks: implications for the geodynamic evolution of the inner accreted terranes of the Canadian Cordillera. *Canadian Journal of Earth Sciences*, **38**, 515–534.

TATSUMI, Y. & EGGINS, S. 1995. *Subduction Zone Magmatism.* Blackwell Publishing, Massachusetts, USA.

WERNICKE, B. & KLEPACKI, D. W. 1988. Escape hypothesis for the Stikine block. *Geology*, **16**, 461–464.

WHITE, R. S., MCKENZIE, D. & O'NIONS, R. K. 1992. Oceanic crustal thickness from seismic measurements and REE inversions. *Journal of Geophysical Research*, **97**, 19683–19715.

WINCHESTER, J. A. & FLOYD, P. A. 1977. Geochemical discrimination of different magma series and their differentiation products using immobile elements. *Chemical Geology*, **20**, 325–343.

ZINDLER, A. & HART, S. R. 1986. Chemical geodynamics. *Annual Review of Earth and Planetary Sciences*, **14**, 493–571.

# Plate tectonics of the Alpine realm

GÉRARD M. STAMPFLI* & CYRIL HOCHARD

*Institute of Geology and Palaeontology, University of Lausanne, Anthropole,
CH 1015 Lausanne, Switzerland*

*Corresponding author (e-mail: gerard.stampfli@unil.ch)*

**Abstract:** New field data on the East Mediterranean domain suggest that this oceanic basin belonged to the larger Neotethyan oceanic system that opened in Permian times. A Greater Apulia domain existed in Mesozoic times, including the autochthonous units of Greece and SW Turkey. It also included a united Adria and Apulia microplate since Early Jurassic times. This key information implies that a new post-Variscan continental fit for the western Tethyan area is necessary, where the relationships between the Adriatic, Apulian and Iberian plates are defined with greater confidence. To construct a reliable palinspastic model of the Alpine realm, plate tectonic constraints must be taken into consideration in order to assess the magnitude of lateral displacements. For most of the plates and their different terranes, differential transport on the scale of thousands of kilometres can be demonstrated. This plate tectonic framework allows a better geodynamic scenario for the formation of the Alpine chain to be proposed, where the western and eastern transects have experienced contrasting geological evolutions. The eastern Alps–Carpathians domain evolved from the north-directed roll-back of the Maliac–Meliata slab and translation of the Meliata suture and Austroalpine domain into the Alpine domain. In the western Alps, the changing African plate boundary in space and time defined the interaction between the Iberian–Briançonnais plate and the Austroalpine accretionary wedge.

## Introduction

To construct a plate tectonic scenario for the Alpine domain is not an easy task, as the wealth of information on these regions cannot be grasped, nor properly cited, in a single publication. Therefore here we try to improve on our former attempts at reconstructing one of the most complicated areas of the planet (Fig. 1), by applying plate tectonic concepts as much as possible, and by working on a large enough scale to integrate the Alps into the geodynamic framework of the western Tethyan area, and also integrating information from the Atlantic domain, such as magnetic anomalies (Fig. 2).

Our approach is totally 'non-fixist' – meaning that if two terranes are now juxtaposed they most likely were not so before, and we would like to dedicate this paper to H. Schardt who, more than a century ago (e.g. Schardt 1889, 1898, 1900, 1907), proposed that the Briançonnais domain of the western Prealps was an exotic terrane. Through his 1900 paper 'encore les régions exotiques' (again the exotic regions) one can see that his 'non-fixist' ideas were strongly rejected at that epoch, we expect present geologists will be more open-minded. Schardt's proposal was finally proven correct, and the Briançonnais terrane is a good example of margin duplication in an orogen (Frisch 1979; Stampfli 1993; Stampfli *et al.* 2002).

For a long time, the Tethys was considered as a large and single oceanic space, mostly of Mesozoic age, located between Gondwana and Eurasia. Already in the 1940s and 1950s, a distinction between a Palaeo- and a Neo-Tethys appeared (see references in Sengör 1985; Stampfli & Kozur 2006) and it was recognized that the latter comprised marine Permian and younger strata, whereas the former opened during the Early Palaeozoic. Stöcklin (1968, 1974), following extensive field work in Iran, gave a formal definition of these two large oceanic entities, the Neotethys becoming a Permian to Cretaceous peri-Gondwanan ocean (whose suture was between Iran and Arabia), whereas the Palaeotethys suture was located just north of Iran, thus between the Cimmerian terranes (sensu Sengör 1979) and Eurasia. In that sense, the Palaeotethys separates the Variscan domain from late Palaeozoic Gondwana-derived terranes.

Besides the two large Palaeotethyan and Neotethyan oceanic domains (one replacing the other during the Triassic) many oceanic back-arc type oceans opened just north of the Palaeotethys subduction zone. They are sometimes erroneously considered as Neotethyan because of their Triassic to Jurassic age, but most of these had no direct connection (either geographical or geological) with the peri-Gondwanan Neotethys ocean, and should therefore be called by their local names (e.g. Meliata, Küre, Maliac, Pindos, Huglu,

*From*: MURPHY, J. B., KEPPIE, J. D. & HYNES, A. J. (eds) *Ancient Orogens and Modern Analogues.*
Geological Society, London, Special Publications, **327**, 89–111.
DOI: 10.1144/SP327.6   0305-8719/09/$15.00 © The Geological Society of London 2009.

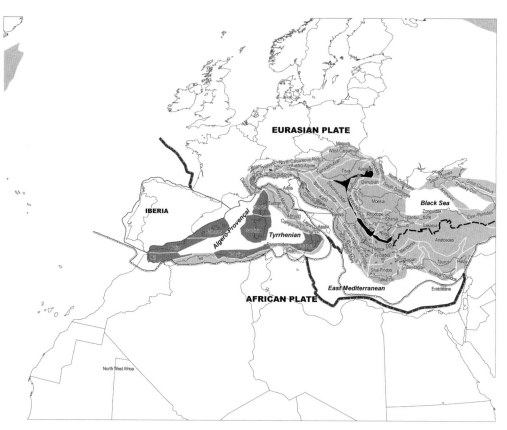

**Fig. 1.** Western Tethys terrane map. Line with triangle marks the limit between the African and Eurasian plates, grey areas are terranes implied in the Alpine orogenic events, darker grey terranes represent fragments of the former Iberian plate. Black area represents the Vardar suture zone, black dash line the Izmir Ankara suture zone. Thick lines with ticks represent the passive margin of respectively the East Mediterranean basin and the Gulf of Biscay.

Vardar) (Stampfli & Kozur 2006). During the break-up of Pangaea, another relatively long, if not wide, oceanic domain appeared, consisting of the Central Atlantic and its eastern extension in the Alpine–Carpathian domain. This Jurassic ocean was named 'Alpine Tethys' (Favre & Stampfli 1992), in order to mark the difference between this relatively northerly ocean (opening in the Variscan domain) and the peri-Gondwanan Neotethys. This Alpine Tethys is made of oceanic segments represented by the Ligurian, Piemont and Penninic/Vahic ophiolitic sequences (e.g. Lagabrielle *et al.* 1984; Liati *et al.* 2005).

The resulting picture of the Tethys realm in Jurassic time is, therefore, quite complex, and made of numerous small oceans and a large peri-Gondwanan Neotethys. Further complexity arose during the convergence stages, as many of these oceanic realms gave birth to new back-arc basins, especially around Turkey and Iran (Moix *et al.* 2008). These are, in most cases, at the origin of the many

ophiolitic belts found in the Tethyan realm, whereas older oceanic domains totally disappeared without leaving large remnants of their sea floors (Stampfli & Borel 2002).

## The East Mediterranean Neotethys connection

In the light of the large amount of new data provided by the CROP Atlas (Finetti 2005), the geodynamic evolution of the Adria and Apulia micro-continents has been recently redefined (Stampfli 2005). One of the main issues is the age of the East Mediterranean–Ionian sea basin, and the nature of the sea floor in this area. This, in turn, influences Mesozoic continental re-assembly models.

This problem was already reviewed in some detail in previous publications (Stampfli 2000; Stampfli *et al.* 2001*b*); depending on the authors, the East Mediterranean–Ionian sea basin is regarded

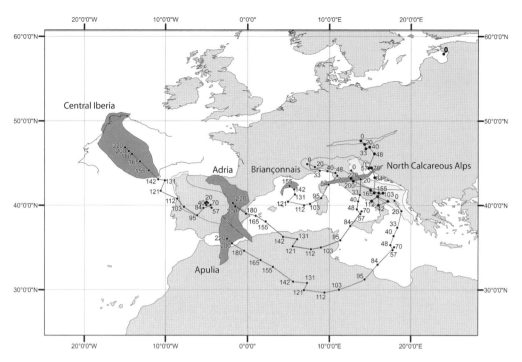

**Fig. 2.** Wander path of some terranes involved in the Alpine orogen. Central Iberia, Apulia, Adria, the western part of the Briançonnais and the North Calcareous Alps are shown in dark grey in their 220 Ma position, and in light grey in their present-day position. Their related velocities are shown in Table 1.

as already opening in the Late Palaeozoic (Vai 1994) or as late as the Cretaceous (e.g. Dercourt *et al.* 1985, 1993). Most people would regard this ocean as opening in the Late Triassic or Early Jurassic (e.g. Garfunkel & Derin 1984; Sengör *et al.* 1984; Robertson *et al.* 1996) and therefore possibly related to the Alpine Tethys–Central Atlantic opening.

A new interpretation (Stampfli *et al.* 2001*b*), showing that the East Mediterranean domain corresponded to a westward extension of the Permian Neotethys, is supported by a large volume of geological and geophysical data. We proposed a Middle to Late Permian onset of sea-floor spreading in the eastern Mediterranean basin, concomitant with the opening of the Neotethys eastward, and the northward drift of the Cimmerian continents since late Early Permian. This model also implies a late closure of the Palaeotethys (Middle to Late Triassic) on a Mediterranean transect (Stampfli *et al.* 2003; Stampfli & Kozur 2006).

Thus, the original position of Apulia with respect to Africa can be well determined by closing the Ionian sea by a *c.* 40° rotation of Apulia around a point located north of Tunisia (Finetti 2005). This represents the total rotation of Apulia from Late Carboniferous to Late Triassic, but a more

complicated scheme should be envisaged when considering that Apulia belonged for a while to the drifting Cimmerian continent, whose rotation point changed in time. Therefore, a slightly different placing and rotation of Apulia is shown in the following reconstructions.

## The Apulia–Adria problem and the Late Triassic fit

The late Triassic reconstruction shown in Figure 3 incorporates all the terranes/geological elements shown on the present-day map (Fig. 1). The main problem regarding this model is the position of Adria and Apulia. The present-day shape of these two terranes (present-day Italy) cannot be fitted between Iberia and Africa without some important deformation in classical fits and remains a major problem at the centre of the western Tethyan realm (e.g. Wortmann *et al.* 2001; Schettino & Scotese 2002).

The continuity between the active subduction zone under Greece and the outer Dinarides (Wortel & Spakman 1992; de Jonge *et al.* 1994) shows that there is a possible plate limit between Apulia s.l. and the autochthonous terrane of

**Table 1.** *Velocities and distances to rotation poles of peri-Alpine terranes (see the paths in Fig. 2). Ages are expressed in Ma, and velocities correspond to the centroids velocities and are expressed in cm/a*

| Initial age | Final age | Adria | | Apulia | | Central Iberia | | North Calcarous Alps | | West Briançonnais | |
|---|---|---|---|---|---|---|---|---|---|---|---|
| | | Velocity (cm/a) | Distance to pole (°) | Velocity (cm/a) | Distance to pole (°) | Velocity (cm/a) | Distance to pole (°) | Velocity (cm/a) | Distance to pole (°) | Velocity (cm/a) | Distance to pole (°) |
| 220 | 200 | 0.26 | 27.71 | 0.44 | 31.92 | 0.16 | 32.58 | 0.52 | 27.48 | 0.00 | 2.58 |
| 200 | 180 | 0.75 | 28.15 | 0.84 | 31.79 | 0.00 | 2.86 | 0.16 | 16.86 | 0.00 | 2.58 |
| 180 | 165 | 0.70 | 36.57 | 1.53 | 29.68 | 0.76 | 33.57 | 0.64 | 52.10 | 0.00 | 42.54 |
| 165 | 155 | 1.33 | 25.06 | 1.53 | 29.18 | 0.75 | 36.80 | 0.89 | 16.44 | 0.00 | 42.54 |
| 155 | 142 | 2.16 | 34.30 | 2.35 | 37.65 | 2.26 | 38.86 | 1.23 | 8.11 | 0.52 | 14.30 |
| 142 | 131 | 1.38 | 32.06 | 1.51 | 35.08 | 0.62 | 5.46 | 0.98 | 40.73 | 0.58 | 11.03 |
| 131 | 121 | 1.08 | 29.94 | 0.97 | 26.68 | 1.48 | 40.58 | 1.26 | 59.34 | 1.18 | 30.71 |
| 121 | 112 | 2.33 | 26.24 | 2.73 | 31.14 | 2.24 | 25.06 | 0.89 | 9.15 | 1.92 | 21.30 |
| 112 | 103 | 1.36 | 10.96 | 2.00 | 16.27 | 1.37 | 11.04 | 1.60 | 17.50 | 0.70 | 5.63 |
| 103 | 95 | 2.89 | 12.92 | 4.05 | 18.25 | 2.29 | 10.20 | 1.00 | 13.78 | 1.73 | 7.69 |
| 95 | 84 | 1.98 | 28.64 | 2.28 | 33.39 | 1.10 | 15.35 | 1.83 | 51.40 | 1.73 | 24.65 |
| 84 | 70 | 1.63 | 17.62 | 2.07 | 22.58 | 0.59 | 6.29 | 0.96 | 12.29 | 1.13 | 21.55 |
| 70 | 57 | 0.25 | 66.92 | 0.24 | 62.41 | 0.27 | 83.39 | 0.23 | 5.04 | 0.00 | 44.67 |
| 57 | 48 | 0.73 | 35.00 | 0.75 | 35.71 | 0.71 | 23.91 | 2.25 | 19.27 | 0.41 | 18.71 |
| 48 | 40 | 1.45 | 23.15 | 1.69 | 27.32 | 0.37 | 6.64 | 1.18 | 6.09 | 1.18 | 8.02 |
| 40 | 33 | 1.42 | 21.43 | 1.69 | 25.83 | 0.17 | 6.09 | 0.58 | 3.47 | 0.88 | 1.74 |
| 33 | 20 | 1.45 | 27.12 | 1.65 | 31.19 | 0.07 | 9.93 | 0.50 | 6.74 | 0.55 | 2.75 |
| 20 | 0 | 0.74 | 36.68 | 0.76 | 37.89 | 0.00 | 40.57 | 0.27 | 4.69 | 0.40 | 1.58 |

Greece. However, we regard this feature as recent and as having no bearing on the fact that a Greater Apulia plate existed in Mesozoic times, in which all the autochthonous portion of Greece was included (PIM, Tor), as well as the Bey–Daglari (Bdg) of SW Turkey, representing the northern margin of the East Mediterranean basin (Moix *et al.* 2008). The apparent present plate limit (between Ap and PIM, Fig. 1) comes from the fact that the Hellenic orogenic/accretionary wedge is oblique with respect to former palaeogeographic domains. It is therefore still colliding with Apulia on an Albanides transect while still subducting the East Mediterranean sea floor on a Greek transect.

The CROP seismic lines through the Adriatic domain (Finetti 2005) clearly show that there is no major tectonic accident cutting Italy into two units, at least since the Jurassic. Thus, Italy must have reached its present configuration between the Triassic and Middle Jurassic. Palaeomagnetic data show that the Apulian plate s.l. (Italy) suffered relatively little rotation in regard to Africa since the Triassic (e.g. Channell 1992, 1996; Muttoni *et al.* 2001).

Our basic hypothesis is that the Apulian part of Italy was definitely an African promontory (Argnani 2002) from Middle Triassic to Recent times, without much displacement with respect to Africa, and that the Adriatic and Apulian microplates were welded in an Eocimmerian collision phase during the Middle–Late Triassic, when both units became part of the African plate.

The need to cut Italy into two microplates comes from the fact that in a Triassic Pangaean reconstruction, there is very little room to insert the present form of Italy in its proper place. In order to solve this dilemma, we had to reconsider the fit of Iberia with Europe, as well as the size and position of the Alboran domain microplates. We made a much tighter fit of these elements with Europe, following similar previous proposals (Srivastava & Tapscott 1986; Srivastava *et al.* 1990). Still, there was not enough room to put the entire present length of Italy in its proper position. The conclusion was that Italy was shorter in Late Triassic times than it is now (or already in the Late Jurassic). A few hundred kilometres were gained through major phases of rifting affecting Greater Apulia since the Triassic and related to the break-up of Pangaea and the opening of the Alpine Tethys [Alpine Lombardian basin in Italy and Ionian rift system affecting the Hellenic domain (Stampfli 2005)].

The reason for widespread Jurassic rifting affecting Italy was that for the Pangaean break-up to succeed, the Atlantic rift system had to join an active plate limit located far to the east in the Neotethyan domain. Thus, all possible ways to break through the Alpine–Mediterranean lithosphere were tried and many resulted in aborted rifts. This pervasive Jurassic rifting finally gave birth, in the Alps, to passive margins, flanked by aborted rifts that became rim basins (e.g. Subbriançonnais, Helvetic-Dauphinois domains, Lombardian basin, Subbetic basin, Magura basin), and a narrow oceanic strip dominated by mantle denudation (Stampfli & Marchant 1997; Rampone & Piccardo 2000). This can be regarded as forced rifting through an already thinned lithosphere, and effectively all this rifting took place under water, with little isostatic/thermal rebound of the rift shoulders.

Thus, the final pre-collisional length and geometry of the Adriatic plate would have been established by Middle–Late Jurassic times only (Figs 4 & 5). Deformation of this plate interior during Alpine times also took place, but only at a small scale as shown by the seismic data, and despite the fact that most of its border was strongly involved in subduction and/or crustal shortening.

The larger Apulia domain had already gone through major rifting phases in the Permian and Triassic, even leading to the opening of small oceans between Greater Apulia and the Austroalpine (AA) domain. This finally led to the subduction of large amounts of continental crust material due to the negative buoyancy of former rift zones and attenuated margins, and regions on which large-scale ophiolite obduction had taken place. On a Hellinides transect this was calculated to be in the order of 900 km of subducted sub-upper-crust continental lithosphere (van Hinsbergen *et al.* 2005), and from our reconstructions a similar amount can be calculated through a Dinarides–Balkan transect (e.g. compare Figs 3 & 12).

## Construction of the models

The reconstructions shown in Figures 3 to 12 are based on a tight pre-Pangaea break-up Permian fit as explained above. From the Early Jurassic onward they are based on magnetic anomalies from the Central Atlantic. Plate tectonic concepts have been systematically applied to our palinspatic models of the western Tethys, moving away from pure continental drift models, not constrained by plate limits, to produce a model which is increasingly self-constrained. In this approach, first explained in Stampfli & Borel 2002, inter-dependent reconstructions are created from the past to the present. Except during collisions, plates are moved step by step, as single rigid entities. The only evolving elements are the plate boundaries, which are preserved and follow a consistent geodynamic evolution through time and an interconnected network through space. Hence, lithospheric plates are constructed by adding to, or removing oceanic

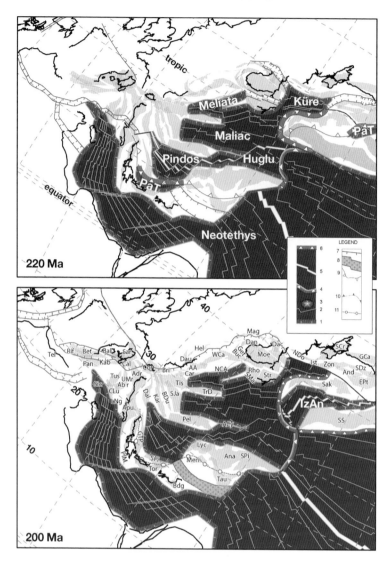

**Fig. 3.** 220–200 Ma. *Late Triassic.* As the Palaeotethys (PaT) subduction reached its final stage, slab roll-back along its northern margin accelerated and was marked by the opening of successive oceanic back-arc basins before the final closure of Palaeotethys in late Carnian times. Extension affected southwestern Eurasia and north Africa. A possible link with the north Atlantic rift system existed through the Pyrenean rift. The East Mediterranean part of Neotethys ceased spreading; on its northern margin, Greater Apulia (LNg, Ap, PIM, Tor, Bdg, Tau) represents at this time the westernmost part of the Cimmerian terranes, detached now from the Iranian blocks (SS). *Triassic–Jurassic boundary.* At this time Palaeotethys was completely closed. The Central Atlantic rift widened but had difficulty finding a way to link with a plate limit to the east. The closure of Palaeotethys south of the Küre basin generated the southward subduction of the latter. Slab roll-back, both in Küre and the Neotethys, allowed the opening of the Izmir-Ankara (IzAn) ocean. See Figure 1 for abbreviation. All reconstructions are in spherical equidistant projection, centred 20E20N. *Symbols:* 1: passive margin; 2: magnetic or synthetic anomalies; 3: seamount; 4: intraoceanic subduction; 5: mid-ocean ridge; 6: active margin; 7: active rift; 8: inactive rift (basin); 9: collision zone; 10: thrust; 11: suture. Oceanic lithosphere in black. Abbreviations for Figures 3 to 12: AA, Austro–Alpine; Abr, Abruzzi; ACy, Attica–Cyclades; Adr, Adria; Ana, Anatolides; And, Andrusov; AnT, Antalya; APr, Algero–Provençal; Ap, Apulia; Apu, Apuseni; Bal, Baleares; Bdg, Beydaglari; BDu, Bosnia–Durmitor; Bet, Betic; Big, Biga; Bri, Briançonnais; BS, Black Sea; Buc, Bucovinian; Bud, Budva; Cal, Calabria; Car, Carnic; CLu, Campania–Lucania; Cor, Corsica; Dac, Dacides; Dal, Dalmatian; Dan, Danubian; Dau, Dauphinois; DoN, Dobrogea North; DoS, Dobrogea South; ECa, East-Carpathian; EPt, East-Pontides; Era, Eratostene; GCa, Great Caucasus; Get, Getic; Gos, Gosau; GTP, Gavrovo-Tripolitza-Pindos; Hat, Hatay; Hel,

material (symbolized by synthetic isochrones) from major continents and terranes.

In recent years we changed our tools and moved into GIS softwares and built a geodynamic database to support the reconstructions, and the model was, and still is, extended to the whole globe (Hochard 2008). An example of this new approach can be found in Ferrari *et al.* (2008).

Most geodynamic/geological constraints and data used for the reconstructions can be found in our previous publications (Stampfli 2000; Stampfli *et al.* 2001*a*, *b*; Stampfli & Borel 2002, 2004). A new global terrane, or rather geological elements map, was established following the approach outlined in Ferrari *et al.* (2008). Part of this map is presented in Figure 1, where only the elements displaced during the Alpine orogenic event are shown. The present-day outlines of the elements should be regarded as geographic markers, their shape having little to do with their original shape, generally most larger, excepted for elements having gone through large-scale Tertiary extension, such has Corsica or Sardinia, or the Cyclades elements. Due to the GIS database, many geodynamic/kinematic information can be derived from the reconstructions; one of the most interesting is the velocity map of the moving terranes, and their wander paths in respect to a fixed Europe (Fig. 2), which can be established here with great precision. This quantitative information can then be confronted to major structural patterns and kinematic indicators in the field; they show the changing direction and speed of convergence in space and time.

## The geodynamic evolution of the larger Alpine area

We shall review some of the major steps of the peri-Alpine evolution used to constrain the reconstructions; Palaeotethys ocean evolution was recently reviewed in Stampfli and Kozur (2006) and is not repeated here. We also do not discuss alternative models, quite numerous, and not often taking into consideration plate tectonics principles, but always very useful for the geological constraints they offer.

## The Jurassic ocean: Alpine Tethys, Central Atlantic and Vardar (Figs 3, 4 & 5)

The results of field-work on the Canary Islands and in Morocco (Favre *et al.* 1991; Favre & Stampfli 1992; Steiner *et al.* 1998) indicate that the onset of sea-floor spreading in the northern part of the Central Atlantic occurred in the Toarcian. Similar subsidence patterns between this region and the Lombardian basin (Stampfli 2000) led us to propose a direct connection between these areas. The Lombardian basin aborted in Middle Jurassic times (Bertotti *et al.* 1993) as it could not link up with the nearby oceanic Meliata-Maliac domains whose already cold lithosphere was rheologically considerably stronger than the surrounding continental areas. Therefore, the Alpine Tethys rift opened to the north of the Meliata basin, separating Adria and the Austro-Carpathian domain from Europe (Bernoulli 1981). Thermal subsidence of areas flanking the Alpine Tethys commenced in the Aalenian in the west (Briançonnais margin: Stampfli & Marchant 1997; Stampfli *et al.* 1998, 2002) and in the Bajocian further to the east (Helvetic and Austroalpine margin: Froitzheim & Manatschal 1996; Bill *et al.* 1997). The Alpine Tethys ocean spreading was considerably delayed with respect to the Central Atlantic; very slow spreading gave birth to a limited amount of oceanic crust, the oceanic area being dominated by continental mantle denudation. A larger transform Maghrebide ocean linked the central Atlantic and the Alpine Tethys, and was also characterized by delays in thermal subsidence (e.g. Rif area, Favre 1995; Stampfli 2000).

Within the Alpine domain, there is a fundamental difference between the AA–Carpathian and Western Alps systems. The AA–Carpathian evolution was rooted in the dynamics of the Triassic back-arc basins located to the south (Meliata–Maliac domain). These back-arc basins were shortened in conjunction with the opening of the Central Atlantic and rotation of Africa with respect to Europe. Subsequent slab roll-back of the subducting Küre, Maliac–Meliata oceanic lithosphere induced opening of the Vardar back-arc ocean, which by Late Jurassic times had completely replaced the pre-existing oceanic basins (Fig. 5).

**Fig. 3.** (*Continued*) Helvetic; Ion, Ionian; Ist, Istanbul; Kab, Kabylies; Kar, Karst; LNg, Lagonegro; Lig, Ligurian ocean; Lom, Lombardian; Lyc, Lycian; Mag, Magura; Men, Menderes; Moe, Moesia; NCA, North Calcareous Alps; NDo, North Dobrogea; Pan, Panormides; Pel, Pelagonia; Pen, Penninic, Vahic ocean; Pie, Piemontais ocean; PIM, Paxi-Ionian-Mani; Pyr, Pyrenean ocean; Rho, Rhodope; Rif, Rif; SA, South Alpine; Sak, Sakarya; Sar, Sardinia; SCr, South Crimea; SDz, Shatsky–Dzir; Sic, Sicani; SJa, Slavonia–Jadar; SMa, Serbo–Macedonian; SPi, Sitia–Pindos; Sre, Srednogorie; SS, Sanandaj–Sirjan; Str, Stranja; SBe, Sub–Betic; Tau, Taurus; Tel, Tell; Tis, Tisia; Tor, Talea–Ori; TrD, Transdanubian; Tro, Troodos; Tus, Tuscan; Tyr, Tyrrhenian; UMr, Umbria–Marches; WCa, West Carpathian; Zon, Zonguldak–Küre.

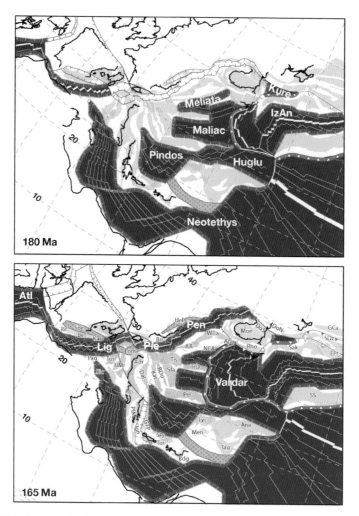

**Fig. 4.** 180–165 Ma. *Toarcian; Bathonian*. Spreading is now active in the Central Atlantic and the segment of the Alpine Tethys located south of Iberia (Lig). Eastward in the Carpathians domain, the Alpine Tethys successfully opened in the Bajocian as it was able to connect with a plate limit through the Dobrogean transform (DoS, DoN) and from there to the active subduction zone around the Sakarya plate. The Izmir–Ankara ocean is a back-arc of the Neotethys, whereas opening of the Vardar corresponds to subduction progradation from the Küre domain toward the Maliac domain. Accelerating roll-back of the Maliac sea floor generated northwestward spreading of the Vardar basin in a scenario of intra-oceanic subduction. Closure of the Küre basin was entering a collisional stage around the Rhodope promontory. Around the Alpine Tethys, aborted rifts become rim basins (sub-Betic, Helvetic–Vocontian, Lombardian) and rifting is still active from Italy to Greece (Ionian zone).

Continued rotation of Africa provoked ridge failure in the Vardar and large-scale Late Jurassic ophiolitic obduction onto the Dinaride–Hellenide passive margin of the Pelagonian terrane (e.g. Laubscher & Bernoulli 1977; Dercourt *et al.* 1986). Roll-back of the oceanic Maliac-Meliata slab was a centrifugal phenomenon that controlled successive collision of the Vardar arc with all passive margins surrounding the Meliata–Maliac basin. Accompanying the obduction on the western Dinaride margin in Late

Jurassic times, the northeastern part of the Vardar arc-trench system collided with the northern Meliata passive margin (the Northern Calcareous Alps: NCA, Fig. 4) (Bernoulli 1981), the Western Carpathian domain (Csontos & Vörös 2004) and the Rhodope. In the latter, remnants of the Vardar arc are found in northern Greece and Bulgaria as tectonic klippen (Bonev & Stampfli 2003, 2008). Closure of the Balkan rift system between Moesia and the Rhodope (Figs 5 & 6) controlled

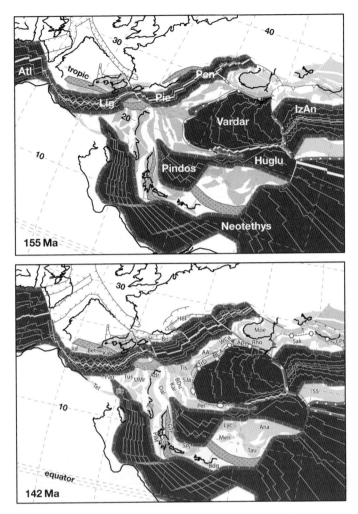

**Fig. 5.** 155–142 Ma. *Kimmeridgian; Berriasian*. Spreading in the Alpine Tethyan ocean now reached the Carpathian domain. In the process, Moesia was detached from Europe by only a few hundreds of kilometres, the point of rotation of Gondwana being located close to Moesia. Northward extension of the Central Atlantic triggered active rifting between Iberia and Newfoundland, and the northern limit of Iberia was affected by rifting extending eastward into the Briançonnais domain, through the Pyrenees and Provence. Extension is affecting also the Helvetic-Dauphinois rim basin. The Küre ocean was totally closed; collision of its arc-trench system with the Rhodope was causing the first phases of the Balkan orogeny, accompanied by inversion of former rift zones. Roll-back of the Meliata–Maliac slab allowed rapid westward expansion of the Vardar back-arc basin, its arc trench system colliding with the Pelagonian and Dinaric (SJa,Tis) landmass. This is accompanied by east–west shortening in the Maliac–Vardar domain due also to the anti-clockwise rotation of Gondwana with respect to Europe. We regard this event as creating a change in spreading direction in the western Neotethyan domain, at the origin of the mid-ocean ridge failure, south of the Sanandaj–Sirjan block (SS).

subsequent development of the Balkan orogen, accompanied by large-scale Early Cretaceous northward nappe emplacement and metamorphism (Georgiev *et al.* 2001). This circum-Vardar orogenic event commenced in the Middle Jurassic and was sealed by Albian to Cenomanian molasse-type sediments in the Balkans (Georgiev *et al.* 2001), and mid-to late Cretaceous platforms in the Hellenides–Dinarides.

Along the NCA margin, elements of the AA micro-continent were scraped off and incorporated into the accretionary wedge, to form the different internal units of the AA–Carpathian orogen (Kozur 1991; Plasienka 1996; Faupl & Wagreich 1999; Csontos & Vörös 2004). This event was accompanied by Early Cretaceous HP–LT metamorphism (e.g. Thöni & Jagoutz 1992; Ivan 2002). Subsequently, the enlarged AA orogenic

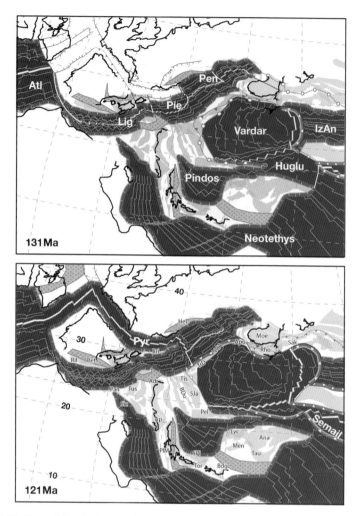

**Fig. 6.** 131–121 Ma. *Hauterivian; Aptian*. Accelerating anti-clockwise rotation of Gondwana was responsible for the obduction of part of the Vardar mid-ocean ridge system onto the Pelagonia, Dinaride (SJa) and Tisia blocks. Collision of the Vardar arc-trench system with the AA block was taking place at this time, detaching the future NCA domain from its basement (internal AA). Collision of the Vardar arc-trench system continued also in the Balkan, where parts of the Rhodope cover and basement were thrust northward . The major changes affecting the Neotethyan domain brought to an end the opening of the Izmir–Ankara back-arc basin system. The Izmir–Ankara slab started retreating eastward, allowing the opening of a new supra-subduction spreading centre. In the western Neotethys, ridge failure generated a new intra-oceanic subduction zone along the former spreading centre; this new oceanic domain will eventually obduct onto Arabia (e.g. Semail ophiolites). The Iberian plate was totally detached from Laurasia, whereas spreading stopped in the Ligurian-Piemont part of the Alpine Tethys. In the eastern Penninic segment of the Alpine Tethys, subduction progradation brought the exotic AA terranes onto the Alpine Tethys sea floor. Orogenic processes were soon to come to an end in the Balkan (sealed by Albian molasses), whereas the Vardar intra-oceanic system extended southward and eastward.

wedge began to move over the eastern segment of the Alpine Tethys (Figs 6 & 7) (Penninic–Vahic ocean), to finally collide with its northern passive margin (Helvetic domain s.l., Magura rim basin) thus forming the present Eastern Alps and Carpathian orogen (Wortel & Spakman 1993). This process involved continuous slab roll-back that

had commenced during the Carnian in the Küre domain in Turkey (Stampfli & Kozur 2006) and continued into the Neogene period in the Eastern Carpathians.

The Cretaceous AA accretionary wedge was affected by large-scale collapse in the Late Cretaceous (Figs 8 & 9), accompanied by the development

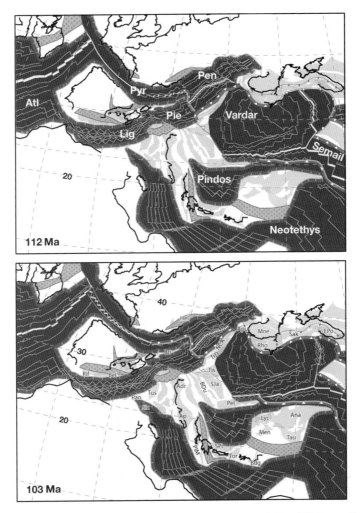

**Fig. 7.** 112–103 Ma. *Albian*. The absence of magnetic anomalies between the M0 and C34 anomalies reduces the constraints on the plate tectonics evolution during the intervening 37 Ma. However, the reconstructions are strongly constrained by the transform movement between the African and Indian plates during that time, leaving very little place for speculation. The Pyrenean 'ocean' opened and closed rapidly due to the rotation of Africa. The Briançonnais peninsula becomes an upper plate promontory slowly advancing onto the Penninic ocean. A flexural basin developed along the northern margin of the Izmir–Ankara ocean, with emplacement of ophiolitic mélanges on the Sakarya–East Pontides domain, preceding subduction reversal. The southern Vardar and Semail intra-oceanic back-arc system was expanding southward following slab retreat of the Huglu and Neotethys oceans.

of the Gosau basins (e.g. Faupl & Wagreich 1999). Before that, an obstacle prevented such a collapse, and we propose that spreading lasted in the Penninic–Vahic ocean until mid-Cretaceous times, the collapse of the accretionary wedge being only possible when the mid-oceanic ridge had been subducted. Recently dated gabbros in the Monte Rosa nappe have given Cenomanian ages (Liati & Froitzheim 2006). These gabbros are interpreted by these authors as belonging to the 'Valais ocean', but in view of their structural position, and

the necessary east–west convergence between the Iberian plate and the Penninic ocean (Figs 7 & 8), it is more likely that these gabbros belong to the latter and were underplated below the Piemont/Penninic Zermatt–Saas ophiolites.

*Cretaceous oceans: North Atlantic and the Pyrenean domain*

From magnetic anomalies in the Atlantic domain, at least from the Aptian (M0 magnetic anomaly,

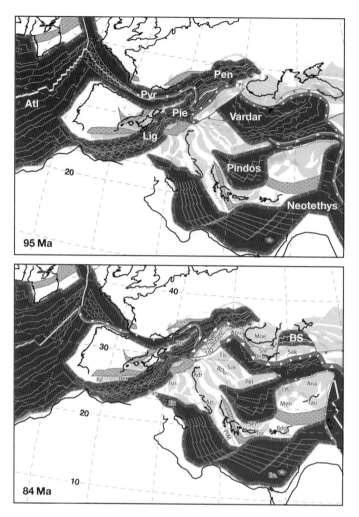

**Fig. 8.** 95–84 Ma. *Cenomanian; Santonian.* The northern limit of the African plate in the Pyrenees became a zone of convergence/subduction, partly extending into the Biscay–Atlantic domain, and generating large-scale inversion from the Pyrenees to the western Alps. The Penninic mid-ocean ridge is subducted and the supra-subduction Gosau rift started to open within the Austroalpine accretionary wedge. East–west shortening in the Alpine-Vardar region brought Adria behind the Austroalpine prism, whereas the north–eastward subducting Vardar remnant ocean generated an active margin setting in the Balkan, accompanied by the opening of the Srednogorie (Sre) and western Black Sea (BS) back-arc basins. The southern Vardar ridge obducted southward on the Anatolide block, and linked eastward with the Troodos–Semail obduction system. Finally, the Anatolian-Tauric plate was nearly totally covered by an ophiolitic-type mélange. The accelerated rotation of Africa narrowed down significantly the Neotethyan domain south of the Semail ocean. The Eratostene seamount could be related to volcanic activity in the Levant during the Cretaceous.

Fig. 7) to the Maastrichtian (Fig. 9) (Stampfli & Borel 2002) and most likely up to the Thanetian (anomaly C25, Fig. 9), it is observed that Iberia rotated with Africa and Apulia–Adria, without appreciable North–South shortening (Fig. 2); it also implies that spreading stopped in the western branch of the Alpine Tethys. Starting in the Early Cretaceous (Figs 6 & 7) the Pangaean break-up had increasing difficulty in linking up eastward with another plate boundary. From divergent, the

African plate limit in this region had become convergent, due to intra-oceanic subduction in the Neotethys and shortening affecting the Vardar region. Also, the African plate was now separated from South America and India and started to rotate more on itself. As a result, the Pyrenean-Biscay ocean opened, with a rifting phase starting in the Oxfordian, and the onset of ocean spreading in the Aptian for the Portuguese–Galician ocean and Pyrenean area, and in the Albian for the Gulf of

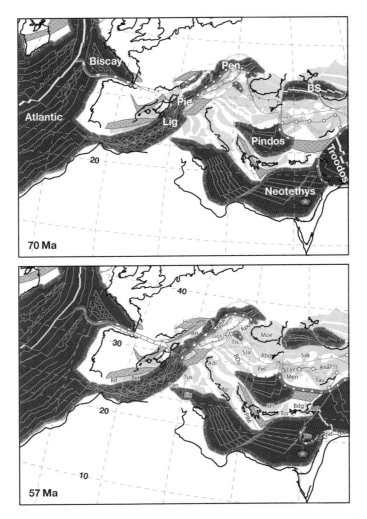

**Fig. 9.** 70–57 Ma. *Maastrichtian; Thanetian.* In the Alps, the northern sinistral Adriatic plate boundary was extending westward into the Piemont ocean, defining a temporary Corsica-Briançonnais plate, which started subducting southward beneath Adria. This was also allowing the AA prism to collapse westward and the supra-subduction Gosau rift to continue to open. North–south shortening was taking place along the northern boundary of the Iberian plate before a slight phase of extension in the Palaeocene. East–west shortening was still very active in the Alpine and Vardar domains, passing into continental subduction. Roll-back of the remnant oceanic slab allows the opening of the narrow east Black Sea back-arc basin. Counter-clockwise rotation of Africa was responsible for the obduction of the Semail ocean on the Arabia margin (from Oman to Syria, Hatay), whereas intra-oceanic subduction persisted in front of the Troodos plate.

Biscay, as clearly shown by peri-Iberian subsidence curves (Borel & Stampfli 1999; Stampfli *et al.* 2002) and a clear mid-Cretaceous thermal event in the Pyrenees (e.g. between 112 and 97 Ma; Schärer *et al.* 1999).

By the end of the Santonian (84 Ma, Fig. 8), the break-up between North America and Greenland took place, and in the Campanian the Biscay spreading aborted (Olivet 1996). Closing of the Pyrenean domain already began during the opening of the

Gulf of Biscay (Vergés & Garcia-Senz 2001), due to the accelerated rotation of the Iberian plate together with Africa (Fig. 8).

It is uncertain how wide the Pyrenean 'ocean' was, and whether it was limited to mantle denudation as indicated by the Palaeozoic lherzolites at Lherz (Fabries *et al.* 1998; Lagabrielle & Bodinier 2008). In order to control the geometry of the plate limit, a tentative ridge is put on the reconstructions for the Pyrenean 'ocean', but as was the case in large parts of the Alpine Tethys (see above), we think that only mantle denudation took place in this domain. A minimum extension of 60–80 km can be calculated from restored cross-sections through the Pyrenees (Vergés & Garcia-Senz 2001). However, from our reconstructions we come to a larger amount (*c.* 200 km), similar to the present northern Red Sea, where indeed no sea-floor spreading has yet taken place. Obviously, most of the distal margins have been subducted/eroded in the Pyrenean domain, as is the case in the Alps.

The rotation of Iberia finally placed the Briançonnais peninsula, by intra-oceanic subduction of the Piemont/Penninic ocean, beside the distal Helvetic margin. This created a repetition of the European margin in the western Alps domain (Figs 8 & 9) (Ringgenberg *et al.* 2002). The space between these two similar margin segments is generally referred to as the Valais 'ocean', a domain that formerly belonged to the Piemont (Alpine Tethys) ocean, trapped by the eastward displacement of the Iberian–Briançonnais block during the opening of the North Atlantic. The Middle Jurassic age (161 Ma) of potential parts of the Valais (Piemont) sea floor from the Misox zone was recently confirmed (Liati *et al.* 2003). As in the Ligurian and Piemont domains, where mantle rocks associated to Alpine ophiolites have shown Permian and even Precambrian ages (Rampone & Piccardo 2000), the Versoyen magmatic complex in the internal Valais domain also has Permian ages, but was clearly reheated during the Cretaceous (110–100 Ma). But it is not yet clear if this complex moved with the Briançonnais, or if it belonged to the toe of the Helvetic margin (Fügenschuh *et al.* 1999). The presence of the Cretaceous thermal event there would place them as part of the northern Briançonnais margin. Other gabbros in the Monte Rosa nappe have been dated as Cenomanian (Liati & Froitzheim 2006), but as discussed above, we would rather see them as pertaining to the Penninic ocean.

## Along strike shortening

A relatively large remnant of the Vardar ocean subducted northeastward under Moesia during the Late Cretaceous, as evidenced by the large Srednogorie volcanic arc of the Balkans and the Late Cretaceous opening of the Black Sea (Nikishin *et al.* 2003), representing the third generation of back-arc opening in that region. This subduction zone and its extension eastward up to Iran, represents a major slab pull force moving Africa in that direction until the closure of the Vardar in the Late Maastrichtian–Palaeocene. This is clearly confirmed by the change of convergence direction and velocity decrease of the African plate at that time (Fig. 2), also related to the synchronous peri-Arabian

ophiolitic obduction (Pillevuit *et al.* 1997) and slab detachment.

This NE directed subduction of the African plate in the Cretaceous, brought Africa–Iberia far to the east, inducing a relative westward escape of the AA wedge into the Piemont oceanic corridor. The southern margin of the Piemont ocean (Southern Alps–AA domain of the western Alps) was affected by tectonic movements since the Coniacian–Santonian, as evidenced by the onset of flysch deposition in the Piemont (Gets and Dranse flysch, Caron *et al.* 1989; Ligurian, Argnani *et al.* 2006 and Lombardian basins, Bernoulli & Winkler 1990). We relate such tectonism to large-scale sinistral strike-slip movements that affected the boundary between Adria and the Alpine domain.

These very large-scale lateral movements are well known also in the eastern Alps (Trümpy 1988, 1992) and finally placed Adria–Tizia south of the AA accretionary prism. In this process, the Piemont oceanic lithosphere was progressively detached from the northern margin of Adria and subducted beneath it. Frontal pieces of the western Adria–AA margin were dragged into the subduction zone, as evidenced by the Late Cretaceous–Early Palaeocene HP–LT metamorphism recorded in the Sesia domain (Oberhänsli *et al.* 1985; Rubatto 1998).

## The Pyrenean cycle

The 'Valais trough', as recognized in the western Alps today, is actually the remnant of trapped Piemont sea floor and of the toe of the Helvetic margin (see above, Stampfli *et al.* 2002). The 'Valais ocean' (as defined in Stampfli 1993) was located south of France and we refer to it here as the 'Pyrenean ocean' to avoid confusion between Valais ocean and Valais trough. No direct traces of this ocean have been found so far because its suture was located exactly where the Algero–Provençal ocean re-opened in Oligo–Miocene times (Roca 2001). A large part of the southern margin of the Pyrenean ocean was the Briançonnais peninsula (Figs 5 & 6); its northern margin was the Corbières–Provençal domain from the Pyrenees to the Maures-Estérel massifs. Elements of the Pyrenean margin of the Briançonnais are found in the Galibier region of the French Alps (Toury 1985), well known for its Late Jurassic Brèche du Télégraphe. Recent investigation there (Luzieux & Ferrari 2002) showed that a pull-apart type basin rapidly deepened under the continental crust deformation (CCD) in Late Jurassic times. This area is regarded as the most external Briançonnais element known so far, unless the Versoyen complex is also seen as a part of this margin (see above). Its conjugate northern Provençal margin area is characterized by important erosion during the Oxfordian (rift shoulder uplift) and the development of Albian basins deepening southward towards the ocean, followed by the accumulation of thousands of metres of upper Cretaceous clastics in a northward migrating fore-deep type basin (Debrand-Passard & Courbouleix 1984). This records a southward closure of the Pyrenean 'ocean' on a Provençal transect, whereas a northward subduction is usually proposed on a Pyrenean transect (Vergés & Garcia-Senz 2001), also seen in Sardinia (Barca & Costamagna 1997, 2000), but again a southward subduction took place westward

in the Biscay ocean (Olivet 1996). We propose here a uniform southward subduction during the late Cretaceous Pyrenean phase (Figs 8 & 9), replaced in the Pyrenees by northward subduction of the Iberian indenter during the second Eocene Pyrenean phase (Fig. 10). This dominating last event gave to the Pyrenees its final lithospheric structure, as seen on the ECORS transect. In between the two phases, and according to the Atlantic magnetic anomalies, the Palaeocene position of Iberia seems to retreat slightly (10–20 km) from Europe. This has created some large-scale extension superimposed on the Late Cretaceous orogenic

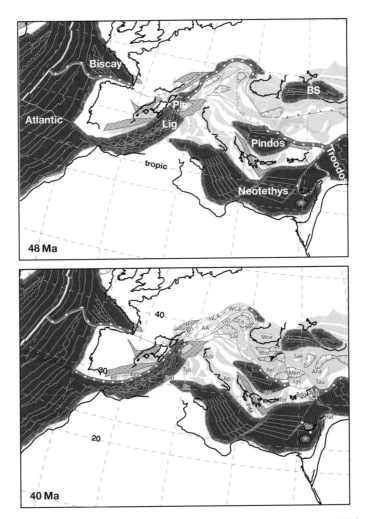

**Fig. 10.** 48–40 Ma. *Lutetian-Bartonian.* The Briançonnais terrane was entering into collision with the AA. The latter has closed a large part of the Alpine Tethys. Behind the prism, extension generated large rift zones, such as on the Tisia terrane (Pannonian basin). Similarly, the subduction of the Pindos and Troodos oceanic slabs, triggered the opening of numerous rifts and associated core-complexes in Turkey, in the Cyclades and in the Balkans. In the Pyrenees, retro-wedge thrusting is now very active; collision of the Iberian and European domains triggered the subduction of the remnant Alpine Tethys south of Iberia.

wedge and is responsible for the deposition of the Palaeocene red beds found all along the Pyrenean chain (Bilotte & Canérot 2006).

The Pyrenean orogen can be followed from the present Pyrenees eastward to southern France (Provence), and continues in the Alps in the form of a large-scale uplift of the Helvetic margin and local inversion of the Jurassic tilted blocks, well expressed by the deposits of the Niesen Flysch (Ackermann 1986) (mainly Maastrichtian) and Meilleret Flysch (Middle Eocene) (Homewood 1974), sedimented on a structured Mesozoic basement. Recent investigations have shown that similar turbiditic deposits of Late Cretaceous age are also found in ultra-Dauphinois units of the French Alps, such as the Pelat units (De Paoli & Thum 2008) and the already known Quermoz unit (Homewood *et al.* 1984). These so-called flysches clearly predate the Alpine syn-collisional event in the Helvetic domain, characterized by the deposition of the classical Alpine flysch sequences not before the earliest Oligocene (Kempf & Pfiffner 2004). These Cretaceous turbiditic deposits clearly point to an orogenic event affecting the distal Dauphinois–Helvetic margin and were deposited in the eastern prolongation of the Provençal foreland basin. On a Provence transect, the closure of the remnant Pyrenean basin certainly took place at the end of the Cretaceous phase (Figs 8 & 9), whereas in the western Alps, this closure was not total, as witnessed by the continuing deposit of coarse clastic turbidites in the Valais trough up to the Late Eocene (Bagnoud *et al.* 1998). This suggests the presence of a subduction/inversion zone between the Valais basin and the Niesen–Quermoz–Pelat foreland basin, both basins being separated by a relief where basement was outcropping (the so-called Tarine cordillera).

Eastward, the uplift of the Bohemian massif (e.g. Tanner *et al.* 1998) was certainly related to the same late Cretaceous inversion event and triggered the deposition of the Rheno–Danubian turbiditic sequence. Along the European margin it is interesting to note the similarity of facies between these Rheno–Danubian turbiditic deposits and the Valais trough sequence (the Valais trilogy) from Albian to Late Cretaceous (Stampfli 1993, and references therein). These deep-water clastic facies were located along the toe of the European margin, often referred to as the North Penninic basin (Figs 7 & 8). The presence of contourites and strong and changing current directions along the basin (Hesse 1974) suggests a connection with major oceanic domains. So the Valais trough, together with the Pyrenean–Biscay ocean, must be regarded as connections between the Eastern Alpine Tethyan realm and the north Atlantic ocean during the Cretaceous.

## The Alpine cycle

After several phases of rifting, as described above, and a new thermal subsidence stage developing during the Cretaceous, the Alpine region entered a phase of convergence between the African and European plate (Fig. 11). The tectonic evolution of the western Alps started with the formation of an accretionary prism related to the closure of the Alpine Tethys where different geological objects, corresponding to different stages of accretion, can be recognized:

- the Adriatic back-stop, comprising an aborted Jurassic rifted basin (Lombardian basin);
- the oceanic accretionary prism of the Piemont/Penninic ocean (the western Alps portion of the Alpine Tethys), including crustal elements from the former toe of the southern passive margin (lower AA elements);
- accreted material of the Briançonnais terrane derived from the Iberian plate; and
- accreted material of the Valais trough, representing the toe of the European passive margin; and
- accreted material of the former European continental margin and rim basin (Dauphinois–Helvetic domain).

In time, one passes from the oceanic accretionary prism to the formation of the orogenic wedge (Escher *et al.* 1996; Pfiffner *et al.* 1997; Ford *et al.* 2006) that we place after the detachment or delamination of the subducting slab in the Early Oligocene (e.g. Stampfli & Marchant 1995; von Blanckenburg & Davies 1995). The resulting heat flux allowed some more units to be detached from the European continental slab, triggering large-scale subduction of continental material (Marchant & Stampfli 1997; Ford *et al.* 2006) and Oligocene to Pliocene overthrusting of the most external units, such as:

- the external Variscan massifs and their cover;
- the Subalpine fold and thrust belts;
- the North Alpine foreland basin (molassic basin); and
- the Jura mountains fold belt.

These late events were accompanied by retro-wedge thrusting of the southern Alps on the Po Plain, a flexural basin located above the former Jurassic Lombardian basin. The retro-wedge becomes wider eastward, forming the Southern Alps thrust belt (Schönlaub & Histon 2000; Schmid *et al.* 2004). This younger S-verging thrust system linked two north-dipping subduction systems: the Adriatic–Dinaride continental subduction to the east, and the subduction of the remnant Alpine Tethys under the Iberian plate in the west (Figs 10 & 12). In the Adriatic–Dinaride subduction zone, a large part of the northern Apulian promontory was subducted under the AA domain (Schmid *et al.* 2004; Kissling

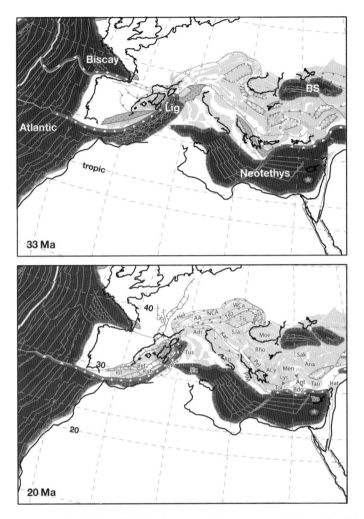

**Fig. 11.** 33–20 Ma. *Rupelian; Burdigalian.* The Alps have entered the main orogenic phase. The Briançonnais was subducted; parts of its cover was detached and accreted to the orogenic prism. Slab roll-back in the Carpathians allowed the Pannonian basin to enlarge and many other rifts and core-complexes were still active in the Balkans, Greece and Turkey until the Pindos was totally closed. Then a collision of the Hellenic orogenic prism took place with the Greater Apulia block (PIM) and triggered the subduction of the East Mediterranean basin (eastern part). Nowadays, this subduction front has reached the Apulian block (Fig. 1). On the Iberian side, the collapse of the active margin onto the remnant Alpine Tethys segment has induced large-scale rifting in the upper plate and the detachment of the Corso–Sardinian block. This rift system link northward with the Bresse–Rhine graben system. A nearly continuous south-vergent subduction/collision front runs from the southern border of Iberia to Turkey, and the north-vergent Alpine prism will progressively become less active.

*et al.* 2006), whereas the northwestern corner of Adria developed as an indenter (Handy *et al.* 2005).

Since the Miocene, the Alps and Carpathian wedges moved mainly due to remnant roll-back of the European plate. The weakly buoyant lithosphere under the rim basins present in that margin allowed further subduction to take place. However, in the western Alps, the Adriatic indenter was pushed against the orogenic prism as it was attached to the still converging African plate, creating maximum shortening in the western Alps.

As we have seen above, the southward subduction of the Alpine Tethys ocean is related to the history of the Meliata–Maliac and Vardar domain, and was inherited in the western Alps from the pre-existing northward vergence of the AA accretionary wedge. This northward vergence is unique in the whole Alpine and Tethyan domain, where most orogens are south vergent.

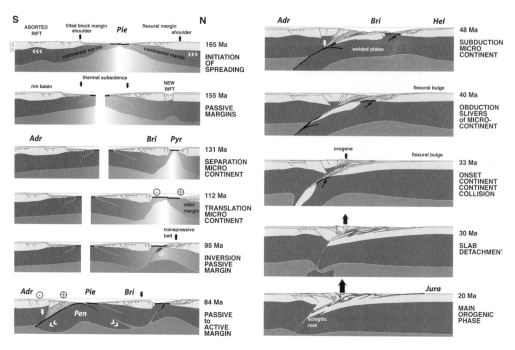

**Fig. 12.** North–south palinspatic cross-sections through the western Alpine segment. The Briançonnais block (Bri, on the reconstruction) is fixed; the other segments move in and out of the picture. Arrows show major uplift and subsidence events.

The change in vergence of Alpine Tethys subduction is found at the connecting region between the Alps and the northern Apennines, south of Corsica. It must be emphasized here that the Penninic Alpine accretionary prism is older (Late Cretaceous–Eocene) than the Apenninic one (Oligocene–Pliocene); actually one started when the other one stopped. During this process, the Apenninic prism re-mobilized parts of the Piemont–Penninic prism as exotic elements (e.g. the Bracco ophiolitic ridge, Elter *et al.* 1966; Hoogenduijn Strating 1991). The Alpine prism collided with the Iberian plate in Corsica (Malavieille *et al.* 1998) in Eocene times (Figs 10–12). The remnant Alpine Tethys oceanic domain (Ligurian basin) south of the Iberian plate started to subduct northward in the Late Eocene (producing HP metamorphism dated at 25–21 Ma, Michard *et al.* 2006) in an ongoing process of shortening between Europe and Africa and following continent-continent collision in the Pyrenees.

On the Miocene reconstruction (Fig. 12), nearly all of this remnant ocean is already subducted. This SE-directed roll-back of the Tethyan slab triggered the opening of the Algero–Provençal back-arc basin starting in the Oligocene (Fig. 12). Thus, the Oligocene volcanism of Sardinia can be regarded as subduction related (e.g. Monaghan 2001). The

Alpine accretionary prism stranded in the eastern border of Corsica started to collapse backward (eastward), following the build-up of the new Apenninic accretionary prism; these Piemont–Ligurian units are now found in an upper structural position, whereas they were in a lower one during the Alpine collision. The northern and southern Apenninic prisms collapsed into pre-existing depressions, the Lombardian rift in the north and the Ionian sea oceanic corridor (the western most part of the East Mediterranean Neotethys ocean) in the south, with its still active back-arc opening (Savelli 2000; Argnani & Savelli 2001). In the eastern part of the East Mediterranean basin, subduction had already started in the late Eocene (Fig. 12); it then prograded westward until its present day position in Greece (Fig. 1).

## Conclusions

The proposed reconstructions require a tight fit of the Iberian–Alboran microplates with respect to North America and Europe, and also imply a long lasting rifting phase in the Atlantic regions now separating these domains. The final position of the Apulia–Adria plate and its geometry close to the present-day one, was reached through Jurassic rifting events, which finally gave birth to the narrow Alpine Tethys. The latter is dominated by stretching of the lithospheric mantle of its borders. On its western side, the Alpine area is largely influenced by the opening Atlantic ocean, and the northward extension of the latter in Early Cretaceous times. On its eastern side, the Alpine domain is dominated by the evolution of the supra-subduction Vardar ocean. The latter collided with its border in latest Jurassic–early Cretaceous times. Its northern border consisted of the AA micro-continent, not buoyant enough to resist subduction. Large portions of its crust and cover were underplated and transported northward onto the eastern Alpine Tethys (Penninic–Vahic ocean).

It is then clear that two specific geodynamic scenarios presided at the evolution of the Alpine belt. The western branch was affected by a second phase of rifting, detaching the Iberian plate and its Briançonnais promontory from Europe, finally duplicating the southern European margin in the western Alps. The eastern branch is dominated by an advancing large accretionary prism that included the AA blocks. The prism was collapsing forward but also westward into the oceanic Piemont part of the western branch. This was made possible by large-scale east–west shortening, related to the kinematics of the African plate with regards to Europe (Fig. 2). Thus, the Vardarian oceanic back-stop of the AA prism was replaced in time by the continental larger Adriatic back-stop. The Piemont oceanic space was, at the same time, forced to subduct southward in front of the advancing AA prism. At the close of the Cretaceous, all the major elements are in place, and after the Palaeocene lull, following the change of wander path of Africa, they were thrust over each other to form the Alpine orogenic wedge.

If this wedge is tectonically relatively cylindrical, its components are not closely related. The Alpine orogen is a clear example of juxtaposition of far travelled tectonic elements and cannot be understood from a 'fixist' point of view.

The Swiss National Funds supported C. Hochard PhD thesis (grants n. 2000-068015 and 200020-109549). M. Ford and A. Argnani thoroughly reviewed this manuscript, their suggestions and corrections are warmly acknowledged. B. Murphy strongly encouraged us to publish this work, we thank him for his patience.

# References

ACKERMANN, A. 1986. Le flysch de la nappe du Niesen. *Eclogae geologicae Helvetiae*, **79**(3), 641–684.

ARGNANI, A. 2002. The Northern Apennines and the kinematics of Africa – Europe Convergence. *Bolletino della Società Geologica Italiana, Volume Speciale*, **1**, 47–60.

ARGNANI, A. & SAVELLI, C. 2001. Magmatic signature of episodic back-arc rifting in the southern Tyrrhenian Sea. *In*: ZIEGLER, P. A., CAVAZZA, W., ROBERTSON, A. H. F. & CRASQUIN-SOLEAU, S. (eds) *PeriTethys Memoir 6: Peritethyan Rift/wrench Basins and Passive Margins, IGCP 369*. Mémoires du Museum Nationale d'Histoire Naturelle, **186**, 735–754.

ARGNANI, A., FONTANA, D., STAFANI, C. & ZUFFA, G. G. 2006. Palaeogeography of the Upper Cretaceous-Eocene carbonate turbidites of the Northern Apennines from provenance studies. *In*: MORATTI, G. & CHALOUAN, A. (eds) *Tectonics of the Western Mediterranean and North Africa*. Geological Society, London, Special Publications, **262**, 259–275.

BAGNOUD, A., WERNLI, R. & SARTORI, M. 1998. Découverte de foraminifères planctoniques dans la zone de Sion-Courmayeur à Sion. *Eclogae geologicae Helvetiae*, **91**(3), 421–429.

BARCA, S. & COSTAMAGNA, L. C. 1997. Compressive 'Alpine' tectonics in Western Sardinia (Italy): geodynamic consequences. *Comptes Rendus Académie des Sciences*, **325**, 791–797.

BARCA, S. & COSTAMAGNA, L. C. 2000. Il Bacino paleogenico del Sulcis-Iglesiente (Sardegna SW): nuovi dati stratigraphico-strutturali per un modello geodinamico nell'ambito dell'orogenesi pirenaica. *Bollettino della Società geologica Italiana*, **119**, 497–515.

BERNOULLI, D. 1981. Ancient continental margins of the Tethyan Ocean. *In*: BALLY, A. W., WATTS, A. B., GROW, J. A., MANSPEIZER, W., BERNOULLI, D., SCHREIBER, C. & HUNT, J. M. (eds) *Geology of passive continental margins; history, structure and sedimentologic record (with special emphasis on the Atlantic margin)*. American Association of Petroleum Geologists, Education course note series, **19/5**, 1–36.

BERNOULLI, D. & WINKLER, W. 1990. Heavy mineral assemblages from Upper Cretaceous South Alpine and Austro-alpine flysch sequences (N-Italy and S-Switzerland): source terranes and paleotectonic implications. *Eclogae geologicae Helvetiae*, **83**(2), 287–310.

BERTOTTI, G., PICOTTI, V., BERNOULLI, D. & CASTELLARIN, A. 1993. From rifting to drifting: tectonic evolution of the South-Alpine upper-crust from the Triassic to the Early Cretaceous. *Sedimentary Geology*, **86**, 53–76.

BILL, M., BUSSY, F., COSCA, M., MASSON, H. & HUNZIKER, J. C. 1997. High precision U–Pb and 40Ar/39Ar dating of an Alpine ophiolite (Gets nappe, French Alps). *Eclogae geologicae Helvetiae*, **90**, 43–54.

BILOTTE, M. & CANEROT, J. 2006. Rôles respectifs des tectoniques fini-crétacée et éocène dans la partie orientale de la chaîne des Pyrénées. Le 'Garumnien' de Cucugnan et ses relations avec le Chevauchement Frontal Nord-Pyrénéen (Corbières méridionales, France). *Eclogae geologicae Helvetiae*, **99**(1), 17–27.

BONEV, N. & STAMPFLI, G. M. 2003. New structural and petrologic data on Mesozoic schists in the Rhodope (Bulgaria): geodynamic implications. *C. R. Geoscience*, **335**, 691–699.

BONEV, N. & STAMPFLI, G. M. 2008. Petrology, geochemistry and geodynamic implications of Jurassic island arc magmatism as revealed by mafic volcanic rocks in the Mesozoic low-grade sequence, eastern Rhodope, Bulgaria. *Lithos*, **100**, 210–233.

BOREL, G. & STAMPFLI, G. M. 1999. Geodynamic evolution of the Briançonnais domain and the individualisation of the Iberian plate. 4th workshop Alpine Geol. Studies, Tübingen, **52**, 13–14.

CARON, C., HOMEWOOD, P. & WILDI, W. 1989. The original Swiss Flysch: a reappraisal of the type deposits in the Swiss Prealps. *Earth-Science Reviews*, **26**, 1–45.

CHANNELL, J. E. T. 1992. Paleomagnetic data from Umbria (Italy): implications for the rotation of Adria and Mesozoic apparent polar wander paths. *Tectonophysics*, **216**, 365–378.

CHANNELL, J. E. T. 1996. Paleomagnetism and Paleogeography of Adria. *In*: MORRIS, A. & TARLING, D. H. (eds) *Paleomagnetism and Tectonics of the Mediterranean Region*. Geological Society, London, Special Publications, 119–132.

CSONTOS, L. & VÖRÖS, A. 2004. Mesozoic plate tectonic reconstruction of the Carpathian region. *Palaeoclimatology, Palaeoecology, Palaeogeography*, **210**, 1–56.

DE JONGE, M. R., WORTEL, M. J. R. & SPAKMAN, W. 1994. Regional scale tectonic evolution and the seismic velocity structure of the lithosphere and upper mantle: the Mediterranean region. *Journal of Geophysical Research*, **99**(B6), 12091–12108.

DE PAOLI, R. & THUM, L. 2008. *Implications géodynamiques de l'étude pétrographique des flyschs du Crétacé supérieur du Piolit, du Pelat et de Baiardo (Alpes occidentales)*. MSc, University of Lausanne, Switzerland, 105.

DEBRAND-PASSARD, S. & COURBOULEIX, S. 1984. Synthèse géologique du S-E de la France. *Mémoires du Bureau de Recherches Géologiques et Minières Orléans*, **126**, 1–28.

DERCOURT, J., RICOU, L. E. & VRIELINCK, B. 1993. *Atlas Tethys, paleoenvironmental maps*. Gauthier-Villars, Paris, 307.

DERCOURT, J., ZONENSHAIN, L. P. *ET AL*. 1985. Présentation des 9 cartes paléogéographiques au 1/20 000 000 s'étendant de l'Atlantique au Pamir pour la période du Lias à l'Actuel. *Bulletin de la Société géologique de France*, **8**(T I, No 5), 637–652.

DERCOURT, J., ZONENSHAIN, L. P. *ET AL*. 1986. Geological evolution of the Tethys from the Atlantic to the Pamirs since the Lias. *Tectonophysics*, **123**, 241–315.

ELTER, G. P. E., STURANI, C. & WEIDMANN, M. 1966. Sur la prolongation du domaine ligure de l'Apennin dans le Montferrat et les Alpes et sur l'origine de la nappe de la Simme s.l. des Préalpes romandes et chablaisiennes. *Archives des Sciences, Genève*, **19**(3), 279–377.

ESCHER, A., HUNZIKER, J. C., MARTHALER, M., MASSON, H., SARTORI, M. & STECK, A. 1996. Geologic framework and structural evolution of the Western Swiss Alps. *In*: PFIFFNER, O. A., LEHNER, P., HEITZMANN, P., MÜLLER, S. & STECK, A. (eds) *Deep structure of Switzerlannd – Results from the National Research Program 20 (NRP20)*. Birkhäuser, Basel, 205–221.

FABRIES, J., LORAND, J. P. & BODINIER, J. L. 1998. Petrogenetic evolution of orogenic lherzolite massifs in the central and western Pyrenees. *Tectonophysics*, **292**(1–2), 145–167.

FAUPL, P. & WAGREICH, M. 1999. Late Jurasic to Eocene Palaeogeography and geodynamic evolution of the Eastern Alps. *Mitteilungen der Österreichischen Geologischen Gesellschaft*, **92**, 79–94.

FAVRE, P. 1995. Analyse quantitative du rifting et de la relaxation thermique de la partie occidentale de la marge transformante nord-africaine: le Rif externe (Maroc). *Geodinamica Acta*, **8**(2), 59–81.

FAVRE, P. & STAMPFLI, G. M. 1992. From rifting to passive margin: the example of the Red Sea, Central Atlantic and Alpine Tethys. *Tectonophysics*, **215**, 69–97.

FAVRE, P., STAMPFLI, G. & WILDI, W. 1991. Jurassic sedimentary record and tectonic evolution of the north-western corner of Africa. *Paleogeography, Paleoecolology, Paleoclimatology*, **87**, 53–73.

FERRARI, O. M., HOCHARD, C. & STAMPFLI, G. M. 2008. An alternative plate tectonic model for the Palaeozoic–Early Mesozoic: palaeotethyan evolution of Southeast Asia (Northern Thailand–Burma). *Tectonophysics*, **451**, 346–365.

FINETTI, I. R. 2005. *CROP Project: deep seismic exploration of the central Mediterranean and Italy*. Elsevier, Amsterdam, Boston.

FORD, M., DUCHENE, S., GASQUET, D. & VANDERHAEGHE, O. 2006. Two-phase orogenic convergence in the external and internal SW Alps. *Journal of the Geological Society, London*, **163**(5), 815–826.

FRISCH, W. 1979. Tectonic progradation and plate tectonic evolution of the Alps. *Tectonophysics*, **60**, 121–139.

FROITZHEIM, N. & MANATSCHAL, G. 1996. Kinematics of Jurassic rifting, mantle exhumation, and passive-margin formation in the Austroalpine and Penninic nappes (Eastern Switzerland). *Geological Society of America Bulletin*, **108**(9), 1120–1133.

FÜGENSCHUH, B., LOPRIENO, A., CERIANI, S. & SCHMID, S. M. 1999. Structural analysis of the Subbriançonnais and Valais units in the area of Moutiers (Savoy, Western Alps): paleogeographic and tectonic consequences. *International Journal of Earth Sciences*, **88**, 201–218.

GARFUNKEL, Z. & DERIN, B. 1984. Permian-early Mesozoic tectonism and continental margin formation in Israel and its implications for the history of the Eastern Mediterranean. *In*: DIXON, J. E. & ROBERTSON, A. H. F. (eds) *The Geological Evolution of the Eastern Mediterranean*. Geological Society, London, Special Publications, **17**, 187–201.

GEORGIEV, G., DABOVSKI, C. & STANISHEVA-VASSILEVA, G. 2001. East Srednogorie-Balkan rift zone. *In*: ZIEGLER, P. A., CAVAZZA, W., ROBERTSON, A. H. F. & CRASQUIN-SOLEAU, S. (eds) *Peri-Tethys memoir 6: Peritethyan rift/wrench basins and passive margins, IGCP 369*. Mémoires du Museum Nationale d'Histoire Naturelle, **186**, 259–294.

HANDY, M., BABIST, J., WAGNER, R., ROSENBERG, C. & KONRAD, M. 2005. Decoupling and its relation to strain partitioning in continental lithosphere; insight from the Periadriatic Fault system (European Alps). *In*: GAPAIS, D., BRUN, J.-P. & COBBOLD, P. R. (eds) *Deformation mechanisms, rheology and tectonics; from minerals to the lithosphere*. Geological Society, London, Special Publications, **243**, 249–276.

HESSE, R. 1974. Long-distance continuity of turbidites: possible evidence for an early Cretaceous trench abyssal plain in the East Alps. *Geological Society of America Bulletin*, **85**, 859–870.

HOCHARD, C. 2008. *GIS and Geodatbases applications to global scale plate tectonics modelling*. PhD thesis, University of Lausanne, Switzerland.

HOMEWOOD, P. 1974. Le flysch du Meilleret (Préalpes romandes) et ses relations avec les unités l'encadrant. *Eclogae geologicae Helvetiae*, **67**(2), 349–401.

HOMEWOOD, P., ACKERMANN, T., ANTOINE, P. & BARBIER, R. 1984. Sur l'origine de la nappe du Niesen et la limite entre les zones ultrahelvétiques et valaisanne. *Comptes Rendus des Séances de l'Académie des Sciences de Paris, série II*, **299**(15), 1055–1059.

HOOGENDUIJN STRATING, E. H. 1991. The evolution of the Piemonte-Ligurian ocean, a structural study of ophiolite complexes in Liguria (NW Italy). *Geologica ultraiectina*, **74**, 1–127.

IVAN, P. 2002. Relict magmatic minerals and textures in the HP/LT metamorphosed oceanic rocks of the Triassic-Jurassic Meliata Ocean, Inner Western Carpathians. *Slovak Geological Magazine*, **8**(2), 109–122.

KEMPF, O. & PFIFFNER, A. 2004. Early Tertiary evolution of the North Alpine foreland basin of the Swiss Alps and adjoining areas. *Basin Research*, **16**(4), 549–567.

KISSLING, R., SCHMID, S. M., LIPPITSCH, R., ANSORGE, J. & FUEGENSCHUH, B. 2006. Lithosphere structure and tectonic evolution of the Alpine Arc; new evidence from high-resolution teleseismic tomography. *In*: GEE, D. G. & STEPHENSON, R. (eds) *European lithosphere dynamics*. Geological Society, London, Memoirs, **32**, 129–145.

KOZUR, H. 1991. The geological evolution at the western end of the Cimmerian ocean in the Western Carpathians and Eastern Alps. *Zentralblatt für Geologie und Paläontologie Teil I*, **1**(H.1), 99–121.

LAGABRIELLE, Y. & BODINIER, J. L. 2008. Submarine reworking of exhumed subcontinental mantle rocks: field evidence from the Lherz peridotites, French Pyrenees. *Terra Nova*, **20**, 11–21.

LAGABRIELLE, Y., POLINO, R. *ET AL.* 1984. Les témoins d'une tectonique intraocéanique dans le domaine téthysien: analyse des rapports entre les ophiolites et leurs couvertures métasédimentaires dans la zone piémontaise des Alpes franco-italiennes. *Ofioliti*, **9**(1), 67–88.

LAUBSCHER, H. P. & BERNOULLI, D. 1977. Mediterranean and Tethys. *In*: NAIRN, A., KANES, W. & STEHLI, F. (eds) *The Ocean Basins and Margins*. Plenum Press, New York, 1–28.

LIATI, A. & FROITZHEIM, N. 2006. Assessing the Valais ocean, Western Alps: U–Pb SHRIMP zircon geochronology of eclogite in the Balma unit, on top of the Monte Rosa nappe. *European Journal of Mineralogy*, **18**, 299–308.

LIATI, A., GEBAUER, D., FROITZHEIM, N. & FANNING, M. 2003. Origin and geodynamic significance of meta-basic rocks from the Antrona ophiolites (western Alps): new insights from SHRIMP-dating. EGS-AGU-EUG joint assembly, Nice, **EAE03-A**, 12648.

LIATI, A., FROITZHEIM, N. & FANNING, M. 2005. Jurassic ophiolites within the Valais Domain of the Western

and Central Alps; geochronological evidence for re-rifting of oceanic crust. *Contributions to Mineralogy and Petrology*, **149**(4), 446–461.

LUZIEUX, L. & FERRARI, O. M. 2002. *Etude géologique et minéralogique du secteur du Col du Galibier (Savoie, France): implications géodynamiques*. MSc thesis, University of Lausanne, Lausanne.

MALAVIEILLE, J., CHEMENDA, A. & LARROQUE, C. 1998. Evolutionary model for the Alpine Corsica: mechanism for ophiolite emplacement and exhumation of high-pressure rocks. *Terra Nova*, **10**, 317–322.

MARCHANT, R. H. & STAMPFLI, G. M. 1997. Subduction of continental crust in the Western Alps. *Tectonophysics*, **269**(3–4), 217–235.

MICHARD, A., NEGRO, F., SADDIQI, O., BOUY-BAOUENE, M. L., CHALOUAN, A., MONTIGNY, R. & GOFFÉ, B. 2006. Pressure–temperature–time constraints on the Maghrebide mountain building: evidence from the Rif–Betic transect (Morocco, Spain), Algerian correlations, and geodynamic implications. *Comptes Rendus Geosciences*, **338**(1–2), 92–114.

MOIX, P., BECCALETTO, L., KOZUR, H. W., HOCHARD, C., ROSSELET, F. & STAMPFLI, G. M. 2008. A new classification of the Turkish terranes and sutures and its implication for the paleotectonic history of the region. *Tectonophysics*, **451**, 7–39.

MONAGHAN, A. 2001. Coeval extension, sedimentation and volcanism along the Cainozoic rift system of Sardinia. *In*: ZIEGLER, P. A., CAVAZZA, W., ROBERTSON, A. H. F. & CRASQUIN-SOLEAU, S. (eds) *PeriTethys memoir 6: Perithethyan rift/wrench basins and passive margins, IGCP 369*. Mémoires du Museum Nationale d'Histoire Naturelle, **186**, 707–734.

MUTTONI, G., GARZANTI, E., ALFONSI, L., CIRILLI, S. & GERMANI, D. 2001. Motion of Africa and Adria since the Permian: paleomagnetic and paleoclimatic constraints from northern Libya. *Earth Planetary Sciences Letter*, **192**, 159–174.

NIKISHIN, A. M., KOROTAEV, M. V., ERSHOV, A. V. & BRUNET, M. F. 2003. The Black-Sea basin: tectonic history and Neogene-Quaternary rapid subsidence modelling. *In*: BRUNET, M. F. & CLOETINGH, S. (eds) *Integrated Peri-Tethyan Basins Studies*. Sedimentary Geology, **156**, 149–168.

OBERHÄNSLI, R., HUNZIKER, J. C., MARTINOTTI, G. & STEN, W. B. 1985. Geochemistry, geochronology and petrology of Monte Mucrone: an example of eo-alpine eclogitisation of Permian granitoïds in the Sesia Lanzo zone, western Alps, Italy. *Chemical Geology*, **52**, 165–184.

OLIVET, J.-L. 1996. Lacinématique de la plaque Ibérique. *Bulletin des Centres de recherches exploration-production Elf-Aquitaine*, **20**, 131–195.

PFIFFNER, O. A., LEHNER, P., HEITZMAN, P. Z., MUELLER, S. & STECK, A. 1997. *Deep structure of the Swiss Alps – Results from NRP 20*. Birkhäuser AG., Basel, 380.

PILLEVUIT, A., MARCOUX, J., STAMPFLI, G. M. & BAUD, A. 1997. The Oman exotics: a key to the understanding of the Neotethyan geodynamic evolution. *Geodinamica Acta*, **10**(5), 209–238.

PLASIENKA, D. 1996. Mid-Cretaceous (120–80 Ma) orogenic processes in the central Western Carpathians:
brief review and interpretation of data. *Slovak Geological Magazine*, **3–4**, 319–324.

RAMPONE, E. & PICCARDO, G. 2000. The
ophiolite-oceanic lithosphere analogue: new insights
from the northern Apennines (Italy). *In*: DILEK, Y.,
MOORES, E. M., ELTHON, D. & NICOLAS, A. (eds)
*Ophiolites and oceanic crust: new insights from field
studies and the ocean drilling project*. Geological
Society of America Special Paper, **349**, 21–34.

RINGGENBERG, Y., TOMASSI, A. & STAMPFLI, G. M.
2002. The Jurassic sequence of the Niesen nappe in
the region of Le Sépey-La Forclaz (Switzerland):
witness of the Piemont rifting in the Helvetic paleogeographic domain. *Bulletin de Géologie de l'Université
de Lausanne*, **87**(4), 353–372.

ROBERTSON, A. H. F., DIXON, J. E. *ET AL.* 1996. Alternative tectonic models for the Late Palaeozoic-Early
Tertiary development of Tethys in the Eastern Mediterranean region. *In*: MORRIS, A. & TARLING, D. H.
(eds) *Palaeomagnetism and Tectonics of the Eastern
Mediterranean Region*. Geological Society, London,
Special Publications, **105**, 239–263.

ROCA, E. 2001. The Northwest Mediterranean Basin
(Valencia trough, Gulf of Lions and Liguro-Provençal
baisns): structure and geodynamic evolution. *In*:
ZIEGLER, P. A., CAVAZZA, W., ROBERTSON,
A. H. F. & CRASQUIN-SOLEAU, S. (eds) *PeriTethys
memoir 6: Peritethyan rift/wrench basins and
passive margins, IGCP 369*. Mémoires du Museum
Nationale d'Histoire Naturelle, **186**, 671–706.

RUBATTO, D. 1998. *Dating of pre-Alpine magmatism,
Jurassic ophiolites and Alpine subductions in the
Westen Alps*. ETHZ, Zürich, 1–173.

SAVELLI, C. 2000. Subduction-related episodes of K'alkaline magmatism (15–0.1 Ma) and geodynamic
implications in the north Tyrrhenian – central Italy
region: a review. *Journal of Geodynamics*, **30**,
575–591.

SCHARDT, H. 1889. Sur l'origine des Prealpes Romandes.
*Eclogae Geologicae Helvetiae*, **4**, 129–142.

SCHARDT, H. 1898. Les régions exotiques du versant Nord
des Alpes Suisse. Préalpes du Chablais et du Stockhorn
et les Klippes. *Bulletin de la Société Vaudoise des
Sciences Naturelles*, **34**(128), 113–219.

SCHARDT, H. 1900. Encore les régions exotiques. Répli-
que aux attaques de M. Emile Haug. *Bulletin de la
Société Vaudoise des Sciences Naturelles*, **36**,
147–169.

SCHARDT, H. 1907. Les vues modernes sur la tectonique et
l'origine de la chaîne des Alpes. *Archives de physique
et des sciences naturelles de Genève*, **4**(23), 356–385/
483–496.

SCHÄRER, U., DE PARSEVAL, P., POLVE, M. & DE SAINT
BLANQUAT, M. 1999. Formation of the Trimouns talc-
chlorite deposit (Pyrenees) from persistent hydrothermal activity between 112 and 97 Ma. *Terra Nova*,
**11**(1), 30–37.

SCHETTINO, A. & SCOTESE, C. 2002. Global kinematic
constraints to the tectonic history of the Mediterranean
region and surrounding areas during the Jurassic
and Cretaceous. *In*: ROSENBAUM, G. A. & LISTER,
G. S. (eds) *Reconstruction of the evolution of the

*Alpine-Himalayan Orogen*. Journal of the Virtual
Explorer, **8**, 149–168.

SCHMID, S. M., FUEGENSCHUH, B., KISSLING, E. &
SCHUSTER, R. 2004. Transects IV, V and VI; the
Alps and their forelands. *In*: CAVAZZA, W., ROURE,
F., SPAKMAN, W., STAMPFLI, G. M. & ZIEGLER, P.
(eds) *The TRANSMED Atlas: The Mediterranean
Region from Crust to Mantle*. Springer Verlag,
Germany, unpaginated.

SCHÖNLAUB, H. P. & HISTON, K. 2000. The Palaeozoic
evolution of the Southern Alps. *Mitteilungen der österreichischen geologischen Gesellschaft*, **92**, 15–34.

SENGÖR, A. M. C. 1979. Mid-Mesozoic closure of Permo-
Triassic Tethys and its implications. *Nature*, **279**,
590–593.

SENGÖR, A. M. C. 1985. The story of Tethys: how many
wives did Okeanos have. *Episodes*, **8**(1), 3–12.

SENGÖR, A. M. C., YILMAZ, Y. & SUNGURLU, O. 1984.
Tectonics of the Mediterranean Cimmerides: nature
and evolution of the western termination of Paleo-
Tethys. *In*: DIXON, J. E. & ROBERTSON, A. H. F.
(eds) *The Geological Evolution of the Eastern Mediterranean*. Geological Society, London, Special Publications, **17**, 77–112.

SRIVASTAVA, S. P. & TAPSCOTT, C. R. 1986. Plate kinematics of the North Atlantic. *In*: VOGT, P. R. &
TUCHOLKE, B. E. (eds) *The Western North Atlantic
region*. Geological Society of America, **M**, 379–404.

SRIVASTAVA, S. P., ROEST, W. R., KOVACS, L. C.,
OAKAY, G., LÉVESQUE, S., VERHOEF, J. &
MACNAB, R. 1990. Motion of Iberia since the Late
Jurassic: results from detailed aeromagnetic measure-
ments in the Newfoundland Basin. *Tectonophysics*,
**184**, 229–260.

STAMPFLI, G. M. 1993. Le Briançonnais, terrain exotique
dans les Alpes? *Eclogae geologicae Helvetiae*, **86**(1),
1–45.

STAMPFLI, G. M. 2000. Tethyan oceans. *In*: BOZKURT,
E., WINCHESTER, J. A. & PIPER, J. D. A. (eds) *Tectonics and magmatism in Turkey and surrounding
area*. Geological Society, London, Special Publications, **173**, 1–23.

STAMPFLI, G. M. 2005. Plate tectonic of the Apulia-Adria
microcontinents. *In*: FINETTI, I. R. (ed.) *CROP
PROJECT deep seismic exploration of the Mediterranean and Italy*. Elsevier, The Netherlands, 747–766.

STAMPFLI, G. M. & MARCHANT, R. H. 1995. Plate configuration and kinematics in the Alpine region. *Accademia Nazionale delle Scienze, Scritti e Documenti*, **14**
(Atti del congresso 'Rapporti tra Alpi e Appennino'),
147–166.

STAMPFLI, G. M. & MARCHANT, R. H. 1997. Geody-
namic evolution of the Tethyan margins of the
Western Alps. *In*: PFIFFNER, O. A., LEHNER, P.,
HEITZMAN, P. Z., MUELLER, S. & STECK, A. (eds)
*Deep structure of the Swiss Alps – Results from NRP
20*. Birkhaüser, Basel, 223–239.

STAMPFLI, G. M. & BOREL, G. D. 2002. A plate tectonic
model for the Paleozoic and Mesozoic constrained by
dynamic plate boundaries and restored synthetic
oceanic isochrons. *Earth and Planetary Science
Letters*, **196**, 17–33.

STAMPFLI, G. M. & BOREL, G. D. 2004. The
TRANSMED transects in space and time: constraints

on the Paleotectonic evolution of the Mediterranean Domain. *In*: CAVAZZA, W., ROURE, F., SPAKMAN, W., STAMPFLI, G. M. & ZIEGLER, P. (eds) *The TRANSMED Atlas: The Mediterranean Region from Crust to Mantle*. Springer Verlag, Germany, 53–80.

STAMPFLI, G. M. & KOZUR, H. 2006. Europe from the Variscan to the Alpine cycles. *In*: GEE, D. G. & STEPHENSON, R. (eds) *European Lithosphere Dynamics*. Geological Society, London, Memoirs, **32**, 57–82.

STAMPFLI, G. M., MOSAR, J., MARCHANT, R., MARQUER, D., BAUDIN, T. & BOREL, G. 1998. Subduction and obduction processes in the western Alps. *In*: VAUCHEZ, A. & MEISSNER, R. (eds) *Continents and Their Mantle Roots*. Tectonophysics, **296**(1–2), 159–204.

STAMPFLI, G. M., BOREL, G., CAVAZZA, W., MOSAR, J. & ZIEGLER, P. A. 2001*a*. *The Paleotectonic Atlas of the Peritethyan Domain*. CD ROM; European Geophysical Society.

STAMPFLI, G. M., MOSAR, J., FAVRE, P., PILLEVUIT, A. & VANNAY, J.-C. 2001*b*. Permo-Mesozoic evolution of the western Tethyan realm: the Neotethys/East-Mediterranean connection. *In*: ZIEGLER, P. A., CAVAZZA, W., ROBERTSON, A. H. F. & CRASQUIN-SOLEAU, S. (eds) *PeriTethys memoir 6: Peritethyan rift/wrench basins and passive margins, IGCP 369*. Mémoires du Museum Nationale d'Histoire Naturelle, **186**, 51–108.

STAMPFLI, G. M., BOREL, G., MARCHANT, R. & MOSAR, J. 2002. Western Alps geological constraints on western Tethyan reconstructions. *Journal Virtual Explorer*, **8**, 77–106.

STAMPFLI, G. M., VAVASSIS, I., DE BONO, A., ROSSELET, F., MATTI, B. & BELLINI, M. 2003. Remnants of the Palaeotethys oceanic suture-zone in the western Tethyan area. *In*: CASSINIS, G. & DECANDIA, F. A. (eds) *Stratigraphic and Structural Evolution on the Late Carboniferous to Triassic Continental and Marine Successions in Tuscany (Italy): Regional Reports and General Correlation*. Bolletino della Società Geologica Italiana, Volume speciale, **2**, 1–24.

STEINER, C. W., HOBSON, A., FAVRE, P., STAMPFLI, G. M. & HERNANDEZ, J. 1998. The Mesozoic sequence of Fuerteventura (Canary islands): witness of an Early to Middle Jurassic sea-floor spreading in the Central Atlantic. *Geological Society of America Bulletin*, **110**(10), 1304–1317.

STÖCKLIN, J. 1968. Structural history and tectonics of Iran: a review. *American Association of Petroleum Geologists Bulletin*, **52**(7), 1229–1258.

STÖCKLIN, J. 1974. Possible ancient continental margin in Iran. *In*: BURK, C. A. & DRAKE, C. L. (eds) *The Geology of Continental Margins*. Springer Verlag, Germany, 873–887.

TANNER, D. C., BEHRMANN, J. H., ONCKEN, O. & WEBER, K. 1998. Three-dimensional retro-modelling of transpression on a linked fault system; the Upper Cretaceous deformation on the western border of the Bohemian Massif, Germany. Geological Society, London, Special Publications, **135**, 275–287.

THÖNI, M. & JAGOUTZ, E. 1992. Some new aspects of dating eclogites in orogenic belts: Sm–Nd, Rb–Sr and Pb–Pb isotopic results from the Austroalpine Saualpe and Koralpe type-locality (Carinthia/Styria, SE Austria). *Geochimica Cosmochimica Acta*, **56**, 347–368.

TOURY, A. 1985. *Etude géologique de la Vallée de la Valloirette entre Valloire, le Col du Galibier et les Aiguilles d'Arves (Alpes occidentales – Savoie)*. Savoie University, Chambéry, 207.

TRÜMPY, R. 1988. A possible Jurassic-Cretaceous transform system in the Alps and the Carpathians. *Geological Society of America Special Paper*, **218**, 93–109.

TRÜMPY, R. 1992. Ostalpen und Westalpen-Verbindendes und Trennendes. *Journal of Geology*, **135**(4), 875–882.

VAI, G. B. 1994. Crustal evolution and basement elements in the Italian area: paleogeography and characterization. *Bollettino di Geofisica Teorica ed Applicata*, **36**(141–144), 411–434.

VAN HINSBERGEN, D. J. J., HAFKENSCHEID, E., SPAKMAN, W., MEULENKAMP, J. E. & WORTEL, R. 2005. Nappe stacking resulting from subduction of oceanic and continental lithosphere below Greece. *Geology*, **33**(4), 325–328.

VERGÉS, J. & GARCIA-SENZ, J. 2001. Mesozoic evolution and Cainozoic inversion of the Pyrenean Rift. *In*: ZIEGLER, P. A., CAVAZZA, W., ROBERTSON, A. H. F. & CRASQUIN-SOLEAU, S. (eds) *PeriTethys memoir 6: Peritethyan rift/wrench basins and passive margins, IGCP 369*. Mémoires du Museum Nationale d'Histoire Naturelle, **186**, 187–212.

VON BLANCKENBURG, F. & DAVIES, J. H. 1995. Slab breakoff: a model for syncollisional magmatism and tectonics in the Alps. *Tectonics*, **14**, 120–131.

WORTEL, M. J. R. & SPAKMAN, W. 1992. Structure and dynamics of subducted lithosphere in the Mediterranean region. *Proceedings of the koninklijke nederlandse Akademie van Wetenschappen. Series B: Palaeontology, Geology, Physics, Chemistry, Anthropology*, **95**(3), 325–347.

WORTEL, M. J. R. & SPAKMAN, W. 1993. The dynamic evolution of the Apenninic-Calabrian, Helvetic, and Carpathian arcs: an unifying approach. *Terra Abstracts*, **5**, 97.

WORTMANN, U. G., WEISSERT, H., FUNK, H. & HAUCK, J. 2001. Alpine plate kinematics revisited: the Adria Problem. *Tectonics*, **20**, 134–147.

# The Calabrian Orocline: buckling of a previously more linear orogen

STEPHEN T. JOHNSTON[1]* & STEFANO MAZZOLI[2]

[1]*School of Earth & Ocean Sciences, University of Victoria, PO Box 3055
STN CSC, Victoria, British Columbia, Canada*

[2]*Dipartimento Scienze della Terra, Università di Napoli 'Federico II', Largo San
Marcellino 10, 10138 Napoli (NA), Italy*

*\*Corresponding author (e-mail: stj@uvic.ca)*

**Abstract:** Recent structural studies of the Apennines and the Calabrian orocline and a compilation of structural, stratigraphic, GPS and palaeomagnetic data from the central and western Mediterranean region show that beginning in the Late Miocene a N–S trending ribbon continent that had been previously deformed, and which we now recognize as the Apennine–Sicilian thrust belt, buckled eastward in response to northward movement of Africa relative to stable Europe. A simple geometric model is consistent with available data and shows how eastward buckling of an originally north–south continental beam explains: (1) opening of the Tyrrhenian Sea basin from 7–2 Ma, at which point sea-floor spreading ceases and the basin begins to shrink by southward subduction beneath Sicily; (2) the coeval development of an east-verging fold-and-thrust belt along the length of the Apennine–Sicilian belt in response to overthrusting of the autochthon to the east, followed by extension beginning at 1 Ma as the Apennine portion of the beam begins to retreat to the SW; and (3) subduction of continental and oceanic lithosphere east of the buckling beam into a trench that migrates eastward through time due to 'push back' by the buckling upper plate.

## Introduction

The Tyrrhenian Sea (Fig. 1) is amongst the world's youngest oceanic basins, having opened at rates of 60–100 km/Ma over the past 7–9 my (Rosenbaum & Lister 2004). The opening of the Tyrrhenian Sea region was coeval with significant crustal shortening in the Apennine and Sicilian thrust belts that bound the sea to the east and south, respectively (Fig. 1). Together these mountain chains define a highly arcuate continental orogen, the bend of which is commonly referred to as the Calabrian Orocline. There is concensus that the Calabrian orocline developed after 10 Ma, and is a tectonic feature that developed as a result of strain of an originally more linear orogen (Cavazza et al. 2004). However, the tectonic setting that gave rise to the bend, its 3D geometry, and the role the bend played in the opening of the Tyrrhenian Sea remain matters of debate (Mantovani et al. 1996; Tapponnier 1977). For instance, models of the bend as a thin-skinned feature, limited to the allochthonous Apennine–Sicilian thrust sheets and not present at deeper crustal or lithospheric levels (Eldredge et al. 1985), provides no explanation for the coeval formation of the bend and opening of the Tyrrhenian Sea. Models of the bend as a product of roll-back of a west- to north-dipping subduction zone along the east to south margin (from north to south,

respectively) of the Apennine–Sicilian peninsula (Kastens et al. 1988) provides no explanation for formation of the intervening Apennine–Sicilian fold-and-thrust belt coeval with opening of the Tyrrhenian Sea. In addition, bend formation in response to roll-back would require significant extension parallel to the length of the Apennine–Sicilian peninsula (perpendicular to the thrust belt), for which there is no evidence.

We ask 'Can the coeval opening of the Tyrrhenian Sea, shortening in the Apennine–Sicilian fold-and-thrust belt, and formation of the Calabrian orocline, be explained in a single tectonic model?' Toward answering this question we describe the major tectonic elements of the southern Italian region and summarize the neo-tectonic setting of the region. We review the data pertaining to the nature of the Calabrian Orocline, and end with the presentation of, and a discussion regarding, a model of lithospheric-scale orocline development. Our orocline model explains opening and the already initiated closing of the Tyrrhenian Sea, the coeval crustal shortening within the Apennine–Sicilian fold-and-thrust belt, and the recently initiated extension in the Apennine portion of the fold-and-thrust belt. Our model implies that the southeastward retreat of the trench bounding the east side of the Apennine–Sicilian orogen was driven, at least in part, by the eastward buckling

*From*: MURPHY, J. B., KEPPIE, J. D. & HYNES, A. J. (eds) *Ancient Orogens and Modern Analogues.*
Geological Society, London, Special Publications, **327**, 113–125.
DOI: 10.1144/SP327.7   0305-8719/09/$15.00 © The Geological Society of London 2009.

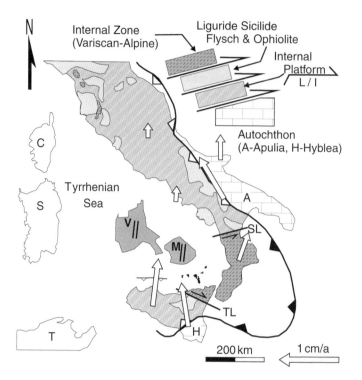

**Fig. 1.** A simplified geological map of the Italian peninsula, Sicily and the Tyrrhenian Sea. Inset cross-section shows schematic relationships between units. Black islands north of Sicily are volcanic islands of the Aeolian magmatic arc. The thick black line with teeth pointing into the hanging wall shows the Apennine (Italy)–Maghrebide (Sicily) thrust front (open teeth) that connect through the Calabrian subduction zone (black teeth) and which continues west through Tunisia (T). Black lines crossing the southernmost Apennines and NE Sicily are the sinistral Sangineto (SL) and dextral Taormina (TL) lines. The Tyrrhenian Sea is locally floored by oceanic crust (grey) that was generated at the Vavilov (V) and Marsili (M) north-trending spreading ridges (parallel lines), and is bound to the west by continental crust of Corsica (C) and Sardinia (S). Sicily is beginning to overthrust the south margin of the Tyrrhenian Sea along a north-verging thrust fault (line with small open teeth pointing into the hanging wall). Large white arrows show modelled crustal velocities based on geodetic data (Oldow *et al.* 2002).

of the Apennine–Sicilian fold-and-thrust belt. In this model, opening of the Tyrrhenian Sea is a product, rather than a cause of orocline formation. Orocline formation has probably been driven by the northward drift of Africa relative to Europe, buckling the intervening Apennine–Sicilian ribbon continent.

# Tectonic elements of southern Italy

The southern Italian region consists of four main tectono-geographic regions: (1) the Apennine–Sicilian fold-and-thrust belt; (2) the Apulian–Hyblean foreland; (3) the Tyrrhenian Sea; and (4) the Calabrian subduction zone. Hence, the Italian peninsula and its continuation through Calabria into Sicily, is divisible into a western to northern Tyrrhenian passive margin; a central Apennine–Sicilian mountain belt, and an eastern

Apulian–Hyblean foreland, all of which have different tectonic histories and which are currently moving with respect to one another (described below). Throughout the following discussion the Apennine geology is first introduced, and then the Sicilian equivalents (Fig. 2).

## The Apennine–Sicilian fold-and-thrust belt

Three main tectono-stratigraphic units comprise the fold-and-thrust belt. These are, in ascending structural order, a Triassic–Palaeogene continental ribbon, referred to as the Internal Platform; an overlying Jurassic–Cretaceous basinal sequence, known as the Ligurides–Sicilides, that include dismembered ophiolite; and an uppermost exotic Alpine continental margin on which sit crystalline Variscan thrust sheets, referred to as the Calabrian–Peloritanian nappes. The narrow (200–400 km wide) continental Internal Platform

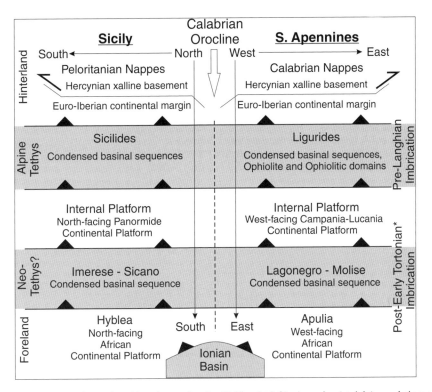

**Fig. 2.** Schematic structural–stratigraphic columns showing Sicilian (at left)–Apennine (at right) correlations around the Calabrian orocline. Continental platform (white) and basinal (grey) units are arranged in structural stacking order with the highest structural levels at the top and the lowest at the bottom. All units are bound by foreland-verging thrust faults (teeth point into the hanging wall). Post-early Tortonian imbrication of basinal strata of the Lagonegro–Immerese–Sicianian basin was preceded by an earlier inversion event of probable Early Miocene age (indicated by *).

was continuous to the north into the northern Apennines and to the south and then west through Sicily and into the Maghrebides of northern Africa (Elter *et al.* 2003). The structural emplacement of the Alpine–Variscan nappes above the Internal Platform along east-south-verging thrust faults was facilitated by Langhian (14–16 Ma) closure of the intervening Liguride-Sicilide (Alpine–Tethys) basin (Stampfli & Borel 2002) (Fig. 2).

To the east and south, the Internal Platform passes out into a condensed Triassic and locally Permian–Eocene basinal sequence of chert and deep-water shales of the Lagonegro–Imerese–Sicianian basin. Although the Ionian Sea has been interpreted as a preserved remnant of the Neotethys (Stampfli & Borel 2002), two observations argue against such a correlation: (1) the Lagonegro–Imerese–Sicianian basinal strata today lies west and north of, is thrust over, and is separated from the Ionian, by shelf and platformal strata of the continental Apulian–Hyblean foreland, making it unlikely that the two basins were continuous; and (2) closure of the Lagonegro–Imerese–Sicianian

basin in the Neogene included removal of the basement beneath the basin, while the Ionian basin remained open and intact. The presence of one or more straits through which the basins may have been linked, for example between Apulia and Hyblea (Gattacceca & Speranza 2002; Stampfli & Borel 2004), cannot be ruled out.

Closure of the Lagonegro–Imerese–Sicianian basin was complicated and included pre-Tortonian (>12 Ma) imbrication (Mazzoli *et al.* 2001; Shiner *et al.* 2004). Post-Tortonian (<7 Ma) NE–S-verging thrust faults resulted in overthrusting of the Lagonegro–Imerese–Sicianian basinal sequences by the Internal Platform and its structurally overlying units, imbrication of the basinal sequence and the removal of its transitional to oceanic lithospheric basement, and structural emplacement of the Internal Platform above continental Apulia–Hyblea foreland (Cello & Mazzoli 1999). These thrust faults steepen to the west, root into the sub-crustal lithospheric mantle, and offset the Moho (Barchi *et al.* 1998; Butler *et al.* 2004). Tomographic imaging shows a discontinuous, steeply west-dipping, relatively low velocity slab beneath the Apennines (Lucente *et al.* 1999; Spakman & Wortel 2004), which likely consists of transitional and entrained continental lithosphere (Faccenna *et al.* 2004).

Overthrusting of the Apulia–Hyblea ended after the Miocene, probably in the Early Pleistocene (Hippolyte *et al.* 1994; Patacca & Scandone 2001). A new tectonic regime, characterized by NE–SW extension (Cello *et al.* 1982; Hippolyte *et al.* 1994; Montone *et al.* 1999) was established within the mountain chain and the adjacent foothills to the east. The result has been the recent and ongoing development of extensional and related transcurrent faults that post-date and dissect the thrust belt (Butler *et al.* 2004; Cello *et al.* 1982).

The regional strike of Apennine–Sicilian nappes and thrust faults varies smoothly around the Calabrian arc (Fig. 3). As demonstrated by Eldredge *et al.* (1985), palaeomagnetic declinations vary around the bend, implying counter-clockwise rotation of the southern Apennines of up to 56°, and clockwise rotation of Sicily of up to 140°, but with a significant scatter. In addition to varying around the Calabrian bend, vertical axis rotations increase structurally up-section, such that the structurally highest thrust sheets exhibit the greatest rotations (Channell *et al.* 1980). Subsequent palaeomagnetic studies (Gattacceca & Speranza 2002; Rosenbaum & Lister 2004; Speranza *et al.* 1998, 1999, 2003) have served to confirm the results of the early palaeomagnetic studies. Rotations attributable to landslides have, at least locally, significantly complicated the story (Catalano *et al.* 1984; Grasso *et al.* 1983; Incoronato & Nardi 1989). Significant rotation of Pleistocene rocks (Scheepers & Langereis 1994) indicates that bend

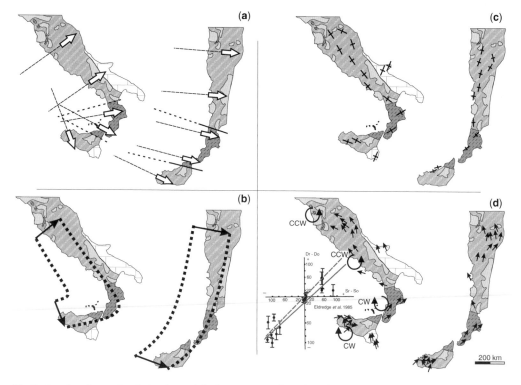

**Fig. 3.** A series of maps (**a–d**) showing, at left, the current configuration of southern Italy and, at right, maps depicting the palaeogeography at 10 Ma in which the Calabrian orocline has been restored according to available palaeomagnetic data. Unit fills are as in Figure 1. (a) Palinspastic restoration (dashed lines) of Oligocene–Miocene thrust sheets according to their vergence directions (arrows) results in a significant overlap of the restored southern Apennine and Sicilian thrust sheets – presenting a significant space problem (at left) that is ameliorated by undoing of the Calabrian orocline (at right). (b) Propagation of thrust sheets toward the foreland requires >225% extension around the Calabrian orocline (at left – note increase in length of dashed line as it propagates to the south and east), whereas pre-oroclinal thrusting does not require extension (at right). (c) The regional strike of structures, which varies smoothly around the orocline (at left) restores to a linear belt of structures (at right). (d) Palaeomagnetic directions vary as a function of change in strike [see inset at left from (Eldredge *et al.* 1985)] consistent with oroclinal bending. Declinations point uniformly north upon restoration of the orocline (at right).

formation is ongoing. Contrary to the palaeo-magnetic data, geodetic data imply fairly uniform northward movement, at rates of $>5$ mm a$^{-1}$, of the Apennine–Sicilian thrust belt (also referred to as Apulia) relative to a fixed Eurasian block around the length of the Calabrian orocline (Fig. 1) (Jimenez-Munt *et al.* 2003; Mattei *et al.* 2007; Nocquet & Calais 2003; Oldow *et al.* 2002).

## The Apulian–Hyblean Foreland

The foreland to the Apennine–Sicilian fold-and-thrust belt consists of a west- to north-facing continental margin sequence of Permian to Pleistocene age, consisting largely of platformal carbonates (Fig. 2). Unlike the Internal Platform of the Apennine–Sicilian fold-and-thrust belt, the foreland carbonate platform was little affected by Neogene tectonism, and is first overrun by orogenic flysch deposits in the Late Miocene. Apulia–Hyblea form part of a coherent continental lithospheric block referred to as Adria (Channell & Horvath 1976). Adria may be a northern promontory of Africa (Channell *et al.* 1979; McKenzie 1972) or a microplate that originated between Africa and Europe (Anderson & Jackson 1987; Dercourt *et al.* 1986). GPS data indicate that the Apulia–Hyblea portion of Adria is currently moving with and appears to be a part of the African continent (Nocquet & Calais 2003; Oldow *et al.* 2002).

Despite now being separated by Calabria, continuity of the Apulian–Hyblean continental margin beneath the Calabrian nappes is suggested by their shared stratigraphy and tectonic evolution (Catalano *et al.* 1976). Palaeomagnetic studies of Neogene strata confirm that Apulia has not rotated with respect to Hyblea, consistent with these platforms having developed on, and forming part of a coherent Adria lithospheric block (Besse *et al.* 1984; Eldredge *et al.* 1985; Scheepers 1992). Interpretations involving the presence of a strait floored by Ionian oceanic lithosphere separating Apulia and Hyblea require that subduction of the oceanic lithosphere underlying the strait occurred without resulting in rotation or translation of Apulia relative to Hyblea (Rosenbaum *et al.* 2002) and are inconsistent with Apulia–Hyblea moving coherently with and forming a coherent continuation of the African continent. Post-Tortonian closure of the Lagonegro–Imerese–Sicanian basin led to overthrusting of the Apulian–Hyblean foreland by the Internal Platform along NE- to S-verging thrust faults.

## The Tyrrhenian Sea

Tortonian rift deposits fringing the sea imply initial opening of the basin at 8–9 Ma and indicate that the sea is the youngest in the Mediterranean region (Sartori 2003). Crust underlying the basin varies between 'transitional' (15-km thick) to true oceanic crust (6-km thick) (Sartori 2003). The north-tapering, triangular shape of the basin implies that it opened about a pole of rotation to the north, consistent with the restricted development of oceanic crust in the southern Tyrrhenian sea (the region farthest from the Euler pole), the north–south trend of spreading ridges (e.g. the Vavilov and Marsili ridges) and the east–west strike of transform faults (Fig. 1) (Marani & Gamberi 2004). The basin is characterized by a positive Bouguer gravity anomaly and a geoid high (Cavazza *et al.* 2004) consistent with its relative youth. Tomographic studies imply that anomolously hot asthenosphere extends to depths of 300 km (Spakman & Wortel 2004). There is, however, no active spreading within the basin. The westerly Vavilov ridge was active from 7–3.5 Ma; spreading along the younger and more easterly Marsilli ridge started at about 2 Ma and ceased by 1 Ma (Kastens & others 1988). Eastward-younging of the basin is consistent with an eastward increase in heatflow to relatively high values ($>120$ mWm$^{-2}$) (Mongelli *et al.* 2004).

First motion analyses of earthquake data and direct stress measurements (Jimenez-Munt *et al.* 2003) imply convergence between and dextral, transpressional overthrusting of the southern Tyrrhenian Sea by the north coast of Sicily (Anderson & Jackson 1987). Actively developing NW-trending grabens that are rotating clockwise characterize the north coast of Sicily (Guarnieri 2004; Somma 2006). The southern margin of the Tyrrhenian sea is being overthrust along a seismically active, north-verging thrust system which roots beneath Sicily (Giunta *et al.* 2003). Regional compilations of earthquake seismic data indicate southeastward subduction of the southern Tyrrhenian Sea lithosphere beneath Sicily and Calabria (Anderson & Jackson 1987; Neri *et al.* 1996).

## Calabrian Subduction Zone

The marine region SE of Calabria is characterized by an accretionary prism bound to the SE by a poorly defined trench, SE of which lies Permian or younger oceanic crust of the Ionian Sea (Fig. 2) (Stampfli 2000). These observations indicate that the Ionian oceanic crust is being consumed in a subduction zone that dips NW beneath Calabria. The volcanic Aeolian Islands, which lie north of Sicily and Calabria, are inferred to be an island arc developed above the dewatering, down-going Ionian oceanic slab (Fig. 1) (Argnani & Savelli 1999). Tomographic studies indicate the presence of a discontinuous, steeply dipping, high velocity zone that dips down to depths of 500 km beneath

Calabria. The upper portions of the high velocity zone appears continuous with the subducting Ionian slab (Lucente *et al.* 1999; Spakman & Wortel 2004), implying that what is being imaged is, at least in part, the Ionian oceanic lithosphere. Moderate to deep earthquakes occurring within this high velocity zone are consistent with a subduction setting (Wortel & Spakman 2000).

Subduction beneath Calabria began at 7 Ma, after a 7 my period of tectonic quiescence marked by a 14–7 Ma magmatic gap that characterizes the region (Argnani & Savelli 1999; Cavazza *et al.* 2004). Low volumes of arc magmas, characterized by the eruption of andesite to rhyolite and the intrusion of granodiorite to monzogranite, occurs from 7 Ma onwards (Cavazza *et al.* 2004; Savelli 2002), presumably providing a record of ongoing subduction throughout this interval. Subduction has consumed a 500 km-wide strip of lithosphere, as indicated by the 500 km length of the tomographically-imaged subducted slab (Lucente *et al.* 1999; Spakman & Wortel 2004). The southern Tyrrhenian Sea is 600 km wide, implying that there is a one-to-one relationship between the amount of lithosphere subducted and the width of the Tyrrhenian Sea. Hence eastward retreat of the subduction zone was accommodated by the opening of the Tyrrhenian Sea in its wake.

## The calabrian orocline

Carey was the first to suggest that the continuity of the geological belts of the Apennine–Sicilian fold-and-thrust belt around the Calabrian bend implies that the bend is an orocline, that the geological belts were formerly more linear in map-view, and that their current geometry is the result of a tectonic rotation (Carey 1955). Several other lines of evidence support an oroclinal origin for the bend (Catalano *et al.* 1976) (Fig. 3). Oligocene and Miocene thrust faults around the Calabrian orocline verge E–NE in the Apennines and south in Sicily. Palinspastic restoration of these thrust sheets results in a significant space problem as the restored Apennine and Sicilian units exhibit considerable overlap (Fig. 3). This space problem can also be thought of as a line length problem: propagation of thrust sheets toward the foreland around the length of the Calabrian orocline should have resulted in enormous (>200%) extension of the thrust sheets along normal faults oriented perpendicular to the thrust faults (Fig. 3). On the contrary, extension is restricted to the Apennines, and neotectonic studies imply continued strike-parallel compression around the Calabrian bend (Di Bucci & Mazzoli 2002; Di Bucci *et al.* 2003).

Palaeomagnetic declination varies as a function of structural strike, implying that bending of an originally more linear belt was responsible for the orocline formation (Fig. 3) (Eldredge *et al.* 1985). However, the increase in vertical axis rotations structurally up-section led to concensus that the orocline was thin-skinned, that the vertical axis rotations were confined to thin allochthonous thrust sheets, and that the rotations occurred during thrust sheet emplacement. Timing constraints imply that orocline formation commenced in the Tortonian, coeval with the initiation of opening of the Tyrrhenian Sea (Rosenbaum & Lister 2004; Speranza *et al.* 1999) and with thick-skinned crustal shortening along the length of the Apennine–Sicilian fold-and-thrust belt.

## A tectonic model

We assume that the Calabrian bend is an orocline that developed in response to lithospheric-scale buckling of a formerly linear portion of a ribbon continent about a vertical axis of rotation. A model of an eastward-buckling beam (Fig. 4) predicts rapid extension behind (west of) the beam giving rise to a significant basin, representing the Tyrrhenian Sea. Basin formation behind the buckling beam is coeval with strike-normal compression within the beam as it migrates east over, and collides with, autochthonous foreland. A model of buckling requires counter-clockwise rotation of the Apennines and clockwise rotation of the thrust belt in Sicily, consistent with available GPS and palaeomagnetic data. Our model predicts that the maximum areal extent of the Tyrrhenian Sea was achieved at *c.* 2 Ma (Fig. 4), consistent with the termination of spreading along the Marselli ridge, the youngest spreading ridge in the basin, at 2 Ma. The subsequent and increasingly rapid reduction in the area of the Tyrrhenian Sea predicted by our model explains the recent initiation of southward subduction of the southern Tyrrhenian Sea lithosphere beneath Sicily and Calabria. The Early Pliestocene switch from crustal shortening to extension along the length of the Apennines is explained as a result of the predicted southwestward retreat of the Apennines after reaching a maximum easterly position at about 2 Ma (Fig. 4). In our model, orocline formation and the related opening of the Tyrrhenian Sea is driven by north–south shortening and requires convergence between autochthonous Africa and Europe since about 9 Ma (Fig. 5).

Map-view palinspastic restoration of the continental ribbon formed by the Apennine–Sicilian fold-and-thrust belt into a north–south trending linear belt (Figs 4 & 5) removes the space and line length problems associated with the fold-and-thrust belt by restoring the Apennine and Sicilian thrust faults into a continuous north–south trending, east-verging belt (Figs 4 & 5). Unbuckling accomplished

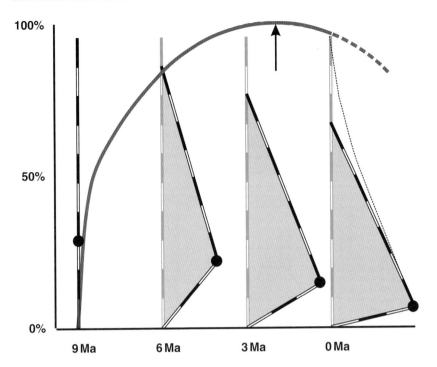

**Fig. 4.** A geometric model showing how buckling of an originally linear beam can explain the origin and evolution of a basin and the coeval development of a spatially-related orocline. The model is organized to represent buckling of an originally linear Apennine–Sicilian continental ribbon, represented at left by the black and white bar. Throughout the model, the buckling bar is pinned at the bottom, with a fixed hinge point (black circle) separating a short lower (southerly) portion of the beam, representing the Sicilian–Calabrian portion of the Calabrian orocline, from a longer upper (northerly) portion, representing the Apennines. A grey and white unbuckled beam is shown in each panel for reference. Buckling is constrained to occur by eastward migration of the hinge, is accomplished by southward migration of the upper (northern) most point of the beam at a constant rate and is assumed to have begun at 9 Ma. The developing buckle is shown in the three panels at right, at 3 Ma intervals. The grey region behind the eastwardly migrating beam represents the Tyrrhenian Sea. The relative area of the basin, from 0–100% (see axis at left) through time is indicated by the solid grey line (dashed line shows projected future area of basin). The arrow indicates the point at which the basin reaches its maximum extent (at c. 2 Ma). By 1 Ma, the basin begins to shrink noticeably. The last panel, at right, represents the current geometry of the Calabrian orocline and Tyrrhenian Sea and requires 30% north–south shortening of the system. The dashed line shows the maximum easterly extent of the buckling beam, also achieved at c. 2 Ma, requiring significant post-2 Ma southwestwardly retreat of the northern (Apennine) portion of the buckling beam.

by undoing the palaeomagnetically constrained clockwise rotations of Sicily–Calabria and the counter-clockwise rotations of the Apennines yields an almost linear peninsula, consistent with our model. The remaining variability in the restored declinations may be attributable to the presence of additional minor bends which have not been restored in our model. Small-scale local rotations adjacent to individual faults may also contribute to the variability. Land-sliding (Incoronato & Nardi 1989) is likely an important factor in contributing to local palaeomagnetic variability. Unbuckling closes the Tyrrhenian Sea and results in juxtaposition of the Apennine–Sicilian thrust belt and the Corsica–Sardinia peninsula, and restores the Sicilian portion of the ribbon continent 600 km to the west

and 100 km to the south relative to northern Italy (Fig. 5).

Eastward migration of the buckling continental beam formed by the Apennine–Sicilian belt requires that autochthonous foreland lithosphere be removed by subduction or dramatically thickened. Ongoing subduction of Ionian oceanic lithosphere beneath Calabria accounts for the foreland lithosphere overrode by the Calabrian–Sicilian southern portion of the orocline. The discontinuous slab imaged beneath the Apennines (Lucente *et al.* 1999) probably constitutes the transitional continental lithosphere that formed the basement to the Lagonegro basin and which was subsequently subducted beneath the encroaching Apennine peninsula.

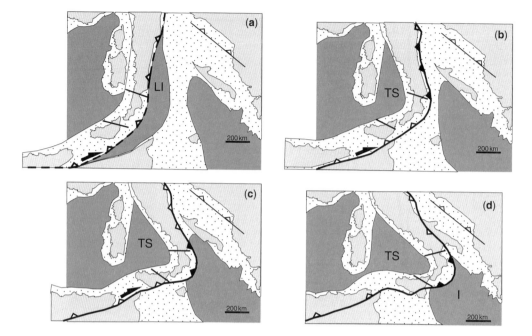

**Fig. 5.** Palaeogeographic evolution of the Apennine–Sicilian belt in response to development of the Calabrian Orocline from 10 Ma (**a**) through 6 Ma (**b**) 3 Ma (**c**) to present (**d**). Present-day continental outlines (grey) are shown with the full extent of continental crust shown in stipple. Oceanic to transitional crust in dark grey. Africa moves north at 1 cm/a throughout this period. (a) Pre-Calabrian orocline palaeogeography with the Apennine–Maghrebide belt restored to a linear geometry adjacent to Corsica-Sardinia and separated to the east from the Apulia–Hyblea continental ribbon by the oceanic to transitional Lagonegro–Imerese–Sicanian (LI) basin. Black lines across the southernmost Apennines and Sicily are the Sangineto and Taormina lines, respectively. Dashed line with open teeth pointing into the upper plate shows incipient thrust fault/subduction zone that develops as the Apennine–Maghrebide belt buckles. In this and subsequent frames, African–European convergence in the Dinarides is indicated by a thrust fault along the eastern margin of the Adriatic Sea. (b) By 6 Ma, eastward buckling Apennine–Maghrebide belt, accommodated by the counter-clockwise rotating Apennines and the clockwise rotating Maghrebides, has subducted the Lagonegro–Imerese–Sicanian basin (black teeth indicating where subduction occurred), with off-scraped basinal sediments having been imbricated beneath the Internal Platform and thrust over the Apulia–Hyblea continental ribbon, and opened the incipient Tyrrhenian Sea (TS). (c) Continued buckling results in the Apennine–Maghrebide belt overthrusting oceanic crust of the Ionian sea SE of Apulia–Hyblea, initating subduction of Ionian (I) oceanic crust (black teeth) with along-strike fold-and-thrust belt development (open teeth) and further opening the TS. (d) Ongoing buckling of the Apennine–Maghrebide belt with continued fold-and-thrust belt development (open teeth) and Ionian subduction (black teeth). Sicily has started to rotate over the southern TS (north-verging thrust fault with small black teeth pointing into hanging wall).

In contrast to the Apennines and the Calabrian hinge region of the orocline, the E–NE displacement of Sicily and the Maghrebides is primarily accomplished by strike-slip displacement. Little or no subduction beneath the southern side of this portion of the belt is required (Fig. 5), consistent with the lack of a subducted slab imaged beneath this region, the paucity of arc magmatism, and the preponderance of Messinian and younger dextral transpressional deformation. An implication of our model is that the Tell thrust front of northern Africa, which bounds the Maghrebides to the south, is probably largely a dextral strike-slip boundary which has been over-ridden by thrust faults with minor (in comparison) dip-slip displacement.

## Discussion

We explain the opening and already ongoing destruction of the Tyrrhenian Sea, compression across the Apennine–Sicilian belt, as well as the recent switch to extension along the Apennines, as a consequence of formation of the Calabrian Orocline. Our model entails the buckling of an originally linear ribbon continent. Such a model implies that the orocline is of lithospheric scale, involving the crust and some of the lithospheric mantle. The buckling beam is bound to the east and south by the steeply west- and north-dipping subduction surface, and to the west and north by the Tyrrhenian lithosphere, and is therefore somewhat wedge-shaped (narrowing at depth) in profile. Buckling of a lithospheric-scale beam is further implied by the subduction of the lithosphere lying east of the eastward-migrating buckle; the opening and growth of an oceanic basin behind it to the west; and by deep reflection seismic and structural studies demonstrating that fold-and-thrust belt deformation coeval with orocline formation was thick-skinned, involving thrust faults that rooted steeply to the west into the sub-crustal mantle, and which cut and offset the Moho. Removal of lithosphere east of the eastward buckling beam was accommodated by the subduction of oceanic lithosphere to the south and transitional lithosphere to the east. Eastward migration of the trench was a consequence of the eastward buckling, upper plate, Apennine–Sicilian beam and hence trench retreat was at least in part a product of 'push back'. In other words, trench retreat was not a passive product of sinking of the lower plate (roll-back), but was an active product of buckling of a beam within the upper plate about a vertical axis of rotation (push back).

Is it probably not reasonable to consider a continental ribbon as a strong beam relative to adjacent oceanic crust. What facilitated buckling of the Apennine–Sicilide beam was the presence of a subduction zone along the length of its eastern margin. Hence the buckling Apennine–Sicilide beam need only have been stronger than the subduction interface bounding the east-margin of the beam. Whether a buckling continental beam is capable of initiating subduction remains unknown. The Lagonegro–Imerese–Sicanian basin east of the Apennine–Sicilide beam had already been deformed prior to the initiation of orocline formation. Subduction may, therefore, have initiated along pre-existing faults.

As with any fold, determining why the hinge developed where it did, remains speculative. Smaller coeval, parasitic bends are present along the length of the Apennine–Sicilide beam, including the Umbrian arc. It may be that bending initially resulted in a number of smaller bends continuous with one another, with much of the strain subsequently being localized within the Calabrian bend. Alternatively, it may be that an external factor dictated the location of the hinge. For example, an oceanic strait separating Apulia from Hyblia may have facilitated bending in the Calabrian region, and would explain the subsequent migration of the subduction zone along which the Lagonegro–Imerese–Sicanian basin was consumed into the Ionian basin.

The Calabrian Orocline has previously been interpreted as a thin-skinned feature (Eldredge *et al.* 1985), because palaeomagnetic and structural data indicate that the structurally highest Apennine–Sicilian thrust sheets are the most rotated. The larger rotations recorded by the structurally highest thrust sheets can, however, be explained as a result of thick-skinned orocline formation (Fig. 6). The line length around the outside, foreland-portion of an orocline significantly

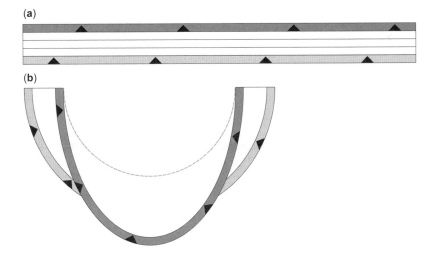

**Fig. 6.** (a) Map-view of a linear lithospheric beam. Light-grey strip indicates the foreland-side of the beam, grey the hinterland side. (b) The same beam buckled into an orocline that has flexed out toward the foreland. The ends of the beam are kept fixed during orocline formation. Because of the excess line length around the inside of the orocline, the hinterland-portion of the buckling beam (grey) is thrust out over the foreland (light grey). Note that the hinterland derived thrust sheet forms the structurally highest element in the orocline, and is characterized throughout by the largest rotations.

exceeds the distance around its hinterland portion – the wider the buckling beam, the greater the outer to inner arc line length discrepancy. The result is that a strip of lithosphere along the inside (hinterland) portion of a buckling beam will significantly exceed the line length of the inside arc of the orocline. A likely manifestation of this excess line length is that lithosphere along the inside of an orocline will be thrust out over its more external foreland regions (Fig. 6), forming the structurally highest thrust sheets. These structurally high thrust sheets will be everywhere more rotated and display a greater arc curvature than the structurally lower foreland portions of the orocline (Fig. 6). Thus the observed up-structural section increase in the magnitude of rotations seen in structural and palaeomagnetic data is a predicted consequence of thick-skinned orocline formation. Our explanation for the up-section increase in rotation requires that this should decrease away from the hinge region of the orocline. Hence our model is testable.

Geodetic data imply northward motion of the Apennines and correlative Sicilian crust relative to a fixed, stable central Europe, whereas palaeomagnetic data imply ongoing opposing senses of rotation. This apparent dichotomy is, however, consistent with a model of orocline formation through buckling of a linear ribbon continent. In our model, the southern Apennines and Sicily–Calabria move north relative to autochthonous stable Europe throughout development of the Calabrian Orocline, despite strongly opposing counter-clockwise and clockwise rotations, respectively (Fig. 5). Hence interpreting the Calabrian Orocline as the result of buckling resolves the dichotomy of uniform convergence of the Apennines and Sicily–Calabria with stable Europe coeval with opposing senses of rotation.

Stratigraphy and geological structures of the Apennine–Sicilian belt are continuous to the west into the Maghrebian chain of northern African, and there continuously on through the Tell Mountains of Algeria, into the Rif Mountains of

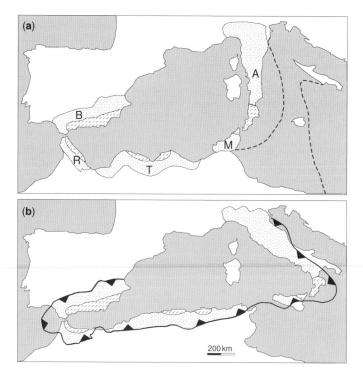

**Fig. 7.** Palaeogeographic evolution of the western Mediterranean from 10 Ma (**a**) to present (**b**). The Western Mediterranean ribbon continent is stippled and includes earlier emplaced nappes of Variscan affinity (wave stipple) that structurally overly the Internal Platform (spot stipple), and can be traced westward from the Apennines (A) through the Maghrebides (M) of Sicily, the Tell (T), the Rif (R) and the Betic (B) mountains. (a) With Africa restored > 100 km to the south, the ribbon continent is characterized by a convex to the south orocline whose apex is located in the Tell region. Dashed lines indicate approximate boundaries of continental lithosphere of the Apulia–Hyblean foreland. (b) Northward displacement of Africa flattens the Tell orocline, tightens the Betic-Rif orocline, and causes escape of the ribbon continent to the east, giving rise to the Calabrian orocline.

Morocco and around a bend, the Gibraltar Orocline, similar in scale to the Calabrian orocline into the Betic mountains of southern Spain (Fig. 7) (Faccenna *et al.* 2004; Rosenbaum *et al.* 2002). Hence the Apennine–Sicilian belt is part of a much longer Western Mediterranean ribbon continent that continues SW into northern Africa (Fig. 7). Structural continuity along the length of this ribbon continent provides a possible mechanism for producing the rapid eastward translation of the buckling Apennine–Maghrebian portion of the continental ribbon. We postulate that prior to 10 Ma the ribbon continent was characterized by a convex to the south geometry with one or more bends in the Tell portion of the ribbon continent (Fig. 7). If the ribbon continent was pinned or fixed in the Gibraltar region to the west, and in the southern Alps to the east, then slow northward motion of Africa would result in rapid easterly-directed escape of the ribbon continent, as proposed by Mantovani *et al.* (1996) and could explain the brisk rate of opening of the Tyrrhenian Sea behind the buckling ribbon continent. A model of easterly escape of the ribbon continent therefore predicts that the Tell portion of the ribbon was formerly characterized by a far more arcuate geometry that has been progressively flattened by the northward motion of Africa over the past 10 Ma (Fig. 7). In addition, our model suggests that the Tell thrust front .is largely a dextral strike-slip boundary. Both these predictions, unbending of the Tell, and largely strike-slip motion between the ribbon continent and Africa, are testable through palaeomagnetic and structural studies.

## Conclusions

Opening of the Tyrrhenian Sea was a direct result of formation of the Calabrian Orocline. Buckling of the Apennine–Maghrebide peninsula was facilitated by over-riding and subducting lithosphere to the east, and the opening of a basin, now the Tyrrhenian Sea, to the west. Hence the Calabrian Orocline is a thick-skinned tectonic feature attributable to an Oroclinal Orogeny that involved buckling of a formerly linear ribbon continent. A model of Oroclinal Orogeny for the origin of the Calabrian Orocline explains: (1) the rapid opening of the Tyrrhenian Sea; (2) the cessation of sea-floor spreading within the basin at 2 Ma; (3) the subsequent and ongoing reduction in the size of the basin by southward subduction of Tyrrhenian Sea lithosphere beneath Sicily opening; and (4) the northward displacement of Sicily–Calabria and the Apennines, despite their respective opposed clockwise and counterclockwise rotations. The Apennine–Maghrebide peninsula is part of a larger Western Mediterranean ribbon

continent. Our model predicts that eastward buckling of the Apennine–Maghrebide peninsula occurred as a result of the northward encroachment of Africa on, and subsequent straightening of, the originally convex to the south Tell portion of the western Mediterranean ribbon continent. Consequently we are faced with the rather surprising conclusion that the Calabrian Orocline owes its existence to the unbending of a Tellian orocline.

The University of Lausanne and the Geological Museum of Lausanne generously made available office space and research facilities to STJ. G. Borel and G. Stampfli are thanked for sharing their expertise on Tethyan geology. This research was funded by an NSERC Discovery Grant, and by a research fellowship from the University of Lausanne to STJ. R. Van der Voo, R. Psyklywec and G. Borel are thanked for their in depth reviews that significantly improved our manuscript. B. Murphy's editorial efforts and encouragement were most appreciated. This is a contribution to IGCP Project 453 Ancient Orogens and Modern Analogues.

## References

ANDERSON, H. & JACKSON, J. 1987. Active tectonics of the Adriatic region. *Royal Astronomical Society Geophysical Journal*, **91**, 937–983.

ARGNANI, A. & SAVELLI, C. 1999. Cenozoic volcanism and tectonics in the southern Tyrrhenian sea: space-time distribution and geodynamic significance. *Journal of Geodynamics*, **27**, 409–432.

BARCHI, R., MINELLI, G. & PIALLI, G. 1998. The CROP 03 profile: a synthesis of results on deep structures of the Northern Apennines. *Memorie della Società Geologica Italiana*, **52**.

BESSE, J., POZZI, J.-P., MASCLE, G. & FEINBERG, H. 1984. Palaeomagnetic study of Sicily: consequences for the deformation of Italian and African margins over the last 100 million years. *Earth and Planetary Science Letters*, **67**, 377–390.

BUTLER, R. W. H., MAZZOLI, S. *ET AL.* 2004. Applying thick-skinned tectonic models to the Apennine thrust belt of Italy – Limitations and implications. *In*: MCCLAY, K. R. (ed.) *Thrust Tectonics and Hydrocarbon Systems*. Volume Memoir 82. American Association of Petroleum Geology, 647–667.

CAREY, S. W. 1955. The Orocline concept in geotectonics – Part 1. *Proceedings of the Royal Society of Tasmania*, **89**, 255–288.

CATALANO, R., CHANNELL, J. E. T., D'ARGENIO, B. & NAPOLEONE, G. 1976. Mesozoic palaeogeography of the southern Apennines and Sicily. *Memoir Society Geological Italy*, **15**, 95–118.

CATALANO, R., DARGENIO, B., GREGOR, C. B., NAIRN, A. E. M. & NARDI, G. 1984. The Mesozoic Volcanics of Western Sicily. *Geologische Rundschau*, **73**, 577–598.

CAVAZZA, W., ROURE, F. & ZIEGLER, P. A. 2004. The Mediterranean area and the surrounding regions; active processes, remnants of former Tethyan oceans and related thrust belts. *In*: CAVAZZA, W., ROURE,

F., SPAKMAN, W., STAMPFLI, G. M. & ZIEGLER, P. A. (eds) *The TRANSMED Atlas; the Mediterranean Region from Crust to Mantle; Geological and Geophysical Framework of the Mediterranean and the Surrounding Areas.* Berlin, Springer-Verlag, 1–29.

CELLO, G. & MAZZOLI, S. 1999. Apennine tectonics in southern Italy: a review. *Journal of Geodynamics,* **27,** 191–211.

CELLO, G., GUERRA, I., TORTORICI, L., TURCO, E. & SCARPA, R. 1982. Geometry of the neotectonic stress-field in southern Italy – geological and seismological evidence. *Journal of Structural Geology,* **4,** 385–393.

CHANNELL, J. E. T. & HORVATH, F. 1976. The African-Adriatic promontory as a paleogeographical premise for Alpine orogeny and plate movements in the Carpatho-Balkan region. *Tectonophysics,* **35,** 71–110.

CHANNELL, J. E. T., D'ARGENIO, B. & HORVATH, F. 1979. Adria, the African promontory in Mesozoic Mediterranean palaeogeography. *Earth-Science Reviews,* **15,** 213–292.

CHANNELL, J. E. T., CATALANO, R. & DARGENIO, B. 1980. Paleomagnetism and deformation of the Mesozoic Continental-Margin in Sicily. *Tectonophysics,* **61,** 391–407.

DERCOURT, J., ZONENSHAIN, L. P. *ET AL.* 1986. Geological evolution of the Tethys belt from the Atlantic to the Pamirs since the Liassic. *Tectonophysics,* **123,** 241–315.

DI BUCCI, D. & MAZZOLI, S. 2002. Active tectonics of the Northern Apennines and Adria geodynamics: new data and a discussion. *Journal of Geodynamics,* **34,** 687–707.

DI BUCCI, D., MAZZOLI, S., NESCI, O., SAVELLI, D., TRAMONTANA, M., DE DONATIS, M. & BORRACCINI, F. 2003. Active deformation in the frontal part of the Northern Apennines: insights from the lower Metauro River basin area (northern Marche, Italy) and adjacent Adriatic off shore. *Journal of Geodynamics,* **36,** 213–238.

ELDREDGE, S., BACHTADSE, V. & VAN DER VOO, R. 1985. Paleomagnetism and the orocline hypothesis. *Tectonophysics,* **119,** 153–179.

ELTER, P., GRASSO, M., PAROTTO, M. & VEZZANI, L. 2003. Structural setting of the Apennine-Maghrebian thrust belt. *Episodes,* **26,** 205–211.

FACCENNA, C., PIROMALLO, C., CRESPO-BLANC, A., JOLIVET, L. & ROSSETTI, F. 2004. Lateral slab deformation and the origin of the western Mediterranean arcs. *Tectonics,* **23,** TC1012.

GATTACCECA, J. & SPERANZA, F. 2002. Paleomagnetism of Jurassic to Miocene sediments from the Apenninic carbonate platform (southern Apennines, Italy): evidence for a 60° counterclockwise Miocene rotation. *Earth and Planetary Science Letters,* **201,** 19–34.

GIUNTA, G., NIGRO, F. *ET AL.* 2003. Il terremoto di Palermo (6 settembre 2002) nel quadro della tettonica attiva della Sicilia nord occidentale e del Tirreno meridionale. *Studi Geologici Camerti, Nuova Serie,* **1/2003,** 57–68.

GRASSO, M., LENTINI, F., NAIRN, A. E. M. & VIGLIOTTI, L. 1983. A Geological and Palaeomagnetic Study of the Hyblean Volcanic-Rocks, Sicily. *Tectonophysics,* **98,** 271–295.

GUARNIERI, P. 2004. Structural evidence for deformation by block rotation in the context of transpressive tectonics, northwestern Sicily (Italy). *Journal of Structural Geology,* **26,** 207–219.

HIPPOLYTE, J. C., ANGELIER, J. & ROURE, F. 1994. A major geodynamic change revealed by Quaternary stress patterns in the southern Apennines (Italy). *Tectonophysics,* **230,** 199–210.

INCORONATO, A. & NARDI, G. 1989. Palaeomagnetic evidences for a peri-Tyrrhenian orocline. *In:* BORIANI, A., BONAFEDE, M. & VAI, G. B. (eds) *The Lithosphere in Italy.* Advances in Earth Science Research Roma, Accademia Nazionale dei Lincei, 217–227.

JIMENEZ-MUNT, I., SABADINI, R., GARDA, A. & BIANCO, G. 2003. Active deformation in the Mediterranean from Gibraltar to Anatolia inforred from numerical modeling and geodetic and seismological data. *Journal of Geophysical Research,* **108,** doi: 10: 1029/2001IB001544.

KASTENS, K. *ET AL.* 1988. ODP Leg 107 in the Tyrrhenian Sea: insights into passive margin and back-arc basin evolution. *Geological Society of America Bulletin,* **100,** 1140–1156.

LUCENTE, F. P., CHIARABBA, C., CIMINI, G. B. & GIARDINI, D. 1999. Tomographic constraints on the geodynamic evolution of the Italian region. *Journal of Geophysical Research-Solid Earth,* **104,** 20 307–20 327.

MANTOVANI, E., ALBARELLO, D., TAMBURELLI, C. & BABBUCCI, D. 1996. Evolution of the tyrrhenian basin and surrounding regions as a result of the Africa-Eurasia convergence. *Journal of Geodynamics,* **21,** 35–72.

MARANI, M. P. & GAMBERI, F. 2004. Distribution and nature of submarine volcanic landforms in the Tyrrhenian Sea: the arc vs the backarc. *Memoir Carta Geologica D'Italia,* **64,** 109–126.

MATTEI, M., CIFELLI, F. & D'AGOSTINO, N. 2007. The evolution of the Calabrian Arc: evidence from palaeomagnetic and GPS observations. *Earth & Planetary Science Letters,* **263,** 259–274.

MAZZOLI, S., BARKHAM, S., CELLO, G., GAMBINI, R., MATTIONI, L., SHINER, P. & TONDI, E. 2001. Reconstruction of continental margin architecture deformed by the contraction of the Lagonegro Basin, southern Apennines, Italy. *Journal of the Geological Society, London,* **158,** 309–319.

MCKENZIE, D. 1972. Active tectonics of the Mediterranean region. *Royal Astronomical Society Geophysical Journal,* **30,** 109–185.

MONGELLI, F., ZITO, G., DE LORENZO, S. & DOGLIONI, C. 2004. Geodynamic interpretation of the heat flow in the Tyrrhenian Sea. *Carta Geologica D'Italia,* **64,** 71–82.

MONTONE, P., AMATO, A. & PONDRELLI, S. 1999. Active stress map of Italy. *Journal of Geophysical Research-Solid Earth,* **104**(B11), 25 595–25 610.

NERI, G., CACCAMO, D., COCINA, O. & MONTALTO, A. 1996. Geodynaimc implications of earthquake data in the southern Tyrrhenian sea. *Tectonophysics,* **258,** 223–249.

NOCQUET, J.-M. & CALAIS, E. 2003. Crustal velocity field of western Europe from permanent GPS array solutions, 1996–2001. *Geophysical Journal International*, **154**, 72–88.

OLDOW, J. S., FERRANTI, L. *ET AL.* 2002. Active fragmentation of Adria, the north African promontory, central Mediterranean orogen. *Geology*, **30**, 779–782.

PATACCA, E. & SCANDONE, P. 2001. Late thrust propagation and sedimentary response in the thrust-belt-foredeep system of the Southern Apennines (Pliocene-Pleistocene). *In*: VAI, G. B. & MARTINI, I. P. (eds) *Anatomy of an Orogen: the Apennines and Adjacent Mediterranean Basins*. London, Kluwer Academic Publishers, 401–440.

ROSENBAUM, G. & LISTER, G. S. 2004. Neogene and Quaternary rollback evolution of the Tyrrhenian Sea, the Apennines, and the Sicilian Maghrebides. *Tectonics*, **23**.

ROSENBAUM, G., LISTER, G. S. & DUBOZ, C. 2002. Reconstruction of the tectonic evolution of the western mediterranean since the Oligocene. *Journal of the Virtual Explorer*, **8**, 105–124.

SARTORI, R. 2003. The Tyrrhenian back-arc basin and subduction of the Ionian lithosphere. *Episodes*, **26**, 217–221.

SAVELLI, C. 2002. Time-space distribution of magmatic activity in the western mediterranean and peripheral orogens during the past 30 Ma (a stimulus to geodynamic considerations). *Journal of Geodynamics*, **34**, 99–126.

SCHEEPERS, P. J. J. 1992. No tectonic rotation for the Apulia-Gargano foreland in the Pleistocene. *Geophysical Research Letters*, **19**, 2275–2278.

SCHEEPERS, P. J. J. & LANGEREIS, C. G. 1994. Palaeomagnetic evidence for counter-clockwise rotations in the southern Apennines fold-and-thrust belt during the Late Pliocene and middle Pleistocene. *Tectonophysics*, **239**, 43–59.

SHINER, P., BECCACINI, A. & MAZZOLI, S. 2004. Thin-skinned versus thick-skinned structural models for Apulian carbonate reservoirs: constraints from the Val d'Agri Fields, S Apennines, Italy. *Marine and Petroleum Geology*, **21**, 805–827, doi: 10.1016/jmarpetgeo.2003.11.020.

SOMMA, R. 2006. The south-western side of the Calabrian Arc (Peloritani Mountains): geological, structural and AMS evidence for passive clockwise rotations. *Journal of Geodynamics*, **41**, 422–439.

SPAKMAN, W. & WORTEL, R. 2004. A tomographic view on western Mediterranean geodynamics. *In*: CAVAZZA, W., ROURE, F., SPAKMAN, W., STAMPFLI, G. M. & ZIEGLER, P. A. (eds) *The Transmed Atlas: The Mediterranean Region from Crust to Mantle*. Berlin, Springer, 31–52.

SPERANZA, F., MATTEI, M., NASO, G., DI BUCCI, D. & CORRADO, S. 1998. Neogene–Quaternary evolution of the central Apennine orogenic system (Italy): a structural and palaeomagnetic approach in the Molise region. *Tectonophysics*, **299**, 143–157.

SPERANZA, F., MANISCALCO, R., MATTEI, M., STEFANO, D. & BUTLER, R. W. H. 1999. Timing and magnitude of rotations in the frontal thrust systems of southwestern Sicily. *Tectonics*, **18**, 1178–1197.

SPERANZA, F., MANISCALCO, R. & GRASSO, M. 2003. Pattern of orogenic rotations in central-eastern Sicily: implications for the timing of spreading in the Tyrrhenian Sea. *Journal of Geological Society, London*, **160**, 183–195.

STAMPFLI, G. M. 2000. Tethyan Oceans. *In*: BOZKURT, E., WINCHESTER, J. A. & PIPER, J. D. A. (eds) *Tectonics and magmatism in Turkey and the surrounding area*. Geological Society, London, Special Publications, **173**, 1–23.

STAMPFLI, G. M. & BOREL, G. D. 2002. A plate tectonic model for the Paleozoic and Mesozoic constrained by dynamic plate boundaries and restored synthetic oceanic isochrons. *Earth and Planetary Science Letters*, **196**, 17–33.

STAMPFLI, G. M. & BOREL, G. D. 2004. The TRANSMED Transects in Space and Time: constraints on the paleotectonic evolution of the Mediterranean domain. *In*: CAVAZZA, W., ROURE, F. M., SPAKMAN, W., STAMPFLI, G. M. & ZIEGLER, P. A. (eds) *The TRANSMED Atlas: The Mediterranean Region from Crust to Mantle*. Berlin, Springer, 53–80.

TAPPONNIER, P. 1977. Evolution tectonique du système alpin en Méditerranée: poinçonnement et écrasement rigide-plastique. *Bulletin Geological Society France*, **7**, 437–460.

WORTEL, J. R. & SPAKMAN, W. 2000. Subduction and slab detachment in the Mediterranean – Carpathian Region. *Science*, **290**, 1910–1917.

# Seismic structure, crustal architecture and tectonic evolution of the Anatolian–African Plate Boundary and the Cenozoic Orogenic Belts in the Eastern Mediterranean Region

YILDIRIM DILEK[1*] & ERIC SANDVOL[2]

[1]*Department of Geology, Miami University, Oxford, OH 45056, USA*

[2]*Department of Geology Sciences, University of Missouri-Columbia, Columbia, MO 65211, USA*

*Corresponding author (e-mail: dileky@muohio.edu)*

**Abstract:** The modern Anatolian–African plate boundary is represented by a north-dipping subduction zone that has been part of a broad domain of regional convergence between Eurasia and Afro–Arabia since the latest Mesozoic. A series of collisions between Gondwana-derived ribbon continents and trench-roll-back systems in the Tethyan realm produced nearly East–West-trending, subparallel mountain belts with high elevation and thick orogenic crust in this region. Ophiolite emplacement, terrane stacking, high-P and Barrovian metamorphism, and crustal thickening occurred during the accretion of these microcontinents into the upper plates of Tethyan subduction roll-back systems during the Late Cretaceous–Early Eocene. Continued convergence and oceanic lithospheric subduction within the Tethyan realm were punctuated by slab breakoff events following the microcontinental accretion episodes. Slab breakoff resulted in asthenospheric upwelling and partial melting, which facilitated post-collisional magmatism along and across the suture zones. Resumed subduction and slab roll-back-induced upper plate extension triggered a tectonic collapse of the thermally weakened orogenic crust in Anatolia in the late Oligocene–Miocene. This extensional phase resulted in exhumation of high-P rocks and medium- to lower-crustal material leading to the formation of metamorphic core complexes in the hinterland of the young collision zones. The geochemical character of the attendant magmatism has progressed from initial shoshonitic and high-K calc-alkaline to calc-alkaline and alkaline affinities through time, as more asthenosphere-derived melts found their way to the surface with insignificant degrees of crustal contamination. The occurrence of discrete high-velocity bodies in the mantle beneath Anatolia, as deduced from lithospheric seismic velocity data, supports our Tethyan slab breakoff interpretations. Pn velocity and Sn attenuation tomography models indicate that the uppermost mantle is anomalously hot and thin, consistent with the existence of a shallow asthenosphere beneath the collapsing Anatolian orogenic belts and widespread volcanism in this region. The sharp, north-pointing cusp (Isparta Angle) between the Hellenic and Cyprus trenches along the modern Anatolian–African plate boundary corresponds to a subduction-transform edge propagator (STEP) fault, which is an artifact of a slab tear within the downgoing African lithosphere.

## Introduction

The present-day geodynamics of the eastern Mediterranean region is controlled by the relative motions of three major plates, Eurasia, Africa and Arabia, and much of the resulting deformation occurs at their boundaries (Fig. 1; Westaway 1994; Jolivet & Faccenna 2000; McClusky *et al.* 2000; Doglioni *et al.* 2002; Dilek 2006; Reilinger *et al.* 2006). The convergence rate between Africa and Eurasia is >40 mm a$^{-1}$ across the Hellenic Trench but decreases to <10 mm a$^{-1}$ across the Cyprus trench to the east (McClusky *et al.* 2000; Doglioni *et al.* 2002; Wdowinski *et al.* 2006) as a result of the subduction of the Eratosthenes seamount beneath Cyprus (Robertson 1998). The Arabia–Eurasia convergence across the Bitlis–Zagros

suture zone has been estimated to be *c*. 16 mm a$^{-1}$ based on global positioning system measurements of present-day central movements in this collision zone (Reilinger *et al.* 1997, 2006). These differential northward motions of Africa (<10 mm a$^{-1}$) and Arabia (16 mm a$^{-1}$) with respect to Eurasia are accommodated along the sinistral Dead Sea fault zone (Fig. 1b). The Anatolian microplate north of these convergent plate boundaries is moving SW with respect to Eurasia (Fig. 1a) at *c*. 30 mm a$^{-1}$ along the North and East Anatolian fault zones (Reilinger *et al.* 1997) and is undergoing complex internal deformation via mainly strike-slip and normal faulting. This deformation has resulted in the extensional collapse of the young orogenic crust, which developed during a series of collisional events in the region (Dewey *et al.* 1986; Dilek &

*From*: Murphy, J. B., Keppie, J. D. & Hynes, A. J. (eds) *Ancient Orogens and Modern Analogues.*
Geological Society, London, Special Publications, **327**, 127–160.
DOI: 10.1144/SP327.8   0305-8719/09/$15.00 © The Geological Society of London 2009.

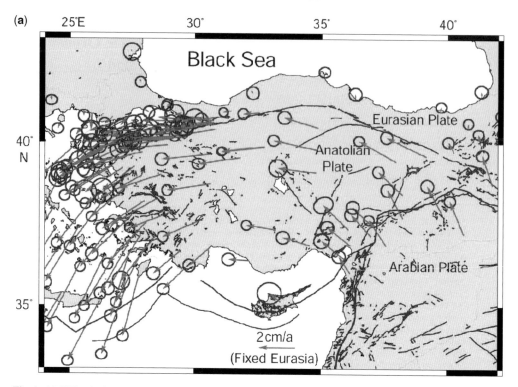

**Fig. 1.** (**a**) GPS velocity vectors for Anatolia and the Aegean Sea plotted in a Eurasia fixed reference frame (modified from Reilinger *et al.* 1997). The major mapped faults for western Anatolia are shown as lines and circles. (**b**) Tectonic map of the Aegean and eastern Mediterranean region showing the main plate boundaries, major suture zones, fault systems and tectonic units. Thick, white arrows depict the direction and magnitude (mm a$^{-1}$) of plate convergence; grey arrows mark the direction of extension (Miocene–Recent). Orange and purple delineate Eurasian and African plate affinities, respectively. Key to lettering: BF, Burdur fault; CACC, Central Anatolian Crystalline Complex; DKF, Datça–Kale fault (part of the SW Anatolian Shear Zone); EAFZ, East Anatolian fault zone; EF, Ecemis fault; EKP, Erzurum–Kars Plateau; IASZ, Izmir–Ankara suture zone; IPS, Intra–Pontide suture zone; ITS, Inner–Tauride suture; KF, Kefalonia fault; KOTJ, Karliova triple junction; MM, Menderes massif; MS, Marmara Sea; MTR, Maras triple junction; NAFZ, North Anatolian fault zone; OF, Ovacik fault; PSF, Pampak–Sevan fault; TF, Tutak fault; TGF, Tuzgölü fault; TIP, Turkish–Iranian plateau (modified from Dilek 2006).

Moores 1990; Yilmaz 1990), giving way to the formation of metamorphic core complexes and intracontinental basins (Bozkurt & Park 1994; Dilek & Whitney 2000; Jolivet & Faccenna 2000; Okay & Satir 2000; Doglioni *et al.* 2002; Ring & Layer 2003). Extensional deformation of the young orogenic belts has been accompanied by magmatism with varying geochemical finger-prints. The cause-effect relations of the spatial and temporal interplay between post-collisional extension and magmatism in the eastern Mediter-ranean region have been a subject of intense scrutiny and interdisciplinary research over the last twenty years.

The modern collision zone between the Anatolian and African plates is an excellent natu-ral laboratory to study the last stages of subduction, subduction roll-back processes, and accretionary

events prior to the onset of continental collision. The 1000 km-long convergent plate boundary between these two plates comprises two separate arcs: the Hellenic and the Cyprean (Fig. 1). The intersection of these two subduction zones occurs in a sharp bend, the Isparta Angle (IA), in the Tauride block. The Hellenic arc is characterized by a relatively steep, retreating subduction, where-as the Cyprean arc appears to involve a shallow subduction with two major seamounts (the Era-tosthenes and Anixamander) impinging on the trench (Kempler & Ben-Avraham 1987; Zitter *et al.* 2003).

In this paper we examine the Cenozoic evolution of the African–Anatolian plate boundary utilizing seismic tomography, and use our observations and interpretations from this modern subduction-collision driven plate boundary to derive

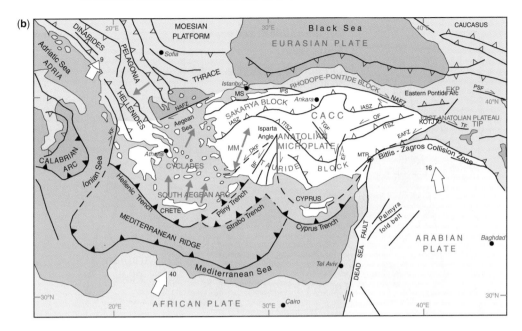

**Fig. 1.** (*Continued*)

conclusions for the geodynamic evolution and mantle dynamics of the young (Cenozoic) orogenic belts in Anatolia. Our tomographic models of the upper mantle beneath Anatolia and the Aegean Sea support the existence of two discrete high-velocity regions. These may be indicative of two separate slab breakoff events that have occurred during the subduction of the Tethyan oceanic litho-sphere. We use our own geological observations and data, and the extant literature to re-interpret the collision-driven tectonic evolution of the western, central and eastern Anatolian orogenic belts and to provide an overview of their crustal architecture.

## Seismic structure of the collision zones in Anatolia and its environs

### Lithospheric seismic velocity structure

Seismic images of the mantle provide important information on the current state of the lithosphere and asthenosphere in the eastern Mediterranean region. In particular, the location of possible pre-served slabs beneath the Anatolian plate could prove useful in understanding the evolution of the African–Anatolian plate boundary and the geody-namics of the Cenozoic orogenic belts in the region. In many models, however, the uppermost mantle picture is not well resolved in regions lacking fairly dense station coverage. This is cer-tainly true for much of western and central Anatolia.

Pn tomography, however, offers an important snapshot of the lithospheric mantle seismic velocity structure, even in regions with sparse station cover-age. Similarly, seismic phase attenuation can also indicate the state of the lithospheric mantle. Speci-fically, the regional seismic phase Sn is very sensi-tive to lithospheric mantle temperature anomalies (e.g. Molnar & Oliver 1969; Rodgers *et al.* 1995; Sandvol *et al.* 2001). Sn attenuation and Pn velocity are, therefore, two complementary and independent measures of the state of the uppermost mantle. Using these two models we find that the majority of the Anatolian plate is underlain by highly attenu-ating and seismically slow material (Figs 2 & 3). The two images from these independent datasets confirm that the Anatolian lithosphere is hot and most probably relatively thin.

Pn velocity tomography with an anisotropy component shows two scales of low Pn velocity anomalies (Fig. 2). First, a broader scale (*c.* 500 km) low (<7.8 km/s) Pn velocity anomaly underlies northwestern Iran, eastern Anatolia, the Caucasus, most of the Anatolian plate, and the northern Aegean Sea. These broad-scale low Pn velocity anomalies occupy regions within the Eurasian side of the Eurasia–Arabia collision zone. In central Iran, fast Pn velocities appear to extend beyond the Zagros suture line, while in northwestern Iran and eastern Turkey the high Pn velocities are limited to the region immediately south of the Bitlis–Zagros suture zone and the Tauride block.

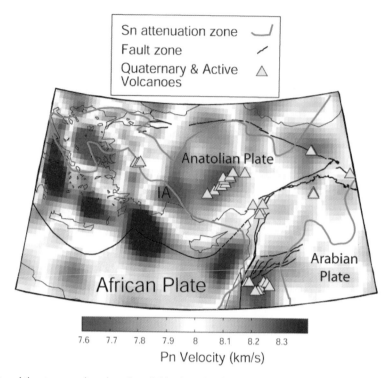

**Fig. 2.** Pn travel time tomography using all available phase data in the Middle East (Al-Lazki *et al.* 2004). The green line is the approximate boundary of a zone of Sn blockage (Fig. 3). Note the anomalously low Pn velocities along southern Turkey especially in the vicinity of the Isparta Angle (IA).

The Sn attenuation tomography model reveals regions of blocked and attenuated Sn (Fig. 3; Gök *et al.* 2000, 2003). In the southern Aegean Sea, a volcanic arc is present north of Crete and parallel to the Hellenic arc and Sn is attenuated. In the northern Aegean Sea, there is a high attenuating Sn zone with low upper mantle velocities (Al-Lazki *et al.* 2003) and Sn is partially attenuated or inefficient in northern Greece. Conversely, there is efficient Sn throughout most of Greece. In the Black Sea there is a very abrupt transition from efficient to blocked Sn regions. Efficient Sn is observed for paths within the eastern Arabian plate and the Zagros fold-and-thrust belt. An efficient zone is also seen at the northern part of the fold-and-thrust belt in the northwestern corner of the Iranian plateau. However, this region might be distorted because of smearing (Fig. 2). The Pn velocity tomography of Al-Lazki *et al.* (2003) shows higher Pn velocities for a smaller portion of the same region. Clearly, Sn is blocked throughout the Anatolian plateau, whereas it propagates efficiently within the eastern part of the Bitlis–Zagros suture zone. Sn also propagates efficiently through the Black Sea and the Mediterranean Sea. In general there is a strong correlation between Quaternary volcanism

and Sn blockage. Throughout the entire Anatolian plateau and northern Arabia, young basaltic volcanism correlates well with regions of high Sn attenuation (Fig. 3).

In the easternmost portion of the model, inefficient Sn is observed in the Greater Caucasus. Throughout most of the Lesser Caucasus, Sn is highly attenuated. Sn is observed as efficient in the western section of the Pontides and inefficient in the eastern Pontides. In western Anatolia, there are some inefficient Sn regions, unlike the regions of Sn blockage observed in eastern Anatolia.

The results of Sn attenuation tomography are consistent with those of Pn tomography, both indicating that the uppermost mantle beneath much of Anatolia is anomalously hot and thin. In western Anatolia, we observe complicated patterns of Sn attenuation, except for the Aegean Sea volcanic arc. A region of clear and distinctly higher Sn attenuation is observed north of the Hellenic trench that is an area of active arc-backarc extension above the northward subducting African plate. Numerous studies have found low velocities in the upper mantle that are consistent with the Sn efficiency results (e.g. Spakman *et al.* 1993; Alessandrini *et al.* 1997; Piromallo & Morelli 2003). The

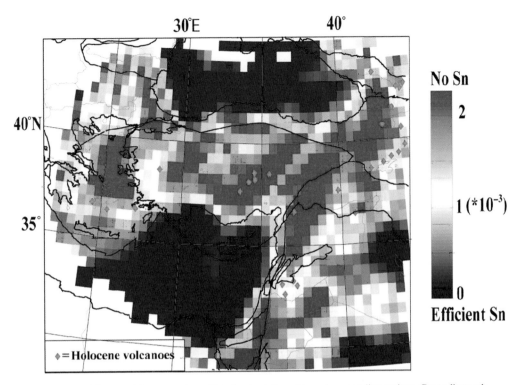

**Fig. 3.** Tomographic map of Sn attenuation within the Anatolian plate and surrounding regions. Green diamonds indicate Holocene volcanoes. Note good correlation between the Pn tomographic image (Fig. 2) and Sn attenuation, both indicating that the lithosphere is thin and hot throughout most of the Anatolian plate (Gök *et al.* 2003).

partial attenuation of Sn in the northern part of the Aegean Sea is related to active extensional deformation in the backarc setting. The continental shortening in NW Greece and Albania does not allow the rotation of the western margin of the region and leads to east–west shortening and north–south extension as the southern Aegean margin moves towards the Hellenic trench. The efficient Sn zone in part of Greece might be related to the subducting slab.

## Upper mantle and transition zone

There have been many tomographic P-wave, S-wave and bulk sound velocity models constructed for the upper mantle of the eastern Mediterranean region (Spakman *et al.* 1993; Bijwaard & Spakman 2000; Kárason & Van Der Hilst *et al.* 2001; Wortel & Spakman 1992, 2000; Widiyantoro *et al.* 2004). Features common to many of these models support their reliability. For example, along the Hellenic trench, the subducted slab of largely oceanic lithosphere of the African plate stands out as a relatively steeply dipping structure with a strong high-velocity P-wave (Fig. 4). This anomaly

extends to depths that exceed the intermediate depth seismicity in the region, a result that concurs with the previous images of deep subduction in this region (e.g. Spakman *et al.* 1988, 1993; Wortel & Spakman 1992). In contrast, the image of the Cyprean slab velocity anomaly is much weaker in most tomographic models (e.g. Widiyantoro *et al.* 2004; Fig. 5). This largely aseismic structure is located near the Vrancea seismic zone and may be related to an earlier episode in the multi-stage closing history of the Tethyan seaways (Dilek *et al.* 1999*a*, 2008). This structure is, however, less obvious in the Kárason & Van der Hilst (2001) model, and further investigation is necessary.

We can use the upper mantle velocity structure to infer the number of slab breakoff events that have occurred in the Neogene. Beneath the Eastern Mediterranean, Aegean Sea and Anatolian plate, the seismic velocity structure of the mantle transition zone is characterized by a number of discrete high-velocity bodies. Current global and regional models for the Eastern Mediterranean clearly show the continuation of subducting ocean lithosphere into the lower mantle along the Hellenic trench and its flattening out at the 660 km discontinuity. Many

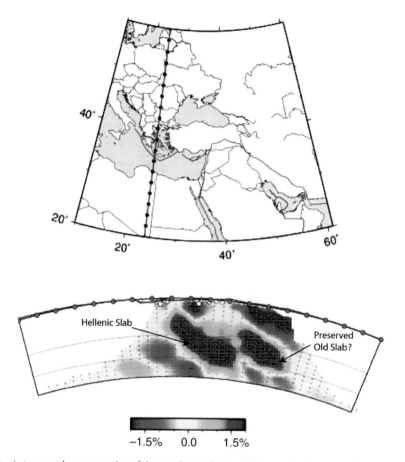

**Fig. 4.** Seismic tomography cross-section of the mantle to a depth of 800 km in the Aegean region mantle taken perpendicular to the Hellenic trench. Colours display the percentage deviation of seismic wave speed. Negative (positive) anomalies likely represent predominantly higher (lower) temperatures than average. Dashed lateral lines in the section depict the mantle discontinuities at 410 km and 660 km depths. See text for discussion.

(not all) of the newer tomographic models, however, show that there are two high-velocity bodies in the mantle transition zone beneath the Aegean Sea (e.g. Kárason 2002; Widiyantoro *et al.* 2004; Fig. 4).

Pn and Sn tomographic models provide no evidence for an attached slab along any of the suture zones within the Anatolian plate. Nor do they provide evidence of a very shallow/flat slab underlying southern Anatolia, as suggested by some researchers (i.e. Doglioni *et al.* 2002; Agostini *et al.* 2007). Many teleseismic velocity models show a relatively shallow high-velocity anomaly underlying much of central and, in some cases, northern Anatolia. This is in contrast to the uppermost mantle images that suggest that the Anatolian lithosphere is hot and possibly thin. This might suggest that the flat or shallow slab of the Cyprean subduction zone may not be in direct contact with the Anatolian lithosphere, and may extend almost

to the North Anatolian Fault Zone (Fig. 5). It still remains unclear as to what role the subducting Cyprean lithosphere had in the very recent uplift of the high topography in the Cyprean back arc (i.e. the Tauride block).

### The Isparta Angle

The Isparta Angle (IA) occurs at the intersection between the Cyprean (east) and Hellenic (west) arcs in the western end of the Tauride block (Fig. 1b). Palaeomagnetic investigations suggest that there has been very little rotation of the IA in the last 10 my (Tatar *et al.* 2002). Seismogenic deformation along the Hellenic arc indicates a transition from compressional to normal faulting near the Isparta Angle. On the other hand, the few available focal mechanisms along the westernmost Cyprean arc are more consistent with thrust

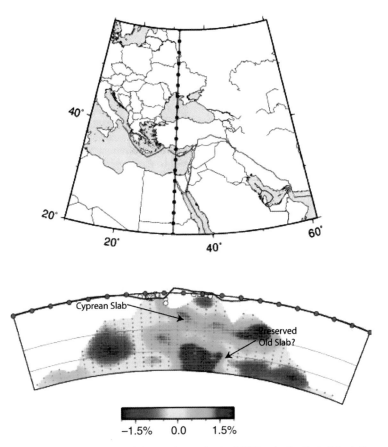

**Fig. 5.** Seismic tomography cross-section of the mantle to a depth of 800 km in the Central Anatolian region mantle taken perpendicular to the Cyprus trench. Colours display the percentage deviation of seismic wave speed. Negative (positive) anomalies likely represent predominantly higher (lower) temperatures than average. Dashed lateral lines in the section depict the mantle discontinuities at 410 km and 660 km depths. See text for discussion.

faulting. The GPS velocity vectors of McClusky *et al.* (2000) suggest that the lithosphere within the IA is moving independently of the rest of the Anatolian plate and that it may be attached to the African plate or to a piece of it (Barka & Reilinger 1997; McClusky *et al.* 2000).

The sharp cusp between the Hellenic and Cyprus trenches (Fig. 1b) and the significant differences in the convergence velocities of the African lithosphere at these trenches (*c.* 40 vs <10 mm a$^{-1}$ at the Hellenic and Cyprus trenches, respectively) are likely to have resulted in a lithospheric tear in the downgoing African plate that allowed the asthenospheric mantle to rise beneath SW Anatolia (Fig. 6; Doglioni *et al.* 2002; Agostini *et al.* 2007; Dilek & Altunkaynak 2008). This scenario is analogous to lithospheric tearing at subduction-transform edge propagator (STEP) faults described by Govers & Wortel (2005) from the Ionian and Calabrian arcs, the New Hebrides trench, the southern edge of the

Lesser Antilles trench, and the northern end of the South Sandwich trench. In all these cases, STEP faults propagate in a direction opposite to the subduction direction, and asthenospheric upwelling occurs behind and beneath their propagating tips. This upwelling induces decompressional melting of shallow asthenosphere, leading to linearly distributed alkaline magmatism younging in the direction of tear propagation.

The North–South-trending potassic and ultrapotassic volcanic fields stretching from the Kirka and Afyon-Suhut region in the north to the Isparta–Gölcük area in the south (Fig. 7) shows an age progression from 21–17 to 4.6–4.0 Ma (Yagmurlu *et al.* 1997; Alici *et al.* 1998; Savaşçin & Oyman 1998; Francalanci *et al.* 2000; Kumral *et al.* 2006; Çoban & Flower 2006; Dilek & Altunkaynak 2007). This distribution of the potassic and ultrapotassic volcanic rocks in SW Anatolia is consistent with a progressive migration of their melt source

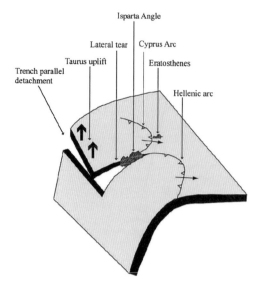

**Fig. 6.** A cartoon illustrating the development of a slab tear or subduction transform edge propagator (STEP) fault beneath the Tauride block and the positions of subducting African lithosphere beneath the Cyprean and Hellenic arcs (after Barka & Reilinger 1997). Evidence for a southward increase in the age of volcanism is consistent with model (Yagmurlu *et al.* 1997; Savaşçin & Oyman 1998; Dilek & Altunkaynak 2008). Also, the presence of ignimbrites is consistent with asthenospheric upwelling leading to widespread crustal melting.

towards the south and supports a STEP model for their origin (Fig. 6; Dilek & Altunkaynak 2008). Asthenospheric low velocities detected through Pn tomographic imaging in this region (Al-Lazki *et al.* 2004) also support the existence of shallow asthenosphere beneath the IA at present.

## Regional geology of western Anatolia

The Aegean province is situated in the upper plate of a north-dipping subduction zone at the Hellenic trench (Fig. 1b) and is considered to have evolved as a backarc environment above this subduction zone (Le Pichon & Angelier 1979; Jolivet 2001; Faccenna *et al.* 2003; van Hinsbergen *et al.* 2005). The slab retreat rate of the subducting African lithosphere has been larger than the absolute velocity of the Eurasian upper plate, causing net *c*. North–South extension in the Aegean region since the early Miocene (Jolivet *et al.* 1994; Jolivet & Faccenna 2000; Faccenna *et al.* 2003; Ring & Layer 2003; Dilek & Altunkaynak 2008). Since then, the thrust front associated with this subduction zone and its slab retreat has migrated from the Hellenic trench (south of Crete) to the south of the Mediterranean Ridge (Jolivet & Faccenna 2000; Le Pichon *et al.* 2003). Backarc extension in the Aegean region appears to have started at *c*. 25 Ma, long before the onset of the Arabian collision-driven southwestward displacement of the Anatolian microplate in the late Miocene (Barka & Reilinger 1997; Jolivet & Faccenna 2000).

The continental crust making up the upper plate of the Hellenic subduction zone south of the North Anatolian fault is composed of the Sakarya and the Anatolide–Tauride continental blocks (Fig. 7). These two microcontinents are separated by the Izmir–Ankara suture zone, which is marked by Tethyan ophiolites and associated tectonic units. The basement of the Anatolide block is composed mainly of the Menderes massif, which is intruded and overlain by Cenozoic granitoid plutons and extrusive rocks; Lower Miocene and younger sedimentary rocks of a series of extensional basins overlie the high-grade metamorphic rocks of the Menderes massif (Bozkurt 2003; Purvis & Robertson 2004; Oner & Dilek 2007).

### Sakarya continent

The Sakarya continent consists of a Palaeozoic crystalline basement with its Permo–Carboniferous sedimentary cover and Permo–Triassic ophiolitic and rift or accretionary-type mélange units (Karakaya complex) that collectively form a composite continental block (Tekeli 1981; Okay *et al.* 1996). The Sakarya continental rocks and the ophiolitic units of the Izmir–Ankara suture zone (IASZ) are

---

**Fig. 7.** Geological map of western Anatolia and the eastern Aegean region, showing the distribution of suture zones, ophiolite complexes, major Cenozoic igneous provinces discussed in this paper, and the salient fault systems. Menderes and Kazdag (KDM) massifs represent metamorphic core complexes with exhumed lower continental crust. Izmir–Ankara suture zone (IASZ) marks the collision front between the Sakarya continental block to the north and the Anatolide–Tauride block to the south. The Intra–Pontide suture zone (IPSZ) marks the collision front between the Sakarya continent to the south and the Rhodope–Pontide block to the north. The Eocene and Oligo–Miocene granitoids (shown in red) represent the post-collisional (post-Eocene) and/or extensional magmatism in the region. Key to lettering of these granitoids: AG, Alasehir; BG, Baklan; CGD, Çataldag; EP, Egrigöz; GBG, Göynükbelen; GYG, Gürgenyayla; IGD, Ilica; KG, Kozak; KOP, Koyunoba; OGD, Orhaneli; TG, Turgutlu; TGD, Topuk; SG, Salihli. Much of western Anatolia is covered by Cenozoic volcanic rocks intercalated with terrestrial deposits. Key to lettering of major fault systems: AF, Acigöl fault; BFZ, Burdur fault zone; DF, Datça fault; IASZ, Izmir–Ankara suture zone; IPSZ, Intra–Pontide suture zone; KF, Kale fault; NAFZ, North Anatolian fault zone; SWASZ, SW Anatolian shear zone.

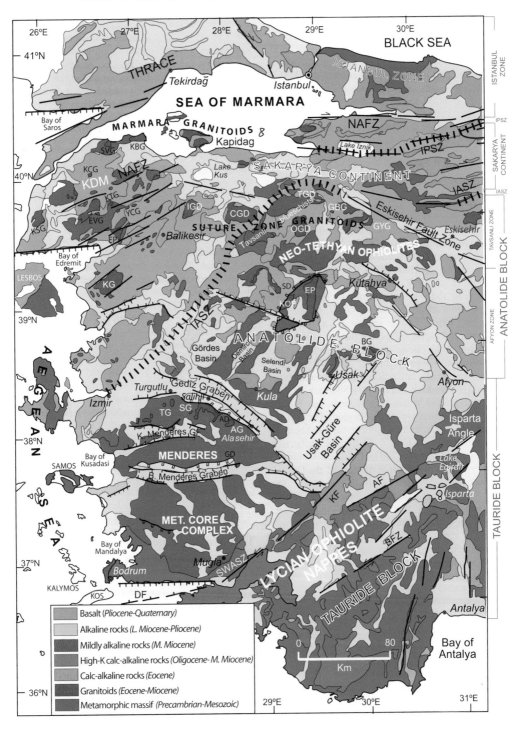

**Fig. 7.** *Continued.*

intruded by a series of east–west trending Eocene and Oligo–Miocene granitoid plutons (Fig. 7; Bingöl *et al.* 1982, 1994; Delaloye & Bingöl 2000; Altunkaynak 2007). The Kazdag massif within the western part of the Sakarya continent (KDM in Fig. 7) represents a metamorphic core complex, which is inferred to have been exhumed from a depth of *c.* 14 km along a north-dipping mylonitic shear zone starting at *c.* 24 Ma (Okay & Satir 2000).

## Izmir–Ankara Suture Zone (IASZ) and Tethyan Ophiolites

The Izmir–Ankara Suture Zone south of the Sakarya continent includes dismembered Tethyan ophiolites, high-pressure low-temperature (HP–LT) blueschist-bearing rocks, and flysch deposits mainly occurring in south-directed thrust sheets (Fig. 7; Önen & Hall 1993; Okay *et al.* 1998; Sherlock *et al.* 1999). Late-stage diabasic dykes crosscutting the ophiolitic units in the Kütahya area are dated at *c.* 92–90 Ma ($^{40}Ar/^{39}Ar$ hornblende ages; Önen 2003), indicating a minimum late Cretaceous igneous age for the ophiolites. The blueschist rocks along the suture zone in the Tavsanli area have yielded $^{40}Ar/^{39}Ar$ cooling ages (phengite crystallization during exhumation) of $79.7 \pm 1.6$–$82.8 \pm 1.7$ Ma (Sherlock *et al.* 1999), suggesting a latest Cretaceous age for the HP–LT metamorphism in the region. The Lycian nappes and ophiolites that occur farther south in the Tauride block (Fig. 7; Collins & Robertson 2003; Ring & Layer 2003; Çelik & Chiaradia 2008) represent the tectonic outliers of the Cretaceous oceanic crust derived from the IASZ. The Lycian nappes are inferred to have once covered the entire Anatolide belt in western Anatolia, and to have been later removed as a result of tectonic uplift and erosion associated with exhumation of the Menderes core complex during the late Cenozoic (Ring & Layer 2003; Thomson & Ring 2006; Dilek & Altunkaynak 2007).

## Menderes massif

The Menderes massif represents a core complex composed of high-grade metamorphic rocks of Pan-African affinity that are intruded by syn-kinematic Miocene granitoid plutons (Hetzel & Reischmann 1996; Bozkurt & Satir 2000; Bozkurt 2004; Gessner *et al.* 2004). Rimmelé *et al.* (2003) estimated the *P–T* conditions of the metamorphic peak for the Menderes massif rocks at >10 kbar and >440 °C. The main episode of metamorphism is inferred to have resulted from the burial regime associated with the emplacement of the Lycian nappes and ophiolitic thrust sheets (Yilmaz 2002).

Imbricate stacking of the Menderes massif beneath the Lycian nappes and ophiolitic thrust sheets appears to have migrated southwards throughout the Palaeocene–Middle Eocene (Özer *et al.* 2001; Candan *et al.* 2005). Unroofing and exhumation of the Menderes massif may have started as early as the Oligocene (25–21 Ma) based on the cooling ages of syn-extensional granitoid intrusions that crosscut the metamorphic rocks (Bozkurt & Satir 2000; Catlos & Çemen 2005; Ring & Collins 2005; Thomson & Ring 2006). This timing may signal the onset of post-collisional tectonic extension in the Aegean region.

## Tauride Block in Western Anatolia

The Tauride Block south of the Menderes core complex consists of Precambrian and Cambro–Ordovician to lower Cretaceous carbonate rocks intercalated with volcano-sedimentary and epiclastic rocks (Ricou *et al.* 1975; Demirtasli *et al.* 1984; Özgül 1984; Gürsu *et al.* 2004). These rocks are tectonically overlain by Tethyan ophiolites (i.e. Lycian, Beysehir, Alihoca and Aladag ophiolites) along south-directed thrust sheets (Dilek *et al.* 1999*a*; Collins & Robertson 2003; Çelik & Chiaradia 2008; Elitok & Drüppel 2008). Underthrusting of the Tauride carbonate platform beneath the Tethyan oceanic crust and its partial subduction at a north-dipping subduction zone in the Inner-Tauride Ocean resulted in HP–LT metamorphism (Dilek & Whitney 1997; Okay *et al.* 1998). Continued convergence caused crustal imbrication and thickening within the platform and resulted in the development of several major overthrusts throughout the Tauride block (Demirtasli *et al.* 1984; Dilek *et al.* 1999*b*).

## Tectonic evolution of the western Anatolian orogenic belt

Emplacement of the Cretaceous ophiolites onto the Anatolide–Tauride block along the Izmir–Ankara suture zone and partial subduction of the continental edge that led to its HP–LT metamorphism (Sherlock *et al.* 1999; Okay 2002; Ring & Layer 2003; Ring *et al.* 2003) in the late Cretaceous mark the initial stages of collision tectonics in western Anatolia (Fig. 8). The terminal closure of the Northern Tethyan seaway resulted in the collision of the Sakarya continent with the Anatolide–Tauride block during the early Eocene that, in turn, caused regional deformation, metamorphism and crustal thickening. This Barrovian-type, collision-driven regional metamorphism was responsible for the development of high-grade rocks in the Kazdag and Menderes metamorphic

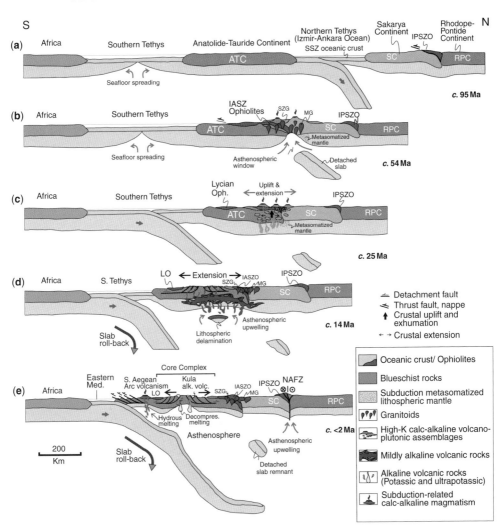

**Fig. 8.** Late Mesozoic–Cenozoic geodynamic evolution of the western Anatolian orogenic belt as a result of collisional and extensional processes in the upper plate of north-dipping subduction zone(s) within the Tethyan realm. See text for discussion.

massifs. Resistance of the buoyant Anatolide–Tauride continental crust to subduction and consequent arrest of the north-dipping subduction zone resulted in isostatic uplift of its partially subducted passive margin, exhumation of high-P rocks, and rapid denudation of upper crustal rocks, leading to widespread flysch formation during the early to middle Eocene (Ring *et al.* 2003, and references therein).

With continued continental collision the leading edge of the subducted Tethyan slab possibly broke off from the rest of the continental lithosphere, resulting in the development of an asthenospheric window (Fig. 8). Slab detachment and breakoff is

a natural response to the gravitational settling of subducted lithosphere in continental collision zones, as a result of a decrease in the subduction rate caused by the positive buoyancy of partially subducted continental lithosphere (Davies & von Blackenburg 1995; von Blackenburg & Davies 1995; Wortel & Spakman 2000; Gerya *et al.* 2004). Our seismic tomography model (Fig. 4) shows the existence of a second high-velocity (cold) slab near the 660 km discontinuity in the lower mantle north of the Hellenic slab, which we interpret as a detached Tethyan slab dipping beneath the western Anatolian orogenic belt. This implies a punctuated evolution of Tethyan subduction zones in the eastern Mediterranean, rather than the single subduction zone hypothesized for much of the Mesozoic and the entire Cenozoic in some recent geodynamic models (i.e. van Hinsbergen *et al.* 2005; Jolivet & Brun 2008).

As the downgoing oceanic plate breaks off, the asthenosphere rises rapidly and is juxtaposed against the thickened mantle lithosphere in the collision zone (Fig. 8, *c.* 54 Ma). Conductive heating of this overriding lithosphere results in melting of the metasomatized and hydrated layers, producing potassic, calc-alkaline magmas. Crustal melting at shallow depths, induced by asthenospheric upwelling, causes granitic/rhyolitic magmatism. Hence, the middle to late Eocene emplacement of widespread granitoid plutons in northwestern Anatolia, mainly through the IASZ and the Sakarya continent (Orhaneli, Topuk, Göynükbelen, etc.), is thought to be a direct result of the heat flux through this window and the associated thermal perturbation that caused melting of the metasomatized continental lithospheric mantle (Dilek & Altunkaynak 2007). Major and trace element compositions of I-type granitoid rocks in NW Anatolia, dated at *c.* 54–35 Ma ($^{40}Ar/^{39}Ar$ hornblende and mica separates, and SHRIMP zircon dating; Dilek & Altunkaynak, unpublished data) indicate their origin by melting of a subduction-modified mantle source that had been enriched in mobile incompatible elements (Altunkaynak 2004, 2007; Köprübasi & Aldanmaz 2004), a signature that is consistent with post-collisional magmatism driven by slab breakoff in other orogenic belts (Schliestedt *et al.* 1987; Hansmann & Oberli 1991; Davies & von Blackenburg 1995; Nemcok *et al.* 1998).

Geological evidence in support of the inferred slab breakoff magmatism in western Anatolia includes: (1) the linear distribution of the plutons in a narrow belt straddling the IASZ where ophiolitic and high-P blueschist rocks are exposed (Figs 7 & 8) – a pattern that suggests a focused heat source likely derived from an asthenospheric window; and (2) the attempted subduction of Anatolide–Tauride continental crust to a depth of ≥80 km (Okay *et al.* 1998), evidenced by the latest Cretaceous blueschist rocks of the Tavsanli zone, that is likely to have clogged the subduction zone and caused the detachment of the sinking Tethyan oceanic lithosphere.

Continued collision of the Sakarya and Anatolide–Tauride continental blocks led to the development of thick orogenic crust, orogen-wide burial metamorphism, and anatectic melting of the lower crust (*c.* 25 Ma, Fig. 8). This episode of post-collisional magmatism coincided with bimodal volcanism and a widespread ignimbrite flare-up in western Anatolia (Ercan *et al.* 1985; Yilmaz *et al.* 2001), and caused thermal weakening of the crust in the western Anatolian orogenic belt that led to

its extensional collapse. The Kazdag core complex in NW Anatolia (Figs 7 & 8) began its initial exhumation in the latest Oligocene–Early Miocene (Okay & Satir 2000) and the Menderes core complex in central western Anatolia (Fig. 7) underwent its exhumation in the earliest Miocene (Isik *et al.* 2004; Thomson & Ring 2006; Bozkurt 2007). Some of the collision-generated thrust faults may have been reactivated during this time as crustal-scale low-angle detachment faults (e.g. Simav detachment fault, SW Anatolian shear zone), facilitating regional extension (Thomson & Ring 2006; Çemen *et al.* 2006).

Starting in the middle Miocene, both lithospheric and asthenospheric mantle melts were involved in the evolution of bimodal volcanic rocks in western Anatolia with the lithospheric input diminishing in time (Akay & Erdogan 2004; Aldanmaz *et al.* 2000, 2006). The timing of this magmatism coincides with widespread lower crustal exhumation and tectonic extension across the Aegean region (*c.* 14 Ma in Fig. 8). This extensional phase and the attendant mildly alkaline volcanism are attributed to thermal relaxation associated with possible delamination of the subcontinental lithospheric mantle beneath the northwestern Anatolian orogenic belt (Altunkaynak & Dilek 2006; Dilek & Altunkaynak 2007, and references therein). Lithospheric delamination may have been triggered by peeling of the base of the subcontinental lithosphere as a result of slab roll-back at the Hellenic trench.

During the advanced stages of extensional tectonism in the late Miocene–Quaternary, the development of regional graben systems (i.e. Gediz, Büyük Menderes, Fig. 7) further attenuated the continental lithosphere beneath the region. This extensional phase was accompanied by upwelling of the asthenospheric mantle and consequent decompressional melting (Fig. 8). Lithospheric-scale extensional fault systems acted as natural conduits for the transport of uncontaminated alkaline magmas to the surface (Richardson-Bunbury 1996; Seyitoglu *et al.* 1997; Alici *et al.* 2002). Asthenospheric flow in the region following the late Miocene (<10 Ma) may also have been driven in part by the extrusion tectonics caused by the Arabian collision in the east, as suggested by the parallelism of the SW-oriented shear wave splitting fast polarization direction in the mantle with the motion of the Anatolian plate (Sandvol *et al.* 2003b; Russo *et al.* 2001). This SW-directed lower mantle flow beneath Anatolia could have played a significant role in triggering intra-plate deformation via extension and strike-slip faulting parallel to the flow direction and horizontal mantle thermal anomalies, which would have facilitated melting and associated basaltic volcanism. This lateral asthenospheric flow may also have resulted in the interaction of different compositional end-members, contributing to the mantle heterogeneity beneath western Anatolia. Lateral displacement of the asthenosphere as a result of the extrusion of collision-entrapped ductile mantle beneath Asia and SE Asia has been similarly identified as the cause of post-collisional high-K volcanism in Tibet and Indo–China during the late Cenozoic (Liu et al. 2004; Williams et al. 2004; Mo et al. 2006).

The apparent SW propagation of Cenozoic tectonic extension and magmatism through time is likely to have been the combined result of slab roll-back associated with the subduction of the Southern Tethys ocean floor at the Hellenic trench and the thermally induced collapse of the western Anatolian orogenic belt (Fig. 8). The thermal input and melt sources were likely provided first by slab breakoff-generated asthenospheric flow, then by litho-spheric delamination-related asthenospheric flow and finally by tectonic extension-driven upward asthenospheric flow and collision-induced (Arabia-Eurasia collision) lateral (westward) mantle flow (Innocenti et al. 2005; Dilek & Altunkaynak 2007). Since the late Miocene, subduction zone magmatism related to the retreating Hellenic trench has been responsible for the progressive southward migration of the South Aegean Arc (Fig. 8; Pe-Piper & Piper 2006). The exhumation of high-P rocks in the Cyclades was likely driven by upper plate extension and channel flow associated with this subduction (Jolivet et al. 2003; Ring & Layer 2003).

## Regional geology of central Anatolia

Much of central Anatolia is occupied by the Central Anatolian Crystalline Complex (CACC), which consists of the Kirsehir, Akdag and Nigde metamorphic massifs, dismembered Tethyan ophiolites, and felsic to mafic plutons ranging in age from the late Cretaceous to the Miocene (Fig. 9; Güleç 1994; Boztug 2000; Düzgören-Aydin 2000; Kadioglu et al. 2003, 2006; Ilbeyli 2004; Ilbeyli et al. 2004). Several curvilinear sedimentary basins (Tuzgölü, Ulukisla and Sivas), which initially evolved as peripheral foreland and/or forearc basins in the late Cretaceous, delimit the CACC in the west, the east and the south. The south–central part of the CACC includes the Cappadocian volcanic province, containing Upper Miocene to Quaternary volcanic-volcaniclastic rocks and polygenetic volcanic centers (Toprak et al. 1994; Dilek et al. 1999b).

### Metamorphic massifs

Protoliths of the high-grade rocks in the metamorphic massifs of the CACC were Palaeozoic to Mesozoic pelitic sediments with local mafic lava units and hypabyssal rocks (Dilek & Whitney 2000, and references therein). These sedimentary units underwent regional metamorphism as a result of burial beneath the Tethyan ophiolites derived from the Izmir–Ankara–Erzincan suture zone to the north (Seymen 1984; Göncüoglu et al. 1991; Whitney & Dilek 1998). Subsequent collisional events in the latest Mesozoic and early Cenozoic resulted in further crustal imbrication and thickening of crystalline rocks in the CACC (Dilek et al. 1999b). Three main massifs (Kirsehir, Akdag and Nigde; Fig. 9) have been delineated in the CACC based on different $P$–$T$–$t$ paths of their protoliths.

The Kirsehir massif in the NW part of the CACC consists of marble and calcsilicate rocks inter-layered with metapelitic/psammitic schist, amphi-bolite ($\pm$ garnet), and quartzite (Whitney et al. 2001). The metamorphic grade appears to increase from 450 °C (garnet zone) in the SE to c. 750 °C (sillimanite zone) in the NW within the massif. The pressure corresponding to peak temperatures for the garnet zone rocks is estimated to be 2.5–4 kbar based on the structural position of these rocks (Whitney et al. 2001). However, pressures of c. 6 kbar are estimated from the first appearance of garnet in amphibolite-facies rocks in the NW part of the massif.

The Akdag massif represents the largest coherent block of metamorphic rocks in the NE part of the CACC (Fig. 9), metamorphic grade of which ranges from chlorite zone to sillimanite–K-feldspar zone. The highest grade rocks occur in an elongate NE–SW-trending belt along the central axis of the Akdag massif (Whitney et al. 2001) containing abundant garnet–muscovite–quartz gneiss (Whitney et al. 2001). Various thermometry and barometry applications using different equilibrium phase assemblages in the Akdag metamorphic rocks have revealed peak pressures ranging from $5 \pm 1$–$8 \pm 1$ kbar and peak temperatures ranging from 550–600 to 660–675 °C (Whitney et al. 2001).

The Nigde massif in the southern part of the CACC (Fig. 9) is exposed in a structural dome that has been interpreted as a Cordilleran-type meta-morphic core complex (Whitney & Dilek 1997). A gently (c. 30°) S-dipping detachment fault bounding the Nigde massif along its southern edge juxtaposes multiply deformed marble, quartzite and schist in the footwall from clastic sedimentary rocks of the Ulukisla basin in the hanging wall. The central part of the Nigde massif consists predominantly of upper amphibolite facies metasedimentary rocks and the peraluminous Uçkapili granite. These meta-pelitic rocks record an earlier episode of regional, medium-pressure/high temperature metamorphism (5–6 kbar and T > 700 °C), and a younger epi-sode of low-pressure/high temperature metamorph-ism (730–770 °C) (Whitney & Dilek 1998; Fayon et al. 2001). The earlier event was most likely

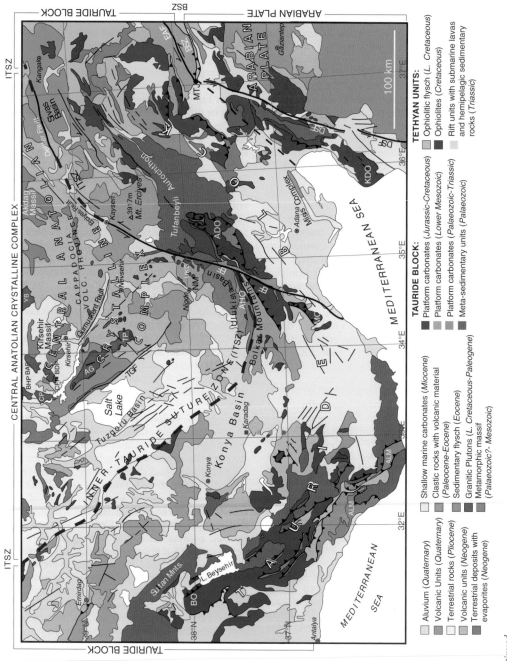

**Fig. 9.** *Continued.*

related to burial and heating after emplacement of the Tethyan ophiolites southward onto the CACC during the late Cretaceous and during the subsequent collision between the CACC and Tauride blocks, whereas the younger event was associated with shallow intrusion of the Uçkapili granite, which crosscuts the metamorphic and structural fabrics in the Nigde massif.

The $P-T-t$ paths of the Akdag and Kirsehir massifs indicate that the northern part of the CACC where they are now exposed was deformed, metamorphosed and unroofed during tectonic events associated with the collision of the CACC with a south-facing intraoceanic arc in the northern branch of the Tethys, followed by a collision with the Pontide arc to the north in the early to mid-Eocene. The sedimentary rocks of the Akdag and Kirsehir massifs were buried to moderate depths ($c.$ 20 km) beneath the Tethyan ophiolite nappes derived from the Izmir–Ankara–Erzincan suture zone and were thickened via folding and thrust faulting. The existence of kyanite in the Akdag massif rocks suggests that the eastern part of the CACC either experienced greater degrees of crustal thickening or was exhumed from deeper crustal levels than the rest of the massifs in the CACC. The northern part of the CACC was slowly exhumed via erosion by the Eocene (Fayon $et$ $al.$ 2001).

The SW part of the CACC experienced relatively high-temperature metamorphism associated with extensive Andean-type arc magmatism at relatively shallow crustal levels that are represented by the CACC plutons (see below). The southern part of the CACC, where the Nigde massif is now exposed, experienced Barrovian metamorphism at mid-crustal pressures ($c.$ 5–6 kbar) and at high temperatures ($>700$ °C), possibly associated with orogenic crustal thickening (Whitney & Dilek 1998). The Nigde core complex was exhumed to $<2$ km depth mainly by tectonic unroofing along low-angle detachment faults. Apatite fission track ages from the Nigde rocks range from $c.$ 9–12 Ma and indicate slow to moderate cooling via exhumation at rates of 30–8 °C/Ma (Fayon $et$ $al.$ 2001). Thus, different unroofing mechanisms appear to have affected the CACC since the early Oligocene:

the northern CACC undergoing erosional unroofing nearly 20 Ma before the tectonic exhumation of the southern CACC starting in the late Miocene.

## Inner-Tauride suture zone (ITSZ) and Tethyan ophiolites

Discontinuous exposures of the Tethyan ophiolites and mélanges define two major suture zones surrounding the CACC in the north and the south. The Izmir–Ankara–Erzincan suture zone to the north and the Inner-Tauride suture (ITSZ) zone to the south (Fig. 1b) mark the obliteration sites of the Tethyan seaways, which had evolved between the Gondwana-derived continental fragments (Sengör & Yilmaz 1981; Robertson & Dixon 1984; Dilek & Moores 1990). Subduction of the Tethyan oceanic lithosphere that evolved in these seaways resulted in the development of incipient arc-forearc complexes that subsequently formed the ophiolites, in the mantle heterogeneity beneath the continental fragments, and eventually, in continental collisions in the Eocene (Dilek & Flower 2003).

Ophiolite complexes within the Izmir–Ankara–Erzincan suture zone include serpentinized upper mantle peridotites and gabbros that are crosscut by dolerite and plagiogranite dikes and overlain by pillow lavas (Tankut $et$ $al.$ 1998; Dilek & Thy 2006). Both dolerite and plagiogranite dykes show negative Ta–Nb patterns typical of arc-related petrogenesis and zircon ages of $c.$ $179 \pm 15$ Ma, indicating that the ophiolitic basement in the Izmir–Ankara–Erzincan suture zone is at least early Jurassic in age or older (Dilek & Thy 2006). The Inner-Tauride ophiolites to the south (i.e. Alihoca, Aladag, Mersin) consist mainly of tectonized harzburgites, mafic-ultramafic cumulates and gabbros, and commonly lack sheeted dykes and extrusive rocks (Parlak $et$ $al.$ 1996, 2002; Dilek $et$ $al.$ 1999$a$). They include thin ($c.$ 200 m) thrust sheets of metamorphic sole rocks beneath them, and both the ophiolitic units and the sole rocks are intruded by mafic dyke swarms composed of basaltic to andesitic rocks with island arc tholeiite (IAT)

---

**Fig. 9.** Geological map of the central Anatolian region, showing the distribution of Tethyan ophiolites, suture zones, major tectonostratigraphic units within the Tauride block and the Central Anatolian Crystalline Complex (CACC) (including metamorphic massifs and major plutons), and major faults systems. Key to lettering of major granitoid plutons in the CACC: AG, Agaçören granitoid; BAP, Bayindir pluton; BDP, Baranadag pluton; BHP, Behrek Dag pluton; CFP, Cefalik pluton; CP, Çelebi pluton; TP, Terlemez pluton; UKG, Uçkapili granite; YB, Yozgat batholith. Key to lettering of major fault systems: BSZ, Bitlis suture zone; DSF, Dead Sea Fault; EAF, East Anatolian Fault; EF, Ecemis Fault; ITSZ, Inner-Tauride Suture Zone; TGF, Tuzgölü Fault. Major ophiolites: ADO, Aladag ophiolite; AHO, Alihoca ophiolite; BO, Beysehir ophiolite; KDO, Kizildag ophiolite; MO, Mersin ophiolite. Other symbols: ALM, Alanya massif; MTJ, Maras Triple Junction; NM, Nigde massif; (data are from Dilek & Moores 1990; Dilek $et$ $al.$ 1999$a$, $b$; Kadioglu $et$ $al.$ 2003, 2006). Notice that the ITSZ, marked by a dark-blue, dashed thick line, is truncated and offset by the sinistral Ecemis fault (EF).

affinities. $^{40}Ar/^{39}Ar$ hornblende ages of 92–90 and 90–91 Ma from the metamorphic sole and dyke rocks, respectively, indicate Cenomanian–Turonian ages for the Inner-Tauride ophiolites (Dilek et al. 1999a; Parlak & Delaloye 1999; Çelik et al. 2006). These age brackets suggest that the Tethyan ophiolites rooted in the Inner-Tauride suture zone are possibly younger than those derived from the Izmir–Ankara–Erzincan suture zone farther north.

In addition to the Cenomanian–Turonian ophiolites, the ITSZ is also marked by discontinuous exposures of blueschist-bearing mafic-ultramafic and carbonate rocks along the northern edge of the Tauride block. The occurrence of sodic amphibole-containing metasedimentary and metavolcanic rocks in the Bolkar Mountains region (Blumenthal 1956; van der Kaaden 1966; Gianelli et al. 1972; Dilek & Whitney 1997) extends into the Tavsanli zone in NW Anatolia and into the Pinarbasi zone in the Eastern Taurides in East–Central Anatolia. These HP/LT rock assemblages showing counterclockwise $P–T–t$ trajectories of their metamorphic evolution indicate increasing $P/T$ ratio with cooling that was associated with continuous subduction within the Inner-Tauride Ocean (Dilek & Whitney 1997, 2000). Late Cretaceous–Palaeocene calc-alkaline plutonic rocks (Dilek, unpublished data) intruding into the Ulukisla basin strata north of the Bolkar Mountains point out that this subduction activity had continued into the early Cenozoic.

## Syn-collisional CACC plutons

The late Cretaceous plutonic rocks in the CACC were emplaced after obduction of the Cretaceous Tethyan ophiolites, which were derived from the Izmir–Ankara–Erzincan suture zone to the north, and before its collision with the Pontide and Tauride continental blocks during the middle Eocene. Therefore, the magmatic evolution of these plutons preceded the terminal continental collisions in the region (Akiman et al. 1993; Erler & Bayhan 1995; Kadioglu et al. 2006).

The CACC plutons can be grouped into three supersuites based on their field occurrences and distinct differences in their mineral and chemical compositions (Kadioglu et al. 2006). Plutonic rocks of the Granite Supersuite commonly occur in a curvilinear belt along the western edge of the CACC (east of the Salt Lake-Tuzgölü, Fig. 9) and consist of calc-alkaline rocks ranging in composition from tonalite, granodiorite and biotite granite to amphibole biotite granite and biotite-alkali feldspar granite. Plutons of the Monzonite Supersuite (i.e. Terlemez, Cefalik, Baranadag plutons) occur immediately east of the Granite Supersuite plutons

and are composed mainly of subalkaline quartz monzonite and monzonite. Both the Granite and Monzonite Supersuite rocks show enrichment in LILE and depletion in HFSE relative to ocean ridge granites (ORG). They display isotopic and trace element signatures suggesting a crustal component that reflect subduction-influenced source enrichment and crustal contamination resulting from assimilation fractional crystallization (AFC) processes during magma transport through the CACC crust (Bayhan 1987; Aydin et al. 1998; Güleç & Kadioglu 1998; Kadioglu et al. 2003, 2006). There is a progression from high-K calc-alkaline and high-K shoshonitic compositions in the Granite Supersuite to typical shoshonitic compositions in the Monzonite Supersuite rocks.

The Syenite Supersuite represents the youngest phase of plutonism in the late Cretaceous (c. 69 Ma) and generally occurs in the inner part of the CACC. Rocks of this supersuite are composed of silica saturated (quartz syenite and syenite) and silica under-saturated, nepheline and pseudoleucite bearing alkaline rocks, which show more enrichment in LILE and a slight enrichment in HFSE in comparison to the other two supersuites (Boztug et al. 1997; Kadioglu et al. 2006). Isotopic and trace element signatures of the Syenite Supersuite plutons suggest that their magmas were more enriched in within-plate mantle components compared to the Granite and Monzonite Supersuite plutons (Kadioglu et al. 2006). $^{40}Ar/^{39}Ar$ age data from these Granite, Monzonite and Syenite Supersuite plutons yield ages of $77.7 \pm 0.3$, $70 \pm 1.0$, and $69.8 \pm 0.3$ Ma, respectively (Kadioglu et al. 2006). Thus, the subduction zone influence on melt evolution beneath the CACC appears to have decreased rapidly (within c. 7–8 Ma) from the earlier calc-alkaline granitic magmatism to the later alkaline, syenitic magmatism during the latest Cretaceous.

## Tauride block

The Tauride block south of the CACC consists of Palaeozoic to upper Cretaceous carbonate, siliciclastic and volcanic rocks (Özgül 1976; Demirtasli et al. 1984) and represents a ribbon continent rifted off from the northwestern edge of Gondwana (Robertson & Dixon 1984; Garfunkel 1998). The Palaeozoic–Jurassic tectonostratigraphic units in the Tauride block are tightly folded and imbricated along major thrust faults that developed first during the obduction of the Inner-Tauride ophiolites from the north in the late Cretaceous, and subsequently during the collision of the Tauride block with the CACC in the latest Palaeocene–Eocene. The buoyancy of the Tauride continental crust in the lower plate eventually arrested the subduction process and caused the isostatic rebound of the partially subducted platform edge, leading to block-fault uplifting of the Taurides during the latest Cenozoic (Dilek & Whitney 1997, 2000). The entire Tauride block experienced gradual uplift in the footwall of a north-dipping frontal normal fault system along its northern edge starting in the Miocene, and developed as a southward-tilted, asymmetric mega-fault block with a rugged, alpine topography (Dilek *et al.* 1999*b*). Apatite fission track ages of 23.6 ± 1.2 Ma from the 55 Ma-old Horoz granite that is intrusive into the Bolkar carbonates are consistent with this uplift history (Dilek *et al.* 1999*b*).

*Cappadocian volcanic province*

The Cappadocian volcanic province defines a *c.* 300 km-long volcanic belt extending NE–SW across the CACC (Fig. 9). The earliest volcanism in the province started in the mid-Miocene (*c.* 13.5 Ma) and continued into the Quaternary (Pasquaré *et al.* 1988; Ercan *et al.* 1994). The initial volcanic products include 13.5–8.5 Ma andesitic lavas, tuff and ignimbrites in the form of effusive centres and endogeneous domes. This volcanic phase was followed by widespread eruption of ignimbrites, volcanic ash, lapilli and agglomerates between 8.5 and 2.7 Ma (Pasquaré *et al.* 1988; Ercan *et al.* 1994). The most recent phase of volcanism produced central volcanoes oriented parallel to the NE–SW axis of the province, consisting of basaltic andesite, andesite, dacite, rhyodacite and basaltic lavas. Geochemically, these rocks collectively have an A-type granitic melt origin showing post-collisional, within-plate affinities (Innocenti *et al.* 1975; Ercan *et al.* 1994; Toprak *et al.* 1994).

The Cappadocian volcanic province broadly corresponds to a structurally controlled topographic depression filled with Upper Miocene to Pliocene fluvial and lacustrine deposits (Dilek *et al.* 1999*b*). The volcanic edifices appear to have been built at the intersections of major strike-slip fault systems (i.e. Mt. Erciyes volcano) and/or in local graben structures associated with tectonic subsidence and a *c.* NNW–SSE-oriented regional extension during the late Cenozoic (Toprak *et al.* 1994; Dilek *et al.* 1999*b*). These relations indicate that faulting and volcanism in the Cappadocian volcanic province were spatially and temporally associated, and that magma transport and extrusion were in part facilitated by crustal-scale fault systems (Dilek *et al.* 1999*b*).

# Tectonic evolution of the central Anatolian orogenic belt

The southward emplacement of the Tethyan suprasubduction zone ophiolites along the northern edge

of the CACC around 90–85 Ma was facilitated by a subduction zone dipping northward beneath the Pontides and away from the CACC (Tankut *et al.* 1998; Dilek & Whitney 2000; Floyd *et al.* 2000). This subduction zone could not have had any effect on the mantle dynamics and heterogeneity beneath the CACC or on the evolution of the Late Cretaceous magmatism on and across the CACC (as suggested in Boztug 2000; Ilbeyli *et al.* 2004; Köksal *et al.* 2004; Ilbeyli 2005). Instead, the subduction zone that was involved in the evolution of the CACC magmatism was located to the SW, dipping northeastward beneath the CACC (in present coordinate system) and consuming the oceanic lithosphere of the Inner-Tauride Ocean (Fig. 10a; Erdogan *et al.* 1996; Dilek *et al.* 1999*a*; Kadioglu *et al.* 2003). The inferred chemical modification of the mantle wedge beneath the CACC was associated with this ITSZ.

There are several independent lines of evidence for the existence of this Inner-Tauride basin between the CACC and the Tauride block and the northward dipping subduction zone within this Tethyan seaway. The Cenomanian age suprasubduction zone ophiolites in the Tauride belt were derived from the arc-forearc setting in the Inner-Tauride Ocean and were emplaced southward onto the continental edge of the Tauride block during its collision with an intra-oceanic arc-trench system (Dilek *et al.* 1999*a*; Parlak *et al.* 1996). Partial subduction of the Tauride edge beneath the ophiolite nappes resulted in HP/LT metamorphism of the platform carbonates and in the formation of blueschists (Fig. 10b), which are currently exposed discontinuously along the northern periphery of the Tauride belt (Okay 1984; Okay *et al.* 1996, 1998; Dilek & Whitney 1997; Önen 2003). The calculated $P–T$ conditions of metamorphism and $^{40}Ar/^{39}Ar$ ages of phengites and glaucophanes from blueschist rocks in northwestern Anatolia suggest that this HP/LT metamorphism occurred around 88 Ma (Harris *et al.* 1994; Okay *et al.* 1996).

The older peraluminous granitoid plutons (*c.* 100–85 Ma) along the western edge of the CACC represent a magmatic arc complex that developed above a NE-dipping subduction zone within the Inner-Tauride Ocean (Fig. 10b; Görür *et al.* 1984). During the closure of this basin as a result of the subduction of the Tethyan oceanic lithosphere beneath the western edge of the CACC, the melts derived from the metasomatized upper mantle were injected into the continental crust, causing its partial melting. This led to interaction of mantle- and crustal-derived magmas that involved AFC, mixing and mingling which collectively produced the calc-alkaline granitoid plutons (Fig. 10b, c; Erdogan *et al.* 1996; Kadioglu *et al.* 2003; Ilbeyli *et al.* 2004).

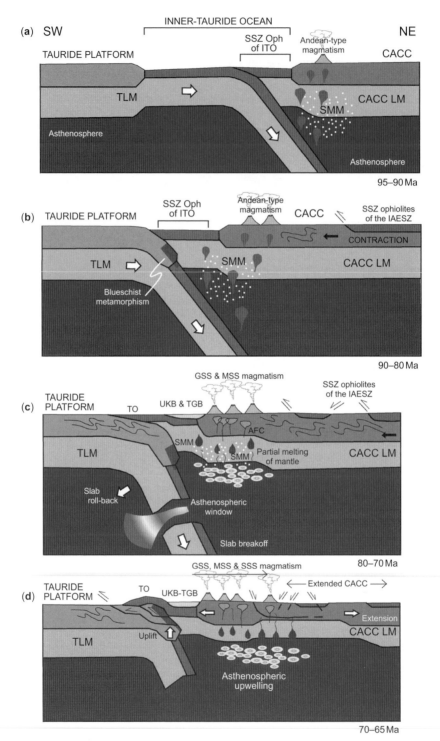

**Fig. 10.** Tectonic model for the evolution of the CACC and the central Anatolian orogenic belts in the late Mesozoic (modified after Kadioglu *et al.* 2006). See text for discussion. Ellipses beneath SMM depict melting in the asthenosphere. Stippled (in white) pattern characterizes subduction-metasomatized mantle. Key to lettering: AFC, Assimilation

Following the demise of the Inner-Tauride oceanic lithosphere at the NE-dipping subduction zone and the emplacement of the incipient arc-forearc ophiolites onto the northern edge of the Tauride block, subduction was arrested by the underplating of the buoyant Tauride continental crust. The leading edge of the subducted Tethyan slab likely broke off from the rest of the Tauride continental lithosphere, resulting in the development of an asthenospheric window (Fig. 10c). The juxtaposition of this asthenospheric heat source against the overlying continental lithosphere may have caused melting of the metasomatized mantle layers, producing the high-K shoshonitic, adakite-like magmas of the monzonitic plutons and then the more-enriched alkaline magmas of the syenitic plutons (Fig. 10d). This process is similar to that inferred for slab breakoff-related collisional magmatism in other orogenic belts (Davies & von Blackenburg 1995, and references therein) and in the early Cenozoic of western Anatolia, as discussed earlier.

As the asthenosphere moved upwards through the window in the slab, partial melting occurred at the boundary between the lithosphere and asthenosphere producing basaltic melts with transitional characteristics between those of calc-alkaline and alkaline basalts. The mantle sources for these primary basaltic melts may have been metasomatized garnet peridotites and/or spinel lherzolites and phlogopite-bearing lherzolites of an upper mantle wedge origin (Conceiçao & Green 2004). Such slab breakoff-related magmatism with similar calc-alkaline to alkaline products has been documented from other orogenic belts, such as the Neogene–Quaternary Carpathian–Pannonian region (Nemcok et al. 1998; Seghedi et al. 2004), the Late Palaeogene Periadriatic–Sava–Vardar magmatic zone in the Dinaride–Hellenide mountain belt (Pamic et al. 2002), the Neogene Maghrebian orogenic belt in northern Africa (Maury et al. 2000) and the late Oligocene–early Miocene central Aegean Sea region (Pe-Piper & Piper 1994).

Thermal perturbation of the continental lithosphere and the alkaline magmatism weakened the orogenic crust, leading to tectonic extension in and across the CACC (Fig. 10d). The results of recent studies of metamorphic massifs and core complexes in central Anatolia suggest that significant crustal extension and unroofing might have occurred within the CACC prior to the Eocene (Gautier et al. 2002).

Strike-slip faulting also played a major role in the late Cenozoic tectonic evolution of the CACC. A series of NE–SW and NW–SE trending strike-slip fault systems crosscut both the crystalline basement rocks and the Palaeogene–Quaternary sedimentary strata and volcanic rock units (Fig. 9). The sinistral Ecemis fault (EF) truncates and offsets the Tauride block and the ITSZ by as much as 100 km, and forms mainly a transtensional fault system with several releasing bends and pull-apart structures. It appears to have accommodated top-to-the south extension and crustal exhumation along the south-dipping detachment fault of the Nigde core complex. It may have, therefore, facilitated the vertical displacement and unroofing of high-grade metamorphic rocks in the eastern part of the Nigde massif during the Oligo–Miocene (Dilek & Whitney 1998). The NW–SE trending Tuzgölü fault (TGF) bounding the Cappadocian volcanic province on the west represents a dextral transpressional fault system (Saroglu et al. 1992). The latest Cretaceous granitoid plutons and their high-grade metamorphic host rocks in the western part of the CACC have been uplifted on the eastern shoulder of the TGF. The Ecemis and Tuzgölü faults together form a triangular-shaped crustal flake within the CACC that has been moving southwards while undergoing rifting and subsidence, which has given way into the development of the Cappadocian volcanic province throughout the latest Cenozoic.

## Regional geology of eastern Anatolia

Much of eastern Turkey is occupied by the East Anatolian High Plateau, which is bounded in the north by the Eastern Pontide arc and in the south by the Bitlis–Pütürge massif (Fig. 11). The mean surface elevation of the plateau is about 2–2.5 km above sea level with scattered Plio–Quaternary volcanic cones over 5 km high (e.g. Mt. Ararat or Mt. Agri). The basement geology of the plateau is composed of ophiolites and ophiolitic mélange, flysch and molasse deposits, and the eastward extension of the Tauride platform carbonates. The Tethyan ophiolites and ophiolitic mélanges, flysch deposits and volcanic arc units collectively constitute the East Anatolian Subduction–Accretion Complex.

**Fig. 10.** (*Continued*) fractional crystallization; CACC, Central Anatolian Crystalline Complex; CACC LM, Central Anatolian Crystalline Complex lithospheric mantle; IAESZ, Izmir–Ankara–Erzincan suture zone; ITO, Inner Tauride Ocean; SMM, Subduction modified mantle; SSZ, Suprasubduction zone; TGB, Tuzgölü basin; TLM, Tauride lithospheric mantle; TO, Tauride ophiolites (including the Aladag, Alihoca and Mersin ophiolites); UKB, Ulukisla basin. GSS, MSS and SSS magmatism represent the Granite, Monzonite and Syenite Supersuites, respectively, of the syncollisional CACC plutons. SSZ ophiolites of the Inner-Tauride Ocean (ITO) include the Beysehir, Alihoca, Aladag, and Mersin ophiolites.

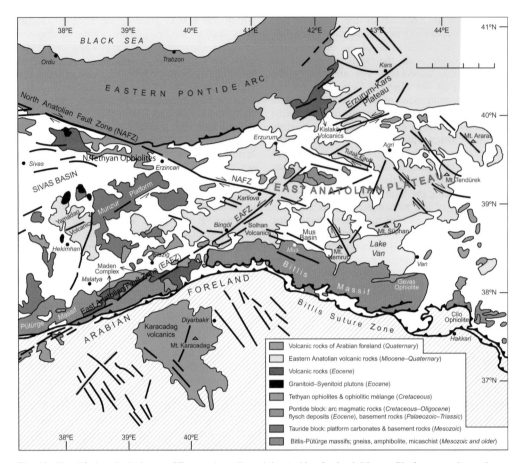

**Fig. 11.** Simplified geological map of Eastern Anatolia and the Arabian foreland. Munzur Platform constitutes the eastern extension of the platform carbonates and basement rocks of the Tauride block. Bitlis–Pütürge massif is a rifted off fragment of the Arabian plate, analogous to the Tauride block. The East Anatolian Plateau is covered by Miocene–Quaternary volcanic rocks; its basement is composed of Tethyan ophiolites and ophiolitic mélanges, flysch and molasses deposits, and platform carbonates of the Tauride block. Key to lettering: EAFZ, East Anatolian fault zone; NAFZ, North Anatolian fault zone.

## Eastern Pontide arc

The Pontide block north of the plateau represents a south-facing early Cretaceous–late Eocene volcano-plutonic arc that developed over a subduction zone dipping northward beneath the Eurasian continental margin (Yilmaz *et al.* 1997). The collision of the Pontide arc with the Tauride microcontinent in the late Eocene terminated the subduction zone magmatism in the Pontide terrane and produced extensive flysch deposits with intense folding in the collision zone (Dewey *et al.* 1986).

## Bitlis–Pütürge massif

The Bitlis–Pütürge massif is a nearly East–West-trending microcontinent that is surrounded on all sides by ophiolitic rocks, mélanges, and volcanic and volcaniclastic rocks of an arc origin. The Sanandaj–Sirjan continental block to the SE in Iran represents the eastern continuation of the Bitlis–Pütürge massif in the peri-Arabian region. The Sanandaj–Sirjan continental block to the SE in Iran represents the eastern continuation of the Bitlis–Pütürge massif in the peri-Arabian region. The Pütürge massif is composed of pre-Triassic gneisses and micaschists, and granitoids (Michard *et al.* 1984; Aktas & Robertson 1990) and is interpreted as a pre-Triassic continental sliver of Afro–Arabian origin, similar to the Bitlis massif to the east. The Pütürge massif and the overlying volcanic and ophiolitic rocks are structurally underlain in the south by an upper Cretaceous–early Tertiary mélange, which is underthrust to the south by foreland sedimentary sequences of the Arabian plate (Fig. 12).

The Bitlis massif is a composite tectonic unit, which is composed of a metamorphic basement

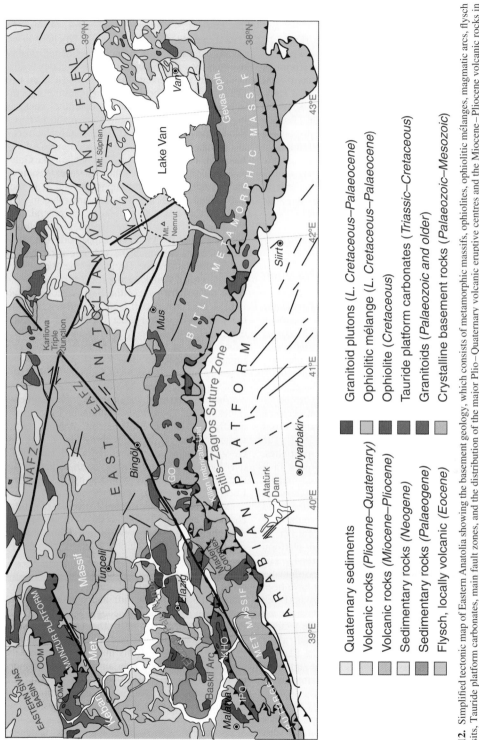

**Fig. 12.** Simplified tectonic map of Eastern Anatolia showing the basement geology, which consists of metamorphic massifs, ophiolites, ophiolitic mélanges, magmatic arcs, flysch deposits, Tauride platform carbonates, main fault zones, and the distribution of the major Plio–Quaternary volcanic eruptive centres and the Miocene–Pliocene volcanic rocks in the East Anatolian High Plateau. EAFZ and NAFZ mark the East and North Anatolian fault zones, respectively. Symbols for the ophiolites: GO, Guleman; IPO, Ispendere; KHO, Kömürhan; OOM, Ovacik ophiolitic mélange.

(Precambrian?), overlying metamorphosed Palaeozoic to Triassic carbonate rocks (Göncüoglu & Turhan 1984; Helvaci & Griffin 1984), and Palaeozoic to late Mesozoic granitoids (Fig. 12). Oberhänsli et al. (2008) recently reported a regionally distributed LT/HP metamorphic overprint in the thermal evolution of the Bitlis massif and suggested that the massif is composed of a stack of nappes formed during the closure of the Southern Tethys between Arabia and the Anatolide–Tauride continental block. The whole massif displays a doubly plunging, multiply folded anticlinorium with overturned limbs both to the north and the south (Dilek & Moores 1990). The relatively youngest thrust faults are south-vergent and synthetic to the Bitlis suture that represents the collision zone between the Arabian and Eurasian plates.

## East Anatolian subduction-accretion complex

The Jurassic (?)–Cretaceous ophiolites underlying the molasse deposits and the Tertiary volcanic cover in the western part of the East Anatolian High Plateau represent the remnants of Mesozoic Tethys and are commonly directed southwards onto the margins of the Eastern Tauride platform and the Pütürge massif (Dilek & Moores 1990). The Maastrichtian Ovacik ophiolitic mélange overlies tectonically the Upper Triassic–Cretaceous Munzur platform (Fig. 12) that is interpreted to be the northeastward extension of the calcareous axis of the Tauride block (Özgül & Tursucu 1984). The Ovacik mélange consists of blocks of serpentinites, metamorphic rocks and pelagic limestones in a fine-grained, phyllitic matrix and is unconformably overlain by Maastrichtian clastic rocks (Özgül & Tursucu 1984). Both the Ovacik mélange and Munzur carbonates are thrust to the south over the Keban metamorphic rocks that consist of Permian to Cretaceous metamorphosed platform carbonates (Fig. 12; Michard et al. 1984). The Keban metamorphics and Munzur carbonates display north-vergent, upright folds that have been subsequently folded by south-vergent folds (Dilek & Moores 1990).

The Keban metamorphic rocks overlie tectonically the Elazig–Palu nappe to the south along south-vergent thrust faults (Fig. 12). The Elazig–Palu nappe consists of calc-alkaline intrusive and extrusive rocks, and overlying Campanian–Maastrichtian volcaniclastic and flysch deposits (Michard et al. 1984; Yazgan 1984; Aktas & Robertson 1990). The Elazig–Palu nappe includes Yazgan's (1984) Baskil magmatic rocks that consist of Coniacian-Santonian granodiorites, tonalites, quartz monzonites, monzodiorites, diorites

and gabbros of an island-arc. These arc rocks intrude the oceanic rocks and mafic extrusives of the Ispendere–Kömürhan ophiolites of Yazgan (1984) and the Guleman nappe of Michard et al. (1984). The Ispendere ophiolite is unconformably overlain by a flysch deposit, which contains fossils of Upper Campanian–Lower Maastrichtian age (Yazgan 1984). The Kömürhan ophiolite structurally beneath the Ispendere ophiolite contains highly deformed, amphibolite-grade mafic, ultramafic and intermediate rocks. The metamorphic rocks are intruded by diorites and granodiorites of the Baskil arc complex (Elazig–Palu nappe). The Guleman ophiolite constitutes the eastern extension of the Ispendere–Kömürhan ophiolite and is in tectonic contact with the Bitlis metamorphic massif. The contact relationships indicate, in general, that the oceanic rocks of the Ispendere–Kömürhan ophiolites and the Guleman nappe form the oceanic basement on which the volcanic arc rocks of the Elazig-Palu nappe (including the Baskil arc) were deposited (Dilek & Moores 1990).

In the eastern part of the East Anatolian High Plateau the Bitlis massif is tectonically overlain to the north and west by the Gevas and Guleman ophiolites, respectively. The Gevas ophiolite, exposed in an east–west trending narrow belt immediately south of Lake Van (Fig. 12), consists of serpentinized ultramafic rocks, cumulate and isotropic gabbros, microgabbros and plagiogranites overlain by extrusive rocks and pelagic sediments (Dilek 1979). These mafic–ultramafic rocks tectonically rest on the Bitlis massif along a south-vergent thrust fault. This thrust fault and the Gevas ophiolite are further deformed and thrust over by the Bitlis massif along north-vergent faults that are depositionally overlain by Palaeocene–Eocene flysch deposits (Dilek 1979).

Mafic–ultramafic rocks and extrusives of similar character crop out farther west, south of the East Anatolian Fault and west of the Bitlis massif, where the Guleman ophiolite overlies tectonically the metamorphosed carbonates of the Bitlis massif (Fig. 11). The Guleman ophiolite includes serpentinized peridotites, banded gabbro, microgabbro, metamorphosed basalt, tuff and agglomerate and radiolarian mudstone (Göncüoglu & Turhan 1984; Aktas & Robertson 1990). The Upper Jurassic–Lower Cretaceous Guleman ophiolite is separated from the underlying Bitlis massif by an intensely mylonitized zone and overlain by an unmetamorphosed Upper Maastrichtian flysch deposit that contains blocks of both the Guleman ophiolite and the Bitlis massif. This depositional relationship constrains the ophiolite emplacement age as pre-late Maastrichtian. The rocks of the Gevas–Guleman ophiolite belt continue farther west and are intruded and overlain by the calc-alkaline intrusives and volcaniclastic rocks of the Elazig–Palu nappe (Ozkaya 1978; Michard *et al.* 1984; Aktas & Robertson 1990).

The Bitlis massif is underlain to the south by a late Cretaceous–early Tertiary tectonic mélange, which directly overlies the foreland deposits of Arabian plate (Fig. 12). The mélange is composed of ophiolitic material tectonically interleaved with hemipelagic and clastic rocks (Aktas & Robertson 1990). In places, however, the Bitlis massif is underlain by thrust sheets of ophiolitic rocks consisting of serpentinized peridotites, gabbro, diabase and basaltic andesites that structurally overlie the tectonic mélange. Both the mélange and the ophiolitic thrust sheets constitute the Killan Imbricate Unit of Aktas and Robertson (1990) that comprises, with the Guleman ophiolite, their Maden complex (Fig. 12). The Maden complex and the overlying Elazig–Palu nappe directly rest on the Arabian platform units along south-vergent thrust faults wherever the Bitlis massif is absent (Dilek & Moores 1990). South of the Bitlis Suture Zone, the lower Palaeozoic to Miocene shelf sequences of the Arabian foreland display a south-vergent fold-and-thrust belt architecture.

East-southeast of the Bitlis massif, south-vergent thrust sheets composed of a complete ophiolite sequence (the Cilo ophiolite) and arc-related calc-alkaline rocks rest tectonically on the Mesozoic Arabian platform (Yilmaz 1985). The Cilo ophiolite contains, from bottom to top, peridotites, cumulate and isotropic gabbros, diorite, quartz diorite, sheeted dykes, pillow and massive lava flows, and ribbon cherts and shales (Yilmaz 1985). The sedimentary rocks associated with the ophiolite give Jurassic to Upper Cretaceous fossil ages; the entire ophiolitic sequence is depositionally overlain by volcanic-pyroclastic rocks and is intruded by granitic–granodioritic intrusions (Fig. 11; Yilmaz 1985). The lower thrust sheet underlying the Cilo ophiolite consists mainly of calc-alkaline lavas, pyroclastics, and blocks of radiolarian chert and recrystallized limestone. These relations suggest that the Cilo ophiolite may constitute the oceanic basement of an ensimatic arc complex represented by the calc-alkaline intrusions and volcanic-pyroclastic rocks.

## Deformation and volcanism in the East Anatolian High Plateau

Emplacement of the Tethyan ophiolites in the late Cretaceous and the subsequent continental collisions in the Eocene and mid-Miocene played a major role in the construction of the East Anatolian High Plateau. North–South shortening and crustal imbrication via thrust faulting and folding in the collision zone took up much of the convergence between Arabia and Eurasia. East–West-oriented thrust faults and folds in the Upper Miocene lavas and pyroclastic rocks of the Solhan volcanic rocks in the Mus basin area (Fig. 11; Yilmaz *et al.* 1987) indicate that crustal shortening continued after the Arabia-Eurasia collision and affected the post-collisional volcanic rocks in the plateau. However, progressive thickening of the crust has been accompanied by major strike-slip faulting on the dextral North Anatolian and the sinistral East Anatolian faults that have accommodated the westward escape of the Anatolian plate at a rate of 0.5 cm a$^{-1}$ (Jackson & McKenzie 1984). NE- and NW-striking conjugate strike-slip faults (sinistral and dextral, respectively) with a significant compressional component also occur within the plateau (i.e. Tutak fault; Fig. 11), and some of these faults are seismically active (Tan & Taymaz 2006).

The widespread volcanism in the Turkish–Iranian Plateau started between 8 and 6 Ma, 4–5 Ma years after the initial collision of Arabia with Eurasia in Serravallian-Tortonian times (Innocenti *et al.* 1982; Yilmaz *et al.* 1987; Pearce *et al.* 1990). In general, volcanism in the southern segment of the plateau is characterized by the construction of stratovolcanoes with significant peaks (i.e. Nemrut, Suphan, Tendürek, Ararat; Fig. 11), whereas in the north it forms an extensive (5000 km$^2$) and relatively flat volcanic field (Erzurum-Kars plateau; Fig. 11) with an average elevation of *c.* 1.5 km above sea level. This volcanic field consists mainly of lava flows intercalated with subordinate ignimbrite units and sedimentary layers giving ages from 6.9 $\pm$ 0.9–1.3 $\pm$ 0.3 Ma (Innocenti *et al.* 1982; Keskin *et al.* 1998). Pleistocene scoriaceous spatter cones locally overlie this lava-ignimbrite sequence. The initial eruptive phase of post-collisional volcanism in the plateau is characterized by basic and intermediate alkaline rocks and was followed by widespread eruptions of andesitic to dacitic calc-alkaline magma during the Pliocene; the last volcanic phase involved the eruption of alkaline and transitional lavas throughout the Plio-Pleistocene and Quaternary (Yilmaz *et al.* 1987). Most of the major stratovolcanoes in the region were built during this last phase of volcanism, which continued until historical times. Alkaline basaltic lavas of the late volcanic phase appear to predominate mainly in the northern part of the plateau, whereas the calc-alkaline rocks of the second major volcanic phase occur most extensively in the south.

The East Anatolian High Plateau has undergone significant uplift since the Arabian collision in the Mid-Miocene (*c.* 13 Ma). The Lower to Middle Miocene fossiliferous marine marl and reefal carbonates around Lake Van indicate that

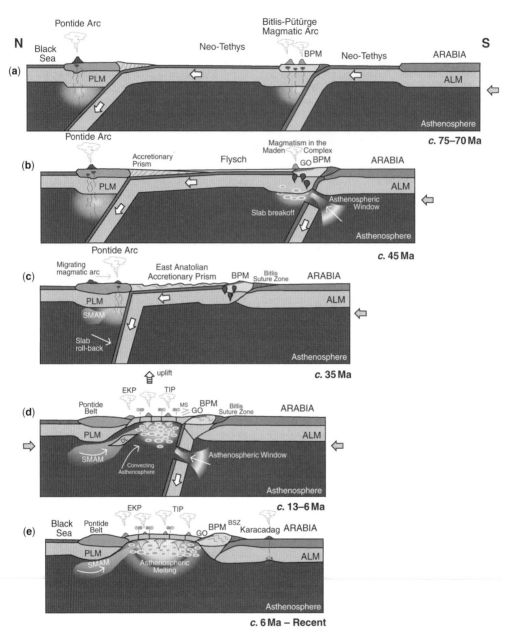

**Fig. 13.** Maastrichtian–Cenozoic geodynamic evolution of the East Anatolian High Plateau and the subduction-accretion complex through subduction and collisional processes in the upper plate of north-dipping subduction zone(s) within the Tethyan realm (data and interpretations are derived in part from Dilek & Moores 1990; Keskin 2003; Sengör *et al.* 2003; and Dilek *et al.* 2010). Guleman ophiolite (GO) together with the Ispendere–Kömürhan ophiolites and Baskil arc units represent backarc and arc oceanic crust tectonically emplaced onto the northern margin of the Bitlis–Pütürge metamorphic massif by the early Eocene. See text for discussion. Key to lettering: ALM, Arabian lithospheric mantle; BPM, Bitlis–Pütürge massif; BSZ, Bitlis suture zone; EKP, Erzurum–Kars plateau; GO, Guleman ophiolite; MS, Mus suture; PLM, Pontide lithospheric mantle; SMAM, Subduction metasomatized asthenospheric mantle; TIP, Turkish–Iranian high plateau.

the area remained under the sea until the Serraval-lian (Gelati 1975). Upper Miocene lacustrine and fluvial sedimentary rocks unconformably overlie these marine sedimentary rocks, suggesting the emergence of land and the onset of terrestrial conditions by the late Miocene (Altinli 1966). A late Miocene–early Pliocene erosional surface trun-cating the fluvial rocks was covered with Pliocene andesitic–dacitic lavas and was subsequently deeply dissected by streams and rivers due to rapid block-uplift of the western and central segments of the plateau (Altinli 1966; Innocenti et al. 1976). These stream valleys were then filled with Pleisto-cene lava flows fed by the alkaline volcanic phase (Erinç 1953).

## Tectonic Evolution of the East Anatolian High Plateau

The late Mesozoic geodynamic evolution of eastern Anatolia was controlled by subduction zone tec-tonics in two separate Tethyan seaways (Fig. 13a). The Northern Tethys seafloor was being consumed at a subduction zone dipping northward beneath the Eastern Pontide arc, and the Black Sea was opening up as a back-arc basin behind this arc around 75–70 Ma (Yilmaz et al. 1997). Subduction of the Southern Tethys seafloor beneath the Tauride microcontinent farther south developed a magmatic arc on the Bitlis–Pütürge microcontinent and the arc-backarc oceanic crust presently represented by the Ispendere–Kömürhan and Guleman ophiolites tectonically overlying the Bitlis–Pütürge massifs. A similar tectonic scenario has been suggested for the Cretaceous arc-ophiolite duo in the Malatya–Maras region farther west in the Tauride block (Parlak 2006).

The collision of the Arabian plate with the Bitlis–Pütürge magmatic arc occurred in the early Eocene (Yilmaz 1993) and produced the mélange and flysch deposits along the Bitlis suture zone (Fig. 13b). This continental collision led to slab breakoff and development of an asthenospheric window, which in turn facilitated partial melting of the subduction-metasomatized lithospheric mantle beneath the Bitlis–Pütürge massifs, pro-ducing the shoshonitic magmatism in the Maden Complex (Fig. 13b; Elmas & Yilmaz 2003).

Continued subduction of the Northern Tethyan seafloor beneath Eurasia farther north and slab stee-pening and roll-back produced southward migrating magmatism in the Eastern Pontide arc during the Eocene–Oligocene, while the subduction-accretion complex widened toward the south (Sengör et al. 2003). As the Tethyan lithosphere continued its subduction beneath the Pontide arc, the widening East Anatolia accretionary complex was shortened

and thickened within the closing basin (Fig. 13c). North–South contraction across the Northern Tethyan realm and vertical thickening of the East Anatolian subduction-accretion complex created a 'tectonic bumper' between the converging Eurasia and Bitlis–Pütürge–Arabia plates that had reached the average thickness of continental crust by the late Oligocene–early Miocene (c. 24 Ma; Sengör et al. 2003). Further steepening and south-ward retreat of the subducting Tethyan lithosphere might have triggered lithospheric delamination beneath the southern margin of the Eastern Pontide arc and the northern part of the East Anato-lian high plateau, resulting in remobilization and partial melting of the subduction-metasomatized asthenospheric mantle (Fig. 13d). This event pro-duced the initial stages of calc-alkaline magmatism in the Erzurum-Kars Plateau by the middle Miocene (Keskin et al. 2006).

Arrival of the Bitlis–Pütürge–Arabia composite continental plate at the trench and the continent-trench collision by c. 13 Ma slowed down and tem-porarily arrested the northward subduction beneath the East Anatolian subduction-accretionary com-plex. However, the continued sinking of the ocea-nic lithosphere in this subduction zone must have caused the detachment of the subducting slab, leading into slab breakoff and development of an asthenospheric window (Fig. 13d). Rising hot asthenosphere beneath the subduction-accretion complex resulted in widespread partial melting both in the upwelling and convecting asthenosphere and in the overlying crust (Fig. 13d; Sengör et al. 2003; Keskin 2003) that produced bimodal volcan-ism throughout the uplifted plateau. Extensive strike-slip and extensional normal faulting in the Turkish–Iranian high plateau (TIP) facilitated the rise and eruption of asthenosphere-derived alkaline olivine basalts at the surface without much conti-nental contamination in the late Miocene–Pliocene (Fig. 13e).

Widespread volcanism across the entire East Anatolian high plateau (>250 km wide) throughout the late Cenozoic and until historic times indicates a significant heat source beneath it, resulting in exten-sive melting. The findings of the recent Eastern Turkey Seismic Experiment (ETSE) and tomo-graphic models have shown the lack of mantle litho-sphere, an average continental crustal thickness (c. 40–45 km), lack of earthquakes deeper than c. 30 km, and very low Pn velocity zones indicating the presence of partially molten material beneath the region (Sandvol et al. 2003a; Al-Lazki et al. 2003; Gök et al. 2003; Zor et al. 2003; Angus et al. 2006). These observations collectively suggest that the East Anatolian high plateau is likely supported by hot asthenospheric mantle, not by overthickened crust (Dewey et al. 1986) or subducted Arabian continental lithosphere (Rotstein & Kafka 1982) as previously inferred.

## Conclusions

The modern Anatolian–African plate boundary is characterized by subduction zone tectonics and is in the initial stages of collision-driven orogenic buildup. The Anatolian microplate itself is made of young orogenic belts (Eocene and younger) that evolved during a series of collisions between Gondwana-derived ribbon continents and trench-roll-back systems within the Tethyan realm. The collision of the Eratosthenes seamount with the Cyprus trench since the late Miocene is a smaller-scale example of this accretionary process and has affected the slab geometry and kinematics of the subducting African lithosphere.

Pn velocity and Sn attenuation tomography results show that the uppermost mantle beneath much of the young orogenic belts in Anatolia is anomalously hot and thin. This is consistent with the surface geology, which is dominantly controlled by strike-slip and extensional tectonics and widespread volcanism in western, central and eastern Turkey. In all these areas, the extension was well under way by the late Oligocene–Miocene, following the main episodes of continental collisions. Pinning of subduction hinge zones by the accreted ribbon continents arrested slab roll-back processes, causing terrane stacking and crustal thickening, and resulted in slab breakoff because of continued convergence of the lithospheric mantle. Slab breakoff-induced asthenospheric upwelling provided the necessary heat and melt to produce the first phases of post-collisional magmatism in these young orogenic belts. Renewed subduction and slab roll-back in the Tethyan realm triggered lithospheric-scale extension in the upper plate, and the thermally weakened orogenic crust started collapsing. These processes resulted in rapid exhumation of recently formed high-pressure metamorphic rocks and in the formation of metamorphic core complexes.

The Cenozoic geodynamic evolution of the western, central and eastern Anatolian orogenic belts indicates that the asthenospheric mantle beneath collision zones responds swiftly to crustal tectonics on time scales of just a few million years. Slab breakoff, lithospheric delamination and slab tearing were common processes that resulted directly from collision-induced events, and caused convective remobilization of the asthenosphere leading to magmatism. Asthenospheric upwelling and partial melting played a major role in a geochemical progression of post-collisional magmatism from initial shoshonitic, calc-alkaline to late-stage alkaline affinities through time.

The collision-driven tectonic evolution of the Anatolian–African plate boundary and the young orogenic belts in the eastern Mediterranean region is typical of the geodynamic development of the Alpine–Himalayan orogenic system. Successive collisions of Gondwana-derived microcontinents with trench-roll-back cycles in the Tethyan realms of the Alpine–Himalayan system caused basin collapse, ophiolite emplacement and continental accretion, producing subparallel mountain belts. Subduction of the Tethyan mantle lithosphere was nearly continuous throughout these accretionary processes, only temporarily punctuated by slab breakoff events.

Part of this study was undertaken during the Eastern Turkey Seismic Experiment (ETSE), which was supported by the NSF under Grant No EAR-9804780; additional support for this experiment was provided by the Bogaziçi University Research Fund under Grant No 99T206. We gratefully acknowledge these grants. The geological fieldwork in Turkey has been funded over the years by the NSF Tectonics Program, Miami University Committee on Faculty Research and Hampton International Initiatives Funds, Scientific & Technical Research Council of Turkey (TUBITAK), and Istanbul Technical University Research Funds. Discussions with S. Altunkaynak, M. Barazangi, E. Bozkurt, C. Genç, R. Gök, C. Helvaci, Y. K. Kadioglu, A. Polat, Y. Savaşçin, F. Yagmurlu, Y. Yilmaz, and E. Zor on various aspects of the Mesozoic–Cenozoic geology, geodynamics and seismic structure of Anatolia have been most helpful to us in developing the ideas and interpretations presented in this paper. Thorough and constructive reviews by D. Nance, B. Murphy and P. T. Robinson helped us improve the paper.

## References

AGOSTINI, S., DOGLIONI, C., INNOCENTI, F., MANETTI, P., TONARINI, S. & SAVAŞÇIN, M. Y. 2007. The transition from subduction-related to intraplate Neogene magmatism in the Western Anatolia and Aegean area. In: BECCALUVA, L., BIANCHINI, G. & WILSON, M. (eds) Cenozoic Volcanism in the Mediterranean Area. Geological Society of America Special Paper, **418**, 1–16.

AKAY, E. & ERDOGAN, B. 2004. Evolution of Neogene calc-alkaline to alkaline volcanism in the Aliaga-Foça region (Western Anatolia, Turkey). Journal of Asian Earth Science, **24**, 367–387.

AKIMAN, O., ERLER, A., GÖNCÜOĞLU, M. C., GULEÇ, N., GEVEN, A., TURELI, T. K. & KADIOGLU, Y. K. 1993. Geochemical characteristics of granitoids along the western margin of the Central Anatolian Crystalline Complex and their tectonic implications. Geological Journal, **28**, 371–382.

AKTAS, G. & ROBERTSON, A. H. F. 1990. Tectonic evolution of the Tethys suture zone in SE Turkey: evidence from the petrology and geochemistry of Late Cretaceous and Middle Eocene extrusives. In: MALPAS, J., MOORES, E. M., PANAYIOTOU, A.

& XENOPHONTOS, C. (eds) *Ophiolites, Oceanic Crustal Analogues, Proceedings of the Symposium 'Troodos 1987'*. The Geological Survey Department, Nicosia, Cyprus, 311–328.

ALDANMAZ, E., PEARCE, J. A., THIRWALL, M. F. & MITCHELL, J. 2000. Petrogenetic evolution of late Cenozoic, post-collision volcanism in western Anatolia, Turkey. *Journal of Volcanology & Geothermal Research*, **102**, 67–95.

ALDANMAZ, E., KÖPRÜBASI, N., GÜRER, Ö. F., KAYMAKÇI, N. & GOURGAUD, A. 2006. Geochemical constraints on the Cenozoic, OIB-type alkaline volcanic rocks of NW Turkey: implications for mantle sources and melting processes. *Lithos*, **86**, 50–76.

ALESSANDRINI, B., BERANZOLI, L., DRAKATOS, G., FALCONE, C., KARANTONIS, G., MELE, F. M. & STAVRAKAKIS, G. 1997. Back arc basins and P-wave crustal velocity in the Ionian and Aegean regions. *Geophysical Research Letters*, **24**, 527–530.

ALICI, P., TEMEL, A., GOURGAUD, A., KIEFFER, G. & GÜNDOGDU, M. N. 1998. Petrology and geochemistry of potassic rocks in the Gölcük area (Isparta, SW Turkey): genesis of enriched alkaline magmas. *Journal of Volcanology and Geothermal Research*, **85**, 423–446.

ALICI, P., TEMEL, A. & GOURGAUD, A. 2002. Pb–Nd–Sr isotope and trace element geochemistry of Quaternary extension-related alkaline volcanism: a case study of Kula region (western Anatolia, Turkey). *Journal of Volcanology & Geothermal Research*, **115**, 487–510.

AL-LAZKI, A. I., SEBER, D., SANDVOL, E., TÜRKELLI, N., MOHAMAD, R. & BARAZANGI, M. 2003. Tomographic Pn velocity and anisotropy structure beneath the Anatolian plateau (eastern Turkey) and the surrounding regions. *Geophysical Research Letters*, **30**(24), 8043, doi: 10.1029/2003GL017391.

AL-LAZKI, A. I., SANDVOL, E., SEBER, D., BARAZANGI, M., TÜRKELLI, N. & MOHAMMAD, R. 2004. Pn tomographic imaging of mantle lid velocity and anisotropy at the junction of the Arabian, Eurasian and African plates. *Geophysical Journal International*, **158**, 1024–1040.

ALTINLI, E. 1966. Dogu ve güneydogu Anadolu'nun jeolojisi (in Turkish). *Bulletin of the Mineral Research & Exploration Institute of Turkey*, **66**, 35–74.

ALTUNKAYNAK, Ş. 2004. *Post collisional multistage magmatism in northwest Anatolia (Turkey): Geochemical and isotopic study of Orhaneli magmatic associations*. International Geological Congress, Florence, Italy, August 20–28, 2004. Abstracts 1298.

ALTUNKAYNAK, Ş. 2007. Collision-driven slab breakoff magmatism in northwestern Anatolia, Turkey. *Journal of Geology*, **115**, 63–82.

ALTUNKAYNAK, Ş. & DILEK, Y. 2006. Timing and nature of postcollisional volcanism in western Anatolia and geodynamic implications. *In*: DILEK, Y. & PAVLIDES, S. (eds) *Postcollisional Tectonics and Magmatism in the Mediterranean Region and Asia*. Geological Society of America Special Paper, **409**, 321–351.

ANGUS, D. A., WILSON, D. C., SANDVOL, E. & NI, J. F. 2006. Lithospheric structure of the Arabian and Eurasian collision zone in eastern Turkey from S-wave receiver functions. *Geophysical Journal International*, **166**, 1335–1346.

AYDIN, N. S., GONCUOGLU, M. C. & ERLER, A. 1998. Latest Cretaceous magmatism in the Central Anatolian Crystalline Complex: review of field, petrographic and geochemical features. *Turkish Journal of Earth Sciences*, **7**, 259–268.

BARKA, A. & REILINGER, R. 1997. Active tectonics of the Eastern Mediterranean region: deduced from GPS, neotectonic and seismicity data. *Annali Di Geofisica*, **XL**, 587–610.

BAYHAN, H. 1987. Cefalıkdag ve Baranadag plütonlarının (Kaman) petrografik ve kimyasal-mineralojik özelikleri. *Jeoloji Mühendisligi*, **30–31**, 11–16 (in Turkish).

BIJWAARD, H. & SPAKMAN, W. 2000. Non-linear global P-wave tomography by iterated linearized inversion. *Geophysical Journal International*, **141**, 71–82.

BINGÖL, E., DELALOYE, M. & ATAMAN, G. 1982. Granitic intrusions in Western Anatolia: A contribution of the geodynamic study of this area. *Eclogae Geologica Helvetica*, **75**, 437–446.

BINGÖL, E., DELALOYE, M. & GENÇ, S. 1994. Magmatism of northwestern Anatolia. *International Volcanological Congress, IAVCEI 1994, Excursion Guide (A3)*, 56.

BLUMENTHAL, M. M. 1956. *Yüksek Bolkardaginin kuzey kenar bölgelerinin ve bati uzantilarinin jeolojisi (Güney Anadolu Toroslari)*. Mineral Research and Exploration Institute of Turkey (MTA). Special Publication Series D, **7**, 153.

BOZKURT, E. 2003. Origin of NE-trending basins in western Turkey. *Geodinamica Acta*, **14**, 61–81.

BOZKURT, E. 2004. Granitoid rocks of the southern Menderes Massif (Southwest Turkey): field evidence for Tertiary magmatism in an extensional shear zone. *International Journal of Earth Sciences*, **93**, 52–71.

BOZKURT, E. 2007. Extensional v. contractional origin for the southern Menderes shear zone, SW Turkey: tectonic and metamorphic implications. *Geological Magazine*, **144**, 191–210.

BOZKURT, E. & PARK, R. G. 1994. Southern Menderes Massif: an incipient metamorphic core complex in western Anatolia, Turkey. *Journal of the Geological Society, London*, **151**, 213–216.

BOZKURT, E. & SATIR, M. 2000. The southern Menderes Massif (western Turkey): geochronology and exhumation history. *Geological Journal*, **35**, 285–296.

BOZTUG, D. 2000. S-I-A-type intrusive associations: geodynamic significance of synchronism between metamorphism and magmatism in Central Anatolia, Turkey. *In*: BOZKURT, E., WINCHESTER, J. A. & PIPER, J. D. A. (eds) *Tectonics and Magmatism in Turkey and the Surrounding Area*. Geological Society, London, Special Publications, **173**, 441–458.

BOZTUG, D., DEBON, F., INAN, S., TUTKUN, Z. S., AVCI, N. & KESGIN, Ö. 1997. Comparative geochemistry of four plutons from the Cretaceous–Palaeogene central Anatolian alkaline province (Divrigi region, Sivas, Turkey). *Turkish Journal of Earth Sciences*, **6**, 96–115.

CANDAN, O., ÇETINKAPLAN, M., OBERHÄNSLI, R., RIMMELÉ, G. & AKAL, C. 2005. Alpine high-P/low-T metamorphism of the Afyon Zone and implications for the metamorphic evolution of Western Anatolia, Turkey. *Lithos*, **84**, 102–124.

CATLOS, E. J. & ÇEMEN, I. 2005. Monazite ages and the evolution of the Menderes Massif, western Turkey. *International Journal of Earth Sciences*, **94**, 204–217.

ÇELIK, Ö. F. & CHIARADIA, M. 2008. Geochemical and petrological aspects of dike intrusions in the Lycian ophiolites (SW Turkey): a case study for the dike emplacement along the Tauride Belt Ophiolites. *International Journal of Earth Sciences*, **97**, 1151–1164, doi: 10.1007/s00531-007-0204-0.

ÇELIK, Ö. F., DELALAOYE, M. & FERAUD, G. 2006. Precise $^{40}Ar-^{39}Ar$ ages from the metamorphic sole rocks of the Tauride Belt Ophiolites, southern Turkey: implications for the rapid cooling history. *Geological Magazine*, **143**, 213–227.

ÇEMEN, I., CATLOS, E. J., GÖGÜS, O. & ÖZERDEM, C. 2006. Postcollisional extensional tectonics and exhumation of the Menderes massif in the Western Anatolia extended terrane. *In*: DILEK, Y. & PAVLIDES, S. (eds) *Postcollisional tectonics and magmatism in the Mediterranean region and Asia*. Geological Society of America Special Paper, **409**, 353–379.

ÇOBAN, H. & FLOWER, M. F. J. 2006. Mineral phase compositions in silica-undersaturated 'leucite' lamproites from the Bucak area, Isparta, SW Turkey. *Lithos*, **89**, 275–299.

COLLINS, A. & ROBERTSON, A. H. F. 2003. Kinematic evidence for Late Mesozoic–Miocene emplacement of the Lycian Allochthon over the Western Anatolide Belt, SW Turkey. *Geological Journal*, **38**, 295–310.

CONCEIÇAO, R. V. & GREEN, D. H. 2004. Derivation of potassic (shoshonitic) magmas by decompression melting of phlogopite + pargasite lherzolite. *Lithos*, **72**, 209–229, doi: 10.1016/j.lithos.2003.09.003.

DAVIS, J. H. & VON BLANCKENBURG, F. 1995. Slab breakoff: a model of lithosphere detachment and its test in the magmatism and deformation of collisional orogens. *Earth and Planetary Science Letters*, **129**, 85–102.

DELALOYE, M. & BINGÖL, E. 2000. Granitoids from western and northwestern Anatolia: geochemistry and modeling of geodynamic evolution. *International Geology Review*, **42**, 241–268.

DEMIRTASLI, E., TURHAN, N., BILGIN, A. Z. & SELIM, M. 1984. Geology of the Bolkar Mountains. *In*: TEKELI, O. & GÖNCÜOGLU, M. C. (eds) *Geology of the Taurus Belt, Proceedings of the International Symposium, Geology of the Taurus Belt, 1984*. Ankara, Mineral Research & Exploration Institute of Turkey (MTA), Ankara, 125–141.

DEWEY, J. F., HEMPTON, M. R., KIDD, W. S. F., SAROGLU, F. & SENGÖR, A. M. C. 1986. Shortening of continental lithosphere: the neotectonics of Eastern Anatolia – a young collision zone. *In*: COWARD, M. P. & RIES, A. C. (eds) *Collision Zone Tectonics*. Geological Society, London, Special Publications, **19**, 3–36.

DILEK, Y. 1979. *Geology of Gevas (Van Province, Turkey) and surrounding areas*. Diploma thesis, Institute of Applied Geology, Faculty of Science, Istanbul University, Turkey, 1–63.

DILEK, Y. 2006. Collision tectonics of the Mediterranean region: causes and consequences. *In*: DILEK, Y. & PAVLIDES, S. (eds) *Postcollisional Tectonics and Magmatism in the Mediterranean Region and Asia*.

Geological Society of America Special Paper, **409**, 1–13.

DILEK, Y. & ALTUNKAYNAK, S. 2007. Cenozoic crustal evolution and mantle dynamics of post-collisional magmatism in western Anatolia. *International Geology Review*, **49**, 431–453.

DILEK, Y. & ALTUNKAYNAK, S. 2008. Geochemical and temporal evolution of Cenozoic magmatism in western Turkey: mantle response to collision, slab breakoff, and lithospheric tearing in an orogenic belt. *In*: VAN HINSBERGEN, D. J. J., EDWARDS, M. A. & GOVERS, R. (eds) *Collision and Collapse at the Africa-Arabia-Eurasia Subduction Zone*. Geological Society, London, Special Publications, **311**, 213–233, doi: 10.1144/SP311.8.

DILEK, Y. & FLOWER, M. F. J. 2003. Arc-trench roll back and forearc accretion: 2. Model template for Albania, Cyprus, and Oman. *In*: DILEK, Y. & ROBINSON, P. T. (eds) *Ophiolites in Earth History*. Geological Society of London Special Publications, **218**, 43–68.

DILEK, Y. & MOORES, E. M. 1990. Regional Tectonics of the Eastern Mediterranean ophiolites. *In*: MALPAS, J., MOORES, E. M., PANAYIOTOU, A. & XENOPHONTOS, C. (eds) *Ophiolites, Oceanic Crustal Analogues, Proceedings of the Symposium 'Troodos 1987'*. The Geological Survey Department, Nicosia, Cyprus, 295–309.

DILEK, Y. & THY, P. 2006. Age and petrogenesis of plagiogranite intrusions in the Ankara mélange, central Turkey. *Island Arc*, **15**, 44–57.

DILEK, Y. & WHITNEY, D. L. 1997. Counterclockwise *P-T-t* trajectory from the metamorphic sole of a Neo-Tethyan ophiolite (Turkey). *Tectonophysics*, **280**, 295–310.

DILEK, Y. & WHITNEY, D. L. 1998. Syn-metamorphic to neotectonic evolution of the Ecemis strike-slip fault zone (Turkey). *EOS Transactions, American Geophysical Union*, **80**, F915.

DILEK, Y. & WHITNEY, D. L. 2000. Cenozoic crustal evolution in central Anatolia: extension, magmatism and landscape development. *In*: PANAYIDES, I., XENOPHONTOS, C. & MALPAS, J. (eds) *Proceedings of the Third International Conference on the Geology of the Eastern Mediterranean*. Geological Survey Department, September 1998, Nicosia, Cyprus, 183–192.

DILEK, Y., THY, P., HACKER, B. & GRUNDVIG, S. 1999a. Structure and petrology of Tauride ophiolites and mafic dike intrusions (Turkey): implications for the Neo-Tethyan ocean. *Bulletin of the Geological Society of America*, **111**, 1192–1216.

DILEK, Y., WHITNEY, D. L. & TEKELI, O. 1999b. Links between tectonic processes and landscape morphology in an alpine collision zone, south-central Turkey. *Annals of Geomorphology (Z. Geomorph. N.F.)*, **118**, 147–164.

DILEK, Y., FURNES, H. & SHALLO, M. 2008. Geochemistry of the Jurassic Mirdita ophiolite (Albania) and the MORB to SSZ evolution of a marginal basin oceanic crust. *Lithos*, **100**, 174–209, doi: 10.1016/j.lithos.2007.06.026.

DILEK, Y., IMAMVERDIYEV, N. & ALTUNKAYNAK, S. 2010. Geochemistry and tectonics of Cenozoic volcanism in the Lesser Caucasus (Azerbaijan) and the Peri-Arabian region: Collision-induced mantle dynamics and its magmatic fingerprint. *International Geology Review*, in press.

DOGLIONI, C., AGOSTINI, S., CRESPI, M., INNOCENTI, F., MANETTI, P., RIGUZZI, F. & SAVAŞÇIN, Y. 2002. On the extension in western Anatolia and the Aegean Sea. *In*: ROSENBAUM, G. & LISTER, G. S. (eds) *Reconstruction of the Evolution of the Alpine-Himalayan Orogen*. Journal of the Virtual Explorer, **8**, 169–183.

DÜZGÖREN-AYDIN, N. S. 2000. Post-collisional granitoid magmatism: case study from the Yozgat Batholith, central Anatolia, Turkey. *Proceedings of the Third International Conference on the Geology of the Eastern Mediterranean*. Geological Survey Department, September 1998, Nicosia, Cyprus, 171–181.

ELITOK, Ö & DRUPPEL, K. 2008. Geochemistry and tectonic significance of metamorphic sole rocks beneath the Beyşehir-Hoyran ophiolite (SW Turkey). *Lithos*, **100**, 322–353.

ELMAS, A. & YILMAZ, Y. 2003. Development of an oblique subduction zone-Tectonic evolution of the Tethys suture zone in southeast Turkey. *International Geology Review*, **45**, 827–840. doi: 10.2747/0020-6814.45.9.827.

ERCAN, T., SATIR, M. *ET AL.* 1985. Bati Anadolu Senozoyik volkanitlerine ait yeni kimyasal, izotopik ve radyometrik verilerin yorumu. *Bulletin of the Geological Society of Turkey*, **28**, 121–136 (in Turkish).

ERCAN, T., TURKECAN, A. & KARABIYIKOGLU, M. 1994. *Neogene and Quaternary Volcanics of Cappadocia*. International Volcanological Congress, Ankara, Turkey, 17–22 September 1994, Post Congress Excursion Guidebook, 1–28.

ERDOGAN, B., AKAY, E. & UGUR, M. S. 1996. Geology of the Yozgat region and evolution of the collisional Çankiri basin. *International Geology Review*, **38**, 788–806.

ERINÇ, S. 1953. Dogu Anadolu Cografyasi. *Istanbul Universitesi Yayini*, 572 (in Turkish).

ERLER, A. & BAYHAN, H. 1995. Orta Anadolu granitoidlerinin genel degerlendirilmesi ve sorunları. *Yerbilimleri*, **17**, 49–67 (in Turkish).

FACCENNA, C., JOLIVET, L., PIROMELLI, C. & MORALLO, A. 2003. Subduction and the depth of convection in the Mediterranean mantle. *Journal of Geophysical Research*, **108**, 2099, doi: 1029/2001JB001690.

FAYON, A. K., WHITNEY, D. L., TEYSSIER, C., GARVER, J. I. & DILEK, Y. 2001. Effects of plate convergence obliquity on timing and mechanisms of exhumation of a mid-crustal terrain, the Central Anatolian Crystalline Complex. *Earth and Planetary Science Letters*, **192**, 191–205.

FLOYD, P. A., GÖNCÜOGLU, M. C., WINCHESTER, J. A. & YALINIZ, K. 2000. Geochemical character and tectonic environment of Neotethyan ophiolitic fragments and metabasites in the Central Anatolian Crystalline Complex, Turkey. *In*: BOZKURT, E., WINCHESTER, J. A. & PIPER, J. D. A. (eds) *Tectonics and Magmatism in Turkey and the Surrounding Area*. Geological Society, London, Special Publications, **173**, 183–202.

FRANCALANCI, L., INNOCENTI, F., MANETTI, P. & SAVAŞÇIN, M. Y. 2000. Neogene alkaline volcanism of the Afyon-Isparta area, Turkey: petrogenesis and geodynamic implications. *Mineralogy & Petrology*, **70**, 285–312.

GARFUNKEL, Z. 1998. Constraints on the origin and history of the Eastern Mediterranean basin. *Tectonophysics*, **298**, 5–35.

GAUTIER, P., BOZKURT, E., HALLOT, E. & DIRIK, K. 2002. Dating the exhumation of a metamorphic dome: geological evidence for pre-Eocene unroofing of the Nigde massif (Central Anatolia, Turkey). *Geological Magazine*, **139**, 559–576.

GELATI, R. 1975. Miocene marine sequence from Lake Van, eastern Turkey. *Rivista italiana di paleontologia e stratigrafia*, **81**, 477–490.

GERYA, T. V., YUEN, D. A. & MARESCH, W. V. 2004. Thermomechanical modeling of slab detachment. *Earth and Planetary Science Letters*, **226**, 101–116.

GESSNER, K., COLLINS, A. S., RING, U. & GÜNGÖR, T. 2004. Structural and thermal history of poly-orogenic basement: U–Pb geochronology of granitoid rocks in the southern Menderes Massif, western Turkey. *Journal of the Geological Society, London*, **161**, 93–101.

GIANELLI, G., PASERINI, P. & SGUAZZONI, G. 1972. Some observations on mafic and ultramafic complexes north of the Bolkardag (Taurus, Turkey). *Bolletino della Societad Geologica Italiana*, **91**, 439–488.

GÖK, R., TÜRKELLI, N., SANDVOL, E., SEBER, D. & BARAZANGI, M. 2000. Regional wave propagation in Turkey and surrounding regions. *Geophysical Research Letters*, **27**(3), 429–432.

GÖK, R., SANDVOL, E., TÜRKELLI, N., SEBER, D. & BARAZANGI, M. 2003. Sn attenuation in the Anatolian and Iranian plateau and surrounding regions. *Geophysical Research Letters*, **30**(24), doi: 10.1029/2003GL018020.

GÖNCÜOGLU, M. C. & TURHAN, N. 1984. Geology of the Bitlis Metamorphic Belt. *In*: TEKELI, O. & GÖNCÜOGLU, M. C. (eds) *Geology of the Taurus Belt*. Proceedings of the International Symposium, Ankara, Turkey, 1983, 237–244.

GÖNCÜOGLU, M. C., TOPRAK, V., KUSCU, I., ERLER, A. & OLGUN, E. 1991. *Orta Anadolu Masifinin batı bölümünün jeolojisi, Bölüm I: Güney Kesim*. TPAO (Turkish Petroleum Corporation) Report No. 2909, 134.

GÖRÜR, N., OKTAY, F., SEYMEN, I. & SENGÖR, A. M. C. 1984. Palaeotectonic evolution of the Tuzgölü basin complex, central Turkey: sedimentary record of a Neo-Tethyan closure. *In*: DIXON, J. E. & ROBERTSON, A. H. F. (eds) *Geological Evolution of the Eastern Mediterranean*. Geological Society, London, Special Publications, **17**, 467–482.

GOVERS, R. & WORTEL, M. J. R. 2005. Lithosphere tearing at STEP faults: response to edges of subduction zones. *Earth and Planetary Science Letters*, **236**, 505–523.

GÜLEÇ, N. 1994, Rb–Sr isotope data from the Agaçören granitoid (East of Tuz Gölü): geochronological and genetical implications. *Turkish Journal of Earth Sciences*, **3**, 39–43.

GÜLEÇ, N. & KADIOGLU, Y. K. 1998. Relative involvement of mantle and crustal components in the Agaçören granitoid (central Anatolia-Turkey): estimates from trace element and Sr-isotope data. *Chemie der Erde – Geochemistry*, **58**, 23–37.

GÜRSU, S., GÖNCÜOGLU, M. C. & BAYHAN, H. 2004. Geology and geochemistry of the Pre-early Cambrian rocks in the Sandikli area: implications for the Pan-African evolution of NW Gondwanaland. *Gondwana Research*, **7**, 923–935.

HANSMANN, W. & OBERLI, F. 1991. Zircon inheritance in an igneous rock suite from the southern Adamello batholith (Italian Alps). *Contributions to Mineralogy and Petrology*, **107**, 501–518.

HARRIS, N. B. W., KELLEY, S. & OKAY, A. I. 1994. Postcollisional magmatism and tectonics in northwest Anatolia. *Contributions to Mineralogy and Petrology*, **117**, 241–252.

HELVACI, C. & GRIFFIN, W. L. 1984. Rb–Sr geochronology of the Bitlis massif, Avnik (Bingöl) area, SE Turkey. *In*: DIXON, J. E. & ROBERTSON, A. H. F. (eds) *Geological Evolution of the Eastern Mediterranean*. Geological Society, London, Special Publications, **17**, 403–413.

HETZEL, R. & REISCHMANN, T. 1996. Intrusion age of Pan-African augen gneisses in the southern Menderes Massif and the age of cooling after Alpine ductile extensional deformation. *Geological Magazine*, **133**, 565–572.

ILBEYLI, N. 2004. Petrographic and geochemical characteristics of the Hamit Alkaline Intrusion in the Central Anatolian Crystalline Complex, Turkey. *Turkish Journal of Earth Sciences*, **13**, 269–286.

ILBEYLI, N. 2005. Mineralogical-geochemical constraints on intrusives in central Anatolia, Turkey: tectonomagmatic evolution and characteristics of mantle source. *Geological Magazine*, **142**, 187–207.

ILBEYLI, N., PEARCE, J. A., THIRWALL, M. F. & MITCHELL, J. G. 2004. Petrogenesis of collision-related plutonics in Central Anatolia, Turkey. *Lithos*, **72**, 163–182.

INNOCENTI, F., MAZZUOULI, G., PASQUARE, F., RADICATI DI BROZOLA, F. & VILLARI, L. 1975. The Neogene calcalkaline volcanism of Central Anatolia: geochronological data on Kayseri-Nigde area. *Geological Magazine*, **112**, 349–360.

INNOCENTI, F., MAZZUOULI, G., PASQUARE, F., RADICATI DI BROZOLA, F. & VILLARI, L. 1976. Evolution of volcanism in the area between the Arabian, Anatolian, and Iranian plates (Lake Van, Eastern Turkey). *Journal of Volcanological and Geothermal Research*, **1**, 103–112.

INNOCENTI, F., MAZZUOULI, G., PASQUARE, F., RADICATI DI BROZOLA, F. & VILLARI, L. 1982. Tertiary and Quaternary volcanism of the Erzurum-Kars area (Eastern Turkey): geochronological data and geodynamic evolution. *Journal of Volcanological and Geothermal Research*, **13**, 223–240.

INNOCENTI, F., AGOSTINI, S., DI VINCENZO, G., DOGLIONI, C., MANETTI, P., SAVAŞÇIN, M. Y. & TONARINI, S. 2005. Neogene and Quaternary volcanism in Western Anatolia: magma sources and geodynamic evolution. *Marine Geology*, **221**, 397–421.

ISIK, V., TEKELI, O. & SEYITOGLU, G. 2004. The $^{40}Ar/^{39}Ar$ age of extensional ductile deformation and granitoid intrusion in the northern Menderes core

complex: implications for the initiation of extensional tectonics in western Turkey. *Journal of Asian Earth Sciences*, **23**, 555–566.

JACKSON, J. & McKENZIE, D. 1984. Active tectonics of the Alpine-Himalayan belt between western Turkey and Pakistan. *Geophysical Journal International*, **77**, 185–264.

JOLIVET, L. 2001. A comparison of geodetic and finite strain pattern in the Aegean, geodynamic implications. *Earth and Planetary Science Letters*, **187**, 95–104.

JOLIVET, L. & BRUN, J.-P. 2008. Cenozoic geodynamic evolution of the Aegean. *International Journal of Earth Sciences*, doi: 10.1007/s00531-008-0366-4.

JOLIVET, L. & FACCENNA, C. 2000. Mediterranean extension and the Africa-Eurasia collision. *Tectonics*, **19**, 1095–1106.

JOLIVET, L., BRUN, J. P., GAUTIER, S., LELLEMAND, S. & PATRIAT, M. 1994. 3-D kinematics of extension in the Aegean from the early Miocene to the present: insight from the ductile crust. *Bulletin de la Societe Geologique de France*, **165**, 195–209.

JOLIVET, L., FACCENNA, C., GOFFÉ, B., BUROV, E. & AGARD, P. 2003. Subduction tectonics and exhumation of high-pressure metamorphic rocks in the Mediterranean orogen. *American Journal of Science*, **303**, 353–409.

KADIOGLU, Y. K., DILEK, Y. & FOLAND, K. A. 2006. Slab breakoff and syncollisional origin of the Late Cretaceous magmatism in the Central Anatolian Crystalline Complex, Turkey. *In*: DILEK, Y. & PAVLIDES, S. (eds) *Postcollisional tectonics and magmatism in the Mediterranean region and Asia*. Geological Society of America Special Paper, **409**, 381–415, doi: 10.1130/2006/2409(19).

KADIOGLU, Y. K., DILEK, Y., GULEÇ, N. & FOLAND, K. A. 2003. Tectonomagmatic evolution of bimodal plutons in the Central Anatolian Crystalline Complex, Turkey. *Journal of Geology*, **111**, 671–690.

KÁRASON, H. 2002. *Constraints on mantle convection from seismic tomography and flow modeling*. PhD thesis, Massachusetts Institute of Technology, Cambridge, USA.

KÁRASON, H. & VAN DER HILST, R. D. 2001. Tomographic imaging of the lowermost mantle with differential times of refracted and defracted core phases (PKP, Pdiff). *Journal of Geophysical Research*, **106**, 6569–6587.

KEMPLER, D. & BEN-AVRAHAM, Z. 1987. The tectonic evolution of the Cyprean Arc. *Annales Tectonicae*, **1**, 58–71.

KESKIN, M. 2003. Magma generation by slab steepening and breakoff beneath a subduction-accretion complex: an alternative model for collision-related volcanism in Eastern Anatolia, Turkey. *Geophysical Research Letters*, **30**, 8046, doi: 10.1029/2003GL018 019.

KESKIN, M., PEARCE, J. A. & MITCHELL, J. G. 1998. Volcano-stratigraphy and geochemistry of collision-related volcanism on the Erzurum-Kars Plateau, North Eastern Turkey. *Journal of Volcanology and Geothermal Research*, **85**, 355–404.

KESKIN, M., PEARCE, J. A., KEMPTON, P. D. & GREENWOOD, P. 2006. Magma-crust interactions and magma plumbing in a postcollisional setting: geochemical evidence from the Erzurum-Kars volcanic plateau, eastern Turkey. *In*: DILEK, Y. & PAVLIDES, S. (eds) *Postcollisional tectonics and magmatism in the Mediterranean region and Asia*. Geological Society of America Special Paper, **409**, 475–505.

KÖKSAL, S., ROMER, R. L., GÖNCÜOGLU, M. C. & TOKSOY-KÖKSAL, F. 2004. Timing of postcollisional H-type to A-type granitic magmatism: U–Pb titanite ages from the Alpine central Anatolian granitoids (Turkey). *International Journal of Earth Sciences (Geologische Rundschau)*, **93**, 974–989.

KÖPRÜBASI, N. & ALDANMAZ, E. 2004. Geochemical constraints on the petrogenesis of Cenozoic I-type granitoids in Northwest Anatolia, Turkey: evidence for magma generation by lithospheric delamination in a post-collisional setting. *International Geology Review*, **46**, 705–729.

KUMRAL, M., ÇOBAN, H., GEDIKOGLU, A. & KILINÇ, A. 2006. Petrology and geochemistry of augite trachytes and porphyritic trachytes from the Gölcük volcanic region, SW Turkey: a case study. *Journal of Asian Earth Sciences*, **27**, 707–716.

LE PICHON, X. & ANGELIER, J. 1979. The Hellenic arc and trench system: a key to the evolution of the Eastern Mediterranean area. *Tectonophysics*, **60**, 1–42.

LE PICHON, X., LALLEMANT, S. J., CHAMOT-ROOKE, N., LEMEUR, D. & PASCAL, G. 2003. The Mediterranean Ridge backstop and the Hellenic nappes. *Marine Geology*, **186**, 111–125.

LIU, M., CUI, X. & LIU, F. 2004. Cenozoic rifting and volcanism in eastern China: a mantle dynamic link to the Indo-Asian collision? *Tectonophysics*, **393**, 29–42, doi: 10.1016/j.tecto.2004.07.029.

MAURY, R. C., FOURCADE, S. ET AL. 2000. Postcollisional Neogene magmatism of the Mediterranean Maghreb margin: a consequence of slab breakoff. *Earth and Planetary Sciences*, **331**, 159–173.

MCCLUSKY, S., BALASSANIAN, S. ET AL. 2000. Global Positioning System constraints on plate kinematics and dynamics in the eastern Mediterranean and Caucasus. *Journal of Geophysical Research*, **105**, 5695–5719.

MICHARD, A., WHITECHURCH, H., RICOU, L.-E., MONTIGNY, R. & YAZGAN, E. 1984. Tauric subduction (Malatya-Elazig provinces) and its bearing on tectonics of the Tethyan realm in Turkey. *In*: DIXON, J. E. & ROBERTSON, A. H. F. (eds) *The Geological Evolution of the Eastern Mediterranean Region*. Geological Society, London, Special Publications, **17**, 361–374.

MO, X., ZHAO, Z. ET AL. 2006. Petrology and geochemistry of postcollisional volcanic rocks from the Tibetan plateau: implications for lithospheric heterogeneity and collision-induced asthenospheric mantle flow. *In*: DILEK, Y. & PAVLIDES, S. (eds) *Postcollisional tectonics and magmatism in the Mediterranean region and Asia*. Geological Society of America Special Paper, **409**, 507–530.

MOLNAR, P. & OLIVER, J. 1969. Lateral variations of attenuation in the upper mantle and discontinuities in the lithosphere. *Journal of Geophysical Research*, **74**, 2648–2682.

NEMCOK, M., POSPISIL, L., LEXA, J. & DONELICK, R. A. 1998. Tertiary subduction and slab break-off model of the Carpathian-Pannonian region. *Tectonophysics*, **295**, 307–340.

OBERHANSLI, R., BOUSQUET, R. & CANDAN, O. 2008. *Bitlis Massif – East Anatolia Plateau connection?* IGC-33, Session T-18, Oslo-Norway, August 2008.

OKAY, A. I. 1984. Distribution and characteristics of the northwest Turkish blueschists. *In*: DIXON, J. E. & ROBERTSON, A. H. F. (eds) *The Geological Evolution of the Eastern Mediterranean Region*. Geological Society, London, Special Publications, **17**, 455–466.

OKAY, A. I. 2002. Stratigraphic and metamorphic inversions in the central Menderes massif. A new structural model. *International Journal of Earth Sciences*, **91**, 173–178.

OKAY, A. I. & SATIR, M. 2000. Coeval plutonism and metamorphism in a latest Oligocene metamorphic core complex in northwest Turkey. *Geological Magazine*, **137**, 495–516.

OKAY, A. I., SATIR, M., MALUSKI, H., SIYAKO, M., MONIE, P., METZGER, R. & AKYÜZ, S. 1996. Paleo- and Neo-Tethyan events in northwest Turkey: geological and geochronological constraints. *In*: YIN, A. & HARRISON, M. T. (eds) *Tectonics of Asia*. Cambridge University Press, 420–441.

OKAY, A. I., HARRIS, N. B. W. & KELLEY, S. P. 1998. Exhumation of blueschists along a Tethyan suture in northwest Turkey. *Tectonophysics*, **285**, 275–299.

ÖNEN, P. 2003. Neotethyan ophiolitic rocks of the Anatolides of NW Turkey and comparison with Tauride ophiolites. *Journal of the Geological Society, London*, **160**, 947–962.

ONER, Z. & DILEK, Y. 2007. Depositional and tectonic evolution of the Late Cenozoic Alasehir supradetachment basin, western Anatolia (Turkey). *Geological Society of America Abstracts with Programs*, **39**(6), 228.

ÖNEN, A. P. & HALL, R. 1993. Ophiolites and related metamorphic rocks from the Kütahya region, northwest Turkey. *Geological Journal*, **28**, 399–412.

ÖZER, S., SÖZBILIR, H., ÖZKAR, I., TOKER, V. & SARI, B. 2001. Stratigraphy of Upper Cretaceous-Palaeogene sequences in the southern and eastern Menderes Massif (Western Turkey). *International Journal of Earth Sciences*, **89**, 852–866.

ÖZGÜL, N. 1976. Some geological aspects of the Taurus orogenic belt (Turkey). *Bulletin of the Geological Society of Turkey*, **19**, 65–78.

ÖZGÜL, N. 1984. Stratigraphy and tectonic evolution of the Central Taurides. *In*: TEKELI, O. & GÖNCÜOĞLU, M. C. (eds) *Geology of the Taurus Belt, Proceedings of the International Symposium on the Geology of the Taurus Belt, 1983*. Ankara, Turkey, Mineral Research and Exploration Institute of Turkey, Ankara, 77–90.

ÖZGÜL, N. & TURSUCU, A. 1984. Stratigraphy of the Mesozoic carbonate sequence of the Munzur Mountains (Eastern Taurides). *In*: TEKELI, O. & GÖNCÜOĞLU, M. C. (eds) *Geology of the Taurus Belt. Proceedings of the International Symposium, Geology of the Taurus Belt, 1984*. Ankara, Mineral Research & Exploration Institute of Turkey (MTA), Ankara, 173–180.

ÖZKAYA, I. 1978. Stratigraphy of Ergani-Maden area, SE Turkey. *Bulletin of the Geological Society of Turkey*, **21**, 129–139.

PAMIC, J., BALEN, D. & HERAK, M. 2002. Origin and geodynamic evolution of Late Paleogene magmatic associations along the Periadriatic-Sava-Vardar magmatic belt. *Geodinamica Acta*, **15**, 209–231.

PARLAK, O. 2006. Geodynamic significance of granitoid magmatism in the southeast Anatolian orogen: geochemical and geochronological evidence from Göksun-Afsin (Kahramanmaras, Turkey) region. *International Journal of Earth Sciences*, **95**, 609–627.

PARLAK, O. & DELALOYE, M. 1999. Precise $^{40}$Ar-$^{39}$Ar ages from the metamorphic sole of the Mersin ophiolite (Southern Turkey). *Tectonophysics*, **301**, 145–158.

PARLAK, O., DELALOYE, M. & BINGÖL, E. 1996. Mineral chemistry of ultramafic and mafic cumulates as an indicator of the arc-related origin of the Mersin ophiolite (southern Turkey). *Geologische Rundschau*, **85**, 647–661.

PARLAK, O., HÖCK, V. & DELALOYE, M. 2002. The suprasubduction zone Pozanti-Karsanti ophiolite, southern Turkey: evidence for high-pressure crystal fractionation of ultramafic cumulates. *Lithos*, **65**, 205–224.

PASQUARÉ, G., POLI, S., VEZZOLI, L. & ZANCHI, A. 1988. Continental arc volcanism and tectonic setting in Central Anatolia, Turkey. *Tectonophysics*, **146**, 217–230.

PEARCE, J. A. & BENDER, J. F. ET AL. 1990. Genesis of collision volcanism in eastern Anatolia, Turkey. *Journal of Volcanology and Geothermal Research*, **44**, 189–229.

PE-PIPER, G. & PIPER, D. J. W. 1994. Miocene magnesian andesites and dacites, Evia, Greece: adakites associated with subducted slab detachment and extension. *Lithos*, **31**, 125–140.

PE-PIPER, G. & PIPER, D. J. W. 2006. Unique features of the Cenozoic igneous rocks of Greece. *In*: DILEK, Y. & PAVLIDES, S. (eds) *Postcollisional Tectonics and Magmatism in the Mediterranean Region and Asia*. Geological Society of America Special Paper, **409**, 259–282.

PIROMALLO, C. & MORELLI, A. 2003. P wave tomography of the mantle under the Alpine-Mediterranean area. *Journal of Geophysical Research*, **108**, 2065, doi: 10.1029/2002JB001757.

PURVIS, M. & ROBERTSON, A. H. F. 2004. A pulsed extension model for the Neogene-Recent E-W-trending Alasehir Graben and the NE-SW-trending Selendi and Gördes Basins, western Turkey. *Tectonophysics*, **391**, 171–201.

REILINGER, R. E., MCCLUSKY, S. C. & ORAL, M. B. 1997. Global positioning system measurements of present-day crustal movements in the Arabia-Africa-Eurasia plate collision zone. *Journal of Geophysical Research*, **102**, 9983–9999.

REILINGER, R. E., MCCLUSKY, S. C. ET AL. 2006. GPS constraints on continental deformation in the Africa-Arabia-Eurasia continental collision zone and implications for the dynamics of plate interactions. *Journal of Geophysical Research*, **111**, V05411, doi: 10.1029/2005JB004051.

RICHARDSON-BUNBURY, J. M. 1996. The Kula volcanic field, western Turkey: the development of a Holocene alkali basalt province and the adjacent normal-faulting graben. *Geological Magazine*, **133**, 275–283.

RICOU, L. E., ARGYRIADIS, I. & MARCOUX, J. 1975. L'axe calcaire du Taurus, un alignement de fenetres arabo-africaines sous des nappes radiolaritiques, ophiolitiques et métamorphiques. *Bulletin de la Société de Geologie de France*, **16**, 107–111.

RIMMELÉ, G., OBERHANSLI, R., GOFFE, B., JOLIVET, L., CANDAN, O. & ÇETINKAPLAN, M. 2003. First evidence of high-pressure metamorphism in the 'Cover Series' of the Southern Menderes Massif. Tectonic and metamorphic implications for the evolution of SW Turkey. *Lithos*, **71**, 19–46.

RING, U. & COLLINS, A. S. 2005. U–Pb SIMS dating of synkinematic granites: timing of core-complex formation in the northern Anatolide belt of western Turkey. *Journal of the Geological Society, London*, **162**, 289–298.

RING, U. & LAYER, P. W. 2003. High-pressure metamorphism in the Aegean, eastern Mediterranean: Underplating and exhumation from the Late Cretaceous until the Miocene to Recent above the retreating Hellenic subduction zone. *Tectonics*, **22**(3), 1022, doi: 10.1029/2001TC001350.

RING, U., JOHNSON, C., HETZEL, R. & GESSNER, K. 2003. Tectonic denudation of a Late Cretaceous-Tertiary collisional belt: regionally symmetric cooling patterns and their relation to extensional faults in the Anatolide belt of western Turkey. *Geological Magazine*, **140**, 421–441.

ROBERTSON, A. H. F. 1998. Tectonic significance of the Eratosthenes Seamount: a continental fragment in the process of collision with a subduction zone in the eastern Mediterranean (Ocean Drilling Program Leg 160). *Tectonophysics*, **298**, 63–82.

ROBERTSON, A. H. F. & DIXON, J. E. 1984. Introduction: aspects of the geological evolution of the eastern Mediterranean: *In*: DIXON, J. E. & ROBERTSON, A. H. F. (eds) *The Geological Evolution of the Eastern Mediterranean*. Geological Society, London, Special Publications, **17**, 1–74.

ROGERS, A. J., NI, J. F. & HEARN, T. M. 1995. Propagation characteristics of short-period Sn And Lg in The Middle East. *Bulletin of The Seismological Society of America*, **87**, 396–413.

ROTSTEIN, Y. & KAFKA, A. L. 1982. Seismotectonics of the southern boundary of Anatolia, eastern Mediterranean region; subduction, collision, and arc jumping. *Journal of Geophysical Research*, **87**, 7694–7706.

RUSSO, R. M., DILEK, Y. & FLOWER, M. F. J. 2001. Collision-driven mantle flow and crustal response in the eastern Mediterranean region during the late Cenozoic. *4th International Turkish Geology Symposium, Work in Progress on the Geology of Turkey and Its Surroundings*. Adana, Turkey, 24–28 September 2001, 98.

SANDVOL, E., AL-DAMEGH, K. ET AL. 2001. Tomographic imaging of Lg and Sn Propagation in the Middle East. *Pure and Applied Geophysics*, **158**, 1121–1163.

SANDVOL, E., TÜRKELLI, N. & BARAZANGI, M. 2003a. The Eastern Turkey Seismic Experiment: the study of a young continent-continent collision. *Geophysical Research Letters*, **30**(24), doi: 10.1029/2003GL018912.

SANDVOL, E., TÜRKELLI, N. *ET AL.* 2003*b*. Shear wave splitting in a young continent-continent collision: an example from eastern Turkey. *Geophysical Research Letters*, **30**(24), doi: 10.1029/2003GL017390.

SAROGLU, F., EMRE, Ö & KUŞÇU, I. 1992. *Active fault map of Turkey* (3 sheets, 1:100,000 scale). General Directorate of Mineral Research and Exploration, MTA, Ankara, Turkey.

SAVAŞÇIN, M. Y. & OYMAN, T. 1998. Tectono-magmatic evolution of alkaline volcanics at the Kirka-Afyon-Isparta structural trend, SW Turkey. *Turkish Journal of Earth Sciences*, **7**, 201–214.

SCHLIESTEDT, M., ALTHERR, R. & MATTHEWS, A. 1987. Evolution of the Cycladic Crystalline Complex: petrology, isotope geochemistry and geochronology. *In*: HELGESON, H. C. (ed.) *Chemical Transport in Metasomatic Processes*. Reidel, Dordrecht, 389–428.

SEGHEDI, I., DOWNES, H. *ET AL.* 2004. Neogene-Quaternary magmatism and geodynamics in the Carpathian-Pannonian region: a synthesis. *Lithos*, **72**, 117–146, doi: 10.1016/j.lithos.2003.08.006.

SENGÖR, A. M. C. & YILMAZ, Y. 1981. Tethyan evolution of Turkey, a plate tectonic approach. *Tectonophysics*, **75**, 181–241.

SENGÖR, A. M. C., ÖZEREN, S., GENÇ, T. & ZOR, E. 2003. East Anatolian high plateau as a mantle-supported, north-south shortened domal structure. *Geophysical Research Letters*, **30**, 8045, doi: 10.1029/2003GL017858.

SEYITOGLU, G., ANDERSON, D., NOWELL, G. & SCOTT, B. 1997. The evolution from Miocene potassic to Quaternary sodic magmatism in western Turkey: implications for enrichment processes in the lithospheric mantle. *Journal of Volcanology and Geothermal Research*, **76**, 127–147.

SEYMEN, I. 1984. Kırsehir metamorfitlerinin jeolojik evrimi: Ketin Simpozyumu. *Türkiye Jeoloji Kurumu Yayını*, 133–148 (in Turkish).

SHERLOCK, S., KELLEY, S. P., INGER, S., HARRIS, N. & OKAY, A. I. 1999. ${}^{40}$Ar-${}^{39}$Ar and Rb–Sr geochronology of high-pressure metamorphism and exhumation history of the Tavsanli Zone, NW Turkey. *Contributions to Mineralogy and Petrology*, **137**, 46–58.

SPAKMAN, W., WORTEL, M. J. R. & VLAAR, N. J. 1988. The Hellenic subduction zone: a tomographic image and its geodynamic implications. *Geophysical Research Letters*, **15**, 60–63.

SPAKMAN, W., VAN DER LEE, S. & VAN DER HILST, R. 1993. Travel-time tomography of the European-Mediterranean mantle down to 1400 km. *Physics of the Earth and Planetary Interiors*, **79**, 3–74.

TAN, O. & TAYMAZ, T. 2006. Active tectonics of the Caucasus: Earthquake source mechanisms and rupture histories obtained from inversion of teleseismic body waveforms. *In*: DILEK, Y. & PAVLIDES, S. (eds) *Postcollisional Tectonics and Magmatism in the Mediterranean Region and Asia*. Geological Society of America Special Paper, **409**, 531–578.

TANKUT, A., DILEK, Y. & ÖNEN, P. 1998. Petrology and geochemistry of the Neo-Tethyan volcanism as revealed in the Ankara Melange, Turkey. *Journal of Volcanological & Geothermal Research*, **85**, 265–284.

TATAR, M., HATZFELD, D., MARTINOD, J., WALPERSDORF, A., GHAFORI-ASHTIANY, M. & CHÉRY, J. 2002. The present day deformation of the central Zagros from GPS measurements. *Geophysical Research Letters*, **29**, 1927, doi: 10.1029/2002GL015427.

TEKELI, O. 1981. Subduction complex of pre-Jurassic age, northern Anatolia, Turkey. *Geology*, **9**, 68–72.

THOMSON, S. N. & RING, U. 2006. Thermochronologic evaluation of postcollision extension in the Anatolide orogen, western Turkey. *Tectonics*, **25**, TC3005, doi: 10.1029/2005TC001833.

TOPRAK, V., KELLER, J. & SCHUMACHER, R. 1994. *Volcano-tectonic features of the Cappadocian Volcanic Province*. International Volcanological Congress, Ankara, Turkey, 17–22 September 1994, Post Congress Excursion Guidebook, 58.

VAN DER KAADEN, G. 1966. The significance and distribution of glaucophane rocks in Turkey. *Bulletin of the Mineral Research and Exploration Institute of Turkey*, **67**, 37–67.

VAN HINSBERGEN, D. J. J., HAFKENSCHEID, E., SPAKMAN, W., MEULENKAMP, J. E. & WORTEL, R. 2005. Nappe stacking resulting from subduction of oceanic and continental lithosphere below Greece. *Geology*, **33**, 325–328.

VON BLANCKENBURG, F. & DAVIES, J. H. 1995. Slab breakoff: a model for syncollisional magmatism and tectonics in the Alps. *Tectonics*, **14**, 120–131.

WDOWINSKI, S., BE-AVRAHAM, Z., ARVIDSSON, R. & EKSTRÖM, G. 2006. Seismotectonics of the Cyprian Arc. *Geophysical Journal International*, **164**, 176–181.

WESTAWAY, B. 1994. Present-day kinematics of the Middle East and eastern Mediterranean. *Journal of Geophysical Research*, **99**, 12071–12090.

WHITNEY, D. L. & DILEK, Y. 1997. Core complex development in central Anatolia, Turkey. *Geology*, **25**, 1023–1026.

WHITNEY, D. L. & DILEK, Y. 1998. Metamorphism during crustal thickening and extension in central Anatolia: the Nigde metamorphic core complex. *Journal of Petrology*, **39**, 1385–1403.

WHITNEY, D. L., TEYSSIER, C., DILEK, Y. & FAYON, A. K. 2001. Metamorphism of the central Anatolian crystalline complex, Turkey: influence of orogen-normal collision vs. wrench-dominated tectonics on *P-T-t*. paths. *Journal of Metamorphic Geology*, **19**, 411–432.

WIDIYANTORO, S., VAN DER HILST, R. D. & WENZEL, F. 2004. Deformation of the Aegean slab in the mantle transition zone. *International Journal of Tomography & Statistics*, **D04**, 1–14.

WILLIAMS, H. M., TURNER, S. P., PEARCE, J. A., KELLEY, S. P. & HARRIS, N. B. W. 2004. Nature of the source regions for post-collisional, potassic magmatism in southern and northern Tibet from geochemical variations and inverse trace element modelling. *Journal of Petrology*, **45**, 555–607.

WORTEL, M. J. R. & SPAKMAN, W. 1992. Structure and dynamics of subducted lithosphere in the Mediterranean region. *Proceedings of the Koninklijke Nederlandse Akademie van Wetenschappen*, **95**, 325–347.

WORTEL, M. J. R. & SPAKMAN, W. 2000. Subduction and slab detachment in the Mediterranean-Carpathian region. *Science*, **290**, 1910–1917.

YAGMURLU, F., SAVAŞÇIN, M. Y. & ERGUN, M. 1997.
Relation of alkaline volcanism and active tectonism
within the evolution of Isparta Angle, SW Turkey.
*Journal of Geology*, **105**, 717–728.

YAZGAN, E. 1984. Geodynamic evolution of the Eastern
Taurus region. *In*: TEKELI, O. & GÖNCÜOĞLU,
M. C. (eds) *Geology of the Taurus Belt*. Proceedings
of the International Symposium, Geology of the
Taurus Belt. 1984, Ankara, Mineral Research &
Exploration Institute of Turkey (MTA), Ankara,
199–208.

YILMAZ, Y. 1985. Geology of the Cilo ophiolite: an
ancient ensimatic island arc fragment on the Arabian
platform, SE Turkey. *Ofioliti*, **10**, 457–484.

YILMAZ, Y. 1990. Comparison of young volcanic associations of western and eastern Anatolia under compressional regime; a review. *Journal of Volcanology and
Geothermal Research*, **44**, 69–87.

YILMAZ, Y. 1993. New evidence and model on the evolution of the Southeast Anatolia orogen. *Geological
Society of America Bulletin*, **105**, 251–271.

YILMAZ, Y. 2002. Tectonic evolution of western Anatolian extensional province during the Neogene and
Quaternary. *Geological Society of America Abstracts
with Programs*, **34**(6), 179.

YILMAZ, Y., SAROGLU, F. & GÜNER, Y. 1987. Initiation
of the neomagmatism in East Anatolia. *Tectonophysics*, **134**, 177–199.

YILMAZ, Y., TÜYSÜZ, O., YIGITBAS, E., GENÇ, S. C. &
SENGÖR, A. M. C. 1997. Geology and tectonic evolution of the Pontides. *In*: ROBINSON, A. G. (ed.)
*Regional and Petroleum Geology of the Black Sea
and Surrounding Region*. AAPG Memoir, **68**, 183–226.

YILMAZ, Y., GENÇ, S. C., KARACIK, Z. & ALTUNKAY-
NAK, S. 2001. Two contrasting magmatic associations
of NW Anatolia and their tectonic significance.
*Journal of Geodynamics*, **31**, 243–271.

ZITTER, T. A. C., WOODSIDE, J. M. & MASCLE, J. 2003.
The Anaximander Mountains: a clue to the tectonics of
Southwest Anatolia. *Geological Journal*, **38**, 375–394,
doi: 10.1002/gj.961.

ZOR, E., SANDVOL, E., GÜRBÜZ, C., TÜRKELLI, N.,
SEBER, D. & BARAZANGI, M. 2003. The crustal structure of the East Anatolian plateau (Turkey) from receiver functions. *Geophysical Research Letters*, **30**(24),
TUR7.1–TUR7.4, doi: 10.1029/2003GL018192.

# The evolution of the Uralian orogen

VICTOR N. PUCHKOV

K. Marx st. 16/2, Institute of Geology, Ufimian Scientific Centre,
Russian Academy of Sciences, 450 000 Ufa, Russia
(e-mail: puchkv@anrb.ru)

**Abstract:** The Uralian orogen is located along the western flank of a huge (>4000 km long) intracontinental Uralo-Mongolian mobile belt. The orogen developed mainly between the Late Devonian and the Late Permian, with a brief resumption of orogenic activity in the Lower Jurassic and Pliocene–Quaternary time. Although its evolution is commonly related to the Variscides of Western Europe, its very distinctive features argue against a simple geodynamic connection. To a first order, the evolution of Uralian orogen shows similarities with the 'Wilson cycle', beginning with epi-continental rifting (Late Cambrian–Lower Ordovician) followed by passive margin (since Middle Ordovician) development, onset of subduction and arc-related magmatism (Late Ordovician) followed by arc–continent collision (Late Devonian in the south and Early Carboniferous in the north) and continent–continent collision (beginning in the mid-Carboniferous). In detail, however, the Uralides preserve a number of rare features. Oceanic (Ordovician to Lower Devonian) and island–arc (Ordovician to Lower Carboniferous) complexes are particularly well preserved as is the foreland belt in the Southern Urals, which exhibits very limited shortening of deformed Mesoproterozoic to Permian sediments. Geophysical studies indicate the presence of 'cold', isostatically equilibrated root. Other characteristic features include a Silurian platinum-rich belt of subduction-related layered plutons, a simultaneous development of orogenic and rift-related magmatism, a succession of collisions that are both diachronous and oblique, and a single dominant stage of transpressive deformation after the Early Carboniferous. The end result is a pronounced bi-vergent structure. The Uralides are also characterized by Meso-Cenozoic post-orogenic stage and plume-related tectonics in Ordovician, Devonian and especially Triassic time. The evolution of the Uralides is consistent with the development and destruction of a Palaeouralian ocean to form part of a giant Uralo-Mongolian orogen, which involved an interaction of cratonic Baltica and Siberia with a young and rheologically weak Kazakhstanian continent. The Uralides are characterized by protracted and recurrent orogenesis, interrupted in the Triassic by tectonothermal activity associated with the Uralo-Siberian superplume.

## Introduction

The last general review of the structural and tectonic evolution of the Uralian orogen was published in English more than 10 years ago (Puchkov 1997). Since then, the stimulus created by the international EUROPROBE *Uralides* Project has resulted in considerable advances in the understanding of the geology of the Urals. However, most publications in English are concerned with the evolution of the Southern and Middle Urals (Brown *et al.* 2006a, 2008, and references therein). This publication provides an overview of the tectonic evolution of the Uralian orogen as a whole, incorporating a wealth of recently published data, and compares this evolution with that of the European Variscides and Mesozoic-Cenozoic orogens.

The Uralian orogen *sensu stricto* partly coincides geographically with the young, neotectonic (Pliocene–Quaternary) Urals mountains, and occurs between three former, Palaeozoic continents: Baltica, Kazakhstania and Siberia (Fig. 1). In the

Early Palaeozoic, these continents were separated by the Palaeouralian and Central Asian oceans. The continents and oceans were partly inherited from the Proterozoic (Vernikovsky *et al.* 2004; Puchkov 2005). The Central Asian (or Palaeo-Asian) ocean existed before the Late Neoproterozoic as the portion of the Palaeo-Pacific Ocean, which surrounded a considerable part of the Siberian continent. A complete separation of this continent from Rodinia by 750 Ma resulted in the birth of the Palaeo-Asian ocean (Li *et al.* 2008).

The E–NE margin of Baltica was modified by the Ediacarian deformational and accretionary events of the Timanian orogeny (Puchkov 1997; Gee *et al.* 2006). A close resemblance of Timanides and Cadomides pointed out by the author (Puchkov 1997) has led to an idea of their immediate lateral connection, supporting a notion of a supercontinent (Pannotia, Dalziel 1992; or Panterra, Puchkov 2000), welded by Ediacarian orogenies. Several variants of such a connection have been suggested, depending on what side of Baltica was thought to

*From*: MURPHY, J. B., KEPPIE, J. D. & HYNES, A. J. (eds) *Ancient Orogens and Modern Analogues.*
Geological Society, London, Special Publications, **327**, 161–195.
DOI: 10.1144/SP327.9   0305-8719/09/$15.00 © The Geological Society of London 2009.

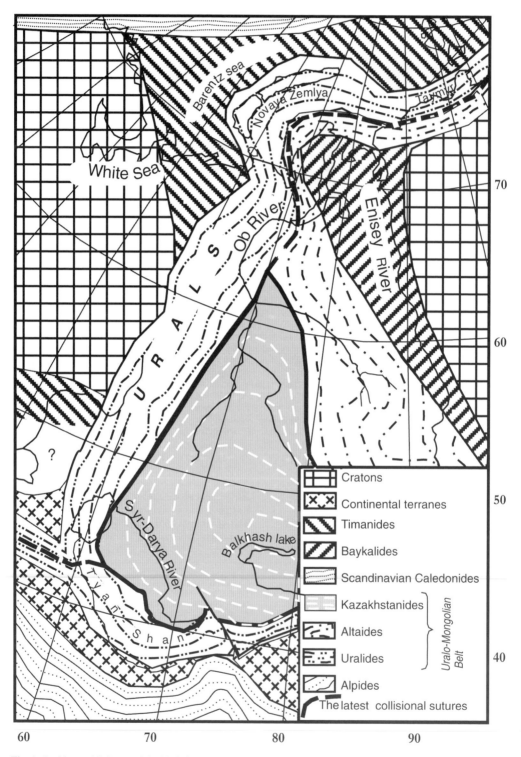

**Fig. 1.** Position and linkages of the Urals in the structure of the Central Eurasia.

be attached to Gondwana (e.g. Linneman *et al.* 1998; Puchkov 2000; Cocks & Torsvik 2006). Unfortunately the APWP of Baltica is still very poorly constrained by existing palaeomagnetic data and additional work is required to distinguish between these hypotheses.

The Timanian/Cadomian orogeny was followed by Late Cambrian–Early Ordovician rifting and passive margin development (Puchkov 2005; Pease *et al.* 2008). These events led to the development of the Palaeouralian ocean, which contained some younger microcontinents (the continent that rifted-away from Baltica is not identified yet). The Mid- to Late Palaeozoic history of the Uralian orogen can be described in terms of formation of an island-arc crust (Late Ordovician–Devonian) and followed by Late Devonian–Early Carboniferous accretion of the island-arcs and other microcontinents with the continental margin of Baltica, which by that time had merged with Laurentia to form Laurussia (Brown *et al.* 1997, 2006*a*–*c*). Kazakhstania formed in the Ordovician–Silurian as a result of subduction-driven accretion of crust around small Neoproterozoic microcontinents (Puchkov 1996*a*). Its collision with the terranes previously accreted to the margins of Laurussia and Siberia started in the Late Bashkirian, after the intervening oceanic crust was completely subducted. This continent–continent collision continued into Permian time, and resulted in the formation of huge foldbelt, variously known as Uralo-Mongolian (Muratov 1979) or Uralo-Okhotskian (Khain 2001) belt, a fundamental event in the assembly of Pangaea. The narrow western flank of the orogen is known as the Uralides, and the more extensive central and eastern flank of the belt is called the Altaides (Sengör *et al.* 1993). However the positioning of the Kazakhstanian continent as a separate Mid-Palaeozoic structure (Figs 1 & 2) suggests that the Kazakhstanides should be excluded from the Altaides.

An attempt to formalize the term 'Uralides' was made by Puchkov (2003), who noted that the northern continuation of the western zones of the Urals is represented by the Pay-Khoy-Novozemelian foldbelt, which was the result of the younger (Early Jurassic) collision between Laurussia and Siberia. More southern parts of the Urals were also deformed by this collision, though the intensity and structural expression are not as prominent.

To the south, Late Palaeozoic collisional structures of the Urals continue in the south-southwestwards vergent fold-and-thrust belt of the Southern Tyan-Shan. Like the Urals, this Late Palaeozoic belt is bordered to the NE and north by the Early Caledonides of Kazakhstania, where the Northern Tyan-Shan occurs (Fig. 1). Attempts to trace Late Palaeozoic structural features from the Uralides

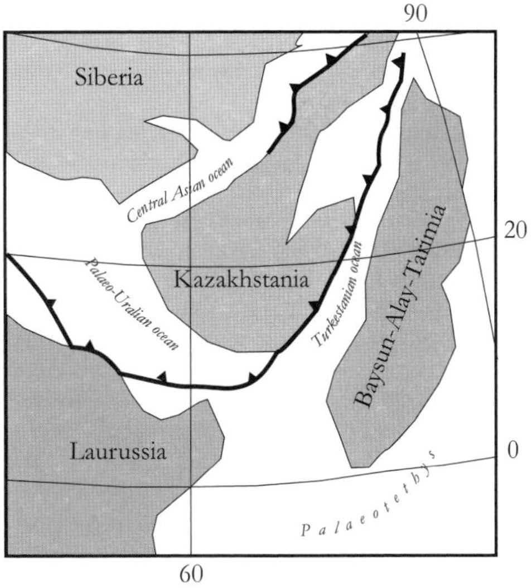

**Fig. 2.** A tentative reconstruction of the Central Eurasia for the Late Devonian (after YUGGEO2002, simplified).

into the Southern Tyan-Shan have been thwarted by the higher degree of shortening and lack of Devonian island arc development in the latter foldbelt. However, palaeocontinental reconstructions and modern structural connections (Figs 1 & 2) suggest a genetic linkage. The evolution of Southern Tyan-Shan is related to the subduction of the Turkestanian ocean (Burtman 2006), that in the Ordovician to Carboniferous time was a direct continuation of the Palaeouralian ocean (Puchkov 2000; Didenko *et al.* 2001; YUGGEO 2002; Biske 2006). The Middle to Late Palaeozoic evolution of the Uralian and Southern Tyan-Shan continental margins has a striking resemblance (Puchkov 1996*a*). The end of orogenic activity in the Southern Tyan-Shan is accompanied by Upper Permian and Lower Triassic continental molasse followed by the development of a stable platform in the whole Tyan-Shan, which lasted until strong intra-plate deformation of the Pliocene–Quaternary age propagated from the Alpine-Himalayan orogenic system, resulting in renewed mountain building (Burtman 2006; Biske 2006; Trifonov 2008).

## Comparison between the Uralides, Variscides and Appalachians

The development of the Ural orogenic belt has traditionally been interpreted to be one of the Late Palaeozoic Variscide (or Hercynide) orogenic belts (Shatsky 1965; International Committee 1982). However, results of geological studies of the last few decades, enhanced by EUROPROBE, have shown such fundamental differences with the

Variscides that this interpretation is no longer considered valid (Brown *et al.* 2006*a*). Instead the Urals is treated as a separate orogenic belt, the main part of the Uralides.

In western Europe, tectonic events leading to the development of the Variscan (also known as Hercynian) orogenic belts (*sensu stricto*) occurred between the Famennian and Late Carboniferous (Fig. 3). Famennian flysch deposits, thought to represent the onset of orogenesis, continued until the end of the Lower Carboniferous, and were followed by molasse deposits which continued until the end of the Carboniferous. Devono-Carboniferous tectonothermal activity accompanied a pre-collisional closure of the Rheic (Saxo-Thuringian) ocean, collisional deformation, granitoid intrusions and metamorphism (including HP–LT type; Franke 2000; Ricken *et al.* 2000). Permian rocks are characterized by rift-related magmatism and the formation of ensialic basins within a wrench regime, and are

followed by Triassic terrigenous and evaporitic deposits in graben structures. The Permian–Triassic events reflect post-orogenic extension (Schwab 1984; Ziegler 1999). The Alleghanian orogeny in the southern Appalachians and Ouachita continued, albeit in a weakened manner, into the Early Permian (Engelder 2007). In the Triassic, the Variscide and Appalachian orogenic belts were characterized by horsts and grabens that strongly influenced the formation of platform oil and gas deposits, especially in the North Sea (Lützner *et al.* 1979; Beutler 1979; Schwab 1984; Ziegler 1999; Khain 2001; Franke 2000).

The Uralides have a much more protracted history of orogenesis than the Variscides, with tectonic events continuing into the Jurassic. In the southern Uralides, syn-orogenic flysch deposition began in the Late Frasnian (Brown *et al.* 2006*b*). Further to the north, collision started in the Visean (Puchkov 2002*a*). This strongly diachronous

| TIME | TECTONIC EVENTS | |
|---|---|---|
| | VARISCIDES | URALIDES |
| Neogene andQuaternary | RIFTING, VOLCANISM | INTRACONTINENTAL COLLISION |
| | BASIN SEDIMENTATION | PENEPLAIN |
| | RIFTING | |
| Palaeogene | | PENEPLAIN FORMATION WITH SEA INGRESSIONS |
| Cretaceous | FORMATION OF PENEPLAIN AND BASINS | |
| Jurassic | | PENEPLAIN FORMATION |
| | | INTRACONTINENTAL COLLISION |
| Triassic | | PENEPLAI AND BASIN FORMATION |
| | RIFTING | RIFTING AND LIP VOLCANISM |
| Permian | WRENCH TECTONICS, INTRAPLATE BI-MODAL VOLCANISM | CONTINENT–CONTINENT COLLISION |
| Carboniferous | CONTINENT–CONTINENT COLLISION | |
| | | SUBDUCTION AND DIACHRONOUS ARC–CONTINENT COLLISION |
| Upper Devonian | INTRAOCEANIC AND MARGINAL S UBDUCTION | |

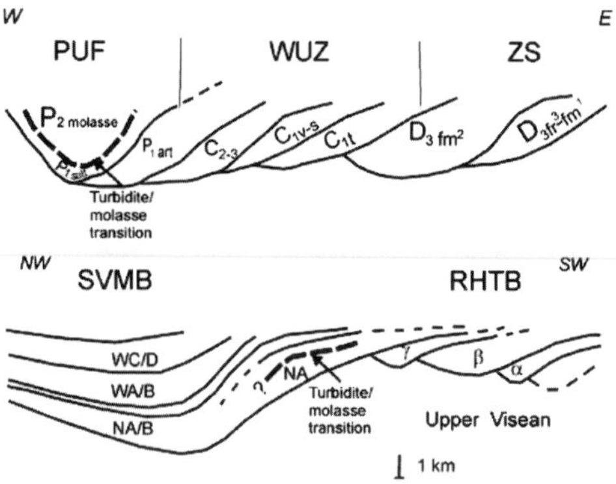

**Fig. 3.** To the left: correlation of orogenic and post-orogenic (intraplate) events in the Variscides and Uralides. To the right: a comparison of idealized sections across the foreland flysch and molasse basins: Preuralian (Southern Urals) after Puchkov (2000) and Central European Variscides, after Ricken *et al.* (2000). PUF, Preuralian foredeep; WUZ, West Uralian zone; ZS, Zilair synclinorium, Stages; art, Asselian to Artinskian stages; v-s, Vizean and Serpukhovian (= *c*. Lower Namurian); t, Tournaisian; fm, Famennian; fr-fm, Frasnian and Famennian; SVMB, Subvariscan molasse basin; RHTB, Renohercynian turbidite basin; WC/D, Westphalian stage (= *c*. the Moscovian stage of the Urals) the upper part; WA/B, Westphalian stage, the lower part; NA/B, Namurian stage (= *c*. Lower Bashkirian and Serpukhovian stages of the Urals); NA, Namurian stage, the lower part (mostly Serpukhovian stage); α, β, γ, the units of the Upper Visean substage. Thick lines, the lower boundary of molasse.

arc-continent collision was accompanied by HP–LT metamorphism (Brown *et al.* 2006*b*; Puchkov 2008, and references therein). In Bashkirian time, continent–continent (Laurussia–Kazakhstania) collision commenced and continued intermittently until the end of the Permian. The early stage of this collision was accompanied by flysch deposition, periodically interrupted by relatively deep-water sediments of a starved type. In the Kungurian (Uppermost Priuralian), evaporite deposition occurred in the southern to northern parts of the Urals, but in its polar area the Kungurian deposits filling the Preuralian foredeep are dominated by flysch and coal-bearing sediments. Late Permian strata are dominated by alluvial to lacustrine molasse deposits and are weakly deformed in the frontal part of the foreland fold-and-thrust belt (Puchkov 2000). At the beginning of the Triassic, tectonic movements connected with Large Igneous Province (LIP) formation led to the accumulation of molasse-like sediments. Voluminous mafic magmatism began in the earliest Triassic, approximately at 250 Ma according to Rb–Sr, Sm–Nd (Anderichev *et al.* 2005) and Ar–Ar (Reichov *et al.* 2007) age data, suggesting a correlation of these events with the contemporaneous events in Siberia and the hypothesis of a single Uralo-Siberian superplume.

The latest collisional events in the Uralides took place in the Early Jurassic, with increasing intensity from the South to the North, and represents the terminal collision between Baltica (Laurussia) and Siberia loose blocks united into Pangaea. The effects of this collision are well exposed in the Pai-Khoy Range, Novaya Zemlya islands and Taymyr.

## Tectonic zones of the Urals (Fig. 4)

The Uralides are divided into several north–south striking structural zones, giving the Urals a general appearance of an approximately linear fold-belt, in contrast to the more strongly oroclinal and more mosaic chains of the European Variscides, Alps or Kazakhstanides (Franke 2000; Khain 2001; Agard & Lemoine, 2005).

The Urals is divided into the following structural zones, which are from west to east (Fig. 4; Puchkov 1997, 2000):

(1)   A – the Preuralian foredeep, which inherited the western part of a bigger and long-living orogenic basin. It is filled mostly by Permian preflysch (deep-water condensed sediments), flysch and molasse.

(2)   B – the West Uralian megazone, predominantly consisting of Palaeozoic shelf and deep-water passive margin sediments. This zone was affected by intense fold-and-thrust

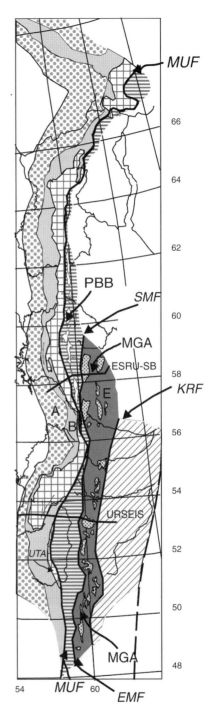

**Fig. 4.** Tectonic zones of the Urals (explanations in the text). Abbreviations: PBB, Platinum-bearing Belt; MGA, Main Granitic Axis; MUF, Main Uralian Fault; EMF, East Magnitogorsk Fault; SMF, Serov–Mauk Fault; KRF, Kartaly ('Troitsk') Fault. URSEIS and ESRU–SB, lines of seismic profiles described in the text.

deformation, and includes klippe containing easterly-derived ophiolites and arc volcanics (Puchkov 2002*a*).

(3)  C – the Central Uralian megazone, where the Precambrian (predominantly Meso- and Neo-proterozoic) crystalline basement of the Urals is exhumed. This basement is traced by geophysical data under the A, B and C megazones. Basically, these megazones were formed as a result of deformation of the continental margin of Baltica, although some allochthons, and partly the Ural-Tau antiform (UTA) were derived from more eastern oceanic structures.

(4)  D – the Tagilo–Magnitogorskian megazone, bordered to the west by serpentinitic mélange within the Main Uralian Fault zone (MUF) and to the east by the East Magnitogorsk Fault (EMF) and Serov-Mauk Fault zones (SMF). This megazone predominantly consists of Ordovician–Lower Carboniferous complexes of oceanic crust and ensimatic island arc, including the Platinum-bearing Belt of layered basic-ultramafic massifs (PBB), overlain by platformal carbonate and rift-related volcanic rocks.

(5)  E – the East Uralian zone, bordered to the west by the East Magnitogorskian mélange zone (EMF) and to the east by the Kartaly (Troitsk) Fault (KRF) (Fig. 4). This zone comprises Proterozoic gneisses and schists overlain by weakly metamorphosed Ordovician to Devonian sedimentary clastic strata and by tectonically emplaced sheets of Palaeozoic (Ordovician–Lower Carboniferous) oceanic and island arc complexes. The East Uralian Zone is intruded by voluminous Late Palaeozoic granite bodies which define the Main Granitic Axis (MGA) of the Urals (Puchkov *et al.* 1986).

(6)  F – the Transuralian zone, the easternmost zone of the Urals has probably an accretionary nature. It contains pre-Carboniferous complexes which preserve a variety of tectonic settings, including Proterozoic blocks of gneisses, crystalline schists and weakly metamorphosed sediments, Ordovician rift (coarse terrigenous and volcanic) and oceanic (ophiolite) deposits, Silurian island-arc complexes and Devonian deep-water deposits overlain unconformably by the Lower Carboniferous suprasubductional volcanogenic strata, which form a post-accretionary overstep complex.

Zones D–F, together with MUF, are traditionally interpreted to comprise of vestiges of the palaeoceanic component of the Urals, relics of the Palaeouralian ocean (Peyve *et al.* 1977; Puchkov

2000). As ophiolites with MORB signatures are poorly preserved in most orogens, the abundance of ophiolites in the eastern zones of the Uralian orogen is a rather anomalous feature, compared to many other orogens.

All megazones are either exposed or are near the Earth's surface only in the Southern Urals. To the north, the easternmost zones are covered by the Mesozoic and Cenozoic strata of the West Siberian basin, and in the Northern and Polar areas only the Preuralian foredeep, West Uralian, Central Uralian and western part of the Tagilo-Magnitogorskian megazone are exposed.

## Structural development of the Urals (Fig. 5)

In general, the Urals comprise the following first-order structural stages: (1) Archaean–Palaeoproterozoic development of cratonic basement; (2) Meso-Neoproterozoic rift and basin development, followed by orogenesis that culminated with the formation of Timanide orogen along the periphery of Baltica; (3) Palaeozoic–Lower Jurassic development of the Uralides; (4) Middle Jurassic–Palaeogene–Miocene (platform); and (5) Pliocene–Quaternary (neo-orogenic) activity which is a far-field effect of Alpine-Himalayan orogenesis. In this paper we focus on the

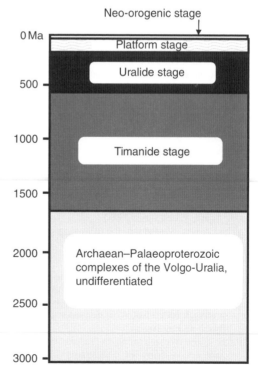

**Fig. 5.** Structural stages of the Urals.

Palaeozoic–Lower Jurassic stage as an example of a full Wilson cycle leading to formation of the Uralide orogen.

## The Uralides (Fig. 6)

The Uralides consist of the following stages of development: (1) rifting of Baltica continental crust, composed of a cratonic crystalline basement and the Neoproterozoic Timanide foldbelt; (2) formation of an oceanic basin and microcontinents; (3) subduction of the oceanic crust and consequent arc generation; (4) arc-continent collision, followed by (5) continent–continent (Laurussia–Kazakhstania and then Laurussia–Siberia) collisions.

### Rifting stage of the Uralides: a precursor of the Palaeouralian ocean

On a global scale, rifting and development of the Palaeouralian ocean episode was preceded by a series of pene-contemporaneous collisions and orogenies (Cadomian, Timanian, Brasilian, Panafrican) associated with the assembly of the supercontinent Pannotia (or Panterra) in the Ediacarian time. The Palaeouralian ocean was formed as a result of a breakup of this supercontinent in the Late Cambrian–Ordovician time (Puchkov 2000). Two other possible scenarios for the origin of the ocean have been suggested. According to Zonenshain *et al.* (1990), a system of rifts formed in the Early Ordovician at the eastern margin of East European continent. Rifting gradually changed to oceanic spreading and the generation of a series of microcontinental fragments (Uvat-Khantymansian, Uraltau, Mugodzharian) which formed adjacent to the boundary between the new-formed Palaeouralian ocean and the older, Asiatian ocean. Alternatively, some researchers (Scarrow *et al.* 2001; Samygin & Ruzhentsev 2003) maintain that Palaeouralian ocean was inherited from the Proterozoic time, implying no distinction between the development of the eastern and western flanks of the Uralo-Mongolian orogenic belt in the Early Palaeozoic. However, there are several strong arguments against the latter interpretation. First, the contrasting orientation of the structural grain of the Timanides and Uralides (especially in the North) support the idea of a continental breakup preceding the formation of the Uralian ophiolites. This interpretation is supported by the pattern of strong NW-trending magnetic anomalies of the Timan-Pechora province, positive anomalies reflecting vast fields of rift and island-arc volcanics, and negative anomalies associated with granites and metamorphic rocks (Fig. 7). These anomalies are truncated by the N–NE-trending magnetic anomalies which correspond to the MUF and palaeo-oceanic structures of the Uralides. Second, the lower age limit of the ophiolites attributed to the Palaeouralian ocean, as determined by recent studies of conodonts, is Arenig–Llandeilian (correlated with two unnamed stages between Tremadocian and Darriwilian (Gradstein *et al.* 2004) and is clearly younger than the Ediacarian–Tremadocian age that would be expected if there was uninterrupted ocean development (Puchkov 2005; Borozdina *et al.* 2004; Borozdina 2006; Smirnov *et al.* 2006). These ages are very different from the relics of the Palaeoasian ocean in Altaides and Kazakhstanides, where Cambrian ophiolites occur. Third, the presence of the Ordovician rift complexes along the margins of the former Baltica continent from one side, and the microcontinent(s) incorporated into the East Uralian and Transuralian zones support the former interpretation (Fig. 8).

A detailed description of the Uralian Early Palaeozoic rift facies in the western slope of the Urals is given in Puchkov (2002*a*, and references therein). Typically, the rift facies consists of Uppermost Cambrian–Tremadocian to Middle Ordovician coarse terrigenous sediments (conglomerates, sandstones, siltstones of very variable thickness, combined with interlayered subalkaline flows and tuffs). They overlie unconformably the crystalline basement and are overlain either by shelf or deepwater facies, reflecting the development of eastern passive continental margin of Baltica. Although the rift facies of the eastern zones of the Urals (Kliuzhina 1985; Snachev *et al.* 2006) resemble them lithologically, their age is restricted to the Middle Ordovician.

Alkaline carbonatite-bearing complexes (mostly miaskites) in the western part of the Middle Urals (Levin *et al.* 1997), originally thought to be a manifestation of this rifting event (Samygin *et al.* 1998; Puchkov 2000) have been re-interpreted to post-date rifting, on the basis of Rb–Sr and U–Pb isotopic data which indicate a latest Ordovician to Silurian age (Puchkov 2006*a*; Nedosekova *et al.* 2006, and references therein). These intrusions are oblique to the Uralian structural grain, and may be analogous to the Early Cretaceous Monteregian alkaline intrusions in eastern Canada or the more or less contemporaneous hot spot tracks of Eastern Brazil (Bell 2001; Cobbold *et al.* 2001). Probable Late Ordovician–Early Silurian plume-related complexes also occur in more northern parts of the Urals (e.g. monzogabbro-syenite-porphyry as indicated by the $447 \pm 8$ Ma U–Pb (SHRIMP) age of the Verkhneserebryansky complex (Petrov 2006) or REE-rich phases of subalkaline granitoids in the North of the Urals, dated as 420–460 Ma by Rb–Sr and U–Pb methods (Udoratina & Larionov 2005).

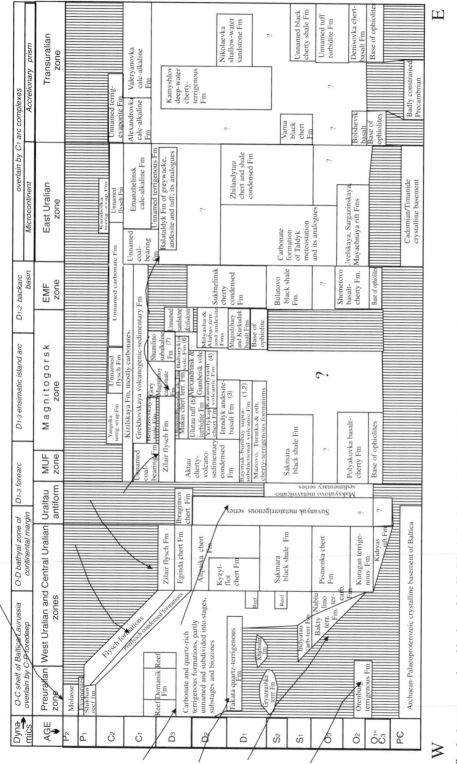

**Fig. 6.** A tectonostratigraphic chart of the Uralides in the Southern Urals. All the formations are tentatively restored to their initial, autochthonous positions. Arrows show a provenance of terrigenous material. (After Maslov *et al.* 2008, strongly modified.)

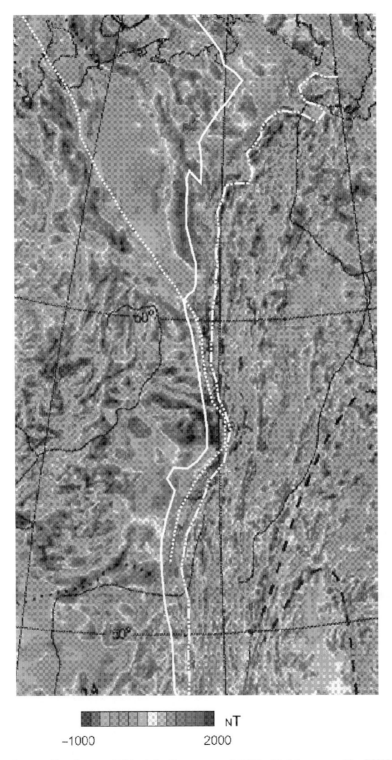

**Fig. 7.** Magnetic anomalies of eastern Baltica (after Jorgensen *et al.* 1995, with data processed by CONOCO Inc., USA). White dotted line, Timanian deformation front; white solid line, Uralide deformation front; white dot-and-dash line, the Main Uralian Fault.

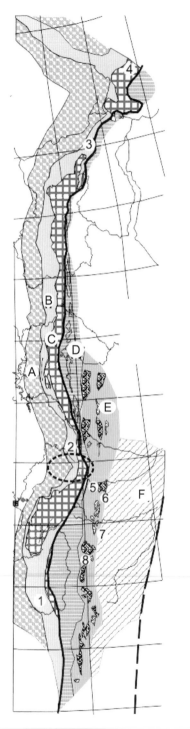

## The origin of the Palaeouralian ocean ophiolites

The aforementioned continental rifting within Baltica led to oceanic spreading and formation of ophiolites. The unusual abundance of ophiolites is a special feature of the Uralian orogen. Several studies report on the petrology, geochemistry, structure and metallogeny of the Uralian ophiolites (e.g. Savelieva 1987; Savelieva & Nesbitt 1996; Saveliev 1997; Melcher *et al.* 1999; Spadea *et al.* 2003; Savelieva *et al.* 2006*a, b*). The general consensus is that the 'ideal' section of the Uralian ophiolites consists of (from top to bottom):

(1) Tholeiitic basalts (mostly pillow lavas) with layers of pelagic sediments (typically cherts containing relics of half-dissolved radiolarians). The age of these basalts, constrained by many occurrences of conodonts, is never older than Arenigian–Llandeilian (see earlier comment). In the Tagil zone, ophiolitic complexes are overlain by Upper Ordovician island-arc formations, whereas in the Magnitogorskian zone, the condensed oceanic sediments overlie Ordovician–Llandoverian basalts, and persist until the onset of island arc magmatism in the Early Devonian;

(2) Dyke-in-dyke sheeted complexes, which are common in the Urals, in contrast to some other orogens (e.g. the Alps);

(3) Alpine-type gabbro;

(4) Banded dunite-wehrlite-clinopyroxenite complexes, interpreted to reflect a fossil MOHO boundary; and

(5) Peridotite complexes, represented by lherzolites, harzburgites and dunites in different proportions and combinations.

However, in detail not all ophiolitic complexes display this simple sequence. For example, Ishkinino, Ivanovka and Dergamysh Ni–Co-rich pyrite deposits in the MUF zone are attributed to ocean-floor black smokers and overlie and partly penetrate peridotitic host-rock, devoid of the several 'standard' members of the ophiolite section. Similar features occur in modern Atlantic thermal fields (e.g. Logachev and Rainbow fields), although their geodynamic setting may not support the direct analogy. The deposits rather belong to the relics of Magnitogorsk forearc (Jonas 2004; Melekesceva 2007).

**Fig. 8.** The position of the Early Palaeozoic rift/plume related complexes. I–VI, tectonic zones of the Urals (correspond to A, B, C, D, E, F in Fig. 4, with the same symbols); 1–8, localities where the Lower Palaeozoic graben formations are developed; 1, Sakmarian;

**Fig. 8.** (*Continued*) 2, Bardym; 3, Lemva; 4, Baydarata; 5, Samar; 6, Sargaza; Uvelka; 8, Mayachnaya. Dashed-line ellipse, location of possible plume-related alkaline complexes Vishnevogorsk and other).

Savelieva *et al.* (2006*a*) classified the Uralian ophiolites into three groups according to their inferred geodynamic setting:

(1) Complete sections of ophiolites (e.g. Kempirsay massif) or their fragments in the south of Magnitogorsk, East Uralian or Denisovka zones), include restite peridotite and overlying succession of plutonic gabbro, parallel diabase dyke complexes and tholeiitic lavas formed in a MOR setting (Savelieva & Nesbitt 1996). However, a supra-subduction geochemical component has been documented in ophiolites in the southern part of the Kempirsay massif (Melcher *et al.* 1999);

(2) Massifs of a lherzolite type representing fairly low depleted lithospheric mantle (e.g. Kraka, Nurali) have a simple evolutionary history consisting of enriched peridotite and dunite associated with less abundant amphibole gabbro (Savelieva 1987). These characteristics are thought to reflect a low degree of partial melting of a mantle diapir, followed by rapid uplift, a scenario typical of rifting that immediately precedes oceanic spreading. Alternatively, Spadea *et al.* (2003) propose a more dynamic history for these massifs, involving re-fertilization of a depleted mantle by basaltic magma, by analogy with Lanzo massif of Alps, which however is also thought to be indicative of pre-spreading rifting (Müntener *et al.* 2005); and

(3) According to the general geodynamic reconstructions of Saveliev (1997), the huge Polar Urals massifs such as Voykar-Synya, Ray-Iz and Syum-Keu, are integrated into a system of allochthons, composed of complexes of two island arcs – Tagil-Schuchya ($O_3$–$S_1$) and Voykar ($S_2$–$D_3$). The restites in these massifs are strongly depleted and preserve evidence of interaction with basaltic magma. The sections consist of multi-phase intrusions of gabbro and diabase, interpreted by Savelieva *et al.* (2006*a*) to reflect the development of a marginal basin when an island arc rifted apart in the Late Silurian-Early Devonian. Therefore this spreading was related to the development of a second island arc. This interpretation, however, may be an over-simplification as the presence of two island arcs suggests the existence of older (Ordovician–Lower Silurian?) ophiolites corresponding to the older of the two arcs. Indeed, Ar–Ar data from the banded complex of Voykar-Synya and Khadata massifs (primary amphibole, fresh plagioclase and clinopyroxene of gabbro) yield 420–490 Ma ages. In addition the Khadata spreading dikes yield a *c.* 423 Ma

age and the Voykar sheeted dykes, described by Remizov (2004) as island arc complex, were dated at 426–444 Ma (Didenko *et al.* 2001). Khain *et al.* (2004) dated zircons from a plagiogranite dyke in the parallel dyke complex of the Voykar-Synya massif at $490 \pm 7$ Ma. The above data indicate that the generation of these ophiolites is probably more complicated than the current geodynamic models purported to explain them.

This complicated scenario is highlighted by recent age data (Gurskaya & Smelova 2003; Savelieva *et al.* 2006*a*; Tessalina *et al.* 2005; Batanova *et al.* 2007; Krasnobaev *et al.* 2008), which yield Neoproterozoic–Ediacarian (536–885 Ma) and some older dates for many of the ultramafic and alpine-type gabbro complexes that were previously regarded as Palaeozoic. These data include Re–Os and Sm–Nd mineral isochrons and U–Pb analyses of zircons. In general, however, most of published age data support also an Ordovician–Lower Devonian age of alteration processes for most of the complexes (summarized by Puchkov 2000 and well illustrated by Krasnobaev *et al.* 2008). Two contrasting explanations have been proposed. According to Tessalina *et al.* (2005), the ultramafic complexes are Neoproterozoic ophiolites and represent relics of the oceanic crust developed during Timanide orogenesis. Alternatively, Puchkov (2006*b*) suggests that the Neoproterozoic dates in the Palaeozoic ophiolites reflect a relict signature preserved in the mantle part of the younger ophiolites (peridotites and partly ex-eclogitic mantle gabbro), notwithstanding the overprinting during subsequent processes of ophiolite formation. The possibility of preservation of ancient mantle zircons and their contamination of younger MOR and island arc volcanics has recently been underlined (Sharkov *et al.* 2004; Bortnikov *et al* 2005; Puchkov *et al.* 2006).

Despite the above complexities, the age of ophiolite basalts determined by conodonts, is never older in the Urals than Arenigian–Llandeilian (see again the earlier comment).

### The passive margin of the continent

Simultaneously with oceanic development, the continental margin of Baltica started to develop by rifting in the Ordovician (Fig. 9). The identity of the conjugate margin to this rift is not known. By the end of the Silurian, Baltica had collided with Laurentia to form Laurussia (Ziegler 1999). The development of the margin is described in detail by Puchkov (2002*a*), and only a general summary is given here. Typically, the succession starts with uppermost Cambrian–Lower Ordovician coarse terrigenous deposits, in some cases accompanied

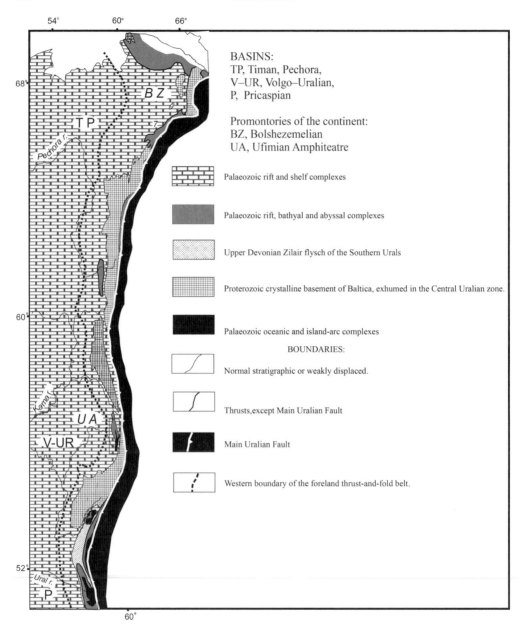

**Fig. 9.** Major structural elements and complexes of Laurussia/Baltica passive margin involved into the Urals (from Puchkov 2002, with minor changes).

by minor volcanic rocks (see above). The margin is classified as a non-volcanic type (in a classification of Geoffroy 2005).

Two facies were established early in the passive margin development – an inner (shelf) and an outer (continental slope, grading to continental rise) (Fig. 6). Generally, shelf sediments are represented by shallow-water carbonates (limestones, dolomitic limestones, dolomites) and terrigenous sediments with west-derived (Smirnov 1957) oligomictic, quartz sandstones. Regressions are marked by barrier reefs at the outer margin of the shelf zone, while the transgressions favour a formation of deep-water, starved basins with condensed facies of marls and oil shales (called 'domanik' in Russia), surrounded by reefs and bioherms.

The outer, continental slope and rise (bathyal) sections consist of thick westerly-derived quartz sandstones, and thin, condensed units consisting of shales, cherts and minor limestones (Puchkov 1979, 2000). Fauna are mostly pelagic: radiolarians, conodonts, graptolites and rare goniatites. The paucity of limestones indicate a transition to abyssal conditions.

The upper strata of the outer facies consist of polymictic, flysch deposits signifying a sharp change of provenance that is connected with the start of orogenesis (Puchkov 1979; Willner et al. 2002, 2004). This change of provenance is diachronous: it is earlier in the east and south of the western slope of the Urals; in the Southern Urals it occurs in the uppermost Frasnian, but in the Polar Urals it starts in the Early Visean.

Puchkov (1979) drew attention to the similarity of these deposits with analogous geodynamic settings in other orogens. For example, eastern Laurentia was bordered by deep-water sediments in the Ordovician that are preserved in the allochthons formed during the generation of the Appalachian orogen. In some cases (e.g. Ouachita), the deep-water facies persisted, like in the Polar Urals, through most of the Palaeozoic (from the lowermost Ordovician until the Carboniferous).

## Subduction of oceanic crust

The Urals is characterized by an exceptionally good preservation of subduction complexes, which permits reconstruction of the development of at least three subduction zones in place and time: Tagil (Late Ordovician–Early Devonian), Magnitogorsk (Early–Late Devonian) and Valerianovka (Tournaisian–Early Bashkirian) (Fig. 10).

### The Late Ordovician–Early Devonian Tagil arc (Fig 10, to the left)

The oldest (Tagil) arc complexes are developed in the Middle, Northern, Cis-Polar and Polar parts of the Tagilo–Magnitogorskian zone. The best sections, well constrained by conodonts, are preserved in the southern part of the Tagil synclinorium (synform). According to recent stratigraphic and petrochemical studies (Narkissova 2005; Borozdina 2006), the oldest ensimatic island-arc succession is predominately represented by the basaltic ($O_3$), basalt-plagiorhyolitic ($O_3$) and basalt–andesite–plagiodacite ($S_1$) volcanic associations. The latter two have calc-alkaline affinities (Narkissova 2005). The overlying Silurian association ($S_1ll_2-S_2w_1$) is represented by flysch consisting of interbedded black cherty siltstones, tuffites and tuffaceous sandstones (arc slope deposits), overlain

by andesites, dacites, basalts and very abundant tuffs. In the Wenlockian, the volcanic rocks are laterally equivalent to reefal limestones (All-Russian Committee 1993). These biohermal deposits persisted until the Pridolian as an unstable, narrow carbonate shelf on the perimeter of the island arc. The above Silurian complexes are substituted laterally by a volcanic association ($S_1ll_3-S_2ld_1$), represented by basalts, andesites and tuffs with rare layers of cherty siltstones. After the Late Ludlovian, the marine conditions partly changed to continental conditions: the Pridolian association is represented by predominant coarse-grained red-coloured polymictic terrigenous-volcanogenic deposits with fragments of the older rocks; such as volcanics, with sublakalic and alkalic basalts being predominant. The island arc succession is terminated by a very specific association ($S_2pr-D_1lh$), preserved in the axial part of the Tagil synform, which resembles the underlying Pridolian association, and includes shoshonitic mafic to intermediate volcanic rocks (Narkissova 2005) and minor flysch-like volcaniclastic deposits.

The volcanism of the Tagil arc evolved from a uniformly tholeiitic affinity to a differentiated calc-alkaline and then to subalkalic shoshonitic affinity, suggesting deeper levels of partial melting in mantle with time, a trend that is opposite to the typical trend in rift and superplume zones (Dobretsov et al. 2001). Geochemically, the volcanics are typical of ensimatic island arcs (Narkissova 2005; Borozdina 2006). Basalts retrieved from the superdeep SG-4 borehole exhibit a distinct Ta–Nb minimum. In general, the volcanics are depleted in Nb,Ta, Zr, Ti, Y and enriched in K, Rb, Ba, Pb relatively to N–MORB. The geochemical trends of contemporaneous volcanics suggest an eastward (in modern co-ordinates) dipping subduction zone (Narkissova 2005).

The Tagil arc is also known for the presence of gabbro-ultramafic massifs composing a gigantic (c. 1000 km) linear, platinum-bearing belt (PBB on Fig. 4). The concentric-zonal massifs consist of dunites, clinopyroxenites, gabbro and plagiogranites, and mafic rocks comprise up to 80% of the belt. Disseminated platinum is hosted by dunites, and industrial deposits are represented mostly by modern (or reworked Meso-Cenozoic) placers. The geodynamic significance of the belt is controversial: models vary from a rift (Efimov 1993) to a supra-subduction zone setting (Ivanov et al. 2006). The age of the belt, determined by several methods as 420–430 Ma, and the similarity of petrogenetic-indicating trace and rare earth elements with island-arc tholeiites (Ivanov et al. 2006) supports the supra-subduction zone model. Such belts are rare in modern arc environments, but possible analogues occur in the northern part

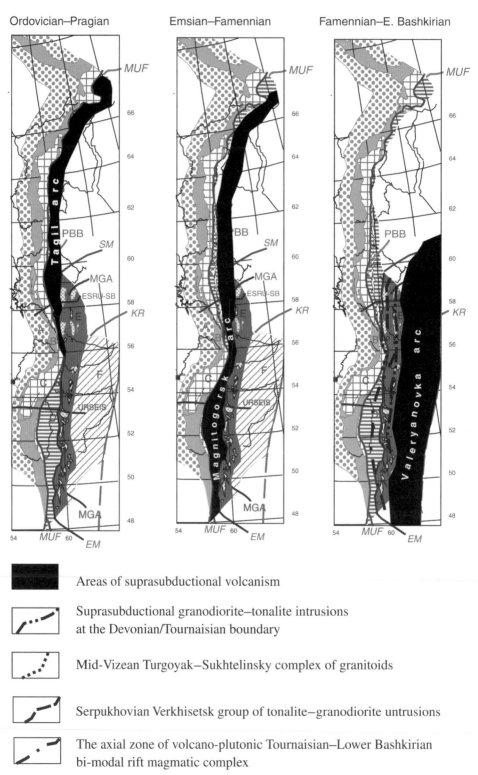

Ordovician–Pragian   Emsian–Famennian   Famennian–E. Bashkirian

**Areas of suprasubductional volcanism**

Suprasubductional granodiorite–tonalite intrusions
at the Devonian/Tournaisian boundary

Mid-Vizean Turgoyak–Sukhtelinsky complex of granitoids

Serpukhovian Verkhisetsk group of tonalite–granodiorite untrusions

The axial zone of volcano-plutonic Tournaisian–Lower Bashkirian
bi-modal rift magmatic complex

**Fig. 10.** The distribution of magmatism of three main stages of subduction: Ordovician–Early Devonian (Tagil): Early Devonian–Famennian (Magnitogorsk); Tournaisian–Early Bashkirian (Valerianovka and contemporaneous to it).

of the circum-Pacific ring, and are known as Alaskan type (Burns 1985).

Ordovician–Early Devonian island-arc volcanism was followed by the development of a relatively stable carbonate shelf which caps the western part of the island arc complexes. After that time the Tagil arc was dismembered and by the Emsian, it was a terrane that accreted to the Magnitogorsk island arc, an event that coincided with the formation of the Magnitogorsk arc itself (see below). The location of the subduction zone changed and the region became characterized by presence of two sub-zones: the western, Petropavlovsk and the eastern, Turyinsk sub-zones.

The Petropavlovsk sub-zone contains Lower to Middle Devonian shallow-water limestones and bauxite, followed in the Late Devonian by deep-water cherty shales and polymictic terrigenous sediments. In the Turyinsk sub-zone, sedimentary strata (shallow-water limestones, shales and cherty shales) are interlayered with andesites, basalts, tuffs and volcanogenic sandstones (All-Russian Committee 1993). Yazeva & Bochkarev (1993) point out that these thick (up to 4–5 km) Devonian volcanic layers occur with comagmatic intrusions in volcano-plutonic complexes. Geochemical parameters (in particular, Rb and Sr contents) indicate that the thickness of the crust was c. 30 km (Yazeva & Bochkarev 1993), which implies that the new arc was ensialic, in contrast with the Ordovician–Lower Devonian ensimatic Tagil island arc.

## Devonian Magnitigorsk arc (Fig. 10, centre)

The development of the Magnitogorsk arc in the Southern Urals was more or less synchronous with the dismemberment of the Tagil arc. The location of the Turinsk zone of Tagil arc along the extension of the Magnitogorsk arc suggests that the Magnitogorsk subduction zone was inherited from the Tagil zone. The Middle–Upper Devonian calc-alkaline complexes can be traced northward to the Polar Urals.

However, in the Southern Urals, island arc development was preceded by a long period of quiescence, expressed by the deposition of deep-water oceanic cherts and carbonaceous cherty shales accompanied by basalts in the Ordovician and Llandoverian. Most of the Silurian is represented by 300 m of distal, condensed cherty shales. They are considered to represent the sedimentary cover of the ophiolites. Non-volcanic sections of the Lower Devonian (Lochkovian–lowermost Emsian) are represented by either deep-water terrigenous chert, argillaceous cherty sediments of Masovo, Turatka, Ishkinino and other formations or bioherm limestones (Artiushkova & Maslov 2003; Fig. 6). This stratigraphic level

includes also olistostromes developed locally with and within serpentinitic mélanges of the Main Uralian fault. The bioherms and olistostromes are local indicators of buckling of the oceanic crust at the onset of subduction and are related to an early, non-volcanic stage of subduction.

Volcanic complexes of the Magnitogorsk arc in the Southern Urals are well studied (Brown et al. 2001, 2006b; Kosarev et al. 2005, 2006, and references therein). The volcanic succession, represented by a characteristic interlayering of tholeiitic basalts, bimodal basalt-rhyolite and regularly differentiated basalt-andesite-dacite-rhyolite series, comprises the following units (the local names are partly shown in Figure 6, and numbered correspondingly):

(1) A bimodal rhyolite-basalt series that overlies a tholeiite-boninite unit (Emsian);
(2) Basalt-andesite-dacite-rhyolitic series (Upper Emsian);
(3) Andesite-basalt series (uppermost Emsian–Lower Eifelian);
(4) Bimodal rhyolite-basalt series (Upper Eifelian);
(5) Basalt-andesite-dacite-rhyolite series (Givetian–Lower Frasnian);
(6) Basalt-andesite formation (Upper Frasnian);
(7) Local shoshonite-absarokite formation (Famennian). In addition, subduction-related 370–350 Ma granitoid intrusions of calc-alkaline affinity are developed in the northern part of Magnitogorsk synclinorium (Bea et al. 2002); and
(8) These intrusions are unconformably overlain by Lower Carboniferous volcanics dominated by tholeiitic basalt in the west and by more widely developed subalkaline bimodal basalt-rhyolite in the east. They are accompanied by a chain of coeval (335–315 Ma, Bea et al. 2002) bimodal gabbro-granitoid intrusions (Magnitogorsk-type plutons). Both volcanic and intrusive bimodal series, according to their field relationships, mineralogy and geochemistry, suggest an extensional or passive within-plate non-arc origin and are probably produced by undepleted lherzolites (Bochkarev & Yazeva 2000; Fershtater et al. 2006). This magmatism may be related to a slab break-off of the Magnitogorsk subduction zone which gave way to a melt from the less depleted, deeper mantle under it (Kosarev et al. 2006).

Notwithstanding differences in composition, magmatic and tectonic affinities, all the Devonian volcanic series share geochemical traits typical of a supra-subduction zone origin, such as negative Nb, Ta, Zr, Hf, Y anomalies, and elevated concentrations of large ion lithophile (LIL) elements

(K, Rb, Ba, Cs) and LREE. They show no signs of contamination by continental crust and are interpreted as ensimatic arc complexes.

There are many parallels with a development of the Tagil arc, including the alkaline trend towards the upper member of the succession. Trace element abundances for contemporaneous volcanic rocks are consistent with an eastward-dipping subduction zone. To the west and east of the main volcanic body of the arc, mostly in mélanges of the MUF and EMF zones and associated alloch-thons, condensed cherty-terrigenous series contem-poraneous to the arc are developed, corresponding to the forearc and backarc basins (Fig. 6).

It looks like the ophiolite basement of the arc is mostly Ordovician in age, except the Mugodzhary section, where the Emsian basalts and cherts of Mugodzhary and Kurkuduk for-mations overlie a large-scale Aktogay sheeted dyke complex, gabbro and serpentinites, composing a Lower Devonian ophiolite (Fig. 6). The ophiolite is tentatively interpreted as a result of a backarc spreading.

## The collision of the Magnitogorsk arc with the passive margin of Laurussia

The collision of the Magnitogorsk arc with the passive margin of Laurussia (former Baltica) has been described in several recent publications (Brown & Puchkov 2004; Brown et al. 2006b, and references therein) and is briefly summarized here (Figs 11 & 12). Since the Early Devonian, an island arc formed within the Uralian palaeocean above an east-dipping subduction zone. Collision of the arc with Laurussia occurred in the Late Devonian and was accompanied by the following events:

(1) scraping-up of the deep-water sediments of the continental passive margin by the rigid wedge (backstop) of the arc and the formation of an accretionary prism;
(2) jamming of the subduction zone followed by a jump in the location of the subduction zone;
(3) slab break-off and opening of a slab window permitting the deeper, more fertile and hotter mantle to produce subalkaline, non-subduction volcanics;
(4) uplift of the buoyant continental part of the slab, exhumation and erosion of UHP(?) and HP–LT metamorphic complexes and their erosion;
(5) formation of the accretionary cordillera of Uraltau antiform (comparable to accretionary avolcanic arc of Indonesia) and two flysch basins flanking both sides of it: forearc and foredeep basins (Fig. 13); and

(6) formation of the suture zone of the Main Uralian Fault, which divides the accretionary prism on the continent from the remnant of the island arc.

In summary, the Southern Urals Magnitogorsk arc was formed in the Early Devonian upon Ordovician–Early Devonian oceanic crust (Puchkov 2000; Snachev et al. 2006) and developed until it collided with the passive continental margin of Laurussia in the Late Devonian. Due to the oblique orientation of the subduction zone relative to the continental margin, collision at this time did not take place along the whole length of the passive margin of Laurussia continent. To the north of the Ufimian promontory, the margin of Laurussia shows no evidence of the Late Devonian arc–continent collision.

Two remarkable Late Devonian events in the Southern Urals can be regarded (along with direct structural data) as important indicators of the tran-sition from subduction to collision. The first is the deposition of the Zilair greywacke flysch formation of the eastern provenance (uppermost Frasnian–Famennian) which overlies Frasnian deep-water and Famennian shallow-water deposits of the continental margin of Laurussia. The second is a culmination of a HP–LT metamorphism at 372–378 Ma that provides additional evidence for the end of subduction and the onset of collision.

As for the metamorphism, its age range in the subduction zone should be broadly contempora-neous with the subduction, and its oldest products should be older than the supra-subductional volcan-ism. However, it is not clear if the products of this early metamorphism were preserved and then exhumed or if they were completely entrained by the slab. The Magnitogorsk arc appears to have these products preserved and then exhumed. In the serpentinitic mélange of MUF, along the margin of the Magnitogorsk arc, garnet pyroxenites (meta-morphic basites) occur. The best documented occur-rence where $P$–$T$ parameters of their origin are determined as 1.5–2 GPa, 800–1200 °C is in the Mindyak peridotite massif of MUF (Pushkarev 2001). Its age was determined by two methods: Sm–Nd isochron is 406–399 Ma (Gaggero et al. 1997); whereas U–Pb analysis of zircons yield an age of 410 ± 5 Ma, which is interpreted as a meta-morphic age (Saveliev et al. 2001). A Pb–Pb analy-sis of zircon cores yield an age of 467 Ma, and is interpreted as a protolith age (Gaggero et al. 1997). The age of garnet pyroxenite from the Bayguskarovo occurrence is 416 ± 6,1 U–Pb SHRIMP (Tretyakov et al. 2008). A series of Ar–Ar age determinations of phengite, whose interpretation depends on mineral dimensions and temperature conditions of equilibration, has been

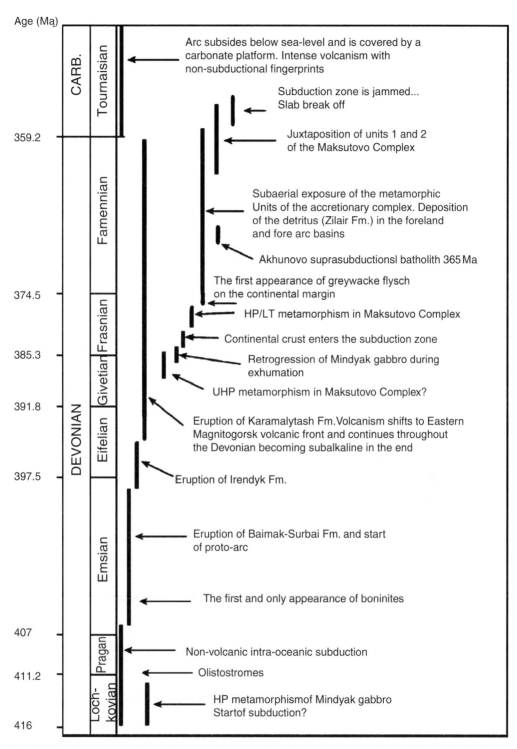

**Fig. 11.** Time/process evolutionary diagram for intra-oceanic subduction and arc-continent collision in the Southern Urals (after Brown *et al.* 2006*b*, with added information given in bold).

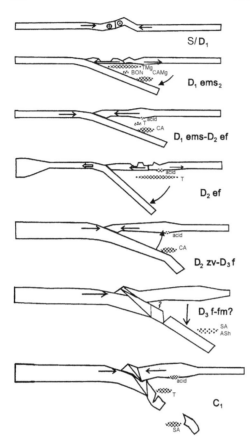

S/D$_1$

D$_1$ ems$_2$

D$_1$ ems-D$_2$ ef

D$_2$ ef

D$_2$ zv-D$_3$ f

D$_3$ f-fm?

C$_1$

**Fig. 12.** A model for development of the Magnitogorsk arc and subduction zone (Kosarev *et al.* 2006, slightly modified). Dotted lenses, supposed zones of melting of initial magmas of different petrogenetic types: T, tholeiitic; BON, boninitic; TMg, tholeiitic magnesial; CA, calc-alkaline; ASh, absarokite-shoshonite; SA, subalkaline. Stages of the Devonian: em, Emsian; ef, Eifelian; gv, Givetian; f, Frasnian; fm, Famennian.

done for a succession of samples across the contact between eclogite and garnet glaucophane schist from Maksiutovo complex. The age range of phengites is from 400 Ma at *c.* 500 °C to *c.* 379 Ma at the final closure temperature of the system (*c.* 370 °C). The Ar–Ar age of glaucophanes from the same sample is 411–389 Ma (Lepesin *et al.* 2006). The peak of Ar–Ar ages obtained from detrital phengites of the Zilair series clusters around 400 Ma (Willner *et al.* 2004). U–Pb SHRIMP dating of zircons from Maksiutovo eclogites yielded 388 ± 4 Ma (Leech & Willingshofer 2004). These older (Lower and Middle Devonian) dates of metamorphism are consistent with the cooling action of the subducting slab, causing the closure of isotopic systems.

The younger ages of the metamorphic rocks cluster around 375–380 Ma and probably are consistent with the general exhumation of the HP–LT metamorphic rocks of the Southern Urals. The rocks are divided into two units, established by Zakharov & Puchkov (1994) and by many later researchers.

The age of the start of general cooling and exhumation (reviewed by Brown *et al.* 2006*b*) for the eclogite facies metamorphism of the lower unit of the Maksutovo Complex is thought to be Frasnian in age, with a mean value of 378 ± 6 Ma according to many isotopic determinations (Matte *et al.* 1993; Lennykh *et al.* 1995; Beane & Connelly 2000; Hetzel & Romer 2000; Glodny *et al.* 1999, 2002). The upper unit was metamorphosed together with the lower unit, suggesting juxtaposition during exhumation at a higher crustal level by 360 ± 8 Ma (Rb–Sr and Ar–Ar methods; Beane & Connelly 2000; Hetzel & Romer 2000).

In the Polar Urals, isotope dating of HP–LT metamorphism of the Marun-Keu complex of eclogites and related metamorphic rocks was reviewed recently by Petrov *et al.* (2005). According to Shatsky *et al.* (2000), the Sm–Nd isotopic analyses of garnet, clinopyroxene and whole rock gave 366 ± 8.5 Ma for the hornblende eclogite and 339 ± 16 Ma for the kyanite eclogite. Rb–Sr whole-rock dating of the eclogites (Glodny *et al.* 1999) gave 358 ± 3 Ma. According to the data of Glodny *et al.* (2003, 2004), the concordant U–Pb age data for the metamorphic zircon domains are between 353 and 362 Ma, coincident with the age of metamorphism as inferred from Rb–Sr internal mineral isochrons (an average value of 355.5 + 1.4 Ma).

The eclogite–glaucophane Nerka-Yu and Parus-Shor complexes in the southernmost Polar Urals yielded 351 ± 3.6 and 352 ± 3.6 Ma ($^{40}$Ar–$^{39}$Ar ages, Ivanov *et al.* 2000). Sm–Nd dating of glaucophane schists of the Salatim belt (Northern Urals) gave 370 ± 35 Ma. Taken together, these dates characterize an Upper Famennian-Middle Tournaisian age of HP–LT metamorphism and the beginning of its exhumation. These data are supported by the Lower Visean age of the oldest known Palaeozoic easterly-derived polymictic sandstones and conglomerates on the continental margin of Laurussia, to the west of the Main Uralian Fault, in the Polar and Northern Urals (Puchkov 2002*a*, and references therein).

Developing the oblique collision model of Puchkov (1996*b*), Ivanov (2001) calculated an average rate of subduction, which led to a gradual northward-shifting collision, of 2.75–2.80 cm. However according to recent data, the diachroneity of events at the end of the Devonian and beginning of the Carboniferous show no gradual south–north

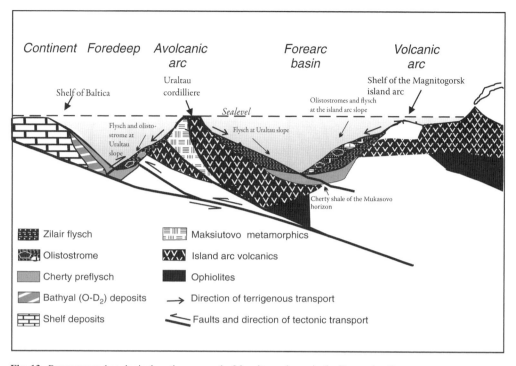

**Fig. 13.** Reconstructed geological section across the Magnitogorsk arc, in the Famennian time.

pattern, implying that behaviour of the subduction zone is more complex.

The collision of the Magnitogorsk arc with Laurussia may have occurred in two discrete stages (Fig. 14). First, collision in the Southern Urals occurred by the Famennian, and a triangular-shaped gap was left between the arc and the continent, similar to the modern Bengal Bay, Northwest Australian Bay or the South China Sea. Second, in the Early Carboniferous, the northern half of the arc was bent to the west and docked to the continental margin. At this stage subduction in the south virtually ceased, but in the north, increasing velocity of subduction resulted in increasing intensity of the HP–LT metamorphism in the same direction (from glaucophane schists of the Salatim belt and Cis-Polar Urals to eclogites of the Polar Urals). In the Middle Urals, these events were immediately followed by intrusion of Turgoyak-Syrostan group of granitoids (335–330 Ma), that was described by Fershtater *et al.* (2006) as 'granitoids connected with tensional structures'. This group can be correlated by the age with the Magnitogorsk within-plate gabbro-granite series (see below).

Unfortunately there is no support for this model from the data on the Early Carboniferous volcanism in the northern part of the Magnitogorsk arc. But the lack of data may be because the eastern limb of this arc is concealed in the Northern to Polar Urals under the Mesozoic–Cenozoic cover of the West Siberian plate.

The above-described Early Carboniferous (Tournaisian–Visean) stage of subduction was followed in the Middle Urals by a Serpukhovian stage, as indicated by the Verkhisetsk chain of granitoids (320 Ma) (Fig. 10), related by Fershtater *et al.* (2006) to another east-dipping subduction zone.

### Early Carboniferous-Bashkirian Valerianovka subduction zone(s) (Fig. 10, to the right)

By the middle of the Lower Carboniferous, the suture zone was established along the whole length of the MUF (Puchkov 2000, 2002*a*). The above-mentioned chain of 335–330 Ma (mid-Visean) massifs (Turgoyak–Syrostan group of granites) intrude the suture zone (Fershtater *et al.* 2006) and therefore post-date the Magnitogorsk subduction, providing an upper age limit for MUF. This conclusion is supported by the age of the Ufaley intrusion (concordant U–Pb, 316 ± 1 Ma, Early Bashkirian), which seals the Main Uralian Fault in the Northern part of the Ufimian promontory (Hetzel & Romer 1999).

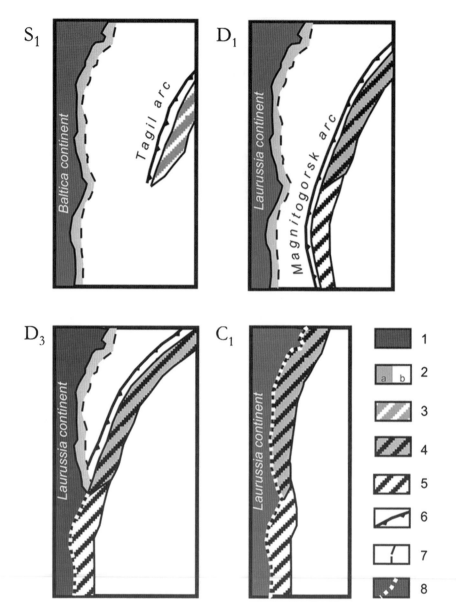

**Fig. 14.** A model for a two-stage Upper Devonian–Lower Carboniferous arc-continent collision in the Urals. 1, continental crust; 2a, transitional crust; 2b, oceanic crust; 3, Tagil island arc; 4 and 5, Magnitogorsk island arc; 4; ensialic (epi-tagilian); 5; ensimatic (Magnitogorsk arc *sensu stricto*); 6, subduction zone; 7, continent–ocean boundary; 8, suture zone of the Main Uralian fault.

With the demise of the Magnitogorskian arc, subduction did not terminate in the Urals as a whole. Ensialic subduction (either island arc or Andean-type or maybe two subduction zones of different type) of uncertain polarity began in the latest Devonian and lasted until the Mid-Bashkirian in the eastern Urals. The Main Granitic Axis of the Urals (Fig. 4) developed first as a chain of

suprasubductional tonalite–granodiorite massifs by the end of the Famennian or the beginning of the Tournaisian (*c.* 360 Ma), when the southern part of the Magnitogorsk subduction zone ceased to operate (Bea *et al.* 2002; Fershtater *et al.* 2006). Simultaneously, immediately to the east, a wide NNE-trending band of calc-alkaline and partly within-plate volcanic rocks and associated plutonic complexes ranging up to mid-Bashkirian in age were formed, suggesting a close affinity with the massifs of the Main Granitic Axis (All-Russian Committee 1993; Tevelev *et al.* 2005; Fershtater *et al.* 2006). The Lower Carboniferous (320 Ma) tonalite-granodiorite massifs occur in the Middle Urals, situated to the east of the Serov-Mauk suture zone (i.e. Verkhisetsk massif and others located to the east of the former Magnitogorsk volcano-plutonic arc).

According to Fershtater *et al.* (2006), increases in $K_2O$ and REE abundances in the granodiorites to the east indicate that subduction had an eastern polarity. However, Kosarev & Puchkov (1999) point out that $K_2O$ concentrations in the Lower Carboniferous volcanic rocks in the eastern Urals increase westward, suggesting a western polarity for the subduction zone. Of the same opinion are Tevelev *et al.* (2005) for the Uralian Lower Carboniferous volcanics, but they propose that the volcanism occurred in a wrench regime and that the easternmost Valeryanovka volcanic band belonged to Kazakhstanides and developed over a separate subduction zone with eastern polarity. Brown *et al.* (2006a) also suggest two oppositely dipping subduction zones, whereas Matte (2006) suggests a westerly dip for several subduction zones. Such a difference in opinion is explained by the complicated nature of the process, involvement of wrench tectonics, and the rather poor exposure of the complexes.

Calc-alkaline volcanic complexes in the Urals abruptly stopped forming by the mid-Bashkirian, signifying the end of a wide-scale subduction of an oceanic crust and transition to a continent-continent-type collision.

## Continent–continent collision and formation of the orogen

### The main Late Bashkirian to Permian stage of collision

The collision between Laurussia and Kazakhstania that resulted in mountain building in the southern and middle Urals has been described recently by Brown *et al.* (2006a), and its main events are summarized here. The external (palaeogeographic and magmatic) expressions of the orogeny, including the northern-to-polar and eastern (epi-Kazakhstanian) regions are emphasized in this synthesis.

By the Mid-Bashkirian, subduction had ceased and collisional processes between Laurussia and Kazakhstania began first as a formation of linear uplifts and basins, documented in the Southern Urals (Puchkov 2000). In the Late Bashkirian and Moscovian, widespread marine flysch were deposited in troughs separated by more slowly subsiding shelf zones and intensely eroded uplifted crustal blocks. As uplift continued, the basins contracted and inverted. By the Kasimovian time, the territory east of MUF was dominated by erosion and subaerial deposits. To the west of MUF, a deep-water foredeep trough was filled by easterly-derived flysch, prograding to the west (Puchkov 2000). A westerly prograding foreland fold-and-thrust belt was developed along the eastern margin of the foredeep, which deformed and uplifted flysch of its eastern limb. The diachroneity of these processes is documented by detailed studies of resedimented conodonts within these strata (Gorozhanina & Pazukhin 2007). In the Gzhelian–Sakmarian, the thrusting and crustal thickening created a hot crustal root in the Southern Urals, which resulted in generation of 305–290 Ma syn-collisional granites of the Main Granitic axis (Fig. 2), followed by 10–15 km of erosion (Fershtater *et al.* 2006). Crustal thickness may reach 65 km (the modern crust thickness of the East Uralian zone is up to 50 km; see below), similar to modern orogens. Tuff layers in deep-water sections of Gzhelian to Lower Kungurian preflysch and distal flysch of Preuralian foredeep may represent volcanic equivalents of this magmatic activity (Davydov *et al.* 2002).

Syn-collisional granite magmatism migrated northward, and is thought to be a manifestation of the oblique, diachronous character of collision (305–290 Ma for the southern Uralian granites, 265 Ma for the Kisegach massif, 250–255 Ma for Murzinka and Adui massifs of the Middle Urals, Fershtater *et al.* 2006). The idea of the transpressive character of orogenic deformation is supported by structural studies (e.g. Pliusnin 1966; Znamensky 2007). A change from thrust-dominated tectonics to sinistral transpression occurred in the Southern Urals (Znamensky 2007), explaining the K-rich concentric-zoned post-tectonic *c.* 283 Ma rift-related granite-monzonite massifs at the northern end of Magnitogorsk synclinorium (Ferstater *et al.* 2006) and the occurrence of *c.* 301–310 Ma lamproite dykes in the Southern Urals (Pribavkin *et al.* 2006).

Post-collisional granite magmatism in each region was followed by uplift and erosion, and the diachroneity of the magmatism is exemplified by the presence of the Late Permian marine sediments with Tethyan fauna in the Southern Urals that is coeval with granite magmatism in the Middle Urals (e.g. Chuvashov *et al.* 1984).

### An interlude: LIP formation and localized rifting

At the Permian–Triassic boundary, the waning effects of orogenesis were overprinted by the formation of the vast Uralo-Siberian LIP, extending from Taymyr in the north to the Kuznetsk and Turgay basins in the south and from the Tunguska basin in the east to the Urals in the west. Volcanism started locally with alkaline basalts and minor rhyolites. Ar–Ar data (Ivanov *et al.* 2005) suggest that the bulk of the volcanism initiated in Siberia at the Permian–Triassic boundary but probably continued for 22–26 Ma, with several short surges. Recent Ar–Ar dates for plagioclase from basalts in the Polar Urals (249.5 $\pm$ 0.7 Ma) and in the east of the Southern Urals (243.3 $\pm$ 0.6 Ma) (Reichov *et al.* 2007) support the simultaneous beginning of the LIP formation over a vast region followed by a more protracted period of reduced magmatism.

In contrast with eastern Siberia, the Early Triassic history of the Urals is dominated by considerable uplift, erosion and formation of thick coarse-grained alluvial to proluvial sediments that resemble orogenic molasse but are attributed to the effects of the Uralo–Siberian distributed rifting and LIP magmatism. Examples include the huge Triassic Koltogorsk–Urengoy graben of Western Siberia, the newly-identified Severososvinsky graben in the subsurface of the Cis-Polar Urals (Ivanov *et al.* 2004), and the eastern parts of the LIP (Kurenkov *et al.* 2002; Ryabov & Grib 2005).

## The Cimmerian orogeny

A short pulse of orogeny occurred at the end of the early Jurassic, and its effects differ along the strike of the Uralides. The Triassic deposits of the Southern Urals are affected by this orogeny only in the Trans-Uralian zone (Chelyabinsk and other graben-like depressions), where Upper Triassic and older deposits are deformed by thrusting (Rasulov 1982), followed by uplift and peneplanation during the Middle and Upper Jurassic, and deposition of Upper Cretaceous marine deposits.

In the Northern Urals, three 'grabens' (Mostovskoi, Volchansky, Bogoslovsk-Veselovsky) (Tuzhikova 1973) containing Upper Triassic coal-bearing sediments were complexly deformed. In the Polar Urals, the Triassic deposits of the foredeep and the Chernyshov and Chernov range are all deformed, and are unconformably overlain by Middle Jurassic strata. However, in the nearby Severososvinsky graben to the east, Triassic and Jurassic deposits are not deformed (Ivanov *et al.* 2004), attesting to the localized nature of Cimmerian orogenic events.

The Pay-Khoy and Novozemelsky ranges were formed in the Cimmerian (Korago *et al.* 1989; Yudin 1994). Cimmerian orogenesis is attributed to a large-scale intra-Pangaean strike-slip faulting, accompanied by block rotation, possibly reflecting lateral escape of Kazakhstania between Laurussia and Siberia and an immediate collision of the latter two (Fig. 1). According to palaeomagnetic data (Kazansky *et al.* 2004), Siberia rotated 30° clockwise between the Triassic and the Late Cretaceous.

## Peneplain formation

Rapid uplift and erosion of the Uralide orogen resulted in the formation of a Cretaceous-Palaeogene peneplain (Papulov 1974; Sigov 1969; Amon 2001), and by the Late Jurassic or Early Cretaceous, there was no topographic barrier dividing Europe and Siberia. Buried river-bed deposits along the eastern slope of the Southern and Middle Urals have a north-eastern direction, as revealed by a shallow prospecting drilling. For the Late Cretaceous and Eocene, the existence of short-lived straits connecting the European and Siberian marine basins is hypothesized.

Along the eastern slope of the Southern and Middle Urals, thin marine sediments occur only during maximal transgressions (in the Late Cretaceous and Middle Eocene); more generally, fluviatile deposits occur. To the north, Upper Jurassic and younger marine sediments occur adjacent to the eastern foothills of the modern Ural mountains. The difference between the southern and northern parts of the modern Urals (as expressed by better exposure of the eastern zones in the south), was probably inherited from this time.

## Neo-orogeny

In the Pliocene–Quaternary time, a modern chain of moderately high mountains was uplifted, forming a natural drainage divide between Europe and Asia. These mountains are formed in an intra-plate setting, having no precursor suture zone. Convergence is indicated by studies which show that maximum stresses are oriented perpendicular to, or at a high angle to, the strike of the belt and by the identification of a zone with anomalously low heat flow (Golovanova 2006). According to Mikhailov *et al.* (2002), mountain building is accompanied by westward-directed thrusting.

The timing of mountain building is controversial. Until recently, there was a consensus that orogenesis began in the Late Oligocene and continued into the Quaternary inclusive (Trifonov 1999; Rozhdestvensky & Zinyakhina 1997) and that ancient peneplains formed in the Triassic–Jurassic were preserved (Sigov 1969; Borisevich 1992).

On the contrary, Puchkov (2002*b*) pointed out that models proposing Late Oligocene uplift of the Urals are inconsistent with: (1) the occurrence of Miocene oligomictic quartz sands and sandstones (Yakhimovich & Andrianova 1959; Kozlov 1976), which indicate stable non-orogenic conditions of weathering, erosion and deposition in both foredeep terrane and the Urals itself. The first appearance of polymictic sediments, indicating more rapid uplift and erosion, is Late Pliocene in age (Verbitskaya 1964); (2) deep Miocene river incision is best explained by a drop of the Caspian sea level, Messinian crisis, as well documented in the Mediterranean (Milanovsky 1963), rather than by crustal uplift; (3) no well-documented Miocene–Early Pleistocene terraces occur in the river valleys of the Urals; (4) no cave deposits older than Middle-Upper Neopleistocene are found; and (5) the velocities of the modern uplift of the Urals surface are 5–7 mm $a^{-1}$, which is an order of magnitude faster than the time needed to build the Urals mountains since Oligocene.

On the other hand, fission-track (Seward et al. 1997; Glasmacher et al. 2002) and unpublished U/Th–He data show that the relief of the axial part of the Southern Urals was not completely stabilized by the Late Cretaceous. Puchkov & Danukalova (2004) demonstrated that the altitudes of the base of shallow marine Upper Cretaceous deposits increase progressively in the direction of the mountain ridge, disappearing at elevations of 500 m. Therefore no Triassic–Jurassic planation surfaces can be preserved in the modern surface. The depth of erosion since the Cretaceous is between 1000 and 2000 m (depending on the thermal gradient), and these numbers are several times greater than the previous estimates.

## The deep structure of the Urals

### The main milestones of the study

Fifteen regional deep seismic survey (DSS) profiles, made between 1961 and 1993 permit the definition of the Moho surface beneath the Urals and demonstrated its layered seismic structure and the anomalous character of its crust. In particular, these surveys suggest the presence of a crustal 'root' under the Tagil–Magnitogorsk zone and a complex compositional transition zone in the lower crust, with $V_p$ velocities between 7.2 and 7.8 km/s (Druzhinin et al. 1976). Puchkov & Svetlakova (1993) placed these results into a plate tectonic context for the first time, by interpreting a DSS profile in the Middle Urals as an indicator of the bi-vergent character of the Uralian orogen.

Reflection profiles made between 1964 and the early 1990s in the Magnitogorsk and Tagil zones (e.g. Menshikov et al. 1983; Sokolov 1992) revealed the inclined reflectors that define synclinoria and anticlinoria, and along-strike variations in the morphology of the Main Uralian fault. In the 1980s, state oil company surveys along the western slope of the Urals (e.g. Skripiy & Yunusov 1989;

Sobornov & Bushuev 1992), combined with drilling, helped to solve some structural problems in the Uralian foreland. In 1993, the commencement of the EUROPROBE Programme 'Uralides', involved acquisition and interpretation of seismic data along two regional profiles (the Southern and Middle Urals) and re-interpretation of some existing shorter profiles. A combined geological and multicomponent geophysical URSEIS-95 project in the Southern Urals (Berzin et al. 1996; Carbonell et al. 1996; Echtler et al. 1996; Knapp et al. 1996; Suleimanov 2006), included a c. 500 km-long seismic reflection line across most of the orogen at a latitude of Kraka and Gebyk massifs. The ESRU-SB profile, ultimately c. 440 km long, crossed the Middle Urals where the 'superdeep' SG-4 borehole (currently c. 5.5 km deep) is located (Kashubin et al. 2006, and references therein).

### The URSEIS profile (Fig. 15)

The interpretation given here is based on combined (vibroseis and explosion) seismic section along the geotraverse profile, after Suleimanov (2006), Spets-Geofizika. The coherency-filtered, depth-migrated vibroseis data by Tryggvason et al. (2001) were also used as an alternative source of information for the upper and middle crust. Along with generally accepted conclusions (Brown et al. 2008, and references therein), the following interpretations contain some latest original inferences of the author.

From 500 km (in the Preuralian foredeep), to the Main Uralian Fault at c. 275 km, the survey characterizes the structure of the foreland fold-and-thrust belt (Fig. 15). From 500–c. 420 km, subhorizontal, moderately coherent reflectivity in the upper 5 km corresponds to weakly deformed Palaeozoic foreland basin (foredeep) and platform margin shelf rocks of Ordovician–Lower Permian age (Brown et al. 2006b). Below this, to approximately 20 km depth, strongly coherent, subhorizontal reflectivity is interpreted to represent undeformed Meso- and Neoproterozoic strata of the SSE prolongation of the Kama-Belsk aulacogen. The base of the reflectivity here is thought to represent the unconformity between undeformed and low-metamorphic Mesoproterozoic strata and the non-reflective Archaean-Palaeoproterozoic crystalline basement (Dianconescu et al. 1998; Echtler et al. 1996). The sedimentary prism, almost 20 km thick, has a convex lens-like shape, consistent with the interpretation that the prism represents an inverted aulacogen. Beneath the crystalline basement, at c. 430 km the Moho is cut by Makarovo normal (?) fault, with up to 5 km of amplitude – probably related to the rift nature of the aulacogen (Fig. 15).

To the east, the upper and middle crust has weak, gently east-dipping reflectivity that, between

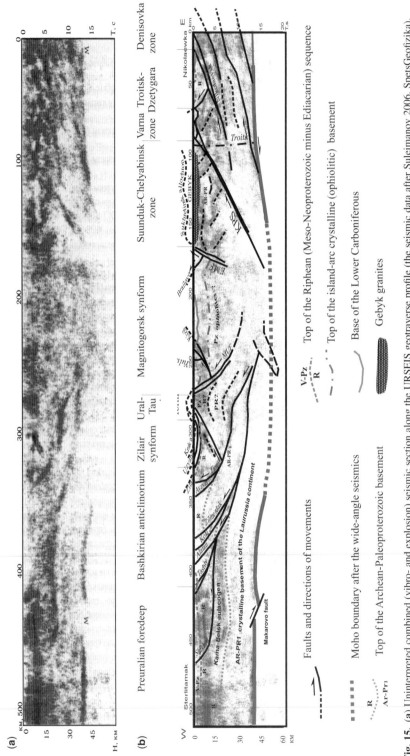

**Fig. 15.** (**a**) Uninterpreted combined (vibro- and explosion) seismic section along the URSEIS geotraverse profile (the seismic data after Suleimanov 2006, SpetsGeofizika). (**b**) Geological interpretation, overlain on the profile. See Figure 4 for location.

420 km and the MUF is concave downward. This reflectivity is associated with the Precambrian rocks in the Bashkirian Anticlinorium which, in its eastern part, was deformed during the Timanide orogeny (Puchkov 2000). The base of the reflectivity is usually interpreted to be the basal detachment contact between Mesoproterozoic strata and the Archaean–Palaeoproterozoic crystalline basement. However, according to structural studies, large anticlines of the central and eastern part of the Bashkirian anticlinoria have detached blocks of the crystalline basement beneath them, close to the surface. Taratash anticline in the north exposes such Precambrian core in the surface.

The lower crust beneath the foreland fold-and-thrust belt is weakly- to non-reflective, though the Moho boundary can be traced by explosion seismic data further to the east, towards the MUF. Close to the MUF, the deep-seated Uraltau antiform is clearly imaged and the MUF fault is traced as a gently concave structure, mainly by a loss of reflections from the Precambrian rocks and assuming that the base of the island-arc complex in the hanging wall of MUF is transparent. The Zilair synform and Uraltau antiform immediately to the west of the MUF form a dynamic couple, with the antiform making a tectonic wedge downthrusted to the west under the antiform.

From the MUF to c. 180 km, the Magnitogorsk arc is almost non-reflective in the upper crust, though the east-vergent Kizil thrust is clearly imaged in the coherency-filtered vibroseis data (Tryggvason et al. 2001) and its interpretation is supported by deep drilling and recent short seismic reflection profiles. The middle and lower crust is relatively transparent. The contact between the Magnitogorsk arc and the East Uralian Zone at c. 180 km (the East Magnitogorsk Fault and suture zone) is imaged by a sharp change from almost transparent crust in the west to coherent, highly reflective, middle crust to the east. In the East Uralian Zone, from c. 180–100 km, the upper crust is nearly transparent down to about 8 km, corresponding to the Gebyk granite. Below this, a series of short east-dipping and subhorizontal reflectors are descending into the middle crust. The lower crust is almost transparent or semi-transparent, except in the east, where a zone with strong west-dipping reflectivity extends downward and westward from the Trans-Uralian Zone (a continuation of Kartaly reflections; see below).

The crust of the Trans-Uralian Zone is imaged as west-dipping, strongly coherent reflectivity called the Kartaly Reflection Sequence (KRS, Fig. 15), which merges with the Moho in a system of thrusts; in this region the Moho appears as a near-horizontal detachment fault. The boundary between the East Uralian and Trans-Uralian zones is thought to be a regional fault called Kartaly or (wrongly) Troitsk fault located immediately to the east of Dzhabyk massif and traced in the SSW and NNE directions where it is interpreted as a wrench fault of a considerable amplitude.

In the western and eastern parts of the profile, the Moho is imaged in the URSEIS combined (vibro- and dynamite) reflection data to a depth of c. 50 km but cannot be traced in the deeper, central portion of the profile. The Moho has been determined from wide-angle data to occur at a maximum depth of 55 km (Carbonell et al. 1998). Although there is some bias between the wide-angle and CDP data, a cloudy reflection under this depth at c. 250 km, can be tentatively interpreted as a wedge of the lower crust protruding into the mantle (compare with a much better imaged wedge of the lower crust in the ESRU-SB profile, see below) (Fig. 16).

## The ESRU-SB profile (Fig. 16)

The latest interpretation of crustal structure of the Middle Urals at latitude 56–62° based on seismic reflection data obtained by yearly installments since 1993 (ESRU-SB profile) was given recently by Kashubin et al. (2006), Rybalka et al. (2006) and Brown et al. (2008). From −100 km in the Pre-ruralian foredeep to c. −25 km in the east, the upper crust has a flat-lying reflectivity, interpreted to represent an undeformed foredeep orogenic basin and platformal deposits (up to c. −65 km). In contrast, to the east, the steeply east-dipping shallow reflectivity of the 'thin-skinned' fold-and-thrust foreland occurs (Fig. 16). Both undeformed and steeply-dipping reflections are underlain by a gently east-dipping zone of reflections at a depth of c. 5–8 km that is interpreted as a low-angle unconformity surface between the platform cover of Ediacarian and Palaeozoic age and the older Neoproterozoic (Upper Riphean), that is transformed in the east into a basal detachment of the fold-and-thrust belt (Brown et al. 2006c). At c. −25 km this reflectivity is abruptly truncated by a series of steep east-dipping concave reflectors corresponding probably to listric-like faults, traced into the middle crust to a depth of 25–30 km. This type of reflectivity persists from a distance mark of 25 km to the MUF zone. The zone is characterized by several pronounced closely-spaced reflectors dipping to the east at angles of 45°–60° between 0 and 10 km, imaging an imbrication zone of the strongly deformed margin of Laurussia continent. From −20–10 km, the steeply east-dipping reflectors represent 'thick-skinned' deformation in the Precambrian-cored Kvarkush anticlinorium and Early Palaeozoic hanging wall.

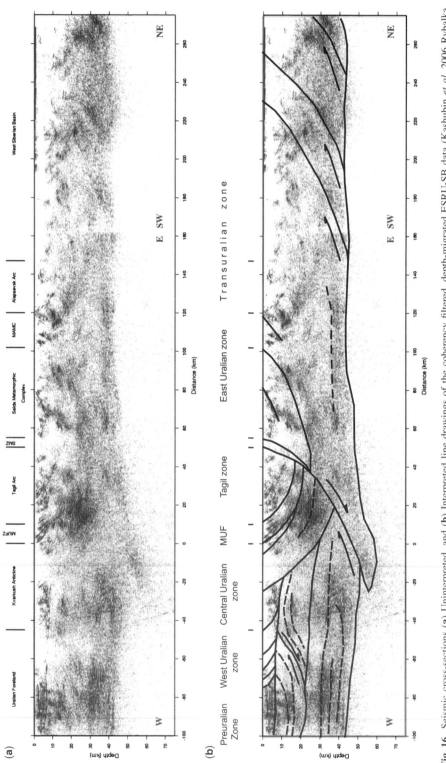

**Fig. 16.** Seismic cross-sections (**a**) Uninterpreted, and (**b**) Interpreted line drawings of the coherency filtered, depth-migrated ESRU-SB data (Kashubin *et al.* 2006 Rybalka *et al.* 2006). See Figure 4 for location. Abbreviations in the Figure A: MUFZ, Main Uralian Fault zone; SMZ, Serov-Mauk Fault zone; MAMC, Murzinka-Aduy metamorphic complex.

Below the undeformed foreland basin and the basal detachment of the fold-and-thrust belt, the middle crust exhibits wave-like concave to convex reflectivity down to approximately 25 km depth. From 25 km to c. 42 km depth, the lower crust is characterized by more coherent and strong, subhorizontal reflectivity. The middle crustal reflectivity probably images the Neoproterozoic and Mesoproterozoic sedimentary rocks, and the lower crustal reflectivity images Archaean–Palaeoproterozoic crystalline basement rocks of the Laurussian margin, though there is a striking difference between the reflectance character of the crystalline basement here and in the URSEIS. The character of crustal reflectivity in the deep part of the section suggests that it is unaffected by the Uralide deformation which also makes a difference between the profiles (Brown et al. 2006c; Kashubin et al. 2006). On the other hand, the reflectivity pattern of the middle crust suggests that at −100 to c. −20 km, the 15 or more kilometre-thick Meso- and Neoproterozoic strata form a large synform that is underthrust by an antiformal tectonic wedge composed of the rocks of the same age. The structure is characteristic of a Timanian foreland deformation, but also resembles the wedge-like relationships between the Uraltau antiform and Zilair synform imaged by URSEIS, though their relationships had been formed by the Uralide orogeny.

The Moho surface is traced here as a gently east-dipping boundary between highly reflective lower crust and almost transparent mantle at a depth of 42–45 km. From about c. 10–50 km, the upper crust of the Tagil arc is imaged as an open synform, thrust to the west over the Meso-Neoproterozoic rocks of Kvarkush aniclinorium. The Tagil synform is asymmetric, and its eastern limb is limited by a serpentinite mélange of the Serov-Mauk fault zone, separating the synform from the Salda metamorphic complex. The mélange zone is transparent and its western boundary can be traced along abrupt truncations of Tagil reflectors, suggesting a westerly dip of the zone at an angle of 60°. The zone can be traced tentatively into the lower crust along weak and diffuse reflectors, changing the steep western dip of the zone to a gentler 30° close to Moho surface.

The Salda metamorphic complex of probable island-arc nature (Rybalka et al. 2006), situated between 55 and 103 km, is characterized by a series of west-dipping reflections, which can be traced under the Tagil synform and Central Uralian zone together with the Serov-Mauk fault. This portion of the crust, correlated with the Salda zone, is wedge-like, protruding down into the mantle under the western slope of the Urals to a depth of 60 km.

The next zone to the east, Murzinka-Adui zone (103–120 km), is represented at the surface by Neoproterozoic metamorphic rocks and Permian granites, and together with the Salda zone belongs to the East Uralian megazone. The character of its reflectivity in the upper crust is incoherent and patchy, and does not permit recognition of its detailed structure. Further east, from 120–180 km within the Trans-Uralian zone, the upper crustal structure is more difficult to interpret owing to both poor surface exposure and the almost complete absence of coherent reflectors in the upper 10-km of the profile.

From c. 180 km to the end of the profile (260 km), the platformal Cretaceous and Cenozoic strata of the West Siberian Basin appear to be characterized by a zone of good subhorizontal reflectivity which thickens to the east up to 1.5 km at 260 km. The details of the profile imaged by Rybalka et al. (2006) show a relief of the Palaeozoic–Cretaceous unconformity surface, which is strongly uneven probably due to pre-Cretaceous grabens, flexures and river-bed incisions. Under the platformal cover, the structure of the Trans-Uralian zone reveals a series of west-dipping reflectors, merging with the Moho at a depth of c. 40 km, in a manner similar to that seen in the KRS of the URSEIS profile, although not as bright.

In general, the Moho is very well defined along the whole profile as a sharp boundary between highly reflective lower crust and an almost transparent mantle. The crust thickens from c. 42–43 km in both the west and east to nearly 60 km beneath the Central Uralian zone. The above-mentioned wedge of the lower crust protruding into the mantle gives here the impression that the eastern limb of the orogen is thrust under the former Laurussian lower crust and Moho.

## Discussion

The development of the Uralides in the Palaeozoic preserves many of the characteristics of a full Wilson cycle. However, if the concept of such a cycle is restricted to a classical 'accordion-type' development, wherein the continent that rifted away returns back to collide (as it was in the case of Rheic ocean), then the development of the Uralides does not conform with the idea. According to palaeomagnetic data and geodynamic reconstructions (e.g. Puchkov 2000; Kurenkov et al. 2002; Svyazhina et al. 2003; Levashova et al. 2003), between Ordovician and Early Permian time, the microcontinents of the East Uralian zone and Kokchetav block of Kazakhstania were transported at least 2000 km from north to south parallel to the margin of Baltica. Kazakhstania as a whole intervened between Siberia and Baltica in the Devonian, and Siberia rotated clockwise at 90° – probably because of the arrival of Kazakhstania. Kazakhstania was squeezed between accreted margins of Siberia and Laurussia, forming a pronounced horseshoe-like orocline (Fig. 1). It looks more like a rock'n'roll dance than an accordion-like motion. On the other hand, only a minority of orogens belongs to the regular 'accordion' type, and therefore we prefer to follow to a more liberal understanding of Wilson cycle.

The rift processes at the beginning of the cycle are demonstrated by a profound difference between the strikes of Timanide structures and the Main Uralian Fault to the north of the Poliud Range (Fig. 7). The accompanying sedimentary deposits are characterized by irregularly distributed coarse-grained polymictic to arkosic sediments unconformably overlying the basement and accompanied by subalkaline basaltic volcanism. Similar formations on the sialic blocks of the East Uralian and Transuralian zones are c. 15–20 Ma younger than those in the Western zones, indicating that the East Uralian microcontinent was not previously detached from the same place where it finally docked.

The suggested Early Palaeozoic plume magmatism described above is considerably younger than the rifting event, and is only indirectly connected, possibly in an analogous manner to the relationship between the transverse chain of the Cretaceous Monteregian alkaline with carbonatite intrusions of eastern North America and the Mesozoic rifts along the margin of the Atlantic Ocean.

Widespread development of ophiolite complexes, almost unprecedented among the Palaeozoic or earlier foldbelts, is one of the most important characteristics of the Uralides. Ophiolites appear to represent anomalous portions of the oceanic tract (Aden and Red sea-type, supra-subduction zone basalts, Lanzo-type mantle blocks) whereas most typical MORB appears to have been almost eliminated by subduction. Island-arc complexes are also widely spread in the Uralides, and subsequent uplift and erosion offers a rare possibility of studying the deeper structure of an island arc, than is available for study in modern island arcs. The internal structure of the arcs exhibits a moderate strain (Brown et al. 2001). According to Alvarez-Marron (2002) the Uralides 'may be seen as a factory for "making" new continental crust in contrast to the Variscides which is a factory for "recycling" existing continental crust'.

One of the possible explanations for the exceptionally good preservation of oceanic complexes in the Uralides is the low rigidity of the Kazakhstanian plate, which became continental crust only in the Silurian (Puchkov 1996a). The deep-seated

deformation of the whole crust with the Moho as a detachment, in the young, eastern limb of the Uralides (see above), absorbed a considerable part of strain. The preservation of ophiolites may depend on strain. In zones of higher strain, ophiolites are squeezed from sutures as allochthonous sheets aided by the formation of rheologically weak serpentinites which may also have acted as a lubricant. This interpretation is supported by experiments that demonstrate the rheological weakness of serpentinite (e.g. Escartin et al. 1997; Hilairet et al. 2007) and by structural studies in the foreland fold-and-thrust belt of the Southern and Middle Urals, where shortening deduced from balanced geological sections is anomalously low (14–17%, Brown et al. 1997). Shortening increases to the north, and in the Mikhailovsk and Serebryansk profiles it is c. 30% (calculated after Brown et al. 2006c). In the Cis-Polar and Polar Urals, however, shortening can be still much greater, judging by the upper section of figure 4 in Puchkov (1997); see also Yudin (1994). This may be explained by the wedging-out of Kazakstania to the north, where two rigid cratons, Laurussia and Siberia, come into contact.

Modern analogies to the arc-continent collision in the Urals include the Indonesian, Taiwan, Tyrrhenian and Greater Antilles arcs. The evolution is similar to that proposed for the Taconian arc in the Appalachian orogen, and the mid-Devonian arc collision with Baltica along the margins of the Rheic Ocean.

Another striking feature of the Uralides, is a long duration and recurrence of orogenic events – from the Devonian until the Early Jurassic, which is comparable to the duration of Appalachian orogenic activity. The oblique, transpressive character and the diachroneity of the continent–continent collision in the Urals is similar to that of the Alleghenian orogeny (Engelder 2007). But the absence of Caledonian collisions, from the Ordovician to the Middle Devonian, is another striking feature of the Uralides.

The characteristics of two regional profiles transecting the orogen, confirm the bi-vergent symmetry of it. Alternatively, uniformly vergent orogens may be part of a bigger, bi-vergent one, that has been dismembered and dispersed by subsequent tectonic events (e.g. Greenland and Scandinavian Caledonides) or its lacking limb overlain by a younger orogen (e.g. a Variscan basement of Alps).

In the western Uralides, where the foreland is cratonic (i.e. ancient and characterized by a thick and rigid lithosphere), the crystalline basement and Moho surface are not affected, or only partly affected, by the main Uralide orogenesis. In the eastern Uralides, however, the Moho is interpreted as a detachment, and the crust is deformed to great depths, an interpretation which supports the idea of a comparative weakness of the Palaeozoic crust and lithosphere as a whole.

In the west, the regional and local profiles show an abrupt transition from 'thin-skinned' – to thick-skinned tectonics along a sharp ramp due to an abrupt change in the plasticity of the rocks. In the east we do not see any 'thin-skinned' tectonics at all. This may reflect the incompleteness of the eastern part of the profiles (URSEIS had been stopped at a state border with Kazakhstan and ESRU – in swamps of Siberia) (Fig. 4). On the other hand, Late Palaeozoic orogenic processes reached much farther to the east than the initial boundary between Uralides and Kazakhstanides (Fig. 1). In the Central Kazakhstanian Caledonides, orogenic processes are documented by Permian deformation of different styles, voluminous Late Palaeozoic syn-orogenic granite and deposition of the contemporaneous molasse in huge intermontane basins such as Chu-Sarysu and Teniz (YUGGEO 2002). Where the Upper Palaeozoic epi-Caledonian sedimentary cover is preserved, as in the Greater Karatau, the east-vergent 'thin-skinned' tectonics was developed during the Upper Carboniferous and Permian (Alekseiev 2008).

## Concluding remarks

Extremely good preservation of oceanic and island-arc complexes and low degree of shortening in the foreland belt are unprecedented in the Palaeozoic orogens and give the Uralides a real individuality. Many other features of the Urals are also rare, such as its well-preserved bi-lateral structure, island-arc related platinum-bearing belt, demonstrative arc-continent collision, diachroneity of collisions, a combination of orogenic and rift-related magmatism in a single stage of transpressive deformation of the lithosphere, and the preservation of a heavy, relatively 'cold', isostatically equilibrated root. On a plate scale, however, the history of the Uralides follows the main stages of a Wilson cycle, modified somewhat by episodes of plume-related tectonics and magmatism in the Early Palaeozoic and Triassic.

Fieldwork campaigns, workshops and informal discussions in more than a decade of co-operation with prominent European and American geologists during the activity of the *EUROPROBE* Commission in the Urals have considerably stimulated the work of the author. The author acknowledges also the importance of his participation in the *IGCP-453* '*Uniformitarianism revisited: a comparison between modern and ancient orogens*', which ultimately inspired him to write this paper. The financial support of the Department of Earth's Sciences of the Russian Academy of Sciences, *Program No. 10 The Central Asian mobile belt: geodynamics and stages of the Earth's crust formation* must be also highly appreciated. The last, but not the least of my expression of gratitude is to the reviewers, V. Ramos and R. Ernst and to B. Murphy, the editor, who helped greatly at the last stages of the work.

## References

AGARD, P. & LEMOINE, M. 2005. *Faces of the Alps: Structure and Geodynamic Evolution*. Commission of the geological map of the world, Paris.

ALEKSEIEV, D. V. 2008. Kinematics of the Variscan Ural and Tian-Shan junction area. *Geologie de la France*, **2**, 75.

ALL-RUSSIAN STRATIGRAPHIC COMMITTEE. 1993. *Stratigraphic Schemes of the Urals: Precambrian and Paleozoic*. Ekaterinburg, 151 scheme (in Russian).

ALVAREZ-MARRON, J. 2002. Tectonic processes during collisional orogenesis from comparison of the Southern Urals with the Central Variscides. *In*: BROWN, D., JUHLIN, C. & PUCHKOV, V. (eds) *Mountain Building in the Uralides: Pangea to Present*. American Geophysical Union, Geophysical Monograph, **132**, 83–100.

AMON, E. O. 2001. The marine equatorial basin of the Uralian region in the Mid- and Late Cretaceous epoch. *The Geology and Geophysics*, **42**, 471–483 (in Russian).

ANDREICHEV, V. L., RONKIN, YU. K., LEPIKHINA, O. P. & LITVINENKO, A. F. 2005. *Rb–Sr and Sm–Nd Isotopic-geochronometric Systems in Basalts of the Polar Cis-Urals*. Syktyvkar, Geoprint (in Russian).

ARTIUSHKOVA, O. V. & MASLOV, V. A. 2003. Lower Devonian (pre-Upper Emsian) deposits of the Magnitogorsk megazone. *Geological Sbornik*, IG USC RAS, **3**, 80–87 (in Russian).

BATANOVA, V. G., BRUEGMANN, G., BELOUSOV, I. A., SAVELIEVA, G. N. & SOBOLEV, A. V. 2007. The processes of melting and migration of melts in mantle based on a study of highly siderophile elements and their isotopes. *In*: GALIMOV, E. M. (ed.) *XVIII Vinogradov Symposium on Geochemistry of Isotopes*. V., GEOKHI, 40–41 (in Russian).

BEA, F., FERSHTATER, G. & MONTERO, P. 2002. Granitoids of the Urals: implications for the evolution of the orogen. *In*: BROWN, D., JUHLIN, C. & PUCHKOV, V. (eds) *Mountain Building in the Uralides: Pangea to Present*. American Geophysical Union, Geophysical Monograph, **132**, 211–232.

BEANE, R. J. & CONNELLY, J. H. 2000. $^{40}Ar-^{39}Ar$, U–Pb and Sm–Nd constraints on the timing of metamorphic events in the Maksyutov Complex, Southern Ural Mountains. *Journal of the Geological Society, London*, **157**, 811–822.

BELL, K. 2001. Carbonatites: relationships to mantle-plume activity. *In*: ERNST, R. E. & BUCHAN, K. L. (eds) *Mantle Plumes: Their Identification Through Time*. Geological Society of America Special Paper, **352**, 267–290.

BERZIN, R., ONCKEN, O., KNAPP, J. H., PEREZ-ESTAUN, A., HISMATULIN, T., YUNUSOV, N. & LIPILIN, A. 1996. Orogenic evolution of the Ural Mountains: results from an integrated seismic experiment. *Science*, **274**, 220–221.

BEUTLER, G. 1979. Verbreitung und Charakter der altkimmerischen Hauptdiskordanz in Mitteleuropa. *Zeitschrift für Geologische Wissenschaften*, **5**, 617–632.

BISKE, YU. S. 2006. *The Foldbelts of the Northern Eurasia. Tyanshan Foldbelt.* St. Petersburg University (in Russian).

BOCHKAREV, V. V. & YAZEVA, R. G. 2000. *Subalkaline Magmatism of the Urals.* Ekaterinburg (in Russian).

BORISEVICH, D. V. 1992. Neotectonics of the Urals. *Geotectonics*, **26**, 41–47 (in Russian).

BOROZDINA, G. N. 2006. *The history of geological development of the Ragil megazone of the Middle and Southern part of the Middle Urals during the Lower Paleozoic.* Candidate of Science thesis, Ekaterinburg, IGG Uralian Branch of Russian Academy (in Russian).

BOROZDINA, G. N., IVANOV, K. S., NASSEDKINA, V. A. & SNIGIREVA, M. P. 2004. On the age and volume of the Shemur formation of the Tagil megazone. *Ezhegodnik-2003*, Institute of Geology and Geophysics, Sverdlovsk, 10–13 (in Russian).

BORTNIKOV, P. S., SAVELIEVA, G. N., MATUKOV, D. I., SERGEEV, S. A., BAREZHNAYA, N. G., LEPEKHINA, E. N. & ANTONOV, A. V. 2005. The age of a zircon from plagiogranites and gabbro after SHRIMP data: pleistocene intrusion in the rift valley of the Mid-Atlantic Ridge, $5°30'6''–5°32'4''$N. *Doklady of Russian Academy of Science, Geology*, **404**, 94–99 (in Russian).

BROWN, D. & PUCHKOV, V. 2004. *Arc-continent Collision in the Uralides.* An IGCP-453 'Uniformitarianism revisited: a comparison between modern and ancient orogens' conference and fieldtrip. Design Poligraph Service, Ufa.

BROWN, D., ALVAREZ-MARRON, J., PEREZ-ESTAUN, A., GOROZHANINA, Y., BARYSHEV, V. & PUCHKOV, V. 1997. Geometric and kinematic evolution of the foreland thrust and fold belt in the southern Urals. *Tectonics*, **16**, 551–562.

BROWN, D., ALVAREZ-MARRON, J., PEREZ-ESTAUN, A., PUCHKOV, V., AYARZA, P. & GOROZHANINA, Y. 2001. Structure and evolution of the Magnitogorsk forearc basin: identifying upper crustal processes during arc-continent collision in the Southern Urals. *Tectonics*, **20**, 364–375.

BROWN, D., PUCHKOV, V. N., ALVAREZ-MARRON, J. & PEREZ-ESTAUN, A. 2006a. Tectonic processes in the Southern and Middle Urals: an overview. *In*: GEE, D. G. & STEPHENSON, R. A. (eds) *European Lithosphere Dynamics*. Geological Society, London, Memoirs, **32**, 407–419.

BROWN, D., SPADEA, P. ET AL. 2006b. Arc-continent collision in the Southern Urals. *Earth-Science Reviews*, **79**, 261–287.

BROWN, D., JUHLIN, C., TRYGGVASON, A., FRIBERG, M., PUCHKOV, V., RYBALKA, A. & PETROV, G. 2006c. Structural architecture of the southern and middle Urals foreland from reflection seismic profiles. *Tectonics*, **25**, 1–12.

BROWN, D., JUHLIN, C. ET AL. 2008. Mountain building processes during continent-continent collision in the Uralides. *Earth-Science Reviews*, **89**, 177–195.

BURNS, L. E. 1985. The Border Ranges ultramafic and mafic complex, south-central Alaska: cumulate fractionates of island-arc volcanics. *Canadian Journal of Earth Sciences*, **22**, 1020–1038.

BURTMAN, V. S. 2006. *Tien Shan and High Asia. Tectonics and Geodynamics in Paleozoic.* Moscow, GEOS (in Russian).

CARBONELL, R., PEREZ-ESTAÚN, A. ET AL. 1996. A crustal root beneath the Urals: Wide-angle seismic evidence. *Science*, **274**, 222–224.

CARBONELL, R., LECERF, D., ITZIN, M., GALLART, J. & BROWN, D. 1998. Mapping the Moho beneath the Southern Urals. *Geophysical Research Letters*, **25**, 4229–4233.

CHUVASHOV, B. I., IVANOVA, R. M. & KOLCHINA, A. N. 1984. *The Upper Paleozoic of the Eastern Slope of the Urals.* Uralian Scientific Centre, Sverdlovsk (in Russian).

COBBOLD, P. R., MEISLING, K. E. & MOUNT, V. S. 2001. Reactivation of obliquely rifted margin, Campos and Santos basins, southeastern Brazil. *AAPG Bulletin*, **85**, 1925–1944.

COCKS, L. R. M. & TORSVIK, T. H. 2006. European geography in a global context from the Vendian to the end of Paleozoic. *In*: GEE, D. G. & STEPHENSON, R. A. (eds) *European Lithosphere Dynamics*. Geological Society, London, Memoirs, **32**, 83–95.

DALZIEL, I. 1992. On the organization of American plates in the Neoproterozoic and the breakout of Laurentia. *GSA Today*, **11**, 237–241.

DAVYDOV, V. I., CHERNYKH, V. V., CHUVASHOV, B. I., NORTHRUP, K. J. & SNYDER, V. S. 2002. Volcanic tuff layers in the Upper Paleozoic of the Southern Urals and prospects of creation of exactly calibrated time scale of the Carboniferous. *In*: CHUVASHOV, B. I. (ed.) *Stratigraphy and Paleogeography of the Carboniferous of Eurasia*. Ekaterinburg, Institute of Geology and Geochemistry, 112–123 (in Russian).

DIANCONESCU, C. C., KNAPP, J. H., BROWN, L. D., STEER, D. N. & STILLER, M. 1998. Precambrian Moho offset and tectonic stability of the East European platform from the URSEIS deep seismic profile. *Geology*, **26**, 211–214.

DIDENKO, A. H., KURENKOV, S. A. ET AL. 2001. *Tectonic History of the Polar Urals*. M. Nauka (in Russian).

DOBRETSOV, N. L., KIRDYASHKIN, A. G. & KIRDYASHKIN, A. 2001. *The Deep Geodynamics.* Novosibirsk, GEO (in Russian).

DRUZHININ, V. S., RYBALKA, V. M. & SOBOLEV, I. D. 1976. *The Connection of the Tectonics and Magmatism with the Deep Structure after DSS Data.* Sverdlovsk (in Russian).

ECHTLER, H. P., STILLER, M. ET AL. 1996. Preserved collisional crustal architecture of the Southern Urals – Vibroseis CMP-profiling. *Science*, **274**, 224–226.

EFIMOV, A. A., EFIMOVA, L. P. & MAEGOV, V. I. 1993. Tectonics of the Platinum-bearing belt of the Urals: relations of complexes and mechanism of structure formation. *Geotectonics*, **3**, 4–46 (in Russian).

ENGELDER, T. 2007. The Alleghanian orogeny of the Appalachian Mountains. *Geologie de la France*, **2**, 94.

ESCARTIN, J., HIRTH, G. & EVANS, B. 1997. Nondilatant brittle deformation of serpentinites: implications for Mohr-Coulomb theory and the strength of faults. *Journal of Geophysical Research*, **102**, B2, 2897.

FERSHTATER, G. B., BEA, F. & MONTERO, P. 2006. Granitoids. *In*: MOROZOV, A. F. (ed.) *Structure and Dynamics of Lithosphere of the Eastern*

*Europe, Issue 2. The Results of the EUROPROBE Research.* Moscow, GEOKART, GEOS, 449–461 (in Russian).

FRANKE, W. 2000. The Mid-European segment of the Variscides: tectonostratigraphic units, terrane boundaries and plate tectonic evolution. *In*: FRANKE, W., HAAK, V., ONCKEN, O. & TANNER, D. (eds) *Orogenic Processes: Quantification and Modelling in the Variscan Belt.* Geological Society, London, Special Publications, **179**, 35–61.

GAGGERO, L., SPADEA, P. & CORTESOGNO, L. 1997. Geochemical investigation of the igneous rocks from the Nurali ophiolite melange zone, Southern Urals. *Tectonophysics*, **276**, 139–161.

GEE, D. G., BOGOLEPOVA, O. K. & LORENZ, H. 2006. The Timanide, Caledonide and Uralide orogen in the Eurasian high Arctic, and relationships to the palaeocontinents Laurentia, Baltica and Siberia European Lithosphere Dynamics. *In*: GEE, D. G. & STEPHENSON, R. A. (eds) *European Lithosphere Dynamics.* Geological Society, London, Memoirs, **32**, 507–520.

GEOFFROY, L. 2005. Volcanic passive margins. *Comptes Rendu Geoscience*, **337**, 1395–1408.

GLASMACHER, U. A., WAGNER, G. A. & PUCHKOV, V. N. 2002. Thermotectonic evolution of the western fold-and-thrust belt, southern Urals, Russia, as revealed by apatite fission-track data. *Tectonophysics*, **354**, 25–48.

GLODNY, J., AUSTRHEIM, H., MONTERO, P. & RUSIN, A. 1999. The Marun-Keu Metamorphic Complex, Polar Urals, Russia Protolith ages, Eclogite facies fluid-rock interaction, and Exhumation History. *EUG–10 Abstracts*, Cambridge Publications, 80.

GLODNY, J., BINGEN, B., AUSTRHEIM, H., MOLINA, J. F. & RUSIN, A. 2002. Precise eclogitization ages deduced from Rb/Sr mineral systematics: the Maksyutov Complex, Southern Urals, Russia. *Geochimica et Cosmochimica Acta*, **66**, 1221–1235.

GLODNY, J., AUSTERHEM, H., MOLINA, J. F., RUSIN, A. & SEWARD, D. 2003. Rb–Sr record of fluid-rock interaction in eclogites: the Marun-Keu complex, Polar Urals, Russia. *Geochimica et Cosmochimica Acta*, **67**, 4353–4371.

GLODNY, J., PEASE, V. L., MONTERO, P., AUSTERHEIM, H. & RUSIN, A. I. 2004. Protolith ages of eclogites, Marun-Keu Complex, Polar Urals, Russia: implications for the pre- and early Uralian evolution of the northeastern European continental margin. *In*: GEE, D. G. & PEASE, V. L. (eds) *The Neoproterozoic Timanide Orogen of Eastern Baltica.* Geological Society, London, Memoirs, **30**, 87–105.

GOLOVANOVA, I. V. 2006. *The Thermal Field of the Southern Urals.* M. Nauka (in Russian).

GOROZHANINA, E. N. & PAZUKHIN, V. N. 2007. The stages of activization of geodynamic processes in the Late Devonian-Mid-Carboniferous time at the western limb of the Zilair synclinorium and their dating by conodonts. *Geolichesky Sbornik*, **6**, Ufa, Institute of Geology, 55–64 (in Russian).

GRADSTEIN, F. M., OGG, J. G., SMITH, A. G., BLEEKER, W. & LOURENS, L. J. 2004. A new Geologic time scale with special reference to Precambrian and Neogene. *Episodes*, **27**, 83–100.

GURSKAYA, L. I. & SMELOVA, L. V. 2003. Platinum-metal ore formation and structure of Syum-Keu

massif (Polar Urals). *Geology of Ore deposits*, **45**, 353–371 (in Russian).

HETZEL, R. & ROMER, R. L. 1999. U–Pb dating of the Verkniy Ufaley intrusion, middle Urals, Russia: a minimum age for subduction and amphibolite facies overprintof the East European continental margin. *Geological Magazine*, **136**, 593–597.

HETZEL, R. & ROMER, R. L. 2000. A moderate exhumation rate for the high-pressure Maksyutov Complex, Southern Urals, Russia. *Geological Journal*, **35**, 327–344.

HILAIRET, N., REYNARD, B., WANG, YANBIN, DANIEL, I., MERKEL, S., NISHIYAMA, N. & PETITGIRARD, S. 2007. High-pressure creep of serpentine, interseismic deformation, and initiation of subduction. *Science*, **318**(5858), 1910–1913.

INTERNATIONAL COMMITTEE OF TECTONIC MAPS. 1982. *Tectonics of Europe and adjacent areas. Variscides, epi-Paleozoic platforms and Alpides. Explanatory note to the International tecnonic map of Europe and adjacent areas, scale 1:2 5000 000.* M. Nauka Publishing House.

IVANOV, A. V., RASSKAZOV, S. V., FEOKTISTOV, G. D., HUAIYU, H. E. & BOVEN, A. 2005. $^{40}Ar–^{39}Ar$ dating of Usol'skii sill in the south-eastern Siberian Traps Large Igneous Province: evidence for long-lived magmatism. *Terra Nova*, **17**, 203–208.

IVANOV, K. S. 2001. The evaluation of paleovelocities of subduction and collision in the origin of the Urals. *Transactions (Doklady) of the Russian Academy of Sciences/Earth Science Section*, **377**, 231–234 (in Russian).

IVANOV, K. S., KARSTEN, L. A. & MALUSKI, G. 2000. The first data on the age of subductional (eclogite-glaucophane) metamorphism in the Polar Urals. *In*: KOROTEEV, V., IVANOV, K. & BOCHKAREV, V. (eds) *Paleozones of Subduction: Nectonics, Magmatism, Metamorphism, Sedimentology.* Ekaterinburg, Urals Branch of Russian Academy of Sciences, 121–128 (in Russian).

IVANOV, K. S., KOROTEEV, V. A. ET AL. 2004. The structure of a conjugation zone between Cis-Urals and West Siberian oil-and-gas-bearing basin. *Lithosphere*, **2**, 108–124 (in Russian).

IVANOV, K. S., SHMELEV, V. R., RONKIN, YU. L., SAVELIEVA, G. N. & PUCHKOV, V. N. 2006. Zonal gabbro-ultramafic complexes. *In*: MOROZOV, A. F. (ed.) *Structure and Dynamics of Lithosphere of the Eastern Europe, Issue 2. The Results of the EUROPROBE Research.* Moscow, GEOKART, GEOS, 437–445 (in Russian).

JONAS, P. 2004. Tectonostratigraphy of oceanic crustal terrains hosting serpentinite-associated massive sulphide deposits in the Main Uralian Fault zone (South Urals). *Freiberger Forschungshefte*, **C498**, Technische Universitat Bergakademie Freiberg.

KASHUBIN, S., JUHLIN, C. ET AL. 2006. Crustal structure of the Middle Urals based on reflection seismic data. *In*: GEE, D. & STEPHENSON, R. (eds) *European Lithosphere Dynamics.* Geological Society, London, Memoirs, **32**, 427–442.

KAZANSKY, A. YU., METELKIN, D. V., BRAGIN, V. YU., MIKHALTSOV, N. V. & KUNGURTSEV, L. V. 2004. Paleomagnetic data on the Mesozoic complexes of a frame of Siberian platform as a reflection of interpolated transcurrent deformations of the Central Asiatic belt. *In*: SKLYAROV, E. M. (ed.) *The Geodynamic Evolution of Lithosphere of the Central Asiatic Belt. From Ocean to Continent.* **2**, Irkutsk, Institute of the Earth Crust, 151–155.

KHAIN, V. E. 2001. *Tectonics of Continents and Oceans (2000).* Nauchnyi Mir, Moscow (in Russian).

KHAIN, E. V., FEDOTOVA, A. A., SALNIKOVA, E. B., KOTOV, A. B. & YAKOVLEVA, S. Z. 2004. New U–Pb data on the age of ophiolites of the Polar Urals and development of the margins of the Paleoasiatian ocean in the Late Precambrian and Early Paleozoic. *In*: KOROTEEV, V. A. (ed.) *Geology and Metallogeny of the Ultramafic-mafic and Granitoid Associations of Foldbelts.* Ekaterinburg, Institute of Geology and Geochemistry, 183–186 (in Russian).

KLIUZHINA, M. L. 1985. *Paleogeography of the Urals in the Ordovician.* M. Nauka (in Russian).

KNAPP, J. H., STEER, D. N. ET AL. 1996. A lithosphere-scale image of the Southern Urals from explosion-source seismic reflection profiling in URSEIS '95. *Science*, **274**, 226–228.

KORAGO, E. A., KOVALEVA, G. N. & TRUFANOV, G. N. 1989. Formations, tectonics and geological history of the Novozemelian Cimmerides. *Geotectonics*, **6**, 40–61 (in Russian).

KOSAREV, A. M. & PUCHKOV, V. N. 1999. The distribution of K, Ti and Zr in the Silurian-Carboniferous volcanic formations of the Southern Urals in connection with the position of the Paleozoic subduction zone. *Ezhegodnik-1977.* Ufa, Institute of Geology, Ufimian Science Centre. 186–191 (in Russian).

KOSAREV, A. M., PUCHKOV, V. N. & SERAVKIN, I. B. 2005. Petrological-geochemical features of the Early Devonian-Eifelian island arc volcanites of the Magnitogorsk zone in a geodynamical context. *Lithosphere*, **4**, 24–40 (in Russian).

KOSAREV, A. M., PUCHKOV, V. N. & SERAVKIN, I. B. 2006. Petrologo-geochemical character of the Middle Devonian–Early Carboniferous island-arc and collisional volcanic rocks of the Magnitogorsk zone in the geodunamical context. *Lithosphere*, **1**, 3–21 (in Russian).

KOZLOV, V. I. 1976. The coal-bearing deposits of the Paleogene and Neogene of the Tirlyan syncline. *In*: YAKHIMOVICH, V. L. (ed.) *The Problems of Stratigraphy and Correlation of Pliocene and Pleistocene Deposits of the Northern and Southern Parts of the Cis-Urals.* Ufa, Bashkirian Branch of the Academy of Sciences, USSR, 210–227 (in Russian).

KRASNOBAEV, A. A., RUSIN, A. I., RUSIN, I. A. & BUSHARINA, S. V. 2008. Zirconology of lherzolite-garnet pyroxenite-dunite complex of the Uzyan Kraka (S.Urals). *In*: KORTOTEEV, V. A. (ed.) *The Structural-material Complexes and Geodynamic Problems of the Precambrian of Phanerozoic Orogens.* Ekaterinburg, Uralian Branch of Russian Academy of Sciences, 58–61 (in Russian).

KURENKOV, S. A., DIDENKO, A. N. & SIMONOV, V. A. 2002. *The Geodynamics of Paleospreading.* M. GEOS. (in Russian).

LEECH, M. L. & WILLINGSHOFER, E. 2004. Thermal modeling of the UHP Maksyutov Complex in the south Urals. *Earth and Planetary Science Letters*, **226**, 85–99.

LENNYKH, V. I., VALISER, P. M., BEANE, R., LEECH, M. & ERNST, W. G. 1995. Petrotectonic evolution of the Makysutov complex, southern Ural Mountains, Russia: implications for ultrahigh-pressure metamorphism. *International Geology Review*, **37**, 584–600.

LEPEZIN, G. G., TRAVIN, A. V., YUDIN, D. S., VOLKOVA, N. V. & KORSAKOV, A. V. 2006. The age and thermal history of the Maksiutovo metamorphic complex (after Ar–Ar data). *Petrology*, **14**, 1–18.

LEVASHOVA, N. M., DEGTYAREV, K. E., BAZHENOV, M. L., COLLINS, A. Q. & VAN DER VOO, R. 2003. Middle Paleozoic paleomagnetism of east Kazakhstan: post-Middle Devonian rotations in a large-scale orocline in the central Ural-Mongol belt. *Tectonophysics*, **377**, 249–268.

LEVIN, V. YA., RONENSON, B. M. & SAMKOV, V. S. 1997. *Alkaline-carbonatite Complexes of the Urals.* Ekaterinburg (in Russian).

LI, Z. X., BOGDANOVA, S. V. ET AL. 2008. Assembly, configuration, and break-up history of Rodinia: a synthesis. *Precambrian Research*, **160**, 179–210.

LINNEMANN, U., GEHMLICH, V., TICHOMIROWA, V. & BUSCHMANN, B. 1998. Introduction to the Pre-Symposium Excursion (part I): the Peri-Gondwanan basement of the Saxothuringian Composite Terrane. Excursion Guide to Saxony, Thüringia, Bohemia. Abstracts. *Schriften des Staatlichen Museums für Mineralogie und Geologie zu Dresden*, 7–13.

LÜTZNER, H., FALK, F. & ELLENBERG, J. 1979. Ubersicht über die Variszische Molasseentwicklung in Mitteleuropa und am Ural. *Zeitschrift für Geologische Wissenschaften*, **9**, 1157–1167.

MASLOV, V. F., ARTIUSHKOVA, O. V., YAKUPOV, R. R. & MAVRINSKAYA, T. M. 2008. The problems of the Lower–Middle Paleozoic stratigraphy of the Southern Urals. *Geologichesky Sbornik*, **7**, 193–204 (in Russian).

MATTE, P. 2006. The Southern Urals: deep subduction, soft collision and weak erosion. *In*: GEE, D. & STEPHENSON, R. (eds) *European Lithosphere Dynamics.* Geological Society, London, Memoirs, **32**, 421–426.

MATTE, P., MALUSKI, H., CABY, R., NICOLAS, A., KEPEZHINSKAS, P. & SOBOLEV, S. 1993. Geodynamic model and $^{39}$Ar/$^{40}$Ar dating for the generation and emplacement of the high-pressure (HP) metamorphic rocks in SW Urals. *Academie de Sciences Comptes Rendus*, **317**, 1667–1674.

MELCHER, F., GRUM, W., THALHAMMER, T. V. & THALHAMMER, O. A. R. 1999. The giant chromite deposits at Kempirsai, Urals: constraints from trace element (PGE, REE) and isotope data. *Mineralium Deposita*, **34**, 250–272.

MELEKESTSEVA, I. YU. 2007. *Heterogenous Co-bearing Massive Sulfide Deposits Associated with Ultramafites of Paleo-island Arc Structures.* M. Nauka (in Russian).

MENSHIKOV, YU. P., KUZNETSOVA, N. V., SHEBUKHOVA, S. V. & NIKISHEVA, G. N. 1983. The tectonics of the northern half of the Magnitogorsk depression after geophysical data. *In*: KHALEVIN, N. (ed.) *The Faults of the Earth's Crust of the Urals and the Methods of Their Study*. Sverdlovsk, Uralian Scientific Centre, 65–78 (in Russian).

MIKHAILOV, V. O., TEVELEV, A. V., BERZIN, R. G., KISELEVA, E. A., SMOLYANINOVA, E. I., SULEIMANOV, A. K. & TIMOSHKINA, E. P. 2002. Constraints on the Neogene-Quaternary Geodynamics of the Southern Urals: comparative study of Neotectonic data and results of Strength and Strain Modelling Along the URSEIS Profile. *In*: BROWN, D., JUHLIN, C. & PUCHKOV, V. (eds) *Mountain Building in the Uralides: Pangea to the Present*. AGU Geophysical Monograph Series, **132**, 273–286.

MILANOVSKI, E. E. 1963. On the paleogeography of the Kaspian basin in the Middle and beginning of the Upper Pliocene. *Bulletin Moscow Society of Nature Studies, Section of Geology*, **38**, 23–26 (in Russian).

MÜNTENER, O., PICCARDO, G. B., POLINO, R. & ZANETTI, A. 2005. Revisiting the Lanzo peridotite (NW Italy): "astenospherization" of ancient mantle lithosphere. *Ofioliti*, **30**, 111–124.

MURATOV, M. V. 1979. The Uralo-Mongolian foldbelt. *In*: MURATOV, M. V. (ed.) *Tectonics of the Uralo-Mongolian Foldbelt*. M. Nauka, 5–11 (in Russian).

NARKISSOVA, V. V. 2005. *Petrochemistry of the Late Ordovician-Early Devonian basaltoids of the Southern part of the Middle Urals (after data on the Uralian superdeep borehole and near-hole area)*. Candidate Dissertation thesis. Moscow University (in Russian).

NEDOSEKOVA, I. L., PRIBAVKIN, S. V., SEROV, P. A. & RONKIN, YU. L. 2006. The isotopic compositions and age of carbonatites of the Ilmeno-Vishnevogorsk alkaline complex. *In*: GALIMOV, E. M. (ed.) *Isotopic Dating of Processes of Ore Formation, Magmatism, Sedimentation and Metamorphism*. M. GEOS, 40–54 (in Russian).

PAPULOV, G. N. 1974. *The Cretaceous Deposits of the Urals*. M. Nauka (in Russian).

PEASE, V., DALY, J. S. *ET AL*. 2008. Baltica in the Cryogenian, 850–630 Ma. *Precambrian Research*, **160**, 46–65.

PETROV, G. A. 2006. *The Geology and Minerageny of the Main Uralian Fault Zone in the Middle Urals*. Ekaterinburg, Uralian Mining University (in Russian).

PETROV, G. A., RONKIN, YU. L., LEPIKHINA, O. P. & POPOVA, O. YU. 2005. High-pressure metamorphism in the Urals – two stages? *Ezhegodnik-2004*. IGG Uralian Branch of RAS, Ekaterinburg. 97–102 (in Russian).

PEYVE, A. V., IVANOV, S. N., NECHEUKHIN, V. M., PERFILYEV, A. S. & PUCHKOV, V. N. 1977. *Tectonics of the Urals. The explanatory notes for the 1:1 000 000-scale tectonic map*. M. Nauka, Moscow (in Russian).

PLIUSNIN, K. P. 1966. Wrench faults of the eastern slope of the Southern Urals. *Geotectonics*, **4**, 57–68 (in Russian).

PRIBAVKIN, S. V., RONKIN, YU. L., TRAVIN, A. V. & PONOMARCHUK, V. A. 2006. The new data on the age of lamproite magmatism of the Urals. *In*: GALIMOV, E. M. (ed.) *III Russian Conference on Isotope Geochronology: Isotopic Dating of Processes*

*Ore Formation, Magmatism, Sedimentation and Metamorphism*. M. GEOS, 123–125.

PUCHKOV, V. N. 1979. *Bathyal Complexes of the Passive Margins of Geosynclines*. M. Nauka, Moscow (in Russian).

PUCHKOV, V. N. 1996a. The Paleozoic geology of Asiatic Russia and adjacent territories. *In*: MOULLADE, M. & NAIRN, A. (ed.) *The Phanerozoic Geology of the World I. The Paleozoic, B*. Elsevier, 1–107.

PUCHKOV, V. N. 1996b. The origin of the Ural-Novozemelian foldbelt as a result of an uneven, oblique collision of continents. *Geotectonics*, **5**, 66–75 (in Russian).

PUCHKOV, V. N. 1997. Structure and geodynamics of the Uralian orogen. *In*: BURG, J.-P. & FORD, M. (eds) *Orogeny Through Time*. Geological Society, London, Special Publications, **121**, 201–236.

PUCHKOV, V. N. 2000. *Paleogeodynamics of the Middle and Southern Urals*. Ufa, Dauria (in Russian).

PUCHKOV, V. N. 2002a. Paleozoic evolution of the East European continental margin involved into the Urals. *In*: BROWN, D., JUHLIN, C. & PUCHKOV, V. (eds) *Mountain Building in the Uralides: Pangea to the Present*. AGU Geophysical Monograph Series, **132**, 9–32.

PUCHKOV, V. N. 2002b. Neotectonics of the Urals. *In*: PUCHKOV, V. N. & DANUKALOVA, G. A. (eds) *Upper Pliocene and Pleistocene of the Southern Urals Regions INQUA-SEQS'02*, Ufa, 70–72.

PUCHKOV, V. N. 2003. Uralides and Timanides: their structural relationship and position in the geological history of the Uralo-Mongolian foldbelt. *Russian Geology and Geophysics*, **44**, 28–39.

PUCHKOV, V. N. 2005. Evolution of lithosphere: from the Pechora ocean to Timanian orogen, from the Paleouralian ocean to Uralian orogen. *In*: LEONOV, YU. G. (ed.) *Problems of Tectonics of the Central Asia*. M., GEOS, 309–342 (in Russian).

PUCHKOV, V. N. 2006a. The outline of the Uralian minerageny. *In*: PYSTIN, M. A. (ed.) *The Problems of Geology and Mineralogy*. Syktyvkar, Geoprint, 195–222 (in Russian).

PUCHKOV, V. N. 2006b. On the age of the Uralian ophiolites. *In*: KOROTEEV, V. A. (ed.) *Ophiolites: Geology, Petrology, Metallogeny and Geodynamics*. Ekaterinburg, IG USC RAS, 121–129 (in Russian).

PUCHKOV, V. N. 2008. The diachroneity of arc-continent collision in the Uralides: subduction jump, oblique subduction or … what? *In*: National Cheng Kung University Tainan. *Arc–continent collision*. Taiwan, 47–51.

PUCHKOV, V. N., RAPOPORT, M. B., FERSHTATER, G. B. & ANANYEVA, E. M. 1986. Tectonic control of the Paleozoic granitoid magmatism in the eastern slope of the Urals. *In*: IVANOV, S. N. (ed.) *Studies of Geology and Metallogeny of the Urals*. UNTs AN SSSR, Sverdlovsk, 85–95 (in Russian).

PUCHKOV, V. N. & SVETLAKOVA, A. N. 1993. The structure of the Urals in the transect of the Troitsk DSS profile. *Transactions (Doklady) of the Russian Academy of Sciences/Earth Science Section*, **333**, 348–351 (in Russian).

PUCHKOV, V. N. & DANUKALOVA, G. A. 2004. New data on the character of tectonic deformations of the

Cretaceous-Paleogene peneplain in the Southern Urals. *In*: PUCHKOV, V. N. (ed.) *Geological Sbornik* **4**. Institute of Geology, Ufimian Geoscience Center, Russian Academy of Science, Ufa, 183–184 (in Russian).

PUCHKOV, V. N., ROSEN, O. M., ZHURAVLEV, D. Z. & BIBIKOVA, E. V. 2006. Contamination of Silurian volcanites of the Tagil synform by Precambrian zircons. *Transactions (Doklady) of the Russian Academy of Sciences/Earth Science Section*, **411**, 1–4 (in Russian).

PUSHKAREV, E. V. 2001. Explosive breccias with inclusiona of mafic and ultramafic high-pressure rocks in the Mindyak lherzolite massif – composition and petrogenetic consequences. *In*: PUCHKOV, V. N. (ed.) *Geology and Prospects of the Raw Materials of Bashkortostan and Adjacent Territories*. Ufa, IG UNC RAS, **1**, 155–168 (in Russian).

RASULOV, A. T. 1982. *The Tectonics of the Early Mesozoic Depressions of the Eastern Slope of the Urals*. Sverdlovsk, Institute of Geology and Geochemistry, Uralian Scientific Centre, USSR (in Russian).

REICHOV, M. K., PRINGLE, M. S. *ET AL.* 2007. New high-precision $^{40}Ar-^{39}Ar$ ages and geochemical data from the greater Siberian large igneous province: The Biggest gets Bigger. *Abstract of AGU meeting*, San Francisco, 425.

REMIZOV, D. N. 2004. *The Island-arc System of the Polar Urals (Petrology and Evolution of Deep-seated Zones)*. Ekaterinburg, Uralian Branch of Russian Academy (in Russian).

RICKEN, W., SCHRADER, S., ONCKEN, O. & PLESCH, A. 2000. Turbidite basin and mass dynamics related to orogenic wedge growth; the Rheno-Hercynian case. *In*: FRANKE, W., HAAK, V., ONCKEN, O. & TANNER, D. (eds) *Orogenic Processes: Quantification and Modelling in the Variscan Belt*. Geological Society, London, Special Publications, **179**, 257–280.

ROZHDESTVENSKY, A. P. & ZINYAKHINA, I. K. 1997. *The Development of the Relief of the Southern Urals in the Cenozoic, the Neogene Period*. Ufa, IG USC RAS (in Russian).

RYABOV, V. V. & GRIB, D. YE. 2005. Multiphase dykes – an example of dissipated spreading in the North of the Siberian platform. *Geology and Geophysics*, **46**, 471–485 (in Russian).

RYBALKA, A. V., PETROV, G. A., KASHUBIN, S. N. & JUHLIN, C. 2006. The Middle Uralian ESRU transsect. *In*: MOROZOV, A. F. (ed.) *Structure and Dynamics of Lithosphere of the Eastern Europe, Issue 2. The Results of the EUROPROBE Research*. Moscow, GEOKART, GEOS, 390–401 (in Russian).

SAVELIEV, A. A. 1997. Ultramafic-gabbro formations in the structure of ophiolites of the Voikar-Synya massif (Polar Urals). *Geotectonics*, **1**, 48–58 (in Russian).

SAVELIEV, A. A., BIBIKOVA, E. V. & SAVELIEVA, G. N. 2001. Garnet pyroxenites of the Mindyak massif, the position and age of formation. *Bulletin MOIP, Section of Geology*, **1**, 22–29.

SAVELIEVA, G. N. 1987. *Gabbro-Ultramafic Complexes of Ophiolites of the Urals and their Analogues in the Modern crust*. M. Nauka (in Russian).

SAVELIEVA, G. N. & NESBITT, R. Q. 1996. A Synthesis of the stratigraphic and tectonic setting of the Uralian ophiolites. *Journal of Geological Society, London*, **153**, 525–537.

SAVELIEVA, G. N., PUCHKOV, V. N. & SPADEA, P. 2006*a*. The ophiolites of the Urals. Structure and dynamics of lithosphere of the Eastern Europe. *In*: MOROZOV, A. F. (ed.) *Structure and Dynamics of Lithosphere of the Eastern Europe, Issue 2. The Results of the EUROPROBE Research*. Moscow, GEOKART, GEOS, 421–436 (in Russian).

SAVELIEVA, G. N., SHISHKIN, M. A., LARIONOV, A. N., SUSLOV, P. V. & BEREZHNAYA, N. G. 2006*b*. Tectono-magmatic events of the Late Vendian in the mantle complexes of ophiolites of the Polar Urals: data of U–Pb dating of zircons from chromites. *In*: KOROTEEV, V. A. (ed.) *Ophiolites: Geology, Petrology, Metallogeny and Geodynamics*. Ekaterinburg, IG USC RAS, 160–164 (in Russian).

SAMYGIN, S. G., KUZNETSOV, N. B., PAVLENKO, T. I. & DEGTYAREV, K. E. 1998. The structure of the Kyshtym-Miass region of the Southern Urals and a problem of conjugation of Magnitogorsk and Tagil complexes. *In*: LEONOV, YU. G (ed.) *The Urals fundamental Problems of Geodynamics and Stratigraphy*. M. Nauka, 73–92 (in Russian).

SAMYGIN, S. G. & RUZHENTSEV, V. S. 2003. The Uralian paleocean: a model of inheritad development. *Transactions (Doklady) of the Russian Academy of Sciences/Earth Science Section*, **392**, 226–229 (in Russian).

SCARROW, J. H., PEASE, V., FLEUTELOT, C. & DUSHIN, V. 2001. The Late Neoproterozoic Enganepe ophiolite, Polar Urals: an extension of the Cadomian arc? *Precambrian Research*, **110**, 255–275.

SCHWAB, M. 1984. The Harz Mountains. *In*: BANKVITZ, P. (ed.) *Sedimentary and tectonic structures in the Saxoturingian and Rhenohercynian Zones*. Central Institute for Physics of the Earth, Potsdam, 34–77.

SENGÖR, A. M. C., NATAL'IN, B. A. & BURTMAN, V. S. 1993. Evolution of the Altaid tectonic collage and Palaeozoic crustal growth in Eurasia. *Nature*, **34** (6435), 299 – 307.

SEWARD, D., PEREZ-ESTAUN, A. & PUCHKOV, V. 1997. Preliminary fission-track results from the southern Urals – Sterlitamak to Magnitogorsk. *Tectonophysics*, **276**, 281–290.

SHARKOV, E. V., BORTNIKOV, N. S. & BOGATIKOV, O. A. 2004. The Mesozoic zircon from gabbro-norites of the axial zone of the Mid-Atlantic Ridge, 6°N (the Basin of Markov). *Transactions (Doklady) of the Russian Academy of Sciences/Earth Science Section*, **396**, 675–679 (in Russian).

SHATSKY, N. S. 1965. On the tectonics of Soviet Union (On 10 years after demise of A. D. Archangelsky. *In*: SHATSKY, V. S. (ed.) *Selected publications*. M. Nauka Publishing House, **IV**, 76–84 (in Russian).

SHATSKY, V. S., SIMONOV, V. A. & YAGOUTZ, E. 2000. New data on the age of eclogites of the Polar Urals. *Doklady of Russian Academy*, **371**, 519–523 (in Russian).

SIGOV, A. P. 1969. *The Mesozoic and Cenozoic Metallogeny of the Urals*. M. Nedra. (in Russian).

SKRIPIY, A. A. & YUNUSOV, H. K. 1989. Tension and compression structures in the Articulation zone of the

Southern Urals and the East European Platform. *Geotectonics*, **23**, 515–522 (in Russian).

SMIRNOV, G. A. 1957. *Materials for Paleogeography of the Urals. The Visean Stage*. Sverdlovsk, Mining and Geology Institute (in Russian).

SMIRNOV, V. N., BOROZDINA, G. N., DESYATNICHENKO, L. I., IVANOV, K. S., MEDVEDEVA, T. YU. & FADEICHEVA, I. F. 2006. On the time of opening of the Uralian paleocean (biostratigraphic and geochemical data). *Russian Geology and Geophysics*, **47**, 755–761 (in Russian).

SNACHEV, A. V., PUCHKOV, V. N., SAVELIEV, D. B. & SNACHEV, V. N. 2006. *Geology of the Aramil-Sukhtelinsk Zone of the Urals*. IG USC RAS, Ufa (in Russian).

SOBORNOV, K. O. & BUSHUEV, F. S. 1992. The kinematics of the conjugation zone of the Northern Urals and the Upper Pechora Basin. *Geotectonics*, **1**, 39–51 (in Russian).

SOKOLOV, V. B. 1992. The structure of the Earth's crust of the Urals. *Geotectonics*, **5**, 3–19 (in Russian).

SPADEA, P., ZANETTI, A. & VANUCCI, R. 2003. Mineral chemistry of ultramafic massifs of the Southern Uralides orogenic belt (Russia) and the petrogenesis of the Lower Palaeozoic ophiolites of the Uralian Ocean. *In*: DILEK, Y. & ROBINSON, P. T. (eds) *Ophiolites in Earth History*. Geological Society, London, Special Publications, **218**, 567–596.

SULEIMANOV, A. K. 2006. The CDP works along the URSEIS profile. *In*: MOROZOV, A. F. (ed.) *Structure and Dynamics of Lithosphere of the Eastern Europe, Issue 2. The Results of the EUROPROBE Research*. Moscow, GEOKART, GEOS, 363–373 (in Russian).

TESSALINA, B. S., BOURDON, G., PUSHKAREV, E., CAPMAS, F., BIRCK, J.-L. & GANNOUN, M. 2005. MinUrals – IPGP contribution. Complex Proterozoic to Paleozoic history of the upper mantle recorded in the Urals lherzolite massifs by Re–Os and Sm–Nd systematics. *In*: CERCAMS-6 Workshop Mineral Deposits of the Urals. London, The Natural History Museum, CD.

TEVELEV, A. L. V., DEGTYAREV, K. E. *ET AL.* 2005. Geodynamic situations of formation of the Carboniferous volcanic complexes of the Southern Urals. *In*: NIKISHIN, A. (ed.) *The Studies on the Regional Tectonics of the Urals, Kazahstan and Tyan-Shan*, **1**. Southern Urals. Moscow, Nauka, 213–247.

TRETYAKOV, A. A., RYAZANTSEV, A. V., KUZNETSOV, N. B. & BELOVA, A. A. 2008. The structural position and geochronological dating of granate ultramafics in the Southern Urals. *In*: KARYAKIN, YU. V. (ed.) *The General and Regional Problems of Tectonics and Geodynamics*, **2**. Moscow, GEOS, 343–349 (in Russian).

TRIFONOV, V. G. 1999. *The neotectonics of the Europe*. Nauchny Mir (in Russian).

TRIFONOV, V. G. 2008. The age and mechanisms of the newest mountain building. *In*: KARYAKIN, YU. V. (ed.) *The General and Regional Problems of Tectonics and Geodynamics*. **2**. Moscow, GEOS, 349–353 (in Russian).

TRYGGVASON, A., BROWN, D. & PEREZ-ESTAUN, A. 2001. Crustal architecture of the southern Uralides from true amplitude processing of the URSEIS vibroseis profile. *Tectonics*, **20**, 1040–1052.

TUZHIKOVA, V. I. 1973. *The History of the Lower Mesozoic Coal Accumulation in the Urals*. M. Nauka (in Russian).

UDORATINA, O. V. & LARIONOV, A. N. 2005. The age of granitoids of Taikeu massif (Polar Urals). *In*: YUSHKIN, N. P. (ed.) *The Structure, Geodynamics and Mineragenic Processes in Lithosphere*. Syktyvkar, Geoprint, 346–349 (in Russian).

VERBITSKAYA, N. P. 1964. *The Regional Specific Features of Neotectonics of the Urals*. Leningrad (in Russian).

VERNIKOVSKY, V. A., VERNIKOVSKAYA, A. A., PEASE, V. L. & GEE, D. G. 2004. Neoproterozoic Orogeny along the margins of Siberia. *In*: GEE, D. G. & PEASE, V. (eds) *The Neoproterozoic Timanide Orogen of Eastern Baltica*. Geological Society, London, Memoirs, **30**, 233–248.

WILLNER, A. P., ERMOLAEVA, T. *ET AL.* 2002. Surface Signals of an Arc-Continent Collision: The Detritus of the Upper Devonian Zilair Formation in the Southern Urals, Russia. *In*: BROWN, D., JUHLIN, C. & PUCHKOV, V. (eds) *Mountain Building in the Uralides: Pangea to the Present*. AGU Geophysical Monograph Series, **132**, 183–209.

WILLNER, A. P., WARTHO, J.-A, KRAMM, U. & PUCHKOV, V. N. 2004. Laser $^{40}Ar–^{39}Ar$ ages of single detrital white mica grains related to the exhumation of Neoproterozoic and Late Devonian high pressure rocks in the Southern Urals (Russia). *Geological Magazine*, **141**, 161–172.

ZAKHAROV, O. A. & PUCHKOV, V. N. 1994. *On the Tectonic Nature of the Maksutovo Complex of the Ural-Tau zone*. Ufimian Science Centre, Ufa (in Russian).

ZIEGLER, P. A. 1999. Evolution of the Arctic-North Atlantic and the Western Tethys. *AAPG Memoir*, **43**, 164–196.

ZNAMENSKY, S. E. 2007. The stages of Late Paleozoic deformations of the Magnitogorsk megazone (Southern Urals). *In*: YUSHKIN, N. P. & SAZONOV, V. N. (eds) *Geodynamics of Formation of Mobile Belts of the Earth*. Ekaterinburg, Institute of Geology and Geochemistry, Uralian Branch of Russian Academy, 108–111 (in Russian).

ZONENSHAIN, L. P., KUZMIN, M. I. & NATAPOV, L. M. 1990. Geology of the USSR, a plate tectonic synthesis. *American Geophysical Union, Geodynamic Series*, **21**, 1–242.

YAKHIMOVICH, V. L. & ANDRIANOVA, O. S. 1959. *The Southern Uralian Brown Coal Basin*. Ufa, Mining-Geological Institute of the Bashkirian Branch of the Academy of Sciences, USSR.

YAZEVA, R. G. & BOCHKAREV, V. V. 1993. Post-collisional magmatism of the Northern Urals. *Geotectonics*, **4**, 56–65 (in Russian).

YUDIN, V. V. 1994. *Orogeny in the Northern Urals and Pai-Khoi*. Nauka, Ekaterinburg (in Russian).

YUGGEO. 2002. *Atlas of Paleogeographic, Structural, Palinspastic and Geoecological Maps of the Central Eurasia*. Almaty (in Russian).

# Timing of dextral strike-slip processes and basement exhumation in the Elbe Zone (Saxo-Thuringian Zone): the final pulse of the Variscan Orogeny in the Bohemian Massif constrained by LA-SF-ICP-MS U–Pb zircon data

M. HOFMANN[1]*, U. LINNEMANN[1], A. GERDES[2], B. ULLRICH[3] & M. SCHAUER[4]

[1]*Staatliche Naturhistorische Sammlungen Dresden, Museum für Mineralogie und Geologie, Königsbrücker Landstraße 159, D-01109 Dresden, Germany*

[2]*Institut für Geowissenschaften, Johann Wolfgang Goethe-Universität, FE Mineralogie, Altenhöferallee 1, 60438 Frankfurt am Main, Germany*

[3]*TU Dresden, Institut für Geotechnik, Lehrstuhl für Angewandte Geologie, Neuffer-Bau, Helmholtzstraße 10, D-01069 Dresden, Germany*

[4]*Am Hexenberg 8, 09224 Grüna, Germany*

*Corresponding author (e-mail: mandy.hofmann@senckenberg.de)*

**Abstract:** The final pulse of the Variscan Orogeny in the northern Bohemian Massif (Saxo-Thuringian Zone) is related to the closure of the Rheic Ocean, which resulted in subduction-related $D_1$-deformation followed by dextral strike-slip activity ($D_2$-deformation, the Elbe Zone). Taken together, these deformation events reflect the amalgamation of Pangaea in central Europe. Lateral extrusion of high-grade metamorphosed rocks from an allochthonous domain (Saxonian Granulitgebirge) and the top–NW-directed transport of these domains (Erzgebirge nappe complex, Saxonian Granulitgebirge) are responsible for these dextral strike-slip movements. Geochronological data presented herein, together with published data, allow the timing of the final pulse of the Variscan Orogeny and related plutonic, volcano-sedimentary and tectonic processes. Marine sedimentation lasted at least until the Tournaisian (357 Ma). Onset of Variscan strike-slip along the Elbe Zone is assumed to be coeval with the beginning of the top–NW-directed lateral extrusion of the Saxonian Granulitgebirge at 342 Ma ($D_2$-deformation). The sigmoidal shape of the Meissen Massif indicates that strike-slip activity was coexistent with intrusion of the pluton at *c.* 334 Ma into the schist belt of the Elbe Zone. In contrast, the intrusion of the Markersbach Granite provides a minimum age of *c.* 327 Ma for the termination of $D_2$ strike-slip activity, because this undeformed pluton cross-cuts all strike-slip related tectonic structures. Geochronological data of an ash bed from the Permo-Carboniferous Döhlen Basin show clearly that post-orogenic sedimentation of Variscan molasse in that area was already active at 305 Ma. This pull-apart basin is a local example of regional Permo-Carboniferous extension within Pangaea. The uplift and denudation of the Variscan basement in the Saxo-Thuringian Zone occurred between *c.* 327–305 Ma.

## Introduction

The Bohemian Massif is the type area of the Variscan Orogeny. The name is derived from an area called by the Romans '*Curia Variscorum*', which is situated adjacent to the city of Hof in the Frankonian Forest (Germany) (Linnemann 2003*a*, and references therein). In the Anglo–American literature, the Variscan Orogeny is also known as the 'Hercynian Orogeny'. Variscan orogenic events are geodynamically linked to Alleghenian processes in the Appalachians (Linnemann *et al.* 2007*b*). Alleghenian–Variscan collisional orogenic activity reflects the closure of the Rheic Ocean and the amalgamation of the supercontinent Pangaea (Rheic

suture, see Fig. 1) (Scotese & Barret 1990). Despite decades of study, a number of important first-order questions remain unanswered, including the precise age of Variscan orogenic processes and their relationship to the closure of the Rheic Ocean. Palaeocontinental reconstructions suggest that convergence of the Rheic Ocean started after the formation of the supercontinent Laurussia at *c.* 420–400 Ma (Romer *et al.* 2003; Sánchez Martínez *et al.* 2007). Northward drift of Gondwana started around the same time and subduction of Rheic oceanic lithosphere terminated at around *c.* 400–370 Ma (Romer *et al.* 2003).

In the Bohemian Massif, continental collision involved subduction of continental crust, which

*From*: Murphy, J. B., Keppie, J. D. & Hynes, A. J. (eds) *Ancient Orogens and Modern Analogues.*
Geological Society, London, Special Publications, **327**, 197–214.
DOI: 10.1144/SP327.10   0305-8719/09/$15.00 © The Geological Society of London 2009.

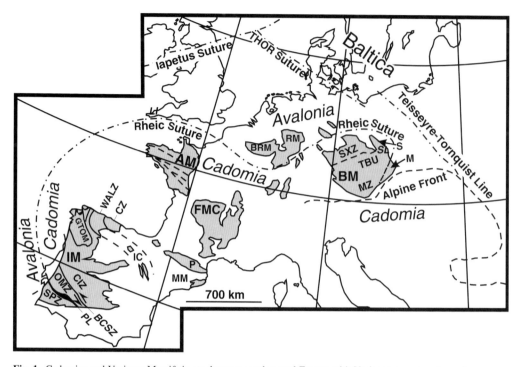

**Fig. 1.** Cadomian and Variscan Massifs in southwestern and central Europe with Variscan oceanic suture of the Rheic Ocean (after Robardet 2002; Linnemann *et al.* 2007*a*, 2008*a*). IM, Iberian Massif; AM, Armorican Massif; FMC, French Massif Central; RM, Rhenish Massif; BRM, Brabant Massif; BM, Bohemian Massif; SPZ, South Portuguese Zone; OMZ, Ossa-Morena Zone; CIZ, Central Iberian Zone; GTOM, Galicia-Trás os Montes Zone; WALZ, West Asturian Leonese Zone; CZ, Cantabrian Zone; PL, Pulo de Lobo oceanic units; IC, Iberian Chains; BCSZ, Badajoz-Cordoba Shear Zone; P-Pyrénées; MM, Maures Massif; SXZ, Saxo-Thuringian Zone; TBU, Teplá-Barrandian Unit; MZ, Moldanubian Zone; S, Sudetes; M, Moravo-Silesian Zone; Li, Lizard Ophiolite; SL, Ślęża ophiolite. Black-oceanic rocks of the Pulo de Lobo suture and ophiolitic units of allochthonous complexes of Galicia.

started at *c.* 370–360 Ma (Kroner *et al.* 2007), as the northern margin of Gondwana underthrusted the southern margin of Laurussia. Complex subduction processes and stacking of Gondwanan continental crust occurred between *c.* 360 and 340 Ma. The final stage of the Variscan Orogeny is characterized by orogen-wide transpressional tectonics, regional HT/LP metamorphism, late orogenic granite intrusions, and the formation of fold-and-thrust belts in the external parts of the orogen (Kroner *et al.* 2007). The related final extrusion of deeper parts of subducted units exhumed rocks of peri-Gondwanan affinity. Their juxtaposition with low-grade assemblages in the upper crust is today an important feature of the Variscan orogen (Kroner *et al.* 2007) that is due to transport directions of allochthonous nappe piles that in part had a large influence on the samples presented in this study.

In order to better characterize the timing of the final pulse of the Variscan Orogeny, we present

herein a number of new U–Pb age determinations of igneous rocks and one ash bed from a major late Variscan shear zone, known as the Elbe Zone. This belt of sheared rocks occurs within the Saxo-Thuringian Zone in the northern part of the Bohemian Massif (Linnemann 1994, 2003*a*; Kroner *et al.* 2007). As field relationships indicate that some plutonic bodies are coeval with shearing, whereas other bodies post-date shearing, our data help constrain the age of D$_2$ strike-slip deformation and therefore constrain the age of Late Variscan tectonic activity in central Europe.

## Geological setting

### The Saxo-Thuringian Zone and the history of the Rheic Ocean

A detailed description of the geology of the Bohemian Massif, and the Saxo-Thuringian Zone in particular, may be found in Franke (2000),

Franke & Żelazniewics (2002), Linnemann (2003*a*, *b*), Kroner *et al.* (2007), Linnemann *et al.* (2008*b*) and Nance *et al.* (2008). The Elbe Zone occurs in the Saxo-Thuringian Zone, which is located in the Bohemian Massif (Kossmat 1927) (Fig. 1). The oldest rocks of the Saxo-Thuringian Zone are *c.* 570–565 Ma sedimentary rocks (Fig. 2), which predominantly consist of turbiditic greywackes and shales, and are inferred to have been deposited in back-arc and retro-arc basins along the northern margin of Gondwana during the Cadomian Orogeny (Linnemann *et al.* 2007*a*).

These sedimentary basins were deformed by arc-continent collision and were intruded by voluminous granitoids at *c.* 540 Ma, which are coeval with a switch from an arc to a transform margin setting, analogous with the modern Eastern Pacific (Linnemann *et al.* 2007*a*). The granitoid intrusions were due to slab break-off of the subducted oceanic plate during the change of the marginal setting (Linnemann *et al.* 2007*a*). The tectonic setting may have been similar to that of the present-day Basin and Range Province adjacent to Baja California and the San Andreas Fault (Nance *et al.* 2002). Cadomian orogenic processes provided the structural heterogeneities which played an important role for the opening of the Rheic Ocean during the Late Cambrian (Linnemann *et al.* 2007*a*, 2008*c*).

Lower to Middle Cambrian strata in the Saxo-Thuringian Zone (Fig. 2) (Elicki 1997) are characterized by carbonates with archaeocyatha, siliciclastic strata and red beds. Lower Ordovician deposits in the Saxothuringian Zone are for example quartzites (Frauenbach and Phycodes groups) that are up to 3000 m thick (e.g. SE-part of the Schwarzburg unit, Fig. 2) and are typical of the Lower Ordovician strata in central and western Europe. In some localities, these deposits overstep Lower–Middle Cambrian strata to overlie directly the Neoproterozoic (Cadomian) basement (Linnemann & Romer 2002). Similar relationships are documented in coeval deposits in the Armorican Massif (NW France) and in different parts of Iberia (e.g. Graindor 1957; Chauvel & Rabu 1988; Chantraine *et al.* 1994; Fernández-Suárez *et al.* 2000; Linnemann *et al.* 2007*a*). These deposits, which are accompanied in some localities by *c.* 485 Ma rift-related bimodal magmatism (Sánchez-García *et al.* 2003), are thought to represent the rift-to-drift transition that heralded the development of the Rheic Ocean and the separation of Avalonia or a related micro-continent or terrane from the Gondwanan margin (Linnemann *et al.* 2007*a*).

After *c.* 480 Ma, the Saxo-Thuringian Zone is characterized by an extended period (Mid-Upper Ordovician–Mid-Devonian) of relative tectonic and magmatic quiescence and shelf sedimentation. Linnemann *et al.* (2007*a*) interpret this to be a passive margin along the northern Gondwanan margin that is flanked to the north by the Rheic Ocean. On the opposing flank of the Rheic Ocean Avalonia collided with Baltica during the closure of the Tornquist Sea at *c.* 450 Ma, which was followed, at *c.* 420 Ma, by the collision of Baltica + Avalonia with Laurentia, which closed the Iapetus Ocean and resulted in the formation of Laurussia (Fig. 3) (Scotese & Barret 1990; Torsvik & Rehnström 2002).

In the Upper Devonian, closure of the Rheic Ocean in central Europe was accompanied by subduction of the Gondwanan margin beneath Laurussia (Kroner *et al.* 2007) (Fig. 3). In the Saxo-Thuringian Zone, this collision was accompanied by deposition of clastic strata until the Lower Carboniferous (Linnemann 2003*b*; Kroner *et al.* 2007).

The Mid-German Crystalline Zone is interpreted as the suture of the Rheic Ocean (Fig. 1). Martínez *et al.* (2007) suggest the closure of the Rheic Ocean occurred in a northward-dipping supra-subduction zone setting because of the general absence of large Silurian-Devonian volcanic arcs on both margins of the ocean.

In the Bohemian Massif and also in other parts of Central and Western Europe, rocks of the Variscan orogen include voluminous Late Devonian–Early Carboniferous high-pressure metamorphic units (allochthonous domains) tectonically juxtaposed with low-grade Neoproterozoic (Cadomian) and Palaeozoic successions (autochthonous domains). The allochthonous domains are surrounded by wrench-and-thrust zones and are interpreted to reflect diachronous subduction of thinned peri-Gondwanan continental crust down to the stability field of metamorphic diamond (*c.* 120 km; Massonne 1998; Kroner *et al.* 2007, and references therein). Top-SE exhumation of high pressure (HP) metamorphic units occurred in a subduction channel ($D_1$ transport) and is preserved in the Saxonian Granulitgebirge complex and in parts of the nappe pile of the Erzgebirge complex, and the nappe complexes of Münchberg, Wildenfels and Frankenberg (Fig. 4) (Kroner *et al.* 2007). Finally, regional dextral transpression, rapid exhumation and the formation and inversion (folding) of Variscan flysch basins occurred during SE–NW oriented $D_2$-transport at *c.* 340–330 Ma (Kroner *et al.* 2007) (Fig. 4). Related extensive Variscan plutonism occurred in the Bohemian Massif between *c.* 334–327 Ma (this paper). The Mid-German Crystalline Zone, the suture of the Rheic Ocean, was formed by large-scale oblique subduction, collision and exhumation tectonics, and strike-slip related slivering (Zeh *et al.* 2005; Kroner *et al.* 2007).

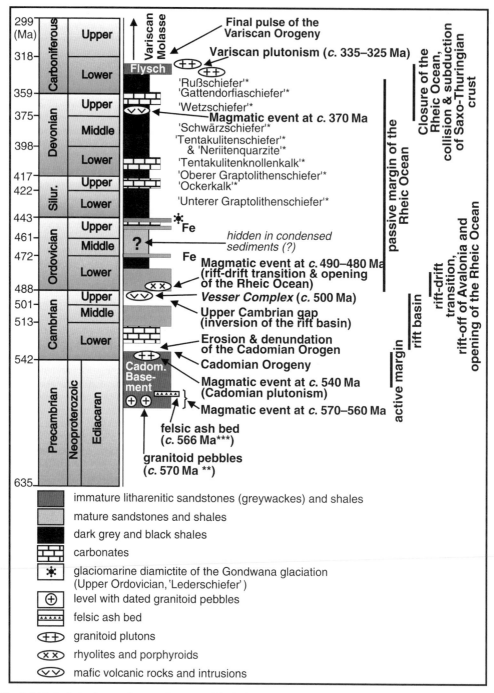

**Fig. 2.** Lithological column and magmatic events of the Cadomian Basement and the overlying Palaeozoic strata of the Thuringian Facies in the Saxo-Thuringian Zone (after Linnemann *et al.* 2004). 1, immature litharenitic sandstones (greywackes) and shales; 2, mature sandstones and shales; 3, dark grey and black shales; 4, carbonates, 5, glaciomarine diamiktite of the Gondwana glaciation in the Late Ordovician (Lederschiefer); 6, level with dated granitoid pebbles; 7, felsic ash bed; 8, granitoid plutons; 9, rhyolites and porphyroids; 10, mafic volcanic rocks and intrusions. *Traditional German terms for lithostratigraphic units. Sources of geochronological data: **Pb–Pb TIMS (Linnemann *et al.* 2000); ***U–Pb SHRIMP (Buschmann *et al.* 2001).

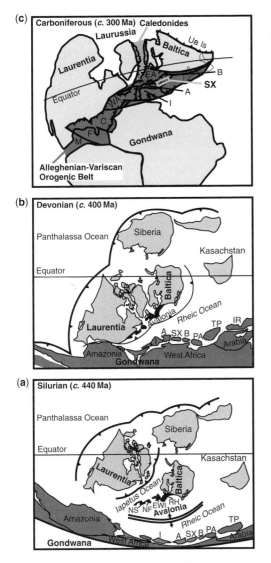

**Fig. 3.** Palaeogeography in (**a**) Lower Silurian (*c.* 440 Ma); (**b**) Lower Devonian (*c.* 400 Ma); and (**c**) Upper Carboniferous (*c.* 300 Ma) (modified after C. R. Scotese: Palaeomap web site: www.scotese.com). A, Armorica (Brittany, Normandy, Massif Central); B, Barrandian; C, Carolina; EA, East Avalonia; EWI, England, Wales, Southern Ireland; F, Florida; I, Iberia; IR, Iran; M, Mexican terranes; NF, New Foundland; NS, Nova Scotia; PA, Basement of the Alps; RH, Rheno-Hercynian; SX, Saxo-Thuringia; TP, 'Turkish plate'; WA, West Avalonia.

## The Elbe Zone and adjoining areas

The Elbe Zone near Dresden (Figs 4 & 5) is a NW–SE striking schist belt that divides the Lausitz Block in the NE from the Erzgebirge nappe complex in the

SW. The geodynamic relationships and the appropriate model for the Saxo-Thuringian Zone during the Variscan Orogeny have been published recently by Kroner *et al.* (2007), including also the division of Saxo-Thuringian units into autochthonous and allochthonous domains. The Lausitz Block belongs to the autochthonous domain of the Saxo-Thuringian Zone (Fig. 4) and was relatively stable during Variscan deformation processes (Fig. 4). The Erzgebirge nappe complex in the SW of the Elbe Zone is part of the allochthonous domain of the Saxo-Thuringian Zone and consists of a nappe pile containing diverse metamorphic rocks ranging from HP/HT to LP/LT units. The Elbe Zone is a schist belt and represents a part of the wrench-and-thrust zone and is situated between the Lausitz Block and the Erzgebirge nappe complex (Fig. 4) (Kroner *et al.* 2007).

According to the model of Kroner *et al.* (2007), the Variscan Orogeny in the Saxo-Thuringian Zone is characterized by two regionally-distributed deformation and tectonic transport events ($D_1$, $D_2$, Fig. 4). The $D_1$-deformation is restricted to the allochthonous domain and is related to a top-SW-directed tectonic transport, which reflects the exhumation of subducted continental crust along a subduction channel during the final stage of the Variscan Orogeny. In the central Bohemian Massif, rock units that escaped from of the subduction channel formed a thick pile of metamorphic rocks. After $D_1$, NW-directed transport of these over-thickened allochthonous domains occurred during a regional $D_2$ event. $D_2$ is characterized by top-NW vergent folding in the southeastern part of the Saxo-Thuringian Zone and a SE-directed back-thrusting in its northwestern part. In contrast to $D_1$, $D_2$ occurs in the allochthonous and the autochthonous domains as well as in the wrench-and-thrust zone (Fig. 4). During $D_2$-transport of the allochthonous nappes of the Erzgebirge complex towards the NW, dextral strike-slip between the Erzgebirge nappe complex and the autochthonous Lausitz Block occurred, which resulted in the formation of the Elbe Zone. Final exhumation of the deeper parts of the allochthonous domain is related to the juxtaposition of the Saxonian Granulitgebirge beneath the wrench-and-thrust zone (e.g. schist belt of the Elbe Zone) and dextral shear between the Erzgebirge nappe complex and the autochthonous domain of the Lausitz Block (Linnemann 1994; Kroner *et al.* 2007). There is broad agreement that the $D_2$-deformation led to the final crustal architecture of the Saxo-Thuringian Zone (Fig. 4) and that the schist belt of the Elbe Zone (Fig. 5) was deformed and overprinted during the dextral strike-slip movements (Linnemann 1994; Mattern 1996; Kroner *et al.* 2007). Estimated distances of strike-slip movement

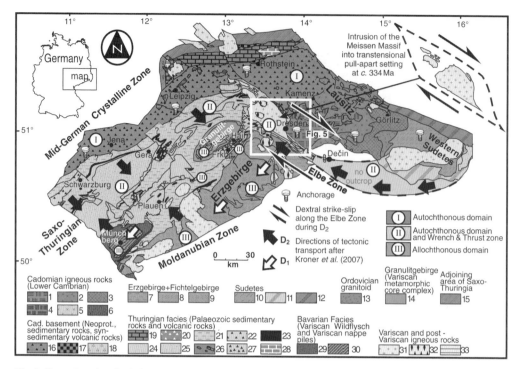

**Fig. 4.** Tectonic and geological map of the Saxo-Thuringian Zone in the NE-part of the Bohemian Massif showing units of Lower Carboniferous and older ages, and the dextral strike-slip movement along the Elbe Zone (modified from Linnemann & Schauer 1999; Linnemann & Romer 2002; Linnemann *et al.* 2007a). Subdivision of the Saxo-Thuringian Zone into autochthonous and allothonous domains and into a wrench-and-thrust Zone is based on Kroner *et al.* (2007). Earlier regional Variscan $D_1$ and later regional $D_2$ transport directions are taken from Kroner *et al.* (2007). Note the anchorage of the autochthonous domain relative to the allochthonous domain and the wrench-and-thrust zone and the resulting dextral strike-slip movements along the Elbe Zone during the regional $D_2$ deformation.

Abbreviations and Numbers: Hain, Hainichen; Frkbg, Frankenberg; 1 to 3, Lausitz granitoids complex; 1, Lausitz anatexite; 2, western Lausitz granitoids; 3, eastern Lausitz granitoids; 4, granitoids of the Leipzig area; 5, granitoids of the Elbe Zone (Dohna & Laas granodiorites); 6, shear zone-related orthogneisses (Grossenhain gneiss); 7, 'red' orthogneisses, anatexites, migmatities (MP–MT unit); 8, gneisses and eclogites with major shear zones (HP–HT unit); 9, phyllites, garnet phyllites and mica schists (HP–LT unit, MP–LT unit and LP–LT unit); 10, low to high grade ortho- and para-rocks; 11, Ordovician, Silurian and Devonian volcano-sedimentary rock complexes; 12, Cambro–Ordovician to Lower Carboniferous volcano-sedimentary rock complexes; 13, Rumburk granite (Lausitz antiform); 14, Granulite and high grade country rocks of the granulite core; 15, southern phyllite zone (Cambro–Ordovician rock complex); 16, greywackes, pelites, cherts, volcanic rocks (Altenfeld Fm., Frohnberg Fm., Lausitz Gr., Leipzig Fm.); 17, greywackes, pelites, cherts, basalts, andesites (Rothstein Fm.); 18, greywackes, pelites, quartzites, basalts (passive margin sequences and tillites of the Clanzschwitz, Rödern and Weesenstein Gr.); 19, carbonates, sandstones, pelites (Lower to Middle Cambrian); 20, quartzites and pelites (Skolithos facies; Collmberg Fm., Hainichen–Otterwisch Fm., Dubrau Fm.; Lower Ordovician, Tremadoc); 21, quartzites, shales, sed. iron ore (Ordovician); 22, Cadomian and Ordovician rocks affected by the dextral Variscan Blumenau shear zone (Schwarzburg antiform); 23, Lower Graptolite Shale ('Unterer Graptolithenschiefer') and 'Ockerkalk' (Silurian); 24, carbonates, sandstones, pelites, diabases (Devonian); 25, greywackes and pelites (Variscan flysch; Tournai,Visean); 26, Variscan wildflysch with large olistolites; 27, Variscan early molasses of Hainichen-Borna (Upper Visean); 28, Variscan early molasses of Doberlug (Upper Visean); 29, olistolithes of Cambrian to Devonian rock complexes within a wildflysch matrix; 30, acid to basic metamorphic rocks of the nappe pile remnants of Münchberg and of the Saxon 'Zwischengebirge' of Wildenfels and Frankenberg; 31, Permo-Carboniferous granitoids; 32, Upper Carboniferous rhyolithes; 33, Upper Carboniferous major granitoids dykes.

are in the range of 80–150 km (see Fig. 4, distance of $D_2$-transport). The schist belt of the Elbe Zone is separated in two parts by the overlying Döhlen Basin, which is one of the several late-Variscan molasse basins formed in Permo-Carboniferous

time (Fig. 5). Against the Erzgebirge nappe complex, the schist belt of the Elbe Zone is bordered by the Mid-Saxon Fault (Fig. 5). A major component of the regional dextral shear occurred along that fault (Pietzsch 1962; Linnemann 1994).

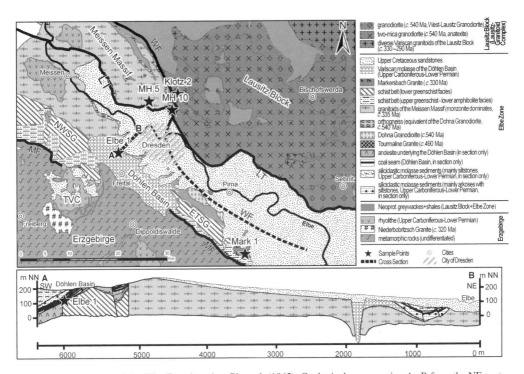

**Fig. 5.** Geological map of the Elbe Zone based on Pietzsch (1962). Geological cross-section A–B from the NE-part of the Döhlen Basin and the Meissen Massif modified after Reichel & Schauer (2007). WF, Westlausitz Fault; LT, Lausitz Thrust; MF, Mid-Saxon Fault; NWSG, Nossen-Wilsdruff Schiefergebirge (Nossen-Wilsdruff schistbelt); ETSG, Elbtalschiefergebirge (Elbtal schistbelt); TVC, Tharandt Volcanic Complex (Upper Carboniferous). In the map sample locations are indicated: Elbe 1, sample of an ash bed within the Döhlen Formation in the Döhlen Basin (Elbe Zone); Klotz 2, sample of a granodiorite from the Lausitz Granitoid Complex (Lausitz Block); MH 5, sample of the monzonite of the Meissen Massif (Elbe Zone); MH 10, sample of a gneiss from the eastern rim of the Meissen Massif (Elbe Zone); Mark 1, sample of a granite near the village Markersbach (Elbe Zone).

Towards the Lausitz Block in the NE, the schist belt of the Elbe Zone is bordered by two major faults, the Westlausitz Fault and the Lausitz Thrust (Fig. 5). The schist belt in the Elbe Zone contains parts of the Neoproterozoic (Cadomian) basement (Linnemann *et al.* 2007*a*) and Lower Ordovician to Lower Carboniferous sedimentary rocks and volcano–sedimentary complexes. The age of the youngest meta-sediments of the schist belt is poorly constrained by badly preserved fossils, but is thought to be Lower Carboniferous (Kurze *et al.* 1992).

The schist belt of the Elbe Zone was intruded by the granitoids of the Meissen Massif (Fig. 5). This complex consists of plutonic, dioritic to mainly monzonitic and granitic rocks, dominated by hornblende-monzonite. The granitoids show a signature typical for I-type granites. The diorite to monzonite intrusions show enriched mantle signatures typical for shoshonitic rocks (Wenzel *et al.* 1997). Nasdala *et al.* (1999) determined ages of $326 \pm 6$ Ma and $330 \pm 5$ Ma (SHRIMP U–Pb) for the time of the monzonite intrusion. The

sigmoidal shape of the Meissen Massif (Fig. 4, inset upper right) and deformation structures along the edges of the plutonic complex suggest an intrusion into an area of tension in the active dextral strike-slip regime along the Elbe Zone (Mattern 1996; Wenzel *et al.* 1997; Kroner *et al.* 2007).

The Markersbach Granite intruded into the southern part of the schist belt (Fig. 5). This alkali feldspar granite clearly intruded after dextral shear along the Elbe Zone was finished because this pluton cross-cuts all the Variscan structures, including faults and cleavage. According to Bonin (2007) this rock with its euhedral dark quartz and pink alkali feldspar, mostly mesoperthitic, and late anhedral mica is a typical A-type granite. Such granites are commonly emplaced in within-plate, continental or oceanic settings, or under transtensional regimes in postorogenic contexts (Bonin 2007).

Adjacent to the southern margin of the Elbe Zone the Tharandt Volcanic Complex (TVC) and the granite of Niederbobritzsch intruded (Fig. 5). As there are components of the TVC and the granite within the earliest *c.* 330 Ma old Variscan molasse of the Saxo-Thuringian Zone (Pietzsch 1962; Gehmlich 2003), both complexes must have been at the surface before 330 Ma. In addition, neither the TVC nor the Niederbobritzsch granite show shear-related structures, what we interpret as both complexes intruded or extruded, respectively, after dextral strike-slip movements along the Elbe Zone finished. It can be inferred that the TVC and the Niederbobritzsch granite may belong to the same magmatic event as the Markersbach granite, marking the age of a final pulse at the very end of the Variscan orogenic processes.

The Döhlen Basin post-dates all Variscan structures of the Elbe Zone and adjoining areas. It originated as a pull-apart basin, 22 × 6 km in dimension, which opened during post-Variscan (Late Carboniferous) strike-slip motion along the Elbe Zone (Reichel & Schauer 2007). It is situated to the SW of Dresden (Fig. 5) and consists of molasse of the Variscan Orogen with a maximum thickness of *c.* 700 m (Reichel & Schauer 2007). The volcano-sedimentary rocks within the basin overlie unconformably the schist belt of the Elbe Zone, the Meissen Massif and the Erzgebirge nappe complex (Figs 4 & 5). Palaeontological data suggest a deposition during latest Carboniferous and Lower Permian times (Roscher & Schneider 2005). Similar molasse deposits are known from the Lausitz Block (Weissig Basin). Thus, the basement rocks of the Lausitz Block, the plutonic rocks of the Meissen Massif, the schist belt of the Elbe Zone and the metamorphic rocks of the Erzgebirge nappe complex were completely exhumed in the Upper Carboniferous.

## Samples and methods

All sample locations are situated in the Elbe Zone and adjoining areas (Saxo-Thuringian Zone). Locations are given in Table 1 and in Figure 5. We collected one sample of a granodiorite from the Lausitz Block (sample Klotz 2) close to the Westlausitz Fault. From the Elbe Zone three samples were taken. Sample MH 10 is an orthogneiss which occurs along the eastern rim of the Meissen Massif. A monzonite (sample MH 05) was taken from the Meissen Massif itself, which shape indicates an intrusion into an active dextral shear setting. Sample Mark 1 is from the Markersbach Granite, which intruded the schist belt of the Elbe Zone and crosscuts all tectonic features. Sample Elbe 1 is a felsic ash layer from coal seam 5 of the Döhlen Formation in the Döhlen Basin (Fig. 6).

From each sample 2–5 kg of rock material was collected. Zircons were separated using heavy liquid, magnetic separation and hand-picking. The grains were subsequently mounted and polished

to expose their centre. Prior to the U–Pb analyses, the internal structures of the zircon grains were investigated by cathodoluminescence (CL). Uranium, thorium and lead isotopes from the zircon grains were analyzed by laser ablation (LA) using a Thermo-Scientific Element 2 SF-ICP-MS (sector field-inductively coupled plasma-mass spectrometer) coupled to a New Wave Research UP-213 ultraviolet laser system at Goethe University Frankfurt (GFU) (Gerdes & Zeh 2006, 2008). Data were acquired in time resolved – peak jumping – pulse counting mode over 810 mass scans, with a 16 s. background measurement followed by 28 s sample ablation. Laser spot-sizes varied from 20 to 30 μm with a typical penetration depth of *c.* 15–20 μm. The signal was tuned for maximum sensitivity for Pb and U while keeping oxide production, monitored as $^{254}UO/^{238}U$, well below 1%. A teardrop-shaped, low volume ($<2.5$ cm$^3$) laser cell was used (Frei & Gerdes 2009 and references therein). This cell enables detection and sequential sampling of heterogeneous grains (e.g. growth zones) during time resolved data acquisition, due to its response time of $<1$ s (time until maximum signal strength was achieved) and wash-out ($<99\%$ of previous signal) time of $<5$ s. With a depth penetration of *c.* 0.6 μms$^{-1}$ and a 0.9 s integration time ($= 15$ mass scans $= 1$ ratio), any significant variation of the Pb/Pb and U/Pb in the μm scale is detectable. Raw data were corrected offline for background signal, common Pb, laser-induced elemental fractionation, instrumental mass discrimination, and time-dependent elemental fractionation of Pb/U using an in-house MS Excel© spreadsheet program (Gerdes & Zeh 2006). A common-Pb correction based on the interference- and background-corrected $^{204}Pb$ signal and a model Pb composition (Stacey & Kramers 1975) was carried out where necessary. The necessity of the correction was judged on whether the corrected $^{207}Pb/^{206}Pb$ lay outside the internal errors of the measured ratios. The interference of $^{204}Hg$ (mean $= 110 \pm 14$ cps; counts per second) on mass 204 was estimated using a $^{204}Hg/^{202}Hg$ ratio of 0.2299 and measured $^{202}Hg$. Laser-induced elemental fractionation and instrumental mass discrimination were corrected by normalization to the reference zircon GJ-1 for each analytical session. Prior to this normalization, the drift in inter-elemental fractionation (Pb/U) during 28 s of sample ablation was corrected for the individual analysis. The correction was done by applying a linear regression through all measured ratios, excluding the outliers ($\pm 2$ standard deviation; 2SD), and using the intercept with the *y*-axis as the initial ratio. The total offset of the measured drift-corrected $^{206}Pb/^{238}U$ ratio from the 'true' ID-TIMS value ($0.0983 \pm 0.0004$; ID-TIMS GFU

**Table 1.** Laser Ablation-SF-ICP-MS U, Pb and Th data of zircon grains from the Cadomian and Variscan basement and from one Upper Carboniferous–Lower Permian ash bed of the Elbe Zone and Lausitz Block, Saxo-Thuringian Zone, Bohemian Massif (co-ordinates: UTM Zone 33)[a]

| Name (spot) | 207Pb[a] (cps) | U[b] (ppm) | Pb[b] (ppm) | Th[b]/U | 206Pb/204Pb | Isotopic ratios[c] 206Pb/238U | 2σ (%) | 207Pb/235U | 2σ (%) | 207Pb/206Pb | 2σ (%) | Rho[d] | Ages 206Pb/238U | ±2σ (Ma) | 207Pb/235U | ±2σ (Ma) | 207Pb/206Pb | ±2σ (Ma) | Conc (%) |
|---|---|---|---|---|---|---|---|---|---|---|---|---|---|---|---|---|---|---|---|
| *Klotz 2 (biotite granodiorite, Lower Cambrian, Lausitz Block, Klotzsche in the north of the city of Dresden, Easting: 54 15448, Northing: 56 66651)* | | | | | | | | | | | | | | | | | | | |
| Klotz 2–1 | 7084 | 180 | 16 | 0.22 | 1890 | 0.08797 | 1.8 | 0.6841 | 5.6 | 0.05640 | 5.3 | 0.31 | 544 | 10 | 529 | 30 | 468 | 118 | 116 |
| Klotz 2–2 | 9057 | 213 | 22 | 0.51 | 2715 | 0.09153 | 1.5 | 0.7334 | 3.4 | 0.05812 | 3.1 | 0.43 | 565 | 8 | 559 | 19 | 534 | 68 | 106 |
| Klotz 2–3 | 8293 | 208 | 18 | 0.18 | 3054 | 0.08790 | 1.7 | 0.7163 | 2.3 | 0.05910 | 1.5 | 0.73 | 543 | 9 | 548 | 12 | 571 | 33 | 95 |
| Klotz 2–4 | 9198 | 206 | 21 | 0.50 | 1023 | 0.08770 | 1.6 | 0.6799 | 4.2 | 0.05623 | 3.9 | 0.39 | 542 | 9 | 527 | 22 | 461 | 85 | 117 |
| Klotz 2–6 | 6745 | 171 | 16 | 0.25 | 3608 | 0.08904 | 1.5 | 0.7069 | 2.4 | 0.05758 | 1.9 | 0.61 | 550 | 8 | 543 | 13 | 514 | 43 | 107 |
| Klotz 2–7 | 3891 | 92 | 8 | 0.31 | 1729 | 0.08679 | 2.0 | 0.6621 | 10 | 0.05533 | 9.8 | 0.20 | 537 | 10 | 516 | 51 | 425 | 218 | 126 |
| Klotz 2–8 | 10124 | 271 | 23 | 0.09 | 57675 | 0.08846 | 1.4 | 0.7237 | 1.5 | 0.05934 | 1.5 | 0.68 | 546 | 7 | 553 | 11 | 580 | 32 | 94 |
| Klotz 2–9 | 8034 | 207 | 17 | 0.09 | 21358 | 0.08855 | 1.5 | 0.7167 | 2.3 | 0.05870 | 1.8 | 0.64 | 547 | 8 | 549 | 13 | 556 | 39 | 98 |
| Klotz 2–10 | 1906 | 46 | 5 | 0.46 | 1315 | 0.09244 | 1.8 | 0.7558 | 3.2 | 0.05929 | 2.6 | 0.56 | 570 | 10 | 572 | 18 | 578 | 57 | 99 |
| Klotz 2–11 | 7868 | 199 | 17 | 0.12 | 4016 | 0.08799 | 1.6 | 0.7066 | 2.4 | 0.05824 | 1.8 | 0.65 | 544 | 9 | 543 | 13 | 539 | 40 | 101 |
| Klotz 2–12 | 7173 | 182 | 15 | 0.13 | 2615 | 0.08650 | 1.9 | 0.6816 | 2.7 | 0.05715 | 2.0 | 0.69 | 535 | 10 | 528 | 14 | 497 | 44 | 108 |
| Klotz 2–13 | 3074 | 77 | 7 | 0.28 | 4466 | 0.09153 | 2.1 | 0.7455 | 3.3 | 0.05907 | 2.6 | 0.61 | 565 | 12 | 566 | 19 | 570 | 57 | 99 |
| Klotz 2–14 | 5322 | 54 | 6 | 6.90 | 177 | 0.08872 | 2.3 | 0.6918 | 8.2 | 0.05655 | 7.9 | 0.28 | 548 | 13 | 534 | 44 | 474 | 175 | 116 |
| Klotz 2–15 | 2929 | 73 | 7 | 0.40 | 1871 | 0.08780 | 1.8 | 0.7016 | 3.5 | 0.05796 | 3.0 | 0.52 | 543 | 10 | 540 | 19 | 528 | 65 | 103 |
| Klotz 2–16 | 2612 | 65 | 6 | 0.50 | 1201 | 0.09243 | 1.8 | 0.7528 | 3.4 | 0.05907 | 2.9 | 0.52 | 570 | 10 | 570 | 19 | 570 | 62 | 100 |
| Klotz 2–17 | 12298 | 314 | 27 | 0.13 | 6924 | 0.08793 | 1.6 | 0.6993 | 2.7 | 0.05769 | 2.2 | 0.59 | 543 | 9 | 538 | 14 | 518 | 47 | 105 |
| Klotz 2–20 | 7134 | 153 | 15 | 0.61 | 948 | 0.08820 | 1.5 | 0.6877 | 4.8 | 0.05655 | 4.6 | 0.30 | 545 | 8 | 531 | 26 | 474 | 102 | 115 |
| Klotz 2–22 | 2939 | 60 | 6 | 0.48 | 860 | 0.09384 | 1.9 | 0.7707 | 3.5 | 0.05957 | 3.0 | 0.55 | 578 | 11 | 580 | 20 | 588 | 64 | 98 |
| Klotz 2–23 | 3617 | 93 | 8 | 0.29 | 3210 | 0.08628 | 2.3 | 0.6804 | 6.4 | 0.05719 | 6.0 | 0.36 | 533 | 12 | 527 | 34 | 499 | 132 | 107 |
| Klotz 2–24 | 4762 | 114 | 10 | 0.16 | 3590 | 0.08818 | 1.7 | 0.7056 | 3.0 | 0.05804 | 3.0 | 0.49 | 545 | 9 | 542 | 19 | 531 | 65 | 103 |
| Klotz 2–25 | 3715 | 91 | 9 | 0.50 | 2142 | 0.09248 | 1.5 | 0.7422 | 3.2 | 0.05821 | 2.8 | 0.48 | 570 | 9 | 564 | 18 | 538 | 61 | 106 |
| Klotz 2–26 | 3337 | 80 | 7 | 0.40 | 1807 | 0.08778 | 1.8 | 0.7083 | 3.1 | 0.05852 | 2.5 | 0.58 | 542 | 10 | 544 | 17 | 549 | 55 | 99 |
| Klotz 2–27 | 3819 | 87 | 8 | 0.37 | 1000 | 0.09271 | 1.5 | 0.7560 | 2.5 | 0.05914 | 2.0 | 0.60 | 572 | 9 | 572 | 14 | 572 | 44 | 100 |
| Klotz 2–28 | 10344 | 153 | 15 | 0.83 | 398 | 0.08641 | 2.0 | 0.7036 | 3.1 | 0.05906 | 2.4 | 0.63 | 534 | 10 | 541 | 17 | 569 | 53 | 94 |
| Klotz 2–29 | 5029 | 115 | 10 | 0.44 | 1302 | 0.08776 | 1.8 | 0.6628 | 6.5 | 0.05477 | 6.2 | 0.28 | 542 | 10 | 516 | 34 | 403 | 140 | 135 |
| Klotz 2–31 | 7148 | 178 | 15 | 0.21 | 1252 | 0.08767 | 1.7 | 0.6965 | 2.9 | 0.05762 | 2.3 | 0.60 | 542 | 9 | 537 | 15 | 515 | 50 | 105 |
| Klotz 2–32 | 23586 | 478 | 43 | 0.54 | 836 | 0.08699 | 1.7 | 0.7086 | 2.8 | 0.05908 | 2.3 | 0.59 | 538 | 9 | 544 | 15 | 570 | 50 | 94 |
| Klotz 2–33 | 4474 | 116 | 11 | 0.43 | 2569 | 0.08861 | 1.4 | 0.7132 | 2.0 | 0.05838 | 2.0 | 0.73 | 547 | 8 | 547 | 11 | 544 | 29 | 101 |
| *MH 05 (monzonite, Carboniferous, Visean, Meissen Massif, Elbe Zone, Boxdorf in the north of the city of Dresden, Easting: 54 09338, Northing: 56 65295)* | | | | | | | | | | | | | | | | | | | |
| MH 05–3 | 5704 | 533 | 33 | 0.92 | 932 | 0.05352 | 2.2 | 0.3979 | 5.2 | 0.05391 | 4.7 | 0.43 | 336 | 7 | 340 | 15 | 367 | 106 | 91 |
| MH 05–7 | 5097 | 570 | 29 | 0.17 | 3545 | 0.05336 | 1.9 | 0.3937 | 2.2 | 0.05351 | 1.2 | 0.85 | 335 | 6 | 337 | 6 | 350 | 26 | 96 |
| MH 05–8 | 20408 | 663 | 55 | 1.55 | 301 | 0.05293 | 2.0 | 0.3855 | 2.8 | 0.05283 | 2.0 | 0.70 | 332 | 6 | 331 | 8 | 321 | 45 | 103 |
| MH 05–11 | 17585 | 897 | 67 | 1.34 | 808 | 0.05262 | 2.2 | 0.3958 | 5.6 | 0.05456 | 5.2 | 0.40 | 331 | 7 | 339 | 16 | 394 | 116 | 84 |
| MH 05–13 | 5792 | 634 | 48 | 1.67 | 2140 | 0.05213 | 1.9 | 0.3810 | 4.4 | 0.05301 | 4.0 | 0.42 | 328 | 6 | 328 | 13 | 329 | 91 | 100 |
| MH 05–18 | 2842 | 315 | 27 | 2.51 | 6167 | 0.05275 | 4.1 | 0.3966 | 5.2 | 0.05453 | 3.3 | 0.79 | 331 | 13 | 339 | 15 | 393 | 72 | 84 |
| MH 05–22 | 6118 | 684 | 39 | 0.65 | 2081 | 0.05464 | 2.5 | 0.3955 | 4.1 | 0.05250 | 3.3 | 0.61 | 343 | 8 | 338 | 12 | 307 | 74 | 112 |
| MH 05–23 | 6671 | 726 | 42 | 0.68 | 1630 | 0.05462 | 2.2 | 0.3979 | 3.5 | 0.05284 | 2.8 | 0.62 | 343 | 7 | 340 | 10 | 322 | 63 | 107 |
| MH 05–24 | 5611 | 700 | 39 | 1.02 | 1765 | 0.05435 | 2.2 | 0.3983 | 4.8 | 0.05316 | 4.3 | 0.45 | 341 | 7 | 340 | 14 | 336 | 98 | 102 |
| MH 05–26 | 3846 | 352 | 20 | 0.38 | 870 | 0.05215 | 2.1 | 0.3778 | 4.6 | 0.05254 | 4.1 | 0.46 | 328 | 7 | 325 | 13 | 309 | 93 | 106 |
| MH 05–29 | 4330 | 962 | 51 | 0.58 | 2038 | 0.05193 | 4.0 | 0.3859 | 5.6 | 0.05390 | 3.9 | 0.72 | 326 | 13 | 331 | 16 | 367 | 88 | 89 |
| MH 05–31 | 6983 | 695 | 43 | 1.15 | 796 | 0.05427 | 3.7 | 0.3949 | 4.9 | 0.05278 | 3.2 | 0.75 | 341 | 12 | 338 | 14 | 319 | 72 | 107 |

*(Continued)*

**Table 1.** *Continued*

| Name (spot) | 207Pb[a] (cps) | U[b] (ppm) | Pb[b] (ppm) | Th[b]/U | Isotopic ratios[c] | | | | | | | Rho[d] | Ages | | | | | | Conc (%) |
|---|---|---|---|---|---|---|---|---|---|---|---|---|---|---|---|---|---|---|---|
| | | | | | 206Pb/204Pb | 206Pb/238U | 2σ (%) | 207Pb/235U | 2σ (%) | 207Pb/206Pb | 2σ (%) | | 206Pb/238U | ±2σ (Ma) | 207Pb/235U | ±2σ (Ma) | 207Pb/206Pb | ±2σ (Ma) | |
| *MH 10 (biotite orthogneiss, Lower Cambrian, Meissen Massif, Elbe Zone, north of the city of Dresden, Easting: 54 14601, Northing: 56 63399)* | | | | | | | | | | | | | | | | | | | |
| MH 10-3 | 8026 | 47 | 29 | 1.62 | 5782 | 0.3953 | 2.8 | 7.452 | 3.2 | 0.1367 | 1.6 | 0.86 | 2148 | 51 | 2167 | 29 | 2186 | 29 | 98 |
| MH 10-4 | 5909 | 306 | 33 | 0.49 | 1304 | 0.08656 | 1.7 | 0.6843 | 4.3 | 0.05734 | 3.9 | 0.40 | 535 | 9 | 529 | 18 | 504 | 86 | 106 |
| MH 10-5 | 7213 | 372 | 38 | 0.24 | 1060 | 0.08742 | 1.6 | 0.6955 | 3.9 | 0.05770 | 3.5 | 0.40 | 540 | 8 | 536 | 16 | 518 | 78 | 104 |
| MH 10-7 | 4626 | 247 | 27 | 0.19 | 3821 | 0.09868 | 1.6 | 0.8205 | 2.1 | 0.06031 | 1.4 | 0.74 | 607 | 8 | 608 | 10 | 615 | 31 | 99 |
| MH 10-8 | 15354 | 529 | 73 | 0.82 | 400 | 0.1005 | 1.8 | 0.8387 | 3.4 | 0.06052 | 2.9 | 0.53 | 617 | 11 | 618 | 16 | 622 | 62 | 99 |
| MH 10-9 | 4089 | 246 | 26 | 0.41 | 2728 | 0.08874 | 1.6 | 0.7198 | 2.8 | 0.05883 | 2.3 | 0.57 | 548 | 8 | 551 | 12 | 561 | 50 | 98 |
| MH 10-10 | 4536 | 201 | 23 | 0.71 | 776 | 0.08776 | 2.4 | 0.6972 | 7.1 | 0.05762 | 6.7 | 0.33 | 542 | 12 | 537 | 30 | 515 | 148 | 105 |
| MH 10-11 | 2655 | 155 | 17 | 0.67 | 4388 | 0.08807 | 2.1 | 0.7184 | 3.5 | 0.05916 | 2.8 | 0.60 | 544 | 11 | 550 | 15 | 573 | 61 | 95 |
| MH 10-12 | 52138 | 200 | 127 | 0.94 | 17630 | 0.4300 | 3.9 | 10.26 | 4.1 | 0.1731 | 1.3 | 0.95 | 2306 | 76 | 2459 | 39 | 2588 | 22 | 89 |
| MH 10-13 | 2687 | 144 | 19 | 1.06 | 4374 | 0.09726 | 2.2 | 0.8000 | 3.4 | 0.05966 | 2.5 | 0.66 | 598 | 13 | 597 | 15 | 591 | 55 | 101 |
| MH 10-14 | 11313 | 564 | 62 | 0.48 | 2972 | 0.08816 | 2.6 | 0.7116 | 3.4 | 0.05854 | 2.2 | 0.76 | 545 | 13 | 546 | 14 | 550 | 48 | 99 |
| MH 10-16 | 9437 | 243 | 25 | 0.12 | 11088 | 0.08811 | 4.3 | 0.6971 | 6.1 | 0.05738 | 4.3 | 0.71 | 544 | 23 | 537 | 26 | 506 | 95 | 108 |
| *Elbe 1 (felsic ash bed, Upper Carboniferous, Döhlen Basin, Elbe Zone, Elbstolln near Dresden, Easting 54 03464, Northing: 56 55772)* | | | | | | | | | | | | | | | | | | | |
| Elbe 1-1 | 2244 | 122 | 6 | 0.45 | 4287 | 0.04760 | 2.1 | 0.3428 | 3.2 | 0.05223 | 3.2 | 0.54 | 300 | 6 | 299 | 10 | 296 | 74 | 101 |
| Elbe 1-4 | 2005 | 102 | 5 | 1.71 | 3333 | 0.04828 | 3.3 | 0.3474 | 6.7 | 0.05219 | 5.8 | 0.50 | 304 | 10 | 303 | 18 | 294 | 133 | 104 |
| Elbe 1-5 | 6732 | 176 | 10 | 0.69 | 4271 | 0.04830 | 2.2 | 0.3384 | 9.6 | 0.05081 | 9.3 | 0.23 | 304 | 7 | 296 | 25 | 232 | 215 | 131 |
| Elbe 1-6 | 4192 | 205 | 10 | 0.44 | 7953 | 0.04889 | 1.5 | 0.3560 | 2.6 | 0.05281 | 2.1 | 0.59 | 308 | 5 | 309 | 7 | 321 | 47 | 96 |
| Elbe 1-9 | 3553 | 136 | 7 | 0.41 | 997 | 0.04841 | 2.5 | 0.3560 | 6.9 | 0.05334 | 6.4 | 0.36 | 305 | 7 | 309 | 18 | 343 | 145 | 89 |
| Elbe 1-11 | 2013 | 96 | 5 | 0.59 | 3788 | 0.04837 | 1.8 | 0.3517 | 3.5 | 0.05273 | 3.0 | 0.51 | 305 | 5 | 306 | 9 | 317 | 69 | 96 |
| *Mark 1 (granite, Lower Carboniferous, Elbe Zone, near Markersbach, Easting 54 29073, Northing: 56 35602)* | | | | | | | | | | | | | | | | | | | |
| Mark 1-10 | 17185 | 273 | 16 | 0.76 | 608 | 0.05068 | 2.3 | 0.3671 | 4.0 | 0.05253 | 3.3 | 0.58 | 319 | 7 | 317 | 11 | 309 | 75 | 103 |
| Mark 1-13 | 79884 | 1190 | 71 | 0.23 | 396 | 0.05333 | 2.3 | 0.3900 | 2.5 | 0.05303 | 1.1 | 0.90 | 335 | 8 | 334 | 7 | 330 | 25 | 101 |
| Mark 1-27 | 609202 | 1985 | 182 | 0.11 | 75 | 0.05328 | 7.2 | 0.3877 | 8.8 | 0.05278 | 4.9 | 0.83 | 335 | 24 | 333 | 25 | 319 | 112 | 105 |
| Mark 1-28 | 3013 | 76 | 4 | 0.57 | 5438 | 0.05264 | 2.2 | 0.3868 | 3.7 | 0.05330 | 2.9 | 0.60 | 331 | 7 | 332 | 10 | 341 | 66 | 97 |
| Mark 1-32 | 249445 | 3010 | 190 | 0.08 | 362 | 0.05304 | 4.8 | 0.3923 | 5.9 | 0.05365 | 3.5 | 0.81 | 333 | 16 | 336 | 17 | 356 | 78 | 93 |
| Mark 1-40 | 114711 | 2228 | 231 | 0.13 | 4157 | 0.05274 | 4.6 | 0.3809 | 4.9 | 0.05238 | 1.5 | 0.95 | 331 | 15 | 328 | 14 | 302 | 34 | 110 |
| Mark 1-41 | 145046 | 3151 | 165 | 0.10 | 1085 | 0.05006 | 3.6 | 0.3614 | 3.8 | 0.05236 | 1.4 | 0.93 | 315 | 11 | 313 | 10 | 301 | 33 | 105 |
| Mark 1-42 | 65241 | 1377 | 84 | 0.25 | 1758 | 0.05369 | 2.1 | 0.3906 | 2.4 | 0.05277 | 1.2 | 0.87 | 337 | 7 | 335 | 7 | 319 | 27 | 106 |
| Mark 1-44 | 333624 | 2607 | 194 | 0.12 | 96 | 0.04918 | 5.7 | 0.3639 | 6.8 | 0.05366 | 3.7 | 0.84 | 309 | 17 | 335 | 19 | 357 | 84 | 87 |
| Mark 1-48 | 87842 | 2117 | 1 | 0.14 | 1579 | 0.05172 | 2.4 | 0.3777 | 2.5 | 0.05296 | 0.7 | 0.95 | 325 | 8 | 325 | 7 | 327 | 17 | 99 |
| Mark 1-51 | 82370 | 904 | 5 | 0.28 | 166 | 0.05156 | 2.6 | 0.3763 | 3.9 | 0.05293 | 2.9 | 0.67 | 324 | 8 | 324 | 11 | 326 | 66 | 99 |
| Mark 1-52 | 102843 | 1295 | 5 | 0.21 | 298 | 0.05001 | 3.5 | 0.3639 | 3.8 | 0.05278 | 1.4 | 0.92 | 315 | 11 | 315 | 10 | 320 | 33 | 98 |
| Mark 1-55 | 259099 | 3073 | 5 | 0.09 | 285 | 0.05318 | 5.1 | 0.3841 | 5.6 | 0.05238 | 2.3 | 0.91 | 334 | 17 | 330 | 16 | 302 | 52 | 111 |

[a] Within-run background-corrected mean 207Pb signal in counts per second.

[b] U and Pb content and Th/U ratio were calculated relative to GJ-1 and are accurate to approximately 10%.

[c] Corrected for background, mass bias, laser induced U–Pb fractionation and common Pb (if detectable, see analytical method) using Stacey & Kramers (1975) model Pb composition. 207Pb/235U calculated using 207Pb/206Pb/(238U/206Pb × 1/137.88). 238U/206Pb errors are propagated by quadratic addition of within-run errors (2 standard error) and the reproducibility of GJ-1 (2 standard deviation). 207Pb/206Pb errors were propagated following Gerdes & Zeh (2008).

[d] Rho is the error correlation defined as err206Pb/238U/err207Pb/235U.
See text for more details.

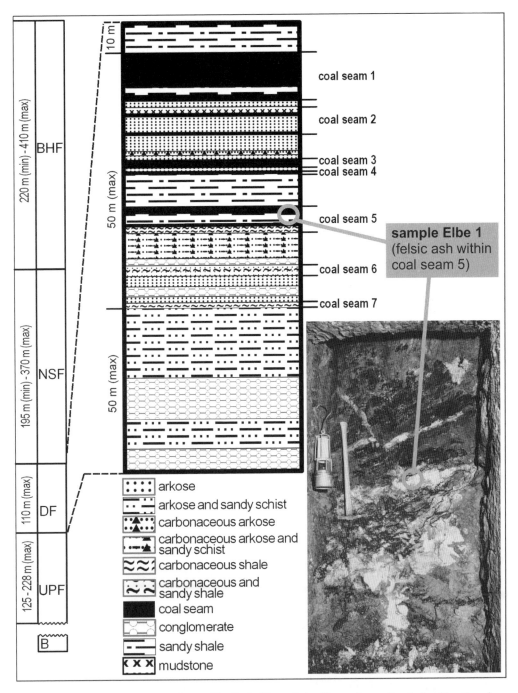

**Fig. 6.** Lithosection (normal profile) of the Döhlen Basin (Upper Carboniferous–Lower Permian) and location of the dated ash bed (sample Elbe 1) (modified after Reichel & Schauer 2007). Sample Elbe 1 was taken within the sequence of coal seam 5. Photograph to the left (by WISMUT GmbH 2008) shows the sample location in the gallery 'Tiefer Elbstolln' at 5872 m. BHF, Bannewitz–Hainsberg Formation; NSF, Niederhäslich–Schweinsdorf Formation; DF, Döhlen Formation; UPF, Unkersdorf–Potschappel Formation; B, Palaeozoic Basement.

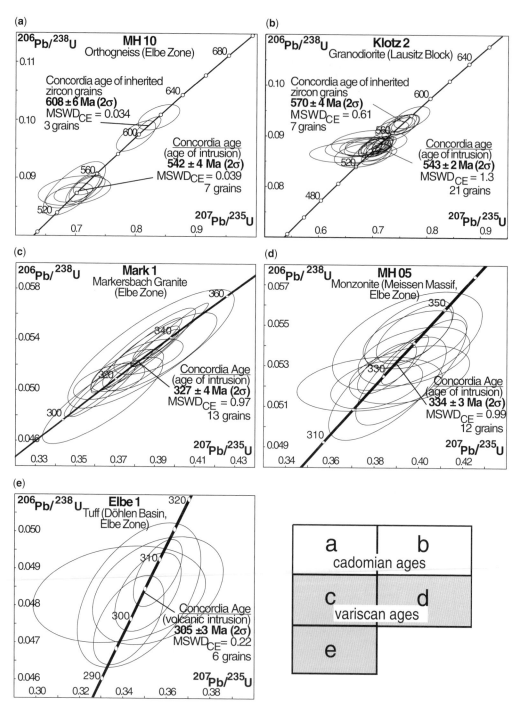

**Fig. 7.** U/Pb concordia plots of zircon grains from samples MH 10, Klotz 2, Mark 1, MH 5 and Elbe 1 (for location of the samples see Figs 5 & 6 and Table 1). Concordia plots (**a**) and (**b**) show the age of the two Cadomian samples (MH 10, Klotz 2), whereas the plots (**c**), (**d**) and (**e**) show the age of the Variscan samples (Mark 1, MH 05, Elbe 1). All data-point error ellipses are $2\sigma$. $MSWD_{C+E}$ = mean squared weighted deviation of concordance and equivalence. (a) and (b): The ages of the younger zircon populations are interpreted as the crystallization age of both Cadomian granitoids ($542 \pm 4$ Ma and $543 \pm 2$ Ma, respectively). The older populations are interpreted in terms of inheritance from the

value) of the analyzed GJ-1 grain was typically around 3–9%. Reported uncertainties ($2\sigma$) of the $^{206}Pb/^{238}U$ ratio were propagated by quadratic addition of the external reproducibility (2 SD %) obtained from the standard zircon GJ-1 (n = 12; 2 SD c. 1.3%) during the analytical session and the within-run precision of each analysis (2 SE %; standard error). In the case of the $^{207}Pb/^{206}Pb$, we used a $^{207}Pb$ signal-dependent uncertainty propagation (Gerdes & Zeh 2008). The accuracy of the method was verified by analyses of reference zircon 91500 (1064.8 ± 4.3 Ma, MSWD of concordance and equivalence = 0.86), Plešovice (337.7 ± 1.6 Ma, $MSWD_{C+E}$ = 0.84) and Temora (416.6 ± 2.5 Ma, $MSWD_{C+E}$ = 0.9).

## Results

Results of the U–Pb geochronology are given in Table 1 and in Figures 7, 8 and 9. For construction of concordia plots and the calculation of concordia ages Isoplot 2.49 (Ludwig 2001) was used. The granodiorite from the Lausitz Block (sample Klotz 2) (Fig. 7b, Table 1) exhibits two distinct zircon populations. The older population yielded a concordia age of 570 ± 4 Ma (7 grains). The younger population gave a concordia age of 543 ± 2 Ma (21 grains), which is interpreted as the crystallization age of this sample. This lowermost Cambrian age of intrusion indicates that the granodiorite is part of the Cadomian basement from the southern part of the Lausitz Block. The older c. 570 Ma population is interpreted to reflect inherited zircon grains consistent with the known age of magmatic activity in the Cadomian arc (e.g. Linnemann et al. 2007a).

The orthogneiss of the Elbe Zone (sample MH 10) (Fig. 7a, Table 1) shows similarities to sample Klotz 2 concerning the concordia age of its youngest zircon population, which yielded an age of 542 ± 4 Ma (7 grains). This age is likewise interpreted to represent the protolith age of the orthogneiss. Three older grains with a concordia age of 608 ± 6 Ma are interpreted as inherited zircons from an older Cadomian magmatic event. Other analyzed zircons from the orthogneiss include a Palaeoproterozoic zircon with a $^{207}Pb/^{206}Pb$ age of 2186 ± 29 Ma and one Neoarchaean zircon with an age of 2588 ± 22 Ma ($^{207}Pb/^{206}Pb$) (Table 1). The crystallization age and the inherited zircon populations suggests that the orthogneiss is part of the Gondwanan Cadomian basement,

which was deformed and metamorphosed during Variscan strike-slip in the Elbe Zone (Figs 4 & 5).

From the monzonite of the Meissen Massif (sample MH 05) 12 grains analyzed were concordant and equivalent with a concordia age of 334 ± 3 Ma (Figs 7d & 8a–b, Table 1). This zircon population is the only one that could be found and is interpreted to reflect the crystallization age of the monzonite. This age is more precise than, but overlaps within error with, the U–Pb SHRIMP age of Nasdala et al. (1999), who published crystallization ages of 326 ± 6 Ma and 330 ± 5 Ma for the monzonite of the Meissen Massif from other localities, and confirms the importance of magmatism of this age in the region.

In the Markersbach Granite (sample Mark 1) 13 grains were analyzed (Figs 7c & 8c–d, Table 1) and gave equivalent and concordant results. These grains represent the only population that could be found in this sample. The calculated concordia age is 327 ± 4 Ma and interpreted to be the crystallization age of the granite.

Six zircons from the felsic ash bed of the coal seam 5 from the Döhlen Basin were analyzed and yielded equivalent and concordant results corresponding to the concordia age of 305 ± 3 Ma (Figs 7e & 8e–f, Table 1). This age is interpreted to date the volcanic eruption that had formed the ash bed. Further, it provides an absolute age for the deposition of coal seam 5 in the Döhlen Formation.

## Discussion and conclusion

Our study identifies the granitoids from the Lausitz Block (sample Klotz 2) and from the Elbe Zone (sample MH 10) as intruded at c. 542–543 Ma. They, therefore, belong to the Cadomian orogenic cycle. According to Linnemann et al. (2007a) granitoids of this age were generated during an interval of high heat flow as a result of slap break-off during the arc–continent-collision of the Cadomian Orogeny. The Cadomian granodiorite from the Lausitz Block (sample Klotz 2) is not deformed by the strike-slip movements because it lays outside the shear zones along the Elbe Zone. In contrast, the Cadomian orthogneiss within the Elbe Zone (sample MH 10) obtained its gneissic foliation from the Variscan strike-slip processes along the Elbe Zone. However, the conditions during this shearing event did not affect the igneous zircon grains in the protoliths.

**Fig. 7.** (*Continued*) activity of the Cadomian arc magmatism at 608 ± 6 Ma and 570 ± 4 Ma, respectively. (c) and (d): Zircon grains of Variscan plutonic rocks of the Elbe Zone. The monzonite of the Meissen Massif (MH 05) results in a concordia age of 334 ± 3 Ma. The Markersbach Granite (sample Mark 1) is slightly younger (327 ± 4 Ma).
(e): Concordia plot showing U–Pb ages of 6 zircons of the ash bed in coal seam 5 of the Döhlen Formation (Elbe 1). The concordia age of 305 ± 3 Ma is interpreted as dating the volcanic eruption that formed the felsic ash bed and thus also reflects the absolute age of sedimentation for coal seam 5 from the Döhlen Formation (Döhlen Basin).

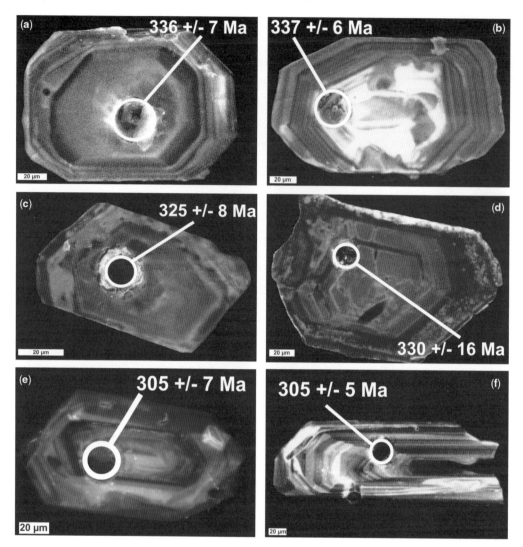

**Fig. 8.** Cathodoluminescence–SEM–Images of representative zircon grains of different samples, which have
been dated in this study to constrain the timing of the strike-slip and the final tectonic procedures during the Variscan
Orogeny. (**a**) and (**b**): Images of zircon grains of sample MH 05 from the Meissen Massif (Elbe Zone) frequently
reveal a uniform or complex zoned core and an wide outer domain with well-developed oscillatory zoning. The
obtained ages from different domains, however, are not distinguishable, suggesting that this structure reflect some
change of the physico-chemical conditions during magma crystallization. The obtained Variscan concordia age of
334 ± 3 Ma for the monzonite is interpreted to date an active strike-slip regime. (**c**) and (**d**): Zircon grains of sample
Mark 1 generally show dark luminescent CL images with less well developed or wide oscillatory zoning. This reflects
their high U content of up to 3150 ppm. Although the U–Pb spots have been carefully selected it is still remarkable
that no Pb-loss has been observed despite of the high U content. It clearly indicates that no later event has affected
this granite. The Variscan concordia age of 327 ± 4 Ma for the granite of Markersbach (Elbe Zone) marks the end of
the strike-slip movements along the Elbe Zone. (**e**) and (**f**): Zircon grains of sample Elbe 1 show well-developed
oscillatory zoning and yielded equivalent and concordant ages. The Variscan concordia age of 305 ± 3 Ma for the
ash bed within coal seam 5 of the Döhlen Formation in the Döhlen Basin (Elbe Zone) is reflecting a time where all
basement units where exhumed and sedimentation already started.

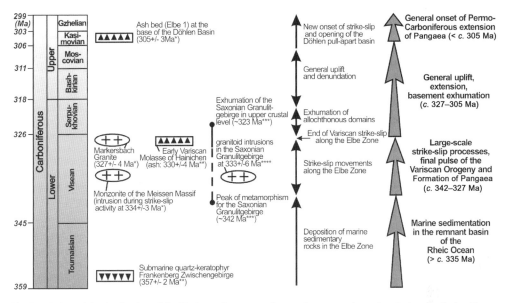

**Fig. 9.** Timing of the final pulse of the Variscan Orogeny and general geotectonic setting during the Carboniferous in the Elbe Zone and adjoining areas (summary) showing all samples with their zircon U/Pb ages presented in this study and additional data from the literature. Source of geochronological data: *, U–Pb LA-SF-ICP-MS (this study); **, Pb–Pb TIMS (Gehmlich 2003); ***, U–Pb TIMS (Romer & Rötzler 2001; Rötzler & Romer 2001); ****, U–Pb SHRIMP (Nasdala *et al.* 1996).

The final pulse of the Variscan Orogeny is responsible for the formation of the Elbe Zone. Due to the $D_2$-deformation (NW-ward transport of the allochthonous units Saxonian Granulitgebirge and Erzgebirge nappe complex) in the northern Bohemian Massif, dextral strike-slip movements along today's Elbe Zone were produced (Fig. 4) (Kroner *et al.* 2007). Our geochronological data allow a precise timing of related plutonic, volcano-sedimentary and tectonic processes (Fig. 9).

Fossils from the sedimentary rocks in the Elbe Zone demonstrate that marine sedimentation occurred until the Lower Carboniferous (Kurze *et al.* 1992). The only published age from marine Lower Carboniferous volcano-sedimentary complexes is a Pb–Pb-age of $357 \pm 2$ Ma from a syn-sedimentary submarine quarz-keratophyre extrusion (Gehmlich 2003) from a tectonostratigraphic unit close to the Elbe Zone (Frankenberg Zwischengebirge) (Fig. 9). This unit is situated between the Saxonian Granulitgebirge and the Erzgebirge nappe complex (Fig. 4). From these data we infer that Lower Carboniferous sedimentation occurs until at least the Tournaisian (Fig. 9).

Peak metamorphism for the Saxonian Granulitgebirge is defined at *c.* 342 Ma (Fig. 9). After that, exhumation of the Saxonian Granulitgebirge into upper crustal level occurred until *c.* 323 Ma (Romer & Rötzler 2001; Rötzler & Romer 2001). According to Kroner *et al* (2007), the beginning of

that exhumation forced dextral strike-slip processes along the Elbe Zone during $D_2$-deformation and NW-ward transport of allochthonous units. The Meissen Massif intruded at *c.* 334 Ma (age of intrusion, Fig. 7d) into a local area of extension in the schist belt of the Elbe Zone during ongoing dextral shear (Kroner *et al.* 2007). The sigmoidal shape of the Meissen Massif demonstrates that strike-slip movements along the Elbe Zone were still active at that time (Fig. 4). Dextral shear had terminated before the Markersbach Granite intruded at $327 \pm 4$ Ma, because the granite cross-cuts all tectonic structures related to the dextral shear. In contrast to the monzonite of the Meissen Massif, the Markersbach Granite itself is undeformed. At about the same time, the first terrestrial molasse (Early Variscan Molasse of Hainichen) was deposited at $330 \pm 4$ Ma along the southern periphery of the Saxonian Granulitgebirge overlying the Frankenberg Zwischengebirge (Gehmlich 2003) (Figs 4 & 9). These events in the range of *c.* 327–330 Ma are related to the final pulse of the Variscan Orogeny in the northern Bohemian Massif (Fig. 9).

For the time span between *c.* 327 Ma and the opening of the Döhlen Basin in Upper Carboniferous times, no significant magmatism is recorded. But during this episode the general uplift and basement exhumation to the palaeosurface occurred. The Döhlen Basin overlies the metamorphic rocks of the Erzgebirge nappe complex, the schist belt of the Elbe Zone and the Meissen Massif (Figs 4 & 5). In addition, equivalents of the Döhlen Basin overlie the Cadomian basement of the Lausitz Block (Weissig Basin).

This study provides the first precise estimate of the commencement of sedimentation in the Döhlen Basin. Available palaeontological data could not resolve the age of the onset of sedimentation. For example, the palaeoflora only allows to constrain the age of deposition to the transition from Upper Carboniferous–Lower Permian. The dated ash bed (sample Elbe 1 305 ± 3 Ma) from coal seam 5 of the Döhlen Formation shows clearly that sedimentation in the Döhlen Basin commenced by the Upper Carboniferous. However, the start of sedimentation in the basin must be earlier, because the underlying Unkerdorf–Potschappel–Formation unconformably overlies the Cadomian basement (Figs 5 & 6).

The time of general uplift, exhumation and denudation of the Variscan basement lasted a maximum of *c*. 22 Ma between the intrusion of the Markersbach Granite at *c*. 327 Ma and the onset of sedimentation in the Döhlen Basin at *c*. 305 Ma (Fig. 9).

This study defines the very end of Variscan orogenic processes in central Europe. These processes where strongly linked to the evolution of the Rheic Ocean, especially to its closure. As the closure of the Rheic started in the west and progressed towards the east (Fig. 3), collision of the dispersed Cadomian blocks with the Old Red Continent (Laurussia) and related subduction processes show chronological differentiations and vary between Early Devonian until Early Carboniferous (Kroner *et al.* 2007). Due to the northward-drift of Gondwana (and Cadomia, see Fig. 3) closure of the Rheic started at *c*. 400 Ma and triggered subduction of thinned Gondwanan continental lithosphere in central Europe (i.e. Bohemian Massif) at *c*. 370 Ma (Linnemann *et al.* 2007*a*; Kroner *et al.* 2007). In the Bohemian Massif (BM, see Fig. 1), the allochthonous domains (e.g. Saxonian Granulitgebirge, Erzgebirge nappe complex, see Fig. 4) mark such parts of subducted continental crust. At least for the northern BM in central Europe, subduction-related $D_1$-deformation lasted until *c*. 340 Ma (Kroner *et al.* 2007). $D_2$-deformation and related dextral strike-slip movements were still active at *c*. 334 Ma represented by the intrusion of the Meissen Massif into the Elbe Zone (this study, Fig. 7d). This means that, for central Europe, it took *c*. 36 Ma from the beginning of first subduction of continental lithosphere (*c*. 370 Ma) up to the point of completed subduction of oceanic and ongoing subduction of thinned continental crust and final orogen-related tectonic processes (*c*. 334 Ma). This $D_2$-deformation was finished at *c*. 327 Ma marked by the intrusion of the Markersbach Granite (this study, Fig. 7c).

The final exhumation and equilibrium of the Saxo-Thuringian crust in the area of the Elbe Zone took place in the time span from *c*. 327 Ma until the onset of Late Carboniferous molasse deposition at *c*. 305 Ma in the Döhlen Basin (this study). From the beginning of subduction of continental crust at *c*. 370 Ma until the opening of the Late Carboniferous molasse basins at *c*. 305 Ma, it took *c*. 65 Ma. When we interpolate that timing to the formation of the whole of Pangaea, we assume that for the formation of the supercontinent, *c*. 65 Ma were needed from the beginning of continental subduction to the start of dispersal represented by general stretching of the lithosphere and opening of molasse basins, as the Döhlen Basin.

This is a contribution to IGCP 453 'Modern and Ancient Orogens' and IGCP 497 ('The Rheic Ocean: Its Origin, Evolution and Correlatives'). The authors benefited from funding by the IGCP 497, by the Ludwig–Reichenbach–Gesellschaft e.V., and the Museum of Mineralogy and Geology (Senckenberg Naturhistorische Sammlungen Dresden). R. Strachan (Portsmouth, UK) and M. F. Pereira (Evora, Portugal) are thanked very much for critical and constructive reviews. Special thanks to B. Murphy for a lot of useful discussions and numerous corrections, not only of the language of the manuscript.

# References

BONIN, B. 2007. A-type granites and related rocks: evolution of a concept, problems and prospects. *Lithos*, **97**, 1–29.

BUSCHMANN, B., NASDALA, L., JONAS, P., LINNEMANN, U. & GEHMLICH, M. 2001. SHRIMP U–Pb dating of tuff-derived and detrital zircons from Cadomian marginal basin fragments (Neoproterozoic) in the northeastern Saxothuringian Zone (Germany). *Neues Jahrbuch Geologie Paläontologie, Monatshefte*, 321–342.

CHANTRAINE, J., AUVRAY, B., BRUN, J. P., CHAUVEL, J. J. & RABU, D. 1994. Introduction. *In*: KEPPIE, J. D. (ed.) *Pre-Mesozoic Geology in France and Related Areas*. Springer, Berlin, 75–80.

CHAUVEL, J. J. & RABU, D. 1988. Brioverian in Central Brittany. *In*: ZOUBEK, V., CONGE, D., KOZOUKHAROV, H. & KRAUTNER, G. (eds) *Precambrian in Younger Fold Belts*. J. Wiley & Sons, Chichester, 462–470.

ELICKI, O. 1997. Biostratigraphic data of the German Cambrian – present state of knowledge. *Freiberger Forschungshefte*, C **466**, 155–165.

FERNÁNDEZ-SUÁREZ, J., GUTIÉRRES-ALONSO, G., JENNER, G. A. & TUBRETT, M. N. 2000. New ideas on the Proterzoic-Early Palaeozoic evolution of NW Iberia: insights from U–Pb detrital zircon ages. *Precambrian Research*, **102**, 185–206.

FRANKE, W. 2000. The mid-European segment of the Variscides: tectonometamorphic units, terrane boundaries and plate tectonic evolution. *In*: FRANKE, W., HAAK, V., ONCKEN, O. & TANNER, D. (eds) *Orogenic processes – Quantification and Modelling in the Variscan Belt of Central Europe*. Geological Society, London, Special Publications, **179**, 35–61.

FRANKE, W. & ŻELAZNIEWICS, A. 2002. *Structure and Evolution of the Bohemian Arc*. Geological Society, London, Special Publications, **201**, 279–293.

FREI, D. & GERDES, A. 2009. Precise and accurate in situ U–Pb dating of zircon with high sample throughput by automated LA-SF-ICP-MS. *Chemical Geology*, **261**, 261–270, doi: 10.1016/j.chemgeo.2008.08.025.

GEHMLICH, M. 2003. Die Cadomiden und Varisziden des Saxothuringischen Terranes – Geochronologie magmatischer Ereignisse. *Freiberger Forschungshefte*, **500**, 1–129.

GERDES, A. & ZEH, A. 2006. Combined U–Pb and Hf isotope LA-(MC)-ICP-MS analyses of detrital zircons: comparison with SHRIMP and new constraints for the provenance and age of an Armorican metasediment in Central Germany. *Earth and Planetary Sciences Letters*, **249**, 47–61.

GERDES, A. & ZEH, A. 2008. Zircon formation versus zircon alteration – New insights from combined U–Pb and Lu–Hf *in-situ* La-ICP-MS analyses of Archean zircons from the Limpopo Belt. *Chemical Geology*, doi: 10.1016/j.chemgeo.2008.07.025.

GRAINDOR, M.-J. 1957. Le Briovérien dans le Nord-Est du massif Armoricain. *Memoires pour servir a l'explication de la Carte Géologique detaillée de la France*. Paris, 1–211.

KOSSMAT, F. 1927. Gliederung des varistischen Gebirgsbaues. *Abhandlungen des Sächsischen Geologischen Landesamtes*, **1**, 1–39.

KRONER, U., HAHN, T., ROMER, R. L. & LINNEMANN, U. 2007. The Variscan orogeny in the Saxo-Thuringian zone-heterogenous overprint of Cadomian/Palaeozoic peri-Gondwana crust. *In*: LINNEMANN, U., NANCE, R. D., KRAFT, P. & ZULAUF, G. (eds) *The Evolution of the Rheic Ocean: From Avalonian-Cadomian Active Margin to Alleghenian-Variscan collision*. Boulder, Colorado, Geological Society of America Special Paper, **423**, 153–172.

KURZE, M., LINNEMANN, U. & TRÖGER, K.-A. 1992. Weesensteiner Gruppe und Altpaläozoikum in der Elbtalzone (Sachsen). *Geotektonische Forschungen*, **77**, 101–167.

LINNEMANN, U.-G. 1994. Geologischer Bau und Strukturentwicklung der südlichen Elbezone. *Abhandlungen des Staatlichen Museums für Mineralogie und Geologie zu Dresden*, **40**, 7–36.

LINNEMANN, U. 2003a. Die Struktureinheiten des Saxothuringikums. *In*: LINNEMANN, U. (ed.) *Das Saxothuringikum*. Geologica Saxonica, **48/49**, 19–28.

LINNEMANN, U. 2003b. Sedimentation und geotektonischer Rahmen der Beckenentwicklung im Saxothuringikum (Neoproterozoikum – Unterkarbon). *In*: LINNEMANN, U. (ed.) *Das Saxothuringikum*. Geologica Saxonica, **48/49**, 71–110.

LINNEMANN, U. & ROMER, R. L. 2002. The Cadomian Orogeny in Saxo-Thuringia, Germany: geochemical and Nd–Sr–Pb isotopic characterisation of marginal basins with constraints to geotectonic setting and provenance. *Tectonophysics*, **352**, 33–64.

LINNEMANN, U. & SCHAUER, M. 1999. Die Entstehung der Elbezone vor dem Hintergrund der cadomischen und variszischen Geschichte des Saxothuringischen

Terranes – Konsequenzen aus einer abgedeckten geologischen Karte. *Zeitschrift für Geologische Wissenschaften*, **27**, 529–561.

LINNEMANN, U., GEHMLICH, M. *ET AL*. 2000. From Cadomian subduction to Early Palaeozoic rifting: the evolution of Saxo-Thuringia at the margin of Gondwana in the light of single zircon geochronology and basin development (Central European Variscides, Germany). London. Geological Society, London, Special Publications, **179**, 131–153.

LINNEMANN, U., MCNAUGHTON, N. J., ROMER, R. L., GEHMLICH, M., DROST, K. & TONK, C. 2004. West African provenance for Saxo-Thuringia (Bohemian Massif): did Armorica ever leave pre-Pangean Gondwana? U–Pb-SHRIMP zircon evidence and the Nd-isotopic record. *International Journal of Earth Sciences*, **93**, 683–705.

LINNEMANN, U., GERDES, A., DROST, K. & BUSCHMANN, B. 2007a. The continuum between Cadomian Orogenesis and opening of the Rheic Ocean: Constraints from LA-ICP-MS U–Pb zircon dating and analysis of plate-tectonic setting (Saxo-Thuringian zone, NE Bohemian massif, Germany). *In*: LINNEMANN, U., NANCE, R. D., KRAFT, P. & ZULAUF, G. (eds) *The Evolution of the Rheic Ocean: From Avalonian-Cadomian Active Margin to Alleghenian-Variscan Collision*. Boulder, Colorado, Geological Society of America Special Paper, **423**, 61–96.

LINNEMANN, U., NANCE, R. D., ZULAUF, G. & KRAFT, P. 2007b. Avalonian-Cadomian Belt, Adjoining Cratons and the Rheic Ocean. *In*: LINNEMANN, U., NANCE, R. D., KRAFT, P. & ZULAUF, G. (eds) *The Evolution of the Rheic Ocean: From Avalonian–Cadomian Active Margin to Alleghenian–Variscan Collision*. Boulder, Colorado, Geological Society of America Special Paper, **423**, 7–8.

LINNEMANN, U., PEREIRA, F., JEFFRIES, T. E., DROST, K. & GERDES, A. 2008a. The Cadomian Orogeny and the opening of the Rheic Ocean: the diacrony of geotectonic processes constrained by LA-ICP-MS U–Pb zircon dating (Ossa-Morena and Saxo-Thuringian Zones, Iberian and Bohemian Massifs). *Tectonophysics*, **461**, 21–43.

LINNEMANN, U., ROMER, R. L. *ET AL*. 2008b. Introduction (Chapter 2: 'The Precambrian'). *In*: MCCANN, T. (ed.) *The Geology of Central Europe*. Geological Society, London, Special Publications, 21–102.

LINNEMANN, U., D'LEMOS, R. *ET AL*. 2008c. Introduction (Chapter 3: 'The Cadomian Orogeny'). *In*: MCCANN, T. (ed.) *The Geology of Central Europe*. Geological Society, London, Special Publications, 103–154.

LUDWIG, K. R. 2001. *Users Manual for Isoplot/Ex rev. 2.49*. Berkeley Geochronology Center Special Publication, **1**, 1–55.

MARTÍNEZ, S. S., ARENAS, R., GARCÍA, F. D., MARTÍNEZ CATALAN, J. R., GÓMEZ-BARREIRO, J. & PEARCE, J. A. 2007. Careón ophiolithe, NW Spain: Suprasubduction zone setting for the youngest Rheic Ocean floor. *Geology*, **35**, 53–56.

MASSONNE, H.-J. 1998. *A new occurrence of microdiamonds in quartzofeldspathic rocks of the Saxonian Erzgebirge, Germany, and their metamorphic evolution*.

7th International Kimberlite Conference, Abstracts, Cape Town, 552–554.

MATTERN, F. 1996. The Elbe Zone at Dresden – a Late Palaeozoic pull-apart intruded shear zone. *Zeitschrift der Deutschen Geologischen Geselschaft*, **147**, 57–80.

NANCE, R. D., MURPHY, J. B. & KEPPIE, J. D. 2002. A Cordilleran model for the evolution of Avalonia. *Tectonophysics*, **352**, 11–31.

NANCE, R. D., BRENDAN, M. *ET AL.* 2008. Neoproterozoic-early Palaeozoic tectonostratigraphy and palaeogeography of the peri-Gondwanan terranes: Amazonian v. West African connections. *In*: ENNIH, H. & LIÉGEOIS, L.-P. (eds) *The Boundaries of the West African Craton*. Geological Society, London, Special Publications, **297**, 345–383.

NASDALA, L., GRUNER, T., NEMCHIN, A. A., PIDGEON, R. T. & TICHOMIROWA, M. 1996. New SCHRIMP ion microprobe measurements on zircons from Saxonian magmatic and metamorphic rocks. *Proceedings of the Freiberg Isotope Colloquium, TU Bergakadamie Freiberg*, 205–214.

NASDALA, L., WENZEL, T., PIDGEON, R. T. & KRONZ, A. 1999. Internal structures and dating of complex zircons from Meissen Massif monzonites, Saxony. *Chemical Geology*, **159**, 331–341.

PIETZSCH, K. 1962. *Geologie von Sachsen*. VEB Deutscher Verlag der Wissenschaften, Berlin.

REICHEL, W. & SCHAUER, M. 2007. *Das Döhlener Becken bei Dresden – Geologie und Bergbau*. Bergbaumonographie. Landesamt für Umwelt und Geologie Sachsen.

ROBARDET, M. 2002. Alternative approach to the Variscan Belt in southwestern Europe: preorogenic paleobiogeographicalo constraints. *In*: MARTINEZ CATALAN, J. R., HATCHER, R. D. J., ARENAS, R. & DIAZ GARCIA, F. (eds) *Variscan-Appalachian Dynamics: The Building of the Late Palaeozoic Basement*. Geological Society of America Special Paper, **364**, 1–15.

ROMER, R. L. & RÖTZLER, J. 2001. *P–T–t* Evolution of Ultrahigh-Temperature Granulites from the Saxon Granulite Massif, Germany. Part II: geochronology. *Journal of Petrology*, **42**(11), 2015–2032.

ROMER, R. L., LINNEMANN, U. & GEHMLICH, M. 2003. Geochronologische und isotopengeochemische Randbedingungen für die cadomische und variszische Orogenese im Saxothuringikum. *In*: LINNEMANN, U.

(ed.) *Das Saxothuringikum*. Geologica Saxonica, **48/49**, 19–28.

ROSCHER, M. & SCHNEIDER, J. W. 2005. An annotated correlation chart for continental Late Pennsylvanian and Permian basins and the marine scale. *In*: LUCAS, S. G. & ZEIGLER, K. E. (eds) *The nonmarine Permian*. New Mexico Museum of Natural History and Science Bulletin, **30**, 282–291.

RÖTZLER, J. & ROMER, R. L. 2001. *P–T–t* Evolution of Ultrahigh-Temperature Granulites from the Saxon Granulite Massif, Germany. Part I: petrology. *Journal of Petrology*, **42/11**, 1995–2013.

SÁNCHEZ-GARCÍA, T., BELLIDO, F. & QUESADA, C. 2003. Geodynamic setting and geochemical signatures of Cambrian-Ordovician rift-related igneous rocks (Ossa-Morena Zone, SW Iberia). *Tectonophysics*, **365**, 233–255.

SÁNCHEZ MARTÍNEZ, S., ARENAS, R., DÍAZ GARCÍA, F., MARTÍNEZ CATALÁN, J. R., GÓMEZ-BARREIRO, J. & PEARCE, J. A. 2007. Careón ophiolite, NW Spain: suprasubduction zone setting for the youngest Rheic Ocean floor. *Geology*, **35**(1), 53–56.

SCOTESE, C. R. & BARRET, S. F. 1990. Gondwana's movement over the South Pole during the Palaeozoic: evidence from lithological indicators of climate. *In*: McKERROW, W. S. & SCOTESE, C. R. (eds) *Palaeozoic Palaeogeography and Biogeography*. Geological Society, London, Memoir, **12**, 75–85.

STACEY, J. S. & KRAMERS, J. D. 1975. Approximation of terrestrial lead isotope evolution by a two-stage model. *Earth and Planetary Science Letters*, **26**, 207–221.

TORSVIK, T. H. & REHNSTRÖM, E. F. 2002. The Tornquist Sea and Baltica-Avalonia docking. *Tectonophysics*, **362**, 67–82.

WENZEL, T., MERTZ, D. F., OBERHÄNSLI, R. & BECKER, T. 1997. Age, geodynamic setting, and mantle enrichment processes of a K-rich intrusion from the Meissen massif (northern Bohemian massif) and implications for related occurrences from the mid-European Hercynian. *Geologische Rundschau*, **86**, 556–570.

ZEH, A., GERDES, A., WILL, T. M. & MILLAR, I. L. 2005. Provenance and Magmatic-Metamorphic Evolution of a Variscan Island-arc Complex: constraints from U–Pb dating, Petrology, and Geospeedometry of the Kyffhäuser Crystalline Complex, Central Germany. *Journal of Petrology*, **46**, 1393–1420.

# Variscan intra-orogenic extensional tectonics in the Ossa–Morena Zone (Évora–Aracena–Lora del Río metamorphic belt, SW Iberian Massif): SHRIMP zircon U–Th–Pb geochronology

M. FRANCISCO PEREIRA[1]*, MARTIM CHICHORRO[2], IAN S. WILLIAMS[3],
JOSÉ B. SILVA[4], CARLOS FERNÁNDEZ[5], MANUEL DÍAZ-AZPÍROZ[6],
ARTURO APRAIZ[7] & ANTONIO CASTRO[8]

[1]*Departamento de Geociências, Centro de Geofísica de Évora, Universidade
de Évora, Apt. 94, 7001-554 Évora, Portugal*

[2]*Centro de Investigação em Ciência e Engenharia Geológica,
Universidade Nova de Lisboa, Portugal*

[3]*Research School of Earth Sciences, The Australian National University,
Canberra, ACT, 0200, Australia*

[4]*Departamento de Geologia, Faculdade de Ciências, Universidade de Lisboa,
Edifício C3, Campo Grande, Lisboa, Portugal*

[5]*Departamento de Geodinámica y Paleontología, Facultad Ciencias Experimentales,
Universidad de Huelva, Campus Carmen, 21071 Huelva, Spain*

[6]*Departamento Ciencias Ambientales, Universidad Pablo Olavide, 41013 Sevilla, Spain*

[7]*Geodinamika Saila, Zientzia eta Teknologia Fak., Euskal Herriko
Unibertsitatea, Apt. 644, 48080 Bilbo, Spain*

[8]*Departamento de Geología, Universidad de Huelva, E21819 La Rábida, Huelva, Spain*

*\*Corresponding author (e-mail: mpereira@uevora.pt)*

**Abstract:** Following a Middle–Late Devonian (*c.* 390–360 Ma) phase of crustal shortening and mountain building, continental extension and onset of high-medium-grade metamorphic terrains occurred in the SW Iberian Massif during the Visean (*c.* 345–326 Ma). The Évora–Aracena–Lora del Río metamorphic belt extends along the Ossa–Morena Zone southern margin from south Portugal through the south of Spain, a distance of 250 km. This major structural domain is characterized by local development of high-temperature–low-pressure metamorphism (*c.* 345–335 Ma) that reached high amphibolite to granulite facies. These high-medium-grade metamorphic terrains consist of strongly sheared Ediacaran and Cambrian–early Ordovician (*c.* 600–480 Ma) protoliths. The dominant structure is a widespread steeply-dipping foliation with a gently-plunging stretching lineation generally oriented parallel to the fold axes. Despite of the wrench nature of this collisional orogen, kinematic indicators of left-lateral shearing are locally compatible with an oblique component of extension. These extensional transcurrent movements associated with pervasive mylonitic foliation (*c.* 345–335 Ma) explain the exhumation of scarce occurrences of eclogites (*c.* 370 Ma). Mafic-intermediate plutonic and hypabyssal rocks (*c.* 355–320 Ma), mainly I-type high-K calc-alkaline diorites, tonalites, granodiorites, gabbros and peraluminous biotite granites, are associated with these metamorphic terrains. Volcanic rocks of the same chemical composition and age are preserved in Tournaisian–Visean (*c.* 350–335 Ma) marine basins dominated by detrital sequences with local development of syn-sedimentary gravitational collapse structures. This study, supported by new U–Pb zircon dating, demonstrates the importance of intra-orogenic transtension in the Gondwana margin during the Early Carboniferous when the Rheic ocean between Laurussia and Gondwana closed, forming the Appalachian and Variscan mountains.

## Introduction

Recent studies of ancient and modern examples of orogenic belts have demonstrated the importance of intra-continental extensional tectonics during the terminal stages of, and immediately after, continental accretion (late- to post-orogenic processes). The main effects of such crustal thinning processes

*From*: MURPHY, J. B., KEPPIE, J. D. & HYNES, A. J. (eds) *Ancient Orogens and Modern Analogues.*
Geological Society, London, Special Publications, **327**, 215–237.
DOI: 10.1144/SP327.11   0305-8719/09/$15.00 © The Geological Society of London 2009.

are distributed over vast areas affected by large-scale orogen-parallel strike-slip tectonics. High heat flow due to asthenospheric upwelling and extensive emplacement of igneous rocks are characteristic of such thermally weakened lithosphere. Ductile deformation on major detachments favours the ascent and emplacement of metamorphic complexes and the rapid exhumation of medium-high-pressure metamorphic rocks (Sandiford & Powell 1986). In the upper crust, strike-slip and normal faulting controls the generation of basins with the development of gravitational collapse structures (Rey *et al.* 2001). The Pangaea amalgamation with the sequential closure of the Rheic ocean and assembly of continental blocks in a multiphase Middle and Late Palaeozoic orogenesis (Variscan Orogeny; Matte 2001) have been highlighted as a major feature of the pre-Permian geodynamic history of the Iberian Massif (e.g. Ribeiro *et al.* 1990). During the last two decades several contrasting plate tectonic models have been proposed to explain the remnants of the Avalonia–Gondwana collisional margin exposed in the western Iberian Massif. Exhaustive searching for the geodynamic significance of tectonothermal and sedimentary events related to the evolution of this Variscan oblique convergence margin (Ossa–Morena Zone/South–Portuguese Zone suture zone) has lead to several conflicting ideas and sparked ongoing controversy (Silva *et al.* 1990; Crespo-Blanc & Orozco 1991; Fonseca & Ribeiro 1993; Giese *et al.* 1994; Castro *et al.* 1996*a*; Matte 2001; Simancas *et al.* 2005; Onèzime *et al.* 2003; Silva & Pereira 2004; Díaz Azpíroz *et al.* 2005, 2006; Pereira *et al.* 2007; Azor *et al.* 2008; Pin *et al.* 2008; Rosas *et al.* 2008). The same happens with the regional geodynamic models that try to link this Variscan suture with the rootless suture of the NW Iberian Massif (Matte 2001; Robardet 2002; Simancas *et al.* 2005; Ribeiro *et al.* 2007; Martinez Catalan *et al.* 2007).

The aim of the present study was to contribute to a better understanding of the geodynamic evolution of this segment of the European Variscan chain during the Visean, a period of poorly-known intra-orogenic extension. Based on structural, geochemical and geochronological data now available from studies of high-grade metamorphic terrains exposed in Portugal (Évora-Chichorro 2006; Pereira *et al.* 2007) and Spain (Aracena-Díaz Azpíroz 2006; Díaz Azpíroz *et al.* 2006; and Lora del Río-Apraiz 1998; Apraiz & Eguiluz 2002). The results obtained are the key to constraining the onset of peak metamorphism and the duration of ductile deformation that accompanied the exhumation of medium to high-pressure rocks, and also the local gravitational collapse recognizable within the Tournaisian–Visean sedimentary basins from the Ossa–Morena Zone (Pereira *et al.* 2006*a*). Our

results have been interpreted through comparisons with NW Iberian Massif correlatives, which help us to better understand the crustal growth processes that led to the amalgamation of Pangaea in the Lower–Middle Mississipian. We discuss the tectonic model of intra-continental extension and onset of high-grade metamorphism that occurred at this time, contemporaneously with large-scale orogen-parallel tectonic (transcurrent) movements in the northern Gondwana margin.

## Geological background

The geological processes related to the formation of Pangaea during the Tournaisian–Visean are recognizable within the SW Iberian Massif regions (Fig. 1), in the western part of the European Variscan chain. The Ossa–Morena Zone experienced extension involving crustal deformation and metamorphism of deep crust, generation of metamorphic complexes, the emplacement of voluminous magmas and the development of marine basins.

The Ossa–Morena Zone is composed of continental crustal rocks with a complex but continuous stratigraphy. The geological record started in the Ediacaran with sedimentation and magmatism related to a magmatic arc (c. 560 Ma; Quesada 1990; Schäfer *et al.* 1993; Eguíluz *et al.* 2000; Pereira *et al.* 2006*b*, 2008), then evolved to a rifting process, active from the Cambrian to the Early Ordovician (c. 540–480 Ma; Liñán & Quesada 1990; Sanchez-Garcia *et al.* 2003; Robardet & Gutierrez-Marco 2004; Chichorro *et al.* in press). The record of tectonothermal processes for the period of approximately 25 Ma in the Middle–Late Devonian (c. 395–360 Ma) that predates the Tournaisian–Visean evolution is poorly known. On contrary, Devonian sedimentation is well-known and includes flysch (Terena syncline; Pereira & Oliveira 1996*a*) and platformal (Valle and Cerrón del Hornillo synclines; Robardet & Gutiérrez Marco 2004) sequences.

The Évora–Aracena–Lora del Río metamorphic belt discontinuously extends along the southern margin of the Ossa–Morena Zone from south Portugal through the south of Spain, a distance of 250 km. The belt includes three major exposures of metamorphic complexes (high-medium-grade metamorphic terrains surrounded by lower-grade metamorphic rocks). These metamorphic complexes are characterized by the development of high-temperature–low-pressure metamorphic conditions that reached upper amphibolite to granulite facies during the Visean (c. 345–335 Ma), affecting strongly sheared Ediacaran and Cambrian–early Ordovician protoliths (c. 540–480 Ma). The main structure consists of a dominant mylonitic foliation with gently plunging stretching lineations generally

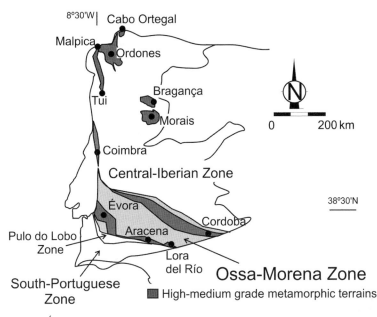

**Fig. 1.** Location of the Évora–Aracena–Lora del Río metamorphic belt in the SW Iberian Massif.

oriented parallel to the fold axes. Kinematic indicators of extension with a coeval left-lateral component (transtension) seem to be locally linked to the exhumation of eclogites.

The recognition of 20-km-wide MORB-derived amphibolites and phyllites of unknown age in between the Ossa–Morena Zone (Gondwana) and South Portuguese Zone (Laurussia) (Bard & Moine 1979; Munhá 1986; Fonseca & Ribeiro 1993; Quesada *et al.* 1994; Castro *et al.* 1999; Díaz Azpíroz *et al.* 2005, 2006) has been used to define a Variscan suture zone (Pulo do Lobo Zone; Quesada 1991; Quesada *et al.* 1994). Evidence for subduction of oceanic crust (Rheic Ocean?) beneath the Ossa–Morena Zone is accepted by the majority of authors but fundamental questions regarding the interactions of deformation, metamorphism and magmatism related to this particular stage of the SW Iberian Massif geodynamic evolution still remain unsolved: Why are there no significant testimonies of early Variscan foliation, regional metamorphism, arc magmatism and fore-arc sedimentation related with the supposed Middle–Upper Devonian subduction/collision (*c.* 390–370 Ma)?

## Évora massif

The Évora massif (Fig. 2), located in the southwesternmost part of the Ossa–Morena Zone, extends from Montemor-o-Novo to Évora (Portugal; Carvalhosa 1983; Pereira *et al.* 2003, 2007). There the lithostratigraphy includes Ediacaran

metagreywackes, micaschists, paragneisses and interbedded black metacherts, amphibolites and felsic gneisses (Série Negra; Carvalhosa 1965), an Early–Middle Cambrian igneous (felsic-dominated) complex with marbles, interbedded felsic and mafic metavolcanics, felsic gneisses and micaschists, and a Late Cambrian–Early Ordovician igneous (mafic-dominated) complex mainly composed of amphibolites with micaschists, quartzites, metatuffs and calc-silicate rocks (Carvalhosa 1999; Chichorro 2006; Pereira *et al.* 2007, 2008; Chichorro *et al.* 2008).

Three main units overprint this lithostratigraphy (Pereira *et al.* 2003, 2007): a southern low- to medium-grade unit (greenschist-upper amphibolite facies; Montemor-o-Novo shear zone) and a northern medium-grade unit (amphibolite facies; Évora medium-grade metamorphic terrains) represent the hanging-wall of a 25–45 km-wide and 75 km-long central high-grade unit (Évora high-grade metamorphic terrains). Orogen-parallel movements were responsible for the partial exhumation of a central high-grade unit with anatectic granitoids, migmatitic gneisses and diatexites (high-amphibolite facies and transitional between the amphibolite and granulite facies). The huge volume of melt produced during uplift and decompression of this high-grade unit is indicated by numerous granitoids emplaced along the boundary with the hanging-wall low- to medium-grade units (Pereira *et al.* 2007).

The structure is characterized by a moderately to steeply dipping mylonitic foliation and a weakly to moderately dipping stretching lineation.

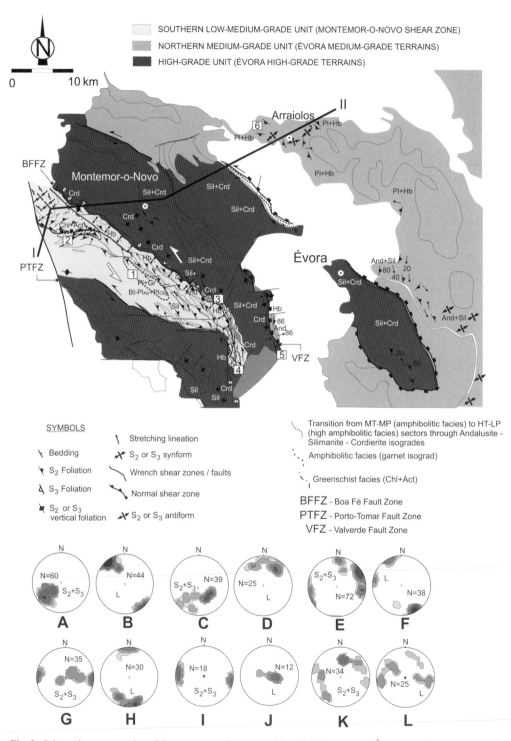

**Fig. 2.** Schematic representation of the structure and metamorphism distribution in the Èvora Massif (adapted from Pereira *et al.* 2007). Location of samples analysed for U–Pb SHRIMP zircon geochronology.

This mylonitic fabric is locally folded with fold axis parallel to the stretching lineation. Different mineral assemblages are found associated with the same fabric, revealing significant variations in the metamorphic conditions without any noticeable change in the regional strain regime. Application of geothermobarometers to garnet-rich amphibolites from the southern low- to medium-grade unit revealed an increase in temperature and pressure from c. 375–425 °C at 4.5–5.25 kbar to c. 475–525 °C at 6.6–7.5 kbar (Pereira et al. 2007). A regional episode of albitization has been interpreted as a consequence of pressure collapse followed by dynamic plagioclase recrystallization associated with transcurrent shearing (Chichorro 2006).

There has been a strong telescoping of the metamorphic gradient, as is typical of extensional regimes. The metamorphic isograds along the major detachment that separates the southern unit from the high-grade unit are telescoped, passing from the biotite to the sillimanite zone over a distance of few hundred metres. Paragneisses, which record the first stages of local melting, differentiation and the highest strain gradient, are characterized by well-developed mylonitic structures and textures (c. 650–750 °C, 2–4 kbar; Chichorro et al. 2004).

The contact of the northern medium-grade unit with the core high-grade unit is marked by an increase of temperature (with local partial melting) and shearing (local mylonitization). Low pressure and medium- to high-temperature metamorphism is characterized by the presence of sillimanite-garnet-cordierite-rich paragneisses (sillimanite zone) and andalusite-rich mica schists (andalusite zone) (Carvalhosa 1999). The highest temperatures were reached within the high-grade unit with diatexites showing mineral paragenesis consisting of plagioclase, K–feldspar, sillimanite and cordierite, and associated with progressive partial melting with dehydration reactions of muscovite and biotite. A regional well-developed late post-mylonitic replacement of biotite-andalusite-silimanite-cordierite by muscovite suggests slow cooling. The subsequent decrease of temperature produced low-grade assemblages with local replacement of biotite by chlorite (Pereira et al. 2007).

## Aracena massif

The Aracena massif (Bard 1969; Florido & Quesada 1984; Crespo-Blanc & Orozco 1991) is 2–10 km wide and extends for 120 km from Aroche to Almadén de la Plata (Spain). This massif (Fig. 3) has been divided into two domains (Castro et al. 1996a, b, 1999; Díaz Azpíroz 2006; Díaz Azpíroz et al. 2006): a southern domain that includes the Acebuches MORB-derived amphibolites (Bard

1969; Dupuy et al. 1979; Quesada et al. 1994; Castro et al. 1996b) dated at c. 340–332 Ma (Azor et al. 2008); and a northern domain separated into a northernmost low- to medium-grade unit (greenschist-amphibolite facies) and a southernmost high-grade unit (upper amphibolite–granulite facies). Both units consist of aluminous and calc-magnesium series derived from Ediacaran (Série Negra) and Cambrian protoliths, respectively. The high-grade unit is composed of pelitic and calc-silicate gneisses, marbles, amphibolites, migmatites and granulites (Ediacaran and Early Cambrian protoliths) that were metamorphosed at high temperature and low pressure. The complex structure of the high-grade unit of the Aracena metamorphic belt is the result of up to four phases of deformation (Díaz Azpíroz 2006). The pervasive phase was responsible for the generation of a crenulation and mylonitic foliation, which is the main planar fabric observed in the field. Structural analysis has revealed that this main foliation was originally sub-horizontal, although it was later folded during the last deformation phases (Díaz Azpíroz 2006). Quartz c-axis fabrics associated with the main foliation show characteristic patterns indicative of axial flattening with a vertical maximum shortening axis (Díaz Azpíroz 2006; Díaz Azpíroz et al. 2006). Chocolate tablet boudinage of calc-silicate rock layers interspersed within marbles supports this conclusion. Locally, shear zones have top-to-the-north and top-to-the-NE displacements. Therefore, the evolution of the high-grade unit of the Aracena metamorphic belt during the second deformation phase is best described as due to the extensional collapse of a previously thickened crust.

Metamorphic isograds are telescoped at the northern boundary of the high-grade unit, suggesting that this could be an original extensional contact (Florindo & Quesada 1984). Extension predated and partly coincided with the temperature peak recorded in the metamorphic assemblages and with migmatization and intrusion of abundant noritic and granitic plutons. Garnet-cordierite-sillimanite-rich migmatites experienced maximum metamorphic conditions estimated at 900 °C and 4–6 kbar (El-Biad 2000; Díaz Azpíroz 2006). Peak temperatures recorded in granulites with cordierite, garnet, spinel, sillimanite and K–feldspar reached 950–975 °C at a pressure of 5 kbar (Patiño Douce et al. 1997; El-Biad 2000). Migmatites have yielded a Visean age of c. 330–320 Ma (Rb–Sr biotite, feldspar; Castro et al. 1999), which is the supposed to be related to the extensional deformation. $^{40}Ar/^{39}Ar$ ages of the Acebuches amphibolites range from c. 340–330 Ma (Fonseca et al. 1993; Castro et al. 1999). The eastern extension of the Aracena metamorphic complex, the Almadén de la Plata core (Abalos

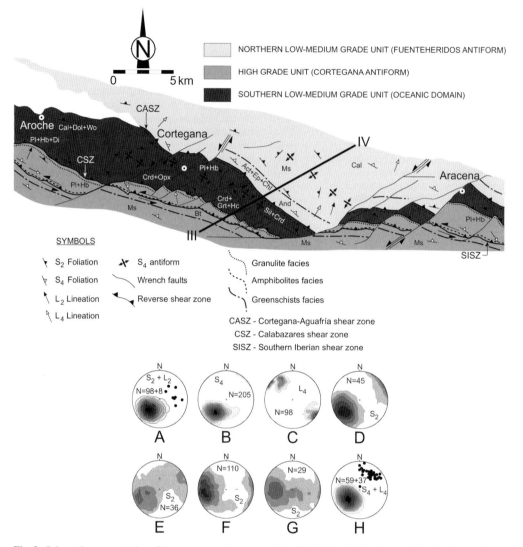

**Fig. 3.** Schematic representation of the structure and metamorphism distribution in the Aracena Massif (adapted from Díaz Azpíroz *et al.* 2006).

*et al.* 1991), also shows a prograde metamorphic evolution reaching in the highest-grade areas low pressure–high temperature mineral parageneses with garnet, cordierite and sillimanite (peak of metamorphism at 600–700 °C and 3–4 kbar).

### Lora del Río massif

The Lora del Río massif, which is approximately 12 km wide and 20 km long, crops out at the southeastern end of the Ossa–Morena Zone, where it is partially covered by Cenozoic sediments of the Guadalquivir basin (north of Seville, Spain; Fabriès 1963; Apraiz 1998; Apraiz & Eguiluz 2002). The

approximate age of extension, as given by metamorphic overgrowths on zircons in high-grade rocks, is Visean (*c.* 340 Ma, SHRIMP U–Pb; Ordóñez-Casado 1998). Three tectonic units that represents three structural levels were here defined (Fig. 4) and attest to a prograde metamorphism under low-pressure conditions (Apraiz & Eguiluz 2002): the upper low-medium grade unit (Los Miradores unit) in the north is represented by felsic tuffs, rhyolites, limestones, gabbros, slates and arkoses, micaschists and pelitic gneisses (greenschist-upper amphibolite facies); the intermediate low-medium grade unit (Hueznar unit) with slates, greywackes, metarhyolites and

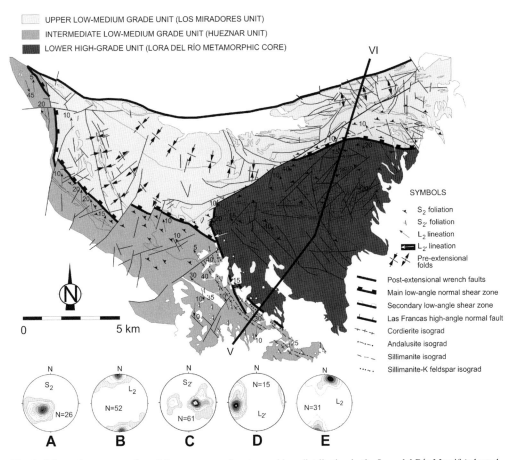

**Fig. 4.** Schematic representation of the structure and metamorphism distribution in the Lora del Río Massif (adapted from Apraiz & Eguiluz 2002).

micaschists (greenschist-upper amphibolite facies); and the lower high-grade unit (Lora del Río metamorphic core) made of metatexites, amphibolites, amphibolitic gneisses, graphite-rich micaschists, diatexites, leucogranulites and anatectic granites (upper amphibolite to granulite facies). Both upper unit and intermediate unit rocks are derived from Early–Middle Cambrian rocks, and the lower unit may also include Ediacaran protoliths. The metamorphism of the upper structural unit increases downwards to high-grade amphibolite facies conditions close to the detachment that marks its limit with the intermediate unit. At the detachment, partial melting is associated with the growth of sillimanite and K–feldspar, indicating that these rocks reached temperatures of $c.$ 650–750 °C and pressures of $c.$ 3–5 kbar (Apraiz 1998). The intermediate unit records maximum metamorphic conditions estimated as $c.$ 650–700 °C and 3–5 kbar (Apraiz 1998). The occurrence of leucogranulites with garnet, kyanite, K–feldspar and rutile in the lower

high-grade unit suggests a prograde metamorphic evolution through the kyanite stability field with the presence of garnet and kyanite (650–750 °C and 10–12 kbar, Apraiz 1998); these relics of higher pressure conditions suggest previous crustal thickening. Stretching lineations strike predominantly N–S, and kinematic criteria indicate a top-to-the-north normal shear sense for the northern boundary of the high-grade unit. Isobaric or decompression heating is interpreted to be coeval with normal slip on the main detachments that bound the lower high-grade unit ($c.$ 750–800 °C and 4–6 kbar, with growth of cordierite and sillimanite; Apraiz 1998).

## Tournaisian-Visean magmatism and marine basins

Diorites, tonalites, granodiorites and gabbros ($c.$ 350–320 Ma) represent I–type, high–K calc-alkaline magmatism (Santos *et al.* 1987, 1990;

Moita *et al.* 2005*a*; Tornos *et al.* 2005) associated with the Évora–Aracena–Lora del Rio metamorphic belt. Less representative norites of boninitic affinity (Castro *et al.* 1996*a*; El Hmidi 2000; Díaz Azpíroz *et al.* 2006) and peraluminous biotite granite are also present (Chichorro 2006; Pereira *et al.* 2008).

The Beja mafic-ultramafic rocks (*c.* 350 Ma; Pin *et al.* 1999, 2008) are banded gabbros with tholeiitic to calc-alkaline affinity, cumulate textures and magmatic foliation (Santos *et al.* 1987). This intrusive episode was followed by another magmatic event at *c.* 330 Ma (Jesus *et al.* 2006) that produced the calc-alkaline Cuba gabbros and diorites.

Further to the north (present co-ordinates), peraluminous I-type, calc-alkaline granites (Arraiolos biotite granite; Chichorro 2006) and tonalites and gabbros (Hospitais tonalites and Alto de São Bento leucogranites, with a crystallization/cooling age of *c.* 320 Ma; Moita 2007) show intrusive relations with the Évora high-grade and medium-grade metamorphic terrains. The mafic-intermediate intrusions are thought to be derived from the mantle based on their isotopic signature (Santos *et al.* 1987; Pin *et al.* 1999; Moita *et al.* 2005*a*; Moita 2007), while the granites contain a mixture of mantle and crustal components.

The same chemical compositions and ages are preserved in Tournaisian–Visean (*c.* 350–335 Ma) volcanic rocks associated with marine sedimentation in a relatively shallow-water platform environment (Santos *et al.* 1987; Chichorro 2006; Pereira *et al.* 2007). These marine basins, dominated by detrital sequences with synsedimentary gravitational collapse structures, are preserved in the southwestern Ossa–Morena Zone in Portugal (Pereira *et al.* 2006*a*) and record an early period of mixed siliciclastic-carbonate platform sedimentation, which was transitional to predominantly siliciclastic deposits. The Toca de Moura volcano-sedimentary complex exposed to the north of the Beja gabbros consists of basalts (pillow-lavas and hyaloclastites), microdiorites, andesites, rhyolites, felsic-intermediate tuffs and shales (Gonçalves 1985; Santos *et al.* 1987; Pereira *et al.* 2006*a*). To the north, the Cabrela volcano-sedimentary complex is exposed. A siliciclastic-carbonate sequence (calcic-turbidites) near the base passes upwards into a predominantly siliciclastic sequence with shales, greywackes and interbedded conglomerates, with associated felsic tuffs, volcanic breccias, rhyodacites, rhyolites and subordinate andesites (Ribeiro 1983; Chichorro 2006).

Shales from both basins contain miospores of late Tournaisian to late Visean age (Pereira *et al.* 2006*a*). Slumps, growth-faults and olistoliths of limestone (of Fammenian and Fransnian ages)

reflect the influence of gravitational processes during sedimentation.

## SHRIMP U–Th–Pb geochronology

### Sample preparation and analytical methods

Five samples were collected from the Évora Massif for U–Pb dating of zircon: two paragneisses from the Ediacaran Série Negra, BSC-1 (Universal Transverse Mercator co-ordinates–29SNC707609) and SEC-1 (29SNC742677); two of the orthogneisses from the Cambrian igneous-sedimentary complexes, ALC-10 (29SNC734568) and VLV-3 (29SNC855655) and one of an anatectic biotite granite ARL-6 (29SNC762664), probably formed by partial melting of the Série Negra sediments and Cambrian igneous-sedimentary complexes. The zircon grains show complex structures, most consisting of a core surrounded by a much younger overgrowth. The ages of the cores and their tectonic implications have been reported by Pereira *et al.* (2008). Here we report the ages of the zircon metamorphic/hydrothermal overgrowths and their significance.

Zircon grains were extracted, then mounted in epoxy resin with zircon standards SL13 (U = 238 ppm) and TEMORA ($^{206}Pb^*/^{238}U$ = 0.06683), using procedures described by Pereira *et al.* (2008). The polished mount was photographed and imaged by SEM cathodoluminescence (CL) to document the internal growth zoning of the grains, prior to SHRIMP analysis at the ANU using a procedure similar to that described by Williams & Claesson (1987). Monazite recovered from VLV-3 was mounted separately with monazite standard WB.T.329 from Thompson Mine, Manitoba ($^{206}Pb^*/^{238}U$ = 0.3152, U ≈ 2100 ppm), and was analysed using a procedure similar to that described by Williams *et al.* (1996). The Th-related isobar which potentially interferes with $^{204}Pb$ was reduced to insignificance by using moderate energy filtering.

The plotted and tabulated analytical uncertainties are 1$\sigma$ precision estimates. Uncertainties in the calculated mean ages are 95% confidence limits (t$\sigma$, where t is the Student's t multiplier) and, for the mean $^{206}Pb/^{238}U$ ages, include the uncertainty in the Pb/U calibration (*c.* 0.3–0.5%). Ages were calculated using the constants recommended by the IUGS Subcommission on Geochronology (Steiger & Jäger 1977). Common Pb corrections assumed a model common Pb composition appropriate to the age of each spot (Cumming & Richards 1975). Best estimates of the individual ages (Inferred ages, Tables 1 & 2) were calculated from the radiogenic $^{206}Pb/^{238}U$ (common Pb correction based on $^{207}Pb$). Wherever

**Table 1.** Zircon SHRIMP data from the paragneisses SEC-1 and BSC-1

| Analysis | Type | $Pb^*$ (ppm) | U (ppm) | Th (ppm) | Th/U | $^{204}Pb$ ppb | $^{204}Pb/^{206}Pb$ | ± | $f_{206}Pb$ | $^{208}Pb^*/^{206}Pb$ | ± | $^{208}Pb^*/^{232}Th$ | ± | $^{206}Pb^*/^{238}U$ | ± | $^{207}Pb^*/^{206}Pb$ | ± | $^{208}/^{232}$ | ± | $^{206}/^{238}$ | ± | $^{207}/^{206}$ | ± | $Inf/Age_§$ | ± |
|---|---|---|---|---|---|---|---|---|---|---|---|---|---|---|---|---|---|---|---|---|---|---|---|---|---|
| *Paragneiss SEC-1* | | | | | | | | | | | | | | | | | | | | | | | | | |
| 1.1 | UZO | 32 | 448 | 108 | 0.24 | 4 | 1.5E-04 | 6.5E-05 | 0.00266 | 0.0856 | 0.0033 | 0.02612 | 0.00113 | 0.07330 | 0.00108 | 0.05692 | 0.00146 | 521.1 | 22.3 | 456.0 | 6.5 | 488.6 | 57.6 | 455.5 | 6.5 |
| 2.1 | UZO | 25 | 523 | 2 | 0.00 | 5 | 5.3E-04 | 1.7E-04 | 0.00373 | – | – | – | – | 0.05219 | 0.00065 | 0.05491 | 0.00154 | – | – | 327.9 | 4.0 | 408.4 | 63.8 | 327.1 | 4.0 |
| 4.1 | UZO | 42 | 812 | 34 | 0.04 | 2 | 5.2E-05 | 2.9E-05 | 0.00095 | 0.0187 | 0.0023 | 0.02501 | 0.00309 | 0.05564 | 0.00050 | 0.05283 | 0.00176 | 499.4 | 61.1 | 349.0 | 3.1 | 321.5 | 77.3 | 349.3 | 3.1 |
| 5.1 | UZO | 15 | 277 | 28 | 0.10 | 6 | 4.3E-04 | 2.2E-04 | 0.00751 | – | – | – | – | 0.05843 | 0.00186 | 0.05144 | 0.00279 | – | – | 366.1 | 11.3 | 260.7 | 129.4 | 367.1 | 11.0 |
| 6.1 | UZO | 27 | 390 | 40 | 0.10 | 32 | 1.0E-03 | 1.9E-04 | 0.02304 | – | – | – | – | 0.07356 | 0.00095 | 0.05573 | 0.00338 | – | – | 457.6 | 5.7 | 441.7 | 141.0 | 457.8 | 5.9 |
| 8.1 | UZO | 21 | 349 | 36 | 0.10 | 16 | 8.9E-04 | 2.0E-04 | 0.01621 | 0.1327 | 0.0108 | 0.07448 | 0.00626 | 0.05847 | 0.00094 | 0.06022 | 0.00360 | 1451.9 | 118.1 | 366.3 | 5.8 | 611.6 | 134.9 | 363.5 | 5.6 |
| 9.1 | UZO | 30 | 546 | 8 | 0.02 | 25 | 1.6E-03 | 3.2E-04 | 0.01605 | – | – | – | – | 0.05997 | 0.00055 | 0.05999 | 0.00204 | – | – | 375.5 | 3.3 | 603.2 | 75.5 | 372.8 | 3.4 |
| 6.2 | UZO | 38 | 668 | 13 | 0.02 | 34 | 2.2E-03 | 3.8E-04 | 0.01715 | – | – | – | – | 0.06228 | 0.00165 | 0.05607 | 0.00209 | – | – | 389.5 | 10.0 | 455.3 | 85.1 | 388.7 | 10.0 |
| 2.2 | UZO | 47 | 958 | 9 | 0.01 | 11 | 1.4E-04 | 9.0E-05 | 0.00436 | – | – | – | – | 0.05446 | 0.00034 | 0.05295 | 0.00127 | – | – | 341.9 | 2.1 | 326.8 | 55.5 | 342.0 | 2.1 |
| 2.3 | UZO | 53 | 951 | 5 | 0.00 | 27 | 9.2E-04 | 2.5E-04 | 0.00973 | – | – | – | – | 0.06160 | 0.00077 | 0.05427 | 0.00113 | – | – | 385.3 | 4.7 | 382.2 | 47.5 | 385.4 | 4.7 |
| 4.2 | UZO | 13 | 297 | 7 | 0.02 | 3 | 2.1E-04 | 1.3E-04 | 0.00473 | – | – | – | – | 0.04959 | 0.00086 | 0.05570 | 0.00236 | – | – | 312.0 | 5.3 | 440.4 | 97.2 | 310.8 | 5.3 |
| 11.1 | UZO | 16 | 337 | 2 | 0.01 | 2 | 8.6E-05 | 1.5E-04 | 0.00224 | – | – | – | – | 0.05147 | 0.00097 | 0.05236 | 0.00158 | – | – | 323.5 | 5.9 | 301.0 | 70.1 | 323.8 | 6.0 |
| *Paragneiss BSC-1* | | | | | | | | | | | | | | | | | | | | | | | | | |
| 1.1 | UZO | 11 | 224 | 4 | 0.02 | 4 | 9.1E-04 | 6.3E-04 | 0.00646 | – | – | – | – | 0.05601 | 0.00060 | 0.05538 | 0.00200 | – | – | 351.3 | 3.6 | 427.5 | 82.5 | 350.5 | 3.7 |
| 3.1 | LUZO | 7 | 153 | 5 | 0.03 | 13 | 3.1E-03 | 5.9E-04 | 0.03270 | – | – | – | – | 0.05363 | 0.00164 | 0.05312 | 0.00382 | – | – | 336.8 | 10.0 | 333.7 | 171.9 | 336.8 | 10.0 |

\*, Radiogenic.

$f$, Correction for common Pb–Fraction of total $^{206}Pb$ ($^{206}Pb_T$) that is common $^{206}Pb$ ($^{206}Pb_c$).

§, Best estimated of the age of the analysed zircon. (See text for explanation.) Uncertainties one standard error. UZO, Unzoned overgrowth; LUZO, Light unzoned overgrowth.

**Table 2.** *Zircon SHRIMP data from the orthogneisses ALC-10 and VLV-3*

| Analysis | Type | Pb* (ppm) | U (ppm) | Th (ppm) | Th/U | $^{204}$Pb/ ppb | $^{204}$Pb/$^{206}$Pb | ± | $f$ $^{206}$Pb | $^{208}$Pb*/$^{206}$Pb | ± | $^{208}$Pb*/$^{232}$Th | ± | $^{206}$Pb*/$^{238}$U | ± | $^{207}$Pb*/$^{206}$Pb | ± | $^{208}$/$^{232}$ | ± | $^{206}$/$^{238}$ | ± | $^{207}$/$^{206}$ | ± | Inf Age§ | ± |
|---|---|---|---|---|---|---|---|---|---|---|---|---|---|---|---|---|---|---|---|---|---|---|---|---|---|
| *Orthogneiss ALC-10* | | | | | | | | | | | | | | | | | | | | | | | | | |
| 1.1 | UZ(r) | 19 | 376 | 15 | 0.04 | 4 | 3.4E-05 | 3.6E-05 | 0.00391 | – | – | – | – | 0.05425 | 0.00078 | 0.05179 | 0.00160 | – | – | 340.6 | 4.7 | 276.2 | 72.4 | 341.2 | 4.8 |
| 2.1 | UZ(r) | 14 | 272 | 10 | 0.04 | 2 | 2.6E-04 | 1.5E-04 | 0.00339 | – | – | – | – | 0.05457 | 0.00111 | 0.05524 | 0.00194 | – | – | 342.5 | 6.8 | 421.8 | 80.5 | 341.7 | 6.8 |
| 3.1 | UZ(r) | 27 | 535 | 10 | 0.02 | 4 | 8.1E-05 | 7.7E-05 | 0.00258 | – | – | – | – | 0.05455 | 0.00049 | 0.05206 | 0.00141 | – | – | 342.4 | 3.0 | 288.2 | 63.2 | 342.9 | 3.0 |
| 4.1 | UZO | 33 | 665 | 28 | 0.04 | 3 | 1.5E-04 | 1.4E-04 | 0.00170 | – | – | – | – | 0.05400 | 0.00052 | 0.05385 | 0.00129 | – | – | 339.1 | 3.2 | 364.6 | 54.8 | 338.8 | 3.2 |
| 5.1 | LUZ(r) | 11 | 230 | 8 | 0.03 | 4 | 4.5E-04 | 2.5E-04 | 0.00658 | – | – | – | – | 0.05244 | 0.00084 | 0.05174 | 0.00263 | – | – | 329.5 | 5.2 | 274.0 | 120.9 | 330.0 | 5.2 |
| *Orthogneiss VLV* | | | | | | | | | | | | | | | | | | | | | | | | | |
| 1.1 | UZO | 87 | 1374 | 334 | 0.24 | 337 | 4.1E-03 | 2.4E-04 | 0.07419 | 0.0860 | 0.0099 | 0.02287 | 0.00266 | 0.06471 | 0.00066 | 0.05744 | 0.00431 | 457.1 | 52.6 | 404.2 | 4.0 | 508.3 | 174.0 | 403.0 | 3.7 |
| 2.1 | UZO | 53 | 722 | 207 | 0.29 | 46 | 7.1E-04 | 1.4E-04 | 0.01760 | – | – | – | – | 0.07462 | 0.00072 | 0.05437 | 0.00208 | – | – | 464.0 | 4.3 | 386.6 | 88.3 | 465.0 | 4.4 |
| 3.1 | UZ | 97 | 1278 | 351 | 0.27 | 6 | 7.1E-05 | 4.2E-05 | 0.00127 | 0.0841 | 0.0023 | 0.02368 | 0.00068 | 0.07737 | 0.00064 | 0.05733 | 0.00108 | 473.1 | 13.4 | 480.4 | 3.8 | 504.3 | 42.0 | 480.0 | 3.8 |
| 4.1 | UZ | 54 | 666 | 219 | 0.33 | 1 | 2.1E-05 | 1.8E-05 | 0.00038 | 0.1066 | 0.0019 | 0.02637 | 0.00068 | 0.08134 | 0.00042 | 0.05837 | 0.00084 | 526.0 | 13.5 | 504.1 | 2.5 | 543.7 | 31.7 | 503.4 | 2.5 |
| 6.1 | UZO | 193 | 3343 | 470 | 0.14 | 1907 | 9.7E-03 | 2.2E-04 | 0.16493 | – | – | – | – | 0.06121 | 0.00164 | 0.05508 | 0.00295 | – | – | 383.0 | 10.0 | 415.6 | 124.5 | 382.7 | 9.9 |
| 7.1 | UZO | 66 | 1317 | 89 | 0.07 | 19 | 3.6E-04 | 8.1E-05 | 0.00548 | – | – | – | – | 0.05460 | 0.00045 | 0.05277 | 0.00127 | – | – | 342.7 | 2.7 | 318.7 | 55.8 | 342.9 | 2.8 |
| 8.1 | UZO | 70 | 1293 | 100 | 0.08 | 14 | 2.6E-04 | 7.2E-05 | 0.00395 | – | – | – | – | 0.05873 | 0.00051 | 0.05502 | 0.00151 | – | – | 367.9 | 3.1 | 413.2 | 62.4 | 367.4 | 3.2 |
| 9.1 | UZO | 48 | 984 | 51 | 0.05 | 3 | 4.0E-05 | 3.6E-05 | 0.00105 | – | – | – | – | 0.05340 | 0.00035 | 0.05405 | 0.00110 | – | – | 335.4 | 2.2 | 373.3 | 46.3 | 335.0 | 2.2 |
| 10.1 | UZO | 70 | 984 | 196 | 0.20 | 16 | 1.3E-04 | 1.0E-04 | 0.00477 | – | – | – | – | 0.07355 | 0.00068 | 0.05511 | 0.00148 | – | – | 457.5 | 4.1 | 416.7 | 61.3 | 458.1 | 4.1 |
| 11.1 | UZO | 56 | 1059 | 105 | 0.10 | 60 | 9.9E-04 | 1.9E-04 | 0.02080 | – | – | – | – | 0.05690 | 0.00040 | 0.05321 | 0.00137 | – | – | 356.8 | 2.5 | 337.9 | 59.5 | 357.0 | 2.5 |
| 12.1 | CZ | 64 | 850 | 244 | 0.29 | 92 | 1.4E-03 | 9.9E-05 | 0.02900 | – | – | – | – | 0.07634 | 0.00047 | 0.05706 | 0.00144 | – | – | 474.2 | 2.8 | 493.9 | 56.6 | 473.9 | 2.8 |
| 13.1 | UZC | 90 | 950 | 926 | 0.97 | 14 | 2.2E-04 | 1.1E-04 | 0.00391 | 0.3011 | 0.0048 | 0.02509 | 0.00044 | 0.08125 | 0.00047 | 0.05624 | 0.00184 | 501.0 | 8.6 | 503.6 | 2.8 | 461.8 | 74.1 | 504.2 | 2.7 |
| 13.2 | LCZC | 9 | 102 | 60 | 0.59 | 43 | 5.0E-03 | 6.1E-04 | 0.09968 | – | – | – | – | 0.08089 | 0.00131 | 0.05728 | 0.00824 | – | – | 501.5 | 7.8 | 502.2 | 352.4 | 501.4 | 7.4 |
| 14.1 | UZC | 85 | 918 | 819 | 0.89 | 11 | 1.7E-04 | 3.2E-05 | 0.00305 | 0.2647 | 0.0023 | 0.02408 | 0.00034 | 0.08112 | 0.00029 | 0.05715 | 0.00072 | 481.0 | 6.7 | 502.8 | 1.7 | 497.4 | 28.1 | 502.9 | 1.7 |
| 14.2 | CZ | 61 | 639 | 698 | 1.09 | 11 | 2.5E-04 | 7.4E-05 | 0.00455 | 0.3287 | 0.0040 | 0.02411 | 0.00036 | 0.08017 | 0.00054 | 0.05627 | 0.00133 | 481.5 | 7.0 | 497.2 | 3.2 | 462.9 | 53.3 | 497.7 | 3.2 |
| 15.1 | UZr | 37 | 460 | 105 | 0.23 | 2 | 5.7E-05 | 3.7E-05 | 0.00102 | 0.0710 | 0.0021 | 0.02591 | 0.00080 | 0.08315 | 0.00056 | 0.05741 | 0.00171 | 517.1 | 15.8 | 514.9 | 3.4 | 507.4 | 66.8 | 515.0 | 3.5 |
| 16.1 | UZC | 86 | 908 | 797 | 0.88 | 10 | 1.5E-04 | 3.1E-05 | 0.00276 | 0.2756 | 0.0052 | 0.02598 | 0.00053 | 0.08278 | 0.00058 | 0.05761 | 0.00075 | 518.3 | 10.4 | 512.7 | 3.5 | 515.1 | 29.0 | 512.7 | 3.5 |

*, Radiogenic.

$f$, Correction for common Pb–Fraction of total $^{206}$Pb ($^{206}$Pb$_{T}$) that is common $^{206}$Pb ($^{206}$Pb$_{C}$).

§, Best estimated of the age of the analysed zircon. (See text for explanation.) Uncertainties one standard error. UZO, Unzoned overgrowth; UZ, Unzoned; CZ, Concentric zoned; UZC, Unzoned core; LCZC, Light concentric zoned core; LUZ, Light unzoned core; (r), Recrystallization front.

possible, the analyses tabulated and plotted were corrected for common Pb using $^{208}$Pb, meaning that radiogenic $^{208}$Pb/$^{206}$Pb and $^{208}$Pb/$^{232}$Th were not independently determined. Some analyses (those with U contents >2500 ppm) required a small matrix correction to the Pb/U ratios, which was applied using the relationship determined by Williams & Hergt (2001).

## SHRIMP U–Th–Pb results

The results of the zircon U–Th–Pb analyses are listed in Tables 1–3 (monazite in Table 4) and CL images of representative zircons in Figures 6 and 7. The data are plotted on a concordia diagrams in Figure 8, and a compilation of all inferred ages is illustrated as a relative probability density distribution (Dodson *et al.* 1988) in Figure 9. The zircon grains are complexly structured, most consisting of a core surrounded by a much younger overgrowth. The ages of the cores and their tectonic implications have been reported by Pereira *et al.* (2008). Here we report the ages of the zircon metamorphic/hydrothermal overgrowths and their significance.

## Série Negra paragneisses

SEC-1 (Fig. 6) was a foliated paragneiss (Ediacaran sedimentary protolith) with a mosaic texture consisting of alternating mm-wide bands of dynamically recrystallized quartz and feldspar, ribbons of polygonal quartz and bands with biotite, quartz, feldspar, sillimanite, zircon, apatite, rutile and opaque minerals (mainly graphite). Most of the detrital zircon grains recovered were partially overgrown by a very thin (<20 μm) nodular layer of weakly luminescent zircon commonly rich in inclusions that have yet to be identified (Fig. 5). Only the thickest of the overgrowths were accessible to analysis using the 10 μm diameter SHRIMP primary probe and accurate targeting to avoid overlap with the host zircons was very difficult. The twelve areas analysed, with a single exception (analysis 1.1) had moderate U contents (280–960 ppm) and low to very low Th/U (0.1–0.004), consistent with growth during metamorphism (Williams & Claesson 1987; Heaman *et al.* 1990). With two exceptions (8.1, 9.1) the analyses were concordant within analytical uncertainty, but there was a very large range in their radiogenic $^{206}$Pb/$^{238}$U apparent ages. Two analyses (1.1, 6.1), both of which probably overlapped the old detrital host grain, gave Ordovician (*c.* 455 Ma) apparent ages that are considered to have no geological significance. The remaining $^{206}$Pb/$^{238}$U ages ranged from *c.* 390–310 Ma, and there was no clustering towards either the upper or lower end of the range

that might help explain the scatter as due to either Pb loss or inheritance of radiogenic Pb. Interpretation of the analyses must remain inconclusive. If Pb loss was the dominant effect, the primary age of the overgrowths is *c.* 380 Ma; if inheritance of radiogenic Pb predominated, the overgrowths might be as young as *c.* 320 Ma. Given that unsupported radiogenic Pb is relatively rare in zircon (Williams *et al.* 1984), an age of *c.* 380 Ma is considered to be the more likely.

BSC-1 was a medium-grained paragneiss (Ediacaran sedimentary protolith) consisting of quartz, biotite, muscovite, plagioclase and feldspar aligned parallel to the regional foliation. Albite porphyroblasts included aligned and folded graphitic inclusions, and had tails of new-grown oligoclase. Blades of muscovite and biotite were oriented parallel to dynamically recrystallized quartz ribbons that defined the foliation. Only 11 zircon grains were recovered from the sample. Most of those consisted of a detrital zircon core partly overgrown by a thin (<10 μm), discontinuous, nodular layer of moderately luminescent, irregularly zoned zircon (Fig. 5). Three overgrowths were thick enough to analyse with a 10 μm probe. Two of those had very similar compositions, moderate U contents (224 and 153 ppm) and very low Th/U (0.02 and 0.03), consistent with growth during metamorphism. The $^{206}$Pb/$^{238}$U ages of these overgrowths were also very similar, *c.* 340 Ma. Their weighted mean age, 348.9 ± 3.5 ($\sigma$) Ma, is dominated by the more precise analysis. With only two analyses, the 95% confidence limits on this result are very large, *c.* 45 Ma. In contrast, the third overgrowth was much thicker (*c.* 20 μm), strongly luminescent and faceted. It had much lower U (9 ppm) and higher Th/U. It also had a much higher age, 748 ± 48 ($\sigma$) Ma. This is interpreted as an overgrowth that was present on a detrital grain, and indeed close inspection showed that it was overgrown in turn by tiny (*c.* 2 μm) nodules of zircon resembling the young overgrowths. The best estimate of the age of metamorphism of paragneiss BSC-1 is 350 ± 45 (95% c.l.) Ma.

## Cambrian orthogneisses

Orthogneiss ALC-10 (Fig. 7) was an igneous rock deformed under upper amphibolite facies metamorphic conditions. It had a foliation defined by aligned individual biotite and amphibole crystals and an alternation of continuous layers of biotite aggregates with bands of heterogeneously dynamically recrystallized quartz and feldspar. Rare fibrolite sillimanite was present. CL imaging revealed that the great majority of the zircon grains had weak concentric growth zoning consistent with precipitation from magma of felsic to intermediate

**Table 3.** *Zircon SHRIMP data from the biotite granite ARL-6*

| Analysis | Type | Pb* (ppm) | U (ppm) | Th (ppm) | Th/U | $^{204}Pb$/ppb | $^{204}Pb/^{206}Pb$ | ± | $f_{206Pb}$ | $^{208}Pb^*/^{206}Pb$ | ± | $^{208}Pb^*/^{232}Th$ | ± | $^{206}Pb^*/^{238}U$ | ± | $^{207}Pb^*/^{206}Pb$ | ± | Apparent age (Ma) 208/232 | ± | 206/238 | ± | 207/206 | ± | Inf Age[§] | ± |
|---|---|---|---|---|---|---|---|---|---|---|---|---|---|---|---|---|---|---|---|---|---|---|---|---|---|
| *Biotite granite ARL-6* | | | | | | | | | | | | | | | | | | | | | | | | | |
| 1.1 | CZIO | 37 | 682 | 212 | 0.31 | 1 | 4.4E-05 | 1.9E-05 | 0.00080 | 0.0946 | 0.0022 | 0.01659 | 0.00041 | 0.05453 | 0.00037 | 0.05305 | 0.00074 | 332.6 | 8.1 | 342.3 | 2.3 | 330.8 | 32.0 | 342.2 | 2.3 |
| 2.1 | CZ | 14 | 257 | 123 | 0.48 | 2 | 1.5E-04 | 6.1E-05 | 0.00277 | 0.1448 | 0.0049 | 0.01608 | 0.00056 | 0.05316 | 0.00041 | 0.04837 | 0.00154 | 322.4 | 11.2 | 333.9 | 2.5 | 117.2 | 76.7 | 334.8 | 2.5 |
| 3.1 | CZIO | 17 | 301 | 166 | 0.55 | 1 | 9.5E-05 | 8.0E-05 | 0.00173 | 0.1705 | 0.0042 | 0.01678 | 0.00044 | 0.05427 | 0.00048 | 0.05258 | 0.00155 | 336.4 | 8.8 | 340.7 | 2.9 | 310.5 | 68.5 | 341.0 | 2.9 |
| 4.1 | DBZ | 108 | 1684 | 1671 | 0.99 | 1 | 1.7E-05 | 1.3E-05 | 0.00032 | 0.3072 | 0.0022 | 0.01698 | 0.00016 | 0.05483 | 0.00026 | 0.05351 | 0.00046 | 340.3 | 3.1 | 344.1 | 1.6 | 350.4 | 19.7 | 344.0 | 1.6 |
| 5.1 | DBZ | 171 | 2777 | 2295 | 0.83 | 4 | 2.8E-05 | 2.2E-05 | 0.00051 | 0.2487 | 0.0019 | 0.01650 | 0.00015 | 0.05481 | 0.00025 | 0.05360 | 0.00049 | 330.7 | 3.1 | 344.0 | 1.5 | 354.2 | 21.0 | 343.9 | 1.5 |
| 7.1 | DBZ | 165 | 2741 | 2684 | 0.98 | 102 | 8.1E-04 | 4.5E-05 | 0.01482 | 0.2750 | 0.0031 | 0.01481 | 0.00019 | 0.05273 | 0.00026 | 0.05365 | 0.00082 | 297.2 | 3.8 | 331.2 | 1.6 | 356.2 | 34.9 | 331.0 | 1.6 |
| 8.2 | CZIO | 26 | 522 | 40 | 0.08 | 0 | 7.4E-06 | 7.9E-06 | 0.00014 | 0.0248 | 0.0012 | 0.01742 | 0.00084 | 0.05319 | 0.00035 | 0.05388 | 0.00083 | 349.1 | 16.7 | 334.1 | 2.2 | 365.9 | 35.0 | 333.8 | 2.2 |
| 13.1 | UZO | 217 | 4281 | 335 | 0.08 | 34 | 1.6E-04 | 1.9E-05 | 0.00299 | 0.0043 | 0.0008 | 0.00307 | 0.00060 | 0.05573 | 0.00017 | 0.05332 | 0.00046 | 62.0 | 12.1 | 349.6 | 1.0 | 342.6 | 19.8 | 349.8 | 1.1 |
| 22.1 | CZO | 17 | 293 | 163 | 0.56 | 2 | 1.5E-04 | 1.0E-04 | 0.00269 | 0.1711 | 0.0058 | 0.01648 | 0.00058 | 0.05367 | 0.00044 | 0.05347 | 0.00185 | 330.3 | 11.5 | 337.0 | 2.7 | 348.7 | 80.3 | 336.9 | 2.6 |
| 23.1 | CZC | 9 | 162 | 87 | 0.54 | 2 | 2.5E-04 | 1.1E-04 | 0.00463 | 0.1636 | 0.0059 | 0.01633 | 0.00071 | 0.05371 | 0.00106 | 0.05031 | 0.00229 | 327.4 | 14.2 | 337.3 | 6.5 | 209.1 | 108.9 | 338.5 | 6.5 |
| 25.1 | CZC | 33 | 633 | 165 | 0.26 | 0 | 3.4E-06 | 5.8E-06 | 0.00006 | 0.0799 | 0.0025 | 0.01623 | 0.00054 | 0.05294 | 0.00052 | 0.05363 | 0.00096 | 325.4 | 10.8 | 332.6 | 3.2 | 355.7 | 40.8 | 332.3 | 3.2 |

*, Radiogenic.

*f*, Correction for common Pb–Fraction of total $^{206}Pb$ ($^{206}Pb_T$) that is common $^{206}Pb$ ($^{206}Pbc$).

§, Best estimated of the age of the analysed zircon. (See text for explanation.) Uncertainties one standard error. CZ, Concentric zoned; CZO, Concentric zoned overgrowth; CZIO, Concentric zoned intermediary overgrowth; UZO, Unzoned overgrowth; CZC, Concentric zoned core; DBZ, Dark banded zoned.

**Table 4.** *Monazite SHRIMP data from the orthogneiss VLV-3*

| Analysis | Pb* (ppm) | U (ppm) | Th (ppm) | Th/U | $^{204}Pb/^{206}Pb$ | ± | $f_{206Pb}$ | $^{208}Pb^*/^{206}Pb$ | ± | $^{208}Pb^*/^{232}Th$ | ± | $^{206}Pb^*/^{238}U$ | ± | $^{207}Pb^*/^{206}Pb$ | ± | Apparent age (Ma) 208/232 | ± | 206/238 | ± | 207/206 | ± | Inf Age[§] | ± |
|---|---|---|---|---|---|---|---|---|---|---|---|---|---|---|---|---|---|---|---|---|---|---|---|
| 1.1 | 703 | 2020 | 44061 | 21.8 | 6.79E-05 | 9.44E-05 | 0.0011 | 6.696 | 0.093 | 0.01591 | 0.00035 | 0.05183 | 0.00082 | 0.0533 | 0.0019 | 319.1 | 7.1 | 325.8 | 5.0 | 339.5 | 80.9 | 325.6 | 5.0 |
| 2.1 | 525 | 1864 | 31780 | 17.0 | 4.46E-05 | 3.64E-05 | 0.0007 | 5.273 | 0.057 | 0.0159 | 0.00037 | 0.05142 | 0.00093 | 0.0513 | 0.0012 | 318.9 | 7.4 | 323.2 | 5.7 | 255.3 | 56.3 | 323.8 | 5.7 |
| 3.1 | 437 | 4281 | 17145 | 4.0 | 2.69E-05 | 2.31E-05 | 0.0004 | 1.256 | 0.013 | 0.01606 | 0.00036 | 0.05122 | 0.00086 | 0.0519 | 0.0013 | 322.0 | 7.1 | 322.0 | 5.3 | 281.3 | 57.7 | 322.3 | 5.3 |
| 4.1 | 316 | 1057 | 19404 | 18.4 | 7.01E-05 | 7.02E-05 | 0.0011 | 5.730 | 0.102 | 0.01587 | 0.00047 | 0.05082 | 0.00109 | 0.0522 | 0.0018 | 318.2 | 9.4 | 319.6 | 6.7 | 294.4 | 80.8 | 319.8 | 6.7 |
| 5.1 | 411 | 1263 | 25473 | 20.2 | 2.00E-05 | 2.00E-05 | 0.0003 | 6.275 | 0.079 | 0.01594 | 0.00037 | 0.05122 | 0.00092 | 0.0521 | 0.0013 | 319.6 | 7.5 | 322.0 | 5.6 | 287.9 | 59.9 | 322.3 | 5.7 |
| 6.1 | 339 | 1043 | 21430 | 20.6 | 1.33E-04 | 7.55E-05 | 0.0021 | 6.291 | 0.084 | 0.01564 | 0.00037 | 0.05109 | 0.00088 | 0.0513 | 0.0020 | 313.7 | 7.5 | 321.2 | 5.4 | 252.1 | 94.5 | 321.7 | 5.4 |

*, Radiogenic.

*f*, Correction for common Pb–Fraction of total $^{206}Pb$ ($^{206}Pb_T$) that is common $^{206}Pb$ ($^{206}Pbc$).

§, Best estimated of the age of the analysed zircon. (See text for explanation.) Uncertainties one standard error.

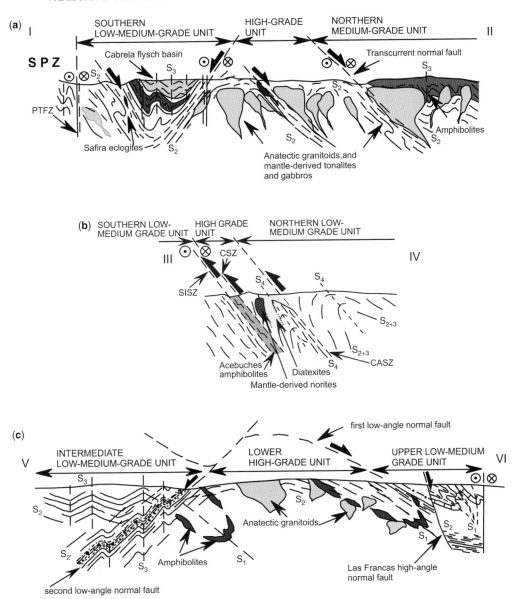

**Fig. 5.** I–II, III–IV and V–VI are cross sections through western (Èvora Massif), middle (Aracena Massif) and eastern (Lora del Rio Massif) sectors of the investigated metamorphic belt of the SW Iberian Massif.

composition. Dating of such grains by Chichorro *et al.* (2008), showed that the igneous protolith was of Cambrian age. Many of the zircon grains had patches, embayments and overgrowths of more luminescent zircon with no visible internal texture. In contrast to the zoned portions of the grains, which had moderate to high Th contents (150–600 ppm) and moderate to high Th/U (0.2–1.2), five luminescent areas analysed for the present study had low to very low Th (28–8 ppm)

and very low Th/U (0.02–0.04). Such textureless areas are a common feature of zircon from high-grade metaigneous rocks, and have been interpreted as the product of thermally-activated solid-state recrystallization (Hoskin & Black 2000). In the present case, the distinctively low Th/U in the recrystallized areas indicates that one of the trace elements preferentially expelled during the recrystallization was Th. The five U–Th–Pb isotopic analyses all yielded concordant Mississippian apparent

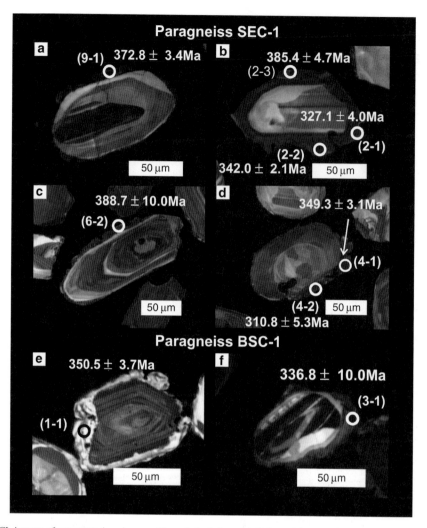

**Fig. 6.** CL images of representative zircons with analytical sites and their resulting ages indicated, from two samples of Ediacaran Serie Negra metasediments from the Evora Massif. Analysis spots and ages are listed in Table 1, first and last columns, respectively.

ages, indicating that another trace element expelled during the recrystallization was radiogenic Pb. In fact, all five $^{206}Pb/^{238}U$ ages were the same within analytical uncertainty, with a well-defined weighted mean age of 339.7 ± 5.5 (95% c.l.) Ma. This is the best estimate of the age of the recrystallization and the metamorphism that caused it.

VLV-3 was a fine- to medium-grained orthogneiss consisting of aligned individual biotite and amphibole crystals, and an alternation of continuous layers of biotite aggregates with bands of heterogeneously dynamically recrystallized quartz and feldspar. Apatite, zircon and monazite were accessory phases. As in ALC-10, most of the zircon occurred

as medium- to coarse-grained (100–200 μm diameter) euhedral to subhedral prisms with simple oscillatory growth zoning, consistent with precipitation from the melt phase of a magma. Many of the grains, however, had complex textures, including convolute zoning and textureless patches, in this case more weakly luminescent than the remainder of the grain. These features are indicative of pervasive solid-state recrystallization. Some grains also had very distinct, nodular, weakly luminescent, weakly zoned or unzoned overgrowths.

U–Th–Pb isotopic analyses of 10 igneous grains showed a range of moderate to high U contents (100–1280 ppm) and moderate to high Th/U

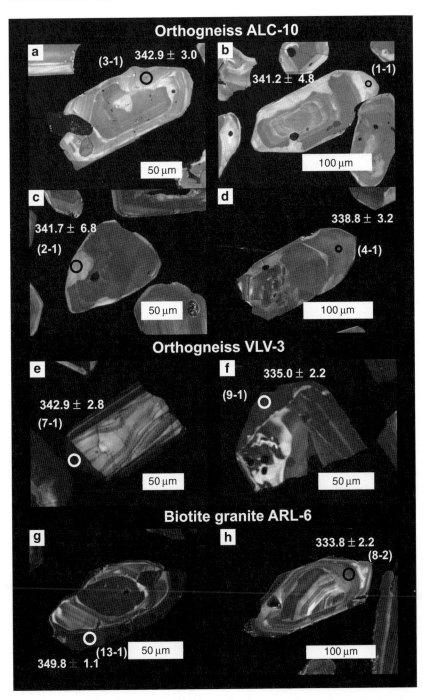

**Fig. 7.** CL images of representative zircons with analytical sites and their resulting ages indicated, from two Cambrian metaigneous rocks and Mississipian biotite granite from the Évora Massif. Analysis spots and ages are listed in Tables 2 and 3, first and last columns, respectively.

(0.22–1.1). The U–Pb isotopic compositions were all concordant within analytical uncertainty, but there was a small range in radiogenic $^{206}Pb/^{238}U$ ages (*c.* 515–475 Ma). Most of the scatter is due to two low values (analyses 3.1 and 12.1). To eliminate the scatter entirely requires the omission of two more analyses, one high and one low; however this is hard to justify. A more objective result is to pool all analyses in the upper group, giving a weighted mean $^{206}Pb/^{238}U$ age of $504.9 \pm 5.4$ (95% c.l.) Ma, the uncertainty taking the minor excess scatter (MSWD = 3.2) into account. The igneous protolith of the orthogneiss was of Cambrian age (Chichorro *et al.* 2008).

Isotopic analyses of 8 overgrowths and recrystallized areas showed a narrower range in U content (720–3340 ppm) and Th/U (0.05–0.28), but a much larger range in $^{206}Pb/^{238}U$ age (*c.* 465–335 Ma). The four 'oldest' areas had relatively high Th/U (0.14–0.28) and the four 'youngest' had low Th/U (0.05–0.10), consistent with the observation in sample ALC-10 that recrystallization was accompanied by the preferential expulsion of Th and Pb. The difference in the case of the zircon from VLV-3 is that the expulsion of radiogenic Pb was incomplete and a partial memory of the primary age was retained. The best estimate of the metamorphic age is provided by the isotopic analyses of six monazite grains, all of which gave the same concordant age within analytical uncertainty, $322.8 \pm 6.8$ (95% c.l.) Ma.

### Mississipian biotite granite

ARL-6 was a medium-grained, weakly foliated anatectic granite composed of plagioclase, K–feldspar, quartz and biotite, with zircon as the main accessory phase. The zircon population was morphologically very heterogeneous. In addition to euhedral grains with simple oscillatory growth zoning, there were large, CL-dark, prismatic grains with indistinct banded zoning and a large number of grains composed of a core surrounded by a weakly zoned or unzoned overgrowth. SHRIMP U–Pb analyses of the cores (Pereira *et al.* 2008) showed that they were a Neoproterozoic–Cambrian inherited component. In contrast, the grains without cores have very similar apparent ages (*c.* 350–330 Ma). Nevertheless, there is a significant range which correlates with grain morphology and composition (Fig. 8). Three dark banded grains and a dark unzoned overgrowth had high U contents (1685–4280 ppm) and a wide range of Th/U (0.07–0.99). These contribute most of the scatter. In contrast, the simply zoned euhedral grains have lower and more uniform U (160–680 ppm) and, with one exception, more uniform Th/U (0.26–0.55), consistent with having crystallized from a

single felsic magma. These grains also all have the same $^{206}Pb/^{238}U$ age within analytical uncertainty (MSWD = 2.1), $337.1 \pm 4.3$ (95% c.l.) Ma. This is the best estimate of the crystallization age of the granite.

## Discussion

### *The significance of different zircon growths and distribution of ages*

The term 'metamorphic zircon' is applied to zircon formed under metamorphic conditions by a range of different postulated processes such as: (1) subsolidus nucleation and crystallization by diffusion of Zr and Si released by metamorphic breakdown reactions of major silicates or other Zr-bearing minerals (Fraser *et al.* 1997); (2) precipitation from Zr-saturated silicate anatectic melts (Vavra *et al.* 1996; Williams *et al.* 1996; Watson 1996), direct crystallization from a zircon-saturated aqueous fluid (hydrothermal fluids) which can be aqueous metamorphic fluids (Williams *et al.* 1996), fluids evolved from a magma during the final stages of crystallization (Hoskin 2005) or low temperature fluids, which precipitate authigenic zircon (Armstrong *et al.* 1995); and (3) solid state annealing (Vavra *et al.* 1996) or recrystallization of protolith zircon (Black *et al.* 1986; Friend & Kinny 1995; Pan 1997; Bowring & Williams 1999).

All the external overgrowths and recrystallized areas that were dated for the present study, whether on detrital zircon grains from Ediacaran paragneisses, on igneous zircon from Cambrian orthogneisses or on inherited zircon from an anatectic granite, have very low Th/U (<0.1). This feature is typical of the zircon precipitated during the metamorphism or partial melting of a peraluminous rock (Williams & Claesson 1987; Heaman *et al.* 1990; Williams 2001). There are significant differences, however, between the zircon precipitated in these different environments. Seven analyses (4 from orthogneiss VLV-3, 1 from orthogneiss ALC-10 and 2 from biotite granite ARL-6) have relatively high U contents (522–4281 ppm) consistent with zircon precipitated from trace element rich partial melts. These overgrowths are very well defined, generally unzoned and pyramidal and tend to have developed preferentially on rapidly growing pyramidal crystal faces.

The two analysed overgrowths on old inherited zircon cores from the biotite granite probably represent the same high temperature processes associated with mylonitization recorded in the Serie Negra paragneisses. These thick new additions resemble the overgrowths developed on detrital zircon during the high temperature/low pressure metamorphism of the Cooma Complex in eastern

Australia that culminated in the production of an anatectic granite (Williams 2001).

The thin, discontinuous, nodular, inclusion rich (spongy) overgrowths developed on the detrital zircon grains from the Serie Negra paragneisses are quite different, and do not appear to be associated with high-grade partial melting. The morphology of the inclusion-rich overgrowths suggests precipitation from intergranular Zr saturated fluid films. Such fluid films, with low Th contents, might be produced during the early stages of partial melting related to prograde high temperature/low pressure metamorphism, or might reflect the inhibition of partial melting, and consequently the inhibition of new zircon growth, by the relatively low $Al_2O_3$ contents ($16\% < Al_2O_3 < 18\%$) of these immature Ediacaran metasediments. This hypothesis suggests a relationship between high-grade solid–solid metamorphic reactions and consequent melting conditions. An alternative is that the new zircon growth was linked to hydrothermal fluid flow driven by temperature gradients around thermal anomalies (Buick & Cartwright 1994). Fluids would have been generated in the Serie Negra paragneisses by the metamorphic biotite dehydration and the melting reaction that produced peritetic cordierite.

Local retrogression of the metamorphic assemblages (also associated with mylonitization) is indicated by intense alteration of cordierite and the replacement of biotite by chlorite (Ribeiro 1994; Chichorro 2006). The presented evidence suggests that hydrothermal fluids may have played a major role in the precipitation of the new zircon. As hydrothermal zircon can be used to date the infilling fluid events and the consequent metasomatic interactions between water and rock (Hoskin 2005), the results obtained here suggest that in the Évora Massif, the transport and deposition of metals in this gold province was probably concomitant with high-grade metamorphic activity, with a maximum remobilization around the age of 340 Ma (Visean).

Four analyses from the orthogneisses were made on cross-cutting mantles with morphologies typical of thermally activated solid-state recrystallization (Black et al. 1986; Varva et al. 1996; Hoskin & Black 2000; Bowring & Williams 1999). The observed difference in Th/U from the igneous grains indicates that the high-grade conditions persisted for an extended period of time.

The distribution of $^{206}Pb/^{238}U$ ages is illustrated on a relative probability diagram (Dodson et al. 1988) in Figure 9. The main peak at 342 Ma contains nearly 32% of the analyses and is statistically significant. This is the principal metamorphic age obtained from several of the samples; $339.7 \pm 5.5$ Ma for the metamorphic imprint on orthogneiss ALC-10

(5 analyses) and $337.1 \pm 4.3$ Ma for the emplacement of biotite granite ARL-6 (7 analyses). These results are slightly but significantly older than the crystallization age of the monazite from orthogneiss VLV-3, $322.8 \pm 6.8$ Ma. This Middle–Upper Mississipian age overlaps the ages obtained for the Hospitais tonalites ($323.5 \pm 5.2$ Ma; Ar–Ar amphibole) and the Alto de São Bento leucogranites ($325 \pm 9$ Ma; Rb–Sr whole rock – biotite) interpreted to be related with crystallization/cooling (Moita 2007).

The results of this study are consistent with a diachronous and variable spatial distribution of multiple zircon growth and recrystallization related to an extended thermal history (discrete thermal pulses) concentrated in the Early Carboniferous, within the interval c. 350–335 Ma (contains 56% of the analyses and is statistically significant). These thermal events are associated with the development of ductile deformation (widespread mylonitic fabrics) related to transcurrent movements.

Finally, the obtained ages of 388–367 Ma from sample SEC-1 seem to be related with metamorphic overgrowths. Apparent ages of $368 \pm 21$ Ma ($2\sigma$, MSWD = 1.6) were also obtained for the Alcáçovas felsic gneisses and interpreted to represent an early Variscan metamorphic overprinting (Cordani et al. 2006). Similar Middle–Upper Devonian ages were found in eclogites and blueschist from Safira and Alvito ($371 \pm 17$ Ma; Sm–Nd, whole-rock– garnet and Ar–Ar, glaucophane; Moita et al. 2005b).

## The significance of Variscan intra-continental extensional tectonics in the SW Iberian Massif

Several tectonic models have been proposed to explain the geological evolution of the Variscan orogeny in the SW Iberian Massif. Bard et al. (1973) proposed a geological scheme with a north-directed (present-day co-ordinates) subduction zone, marked by the exposure of the Ossa–Morena and South Portuguese suture zone. The hypothesis of a SW Iberian Massif suture zone was reinforced by the identification of MORB-derived amphibolites along this boundary (unknown age Beja–Acebuches amphibolites; Bard & Moine 1979; Dupuy et al. 1979; Quesada 1991; Crespo-Blanc & Orozco 1991; Fonseca & Ribeiro 1993; Quesada et al. 1994; Castro et al. 1996b).

The widespread Tournaisian (c. 350 Ma) calc-alkaline igneous rocks from the southwestern Ossa–Morena, which were also interpreted to be arc-related (Beja gabbros and Toca da Moura– Odivelas volcanics; Santos et al. 1987, 1990), favoured this plate convergence tectonic setting.

A more complex structural scenario was proposed by Araújo *et al.* (2005) who postulated the existence of obducted oceanic crust within the Ossa–Morena zone. These ophiolitic nappes, composed of MORB-derived amphibolites and serpentinized peridotites, occur to the north of the supposed suture zone and magmatic arc. According to these authors, such dismembered oceanic rocks were transported on the top of, and/or imbricated in, a tectonic mélange (Moura phyllonitic accretionary complex). This contractional deformation was, by their interpretation, responsible for the exhumation of high- to medium-pressure rocks (Safira–Viana do Alentejo eclogites and Moura blueschists; Araújo *et al.* 2005; Ribeiro *et al.* 2007) during the Middle–Late Devonian (*c.* 390–370 Ma). Meanwhile, Rosas *et al.* (2008) admit extension to explain the exhumation of high-pressure rocks after the earlier episode of contraction deformation.

A major feature of the SW Ossa–Morena Zone is the extensive record of a high-grade thermal event with a prograde history that reached temperatures of crustal anatexis under low-pressure conditions (Abalos *et al.* 1991; Eguiluz *et al.* 2000; Díaz Azpíroz 2006; Díaz Azpíroz *et al.* 2006). This high-temperature/low-pressure metamorphism locally overprinted high- to medium-pressure mineral assemblages (Apraiz 1998; Apraiz & Eguiluz 2002; Chichorro 2006; Pereira *et al.* 2007) observed in small outcrops (metre-scale boudins) with as yet undeciphered relationships to their deformed host rocks.

The Évora–Aracena–Lora del Rio metamorphic belt is a discontinuous linear metamorphic zone characterized by strong ductile shearing associated with high-medium-temperature/low-pressure mineral assemblages with characteristic isobarically cooled *P–T* paths (Fig. 9). The high-grade metamorphism appears to be related to thermal relaxation, extensional unroofing and rapid exhumation due to the activity of major transcurrent faults (*c.* 350–335 Ma). Structural elements formed during ductile conditions show the progressive character of shear deformation (strong mylonitization) at higher and lower temperatures, with orogen-parallel tectonic transport.

Our data emphasise the existence of an important Tournaisian–Visean intra-orogenic extension (transtension) of continental lithosphere in the SW Iberian Massif, with strong across-strike thermal gradients.

A strong heat source probably induced partial melting in the upper mantle and hot material moved upward into shallow levels where melting was promoted in the continental crust. The strong ductile deformation is attributed to the influence of transcurrent shear zones, which in places penetrated down to deep crustal levels. The high-medium

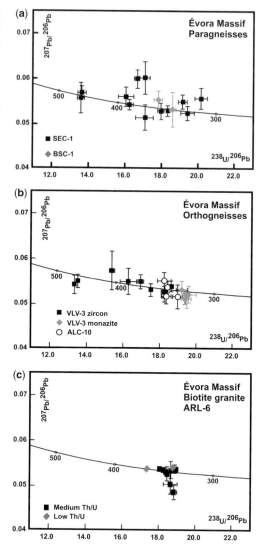

**Fig. 8.** Modified Tera–Wasserburg concordia diagrams for zircon overgrowths from: Série Negra paragneiss, Cambrian orthogneisses and Mississipian biotite granite of the Évora Massif.

pressure rocks surrounded by a strong mylonitic fabric were probably rapidly exhumed along these major structures that gave place to uplift of metamorphic complexes and melt generation and intrusion. The dynamics associated with mantle thermal perturbations and the consequent gravity collapse of the crust still need to be identified.

The thermally anomaly responsible for the Mississippian extensional collapse has been attributed to asthenospheric up-welling apparently in connection with a lithosphere slab-window (Castro *et al.* 1996*a, b*; Díaz Azpíroz *et al.* 2006). Another

(a)  **Variscan intra-continental extensional transcurrent tectonics**

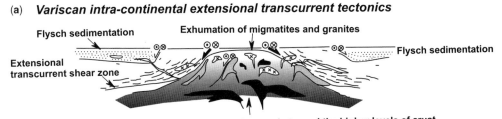

Flysch sedimentation          Exhumation of migmatites and granites

Extensional                                                                    Flysch sedimentation
transcurrent shear zone

Propagation of a large thermal anomaly toward the higher levels of crust
and intrusion of mantle-derived magmas causing voluminous partial melting

(b)  **Pressure–Temperature paths**              (c)  **Ages of metamorphism**

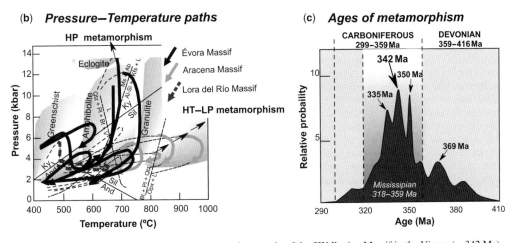

**Fig. 9. (a)** Schematic reconstruction of the geodynamic scenario of the SW Iberian Massif in the Visean (*c.* 342 Ma) before the shortening caused by the dominant Variscan wrench regime (the final result was the steepening of earlier structures as shown in cross-sections of Figs 2, 3 & 4); (**b**) Summary of the *P–T–t* path for the Évora–Aracena–Lora del Río metamorphic belt (data compiled from Apraiz & Eguiluz 2002; Moita *et al.* 2005b; Díaz Azpíroz *et al.* 2006; Pereira *et al.* 2007); (**c**) Relative probability density distribution with compilation of all inferred ages obtained from zircon overgrowths (time scale of Gradstein *et al.* 2004).

interpretation proposes that the heat source to trigger lithosphere extension was a kilometre-scale layered mid-crustal mafic intrusion probably associated with a sub-crustal thermal perturbation akin to a mantle plume (Simancas *et al.* 2003).

Finally, mantle convection slowed down, thermal gradients attenuated, and extension probably stopped at *c.* 320 Ma.

Comparison of our dating results with those obtained for the NW Iberian Massif indicate common geological characteristics.

The NW Iberian Massif (Galicia-Trás-os-Montes Zone; Martinez Catalan *et al.* 2003) preserves relics of the Variscan suture related with the convergence between Gondwana and Laurussia (Martinez Catalán *et al.* 2007). Here the allochthonous complexes of Cabo Ortegal, Órdones, Malpica-Tui, Bragança and Morais constitute a nappe pile of different tectonic units. The basal units are considered to derive from

the Gondwana continental margin involved in the Variscan collision (Martinez Catalán *et al.* 1996). They consist of unknown age metasediments (Cambrian–Ordovician) intruded by early Ordovician felsic and mafic igneous rocks that underwent subduction and high-pressure metamorphism in the course of early Variscan events (Martinez Catalán *et al.* 1996, 2007).

The Malpica–Tui Complex lithostratigraphy and structure looks like that of the Évora Massif. An upper Igneous (mafic dominated)–Sedimentary Complex composed of MORB-derived amphibolites and phyllites with interbedded carbonates and cherts overlies a probably older Igneous (felsic dominated)–Sedimentary Complex with calc-alkaline felsic orthogneisses, schists and paragneisses. Contrary to the Évora Massif here, evidence for early Variscan subduction seems to be well preserved by variably retrograde high-pressure metamorphic fabrics in blueschists and eclogites (Rodriguez 2005). The age of *c.* 370 Ma obtained for the Évora Massif eclogites and blues-chists (Sm–Nd, garnet and Ar–Ar, amphibole, Moita *et al.* 2005*b*) coincide with the eclogite-facies metamorphic event recorded in the Malpica–Tui Complex at 365 ± 1 Ma (Ar–Ar, phengite; Rodriguez *et al.* 2003), interpreted to be related to exhumation. The tangential deformation respon-sible for the emplacement of nappes occurred by a combination of thrusts and normal faults developed under amphibole-facies at 350–340 Ma. The perva-sive greenschist-facies mylonitic fabric parallel with the older ones indicate orogen-parallel tectonic transport until the Visean (330 ± 1 Ma; Ar–Ar, phengite; Rodriguez *et al.* 2003).

Variscan intra-continental extensional tectonics related with orogen-parallel transport seems to be more important than previously recognized in the Iberian Massif. This tectonic context was pro-bably caused by complex interactions between lithospheric plates in convergent settings, when any process affecting subduction/collision rates and/or tectonic consequences of triple junction migration causes collapse of thickened continental crust. Additional influence from sub-lithosphere mantle dynamics cannot be disregarded.

This work is a contribution to the projects: CGL 2004-06808-C04-02-BTE, 'Estudio geoquímico, tectónico y experimental de los procesos de reciclaje cortical y interac-ción manto-corteza' and IGCP 497: 'The Rheic Ocean: Its origin, evolution and correlatives'. The authors thank C. Quesada and J. Allen for their constructive revisions, and P. Moita for providing sample VLV.

# References

ABALOS, B. V., GIL IBARGUCHI, I. & EGUILUZ, L. 1991. Structural and metamorphic evolution of the Almadén de la Plata Core (Seville, Spain) in relation to syn-metamorphic shear between the Ossa–Morena and South Portuguese zones of the Iberian Variscan fold belt. *Tectonophysics*, **191**, 365–387.

APRAIZ, A. 1998. *Geología de los macizos de Lora del Río y Valuengo (Zona de Ossa Morena). Evolución tectonometamórfica y significado geodinámico.* Unpublished PhD dissertation, Universidad del País Vasco.

APRAIZ, A. & EGUILUZ, L. 2002. Hercynian tectono-thermal evolution associated with crustal extension and exhumation of the Lora del Río metamorphic core complex (Ossa–Morena Zone, Iberian Massif, SW Spain). *Geologische Rundschau*, **91**, 76–92.

ARAÚJO, A., FONSECA, P., MUNHÁ, J., MOITA, P., PEDRO, J. & RIBEIRO, A. 2005. The Moura phyllonitic complex: an accretionary complex related with obduc-tion in the southern Iberia Variscan suture. *Geodina-mica Acta*, **18**, 375–388.

ARMSTRONG, R. A., FANNING, C. M., ELDRIDGE, C. S. & FRIMMEL, H. E. 1995. Geochronological and isotopic constraints on provenance and post-depositional alteration of the Witwatersrand Super-group, Extended Abstracts, Centennial Geocongress (1995). *Geological Society of South Africa*, 1086.

AZOR, A., RUBATTO, D., SIMANCAS, J. F., GONZÁLEZ LODEIRO, F., MARTÍNEZ POYATOS, D., MARTÍN PARRA, L. M. & MATAS, J. 2008. Rheic Ocean ophio-litic remnants in southern Iberia questioned by SHRIMP U–Pb zircon ages on the Beja–Acebuches amphibolites. *Tectonics*, **25**, TC5006.

BARD, J. P. 1969. *Le métamorphisme régional progressif des Sierra de Aracena en Andalousie occidentale (Espagne).* Unpublished PhD dissertation, Université de Montpellier, France.

BARD, J. P. & MOINE, B. 1979. Acebuches amphibolites in the Aracena hercynian metamorphic belt (southwest Spain): geochemical variations and basaltic affinities. *Lithos*, **12**, 271–282.

BARD, J. P., CAPDEVILA, R., MATTE, PH. & RIBEIRO, A. 1973. Geotectonic model for the Iberian Variscan Orogen. *Nature*, **241**, 50–52.

BLACK, L. P., WILLIAMS, I. S. & COMPSTON, W. 1986. Four zircon ages from one rock: the history of a 3930 Ma-old granulite from Mount Sones, Enderby Land, Antarctica. *Contributions to Mineralogy and Petrology*, **94**, 427–437.

BOWRING, S. A. & WILLIAMS, I. S. 1999. Priscoan (4.00–4.03 Ga) orthogneisses from northwestern Canada. *Contributions to Mineral Petrology*, **134**, 3–16.

BUICK, I. S. & CARTWRIGHT, I. 1994. Fluid-rock inter-action across a green-schist-to granulite facies tran-sition, Reynolds range, central Australia: implications for regional-scale fluid flow in LP/HT orogenic belts. *Mineralogical Magazine*, **58**, 130–131.

CARVALHOSA, C. 1965. Contribuição para o conheci-mento geológico da região entre Portel e Ficalho (Alentejo). *Memória dos Serviços Geológicos de Portugal*, **11**.

CARVALHOSA, A. 1983. Esquema geológico do Maciço de Évora. *Comunicações dos Serviços Geológicos de Portugal*, **69**, 201–208.

CARVALHOSA, A. 1999. *Carta Geológica de Portugal, Noticia Explicativa da Folha 36-C (Arraiolos).* Insti-tuto Geológico e Mineiro, Lisboa, Portugal, scale 1:50 000.

CASTRO, A., FERNÁNDEZ, C. *ET AL.* 1996*a*. Triple-junction migration during Paleozoic plate conver-gence: the Aracena metamorphic belt, Hercynian massif, Spain. *Geologische Rundschau*, **85**, 180–185.

CASTRO, A., FERNÁNDEZ, C., DE LA ROSA, J. D., MORENO-VENTAS, I. & ROGERS, G. 1996*b*. Signifi-cance of MORB-derived Amphiboles from the Aracena Metamorphic Belt, Southwest Spain. *Journal of Petrology*, **37**, 235–260.

CASTRO, A., FERNÁNDEZ, C., EL-HMIDI, H., EL-BIAD, M., DÍAZ, M., DE LA ROSA, J. & STUART, F. 1999. Age constraints to the relationships between magma-tism, metamorphism and tectonism in the Aracena metamorphic belt, southern Spain. *International Journal of Earth Sciences*, **88**, 26–37.

CHICHORRO, M. 2006. *Estrutura do Sudoeste da Zona de Ossa-Morena: Área de Santiago de Escoural – Cabrela (Zona de Cisalhamento de Montemor-o-Novo,*

*Maciço de Évora).* Unpublished PhD dissertation, Universidade de Évora, Portugal.

CHICHORRO, M., PEREIRA, M. F., APRAIZ, A. & SILVA, J. B. 2004. High temperature–Low pressure tectonites from the Boa Fé Fault Zone (Évora Massif, Ossa Morena Zone, Portugal): evidences for transtensional tectonics. *Geogaceta,* **34,** 43–46.

CHICHORRO, M., PEREIRA, M. F., DÍAZ-AZPIROZ, M., WILLIAMS, I. S., FÉRNANDEZ, C., PIN, C. & SILVA, J. B. 2008. Cambrian ensialic rift-related magmatism in the Ossa–Morena Zone (Évora–Aracena metamorphic belt, SW Iberian Massif): Sm–Nd isotopes and SHRIMP zircon U–Th–Pb geochronology. *Tectonophysics,* **461,** 91–113.

CORDANI, U. G., NUTMAN, A. P., ANDRADE, A. S., SANTOS, J. F., AZEVEDO, M. R., MENDES, M. H. & PINTO, M. S. 2006. New U–Pb SHRIMP zircon ages for pre-variscan orthogneisses from Portugal and their bearing on the evolution of the Ossa–Morena Tectonic Zone. *Anais da Academia Brasileira de Ciencias,* **78**(1), 133–149.

CRESPO-BLANC, A. & OROZCO, M. 1991. The boundary between the Ossa–Morena and South Portuguese zones (southern Iberian Massif): a major suture in the European Hercynian chain. *Geologische Rundschau,* **80,** 691–702.

CUMMING, G. L. & RICHARDS, J. R. 1975. Ore lead isotope ratios in a continuously changing earth. *Earth and Planetary Science Letters,* **28,** 155–171.

DÍAZ AZPIÍROZ, M. 2006. Evolución tectonometamórfica del domínio de alto grado da la banda metamórfica de Aracena, Laboratorio Xeolóxico de Laxe, Coruña. *Nova Terra,* **30,** 858.

DÍAZ AZPÍROZ, M. & FERNÁNDEZ, C. 2005. Kinematic analysis of the southern Iberian shear zone and tectonic evolution of the Acebuches metabasites (SW Variscan Iberian Massif). *Tectonics,* **24,** TC3010, doi: 10.1029/2004TC001682.

DÍAZ AZPÍROZ, M., FERNÁNDEZ, C., CASTRO, A. & EL-BIAD, M. 2006. Tectonometamorphic evolution of the Aracena metamorphic belt (SW Spain) resulting from ridge-trench interaction during Variscan plate convergence. *Tectonics,* **25**(1), TC1001, 10.1029/2004TC001742.

DODSON, M. H., COMPSTON, W., WILLIAMS, I. S. & WILSON, J. F. 1988. A search for ancient detrital zircons in Zimbabwean sediments. *Journal of the Geological Society, London,* **145,** 977–983.

DUPUY, C., DOSTAL, J. & BARD, J. P. 1979. Trace element geochemistry of Paleozoic amphibolites from SW Spain. *Tschermaks Minerologische und Petrographische Mitteilungen,* **26,** 87–93.

EGUILUZ, L., GIL IBARGUCHI, J. I., ABALOS, B. & APRAIZ, A. 2000. Superposed Hercynian and Cadomian orogenic cycles in the Ossa-Morena Zone and related areas of the Iberian Massif. *Geological Society America Bulletin,* **112,** 1398–1413.

EL-BIAD, M. 2000. *Generacíon de granitoides en ambientes geológicamente contrastados del Macizo Ibérico: Limitaciones experimentales entre 2 y 15 kbar.* Unpublished PhD dissertation, Universidad de Huelva, Spain.

FABRIÈS, J. 1963. *Les formations cristallines et métamorphiques du Nord-Est de la Province de Seville (Espagne). Éssai sur le métamorphisme des roches éruptives basiques.* Unpublished PhD dissertation, Université de Nancy, France.

FLORIDO, P. & QUESADA, C. 1984. Estado actual de conocimientos sobre el Macizo de Aracena. *Cuadernos Laboratorio Geológico de Laxe,* **8,** 257–277.

FONSECA, P. & RIBEIRO, A. 1993. Tectonics of the Beja-Acebuches ophiolite: a major suture in the Iberian Variscan Foldbelt. *Geologische Rundschau,* **82,** 440–447.

FRASER, G., ELLIS, D. & EGGINS, S. M. 1997. Zirconium abundance in granulite-facies minerals, with implications for zircon geochronology in high-grade rocks. *Geology,* **25,** 607–610.

FRIEND, C. R. L. & KINNY, P. D. 1995. New evidence for protolith ages of Lewisian granulites, northwest Scotland. *Geology,* **23,** 1027–1030.

GIESE, U., HOEGEN VON, R., HOYMANN, K.-H. & WALTER, R. 1994. The Palaeozoic evolution of the Ossa Morena Zone and its boundary to the South Portuguese Zone in SW Spain: geological constraints and geodynamic interpretation of a suture in the Iberian Variscan orogen. *Neues Jahrbuch fur Geologie und Palaontologie, Abhandlungen,* **192,** 383–412.

GONÇALVES, F. 1985. Contribuição para o conhecimento geológico do complexo volcano-sedimentar da Toca da Moura (Alcácer do Sal). *Memoria da Academia de Ciências de Lisboa,* **26,** 263–267.

HEAMAN, L. M., BOWINS, R. & CROCKET, J. 1990. The chemical composition of igneous zircon studies: implications for geochemical tracer studies. *Geochimica et Cosmochimica Acta,* **54,** 1597–1607.

HOSKIN, P. W. O. 2005. Trace-element composition of hydrothermal zircon and the alteration of Hadean zircon from the Jack Hills, Australia. *Geochimica et Cosmochimica Acta,* **69**(3), 637–648.

HOSKIN, P. W. O. & BLACK, L. P. 2000. Metamorphic zircon formation by solid-state recrystallization of protolith igneous zircon. *Journal of Metamorphic Geology,* **18,** 423–439.

JESUS, A., MUNHÁ, J., MATEUS, A., TASSINARI, C. & NUTMAN, A. P. 2007. The Beja Layered Gabbroic Sequence (Ossa–Morena Zone, Southern Portugal): geochronology and geodynamic implications. *Geodinamica Acta,* **20**(3), 139–157.

LIÑÁN, E. & QUESADA, C. 1990. Part V Ossa-Morena Zone: 2. Stratigraphy. *In*: DALLMEYER, R. D. & MARTÍNEZ-GARCÍA, E. (eds) *Pre-Mesozoic Geology of Iberia.* Springer, Berlin, 229–266.

MARTINEZ-CATALAN, J. R., ARENAS, R. & DÍEZ BALDA, M. A. 2003. Large extensional structures developed during emplacement of a crystalline thrust sheet: the Mondoñedo nappe (NW Spain). *Journal of Structural Geology,* **25,** 1815–1839.

MARTINEZ-CATALAN, J. R., ARENAS, R. *ET AL.* 2007. Space and time in the tectonic evolution of the northwestern Iberian Massif. Implications for the comprehension of the Variscan belt. *In*: HATCHER, R. D., CARLSON, M. P., MCBRIDE, J. H. & MARTINEZ-CATALAN, J. R. (eds) *4-D Framework of Continental Crust.* Geological Society of America Memoir, **200.**

MATTE, P. 2001. The Variscan collage and orogeny (480–290 Ma) and the tectonic definition of the Armorica microplate: a review. *Terra Nova,* **13,** 122–128.

MOITA, P. 2007. *Granitóides no SW da zona de Ossa–Morena (Montemor-o-Novo-Évora). Petrogénese e processos geodinâmicos.* Unpublished PhD thesis, Universidade de Évora.

MOITA, P., SANTOS, J. F. & PEREIRA, M. F. 2005a. Tonalites from the Hospitais Massif (Ossa-Morena Zone, SW Iberian Massif, Portugal) II: geochemistry and petrogenesis. *Geogaceta*, **37**, 55–58.

MOITA, P., FONSECA, P., TASSINARI, C., ARAUJO, A. & PALACIOS, T. 2005b. Phase equilibria and geochronology of Ossa–Morena eclogites. *Actas da XIV Semana de Geoquímica/VII Congresso de geoquímica dos Países de Língua Portuguesa*, **2**, 463–466.

MUNHÁ, J., OLIVEIRA, J. T., RIBEIRO, A., OLIVEIRA, V., QUESADA, C. & KERRICH, R. 1986. Beja–Acebuches Ophiolite: characterization and geodynamic significance. *Maleo*, **2**, 1–3.

ONÈZIME, J., CHARVET, J., FAURE, M., BOURDIER, J. & CHAUVET, A. 2003. A new geodynamic interpretation for the South Portuguese Zone (SW Iberia) and the Iberian Pyrite Belt genesis. *Tectonics*, **22**, 10–27.

ORDOÑEZ-CASADO, B. 1998. *Geochronological studies of the Pre-Mesozoic basement of the Iberian Massif: the Ossa Morena Zone and Allochthonous Complexes within the Central-Iberian Zone.* PhD thesis, ETH (12.940), Zurich.

PAN, Y. 1997. Zircon- and monazite-forming metamorphic reactions at Manitouwadge, Ontario. *Canadian Mineralogist*, **35**, 105–118.

PATIÑO DOUCE, A. E. A., CASTRO, A. & EL-BIAD, M. 1997. *Thermal evolution and tectonic implications of spinel-cordierite granulites from the Aracena metamorphic belt, Southwest Spain.* GAC/MAC Annual meeting, Ottawa, Canada.

PEREIRA, M. F., SILVA, J. B. & CHICHORRO, M. 2003. Internal Structure of the Évora High-grade Terrains and the Montemor-o-Novo Shear Zone (Ossa–Morena Zone, Portugal). *Geogaceta*, **33**, 79–82.

PEREIRA, Z., OLIVEIRA, V. & OLIVEIRA, J. T. 2006a. Palynostratigraphy of the Toca da Moura and Cabrela Complexes, Ossa–Morena Zone, Portugal. Geodynamic implications. *Review of Palaeobotany & Palynology*, **139**, 227–240.

PEREIRA, M. F., CHICHORRO, M., LINNEMANN, U., EGUILUZ, L. & SILVA, J. B. 2006b. Inherited arc signature in Ediacaran and Early Cambrian basins of the Ossa–Morena Zone (Iberian Massif, Portugal): paleogeographic link with European and North African Cadomian correlatives. *Precambrian Research*, **144**, 297–315.

PEREIRA, M. F., SILVA, J. B., CHICHORRO, M., MOITA, P., SANTOS, J. F., APRAIZ, A. & RIBEIRO, C. 2007. Crustal growth and deformational processes in the Northern Gondwana margin: Constraints from the Évora Massif (Ossa-Morena Zone, SW Iberia, Portugal). *In*: LINNEMANN, U., NANCE, R. D., KRAFT, P. & ZULAUF, G. (eds) *The Evolution of the Rheic Ocean: From Avalonian–Cadomian Active Margin to Alleghenian–Variscan Collision.* Special Paper Geological Society of America, **423**, 333–358.

PEREIRA, M. F., CHICHORRO, M., WILLIAMS, I. S. & SILVA, J. B. 2008. Zircon U–Pb geochronology of paragneisses and biotite granites from the SW Iberian Massif (Portugal): Evidence for a paleogeographic link between the Ossa–Morena Ediacaran basins and the West African craton. *Geological Society, London, Special Publications*, **297**, 385–408.

PIN, C., PAQUETTE, J.-L. & FONSECA, P. 1999. 350 Ma (U–Pb zircon) igneous emplacement age and Sm-Nd isotopic study of the Beja gabbroic complex (S Portugal), XV Reunión Geología del Oeste Peninsular, Badajoz, Spain. *Journal of Conference Abstracts*, **4**(3), 1019.

PIN, C., FONSECA, P. E., PAQUETTE, J.-L., CASTRO, P. & MATTE, P. 2008. The ca. 350 Ma Beja Igneous Complex: A record of transcurrent slab break-off in the Southern Iberia Variscan Belt? *Tectonophysics*, **461**, 356–377.

QUESADA, C. 1990. Precambrian successions in SW Iberia: their relationship to 'Cadomian' orogenic events. *In*: D'LEMOS, R. S., STRACHAN, R. A. & TOPLEY, G. G. (eds) *The Cadomian Orogeny.* Geological Society, London, Special Publications, **551**, 353–362.

QUESADA, C. 1991. Geological constraints on the Paleozoic tectonic evolution of tectonostratigraphic terranes in the Iberian Massif. *Tectonophysics*, **185**, 225–245.

QUESADA, C., FONSECA, P. E., MUNHÁ, J. M., OLIVEIRA, J. T. & RIBEIRO, A. 1994. The Beja-Acebuches Ophiolite (southern Iberia Variscan fold belt): geological characterization and geodynamic significance. *Boletín Geológico y Minero*, **105**, 3–49.

REY, P., VANDERHAEGHE, O. & TEYSSIER, C. 2001. Gravitational collapse of the continental crust: definition, regimes and modes. *Tectonophysics*, **342**, 435–449.

RIBEIRO, A. 1983. Relações entre formações do Devónico superior e o Maciço de Évora na região de Cabrela (Vendas Novas). *In*: RIBEIRO, A., GONÇALVES, F., SOARES DE ANDRADE, A. & OLIVEIRA, V. (eds) *Guia das excursões do bordo sudoeste da Zona de Ossa-Morena.* Comunicações dos Serviços Geológicos de Portugal, **69**, 267–269.

RIBEIRO, A., QUESADA, C. & DALLMEYER, R. D. 1990. Geodynamic evolution of the Iberian Massif. *In*: DALLMEYER, R. D. & MARTÍNEZ GARCÍA, E. (eds) *Pre-Mesozoic Geology of Iberia.* Berlin-Heidelberg, Springer-Verlag, 399–409.

RIBEIRO, A., MUNHÁ, J. *ET AL.* 2007. Geodynamic evolution of the SW Europe Variscides. *Tectonics*, **26**, TC6009.

ROBARDET, M. 2002. Alternative approach to the Variscan Belt in southwestern Europe: Preorogenic palaeobiogeographical constraints. *In*: MARTINEZ-CATALAN, J. R., HATCHER, R. D., JR., ARENAS, R. & DIAZ GARCIA, F. (eds) *Variscan–Appalachian Dynamics: The Building of the Late Palaeozoic Basement.* Geological Society of America, Special Papers, **364**, 1–15.

ROBARDET, M. & GUTÍERREZ-MARCO, J. C. 2004. The Ordovician, Silurian and Devonian sedimentary rocks of the Ossa–Morena Zone (SW Iberian Peninsula, Spain). *Journal of Iberian Geology*, **30**, 73–92.

RODRIGUEZ, J. 2005. Recristaizacion y deformacion de litologias supracorticales sometidas a metamorfismo de alta presion (Complejo Malpica-Tuy, NO del Macizo Ibérico). *Laboratório Xeológico de Laxe, Serie Nova Terra*, **29**.

RODRIGUEZ, J., COSCA, M. A., GIL IBARGUCHI, J. I. & DALLMEYER, R. D. 2003. Strain partitioning and preservation of $^{40}Ar/^{39}Ar$ ages during Variscan exhumation of a subducted crust (Malpica-Tuy Complex, SW Spain). *Lithos*, **70**, 111–139.

ROSAS, F. M., MARQUES, F. O., BALLÈVRE, M. & TASSINARI, C. 2008. Geodynamic evolution of the SW Variscides: Orogenic collapse shown by new tectonometamorphic and isotopic data from western Ossa–Morena Zone, SW Iberia. *Tectonics*, **27**, TC6008.

SANCHEZ-GARCIA, T., BELLINDO, F. & QUESADA, C. 2003. Geodynamic setting and geochemical signatures of Cambrian–Ordovician rift-related igneous rocks (Ossa–Morena Zone, SW Iberia). *Tectonophysics*, **365**, 233–255.

SANDIFORD, M. & POWELL, R. 1986. Deep crustal metamorphism during continental extension: modern and ancient examples. *Earth and Planetary Science Letters*, **79**, 151–158.

SANTOS, J. F., MATA, J., GONÇALVES, F. & MUNHÁ, J. M. 1987. Contribuição para o conhecimento geológico-petrológico da região de Santa Suzana: o Complexo Vulcano-sedimentar de Toca da Moura. *Comunicações Serviços Geológicos Portugal*, **73**, 29–40.

SANTOS, J. F., SOARES DE ANDRADE, A. & MUNHÁ, J. M. 1990. Magmatismo orogénico Varisco no limite meridional da Zona de Ossa-Morena. *Comunicações dos Serviços Geológicos de Portugal*, **76**, 91–124.

SCHÄFER, H. J., GEBAUER, D., NÄGLER, T. F. & EGUILUZ, L. 1993. Conventional and ion-microprobe U-Pb dating of detrital zircons of the Tentudía Group (Serie Negra, SW Spain): implications for zircon systematics, stratigraphy, tectonics and the Precambrian/Cambrian boundary. *Contributions to Mineral Petrology*, **113**, 289–299.

SILVA, J. B. & PEREIRA, M. F. 2004. Transcurrent continental tectonics model for the Ossa-Morena Zone Neoproterozoic-Paleozoic evolution, SW Iberian Massif, Portugal. *Geologische Rundschau*, **93**, 886–896.

SILVA, J. B., OLIVEIRA, J. T. & RIBEIRO, A. 1990. South Portuguese Zone, structural outline. *In*: DALLMEYER, R. D. & MARTÍNEZ GARCÍA, E. (eds) *Pre-Mesozoic Geology of Iberia*. Berlin-Heidelberg, Springer-Verlag, 348–362.

SIMANCAS, F., CARBONELL, R. ET AL. 2003. Crustal structure of the transpressional Variscan orogen of SW Iberia: SW Iberia deep seismic reflection profile (IBERSEIS). *Tectonics*, **22**(6), 1062, doi: 10.1029/2002TC001149.

SIMANCAS, J. F., TAHIRI, A., AZOR, A., GONZÁLEZ LODEIRO, F., MARTÍNEZ POYATOS, D. J. & EL HADI, H. 2005. The tectonic frame of the Variscan-Alleghanian orogen in Southern Europe and Northern Africa. *Tectonophysics*, **398**, 181–198.

STEIGER, R. H. & JÄGER, E. 1977. Subcommision on geochronology: convention on the use of decay constants in geo- and cosmochronology. *Earth and Planetary Science Letters*, **36**, 359–362.

TORNOS, F., CASQUET, C. & RELVAS, M. R. S. J. 2005. 4: Transpressional tectonics, lower crust decoupling and intrusion of deep mafic sills: a model for the unsual metallogenesis of SW Iberia. *Ore Geology Reviews*, **27**, 133–163.

VAVRA, G., GEBAUER, D., SCHMID, R. & COMPSTON, W. 1996. Multiple zircon growth and recrystallisation during polyphase Late Carboniferous to Triassic metamorphism in granulites of the Ivrea Zone (Southern Alps): an ion microprobe (SHRIMP) study. *Contributions to Mineralogy and Petrology*, **122**, 337–358.

WATSON, E. B. 1996. Dissolution, growth and survival of zircons during crustal fusion: kinetic principles, geologic models and implications for isotopic inheritance. *Geological Society of America Special Paper*, **315**, 43–56.

WILLIAMS, I. S. 2001. Response of detrital zircon and monazite, and their U-Pb isotopic systems, to regional metamorphism and host-rock partial melting, Cooma Complex, southeastern Australia. *Australian Journal of Earth Sciences*, **48**, 557–580.

WILLIAMS, I. S. & CLAESSON, S. 1987. Isotopic evidence for the Precambrian provenance and Caledonian metamorphism of high grade paragneisses from the Seve Nappes, Scandinavian Caledonides, II Ion microprobe zircon U-Th-Pb. *Contributions to Mineralogy and Petrology*, **97**, 205–217.

WILLIAMS, I. S. & HERGT, J. M. 2001. U–Pb dating of Tasmanian dolerites: a cautionary tale of SHRIMP analysis of high-U zircon. *In*: WOODHEAD, J. D., HERGT, J. M. & NOBLE, W. P. (eds) *Beyond 2000: New frontiers in Isotope Geoscience*. University of Melbourne, Melbourne, 185–188.

WILLIAMS, I. S., BUICK, I. S. & CARTWRIGHT, I. 1996. An extended episode of early Mesoproterozoic metamorphic fluid flow in the Reynolds Range, central Australia. *Journal of Metamorphic Geology*, **14**, 29–47.

WILLIAMS, O. S., COMPSTON, W., BLACK, L. P., IRELAND, T. R. & FOSTER, J. J. 1984. Unsupported radiogenic Pb in zircon: a cause of anomalously high Pb–Pb, U–Pb and Th–Pb ages. *Contributions to Mineralogy and Petrology*, **88**, 322–327.

# Palaeozoic palaeogeography of Mexico: constraints from detrital zircon age data

R. DAMIAN NANCE[1]*, J. DUNCAN KEPPIE[2], BRENT V. MILLER[3],
J. BRENDAN MURPHY[4] & JAROSLAV DOSTAL[5]

[1]*Department of Geological Sciences, 316 Clippinger Laboratories, Ohio University, Athens, Ohio 45701 USA*

[2]*Departamento de Geología Regional, Instituto de Geología, Universidad Nacional Autonoma de México, 04510, México D.F., México*

[3]*Department of Geology and Geophysics, Texas A&M University, College Station, Texas 77843, USA*

[4]*Department of Earth Sciences, St. Francis Xavier University, Antigonish, Nova Scotia, Canada, B2G 2W5*

[5]*Department of Geology, St. Mary's University, Halifax, Nova Scotia, Canada, B3H 3C3*

*\*Corresponding author (e-mail: nance@ohio.edu)*

**Abstract:** Detrital zircon age populations from Palaeozoic sedimentary and metasedimentary rocks in Mexico support palinspastic linkages to the northwestern margin of Gondwana (Amazonia) during the late Proterozoic–Palaeozoic. Age data from: (1) the latest Cambrian-Pennsylvanian cover of the *c.* 1 Ga Oaxacan Complex of southern Mexico; (2) the ?Cambro-Ordovician to Triassic Acatlán Complex of southern Mexico's Mixteca terrane; and (3) the ?Silurian Granjeno Schist of northeastern Mexico's Sierra Madre terrane, collectively suggest Precambrian provenances in: (1) the *c.* 500–650 Ma Brasiliano orogens and *c.* 600–950 Ma Goias magmatic arc of South America, the Pan-African Maya terrane of the Yucatan Peninsula, and/or the *c.* 550–600 Ma basement that potentially underlies parts of the Acatlán Complex; (2) the Oaxaquia terrane or other *c.* 1 Ga basement complexes of the northern Andes; and (3) *c.* 1.4–3.0 Ga cratonic provinces that most closely match those of Amazonia. Exhumation within the Acatlán Complex of *c.* 440–480 Ma granitoids prior to the Late Devonian–early Mississippian, and *c.* 290 Ma granitoids in the early Permian, likely provided additional sources in the Palaeozoic. The detrital age data support the broad correlation of Palaeozoic strata in the Mixteca and Sierra Madre terranes, and suggest that, rather than representing vestiges of Iapetus or earlier oceanic tracts as has previously been proposed, both were deposited along the southern, Gondwanan (Oaxaquia) margin of the Rheic Ocean and were accreted to Laurentia during the assembly of Pangaea in the late Palaeozoic.

## Introduction

The geology of Mexico is dominated by a collage of suspect terranes, the majority of which constitute part of the North American Cordillera having been accreted to the southwestern margin of Laurentia during the Mesozoic–Cenozoic (e.g. Keppie 2004). However, several Mexican terranes were accreted to Laurentia during the late Palaeozoic amalgamation of Pangaea and record Palaeozoic histories that constrain continental reconstructions for Pangaea assembly. These terranes (Fig. 1) include Oaxaquia, a *c.* 1 Ga crustal block that underlies much of central Mexico and is overlain by a thin veneer of unmetamorphosed Palaeozoic strata (Ortega-Gutiérrez *et al.* 1995), and the Mixteca and Sierra Madre terranes, which are dominated by Palaeozoic siliciclastic and oceanic rocks tectonically juxtaposed against the Oaxaquia terrane along major, north–south dextral faults of late Palaeozoic age (e.g. Elías-Herrera & Ortega-Gutiérrez 2002; Dowe *et al.* 2005).

The metasedimentary rocks of the Mixteca and Sierra Madre terranes have long been recognized to preserve an important record of Palaeozoic ocean opening and closure with links to the adjacent Oaxaquia terrane (e.g. Yañez *et al.* 1991; Ramírez-Ramírez 1992). But whether this ocean was Iapetus, the Rheic Ocean, or some other oceanic tract, remains controversial. Yañez *et al.* (1991) proposed that the Acatlán Complex documented a Devonian episode of Laurentia–Gondwana collision, an event they linked to the Acadian belt of the Appalachian orogen. On the basis of revised geochronology,

*From*: MURPHY, J. B., KEPPIE, J. D. & HYNES, A. J. (eds) *Ancient Orogens and Modern Analogues.*
Geological Society, London, Special Publications, **327**, 239–269.
DOI: 10.1144/SP327.12   0305-8719/09/$15.00 © The Geological Society of London 2009.

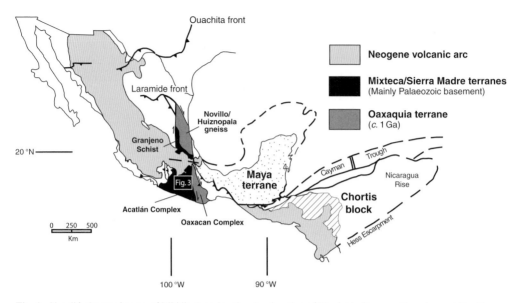

**Fig. 1.** Simplified tectonic map of Middle America showing location of Mexico's Oaxacan Complex and Novillo Gneiss (Oaxaquia terrane), Acatlán Complex (Mixteca terrane) and Granjeno Schist (Sierra Madre terrrane). Figure modified after Keppie (2004).

however, Ortega-Gutiérrez *et al.* (1999) later reassigned this event to the Late Ordovician–early Silurian and proposed that the collision recorded closure of the Iapetus Ocean, a view subsequently developed for the Acatlán Complex by Talavera-Mendoza *et al.* (2005) and Vega-Granillo *et al.* (2007). Ramírez-Ramírez (1992) similarly advocated a marginal basin of the Iapetus Ocean as the depositional setting for the metasedimentary rocks of the Sierra Madre terrane. A linkage to the Iapetus Ocean, however, is not supported by new age data that constrain the high-grade collisional metamorphism of the Acatlán Complex to the latest Devonian–early Mississippian (Middleton *et al.* 2007). Instead, these data have been used to link the Acatlán Complex to the southern (Gondwanan) margin of the Rheic Ocean (e.g. Nance *et al.* 2006, 2007*a*; Keppie *et al.* 2008*a*), a view consistent with the continental reconstructions of Keppie & Ramos (1999).

Discriminating between these mutually exclusive models is essential if the Palaeozoic stratigraphic and tectonothermal history of Mexico is to be placed within a broader plate tectonic framework, and the country's important role in continental reconstructions for the assembly of Pangaea is to be resolved. In this article, we examine the problem from the perspective of sedimentary provenance by using available detrital zircon data from Palaeozoic strata to identify potential cratonic source areas and, thereby, constrain the palinspastic restoration of those terranes that record Palaeozoic sedimentary histories.

## Palaeozoic detrital zircon data

Palaeozoic U–Pb detrital zircon age data are presently available in Mexico from three main sources (Fig. 1): (1) the sedimentary cover of the *c.* 1 Ga Oaxacan Complex in southern Mexico, the largest exposed portion of the Oaxaquia terrane (Gillis *et al.* 2005); (2) metasedimentary rocks of the largely Palaeozoic Acatlán Complex, which forms the basement of southern Mexico's Mixteca terrane (Sánchez-Zavala *et al.* 2004; Talavera-Mendoza *et al.* 2005; Keppie *et al.* 2006*a*; Murphy *et al.* 2006; Vega-Granillo *et al.* 2007; Grodzicki *et al.* 2008; Hinojosa-Prieto *et al.* 2008; Morales-Gámez *et al.* 2008); and (3) the Palaeozoic Granjeno Schist of northeastern Mexico's Sierra Madre terrane, which is tectonically juxtaposed against the second largest exposure of the Oaxaquia terrane, the *c.* 1 Ga Novillo Gneiss (Nance *et al.* 2007*b*).

These ages have been determined for samples of variable size using a variety of protocols and analytical techniques, including conventional, isotope dilution and thermo-ionization mass spectrometry (ID-TIMS), laser ablation-inductively coupled plasma mass spectrometry (LA-ICPMS), and sensitive high-resolution ion microprobe (SHRIMP). For comparative purposes, therefore, the data are best depicted using relative age probability plots (Ludwig 2003) normalized for the sample size. These plots are constructed by: (1) calculating a normal distribution for each analysis based on the reported age and uncertainty; (2)

summing the probability distributions of all acceptable analyses into a single curve; and (3) dividing the area under the curve by the number of analyses (Gehrels *et al.* 2006). The plots were constructed at the same scale from original data using an Excel macro available on the website of the Laser-Chron Center at the University of Arizona (www.geo.arizona.edu/alc). For consistency, the normalized relative age probability plots were constructed using the $^{206}Pb/^{238}U$ age for young ($<1.0$ Ga) zircons and the $^{206}Pb/^{207}Pb$ age for older ($>1.0$ Ga) grains, and analyses with an error $>10\%$, and older ages with $>20\%$ discordance or $>5\%$ reverse discordance, were not used.

## Oaxacan Complex

The Oaxacan Complex of the Oaxaquia terrane (Fig. 1) comprises granulite facies paragneisses and within-plate, rift-related AMCG (anorthosite-mangerite-charnockite-granite) suites that have yielded U–Pb emplacement ages of *c.* 1000–1020 and *c.* 1100–1160 Ma (Keppie *et al.* 2001, 2003; Solari *et al.* 2003), and orthogneisses with a protolith age of $\geq$1350 Ma that were migmatized at *c.* 1106 Ma. These rocks underwent polyphase deformation and granulite facies metamorphism at *c.* 980–1005 Ma, cooling through 500 °C ($^{40}Ar/^{39}Ar$ on hornblende) at *c.* 975 Ma (Solari *et al.* 2003). A shallow-level, calc-alkaline granitoid pluton intruded the northern Oaxacan Complex at *c.* 917 Ma (Ortega-Obregon *et al.* 2003).

Further north, inliers of the Oaxaquia terrane include the Huiznopala and Novillo gneisses (Fig. 1), the latter juxtaposed against the Palaeozoic Granjeno Schist of the Sierra Madre terrane along a north–south dextral shear zone of late Palaeozoic age (Dowe *et al.* 2005). The Huiznopala Gneiss comprises layered paragneisses and arc-related orthogneisses that have yielded ages of *c.* 1150–1200 Ma (Lawlor *et al.* 1999), and an anorthosite-gabbro complex emplaced during granulite facies metamorphism at *c.* 1000 Ma. Pegmatites that give an age of 988 $\pm$ 3 Ma post-date ductile deformation. The Novillo Gneiss likewise comprises metasedimentary rocks intruded by gabbro-anorthosite, granite and amphibolite that have yielded U–Pb emplacement ages of *c.* 1010–1035 and *c.* 1115–1235 Ma (Cameron *et al.* 2004). These rocks underwent polyphase deformation and granulite facies metamorphism at 990 $\pm$ 5 Ma, followed by post-tectonic anorthositic pegmatite emplacement at 978 $\pm$ 13 Ma. At *c.* 546 Ma, the Novillo Gneiss was intruded by a mafic dyke swarm with plume-related geochemical affinities (Keppie *et al.* 2006*b*).

Unmetamorphosed Palaeozoic strata that nonconformably overlie the Oaxacan Complex include three formations that are separated by shear zones (Centeno-Garcia & Keppie 1999): (1) the latest Cambrian–Ordovician Tiñu Formation, which contains shallow-marine fauna of Gondwanan affinity (Robison & Pantoja-Alor 1968; Landing *et al.* 2007); (2) the Mississippian Santiago Formation with midcontinent (US) brachiopod fauna (e.g. Navarro-Santillan *et al.* 2002); and (3) the Pennsylvanian Ixtaltepec Formation. Gillis *et al.* (2005) report ID-TIMS and LA-ICPMS U–Pb detrital zircon ages from each of these three formations (Fig. 2).

*Tiñu Formation.* The latest Cambrian–earliest Ordovician Tiñu Formation comprises up to 200 m of limestone, shale and siltstone interpreted to represent a transgressive sequence of wave-dominated shelf sediments (Sánchez-Zavala *et al.* 1999; Landing *et al.* 2007). The basal unit is a coarse arkosic sandstone, detrital zircons from which yield a normalized relative age population (Fig. 2a) with a series of peaks between *c.* 995 Ma and 1225 Ma, and a single older zircon at *c.* 1450 Ma (Gillis *et al.* 2005). The youngest detrital zircon gives an LA-ICPMS age of *c.* 960 Ma.

*Santiago Formation.* The early Mississippian Santiago Formation comprises some 200 m of brachiopod-rich limestone, shale and minor sandstone (Pantoja-Alor & Robison 1967; Navarro-Santillan *et al.* 2002) interpreted to have been deposited in either a shallow marine (Sánchez-Zavala *et al.* 1999) or deeper shelf–slope (Flores de Dios Gonzalez *et al.* 1998) environment. Detrital zircons from a calcareous sandstone at the base of the formation yield a normalized relative age population (Fig. 2b) with peaks at *c.* 470 and 990 Ma (Gillis *et al.* 2005). The youngest detrital zircon gives an ID-TIMS age of *c.* 460 Ma (mid-Ordovician; Gradstein *et al.* 2004).

*Ixtaltepec Formation.* The fossiliferous early–middle Pennsylvanian Ixtaltepec Formation comprises *c.* 425 m of shale with interbedded limestone and sandstone (Pantoja-Alor & Robison 1967; Flores de Dios Gonzalez *et al.* 1998) that are interpreted to be shallow marine (Sánchez-Zavala *et al.* 1999). Detrital zircons from a sandstone near the base of the formation yield a normalized relative age population (Fig. 2c) with a peak at *c.* 360 Ma and a series of peaks in the interval *c.* 990–1185 Ma (Gillis *et al.* 2005). The youngest detrital zircons give ID-TIMS and LA-ICPMS ages of *c.* 340 Ma (Mississippian; Gradstein *et al.* 2004).

## Acatlán Complex

To the west, the Oaxacan Complex is faulted against the mainly Palaeozoic rocks of the Acatlán Complex

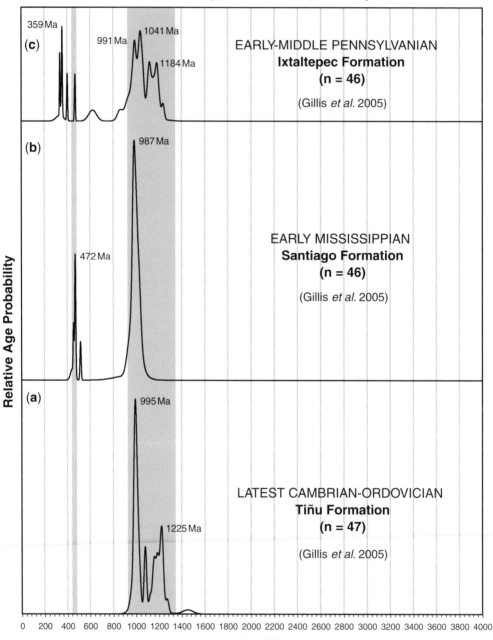

**Fig. 2.** Normalized relative age distribution plots of U–Pb detrital zircon analyses from the sedimentary cover of the Oaxacan Complex: (**a**) latest Cambrian-Ordovician Tiñu Formation; (**b**) early Mississippian Santiago Formation; and (**c**) early–middle Pennsylvanian Ixtaltepec Formation (data from Gillis *et al.* 2005). Shaded bars outline the age ranges characteristic of the Ordovician megacrystic granitoids of the Acatlán Complex (*c.* 440–480 Ma: Sánchez-Zavala *et al.* 2004; Talavera-Mendoza *et al.* 2005; Miller *et al.* 2007; Vega-Granillo *et al.* 2007) and the Mesoproterozoic basement of the Oaxaquia terrane (*c.* 920–1350 Ma: Keppie *et al.* 2001, 2003; Solari *et al.* 2003; Cameron *et al.* 2004). n = sample size.

of the Mixteca terrane (Fig. 1). The north–south Caltepec fault zone that defines the terrane boundary between these two complexes (Fig. 3) is a dextral transpressive structure that was active at *c.* 276 Ma and is overstepped by early Permian (Leonardian) red beds of the Matzitzi Formation (Elías-Herrera & Ortega-Gutiérrez 2002).

The Acatlán Complex is made up of several tectonic assemblages that range in age from ?Cambro–Ordovician to Triassic and record major tectonothermal events of Devono–Carboniferous, Permo–Triasssic and Jurassic age (Fig. 4). For recent reviews, see Nance *et al.* (2006, 2007*a*) and Keppie *et al.* (2008*a*). For the purpose of this paper, however, the principal units of the complex can be grouped into five major packages (Keppie *et al.* 2008*a*): (1) Pre-Silurian high-grade metasedimentary rocks (Piaxtla Suite) that are thought to represent an allochthonous, ?Cambro–Ordovician rift-drift assemblage and include locally eclogitic mafic-ultramafic rocks, serpentinites, blueschists, migmatites and amphibolite facies metasedimentary units (Xayacatlán Formation, Asis Unit and Ixcamilpa blueschists) that are intruded by a suite of highly deformed K–feldspar/quartz megacrystic granitoid bodies (Esperanza Grantoids) of Ordovician (*c.* 480–440 Ma) age (e.g. Talavera-Mendoza *et al.* 2005; Murphy *et al.* 2006; Miller *et al.* 2007; Vega-Granillo *et al.* 2007) and were exhumed in the Late Devonian–early Mississippian (Middleton *et al.* 2007); (2) pre-Silurian low-grade metasedimentary rocks that are thought to represent a ?Cambro–Ordovician rift-passive margin clastic sequence (El Rodeo Formation and Amate, Canoas, Huerta, El Epazote and Las Calaveras units) and are associated with bimodal, rift-related igneous rocks and intruded by Ordovician megacrystic granitoids (e.g. Keppie *et al.* 2007, 2008*b*; Miller *et al.* 2007; Hinojosa-Prieto *et al.* 2008; Grodzicki *et al.* 2008; Morales-Gámez *et al.* 2008; Ramos-Arias *et al.* 2008); (3) Devonian–Carboniferous low-grade metasedimentary rocks that include phyllite, quartzite and minor mafic volcanic rocks (Cosoltepec Formation, Salida and Coatlaco units), the deposition of which was coeval with exhumation of the high pressure rocks (Grodzicki *et al.* 2008; Morales-Gámez *et al.* 2008); (4) Carboniferous–Permian low-grade metasedimentary rocks that form a continental-shallow marine succession dominated by slate, sandstone, conglomerate and limestone (Tecomate, Olinalá and Patlanoaya formations), the deposition of which was coeval with Permo–Triassic arc magmatism (Torres *et al.* 1999; Malone *et al.* 2002) and which are locally overthrust by the broadly coeval (*c.* 288 Ma), arc-related Totoltepec pluton (Yañez *et al.* 1991; Keppie *et al.* 2004*a*); and (5) Permian–Triassic high-grade metasedimentary rocks

(Petlalcingo Suite) that are thought to represent an active margin clastic wedge (Keppie *et al.* 2006*a*) and comprise pelites, psammites and minor mafic units (Magdalena protolith and Chazumba Formation) that were metamorphosed in the amphibolite facies and locally pervasively migmatized in the mid-Jurassic.

*Pre-Silurian high-grade metasedimentary rocks (Piaxtla Suite).* LA-ICPMS U–Pb detrital zircon age data have been obtained from several high-grade metasedimentary units of the Acatlán Complex sampled within the Piaxtla Suite (Fig. 5). High-grade metasedimentary rocks reportedly intruded by the Esperanza Granitoids have been sampled for detrital zircons at Mimilulco (Xayacatlán Formation; Talavera-Mendoz *et al.* 2005), San Francisco de Asis (Asis Lithodeme; Murphy *et al.* 2006) and Santa Cruz Organal (quartzite; Vega-Granillo *et al.* 2007) near the northern margin of the Acatlán Complex (Fig. 3). They have also been sampled at Ixcamilpa (Ixcamilpa blueschist; Talavera-Mendoza *et al.* 2005) near the western margin of the complex.

At Mimilulco, detrital zircons from a chloritoid–phengite–garnet psammitic schist of the Xayacatlán Formation interbedded with retrogressed eclogites yield a normalized relative age population (Fig. 5a) with peaks at *c.* 870, 990, 1120 and 1390 Ma (Talavera-Mendoza *et al.* 2005). At San Francisco de Asis, detrital zircons from a quartz-rich metapsammite of the Asis Lithodeme interlayered with garnet-bearing metapelites and amphibolites yield a normalized population (Fig. 5b) with major peaks at *c.* 920, 1160 and 1200 Ma (Murphy *et al.* 2006). Older zircons give minor peaks at *c.* 1490 and 1555 Ma. At Santa Cruz Organal, detrital zircons from a garnet-phengite quartzite yield age probabilities (Fig. 5c) with a peak at *c.* 720 Ma and a series of peaks between *c.* 945 and 1125 Ma (Vega-Granillo *et al.* 2007). Older zircon peaks occur at *c.* 1450, 1510 and 1675 Ma. At Ixcamilpa, zircons from a chlorite-phengite schist interbedded with blueschists (Ixcamilpa blueschist) yield a very different normalized relative age population (Fig. 5d) with major peaks at *c.* 475, 540 and 600 Ma, a broad spectrum of smaller peaks between *c.* 730 and *c.* 1230 Ma, and older peaks at *c.* 1820 and 1950 Ma (Talavera-Mendoza *et al.* 2005).

The youngest detrital zircons in the four units give ages of *c.* 800, 705, 670 and 450 Ma (Late Ordovician; Gradstein *et al.* 2004), respectively, and place maximum age constraints on the timing of their deposition. The youngest zircon cluster in the Ixcamilpa sample, which Talavera-Mendoza *et al.* (2005) consider to provide a more reliable maximum age constraint, peaks at *c.* 477 Ma

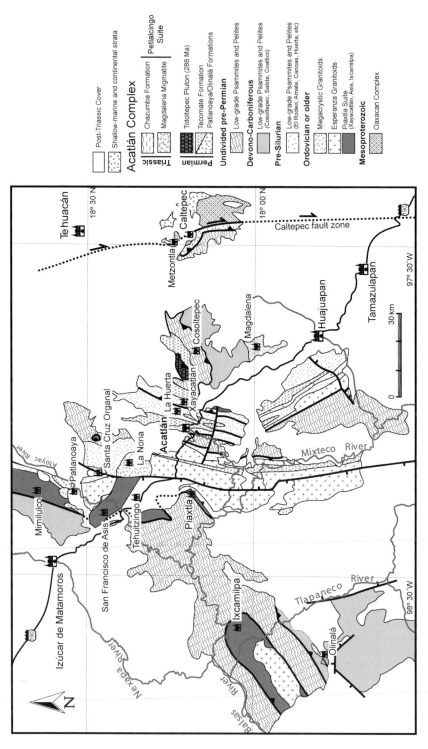

**Fig. 3.** Simplified geological map of the Acatlán Complex, southern Mexico (modified from Ortega-Gutiérrez *et al.* 1999 & Keppie *et al.* 2008*a*). See Figure 1 for location.

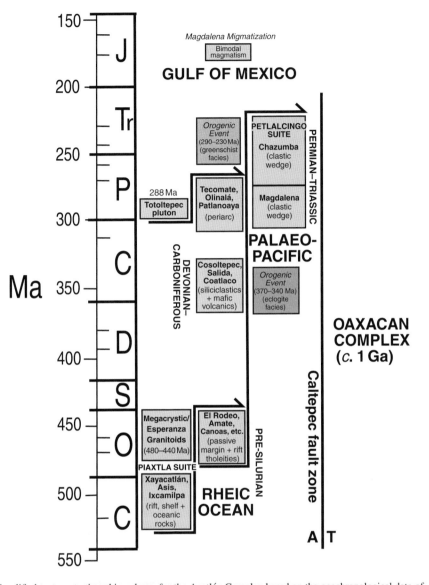

**Fig. 4.** Simplified tectonostratigraphic column for the Acatlán Complex based on the geochronological data of Keppie *et al.* (2004*a*, *b*, 2006*a*), Sánchez-Zavala *et al.* (2004), Talavera-Mendoza *et al.* (2005), Middleton *et al.* (2007), Miller *et al.* (2007), Vega-Granillo *et al.* (2007) and Morales-Gámez *et al.* (2008). Arrows indicate present structural positions of the units but not the nature of their emplacement. Figure modified from Nance *et al.* (2006). C, Cambrian; O, Ordovician; S, Silurian; D, Devonian; C, Carboniferous; P, Permian; Tr, Triassic; J, Jurassic; A, away; T, towards.

(Early Ordovician; Gradstein *et al.* 2004). The significant difference in youngest zircon age between the northern and western units suggests that the former provide constraints on the deposition of the host rock into which the megacrystic granitoids were emplaced, whereas the latter dates a volcaniclastic protolith whose deposition was broadly coeval with granitoid intrusion.

*Ordovician megacrystic granitoids.* ID-TIMS, SHRIMP and LA-ICPMS crystallization and inheritance ages have been determined for 15 megacrystic granitoid bodies. The granitoids have yielded a lower intercept age of $440 \pm 14$ Ma (Ortega-Gutiérrez *et al.* 1999), LA-ICPMS ages ranging from $440 \pm 15$–$478 \pm 5$ Ma (Sánchez-Zavala *et al.* 2004; Talavera-Mendoza *et al.* 2005;

# Pre-Silurian High-Grade Metasedimentary Rocks
## (Piaxtla Suite)

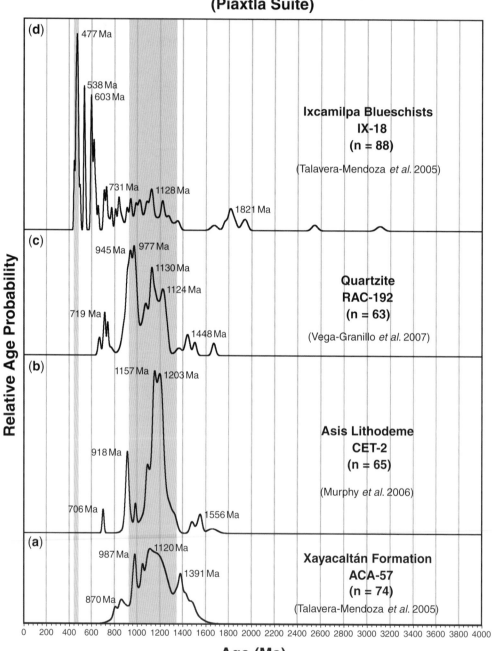

**Fig. 5.** Normalized relative age probability plots of U–Pb detrital zircon analyses from pre-Silurian high-grade metasedimentary units of the Acatlán Complex (Piaxtla Suite): (**a**) Xayacatlán Formation at Mimilulco (Talavera-Mendoza *et al.* 2005); (**b**) Asis Lithodeme at San Francisco de Asis (Murphy *et al.* 2006); (**c**) quartzite from Santa Cruz Organal (Vega-Mendoza *et al.* 2007); and (**d**) Ixcamilpa blueschist at Ixcamilpa (Talavera-Mendoza *et al.* 2005). Shaded bars outline the age ranges characteristic of the Ordovician megacrystic granitoids of the Acatlán Complex (*c.* 440–480 Ma: Sánchez-Zavala *et al.* 2004; Talavera-Mendoza *et al.* 2005; Miller *et al.* 2007; Vega-Granillo *et al.* 2007) and the Mesoproterozoic basement of the Oaxaquia terrane (*c.* 920–1350 Ma: Keppie *et al.* 2001, 2003; Solari *et al.* 2003; Cameron *et al.* 2004). n = sample size.

Vega-Granillo *et al.* 2007), a SHRIMP age of 467 ± 16 Ma, and a single precise ID-TIMS age of 461 ± 2 Ma (Miller *et al.* 2007), all of which are interpreted to date their crystallization. Two additional granitoids were interpreted by Talavera-Mendoza *et al.* (2005) to have Mesoproterozoic (1043 ± 50 and 1149 ± 6 Ma) crystallization ages. However, this interpretation is contentious since the granitoids contain significant Mesoproterozoic inheritance in precisely this age range. LA-ICPMS ages of 464 ± 4 Ma for a granite pegmatite and 452 ± 6 and 447 ± 3 Ma for foliated granite dykes have been reported by Morales-Gámez *et al.* (2008), and a younger LA-ICPMS age of 372 ± 8 Ma has been reported for a leucogranite (Vega-Granillo *et al.* 2007). Collectively, the intrusive age of the granitoids centres at *c.* 460 Ma (Fig. 6a), whereas the normalized relative age population for xenocrystic zircons peaks at *c.* 565, 1015 and 1185 Ma (Fig. 6b).

*Pre-Silurian low-grade metasedimentary rocks.* LA-ICPMS U–Pb detrital zircon age data have been obtained from low-grade, ?Cambro–Ordovician siliciclastic rocks of the Acatlán Complex near the villages of Olinalá (El Rodeo Formation and Canoas Unit) in the southwestern part of the complex, Xayacatlán (Amate Unit) and La Huerta (Huerta Unit) in the east-central part, and near La Noria (Las Calaveras and El Epazote units) in the northern Acatlán Complex (Fig. 3).

The El Rodeo Formation consists of greenschist facies mafic, conglomeratic and volcaniclastic rocks that are intruded by an Early Ordovician (476 ± 8 Ma) granitoid and unconformably overlain by the middle Permian Olinalá Formation. Detrital zircons from a metavolcaniclastic sandstone of the El Rodeo Formation yield a normalized relative age population (Fig. 7a) that defines a broad peak between *c.* 990 and *c.* 1500 Ma centred at *c.* 1170 Ma (Talavera-Mendoza *et al.* 2005). The youngest zircon at *c.* 950 Ma or the youngest zircon cluster at *c.* 988 Ma, and the *c.* 476 Ma age of the intrusive granite, provide maximum and minimum constraints, respectively, on the formation's depositional age.

The Amate Unit comprises low-grade arkoses, psammites and pelites that are intruded by mafic dykes dated at 442 ± 1 Ma (Keppie *et al.* 2008*b*) and granitoid dykes in which the youngest concordant $^{206}$Pb/$^{238}$U zircon ages are 452 ± 6 and 447 ± 3 Ma (Morales-Gámez *et al.* 2008). A feldspathic metapsammite from this unit yielded detrital zircons with a normalized relative age population that shows a series of peaks between *c.* 950 and *c.* 1300 Ma (Fig. 7b). The youngest concordant zircon gives an age of *c.* 900 Ma.

The Canoas Unit is dominated by low-grade psammites and pelites that record three phases of penetrative deformation. Detrital zircons from a sample of metapsammite show a normalized relative age population (Fig. 7c) with peaks at *c.* 960, 1160 and 1510 Ma (Grodzicki *et al.* 2008). The youngest detrital zircon, which records an age of *c.* 453 Ma (Late Ordovician; Gradstein *et al.* 2004), provides a maximum age constraint for the timing of deposition.

The Huerta Unit comprises greenschist facies interbedded psammites and pelites that closely resemble those of the Canoas Unit and likewise record three phases of penetrative deformation (Malone *et al.* 2002). A semipelite from the Huerta Unit yielded detrital zircons that show a normalized relative age population (Fig. 7d) with major peaks at *c.* 510, 570 and 630 Ma, a series of smaller peaks between *c.* 950 and 1220 Ma, and minor peaks at *c.* 1680 and 1995 Ma (Keppie *et al.* 2006*a*). The youngest detrital zircon records an age of *c.* 455 Ma.

The El Epazote and Las Calaveras units are greenschist facies meta-volcanosedimentary assemblages in the northern Acatlán Complex (Hinojosa-Prieto *et al.* 2008). The El Epazote Unit comprises metapelites, metavolcaniclastics and fine-grained metapsammites that are in tectonic contact with megacrystic granitoids, one of which (the La Noria granite) has yielded a weighted mean $^{206}$Pb/$^{238}$U SHRIMP age of 467 ± 16 Ma (Miller *et al.* 2007). The Las Calaveras Unit comprises quartzites and metagreywackes that are intruded by granite. Both units record correlative deformational histories involving three phases of deformation that are also recorded in the Ordovician La Noria granite (Hinojosa-Prieto *et al.* 2008). Detrital zircons from a volcaniclastic epidote-chlorite schist of the El Epazote Unit yield a normalized relative age population with a series of peaks in the intervals *c.* 440–650 and *c.* 975–1280 Ma, and a minor peak at *c.* 1800 Ma (Fig. 7f). Zircons from a metagreywacke of the Las Calaveras Unit yield a normalized relative age probability with major peaks at *c.* 470, 1040 and 1225 Ma, and a minor peak at *c.* 1760 Ma (Fig. 7e). The youngest concordant zircons at *c.* 486 Ma (Early Ordovician; Gradstein *et al.* 2004) and *c.* 452 Ma (Late Ordovician; Gradstein *et al.* 2004), respectively, and youngest zircon clusters at *c.* 506 Ma (late Cambrian; Gradstein *et al.* 2004) and 466 Ma (Middle Ordovician; Gradstein *et al.* 2004), provide maximum limits on the depositional ages of the two units.

*Devonian–Carboniferous low-grade metasedimentary rocks.* LA-ICPMS U–Pb detrital zircon age data have been obtained from low-grade, Devonian–Carboniferous siliciclastic rocks near the village of Mimilulco (Cosoltepec Formation) in the northern

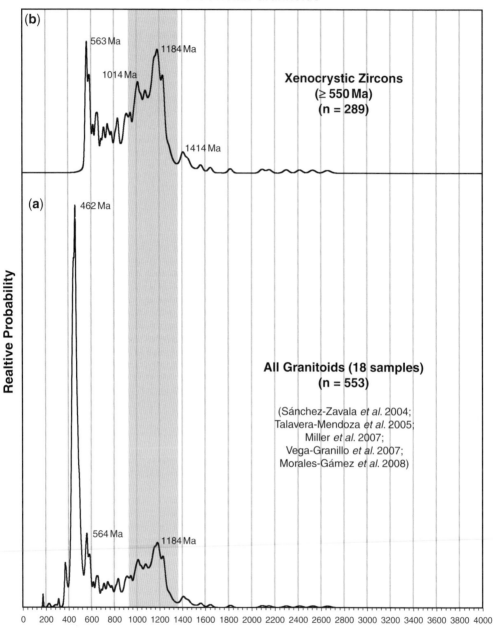

**Fig. 6.** Normalized relative age probability plot of U–Pb zircon analyses from Ordovician granitoids of the Acatlán Complex: (**a**) compilation of all zircon data and (**b**) compilation of xenocrystic zircon analyses (all those with ages ≥550 Ma). Data from Sánchez-Zavala *et al.* (2004), Talavera-Mendoza *et al.* (2005), Miller *et al.* (2007), Vega-Granillo *et al.* (2007) and Morales-Gámez *et al.* (2008). Shaded bar outlines the age range characteristic of the Mesoproterozoic basement of the Oaxaquia terrane (*c.* 920–1350 Ma: Keppie *et al.* 2001, 2003; Solari *et al.* 2003; Cameron *et al.* 2004). n = sample size.

## Pre-Silurian Low-Grade Metasedimentary Rocks

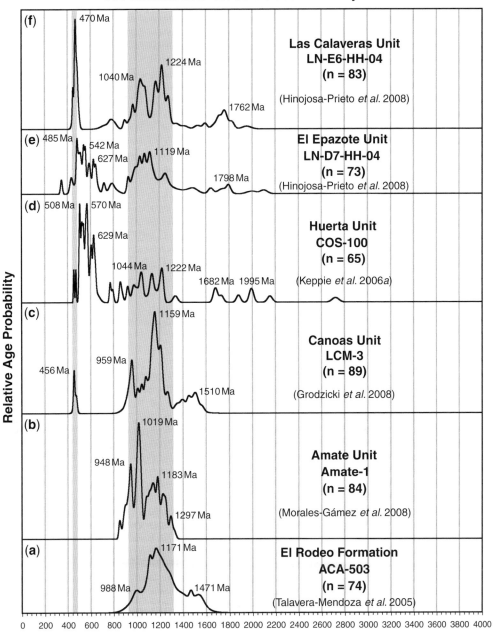

**Fig. 7.** Normalized relative age probability plots of U–Pb detrital zircon analyses from pre-Silurian low-grade metasedimentary units of the Acatlán Complex: (**a**) El Rodeo Formation at Olinalá (Talavera-Mendoza *et al.* 2005); (**b**) Amate Unit at Xayacatlán (Morales-Gámez *et al.* 2008); (**c**) Canoas Unit near Olinalá (Grodzicki *et al.* 2008); (**d**) Huerta Unit at La Huerta (Keppie *et al.* 2006a); (**e**) El Epazote Unit near La Noria (Hinojosa-Prieto *et al.* 2008); and (**f**) Las Calaveras Unit near La Noria (Hinojosa-Prieto *et al.* 2008). Shaded bars outline the age ranges characteristic of the Ordovician megacrystic granitoids of the Acatlán Complex (*c.* 440–480 Ma: Sánchez-Zavala *et al.* 2004; Talavera-Mendoza *et al.* 2005; Miller *et al.* 2007; Vega-Granillo *et al.* 2007) and the Mesoproterozoic basement of the Oaxaquia terrane (*c.* 920–1350 Ma: Keppie *et al.* 2001, 2003; Solari *et al.* 2003; Cameron *et al.* 2004). n = sample size.

Acatlán Complex, at Cosoltepec (Cosoltepec Formation) and Xayacatlán (Salada Unit) in the east-central part, and near Olinalá (Coatlaco Unit) in the southwestern part of the complex (Fig. 3).

The Cosoltepec Formation comprises pelites, quartzites and minor mafic volcanic rocks that record three phases of penetrative deformation (e.g. Malone *et al.* 2002). As originally mapped (Ortega-Gutiérrez 1975), the formation included all low-grade pelite-psammite assemblages with this record of deformation and formed the most widely distributed unit within the complex (Fig. 3). However, detrital zircon data (Talavera-Mendoza *et al.* 2005; Keppie *et al.* 2006*b*; Morales-Gámez *et al.* 2008) have since revealed the existence of polydeformed pelite-psammite units of at least two separate ages. The term Cosoltepec Formation is used here only for pelite-psammite units of Devonian–Carboniferous age, like those of the type-area at Cosoltepec.

Detrital zircons from a quartzite of the Cosoltepec Formation sampled near Mimilulco (Talavera-Mendoza *et al.* 2005) yield a normalized relative age population with peaks at *c.* 340 and 410 Ma, a series of major peaks in the interval *c.* 500–700 Ma, and minor peaks at *c.* 985 and 2190 Ma (Fig. 8a). Zircons from a quartzite of the Cosoltepec Formation sampled in the type-area (Talavera-Mendoza *et al.* 2005) yield a similar population with a peak at *c.* 395 Ma, a series of major peaks in the interval *c.* 540–620 Ma, and a series of minor peaks in the intervals *c.* 1000–1360 and *c.* 1700–1960 Ma (Fig. 8b). Maximum age constraints on the timing of deposition of the formation are provided by the youngest detrital zircons in each sample, which record ages of *c.* 340 Ma (Mississippian; Gradstein *et al.* 2004) and *c.* 375 Ma (late Devonian; Gradstein *et al.* 2004), respectively. The youngest zircon cluster in each sample record an age of *c.* 410 Ma (early Devonian; Gradstein *et al.* 2004).

The Salida Unit consists of greenschist facies psammites, pelites and thin tholeitiic mafic slices tectonically juxtaposed against the pre-Silurian low-grade Amate Unit in the vicinity of Xayacatlán (Fig. 3). Detrital zircons from a metapsammite in this unit (Morales-Gámez *et al.* 2008) yield a normalized relative age distribution with a series of peaks in the interval *c.* 355–625 Ma, a peak at 895 Ma, and a series of minor peaks in the intervals *c.* 1070–1300 and *c.* 1765–1985 Ma (Fig. 8c). An older limit on the time of deposition of the Salada Unit is provided by the *c.* 350 Ma age (early Mississippian; Gradstein *et al.* 2004) of the youngest concordant detrital zircon.

The Coatlaco Unit consists of interbedded quartzite and locally pillowed, greenschist facies meta-basalts that show sub-alkaline tholeitiic affinity and a within-plate geochemical signature (Grodzicki *et al.* 2008). The unit is exposed near Olinalá where it is in tectonic contact with the pre-Silurian low-grade Canoas Unit, but records only two phases of penetrative deformation compared to the three or more phases present in the Canoas Unit, suggesting an originally unconformable relationship. Detrital zircons from a quartzite of the Coatlaco Unit (Grodzicki *et al.* 2008) yield a normalized relative age distribution with peaks in the interval *c.* 310–370 Ma, major peaks at *c.* 535 and 575 Ma, a series of peaks between *c.* 780 and *c.* 1200 Ma, and numerous minor peaks in the broad interval *c.* 1760–2670 Ma (Fig. 8d). The youngest zircon at *c.* 308 Ma (Pennsylvanian; Gradstein *et al.* 2004) and the youngest zircon cluster at *c.* 357 Ma (late Devonian; Gradstein *et al.* 2004) provide maximum constraints on the timing of the unit's deposition.

*Carboniferous–Permian low-grade metasedimentary rocks.* LA-ICPMS U–Pb detrital zircon age data for Carboniferous–Permian units of the Acatlán Complex have been obtained from low-grade siliciclastic rocks near the village of Acatlán (Tecomate Formation) in the centre of the complex and at Olinalá (Olinalá Formation) near its western margin (Fig. 3). The Patlanoaya Formation, which is exposed near the northern margin of the complex, has not been sampled for detrital zircons but comprises fossiliferous shales, sandstones, conglomerates and limestones that range in age from latest Devonian (Strunian) to early Permian (Leonardian) (Vachard *et al.* 2000; Vachard & Flores de Dios 2002).

The Tecomate Formation comprises low-grade, pelitic and volcanosedimentary psammitic rocks, conglomerates and marbles that, in the type section south of the town of Acatlán (Fig. 3), non-conformably overlie a deformed megacrystic granitoid of presumed Ordovician age (Sánchez-Zavala *et al.* 2004). The formation is mildly to strongly deformed, recording up to two phases of penetrative deformation (e.g. Malone *et al.* 2002), and is stratigraphically similar to the essentially undeformed Patlanoaya Formation. Two marble horizons towards the top of the type section have yielded conodonts with age ranges close to the Carboniferous–Permian and early–middle Permian boundaries, respectively (Keppie *et al.* 2004*a*). East of the town of Acatlán, the formation is over-thrust by the Totoltepec pluton, the early Permian ID-TIMS U–Pb age of which ($287 \pm 2$ Ma, Yañez *et al.* 1991; $289 \pm 1$ Ma, Keppie *et al.* 2004*b*) matches the SHRIMP U–Pb ages of granitoid pebbles sampled from a conglomerate of the Tecomate Formation west of Cosoltepec (*c.* 280–310 Ma; Keppie *et al.* 2004*a*).

## Devonian-Carboniferous Low-Grade Metasedimentary Rocks

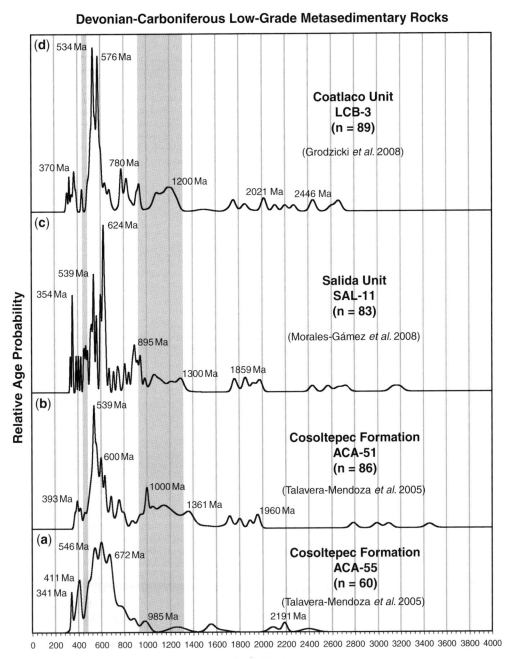

**Fig. 8.** Normalized relative age probability plots of U–Pb detrital zircon analyses from Devonian–Carboniferous low-grade metasedimentary units of the Acatlán Complex: (**a**) Cosoltepec Formation at Mimiluco (Talavera-Mendoza *et al.* 2005); (**b**) Cosoltepec Formation at Cosoltepec (Talavera-Mendoza *et al.* 2005); (**c**) Salida Unit at Xayactalán (Morales-Gámez *et al.* 2008); and (**d**) Coatlaco Unit near Olinalá (Grodzicki *et al.* 2008). Shaded bars outline the age ranges characteristic of the Ordovician megacrystic granitoids of the Acatlán Complex (*c.* 440–480 Ma: Sánchez-Zavala *et al.* 2004; Talavera-Mendoza *et al.* 2005; Miller *et al.* 2007; Vega-Granillo *et al.* 2007) and the Mesoproterozoic basement of the Oaxaquia terrane (*c.* 920–1350 Ma: Keppie *et al.* 2001, 2003; Solari *et al.* 2003; Cameron *et al.* 2004). n = sample size.

Detrital zircon age data are available for seven low-grade recrystallized sandstones of the Tecomate Formation sampled across the type section south of Acatlán (Sánchez-Zavala *et al.* 2004). Collectively, these yield a normalized relative age distribution with a prominent peak at *c.* 465 Ma, a broad Mesoproterozoic population in the interval *c.* 1000–1300 Ma, and a minor peak at *c.* 1520 Ma (Fig. 9a). No systematic variation in detrital age populations exists with stratigraphic level, but individual sandstones are dominated by either Ordovician (*c.* 460–480 Ma) or Mesoproterozoic–early Neoproterozoic (*c.* 900–1300 Ma) zircons, presumably in response to varying drainage patterns. The dominance of Carboniferous–Permian granitoid cobbles in the conglomerate west of Cosoltepec further documents the influence of local drainage on the age populations of the formation's detrital zircons. Older age limits for the deposition of the Tecomate Formation are provided by the youngest concordant zircon at *c.* 450 Ma and the youngest zircon cluster at *c.* 460 Ma (Late Ordovician; Gradstein *et al.* 2004).

The Olinalá Formation is a marine succession in the southwestern Acatlán Complex that, like the Tecomate Formation, unconformably overlies older metasedimentary units. The formation is floored by a massive conglomerate containing quartzite and schist detritus, and includes fossiliferous black shales, estuarine sandstones and locally stromatolitic limestones that have been dated as middle Permian on the basis of ammonoids, brachiopods and fusilinids (González-Arreola *et al.* 1994; Vachard *et al.* 2004).

Detrital zircons from a quartz-rich calcareous sandstone collected from the type section of the Olinalá Formation, east of the town of Olinalá (Talavera-Mendoza *et al.* 2005), yield a normalized relative age distribution with a major peak at *c.* 300 Ma, a minor peak at *c.* 825 Ma, and a broad Mesoproterozoic population in the interval *c.* 1100–1350 Ma (Fig. 9b). The youngest concordant zircon at *c.* 290 Ma and the youngest zircon cluster at *c.* 297 Ma (early Permian; Gradstein *et al.* 2004) provide maximum constraints on the formation's depositional age.

*Permian-Triassic high-grade metasedimentary rocks.* LA-ICPMS U–Pb detrital zircon age data have been obtained from both of the high-grade metasedimentary units of the Permian–Triassic Petlalcingo Suite (Magdalena Migmatite and Chazumba Formation) near the village of Magdalena in the eastern Acatlán Complex (Fig. 3). The Chazumba Formation consists of repeatedly deformed, amphibolite facies psammites and pelites that were metamorphosed during the Jurassic and contain several mafic-ultramafic lenses of Jurassic

(174 ± 1 Ma) age (Keppie *et al.* 2004*b*). The gradationally underlying Magdalena Migmatite is a similar lithological unit with additional calcsilicate and marble lenses, and was pervasively migmatized during the same Jurassic (*c.* 175–170 Ma; Keppie *et al.* 2004*b*) tectonothermal event. Both units are interpreted to represent portions of a clastic wedge deposited ahead of thrust faults developed along the active margin of the palaeo-Pacific during the Permo–Traissic (Keppie *et al.* 2006*a*).

Data are available for two samples of the Magdalena Migmatite palaeosome, both collected south of Magdalena, and three samples of the Chazumba Formation, one collected from the structural base of the succession, near Magdalena, and the others from the structural top to the north (Talavera-Mendoza *et al.* 2005; Keppie *et al.* 2006*a*). The two Magdalena Migmatite samples – a biotite–hornblende schist and a biotite–garnet–amphibole schist – yield normalized relative age populations with peaks (1) in the intervals *c.* 315–370, 500–650, 800–1200 Ma and a *c.* 1865 Ma (Fig. 10a; Talavera-Mendoza *et al.* 2005); and (2) at *c.* 310, 890 and 1000 Ma, in the interval *c.* 1100–1200 Ma, and at *c.* 1590 Ma (Fig. 10b; Keppie *et al.* 2006*a*), respectively. The youngest detrital zircons at *c.* 303 Ma (late Pennsylvanian; Gradstein *et al.* 2004) and *c.* 244 Ma (earliest Middle Triassic; Gradstein *et al.* 2004), respectively, and youngest zircon cluster at *c.* 317 Ma (earliest Pennsylvanian; Gradstein *et al.* 2004), provide maximum limits on the depositional age of the protolith.

The three Chazumba Formation samples – a biotite–sillimanite schist from the base of the succession and two biotite–muscovite–garnet schists from near the top – yield normalized relative age populations with peaks (1) at *c.* 275, 300 and 740 Ma, in the intervals *c.* 850–1000 and 1095–1230 Ma, and at *c.* 1465 Ma (Fig. 10c; Talavera-Mendoza *et al.* 2005); (2) at *c.* 300 Ma, 590 Ma and in the intervals *c.* 900–1000 and 1120–1240 Ma (Fig. 10d; Talavera-Mendoza *et al.* 2005); and (3) at *c.* 240 Ma and in the interval *c.* 925–1130 Ma (Fig. 10e; Keppie *et al.* 2006*a*), respectively. The youngest detrital zircons at *c.* 274 Ma (mid-Permian; Gradstein *et al.* 2004), *c.* 249 Ma (Early Triassic; Gradstein *et al.* 2004) and *c.* 239 Ma (Middle Triassic; Gradstein *et al.* 2004), respectively, and youngest zircon clusters at *c.* 275 Ma (mid-Permian; Gradstein *et al.* 2004) and 301 Ma (late Pennsylvanian; Gradstein *et al.* 2004), provide maximum age constraints on the formation's deposition.

## Granjeno Schist

The Granjeno Schist of northeastern Mexico's Sierra Madre terrane (Fig. 1) comprises repeatedly deformed, pelitic metasedimentary and

## Carboniferous-Permian Low-Grade Metasedimentary Rocks

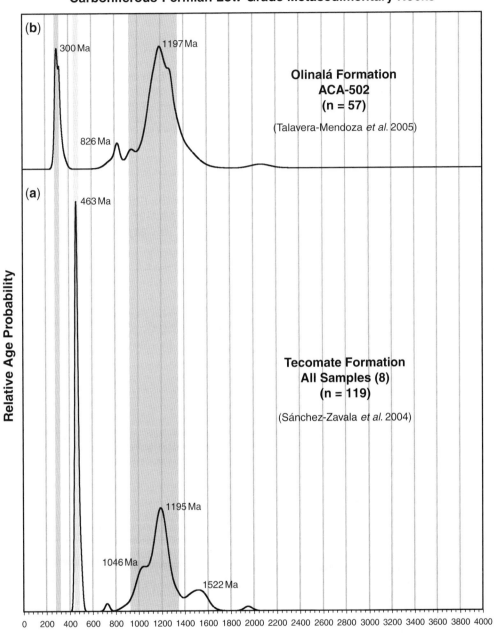

**Fig. 9.** Normalized relative age probability plots of U–Pb detrital zircon analyses from Carboniferous–Permian low-grade metasedimentary rocks of the Acatlán Complex: (**a**) Tecomate Formation south of Acatlán (Sánchez-Zavala *et al.* 2004) and (**b**) Olinalá Formation at Olinalá (Talavera-Mendoza *et al.* 2005). Shaded bars outline the age ranges characteristic of the Totoltepec pluton (280–310 Ma: Yañez *et al.* 1991; Keppie *et al.* 2004*a, b*) and the Ordovician megacrystic granitoids (*c.* 440–480 Ma: Sánchez-Zavala *et al.* 2004; Talavera-Mendoza *et al.* 2005; Miller *et al.* 2007; Vega-Granillo *et al.* 2007) of the Acatlán Complex, and the Mesoproterozoic basement of the Oaxaquia terrane (*c.* 920–1350 Ma: Keppie *et al.* 2001, 2003; Solari *et al.* 2003; Cameron *et al.* 2004). n = sample size.

## Permian-Triassic High-Grade Metasedimentary Rocks

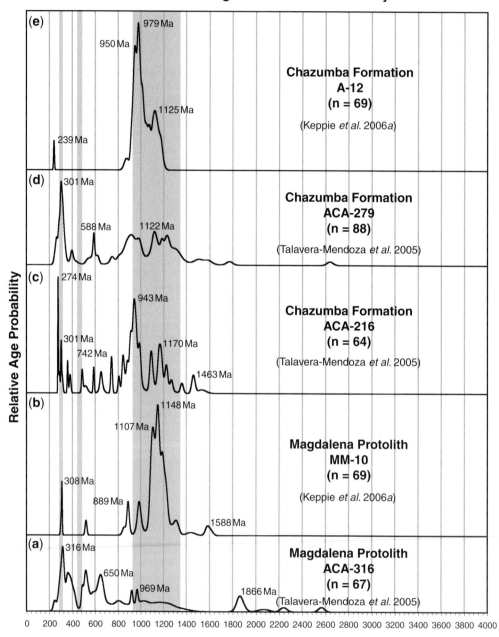

**Fig. 10.** Normalized relative age probability plots of U–Pb detrital zircon analyses from Permian–Triassic high-grade metasedimentary rocks of the Acatlán Complex (Petlalcingo Suite) near Magdalena: (**a**) Magdalena protolith (Talavera-Mendiza *et al.* 2005); (**b**) Magdalena protolith (Keppie *et al.* 2006*a*); (**c**) base of Chazumba Formation (Talavera-Mendiza *et al.* 2005); (**d**) top of Chazumba Formation (Talavera-Mendiza *et al.* 2005); and (**e**) Chazumba Formation (Keppie *et al.* 2006*a*). Shaded bars outline the age ranges characteristic of the Totoltepec pluton (280–310 Ma: Yañez *et al.* 1991; Keppie *et al.* 2004*a, b*) and the Ordovician megacrystic granitoids (*c.* 440–480 Ma: Sánchez-Zavala *et al.* 2004; Talavera-Mendoza *et al.* 2005; Miller *et al.* 2007; Vega-Granillo *et al.* 2007) of the Acatlán Complex, and the Mesoproterozoic basement of the Oaxaquia terrane (*c.* 920–1350 Ma: Keppie *et al.* 2001, 2003; Solari *et al.* 2003; Cameron *et al.* 2004). n = sample size.

## Pre-Permian Low-Grade Metasedimentary Rocks

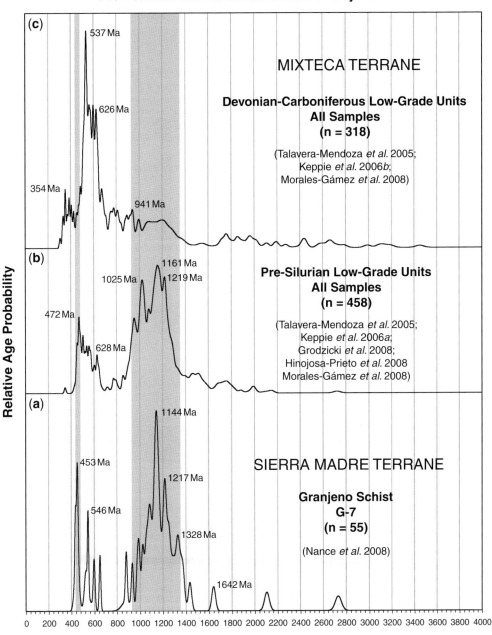

**Fig. 11.** Normalized relative age probability plots of U–Pb detrital zircon analyses (**a**) from the Granjeno Schist of the Sierra Madre terrane (data from Nance *et al.* 2007); (**b**) from all samples of the pre-Silurian low-grade units of the Acatlán Complex (Talavera-Mendoza *et al.* 2005; Keppie *et al.* 2006*a*; Grodzicki *et al.* 2008; Hinojosa-Prieto *et al.* 2008; Morales-Gámez *et al.* 2008); and (**c**) from all samples of the Devonian-Carboniferous low-grade units of the Acatlán Complex (Talavera-Mendoza *et al.* 2005; Grodzicki *et al.* 2008; Morales-Gámez *et al.* 2008). Shaded bars outline the age ranges characteristic of the Ordovician megacrystic granitoids of the Acatlán Complex (*c.* 440–480 Ma: Sánchez-Zavala *et al.* 2004; Talavera-Mendoza *et al.* 2005; Miller *et al.* 2007; Vega-Granillo *et al.* 2007) and the Mesoproterozoic basement of the Oaxaquia terrane (*c.* 920–1350 Ma: Keppie *et al.* 2001, 2003; Solari *et al.* 2003; Cameron *et al.* 2004). n = sample size.

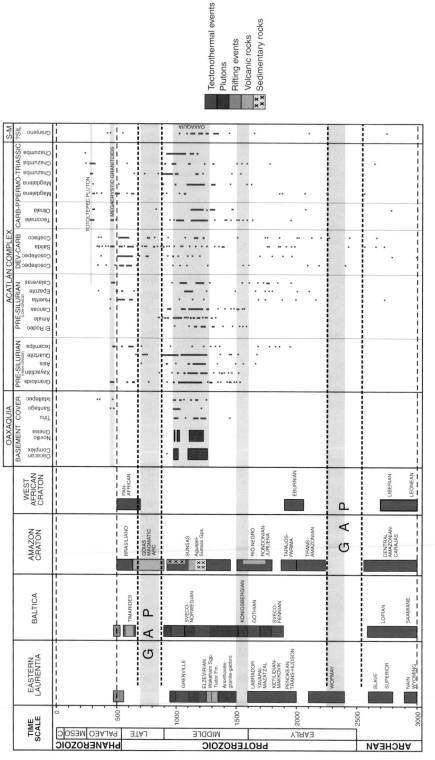

**Fig. 12.** U–Pb detrital zircon ages from (meta)sedimentary rocks of the Oaxaquia, Mixteca (Acatlán Complex) and Sierra Madre (Granjeno Schist) terranes, the Oaxaquia basement (Oaxacan Complex and Novillo Gneiss), and the megacrystic granitoids of the Acatlán Complex (see text for sources) compared to the cratonic age provinces of eastern Laurentia (Cawood *et al.* 2001, 2007), Baltica (Gower *et al.* 1990; Roberts 2003), the Amazon craton (Sadowski & Bettencourt 1996; Tassinari & Macambira 1999; Santos *et al.* 2000; Santos 2003) and the West Africa craton (Rocci *et al.* 1991; Boher *et al.* 1992; Potrel *et al.* 1998; Hirdes & Davis 2002). Duplicate samples (Cosoltepec, Magdalena and Chazumba) are listed in same order as presented in text. Among the detrital zircons, red bars identify zircon age ranges, whereas red dots show ages of individual zircons or zircon pairs.

metavolcaniclastic rocks that enclose lenses of serpentinite-metagabbro. This low-grade assemblage has been correlated with the pre-Silurian units (formerly the Cosoltepec Formation) of the Acatlán Complex on the basis of lithologic similarity (Ortega-Gutiérrez 1978; Ramírez-Ramírez 1992), deformational history (Dowe *et al.* 2005) and detrital zircon ages (Nance *et al.* 2007*b*) and, like the Acatlán Complex, is tectonically juxtaposed against *c.* 1 Ga granulite facies gneisses (Novillo Gneiss), the unmetamorphosed middle Silurian sedimentary cover of which contains fauna of Gondwanan affinity (Boucot *et al.* 1997; Stewart *et al.* 1999).

LA-ICPMS U–Pb detrital zircon age data obtained from a single sample of phyllite from the Granjeno Schist yielded a normalized relative age population (Fig. 11a) with major peaks in the intervals *c.* 450–650 and 880–1330 Ma (Nance *et al.* 2007*b*). The youngest detrital zircon at 433 ± 14 Ma (early Silurian; Gradstein *et al.* 2004) and youngest zircon cluster at *c.* 453 Ma (Late Ordovician; Gradstein *et al.* 2004) indicate a maximum depositional age for the Granjeno Schist.

## Zircon Provenance

### Oaxacan Complex

Deposition of the Tiñu, Santiago and Ixtaltepec formations, which form the sedimentary cover to the Oaxacan Complex, spans an interval of almost 200 Myr (latest Cambrian–middle Pennsylvanian). Yet they are characterized by similar detrital zircon age spectra dominated by zircons of Mesoproterozoic–early Neoproterozoic (*c.* 980–1080 Ma and/or 1100–1240 Ma) age (Fig. 2). These data suggest that the cover sequence records very little change in provenance for much of the Palaeozoic. The most obvious source for these zircons is the underlying basement rocks of the Oaxacan Complex (Gillis *et al.* 2005), the granulite facies gneisses and AMCG suites of which have yielded ages of *c.* 980–1020 and 1100–1265 Ma (Keppie *et al.* 2001, 2003; Solari *et al.* 2003). The Palaeozoic detrital zircon clusters that peak at *c.* 470 Ma in the Santiago and Ixtaltepec formations match the age of the megacrystic granitoids in the Acatlán Complex (*c.* 440–480 Ma) from which they were likely

derived (Gillis *et al.* 2005). Similarly, the youngest (*c.* 340–360 Ma) zircons in the Ixtaltepec Formation record ages that coincide with the high-pressure metamorphism and exhumation of the pre-Silurian high-grade units (Piaxtla Suite) of the Acatlán Complex (*c.* 346 Ma; Middleton *et al.* 2007) and so may also have been derived locally. However, a more distal source in the magmatic arc above the subduction zone from which the Piaxtla Suite was exhumed cannot be excluded.

### Acatlán Complex: Pre-Silurian high-grade metasedimentary rocks

Based on their detrital zircon age populations (Fig. 5), two separate sets of pre-Silurian high-pressure rocks (Piaxtla Suite) have been sampled from the Acatlán Complex – those that are intruded by, and so host, the Ordovician megacrystic granitoids (Xayacatlán Formation and Asis Lithodeme) and those that broadly coincide with granitoid emplacement (Ixcamilpa blueschist). The first set shows normalized relative age probabilities (Fig. 5a–c) that closely match that of the latest Cambrian–Ordovician Tiñu Formation overlying the Oaxacan Complex (Fig. 2a). None of these units contains Palaeozoic zircons and all are dominated by detrital zircons of Mesoproterozoic–early Neoproterozoic age, the likely provenance for which is the *c.* 920–1350 Ma basement of the Oaxacan Complex. In addition, with the exception of the late Neoproterozoic peak, the relative age frequencies of these units closely match the xenocrystic age populations of the megacrystic granitoids (Fig. 6b). These correspondences in zircon ages suggest that the Oaxacan Complex provided an important source for both the sedimentary protoliths of the Piaxtla Suite and the Ordovician granitoid magmas. Potential cratonic sources for those zircons lying outside the age range represented in the Oaxacan Complex are present in eastern Laurentia, Baltica and the Amazon craton but, contrary to the eastern Laurentian provenance proposed by Talavera–Mendoza *et al.* (2005), the presence of both xenocrystic and detrital zircons with ages in the ranges *c.* 700–900 Ma (which is not preserved in eastern Laurentia and Baltica but present in South America) and *c.* 1500–1600 Ma (which is not preserved in eastern Laurentia but present in the

**Fig. 12.** (*Continued*) Purple bars and dots identify xenocrystic zircons in the megacrystic granitoids of the Acatlán Complex. Horizontal light blue bars outline the age ranges characteristic of the Totoltepec pluton (280–310 Ma: Yañez *et al.* 1991; Keppie *et al.* 2004*a, b*) and the Ordovician megacrystic granitoids (*c.* 440–480 Ma: Sánchez-Zavala *et al.* 2004; Talavera-Mendoza *et al.* 2005; Miller *et al.* 2007; Vega-Granillo *et al.* 2007) of the Acatlán Complex, and the Mesoproterozoic basement of the Oaxaquia terrane (*c.* 920–1350 Ma: Keppie *et al.* 2001, 2003; Solari *et al.* 2003; Cameron *et al.* 2004). Horizontal yellow bars highlight *c.* 700–850, 1500–1600 and 2250–2400 Ma age ranges most pertinent to the identification of cratonic source areas. DEV–CARB, Devonian–Carboniferous low-grade units; CARB–P, Carboniferous–Permian low-grade units; ?SIL, possibly Silurian; S–M, Sierra Madre terrane.

Amazon craton) favours an Amazonian source (Fig. 12).

In contrast, the Ixcamilpa blueschist, while containing Mesoproterozoic detrital zircons that match the age range of the Oaxacan Complex (Fig. 5d), is dominated by Palaeozoic and Neoproterozoic zircons with ages that peak at *c*. 475, 540, 600 and 730 Ma (Talavera–Mendoza *et al.* 2005). The youngest of these peaks overlaps the emplacement age range of the Ordovician megacrystic granitoids, the magmatic episode of which the sample's protolith may have been part. However, the remaining clusters cannot presently be sourced in either the Acatlán or Oaxacan complexes unless the presence of Neoproterozoic xenocrystic zircons in the Ordovocian megacrystic granitoids (Fig. 6b) indicates the existence of Neoproterozoic basement beneath the Acatlán Complex that may once have been exposed. The likelihood, however, is that these ages indicate a significant change in provenance.

Rocks of this age are present in Baltica only in the Timanides on its northern margin (e.g. Roberts & Siedlecka 2002), and were largely missing in eastern Laurentia prior to the Silurian arrival of the peri-Gondwanan terranes (e.g. van Staal *et al.* 1998), as indicated by the almost complete absence of detrital zircons of this age in Neoproterozoic–Ordovician strata on the eastern Laurentian margin (Cawood & Nemchin 2001; Cawood *et al.* 2007). If deposition of the Ixcamilpa Unit is as old as *c*. 477 Ma, a local source for the Neoproterozoic–Cambrian detrital zircons may have existed in peri-Gondwanan Avalonia and Carolinia, since these terranes are inferred to have lain adjacent to Oaxaquia until the opening of the Rheic Ocean in the Early Ordovician (Keppie 2004). However, the zircons are considered more likely to have come from either the Pan-African basement of the Maya terrane (Fig. 1) beneath the Yucatan Peninsula (Krogh *et al.* 1993a), or the Brasiliano orogens and Goias magmatic arc of South America (e.g. Brito Neves *et al.* 1999; Pimental *et al.* 2000). Potential sources for the older (Palaeoproterozoic and Archaean) zircons can likewise be found in the age provinces of Amazonia (Fig. 12).

## Acatlán Complex: Pre-Silurian low-grade metasedimentary rocks

The detrital zircon age populations of the pre-Silurian low-grade units of the Acatlán Complex show striking similarities to those of the high-grade metasedimentary rocks, the data again suggesting the presence of two groups of assemblages – those that pre-date the Ordovician megacrystic granitoids (El Rodeo Formation and Amate Unit) and those whose deposition was broadly contemporaneous

with their emplacement (Canoas, Huerta, El Epazote and Las Calaveras units).

The El Rodeo Formation and Amate Unit show normalized relative age probabilities (Fig. 7a–b) that closely match those of the older set of high-grade metasedimentary rocks (Xayacatlán Formation, Asis Lithodeme and Quartzite; Fig. 5a–c), as well as those of the latest Cambrian–Ordovician Tiñu Formation of the Oaxacan Complex (Fig. 2a) and the xenocrystic population within the megacrystic granitoids (Fig. 6b). In each case, the age spectra lack Palaeozoic detrital zircons and are dominated by those of Mesoproterozoic–early Neoproterozoic age like that of the Oaxacan Complex, which likely provided the source. This striking similarity in the detrital zircon signatures suggests that the pre-Silurian siliciclastic units of the Acatlán and Oaxacan complexes are broadly correlative and represent portions of a shelf succession deposited on the margin of Oaxaquia (Xayacatlán Formation, Asis Lithodeme and Quartzite) that were later subjected to varying grades of metamorphism.

With the exception of a cluster of Ordovician (*c*. 450–475 Ma) detrital zircons likely derived from the megacrystic granitoids, the age spectrum of the Canoas Unit (Fig. 7c) strongly resembles those of the El Rodeo Formation and Amate Unit, suggesting a similar, largely local provenance. Older zircons in the Canoas Unit, like those of the El Rodeo Formation, include ages in the range *c*. 1500–1600 Ma, which favours an Amazonian source (Fig. 12).

The detrital zircon signatures of the Huerte, El Epazote and Las Calaveras units (Fig. 7d–f), on the other hand, are closely similar to that of the Ixcamilpa blueschist (Fig. 5d), showing a broad age span dominated by Ordovician (*c*. 450–490 Ma) and/or late Neoproterozoic–Cambrian (*c*. 510–630 Ma) zircons with less conspicuous Mesoproterozoic–early Neoproterozoic populations. Sources for the Ordovician and Mesoproterozoic zircons exist both locally and in eastern Laurentia. However, a late Neoproterozoic–Cambrian source was essentially absent in eastern Laurentia prior to the accretion of the peri-Gondwanan terranes in the Silurian. Hence, likely sources for the three age populations include the Ordovician megacrystic granitoids, the Pan-African Maya terrane and/or the Brasiliano belts of South America, and the Oaxacan Complex, although the possibility of nearby peri-Gondwanan terranes cannot be excluded. In addition, all three units contain a representation of ages in the range *c*. 750–950 Ma, the only viable source for which is the Goias magmatic arc (Fig. 12), which flanks the Amazon craton in the Tocantins Province of Brazil (e.g. Pimental *et al.* 2000). Possible sources for the Palaeoproterozoic and Archaean zircons in these units also exist within the Amazon craton.

## Acatlán Complex: Devonian–Carboniferous low-grade metasedimentary rocks

All of the low-grade Devono–Carboniferous rocks sampled within the Acatlán Complex – the Cosoltepec Formation (Fig. 8a, b), the Salida Unit (Fig. 8c) and the Coatlaco Unit (Fig. 8d) – show similar detrital zircon age spectra (Talavera-Mendoza *et al.* 2005; Morales-Gámez *et al.* 2008), suggesting a similar provenance for each of these assemblages. Like those of the pre-Silurian Ixcamilpa blueschist (Fig. 5d), Huerta Unit (Fig. 7d) and El Epasote Unit (Fig. 7e), each of the Devonian–Carboniferous units shows a spectrum with a broad span of ages that is dominated by late Neoproterozoic (*c.* 540–640 Ma) zircons with only subordinate populations of Mesoproterozoic–early Neoproterozoic age like that of the Oaxacan Complex.

The importance of late Neoproterozoic zircon ages again points to sources in the Maya terrane and/or the Brasiliano belts (Keppie *et al.* 2006*a*), although a local or locally recycled source cannot be excluded given the matching xenocrystic populations in the Ordovician megacrystic granitoids (Fig. 6b). Given the maximum depositional age of these units, however, the abundance of late Neoproterozoic detrital zircons could also reflect increasing proximity to the previously accreted peri-Gondwanan terranes of the eastern Laurentian margin (Grodzicki *et al.* 2008), but the units also contain a significant population of older Neoproterozoic (*c.* 750–900 Ma) zircons, a source for which is absent in Laurentia and present only in the Goias magmatic arc of Amazonia (Fig. 12).

The Coatlaco Unit, which records a structural history like that of the low-grade Permo-Carboniferous units, nevertheless contains detrital zircons whose age spectrum is closely similar to those of the structurally more complex Cosoltepec Formation and Salida Unit, particularly with respect to the broad age-span of the spectrum and the importance of detrital zircons of late Neoproterozoic–Cambrian (*c.* 530–630 Ma) and Neoproterozoic (*c.* 770–940 Ma) age (Fig. 8d), both of which suggest sources in Amazonia (Fig. 12).

Relatively few of the Palaeozoic detrital zircons in the Devonian–Carboniferous units match the age of the megacrystic granitoids, the bulk showing younger (*c.* 340–410 Ma) ages, like those of the Pennsylvanian Ixtaltepec Formation overlying the Oaxacan Complex. These ages broadly coincide with the exhumation of the pre-Silurian high-grade units (Piaxtla Suite) within the Acatlán Complex (*c.* 346 Ma; Middleton *et al.* 2007) and so could have been derived locally. However, subduction of the high-grade units would have been associated with the development of a Devono–Carboniferous magmatic arc, which could have provided an additional source of zircons of this age. No such arc exists within the Acatlán Complex, but evidence of an arc of this age is preserved in the Ouachita orogenic belt on the southern margin of Laurentia (e.g. Viele & Thomas 1989).

## Acatlán Complex: Carboniferous–Permian low-grade metasedimentary rocks

Detrital zircons in the two sampled Carboniferous–Permian low-grade units of the Acatlán Complex – the Tecomate and Olinalá formations (Fig. 9) – share a broad Mesoproterozoic (*c.* 950–1300 Ma) age population most likely derived from the Oaxacan Complex, but differ in their Palaeozoic age populations. The Tecomate Formation is dominated by an Ordovician (*c.* 450–480 Ma) population of zircons that matches the age of the megacrystic granitoids, whereas the Olinalá Formation lacks this age group and contains, instead, a population of Permo-Carboniferous (*c.* 280–310 Ma) zircons that matches the TIMS U–Pb age of the Totoltepec pluton (*c.* 290 Ma; Yañez *et al.* 1991; Keppie *et al.* 2004*a*) and the SHRIMP U–Pb ages of granitoid pebbles sampled from a conglomerate of the Tecomate Formation west of Cosoltepec (*c.* 280–310 Ma; Keppie *et al.* 2004*a*). In addition, detrital zircon age spectra from individual horizons of the Tecomate Formation differ significantly, some being dominated by Mesoproterozoic zircons, whereas others are dominated by zircons of Ordovician age (Sánchez-Zavala *et al.* 2004).

A provenance for the Mesoproterozoic zircons in the granulite facies Oaxacan Complex rather than the megacrystic granitoids, which record inheritance of similar age, is supported by the presence of pebbles of blue-quartz-bearing gneiss in the Tecomate Formation, and the existence in the age-equivalent Patlanoaya Formation of abundant granulite facies grains such as mesoperthite and rutile-bearing quartz (Sánchez-Zavala *et al.* 2004). As these authors suggest, the data imply that the formation was sourced predominantly locally and that variations in drainage pattern played an important role in controlling the local provenance. Given that its deposition post-dates the assembly of Pangaea, older detrital zircons in the Tecomate Formation could have been sourced in either Amazonia, Laurentia or West Africa. However, most show populations with ages in the range *c.* 1500–1600 Ma, which favours an Amazonian source (Fig. 12). The presence of a *c.* 825 Ma population in the Olinalá Formation likewise favours a South American provenance.

Hence, the data suggest that the two formations are not only of similar age, but also shared a similar, largely local provenance sourced in the Oaxacan and Acatlán complexes, with possible minor contributions of South American origin.

## Acatlán Complex: Carboniferous–Permian high-grade metasedimentary rocks

The relative age probabilities of detrital zircons in the Carboniferous–Permian high-grade units of the Acatlán Complex – the Magdalena Migmatite and the Chazumba Formation – show significant variation, even within the same unit. This variation may reflect the creation of new, lithologically complex source areas as a result of the onset of convergent tectonics along the palaeo-Pacific margin.

Detrital zircons from the palaeosome of the Magdalena Migmatite show a broadly similar span of age populations but quite variable normalized relative age frequencies (Fig. 10a, b). Both samples of the protolith contain significant populations of Mesoproterozoic, early Neoproterozoic (*c.* 850–920 Ma), Cambrian (*c.* 520–530 Ma) and late Palaeozoic zircons (*c.* 310–370 Ma), but whereas that sampled by Keppie *et al.* (2006*a*) is dominated by *c.* 1100–1200 Ma ages like those of the older components of the Oaxacan Complex, relatively few zircons of this age occur in the sample collected by Talvera-Mendoza *et al.* (2005). In both cases, the zircons were likely sourced from the Oaxacan Complex or recycled from older units of the Acatlán Complex, although a South American provenance is suggested by the early Neoproterozoic zircons, the only obvious source for which lies in the Goiás magmatic arc of central Brazil (e.g. Pimental *et al.* 2000). The latter sample also contains an important late Neoproterozoic (*c.* 650 Ma) population and shows a broader range of ages that extend into the Palaeoproterozoic. Talavera-Mendoza *et al.* (2005) link the late Palaeozoic, late Neoproterozoic–Cambrian and Palaeoproterozoic zircons to the Alleghanian orogen, the rifting of the Iapetus Ocean and the Trans-Hudson orogen, respectively, and so argue for a Laurentian provenance for these detrital populations. This is a plausible source, given the depositional age of the Magdalena protolith, but better matches for the Cambrian–Neoproterozoic zircons exist within the Ribeira Belt and Borborema Province of South America (e.g. Machado *et al.* 1996; Brito Neves *et al.* 2003) and the late Palaeozoic zircons match the age of locally derived boulders of the Totoltetec granite in the Tecomate Formation (Keppie *et al.* 2004*a*). Sources for the older zircons in both samples similarly match age provinces in both the Amazon craton and Laurentia (Fig. 12).

The detrital zircon populations in the Chazumba Formation resemble those of the Magdalena protolith, and again show similar age ranges but differing normalized relative age frequencies (Fig. 10c). The two samples collected by Talavera-Mendoza *et al.* (2005) show similar spectra dominated by peaks in the Mesoproterozoic (at *c.* 1100–1250 Ma),

Neoproterozoic (at *c.* 590, 740 and 850–1000 Ma) and Late Palaeozoic (at *c.* 275 and 300 Ma). However, almost all the zircons in the sample collected by Keppie *et al.* (2006*a*) fall in the range *c.* 925–1130 Ma, with a single Triassic grain at *c.* 240 Ma. Once again, a likely provenance for the youngest zircon grains lies in the Permo–Triassic magmatic arc (Torres *et al.* 1999) represented by the Totoltepec pluton, whereas those of Mesoproterozoic age were probably sourced from either the Oaxacan or Acatlán complexes. Older detrital zircons are few, but again match age provinces in both Amazonia and eastern Laurentia. However, the Neoproterozoic zircon ages best match sources in the Brasiliano belts and Goiás magmatic arc of South America (Fig. 12).

## Granjeno Schist

Detrital zircons from the Granjeno Schist (Nance *et al.* 2007*b*) show a relative age distribution (Fig. 11a) that resembles those of various units within the Acatlán Complex, with two major populations in the early Neoproterozoic–Mesoproterozoic (*c.* 880–1330 Ma) and late Neoproterozoic–early Palaeozoic (*c.* 450–650 Ma). However, with respect to the pre-Permian low-grade units (Fig. 11b, c), the most striking match is with those of pre-Silurian age (formerly the Cosoltepec Formation) with which the schist has been traditionally correlated (Ortega-Gutiérrez 1978). Despite the slightly younger maximum depositional age of the schist (*c.* 435 Ma), this correlation is further supported by its lithologic association (Ramírez-Ramírez 1992) and deformational history (Dowe *et al.* 2005).

Potential provenances for the Mesoproterozoic detrital zircons exist within the Grenville Belt of southern Laurentia and in belts of similar age in Amazonia. However, the most likely provenance is the adjacent Novillo Gneiss, which has yielded ages of *c.* 990–980, *c.* 1035–1010 and *c.* 1235–1115 Ma (Cameron *et al.* 2004) that are almost identical to those of the Oaxacan Complex (Fig. 12). Sources for the Neoproterzoic detrital zircons can be found in the Maya terrane and in the Brasiliano belts of Amazonia, but also occur in the rifted margin of Laurentia and in the Appalachian peri-Gondwanan terranes, some of which may have been accreted to Laurentia prior to the schist's deposition (e.g. van Staal *et al.* 1998). Sources for the early Palaeozoic zircons exist within the Taconian belt of the Appalachians and the Arequipa–Antofalla terrane on the southwestern margin of Amazonia, but the most likely source is the *c.* 480–440 Ma megacrystic granitoids of the Acatlán Complex (Nance *et al.* 2007*b*). Sources for the older zircons exist in both Amazonia and eastern Laurentia (Fig. 12).

## Palaeogeographic significance

The detrital zircon signatures of Palaeozoic sedimentary and metasedimentary rocks in the Oaxaquia, Mixteca and Sierra Madre terranes show striking similarities despite the wide range in their likely depositional ages (Fig. 12). The most important zircon population is of Mesoproterozoic (*c.* 950–1300 Ma) age. A provenance of this age provided virtually all of the detrital zircons in the sedimentary cover of the Oaxacan Complex from the latest Cambrian to the Pennsylvanian. Such a provenance was also a major source in almost all units of the Acatlán Complex into the Triassic, and dominates the age distribution in the Granjeno Schist. Potential sources for detrital zircons of this age exist within the Grenville Belt of eastern Laurentia (e.g. Tollo *et al.* 2004; Cawood *et al.* 2007), in the 'Grenville' massifs of the northern Andes (e.g. Cediel *et al.* 2003), and in the Sunsas Belt of the Amazon craton (e.g. Boger *et al.* 2005). However, the most likely source is exposed portions of the basement of the Oaxaquia terrane, represented today by the Oaxacan Complex and Novillo Gneiss, available U–Pb ages for which span the interval *c.* 920–1350 Ma (Keppie *et al.* 2003; Cameron *et al.* 2004). That this basement source, which also includes other *c.* 1 Ga inliers in Mexico (Huiznopala Gneiss and Guichicovi Gneiss) with similar age ranges (Lawlor *et al.* 1999; Weber & Köehler 1999), is of Gondwanan rather than Laurentian affinity as indicated by the Gondwanan fauna of the early Palaeozoic cover sequences of both the Oaxacan Complex and the Novillo Gneiss (e.g. Stewart *et al.* 1999; Landing *et al.* 2007). The significant number of studied (meta)sedimentary units that contain Ordovician detrital zircons, which match the *c.* 440–480 Ma age of the megacrystic granitoids of the Acatlán Complex, and Permo–Carboniferous detrital zircons, which match the age of the *c.* 290 Ma Totoltepec pluton, further attests to the importance of local sources throughout the Palaeozoic.

However, the detrital zircon signatures of the (meta)sedimentary rocks also contain important age populations that are not represented by local sources and so bear upon the palaeogeography of the Oaxaquia, Mixteca and Sierra Madre terranes at the time of their deposition. Chief among these are populations of *c.* 500–650 Ma (late Neoproterozoic–Cambrian), *c.* 750–950 Ma (early Neoproterozoic) and >1300 Ma (Mesoproterozoic and older) age.

Detrital zircons of late Neoproterozoic–Cambrian age are absent in the latest Cambrian–Ordovician, Mississippian and Pennsylvanian sedimentary cover of the Oaxacan Complex, in metasedimentary rocks of the Acatlán Complex that pre-date emplacement of the megacrystic granitoids (Xayacatlán Formation and Asis, Quartzite,

El Rodeo, Amate and Canoas units) and in the Permo-Carboniferous Tecomate and Olinalá formations. However, they are present in all other units and constitute an important population in the pre-Silurian Ixcamilpa blueschists and Huerta, El Epazote and Las Calaveras units, the Devonian–Carboniferous low-grade units, and the protolith of the Magdalena Migmatite. Zircons of this age also occur in the Granjeno Schist and as inheritance in the megacrystic granitoids. Potential proximal sources of zircons of this age may exist in the basement of the Acatlán Complex and are available in the Maya terrane of the Yucatan Peninsula (Fig. 1). However, prior to the Jurassic opening of the Gulf of Mexico, this basement was contiguous with that of the Suwannee terrane in Florida (e.g. Pindell *et al.* 2000; Dickinson & Lawson 2001), the accretion of which did not occur until the Alleghanian at *c.* 300 Ma (e.g. Heatherington *et al.* 1996). Hence, this source is unlikely to have been available prior to the late Palaeozoic assembly of Pangaea. More distal sources of such zircons exist in the Brasiliano belts of South America (e.g. Brito Neves *et al.* 2003) and, following the Silurian closure of the Iapetus Ocean (e.g. van Staal *et al.* 1998), in the accreted peri-Gondwanan terranes of eastern Laurentia. In those (meta)sedimentary units whose deposition pre-dates the Silurian, however, a Gondwanan (peri-Amazonian) source would appear to be required.

A peri-Amazonian source is also suggested by detrital zircons of *c.* 750–950 Ma (early–middle Neoproterozoic) age, which are present in virtually all Palaeozoic units of the Mixteca and Sierra Madre terranes (Fig. 12). Provenances of this age range, which post-dates the Grenvillian orogeny (e.g. Rivers 1997), are absent in eastern Laurentia, Baltica and West Africa. However, ages in this range are characteristic of the *c.* 600–930 Ma Goias magmatic arc of central Brazil (Pimental *et al.* 2000).

Likewise, the populations of mid-Proterozoic and older (<1300 Ma) detrital zircons in both the Acatlán Complex and the Granjeno Schist most closely match the age provinces of the Amazon craton (Fig. 12), suggesting an Amazonian provenance. This is indicated by the presence of detrital zircons with ages in the range *c.* 1300–1900 Ma (absent in West Africa), *c.* 1900–2250 Ma ages (absent in Baltica) and *c.* 1500–1600 and 2000–2100 Ma ages (absent in Laurentia), and by the paucity of detrital zircons with ages in the range *c.* 2250–2550 Ma (absent in Amazonia). In support of this, Bream *et al.* (2004) have shown in the southern Appalachians that basements of Amazonian and Laurentian affinity can be distinguished on the basis of zircon inheritance. Whereas zircon inheritance from a Laurentian source is typically no older than *c.* 1300–1350 Ma, inheritance from an Amazonian source is distinguished by the frequency of 1600–2100 Ma and, to a lesser degree, 2700–2900 Ma age ranges, both of which are well represented in the detrital zircon spectra from the Mixteca and Sierra Madre terranes. Cawood *et al.* (2007) have likewise shown that sedimentary basins of Laurentian provenance in the southern Appalachians are essentially devoid of detrital zircons much older than those of Grenville age.

## Discussion

Available detrital zircon data for the Cambrian through Permo–Triassic (meta)sedimentary rocks of the Oaxaquia, Mixteca and Sierra Madre terranes suggests a preponderance of local and/or Amazonian sources throughout the Palaeozoic. This conclusion concurs with that of Gillis *et al.* (2005) for the sedimentary cover of the Oaxacan Complex, with that of Talavera-Mendoza *et al.* (2005) and Keppie *et al.* (2006*a*) for the low-grade Devonian–Carboniferous and high-grade Permian–Triassic units of the Acatlán Complex, and with that of Nance *et al.* (2007*b*) for the Granjeno Schist. However, it is contrary to the view put forward by Talavera-Mendoza *et al.* (2005), and adopted by Vega-Granillo *et al.* (2007), with regard to the provenance of the pre-Silurian high-grade units of the Acatlán Complex. These authors interpret the detrital zircon signatures of the Xayacatlán and Ixcamilpa units to indicate a Laurentian provenance for this component of the Acatlán Complex, and likewise proposed a Laurentian source for the El Rodeo Formation, the depositional age constraints for which are the same as those for the Xayacatlán Formation. Their interpretation is based on the existence in the Xayacatlán and El Rodeo formations of *c.* 800–950 Ma and >1260 Ma age populations, which are rare or absent in the neighbouring Oaxacan Complex, and *c.* 1150–1300 Ma ages, which are rare in Amazonia but common in the southwestern Grenville Province of North America. They also link the Ordovician (*c.* 480 Ma), Neoproterozoic (*c.* 600 and 710 Ma) and Palaeoproterozoic (*c.* 1800 Ma) zircon populations in the Ixcamilpa blueschist to age provinces in eastern Laurentia. Accordingly, for the Acatlán Complex, these authors propose contrasting (Laurentian and Amazonian) provenances for the pre-Silurian high-grade units (Piaxtla Suite) and Devonian–Carboniferous low-grade units (Cosoltepec Formation).

Based on this interpretation, Talavera-Mendoza *et al.* (2005) proposed a complex model for the tectonothermal evolution of the Acatlán Complex in which: (1) eclogite facies metamorphism (Xayacatlán Formation) and rifted arc magmatism (El Rodeo Formation and *c.* 470–480 Ma megacrystic granitoids) are attributed to east-directed subduction beneath a Laurentia-derived arc terrane floored by Grenville basement that is accreted to Laurentia in the early Ordovician; (2) blueschist metamorphism (Ixcamilpa blueschists) and renewed arc magmatism (*c.* 440–460 Ma megacrystic granitoids) are attributed to east-directed subduction beneath an ensimatic arc that is accreted to Laurentia in the late Ordovician–early Silurian; and (3) thrusting of these previously accreted elements over the Cosoltopec Formation (following its deposition at the leading edge of South America) is attributed to west-directed subduction beneath Laurentia and the tectonic transfer of these elements from the upper plate to the lower plate during Laurentia-South America collision in the late Pennsylvanian. The oceanic domains whose closure is recorded by each of these collisional events are not specified, but from the perspective of the eastern Laurentian margin, they would correspond to the Baie Verte seaway, and the Iapetus and Rheic oceans, respectively (van Staal *et al.* 1998; Martínez Catalán *et al.* 2002). According to their model, therefore, the Acatlán Complex records the closure of three separate oceanic tracts.

However, we would argue that the detrital zircon signatures of pre-Silurian units in the Acatlán Complex, and the Ixcamilpa blueschist in particular, are more consistent with local and Amazonian (rather than Laurentian) sources given that: (1) *c.* 800–950 Ma ages are rare in Laurentia but characteristic of the Goias magmatic arc of central Brazil (Pimental *et al.* 2000); (2) *c.* 1150–1300 Ma ages can be found in the Oaxacan Complex (Solari *et al.* 2003); (3) *c.* 480 Ma ages occur in the megacrystic granitoids (Sánchez-Zavala *et al.* 2004; Talavera-Mendoza *et al.* 2005); and (4) *c.* 600 and 700 Ma ages are typical of the peri-Gondwana terranes and rare in Laurentia until the accretion of these terranes in the Silurian (van Staal *et al.* 1998). In addition, several of the pre-Silurian units, including the El Rodeo Formation, contain detrital zircons with ages in the range *c.* 1500–1600 Ma, which can be linked to Amazonia but not to Laurentia (Fig. 12).

If the pre-Silurian high-grade units (Piaxtla Suite) of the Acatlán Complex are of Amazonian rather than Laurentian provenance and their eclogite facies metamorphism, which has been dated as late Devonian–Mississippian (Middleton *et al.* 2007, Elías-Herrera *et al.* 2007), is divorced from the Ordovician emplacement of the megacrystic granitoids, a much simpler model for the Palaeozoic evolution of the Acatlán Complex than that proposed by Talavera-Mendoza *et al.* (2005) can be advanced (Fig. 13). This model (e.g. Nance *et al.* 2006, 2007*a*), which can be used to link the Oaxaquia, Mixteca and Sierra Madre terranes, is analogous to that originally proposed for the Acatlán

# ACATLÁN COMPLEX

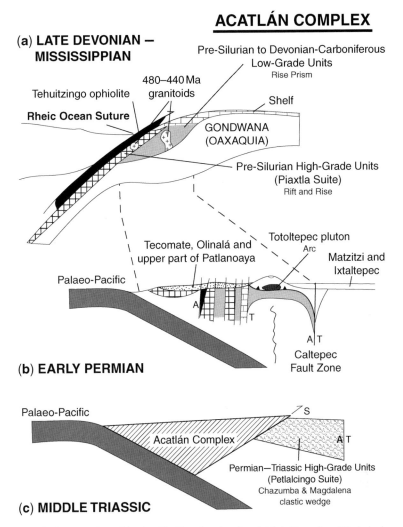

**(a) LATE DEVONIAN – MISSISSIPPIAN**

Pre-Silurian to Devonian-Carboniferous
Low-Grade Units
Rise Prism

480–440 Ma granitoids

Tehuitzingo ophiolite

Shelf

**Rheic Ocean Suture**

GONDWANA
(OAXAQUIA)

Pre-Silurian High-Grade Units
(Piaxtla Suite)
Rift and Rise

Totoltepec pluton
Arc

Tecomate, Olinalá and
upper part of Patlanoaya

Matzitzi and
Ixtaltepec

Palaeo-Pacific

A

T

A T

Caltepec
Fault Zone

**(b) EARLY PERMIAN**

Palaeo-Pacific

S

Acatlán Complex

A T

Permian–Triassic High-Grade Units
(Petlalcingo Suite)
Chazumba & Magdalena
clastic wedge

**(c) MIDDLE TRIASSIC**

**Fig. 13.** Plate tectonic interpretation of the Acatlán Complex showing: (**a**) late Devonian–Mississippian subduction and exhumation of the pre-Silurian high-grade rocks of the Acatlán Complex (Piaxtla Suite), an oceanic and rift-related continental margin assemblage (Tehuitzingo ophiolite from Proenza *et al.* 2004), onto the Gondwanan continental rise deposits of the low-grade pre-Silurian and Devono-Carboniferous pelite-psammite assemblages during closure of the Rheic Ocean between Oaxaquia and Laurentia; (**b**) early Permian development of the Totoltepec arc as a result of subduction of the palaeo-Pacific following the amalgamation of Pangaea, and deposition and deformation of the Carboniferous-Permian low-grade units of the Acatlán Complex coincident with dextral strike-slip faulting that juxtaposed the Acatlán and Oaxacan complexes; and (**c**) Middle Triassic development of the clastic wedge represented by the Permian-Triassic high-grade units of the Acatlán Complex (Petlalcingo Suite) in response to continued subduction of the palaeo-Pacific. S, south; A, away; T, toward (modified from Nance *et al.* 2007*b*).

Complex by Ortega-Gutierrez *et al.* (1999), but views this complex (and the Granjeno Schist), not as a remnant of the Iapetus Ocean, but as a vestige of its immediate successor, the Rheic Ocean, that was subsequently overprinted by convergent tectonics at the margin of the palaeo-Pacific.

Accordingly, the pre-Silurian units, which include quartzites, pelites and mafic rocks of both MORB (Meza-Figueroa *et al.* 2003) and continental

tholeiitic (Keppie *et al.* 2008*b*) affinity, are interpreted to record the development of a rift/passive margin on the southern (Oaxaquia) flank of the Rheic Ocean in the ?Cambro–Ordovician. The broadly coeval megacrystic granitoids geochemically straddle the volcanic arc/within plate fields (Reyes-Salas 2003; Miller *et al.* 2007; Hinojosa-Prieto *et al.* 2008); however, they are synchronous with the intrusion of mafic rocks and an Ordovician,

**(a) Silurian**

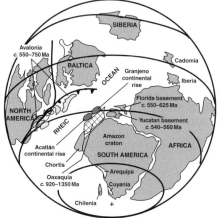

**(b) Devonian–Carboniferous**

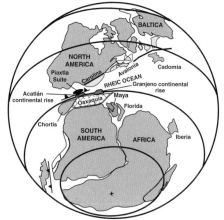

**(c) Permian–Triassic**

**Fig. 14.** Continental reconstructions of the margins of the Rheic Ocean (modified from Keppie 2004 and Keppie *et al.* 2007, 2008*a*) showing locations of

tholeiitic rift-related, dyke swarm (Morales-Gámez *et al.* 2008). Thus the magmatism is bimodal and likely formed a rift-related suite that developed along the northern margin of Gondwana during the opening of the Rheic Ocean. Analogous megacrystic granitoid magmatism is associated with the opening of the Rheic Ocean in the Central Iberian Zone of Spain (e.g. Valverde-Vaquero *et al.* 2000).

The low-grade pelite-psammite units of the Acatlán Complex (Huerta Unit, Cosoltepec Formation and Salida Unit) and the Granjeno Schist, on the other hand, are interpreted to be parts of an Ordovician-Mississippian continental rise prism deposited on oceanic lithosphere along the northern margin of Gondwana, represented by the Oaxaquia terrane, following ocean opening. This interpretation is based on the distal nature of the continental-derived detritus in these units (Dowe *et al.* 2005; Keppie *et al.* 2006*a*) and the presence of MORB and OIB basalts within the low-grade pelite-psammite units of the Acatlán Complex (Keppie *et al.* 2007) and tectonic lenses of serpentinite-metagabbro in the Granjeno Schist (Ramírez-Ramírez 1992), both of which are likely to be of oceanic origin.

The subsequent high-grade metamorphism of pre-Silurian units in the Acatlán Complex and their emplacement onto low-grade pre-Silurian and Devono–Carboniferous rocks is taken to document subduction and exhumation of the leading edge of Gondwana with the closure of the Rheic Ocean in the Late Devonian–Mississippian. Terminal collision at this time is consistent with the appearance of midcontinent (US) brachiopod fauna in the Mississippian rocks of the Oaxaquia

---

**Fig. 14.** (*Continued*) Oaxaquia, the Acatlán Complex (Acatlán continental rise and Piaxtla Suite) and Sierra Madre terrane (Granjeno continental rise) relative to Gondwana (South America–Africa), North America (Laurentia) and Baltica. (**a**) Silurian (*c.* 420 Ma) palaeogeographic map showing location of pre-Silurian low-grade units of the Acatlán Complex (Mixteca terrane) and the Granjeno Schist of the Sierra Madre terrane as parts of a continental rise flanking Oaxaquia and South America; (**b**) Devonian–Carboniferous palaeogeographic map showing subduction and exhumation following the amalgamation of Pangaea of the leading edge of Oaxaquia/Amazonia beneath the southern margin of Lauentia. Loss of Devono–Carboniferous magmatic arc as a result of subduction erosion (Keppie *et al.* 2008*a*) would reverse polarity of subduction; (**c**) Permian–Triassic palaeogeographic map showing formation of a Permo–Triassic magmatic arc and dextral strike-slip fault system, and subsequent development of a clastic wedge, in response to subduction of the palaeo-Pacific following the amalgamation of Pangaea. +, South Pole.

terrane (Stewart *et al.* 1999; Navarro-Santillan *et al.* 2002).

Following the amalgamation of Pangaea and the tectonic juxtapositioning of the Acatlán Complex and Granjeno Schist with the Oaxaquia terrane by dextral strike slip, the development of a Permo–Triassic magmatic arc is recorded in the Acatlán Complex by the emplacement of the Totoltepec pluton and the deposition of the arc-related Tecomate, Olinalá and Patlanoaya formations. This arc is considered to have developed in response to subduction of the palaeo-Pacific on the southwestern (present co-ordinates) flank of Pangaea following Pangaea assembly. Continued convergent tectonics along this margin during the Permo–Triassic led to the development of the Chazumba–Magdalena clastic wedge, and was followed in the Jurassic by hotspot activity and the development of the Magdalena Migmatite coeval with the opening of the Gulf of Mexico.

In this much simpler scenario, the Acatlán Complex and Granjeno Schist were deposited along the southern Rheic margin of the Oaxaquia terrane (Fig. 14a), which lay adjacent to northwestern South America throughout the Palaeozoic as proposed in the continental reconstructions of Keppie & Ramos (1999). Hence, rather than recording the closure of a succession of oceanic tracts as early as the Ordovician, the Mixteca and Sierra Madre terranes are considered to have been accreted to Laurentia only with the late Palaeozoic closure of the Rheic Ocean and the assembly of Pangaea (Fig. 14b). The subsequent development of a dextral strike-slip regime and Permian magmatic arc is considered to record subduction of the palaeo-Pacific, which led to thrusting and the deposition of a Permo–Triassic clastic wedge (Fig. 14c).

This model places the Oaxaquia, Mixteca and Sierra Madre terranes on the lower plate during closure of the Rheic Ocean (Fig. 13a), consistent with the absence of a Devono–Carboniferous arc, and on the upper plate with respect to subduction of the palaeo-Pacific (Fig. 13b, c), consistent with the existence of Permian arc. However, Keppie *et al.* (2008*a*) advocate an alternative explanation for the missing Devono–Carboniferous arc, in which its absence is attributed to subduction erosion beneath the Gondwanan margin. If so, the polarity of subduction would be reversed, placing southern Laurentia on the lower plate consistent with models for the tectonic evolution of the Ouachita orogen (Viele & Thomas 1989).

## Conclusions

Detrital zircon age spectra in the Palaeozoic (meta)-sedimentary rocks of the Oaxaquia, Mixteca and Sierra Madre terranes of Mexico are dominated by Mesoproterozoic (*c.* 950–1300 Ma), late Neoproterozoic–Cambrian (*c.* 500–700 Ma), Ordovician (*c.* 440–480 Ma) and Permo–Carboniferous (*c.* 290 Ma) ages, with additional early–middle Neoproterozoic (*c.* 750–950 Ma) and mid-Proterozoic and older (*c.* 1300–2200 Ma) signatures. These ages are interpreted to indicate sources dominated by local (Oaxaquia terrane and Acatlán Complex) and Gondwanan (Amazonia, Brasiliano and Tocantins) provenances, and support the continental reconstruction of Keppie & Ramos (1999), which places the Oaxaquia terrane adjacent to northwestern South America throughout the Palaeozoic. We therefore consider the Mixteca and Sierra Madre terranes to preserve vestiges, not of the Iapetus Ocean as traditionally advocated, but of the southern margin of the Rheic Ocean. The pre-Silurian sedimentary and magmatic history of the Mixteca terrane is interpreted to record the initial rifting of this ocean, whereas deposition of its pre-Silurian and Devono–Carboniferous pelite-psammite units, and the Granjeno Schist of the Sierra Madre terrane, chronicle the development of a continental rise prism on its Gondwanan (Oaxaquia) margin following ocean opening. Closure of the Rheic Ocean during the Late Devonian–Mississippian is thought to be documented in the Mixteca terrane by the subduction-related eclogite facies metamorphism and exhumation of the pre-Silurian high-grade units of the Acatlán Complex (Piaxtla Suite), while the deposition and subsequent deformation of its Permo–Carboniferous and Permo–Triassic units are interpreted to reflect convergent tectonics on the palaeo-Pacific margin following the amalgamation of Pangaea.

The authors are indebted to P. Cawood and F. McDowell for their constructive reviews, and to G. Gehrels and the University of Arizona LaserChron Center for making available on-line the analysis tool for constructing normalized relative age distribution plots. Funding for this project was provided by NSF grant (EAR-0308105) and an Ohio University Baker Award to RDN, Conacyt and PAPIIT (IN103003) grants to JDK, NSF grant (EAR-0308437) to BVM, and NSERC Discovery and Research Capacity Development grants to JBM and JD. For introducing us to the Acatlán Complex and for continued discussions and support, we are especially indebted to F. Ortega-Gutiérrez. This paper is a contribution to IGCP Projects 453 and 497.

## References

BOGER, S. D., RAETZ, A., GILES, D., ETCHART, E. & FANNING, C. M. 2005. U–Pb age data from the Sunsas region of Eastern Bolivia, evidence for the allochthonous origin of the Paragua block. *Precambrian Research*, **139**, 121–146.

BOHER, M., ABOUCHAMI, W., MICHARD, A. N., ALBAREDE, F. & ARNDT, N. 1992. Crustal growth in West Africa at 2.1 Ga. *Journal of Geophysical Research*, **97**, 345–369.

BOUCOT, A. J., BLODGETT, R. B. & STEWART, J. H. 1997. European Province Late Silurian brachiopods from the Ciudad Victoria area, Tamaulipas, northeastern Mexico. *In*: KLAPPER, G., MURPHY, M. A. & TALENT, J. A. (eds) *Paleozoic Sequence Stratigraphy, Biostratigraphy, and Biogeography: Studies in Honor of J. Granville ('Jess') Johnson*. Geological Society of America Special Paper, **321**, 273–293.

BREAM, B. R., HATCHER, R. D., MILLER, C. F. & FULLAGAR, P. D. 2004. Detrital zircon ages and Nd isotopic data from the southern Appalachian crystalline core, Georgia, South Carolina, North Carolina, and Tennessee: new provenance constraints for part of the Laurentian Margin. *In*: TOLLO, R. P., CORRIVEAU, L., MCLELLAND, J. & BARTHOLOMEW, M. J. (eds) *Proterozoic Evolution of the Grenville Orogen in North America*. Geological Society of America Memoir, **197**, 459–476.

BRITO NEVES, B. B., CAMPOS NETO, M. C. & FUCK, R. 1999. From Rodinia to Western Gondwana: an approach to the Brasiliano – Pan African cycle and orogenic collage. *Episodes*, **22**, 155–199.

BRITO NEVES, B. B., PASSARELLI, C. R., BASEI, M. A. S. & SANTOS, E. J. 2003. *U–Pb zircon ages of some classic granites of the Borborema Province*. *In*: Short Papers – IV South American Symposium on Isotope Geology, Salvador, Brazil, August 24–27, 2003 [CD-ROM], 158–159.

CAMERON, K. L., LOPEZ, R., ORTEGA-GUTIÉRREZ, F., SOLARI, L. A., KEPPIE, J. D. & SCHULZE, C. 2004. U–Pb constraints and Pb isotopic compositions of leached feldspars: constraints on the origin and evolution of Grenvillian rocks from eastern and southern Mexico. *In*: TOLLO, R. P., CORRIVEAU, L., MCLELLAND, J. & BARTHOLOMEW, M. J. (eds) *Proterozoic Tectonic Evolution of the Grenville Orogen in North America*. Geological Society of America Memoir, **197**, 755–770.

CAWOOD, P. A. & NEMCHIN, A. A. 2001. Paleogeographic development of the East Laurentian margin; constraints from U–Pb dating of detrital zircons in the Newfoundland Appalachians. *Geological Society of America Bulletin*, **113**, 1234–1246.

CAWOOD, P. A., MCCAUSLAND, P. J. A. & DUNNING, G. R. 2001. Opening Iapetus: constraints from the Laurentian margin of Newfoundland. *Geological Society of America Bulletin*, **113**, 443–453.

CAWOOD, P. A., NEMCHIN, A. A., STRACHAN, R., PRAVE, T. & KRABBENDAM, M. 2007. Sedimentary basin and detrital zircon record along East Laurentia and Baltica during assembly and breakup of Rodinia. *Journal of the Geological Society, London*, **164**, 257–275.

CEDIEL, F., SHAW, R. P. & CÁCERES, C. 2003. Tectonic assembly of the Northern Andean Block. *In*: BARTOLINI, C., BUFFLER, R. T. & BLICKWEDE, J. (eds) *The Circum-Gulf of Mexico and the Caribbean: Hydrocarbon Habitats, Basin Formation, and Plate Tectonics*. American Association of Petroleum Geologists Memoir, **79**, 815–848.

CENTENO-GARCIA, E. & KEPPIE, J. D. 1999. Latest Paleozoic–early Mesozoic structures in the central

Oaxaca Terrane of southern Mexico: deformation near a triple junction. *Tectonophysics*, **301**, 231–242.

DICKINSON, W. R. & LAWTON, T. F. 2001. Carboniferous to Cretaceous assembly and fragmentation of Mexico. *Geological Society of America Bulletin*, **113**, 1142–1160.

DOWE, D. S., NANCE, R. D., KEPPIE, J. D., CAMERON, K. L., ORTEGA-RIVERA, A., ORTEGA-GUTIÉRREZ, F. & LEE, J. W. K. 2005. Deformational history of the Granjeno Schist, Ciudad Victoria, Mexico: constraints on the closure of the Rheic Ocean? *International Geology Review*, **47**, 920–937.

ELÍAS-HERRERA, M. & ORTEGA-GUTIÉRREZ, F. 2002. Caltepec fault zone: an Early Permian dextral transpressional boundary between the Proterozoic Oaxacan and Paleozoic Acatlán complexes, southern Mexico, and regional implications. *Tectonics*, **21**(3), 10.1029/200TC001278.

ELÍAS-HERRERA, M., MACÍAS-ROMO, C., ORTEGA-GUTIÉRREZ, F., SÁNCHEZ-ZAVALA, J. L., IRIONDO, A. & ORTEGA-RIVERA, A. 2007. Conflicting stratigraphic and geochronologic data from the Acatlán Complex: 'Ordovician' granites intrude metamorphic and sedimentary rocks of Devonian–Permian age. *Eos Transactions of the American Geophysical Union*, **88**(23), Joint Assembly Supplement, Abstract T41A-12.

FLORES DE DIOS GONZALÉZ, L. A., VACHARD, D. & BUITRON-SÁNCHEZ, B. E. 1998. The Tiñu, Santiago Ixtaltepec and Yododene Formations, Oaxaca State: sedimentological, stratigraphic and paleogeographic reinterpretations. *In*: PROGRAM, I. G. C. (ed.) *Laurentia–Gondwanan Connections Before Pangea*. Program and Abstracts. Instituto de Geología, Universidad Nacional Autonoma de México, Oaxaca City, Oaxaca Mexico, 16.

GEHRELS, G., VALENCIA, V. & PULLEN, A. 2006. Detrital zircon geochronology by laser ablation multicollector ICPMS at the Arizona LaserChron Center. *In*: OLSZEWSKI, T. (ed.) *Geochronology: Emerging Opportunities*. Paleontology Society Papers, **12**, 67–76.

GILLIS, R. J., GEHRELS, G. E., RUIZ, J. & FLORES DE DIOS GONZALÉZ, L. A. 2005. Detrital zircon provenance of Cambrian-Ordovician and Carboniferous strata of the Oaxaca terrane, southern Mexico. *Sedimentary Geology*, **182**, 87–100.

GONZÁLEZ-ARREOLA, C., VILLASEÑOR-MARTINEZ, A. B. & CORONA-ESQUIVEL, R. 1994. Permian fauna of the Los Arcos Formation, Municipality of Olinalá, State of Guerrero, Mexico. *Revista Mexicana de Ciencias Geológicas*, **2**(2), 214–221.

GOWER, C. F., RYAN, A. B. & RIVERS, T. 1990. Mid-Proterozoic Laurentia–Baltica: an overview of its geological evolution and summary of the contributions by this volume. *In*: GOWER, C. F., RIVERS, T. & RYAN, A. B. (eds) *Mid-Proterozoic Laurentia–Baltica*. Geological Association of Canada Special Paper, **38**, 1–20.

GRADSTEIN, F. M., OGG, J. G. & SMITH, A. G. 2004. *A Geologic Time Scale*. Cambridge University Press, Cambridge, 1–585.

GRODZICKI, K. R., NANCE, R. D., KEPPIE, J. D., DOSTAL, J. & MURPHY, J. B. 2008. Structural, geochemical and geochronological analysis of metasedimentary and metavolcanic rocks of the Coatlaco area, Acatlán Complex, southern Mexico. *Tectonophysics*, **461**, 311–323.

HEATHERINGTON, A. L., MUELLER, P. A. & NUTMAN, A. P. 1996. Neoproterozoic magmatism in the Suwannee terrane: implications for terrane correlation. *In*: NANCE, R. D. & THOMPSON, M. D. (eds) *Avalonian and Related Peri-Gondwanan Terranes of the Circum-North Atlantic*. Geological Society of America Special Paper, **304**, 257–268.

HINOJOSA-PRIETO, H. R., NANCE, R. D., KEPPIE, J. D., DOSTAL, J., ORTEGA-RIVERA, A. & LEE, J. W. K. 2008. Ordovician and Late Paleozoic-Early Mesozoic tectonothermal history of the La Noria area, northern Acatlán Complex, southern Mexico: record of convergence in the Rheic and paleo-Pacific Oceans. *Tectonophysics*, **461**, 324–342.

HIRDES, W. & DAVIS, D. W. 2002. U–Pb geochronology of Paleoproterozoic rocks in the southern part of the Kedougou-Kéniéba inlier, Senegal, West Africa: evidence for diachronous accretionary development of the Eburnian province. *Precambrian Research*, **118**, 83–99.

KEPPIE, J. D. 2004. Terranes of Mexico revisited: a 1.3 billion year odyssey. *International Geology Review*, **46**, 765–794.

KEPPIE, J. D. & RAMOS, V. S. 1999. Odyssey of terranes in the Iapetus and Rheic Oceans during the Paleozoic. *In*: KEPPIE, J. D. & RAMOS, V. A. (eds) *Laurentia-Gondwana Connections Before Pangea*. Geological Society of America Special Paper, **336**, 267–276.

KEPPIE, J. D., DOSTAL, J., ORTEGA-GUTIÉRREZ, F. & LOPEZ, R. 2001. A Grenvillian arc on the margin of Amazonia: evidence from the southern Oaxacan Complex, southern Mexico. *Precambrian Research*, **112**, 165–181.

KEPPIE, J. D., DOSTAL, J., CAMERON, K. L., SOLARI, L. A., ORTEGA-GUTIÉRREZ, F. & LOPEZ, R. 2003. Geochronology and geochemistry of Grenvillian igneous suites in the northern Oaxacan Complex, southern Mexico: tectonic implications. *Precambrian Research*, **120**, 365–389.

KEPPIE, J. D., SANDBERG, C. A., MILLER, B. V., SÁNCHEZ-ZAVALA, J. L., NANCE, R. D. & POOLE, F. G. 2004a. Implications of latest Pennsylvanian to Middle Permian paleontological and U–Pb SHRIMP data from the Tecomate Formation to re-dating tectonothermal events in the Acatlán Complex, southern Mexico. *International Geology Review*, **46**, 745–753.

KEPPIE, J. D., NANCE, R. D. ET AL. 2004b. Mid-Jurassic tectonothermal event superposed on a Paleozoic geological record in the Acatlán Complex of southern Mexico: hotspot activity during the breakup of Pangea. *Gondwana Research*, **7**, 239–260.

KEPPIE, J. D., NANCE, R. D., FERNÁNDEZ-SUÁREZ, J., STOREY, C. D., JEFFRIES, T. E. & MURPHY, J. B. 2006a. Detrital zircon data from the eastern Mixteca terrane, southern Mexico: evidence for an Ordovician-Mississippian continental rise and a Permo-Triassic clastic wedge adjacent to Oaxaquia. *International Geology Review*, **48**, 97–111.

KEPPIE, J. D., DOSTAL, J., NANCE, R. D., MILLER, B. V., ORTEGA-RIVERA, A. & LEE, J. K. W. 2006b. Circa 546 Ma plume-related dykes in the ~1 Ga Novillo Gneiss (east-central Mexico): evidence of the initial separation of Avalonia. *Precambrian Research*, **147**, 342–353.

KEPPIE, J. D., DOSTAL, J. & ELÍAS-HERRERA, M. 2007. Ordovician-Devonian oceanic basalts in the Cosoltepec Formation, Acatlán Complex, southern Mexico: vestiges of the Rheic Ocean? *In*: LINNEMANN, U., NANCE, R. D., KRAFT, P. & ZULAUF, G. (eds) *The Evolution of the Rheic Ocean: From Avalonian-Cadomian Active Margin to Alleghenian-Variscan Collision*. Geological Society of America Special Paper, **423**, 477–487.

KEPPIE, J. D., DOSTAL, J., MURPHY, J. B. & NANCE, R. D. 2008a. Synthesis and tectonic interpretation of the westernmost Paleozoic Variscan orogen in southern Mexico: from rifted Rheic margin to active Pacific margin. *Tectonophysics*, **461**, 277–290.

KEPPIE, J. D., DOSTAL, J. ET AL. 2008b. Ordovician rift tholeiites in the Acatlán Complex, southern Mexico: evidence of rifting on the southern margin of the Rheic Ocean. *Tectonophysics*, **461**, 130–156.

KROGH, T. E., KAMO, S. L. & BOHOR, B. F. 1993a. Fingerprinting the K/T impact site and determining the time of impact by U/Pb dating of single shocked zircons from distal ejecta. *Earth and Planetary Science Letters*, **119**, 425–429.

LANDING, E., WESTROP, S. R. & KEPPIE, J. D. 2007. Terminal Cambrian and lowest Ordovician succession of Mexican West Gondwana: biotas and sequence stratigraphy of the Tiñu Formation. *Geological Magazine*, **144**, 909–9372.

LAWLOR, P. J., ORTEGA-GUTIÉRREZ, F., CAMERON, K. L., OCHOA-CAMARILLO, H., LOPEZ, R. & SAMPSON, D. E. 1999. U–Pb geochronology, geochemistry, and provenance of the Grenvillian Huiznopala Gneiss of eastern Mexico. *Precambrian Research*, **94**, 73–99.

LUDWIG, K. J. 2003. Isoplot 3.00. *Berkeley Geochronology Center Special Publication*, **4**, 70.

MACHADO, N., VALLADARES, C. S., HEILBRON, M. & VALERIANO, C. M. 1996. U/Pb geochronology of the Central Ribeira Belt. *Precambrian Research*, **79**, 347–361.

MALONE, J. W., NANCE, R. D., KEPPIE, J. D. & DOSTAL, J. 2002. Deformational history of part of the Acatlán Complex: Late Ordovician–Early Silurian and Early Permian orogenesis in southern Mexico. *Journal of South American Earth Sciences*, **15**, 511–524.

MARTÍNEZ CATALÁN, J. R., HATCHER, R. D., JR., ARENAS, R. & DÍAZ GARCÍA, F. (eds) 2002. *Variscan-Appalachian dynamics: The building of the late Paleozoic basement*. Geological Society of Amererica Special Paper, **364**, 312.

MEZA-FIGUEROA, D., RUIZ, J., TALAVERA-MENDOZA, O. & ORTEGA-GUTIERREZ, F. 2003. Tectonometamorphic evolution of the Acatlán Complex eclogites (southern Mexico). *Canadian Journal of Earth Sciences*, **40**, 27–44.

MIDDLETON, M., KEPPIE, J. D., MURPHY, J. B., MILLER, B. V., NANCE, R. D., ORETEGA-RIVERA, A. & LEE,

J. K. W. 2007. *P–T–t* constraints on exumation following subduction in the Rheic Ocean from eclogitic rocks in the Acatlán Complex of southern Mexico. *In*: LINNEMANN, U., NANCE, R. D., KRAFT, P. & ZULAUF, G. (eds) *The Evolution of the Rheic Ocean: From Avalonian–Cadomian active margin to Alleghenian–Variscan collision*. Geological Society of America Special Paper, **423**, 489–509.

MILLER, B. V., DOSTAL, J., KEPPIE, J. D., NANCE, R. D., ORTEGA-RIVERA, A. & LEE, J. W. K. 2007. Ordovician calc-alkaline granitoids in the Acatlán Complex, southern Mexico: geochemical and geochronological data and implications for the tectonics of the Gondwanan margin of the Rheic Ocean. *In*: LINNEMANN, U., NANCE, R. D., ZULAF, G. & KRAFT, P. (eds) *The Evolution of the Rheic Ocean: From Avalonian–Cadomian active margin to Alleghenian–Variscan collision*. Geological Society of America Special Paper, **423**, 465–475.

MORALES-GÁMEZ, M., KEPPIE, J. D. & NORMAN, M. 2008. The Ordovician–Silurian rift-passive margin on the Mexican margin of the Rheic Ocean overlain by Carboniferous-Permian periarc rocks: evidence from the eastern Acatlán Complex, southern Mexico. *Tectonophysics*, **461**, 291–310.

MURPHY, J. B., KEPPIE, J. D., NANCE, R. D., MILLER, B. V., MIDDLETON, M., FERNÁNDEZ-SUAREZ, J. & JEFFRIES, T. E. 2006. Geochemistry and U–Pb protolith ages of eclogitic rocks of the Asís Lithodeme, Piaxtla Suite, Acatlán Complex, southern Mexico: tectonothermal activity along the southern margin of the Rheic Ocean. *Journal of the Geological Society, London*, **163**, 683–695.

NANCE, R. D., MILLER, B. V., KEPPIE, J. D., MURPHY, J. B. & DOSTAL, J. 2006. The Acatlán Complex, southern Mexico: record of Pangea assembly to breakup. *Geology*, **34**, 857–860.

NANCE, R. D., MILLER, B. V., KEPPIE, J. D., MURPHY, J. B. & DOSTAL, J. 2007*a*. Vestige of the Rheic Ocean in North America: the Acatlán Complex of southern México. *In*: LINNEMANN, U., NANCE, R. D., ZULAF, G. & KRAFT, P. (eds) *The Evolution of the Rheic Ocean: From Avalonian–Cadomian active margin to Alleghenian–Variscan collision*. Geological Society of America Special Paper, **423**, 437–452.

NANCE, R. D., FERNÁNDEZ-SUÁREZ, J., KEPPIE, J. D., STOREY, C. & JEFFRIES, T. E. 2007*b*. Provenance of the Granjeno Schist, Ciudad Victoria, México: detrital zircon U–Pb age constraints and implications for the Paleozoic paleogeography of the Rheic Ocean. *In*: LINNEMANN, U., NANCE, R. D., ZULAF, G. & KRAFT, P. (eds) *The Evolution of the Rheic Ocean: From Avalonian-Cadomian active margin to Alleghenian-Variscan collision*. Geological Society of America Special Paper, **423**, 453–464.

NAVARRO-SANTILLAN, D., SOUR-TOVAR, F. & CENTENO-GARCIA, E. 2002. Lower Mississippian (Osagean) brachiopods from the Santiago Formation, Oaxaca, Mexico: stratigraphic and tectonic implications. *Journal of South American Earth Sciences*, **15**, 327–336.

ORTEGA-GUTIÉRREZ, F. 1975. *The Pre-Mesozoic geology of the Acatlán area, south Mexico*. PhD thesis, Leeds University, UK.

ORTEGA-GUTIÉRREZ, F. 1978. El Gneiss Novillo y rocas metamorficas asociadas en los canones del Novillo y la Peregrina, area Ciudad Victoria, Tamaulipas. Universidad Nacional Autónoma de México, Instituto de Geología. *Revista*, **2**, 19–30.

ORTEGA-GUTIÉRREZ, F., RUIZ, J. & CENTENO-GARCÍA, E. 1995. Oaxaquia, a Proterozoic microcontinent accreted to North America during the late Paleozoic. *Geology*, **23**, 1127–1130.

ORTEGA-GUTIÉRREZ, F., ELIAS-HERRERA, M., REYES-SALAS, M., MACIAS-ROMO, C. & LÓPEZ, R. 1999. Late Ordovician-Early Silurian continental collision orogeny in southern Mexico and its bearing on Gondwana-Laurentia connections. *Geology*, **27**, 719–722.

ORTEGA-OBREGÓN, C., KEPPIE, J. D. *ET AL.* 2003. Geochronology and geochemistry of the ~917 Ma, calc-alkaline Etla granitoid pluton (Oaxaca, southern México): evidence of post-Grenvillian subduction along the northern margin of Amazonia. *International Geology Review*, **45**, 596–610.

PANTOJA-ALOR, J. & ROBISON, R. 1967. Paleozoic sedimentary rocks in Oaxaca, Mexico. *Science*, **157**, 1033–1035.

PIMENTAL, M. M., FUCK, R. A. & GIOIA, S. M. C. 2000. The Neoproterozoic Goiás magmatic arc, central Brazil: a review and new Sm–Nd isotopic data. *Revista Brasileira de Geociências*, **30**, 35–39.

PINDELL, J., KENNAN, L. & BARRETT, S. 2000. Putting it all together again. *AAPG Explorer*, **21**(10), 34–37.

POTREL, A., PEUCAT, J. & FANNING, C. M. 1998. Archean crustal evolution of the West African Craton: example of the Amsaga Area (Reguibat Rise). U–Pb and Sm–Nd evidence for crustal growth and recycling. *Precambrian Research*, **90**, 107–117.

PROENZA, J. A., ORTEGA-GUTIÉRREZ, F., CAMPRUBÍ, A., TRITLLA, J., ELÍAS-HERRERA, M. & REYES-SALAS, M. 2004. Paleozoic serpentinite-enclosed chromitites from Tehuitzingo (Acatlán Complex, southern Mexico): a petrological and mineralogical study. *Journal of South America Earth Sciences*, **16**, 649–666.

RAMÍREZ-RAMÍREZ, C. 1992. *Pre-Mesozoic geology of Huizachal-Peregrina Anticlinorium, Ciudad Victoria, Tamaulipas, and Adjacent Parts of Eastern Mexico*. PhD thesis, University of Texas, Austin, 1–317.

RAMOS-ARÍAS, M. A., KEPPIE, J. D., ORTEGA-RIVERA, A. & LEE, J. W. K. 2008. Extensional Late Paleozoic deformation on the western margin of Pangea, Patlanoaya area, Acatlán Complex, southern Mexico. *Tectonophysics*, **461**, 60–76.

REYES-SALAS, A. M. 2003. *Mineralogia y petrologia de los granitoides Esperanza del compljo Acatlán, sur de México*. PhD thesis, Universidad Nacional Autonoma de México, Mexico City.

RIVERS, T. 1997. Lithotectonic elements of the Grenville Province: review and tectonic implications. *Precambrian Research*, **86**, 117–154.

ROBERTS, D. 2003. The Scandinavian Caledonides: event chronology, palaeogeographic settings and likely modern analogues. *Tectonophysics*, **365**, 283–299.

ROBERTS, D. & SIEDLECKA, A. 2002. Timanian orogenic deformation along the northeastern margin of Baltica, northwest Russia and northeast Norway, and

Avalonian–Cadomian connections. *Tectonophysics*, **352**, 169–184.

ROBISON, R. & PANTOJA-ALOR, J. 1968. Tremadocian trilobites from Nochixtlan region, Oaxaca, Mexico. *Journal of Paleontology*, **42**, 767–800.

ROCCI, G., BRONNER, G. & DESCHAMPS, M. 1991. Crystalline basement of the West African craton. *In*: DALLMEYER, R. D. & LÉCORCHÉ, J. P. (eds) *The West African Orogens and Circum-Atlantic Correlatives*. Springer-Verlag, New York, 31–61.

SADOWSKI, G. R. & BETTENCOURT, J. S. 1996. Mesoproterozoic tectonic correlations between eastern Laurentia and the western border of the Amazon craton. *Precambrian Research*, **76**, 213–227.

SÁNCHEZ-ZAVALA, J. L., CENTENO-GARCIA, E. & ORTEGA-GUTIÉRREZ, F. 1999. Review of Paleozoic stratigraphy of Mexico and its role in the Gondwana–Laurentia connections. *In*: RAMOS, V. A. & KEPPIE, J. D. (eds) *Laurentian–Gondwanan Connections Before Pangea*. Geological Society of America Special Paper, **336**, 211–226.

SÁNCHEZ-ZAVALA, J. L., ORTEGA-GUTIÉRREZ, F., KEPPIE, J. D., JENNER, G. A., BELOUSOVA, E. & MACÍAS-ROMO, C. 2004. Ordovician and Mesoproterozoic zircons from the Tecomate Formation and Esperanza granitoids, Acatlán Complex, southern Mexico: local provenance in the Acatlán and Oaxacan complexes. *International Geology Review*, **46**, 1005–1021.

SANTOS, J. O. S. 2003. Geotectônica dos Escudos das Guianas e Brasil-Central. *In*: BIZZI, L. A., SCHOBBENHAUS, C., VIDOTTI, R. M. & GONÇALVES, J. H. (eds) *Geologia, Tectônica e Recursos Minerais do Brasil, Texto, Mapas e SIG*. Brasilia, Seriço Geológico do Brasil – CPRM, 169–226.

SANTOS, J. O. S., HARTMANN, L. A., GAUDETTE, H. E., GROVES, D. I., McNAUGHTON, N. J. & FLETCHER, I. R. 2000. A new understanding of the provinces of the Amazon Craton based on integration of field mapping and U–Pb and Sm–Nd geochronology. *Gondwana Research*, **3**, 453–488.

SOLARI, L. A., KEPPIE, J. D., ORTEGA-GUTIÉRREZ, F., CAMERON, K. L., LOPEZ, R. & HAMES, W. E. 2003. 990 and 1100 Ma Grenvillian tectonothermal events in the northern Oaxacan Complex, southern Mexico: roots of an orogen. *Tectonophysics*, **365**, 257–282.

STEWART, J. H., BLODGETT, R. B., BOUCOT, A. J., CARTER, J. L. & LOPEZ, R. 1999. Exotic Paleozoic strata of Gondwanan provenance near Ciudad Victoria, Tamaulipas, Mexico. *In*: KEPPIE, J. D. & RAMOS, V. A. (eds) *Laurentian–Gondwana Connections Before Pangea*. Geological Society of America Special Paper, **336**, 227–252.

TALAVERA-MENDOZA, O., RUÍZ, J., GEHRELS, G. E., MEZA-FIGUEROA, D. M., VEGA-GRANILLO, R. & CAMPA-URANGA, M. F. 2005. U–Pb geochronology of the Acatlán Complex and implications for the Paleozoic paleogeography and tectonic evolution of southern Mexico. *Earth and Planetary Science Letters*, **235**, 682–699.

TASSINARI, C. C. G. & MACAMBIRA, M. J. B. 1999. Geochronological provinces of the Amazon Craton. *Episodes*, **22**, 174–182.

TOLLO, R. P., CORRIVEAU, L., McLELLAND, J. & BARTHOLOMEW, M. J. (eds) 2004. *Proterozoic tectonic evolution of the Grenville Orogen in North America*. Geological Society of America Memoir, **197**, 1–830.

TORRES, R., RUÍZ, J., PATCHETT, P. J. & GRAJALES-NISHIMURA, J. M. 1999. Permo-Triassic continental arc in eastern Mexico; tectonic implications for reconstructions of southern North America. *In*: BARTOLINI, C., WILSON, J. L. & LAWTON, T. F. (eds) *Mesozoic Sedimentary and Tectonic History of North-central Mexico*. Geological Society of America Special Paper, **340**, 191–196.

VACHARD, D. & FLORES DE DIOS, A. 2002. Discovery of latest Devonian/earliest Mississippian microfossils in San Salvador Patlanoaya (Puebla, Mexico): biogeographic and geodynamic consequences. *Compte Rendu Geoscience*, **334**, 1095–1101.

VACHARD, D., FLORES DE DIOS, A., BUITRÓN, B. E. & GRAJALES, M. 2000. Biostratigraphie par fusulines des calcaires Carbonifères et Permiens de San Salvador Patlanoaya (Puebla, Mexique). *Geobios*, **33**, 5–33.

VACHARD, D., FLORES DE DIOS, A. & BUITRÓN, B. 2004. Guadalupian and Lopingian (Middle and Late Permian) deposits from México and Guatemala, a review with new data. *Geobios*, **37**, 99–115.

VAN STAAL, C. R., DEWEY, J. F., MAC NIOCAILL, C. & McKERROW, W. S. 1998. The Cambrian-Silurian tectonic evolution of the Northern Appalachians and British Caledonides; history of a complex, west and southwest Pacific-type segment of Iapetus. *In*: BLUNDELL, D. & SCOTT, A. C. (eds) *Lyell: The past is the key to the present*. Geological Society, London, Special Publications, **143**, 199–242.

VALVERDE-VAQUERO, P. & DUNNING, G. R. 2000. New U–Pb ages from Early Ordovician magmatism in Central Spain. *Journal of the Geological Society, London*, **157**, 15–26.

VEGA-GRANILLO, R., TALAVERA-MENDOZA, O., DIANA MEZA-FIGUEROA, D., RUIZ, J., GEHRELS, G. E. & LÓPEZ-MARTÍNEZ, M. 2007. Pressure-temperature-time evolution of Paleozoic high-pressure rocks of the Acatlán Complex (southern Mexico): implications for the evolution of the Iapetus and Rheic Oceans. *Geological Society of America Bulletin*, **119**, 1249–1264.

VIELE, G. W. & THOMAS, W. A. 1989. Tectonic Synthesis of the Ouachita Orogenic Belt. *In*: HATCHER, R. D., THOMAS, W. A. & VIELE, G. W. (eds) *The Geology of North America*. The Appalachian–Ouachita Orogen in the United States, **F-2**, 695–728.

WEBER, B. & KÖEHLER, H. 1999. Sm–Nd, Rb–Sr and U–Pb geochronology of a Grenville Terrane in Southern Mexico: origin and geologic history of the Guichicovi Complex. *Precambrian Research*, **96**, 245–262.

YAÑEZ, P., RUIZ, J., PATCHETT, P. J., ORTEGA-GUTIÉRREZ, F. & GEHRELS, G. 1991. Isotopic studies of the Acatlán Complex, southern Mexico: implications for Paleozoic North American tectonics. *Geological Society of America Bulletin*, **103**, 817–828.

# Pre-Carboniferous, episodic accretion-related, orogenesis along the Laurentian margin of the northern Appalachians

CEES R. VAN STAAL[1]*, JOSEPH B. WHALEN[2], PABLO VALVERDE-VAQUERO[3], ALEXANDRE ZAGOREVSKI[2] & NEIL ROGERS[2]

[1]*Geological Survey of Canada, 625 Robson Street, Vancouver, British Columbia, Canada*

[2]*Geological Survey of Canada, 601 Booth Street, Ottawa, Ontario, K1A 0E8, Canada*

[3]*Instituto Geologico y Minero de España (IGME), La Calera 1, Tres Cantos (Madrid), 28760, Spain*

*\*Corresponding author (e-mail: Cees.vanStaal@NRCan-RNCan.gc.ca)*

**Abstract:** During the Early to Middle Palaeozoic, prior to formation of Pangaea, the Canadian and adjacent New England Appalachians evolved as an accretionary orogen. Episodic orogenesis mainly resulted from accretion of four microcontinents or crustal ribbons: Dashwoods, Ganderia, Avalonia and Meguma. Dashwoods is peri-Laurentian, whereas Ganderia, Avalonia and Meguma have Gondwanan provenance. Accretion led to a progressive eastwards (present co-ordinates) migration of the onset of collision-related deformation, metamorphism and magmatism. Voluminous, syn-collisional felsic granitoid-dominated pulses are explained as products of slab-breakoff rather than contemporaneous slab subduction. The four phases of orogenesis associated with accretion of these microcontinents are known as the Taconic, Salinic, Acadian and Neoacadian orogenies, respectively. The Ordovician Taconic orogeny was a composite event comprising three different phases, due to involvement of three peri-Laurentian oceanic and continental terranes. The Taconic orogeny was terminated with an arc–arc collision due to the docking of the active leading edge of Ganderia, the Popelogan–Victoria arc, to an active Laurentian margin (Red Indian Lake arc) during the Late Ordovician (460–450 Ma).

The Salinic orogeny was due to Late Ordovician–Early Silurian (450–423 Ma) closure of the Tetagouche–Exploits backarc basin, which separated the active leading edge of Ganderia from its trailing passive edge, the Gander margin. Salinic closure was initiated following accretion of the active leading edge of Ganderia to Laurentia and stepping back of the west-directed subduction zone behind the accreted Popelogan–Victoria arc. The Salinic orogeny was immediately followed by Late Silurian–Early Devonian accretion of Avalonia (421–400 Ma) and Middle Devonian–Early Carboniferous accretion of Meguma (395–350 Ma), which led to the Acadian and Neoacadian orogenies, respectively. Each accretion took place after stepping-back of the west-dipping subduction zone behind an earlier accreted crustal ribbon, which led to progressive outboard growth of Laurentia. The Acadian orogeny was characterized by a flat-slab setting after the onset of collision, which coincided with rapid southerly palaeolatitudinal motion of Laurentia. Acadian orogenesis preferentially started in the hot and hence, weak backarc region. Subsequently it was characterized by a time-transgressive, hinterland migrating fold-and-thrust belt antithetic to the west-dipping A–subduction zone. The Acadian deformation front appears to have been closely tracked in space by migration of the Acadian magmatic front. Syn-orogenic, Acadian magmatism is interpreted to mainly represent partial melting of subducted fore-arc material and pockets of fluid-fluxed asthenosphere above the flat-slab, in areas where Ganderian's lithosphere was thinned by extension during Silurian subduction of the Acadian oceanic slab. Final Acadian magmatism from 395–c. 375 Ma is tentatively attributed to slab-breakoff.

Neoacadian accretion of Meguma was accommodated by wedging of the leading edge of Laurentia, which at this time was represented by Avalonia. The Neoacadian was devoid of any accompanying arc magmatism, probably because it was characterized by a flat-slab setting throughout its history.

## Introduction

This paper focuses on the style and nature of the relationships between deformation, metamorphism and magmatism generated during the Taconic, Salinic, Acadian and Neoacadian orogenies, mainly by using the critical relationships established in the Canadian, and to a lesser extent, the immediately adjacent parts of the New England segment of the northern Appalachians (Fig. 1). The Alleghenian orogenic overprint in the Canadian segment of the Appalachians is relatively small and we are

*From*: MURPHY, J. B., KEPPIE, J. D. & HYNES, A. J. (eds) *Ancient Orogens and Modern Analogues.*
Geological Society, London, Special Publications, **327**, 271–316.
DOI: 10.1144/SP327.13   0305-8719/09/$15.00 © The Geological Society of London 2009.

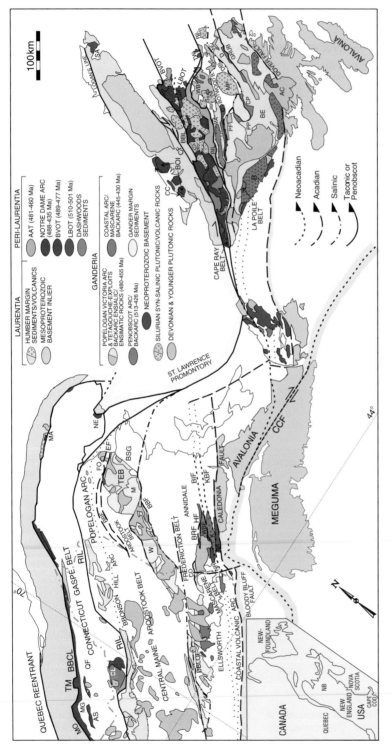

**Fig. 1.** Geology of the Canadian and adjacent New England Appalachians with the geographical distribution of the major tectonic elements discussed in text. A, Arisaig Group; AC, Ackley granite; B, Burgeo batholith; BB, Badger basin; BBF, Bamford Brook fault; BBL, Baie Verte Brompton Line; BVOT, Baie Verte oceanic tract; CCF, Cobequid–Chedabucto fault; CF, Cabot fault; CL, Chain Lakes Massif; CO, Cookson Group; DBL, Dog Bay Line; FO, Fournier Group; F, Fogo Island pluton; GBF, Green Bay fault; GRUB, Gander River ultrabasic belt; HH, Hodges Hill pluton; LBOT, Lushs Bight oceanic tract; MP, Mount Peyton pluton; RBF, Rocky Brook–Millstream fault system; RF, Restigouche fault; RIL, Red Indian Line; SGB, St. George batholith; SM, South Mountain batholith; U, Utopia granite; VA, Victoria arc; TP, Tally Pond Group.

intimately familiar with its geology, some of us having worked here for >25 years. In addition, this segment of the northern Appalachians has relatively large and regionally extensive databases of radiometric and fossil ages as well as high-quality magmatic rock geochemical data, while the grade of regional metamorphism is generally low. Also available is a wide set of seismic reflection and refraction data acquired during Lithoprobe, which imaged the structural architecture at depth and allows its linking with crustal structures mapped at surface (e.g. van der Velden *et al.* 2004, and references therein). Hence, the vital linkages between deformation, metamorphism, magmatism, exhumation and syn-orogenic sedimentation, essential to understand the dynamic processes responsible for orogenesis, are relatively well preserved, yet the constraints provided by these processes rarely have been integrated together. The latter particularly applies to the relationships between magmatism and tectonic processes responsible for the Appalachian orogenic events (see below). The details of the tectonic elements, plate tectonic setting and plate interactions involved have been discussed in several recent papers (e.g. van Staal *et al.* 1998, 2007; van Staal 2005, 2007; Lissenberg *et al.* 2005*b*; Zagorevski *et al.* 2006, 2007*a, c*; Valverde-Vaquero *et al.* 2006; Lin *et al.* 2007; Hibbard *et al.* 2007) and the reader is referred to these for more details.

Although plutonic rocks make up one quarter of the exposure in the Canadian and adjacent New England Appalachians and occur in all the tectonostratigraphic terranes discussed herein, understanding of their petrogenesis generally has been either poor or equivocal (cf. Currie 1995). The subject of their tectonic implications commonly tended to raise more questions than provide answers. In this study, we have attempted to interpret the observed spatial-temporal changes and/or repetitions in plutonic rock compositions within the tectonic context, to a large extent deduced from other less-equivocal lines of geological evidence. This approach not only provides new insights into understanding granitoid petrogenesis in the Appalachians, but it may also represent a 'granitoid geochemical template' that could aid in understanding the tectonic evolution of orogens elsewhere that are also characterized by abundant plutonism but with less well-defined tectonic constraints. The vast majority of the over 5400 analyses employed herein are from plutonic rocks, because our evidence indicates that the most voluminous plutonic episodes occurred in syn- to post-collisional settings, and as such mainly reflect the accretion-related processes and provide information concerning the various crustal- and mantle materials juxtaposed and/or involved during these events. In contrast, much of the volcanic rock record is thought to have formed within arc or back-arc settings prior to accretion (e.g. Swinden *et al.* 1997; van Staal *et al.* 1998; Barr *et al.* 2002). Details concerning spatial-temporal groupings, interpretation methods/approach and data sources are given in Appendix 1. The reader is referred to these data sources for much more comprehensive interpretations of specific portions of our dataset.

A problem encountered in any orogenic research is the well-established classification of orogenies on the basis of time. We know from modern plate tectonics that tectonic events are commonly markedly diachronous and additionally, spatially and kinematically unrelated orogenic events, which overlap in time but not in space, may be juxtaposed fortuitously (e.g. van Staal *et al.* 1998). Hence, as much as possible, we grouped and separated structures, metamorphism and associated magmatic rocks on the basis of their inferred relationship to the specific tectonic events that have been recognised.

The data discussed in this paper indicate that the Appalachians and closely-related British Caledonides (Fig. 1) were an accretionary orogen prior to Alleghenian–Variscan orogenesis. During the early to middle Palaeozoic, the roughly east–west oriented Appalachian Laurentian margin, situated close to the equator, progressively expanded southwards (van Staal *et al.* 1998) due to episodic accretions of ribbons of suprasubduction zone oceanic and continental lithosphere that were present in the Iapetus and to a less extent the Rheic oceans (van Staal *et al.* 1998; Hibbard 2000; van Staal 2005, 2007; van Staal & Hatcher 2010). Hence, large parts of this orogen are polyorogenic (Fig. 2). For the sake of simplicity we will discuss the orientation of large structural features, such as Laurentia's Appalachian margin, in present co-ordinates rather than use Laurentia's true orientation during the Palaeozoic as determined by palaeomagnetism. Accretion generally led to aerially restricted, episodic orogenesis with a progressive eastwards migration of the locus of collision-related deformation, metamorphism and magmatism (Fig. 2). The eastward migration of the locus of orogenesis was followed by uplift and cratonization of the accreted terranes in the orogen's hinterland. This process is particularly evident in Newfoundland by deposition of an unconformable cover of terrestrial, Laurentia-derived clastic rocks (equivalent to the Old Red Sandstone in the British Isles). Deposition of the terrestrial rocks and cratonization of the accreted terranes generally becomes younger eastwards (Fig. 3) and mimics the eastward time-transgressive migration of the site of accretion. This eastward shift is due to progressive stepping back of the subduction zone behind the accreted terranes (see below) and explains the progressive, time-transgressive uplift of the latter. Some of the accreted ribbons were composite

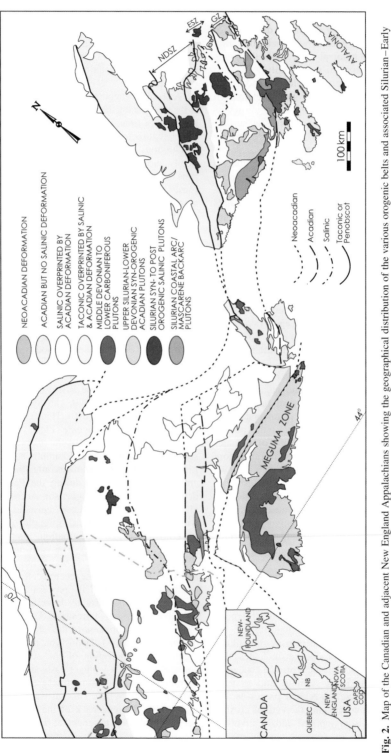

**Fig. 2.** Map of the Canadian and adjacent New England Appalachians showing the geographical distribution of the various orogenic belts and associated Silurian–Early Carboniferous plutonism. Note that the Acadian starts in the part of Ganderia where Salinic orogenesis was absent. Over time it progressively overprints first Salinic and then the Taconic orogenic belts.

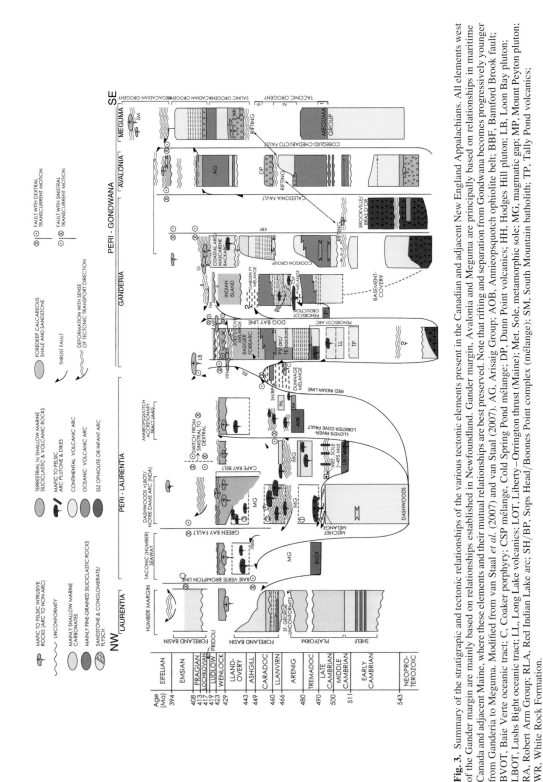

**Fig. 3.** Summary of the stratigrapic and tectonic relationships of the various tectonic elements present in the Canadian and adjacent New England Appalachians. All elements west of the Gander margin are mainly based on relationships established in Newfoundland. Gander margin, Avalonia and Meguma are principally based on relationships in maritime Canada and adjacent Maine, where these elements and their mutual relationships are best preserved. Note that rifting and separation from Gondwana becomes progressively younger from Ganderia to Meguma. Modified from van Staal *et al.* (2007) and van Staal (2007). AG, Arisaig Group; AOB, Annieopsquotch ophiolite belt; BBF, Bamford Brook fault; BVOT, Baie Verte oceanic tract; C, Coaker porphyry; CSP mélange, Cold Spring Pond mélange; DP, Dunn Point volcanics; HH, Hodges Hill pluton; LB, Loon Bay pluton; LBOT, Lushs Bight oceanic tract; LL, Long Lake volcanics; LOT, Liberty–Orrington thrust (Maine); Met. Sole, metamorphic sole; MG, magmatic gap; MP, Mount Peyton pluton; RA, Robert Arm Group; RLA, Red Indian Lake arc; SH/BP, Sops Head/Boones Point complex (mélange); SM, South Mountain batholith; TP, Tally Pond volcanics; WR, White Rock Formation.

due to collision with suprasubduction zone oceanic or continental lithospheric terranes present within Iapetus, prior to their docking with Laurentia. For example, the closure of the Penobscot back-arc basin led to Tremadoc (486–479 Ma) obduction of back-arc ophiolites onto the Gander margin (Fig. 3) and hence to orogenesis, while the latter was still proximal to Gondwana in high southerly latitudes (van Staal 1994; Liss *et al.* 1994; Zagorevski *et al.* 2007*a*). This phase of orogenesis, which took place near the Gondwanan margin far from Laurentia, is not dealt with here.

The accretionary stage of the Appalachian orogen was terminated by the arrival of the bulk of Gondwana during the Carboniferous, which led to formation of the supercontinent Pangaea. The part of the northern Appalachians exposed on land in Canada and northern New England, in contrast to the Appalachians further to the south, largely escaped the penetrative effects of terminal continent–continent collision-related deformation, metamorphism and magmatism. The remnants of the Alleghenian collision zone are mainly situated offshore below the modern Atlantic margin, where they were strongly overprinted by Mesozoic extension accompanying opening of the Atlantic Ocean.

## Northern Appalachian microcontinents

The northern Appalchians comprise the remnants of at least four accreted ribbon-shaped crustal blocks or microcontinents: Dashwoods, Ganderia, Avalonia and Meguma (van Staal 2007). The accretion of these crustal ribbons to Laurentia is the main focus of this paper and hence, below we present a brief overview of the tectonic setting and provenance of these four crustal blocks.

Dashwoods, which is the most inboard crustal block, has a Laurentian provenance. Isotopic data and strong zircon inheritance in Ordovician arc plutons and volcanic rocks suggest it is probably underlain by North American Grenvillian basement (van Staal *et al.* 2007). Fossils and palaeomagnetic data suggest it was never separated by a wide ocean basin from the Humber passive margin of Laurentia (= Humber Zone of Williams 1979) (van Staal *et al.* 1998; Waldron & van Staal 2001, and references therein).

Ganderia, which underlies most of the central core of the northern Appalachians (Fig. 1), was a Late Neoproterozoic–Early Cambrian arc terrane that was probably built upon the Amazonian margin of Gondwana (van Staal *et al.* 1996). The composition and nature of Middle Cambrian volcanic rocks extruded at its trailing edge (White *et al.* 1994; Schultz *et al.* 2008), suggest that Ganderia rifted off Amazonia and became an isolated

microcontinent at *c.* 505 Ma, while its leading edge was still an active margin (511–505 Ma Tally Pond-Long Lake phase of Penobscot arc; Rogers *et al.* 2006) (Figs 1 & 3). The latter suggest that slab rollback forces may have been involved in Ganderia's separation from Gondwana. Slab rollback-induced extension continued and culminated in rifting of the Penobscot arc and opening of a backarc basin at 500–495 Ma, which divided Ganderia into two blocks: an active leading edge (Penobscot arc) and a passive trailing edge. The latter is known as the Gander margin (van Staal 1994). The Penobscot backarc basin temporarily closed between 485 and 479 Ma, which produced the short-lived Penobscot orogeny (Zagorevski *et al.* 2007*a*). Penobscot orogenesis mainly led to obduction of back-arc ophiolites onto the Gander margin (Colman-Sadd *et al.* 1992) and temporary arc shut-off (Fig. 3). After 478 Ma a new arc was resurrected on Ganderia's leading edge, which is known as the Popelogan–Victoria arc (van Staal *et al.* 1998). This arc rifted too, which led to the opening of the wide Tetagouche–Exploits back-arc basin. Ganderia's active leading edge was accreted to Laurentia at 455–450 Ma (van Staal 1994; Zagorevski *et al.* 2007*c*), while the passive trailing edge was accreted at 430–423 Ma, producing the Salinic orogeny (see below).

Avalonia (Fig. 1) is a collage of several fault-bounded Neoproterozoic, largely juvenile arc-related volcanic-sedimentary belts of low metamorphic grade and associated plutonic rocks that experienced a complicated and long-lived Neoproterozoic tectonic history before deposition of an overstepping Cambrian–Early Ordovician shale-rich platformal sedimentary succession (Fig. 3) (O'Brien *et al.* 1996; Landing 1996; Kerr *et al.* 1995). Palaeomagnetic data indicate that Avalonia resided at a high southerly latitude near Gondwana from the Middle Cambrian to the end of the Early Ordovician (Johnson & van der Voo 1986; van der Voo & Johnson 1985; MacNiocaill 2000; Hamilton & Murphy 2004) following a more intermediate latitude position during the Late Neoproterozoic (*c.* 580 Ma; McNamara *et al.* 2001). Fossils also show strong links to Gondwana (e.g. Fortey & Cocks 2003), but previously proposed connections to NW Africa are inconsistent with a wide range of geological arguments (e.g. Landing 1996). This led Murphy *et al.* (2002) to propose an alternative position opposite the Neoproterozoic northern margin of Amazonia, in proximity to but not connected yet with Ganderia (van Staal *et al.* 1996; Rogers *et al.* 2006). Avalonia has a very different geological evolution during the Palaeozoic than Ganderia (Fig. 3), indicating they were two different, unrelated terranes, at least after the onset of the Palaeozoic.

Bimodal volcanic rocks were extruded sporadically during the Middle Cambrian in Avalonia and attributed to transtension (Murphy *et al.* 1996). They are probably related to the onset of extension that culminated in Avalonia's departure from Gondwana during the Early Ordovician at *c.* 475 Ma (Prigmore *et al.* 1997; van Staal *et al.* 1998). This event is characterized by deposition of a transgressive arenitic cover sequence both in Gondwana (Armorican quartzite) and Avalonia (e.g. Stiperstones quartzite in England). Avalonia thus rifted-off Gondwana *c.* 30 Ma after Ganderia, which is an additional argument that Avalonia and Ganderia were two separate microcontinents during the Early Palaeozoic.

Meguma represents the most outboard terrane preserved in the Canadian Appalachians and is exposed on land only in southern Nova Scotia (Fig. 1). However, its regional extent is much larger and its rocks have been traced offshore by an impressive set of geophysical and well data from the southernmost part of the Grand Banks SE of Newfoundland across the Scotian shelf, and the Gulf of Maine to southernmost Cape Cod (see inset Fig. 1) (Hutchinson *et al.* 1988; Keen *et al.* 1991; Pe-Piper & Jansa 1999). Meguma overall had a different Palaeozoic geological evolution than Avalonia (Fig. 3). The oldest exposed rocks in Meguma comprises a thick (<10 km) Cambrian–Early Ordovician turbiditic sandstone–shale sequence, which was deposited on the continental rise and/or slope to outer shelf of a Gondwanan passive margin. These rocks are disconformably overlain by Silurian rift-related bimodal volcanic rocks (442–438 Ma) of the White Rock Formation (Schenk 1997; Keppie & Krogh 2000; MacDonald *et al.* 2002), which may mark the onset of final rifting and departure of Meguma from Gondwana. A combination of detrital zircon, sedimentological and sparse fossil data support an original provenance along the West African continental margin (e.g. Krogh & Keppie 1990; Schenk 1997), possibly in a position separating Avalonia from West Africa (Waldron *et al.* 2009). Fossil evidence suggests that during the Late Silurian, Meguma was close to Avalonia and/or Baltica and probably had moved to Laurentia (Bouyx *et al.* 1997) independently from Gondwana.

Gondwana is generally inferred to have accreted to Laurentia not earlier than the end of the Early Carboniferous.

## Accreted oceanic terranes

In addition to the accreted crustal ribbons, the Appalachian margin of Laurentia grew by accretion of several oceanic terranes (Fig. 3). They are the Cambrian (510–500 Ma) Lushs Bight oceanic tract (LBOT) and correlatives, the Tremadoc (490–483 Ma) Baie Verte oceanic tract (BVOT), the Arenig (480–464 Ma) Annieopsquotch accretionary tract (AAT) and remnants of dismembered Middle Ordovician ophiolitic rocks formed in the Tetagouche–Exploits back-arc basin (van Staal *et al.* 2003; Valverde-Vaquero *et al.* 2006). All accreted oceanic terranes have suprasubduction zone geochemical signatures and either formed during upper plate extension and spreading related to subduction initiation and creation of infant arcs (van Staal *et al.* 1998, 2007; Lissenberg *et al.* 2005*b*) or back-arc basin opening (Rogers & van Staal 2003; Zagorevski *et al.* 2006; Valverde-Vaquero *et al.* 2006). The LBOT, BVOT and AAT formed close to Laurentia and are related to Taconic convergence (Zagorevski *et al.* 2006, 2007*c*; van Staal *et al.* 2007), whereas the Tetagouche–Exploits ophiolitic rocks formed while Ganderia was still at high southerly latitudes near Gondwana (Liss *et al.* 1994).

## Taconic Orogeny

The Taconic orogeny recently has been redefined by van Staal *et al.* (2007) to encompass all the orogenic events that took place in the peri-Laurentian realm between the Late Cambrian and Late Ordovician (495–450 Ma) and the reader is referred to this and related papers (Lissenberg *et al.* 2005*a, b*; Lissenberg & van Staal 2006; Zagorevski *et al.* 2006, 2007*c*) for more detailed information on the tectonic processes responsible for orogenesis. It encompasses three dynamically distinct orogenic events, referred to as Taconic 1, 2 and 3 (Figs 3, 4 & 5).

### Taconic 1

Taconic 1 represents short-lived, Late Cambrian (*c.* 495 Ma) west-directed obduction of an oceanic infant arc terrane, the Lushs Bight oceanic tract (LBOT) onto the peri-Laurentian Dashwoods microcontinent in Newfoundland (Fig. 4) (Waldron & van Staal 2001; van Staal *et al.* 2007). Equivalents of Dashwoods and the LBOT have also been recognized or postulated to exist in the subsurface of southern New England, Maine and Quebec (Laird *et al.* 1993; Karabinos *et al.* 1998; Huot *et al.* 2002; Gerbi *et al.* 2006*a*). Evidence for LBOT's Late Cambrian obduction in Newfoundland is principally indicated by: (1) the *c.* 495 Ma age of the metamorphic aureole beneath the ophiolitic St. Anthony complex (Jamieson 1988; G. Dunning, pers. comm. 2004); (2) the *c.* 493 Ma markedly crustal-contaminated mafic dikes that cut and intruded into metamorphic tectonites of

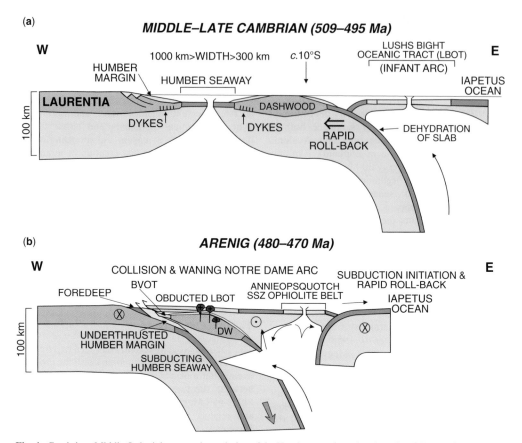

**Fig. 4.** Cambrian–Middle Ordovician tectonic evolution of the Humber margin and outboard peri-Laurentian terranes. Modified from van Staal *et al.* (2007). (**a**) Subduction initiation (using an abandoned spreading ridge, see van Staal *et al.* 2007) and rapid hinge retreat of the east-dipping (present co-ordinates) Dashwoods plate is responsible for formation of the Lushs Bight oceanic tract infant arc terrane. (**b**) Stepping-back of the subduction zone in the Taconic (Humber) Seaway produces the Baie Verte oceanic tract and Notre Dame arc and led to the Taconic 2 arc-continent collision. The onset of collision slowed down convergence in the Taconic seaway, which is thought to have been responsible for initiation of west-directed subduction outboard of Dashwoods. The latter led to formation of the Annieopsquotch ophiolite belt (Lissenberg *et al.* 2005*b*).

the LBOT late synkinematically (Szybinski 1995; Swinden *et al.* 1997); and (3) the *c.* 488 Ma age of the ensialic Cape Ray granodiorite (Dube *et al.* 1996; Whalen *et al.* 1997*b*), which cuts the ophiolitic Long Range ultramafic–mafic complex, a correlative of the LBOT in southern Newfoundland, and underlying Mischief mélange (Hall & van Staal 1999). The mélange (Fig. 3) contains large knockers of gabbroic and serpentinized ultramafic rocks, including dunite and harzburgite, in a predominantly pelitic to semi-pelitic matrix that generally lacks any coherent stratification (Fig. 6a). The zone of mélange can be traced at least for *c.* 100 km along strike (Fox & van Berkel 1988; Brem *et al.* 2007). The mélange was strongly deformed, metamorphosed and injected by Ordovician plutonic rocks of the 1st and 2nd stages of the

Notre Dame arc (see below) after its assembly with the overlying ophiolitic rocks during Taconic 2 (Fig. 4).

Evidence for ductile deformation and metamorphism related to Taconic 1 is relatively cryptic. Szybinski (1995) documented *c.* 493 Ma dykes that had intruded late syn-kinematically into chlorite-rich shear zones. In addition, foliated and flattened Cambrian basalts of the Sleepy Cove Group of the LBOT in Notre Dame Bay were cut by the *c.* 507 Ma Twillingate trondhjemite (Elliot *et al.* 1991; Dewey 2002), which itself was also foliated before the emplacement of the *c.* 473 Ma crosscutting Moreton's Harbour dikes (Williams *et al.* 1976), indicating that the LBOT locally has preserved evidence for a complex structural history prior to and during its accretion to Dashwoods.

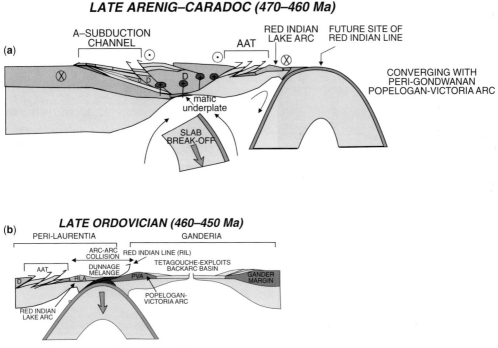

**Fig. 5. (a)** Full-scale Middle Ordovician collision of Dashwoods with the Humber margin (Taconic 2). This phase of the collision led to significant thickening of Dashwoods and east-directed thrusting at its boundary with the Annieopsquotch ophiolite belt (AOB). The latter led to closure of the AOB, accretion of parts of the Red Indian Lake arc and hence formation of the AAT. **(b)** Late Ordovocian Taconic 3 arc–arc collision (Zagorevski *et al.* 2007c). Accretion of the Popelogan–Victoria arc (leading edge of Ganderia) closed the main Iapetan oceanic tract.

In northern Vermont, barroisite in garnet amphibolite that forms part of an ophiolitic sliver at Belvidere Mountain yielded $^{39}Ar/^{40}Ar$-ages of 505–490 Ma (Laird *et al.* 1993), suggesting that this phase of relatively high-pressure metamorphism and ophiolite emplacement may also be related to Taconic 1. The much localized nature of ductile deformation and metamorphism associated with this oceanic infant arc–microcontinent collision, suggests that Taconic 1 was a relatively soft event, mainly characterized by mélange development beneath the overriding infant arc and development of low-temperature shear zones. Ductile deformation and metamorphism were probably mainly restricted to a relatively narrow subduction channel at depth, remnants of which appears to have been rarely exhumed and/or preserved.

*Taconic 2*

Taconic 2 is the main Ordovician orogenic phase in the Appalachians and was due to dextral oblique collision of an Early Ordovician west-facing (present co-ordinates) peri-Laurentian arc (Notre Dame arc in Newfoundland; Ascot arc in Quebec), containing both ensimatic and ensialic segments, with the Humber margin and obduction of suprasubduction zone oceanic lithosphere of the intervening Taconic (Humber in Newfoundland) seaway (Fig. 4) (Waldron & van Staal 2001; van Staal *et al.* 2007; Hibbard *et al.* 2007; Brem *et al.* 2007). The continental part of the arc was built upon Dashwoods in Newfoundland and upon equivalent continental ribbons elsewhere (e.g. Chain Lakes massif in southern Quebec and adjacent Maine) (Karabinos *et al.* 1998; Gerbi *et al.* 2006a). Lower Ordovician arc plutons, formed during east-directed subduction of the Taconic seaway, intrude into continental basement and its LBOT suprastructure in Newfoundland (Hall & van Staal 1999; van Staal *et al.* 2007), probably severely masking structures formed during Taconic 1. Closure of the Taconic seaway was initiated following choking of the Taconic 1 subduction zone after entrance of Dashwoods into the trench (Fig. 4), which forced subduction to step

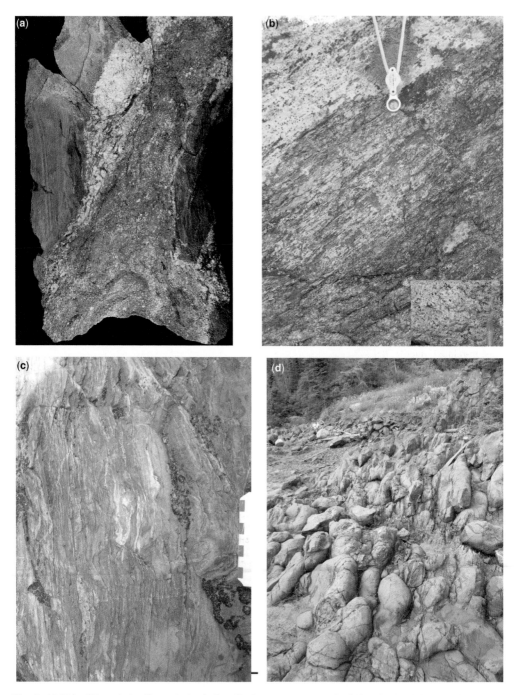

**Fig. 6.** (**a**) Slab of Taconic 1 mélange (*c.* 1 m in length), that occurs structurally below the Cambrian ophiolitic Long Range ultramafic–mafic complex in southwestern Newfoundland. Slab contains a clast of sandstone (left) and gabbro (right). Pelitic to semi-pelitic matrix was deformed and metamorphosed to sillimanite grade, and injected by *c.* 488 Ma and younger granitoid veins. (**b**) Amphibolite facies ultramylonite of granodiorite and tonalite along the moderately to shallowly NW-dipping Hungry Mountain thrust. Hand lens for scale. Thrust was active during Middle Ordovician magmatism (van Staal *et al.* 2007). Inset shows protolith of mylonite. (**c**) Amphibolite tectonite with pods (red) of relict and partly retrogressed eclogite in thrust panel at the base of the Hungry Mountain thrust zone. Scale at right of photograph. (**d**) Panel of tilted and weakly foliated Ordovician pillow basalt in the AAT, structurally below the Hungry Mountain–Lobster Cove thrust system. Rocks locally contain pumpellyite.

back in the remaining oceanic seaway that separated Dashwoods from Laurentia (van Staal *et al.* 2007). Remnants of the Taconic 2-related arc can be traced along the length of the orogen from the southern Appalachians into the British Isles, where its accretion is known as the Grampian orogeny (Dewey & Shackleton 1984; van Staal *et al.* 1998).

Taconic 2-related arc magmatism is most extensively exposed in Newfoundland, where it is known as the 1st stage of the Notre Dame arc (van Staal *et al.* 2007), which lasted from 489–475 Ma. Arc magmatism in Dashwoods has a continental signature, although the degree of contamination by continental crust varies along its length, and comprises both volcanic and plutonic rocks. Geochemical plots (Fig. 7a, b), together with extended element plots, showing well-developed negative Nb anomalies and relatively flat REE patterns (van Staal *et al.* 2007, their fig. 13) indicate that the mafic volcanic and plutonic rocks exhibit island arc tholeiite characteristics.

Stratigraphic and other age constraints suggest that arc–continent collision had started, with little diachroneity between 480 and 470 Ma along the full length of the mountain chain, with most collision-related ductile deformation finished by *c.* 455 Ma (Friedrich *et al.* 1999; Chew *et al.* 2003; Castonguay *et al.* 2001, 2007; van Staal *et al.* 2007). Collision-related deformation, mainly accommodated by west-directed thrusting and folding (e.g. Tremblay & Pinet 1994; Waldron *et al.* 2003), locally seems to have continued until 450–445 Ma in parts of New England (e.g. Ratcliffe *et al.* 1998, 1999) and Quebec (Sacks *et al.* 2004), possibly due to convergence driven by subduction rollback of old oceanic lithosphere trapped in re-entrants in the Laurentian margin. Blueschists and eclogites in the Piedmont of the southern Appalachians, Vermont, southern Quebec and the west of Ireland, which mainly involved rocks of the continental margin, yielded ages between 470 and 459 Ma (Laird *et al.* 1993; Chew *et al.* 2003; Miller *et al.* 2006). These high-pressure metamorphic rocks thus formed during A–subduction of the leading edge of the Laurentian margin to blueschist and locally eclogite facies depths after the start of arc–continent collision. Combined, the available age data indicate little diachroneity in the time of collision along the full length of the Laurentia's Appalachian margin.

Taconic 2 led to intense ductile deformation and metamorphism, locally reaching granulite facies, in rocks involved in arc thickening and tectonic underplating in Newfoundland (van Staal *et al.* 2007) and New England (Karabinos *et al.* 1998; Gerbi *et al.* 2006*a*). Thickening of the Notre Dame arc is manifested by the rapid burial of Lower Ordovician (489–475 Ma) arc volcanic and associated sedimentary rocks to high pressures and temperatures, for example, garnet–clinopyroxene granulite facies conditions at 467–462 Ma (Pehrsson *et al.* 2003; van Staal *et al.* 2007). Such burial of upper plate rocks is explained by widening of the subduction channel towards the arc hinterland, such that progressively more of the channel's hanging wall became dragged down and buried (Fig. 8). Deformation of the upper plate also led to SE-directed thrusting at the boundary (Fig. 6b) of Dashwoods with the Annieopsquotch accretionary tract (Figs 3 & 4). Tectonic transport in Dashwoods thus was accommodated by bi-vergent thrusting (Thurlow *et al.* 1992; Lissenberg *et al.* 2005*a*; Lissenberg & van Staal 2006). Deformation led to development of a strong metamorphic foliation ($S_1$) in the buried arc volcanic and sedimentary rocks, marked by alignment of minerals such as sillimanite, biotite, hornblende and gedrite. $S_1$ was refolded into steep orientations by upright or steeply inclined, tight to isoclinal $F_2$ folds forming a composite $S_1/S_2$ transposition foliation during terminal collision. The interlimb angle of $F_2$ folds appears to become tighter and the $S_2$ transposition foliation better developed on approaching the now subvertical Baie Verte–Brompton Line (= Taconic suture, Williams & St. Julien 1982) in central Newfoundland (van der Velden *et al.* 2004). Dashwoods rocks in the high strain zone, immediately east of the suture, locally show marked retrogression to greenschist facies phyllonites and were intruded by syn-tectonic 459–455 Ma S-type muscovite granite and pegmatite (Brem *et al.* 2007). Such granitic rocks are very rare or absent elsewhere in Dashwoods and its Notre Dame arc and are obviously spatially and temporally associated with the final increments of Taconic deformation localized in the sheared rocks present immediately east of the boundary with the adjacent Humber margin in central Newfoundland. The S-type melts probably formed by melting of underthrusted Laurentian sediments, consistent with the presence of xenocrystic Grenville and older zircons typical of Laurentia, and were channelled into a progressively narrowing and steepening shear zone that must have accommodated significant uplift of the arc after peak metamorphism. In many places the fault zone also accommodated significant dextral transcurrent translation (Brem 2007; Brem *et al.* 2007). Investigations in southern Quebec (e.g. Castonguay *et al.* 2001, 2007) and Gaspé (Sacks *et al.* 2004; Pinchivy *et al.* 2003) revealed a comparable Taconic movement picture involving both orthogonal west-directed thrusting and oblique–dextral tectonic transport near the Baie Verte–Brompton line.

Although evidence for ophiolite obduction, Taconic thrusting and mélange formation is

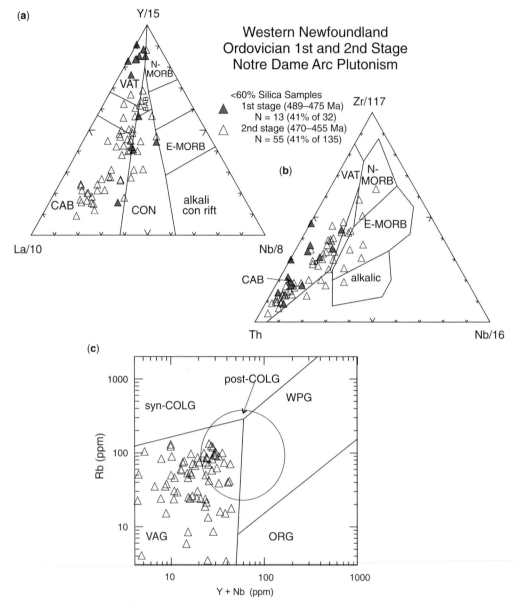

**Fig. 7.** 1st (489–475 Ma) and 2nd stage (470–455 Ma) Notre Dame Arc plutonism in western Newfoundland. Mafic samples (<60 wt% $SiO_2$) plotted on: (**a**) La–Y–Nb diagram (Cabanis & Lecolle 1989); and (**b**) Th–Zr–Nb diagram (Wood 1980). Felsic samples (>60 wt% $SiO_2$) plotted on: (**c**) Rb–Y + Nb diagram (Pearce 1996) for 2nd stage granitoids (1st stage granitoids were omitted because most exhibited evidence of Rb-mobility, see van Staal *et al.* 2007). Beside the diagrams, information on number (N) of samples plotted, sample total (mafic + felsic) and % samples plotted in diagram represent of group total is given. Abbreviations: CAB, calc-alkaline basalt; VAT, volcanic arc tholeiite; N- and E-MORB, normal and enriched mid-ocean-ridge basalt; OIB, ocean island basalt; BAB, backarc basin basalt; CON, continental tholeiite; VAG, volcanic-arc granite; syn-, post-COLG, syn-/post-collisional granite; WPG, within-plate granite; and ORG, ocean-ridge granite. See Appendix 1 for data sources.

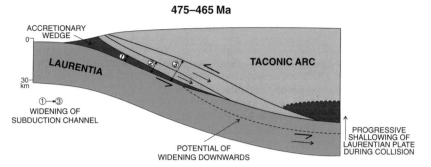

**Fig. 8.** Proposed tectonic evolution of the A–subduction channel beneath Dashwoods. Progressive shallowing of the Humber margin slab caused progressive widening of the subduction channel, which led to underthrusting of upper plate, Notre Dame arc (phase 1) rocks to great depths. The widening of subduction channel may also have involved rocks of the down-going plate, producing a complex kinematic system of sliding lower and upper plate rocks.

widespread in the rocks of the Laurentian margin (e.g. Tremblay & Pinet 1994; Waldron *et al.* 1998, 2003), dating using both the $^{39}Ar/^{40}Ar$ and U–Pb methods (Cawood *et al.* 1994; Castonguay *et al.* 2001, Brem 2007) revealed surprisingly little evidence for Taconic metamorphism. Most metamorphism is either Salinic or Acadian. Castonguay *et al.* (2007) suggested that the scarcity of Taconic ages west of the Baie Verte–Brompton line in southern Quebec was mainly due to nearly complete recrystallization and resetting of Taconic micas during superimposed Silurian (Salinic) and Devonian (Acadian) orogenesis in Quebec. On the other hand, Brem (2007) replicated the results of Cawood *et al.* (1994) in the metamorphic rocks of the Humber margin of central Newfoundland (internal Humber Zone of Williams 1979) and failed to detect any evidence for Taconic metamorphism, despite careful *in-situ* dating of tiny zircon and monazite inclusions in garnets grown in Salinic amphibolite-facies pelitic rocks immediately west of the Baie Verte–Brompton line. Taconic collision-related metamorphism thus seems to have been restricted mainly to rocks of the overriding plate and underplated Laurentian sediments, but appears generally weak or absent in metamorphic tectonites developed in rocks of the immediately adjacent Humber margin (van Staal *et al.* 2007). The lack of evidence for this event in the adjoining metamorphic tectonites of the Humber margin is puzzling, particularly considering the relatively large database of radiometric ages. Hence, it is unlikely to be due to inadequate dating. One possible solution is that Taconically thickened and deeply buried parts of the Humber margin were later (post-peak Taconic metamorphism) overthrust and hidden beneath rocks of the Notre Dame arc and/or adjacent fore-arc ophiolites of the Baie Verte oceanic tract (BVOT; van Staal *et al.* 2007). It is in such rocks, exposed in the

suture zone of the Baie Verte area of northern Newfoundland and structural windows east of the suture in southern Quebec that van Staal *et al.* (2009) and Castonguay *et al.* (2001, 2007) respectively measured Taconic metamorphic ages. Alternatively, large strike-slip movements could have excised segments with high-grade Taconic metamorphism and juxtaposed parts of the Laurentian margin that structurally were never deeply buried with high-grade upper plate rocks. For example, rocks deposited in first- or second-order re-entrants in the Laurentian margin may have escaped deep Taconic-related structural burial, because A–subduction ceased or slowed down significantly shortly after arrival of promontories at the trench. Such early choking of the A–subduction channel was inferred to be the main cause for initiation of west-directed subduction at *c*. 480 Ma, immediately outboard of the Notre Dame arc in the Iapetus Ocean (van Staal *et al.* 2007). This led to formation of the 480–473 Ma suprasubduction zone oceanic lithosphere (e.g. Annieopsquotch and Lloyds River ophiolites, Lissenberg *et al.* 2005*b*; Zagorevski *et al.* 2006) (Fig. 4b) and associated arc sequences, such as the Roberts Arm Group in Newfoundland and Boil Mountain complex in Maine (Gerbi *et al.* 2006*b*). The Annieopsquotch and Lloyds River ophiolites were partially subducted westwards beneath Dashwoods shortly after their formation at *c*. 468 Ma (Lissenberg *et al.* 2005*b*). This event established the east-facing Annieopsquotch accretionary tract (Fig. 5a) (AAT, van Staal *et al.* 1998; van der Velden *et al.* 2004). The AAT progressively expanded eastwards over time due to accretion and underplating of oceanic and arc lithosphere that was originally situated slightly outboard of the Dashwoods microcontinent (Zagorevski *et al.* 2006, 2007*b*, *c*). Amphibolite and eclogite assemblages in the most inboard panels of the AAT (Fig. 6c) indicate that these rocks were subducted to relatively deep depths (Lissenberg & van Staal 2006; Zagorevski *et al.* 2007*d*).

Middle Ordovician Taconic 2 deformation and metamorphism were accompanied by voluminous granitoid magmatism (van Staal *et al.* 2007). This is the 2nd stage of Notre Dame arc magmatism (Figs 3 & 7b). Its mafic end-members exhibit mainly calc-alkaline to arc tholeiite characteristics but also include non-arc-like N- and E-MORB compositions (Fig. 7a, b). Almost equal proportions of mafic and felsic compositions attest to the major role played by mantle-derived magmas in this 2nd stage of Notre Dame arc magmatism. Felsic compositions exhibit volcanic-arc granite (VAG) signatures, plotting mainly to the left or below the area of overlap between VAG and post-collisional granites (post-COLG) (Fig. 7c). Such low-Rb and Y + Nb contents indicate derivation from mainly primitive depleted sources. However, the range of $\varepsilon_{Nd}(T)$ values ($-14$ to $+1$) and infracrustal $\delta^{18}O$ (VSMOW) values ($+5$ to $+10$‰) exhibited by this magmatism (Whalen *et al.* 1997*a*) indicate it formed within Precambrian Laurentian crust. Characteristics of 2nd stage Notre Dame arc magmatism, such as the presence of non-arc-like athenospheric mantle-derived magmas, support a model in which progressive removal of the bulk of the arc's mantle lithosphere during the final stages of collision and breakoff of the oceanic part of the downgoing Taconic seaway slab induced rapid upwelling of the asthenosphere, which caused large-scale melting of the arc's infrastructure (Fig. 5a). Melting led to a relatively short-lived (464–461 Ma), but a very voluminous tonalite bloom, which overlapped with $D_2$ (van Staal *et al.* 2007) and was rapidly followed by uplift and exhumation of the arc's infrastructure. The latter was already exhumed and exposed at the surface by the end of the Caradoc (*c.* 453 Ma). Collision-related Taconic magmatism thus was mainly confined to the site of the original arc.

## Taconic 3

Rocks involved in Taconic 3 mainly occur in a narrow belt situated along the Red Indian Line (RIL, Williams *et al.* 1988), which is the principal Iapetan suture in the northern Appalachians (Figs 1, 3 & 5), along which rocks with Laurentian and Gondwanan provenances have been juxtaposed. The Red Indian Line (RIL) is largely covered in New Brunswick and New England by younger rocks; hence the nature of Taconic 3 is largely based on our research in Newfoundland. Rocks affected by Taconic 3 were deformed and metamorphosed to low grade conditions as a result of a Caradoc (460–450 Ma) collision between an east-facing Middle Ordovician peri-Laurentian (RIL

arc), previously accreted to Dashwoods during Taconic 2 during closure of the Lloyds River backarc basin (Lissenberg *et al.* 2005*b*; Zagorevski *et al.* 2006) with a west-facing peri-Gondwanan arc (Popelogan–Victoria arc), which was built on the leading edge of Ganderia (van Staal *et al.* 1998; Zagorevski *et al.* 2007*a*, *c*) (Fig. 5b). Taconic 3 led to assembly of arc volcanic and associated sedimentary rocks into a series of west-dipping structural panels at the leading edge of the AAT. The structures and other evidence supporting these two processes are described in more detail by Zagorevski *et al.* (2007*b*, *c*). The presence of Laurentian-derived detrital zircon populations and Notre Dame arc tonalite clasts in the late Ordovician (450–440 Ma) sandstones (Fig. 9a) overlying the Victoria arc in Newfoundland (McNicoll *et al.* 2001; O'Brien 2003) confirmed that these two opposing arcs were assembled at this time and hence the main tract of the Iapetus Ocean had closed by this stage.

Mélanges (Figs 3 & 9b; Dunnage and Sops Head-Boones Point mélanges), west-dipping, brittle-ductile, sinistral oblique reverse faults, tight overturned folds and localized low-grade metamorphism with stilpnomelane- and/or phengite-bearing greenschist, pumpellyite–actinolite and prehnite-pumpellyite facies assemblages (Fig. 6d) characterize the style of tectonometamorphism of the rocks accreted as a result of Taconic 3 in the AAT (e.g. Bostock 1988; Zagorevski *et al.* 2006, 2007*b*, *c*). Such metamorphic assemblages, particularly the presence of pumpellyite and phengite, suggest relatively low geothermal gradients typical of accretionary complexes formed above subduction zones.

## Salinic orogeny

The Salinic orogeny has only recently been recognized as a significant tectonic event, although its imprint on the geology of the Ganderian core of the northern Appalachians had been recognized as early as the 1960s (e.g. Boucot 1962). Structural investigations in conjunction with extensive radiometric age dating of critical rock units highlighted the regional importance of this orogenic event, which has been documented in Newfoundland (Dunning *et al.* 1990), New Brunswick (van Staal & de Roo 1995) and Maine (West *et al.* 1992; Hibbard 1994). Salinic orogenesis is kinematically and dynamically distinct from the predominantly Early–Middle Devonian Acadian orogeny (van Staal 2007). Structural evidence suggests that Salinic orogenesis was due to sinistral oblique convergence, whereas the obliquity of the Acadian appears predominantly dextral (Holdsworth 1994;

**Fig. 9.** (**a**) Poorly sorted matrix-supported conglomerate (olistostrome) in the Upper Ordovician Point Leamington Formation, Badger Group, a few hundred metres south of the Red Indian Line, SE of Cottrel's Cove, Notre Dame Bay, Newfoundland. Inset: detail of a tonalite cobble clast typical of the Notre Dame arc, indicating that the latter was uplifted due to Taconic 3 arc–arc collision and supplied detritus to the Badger basin. The latter in part is a new, post-Taconic basin associated with the downgoing Salinic slab (oceanic lithosphere of the Tetagouche–Exploits basin). (a) Zagorevski for scale. (**b**) Dunnage mélange with clasts of basalt and sandstone in a scaly cleaved shaly matrix. Hammer for scale.

Hibbard 1994; van Staal 1994; van Staal & de Roo 1995).

Salinic orogenesis was mainly due to a mid-Silurian (430–422 Ma) collision between the Gander margin and composite Laurentia following terminal closure of the intervening, wide Teta-gouche–Exploits backarc basin (Fig. 10a). This basin separated its passive Gander margin (Fig. 3) from Ganderia's leading edge, the Popelogan–Victoria arc (van Staal 1994, 2007). Ganderia had already accreted to Baltica during the Ashgill Shel-vian orogeny at its eastern extremity in the British

Caledonides; hence, the coeval Scandian and Salinic orgenies are both dynamically related to collision between Laurentia and an enlarged and composite Baltica (Valverde-Vaquero *et al.* 2006; van Staal & Hatcher 2009). The Tetagouche–Exploits back–arc basin had started to open during the Early Ordovician (*c.* 475 Ma) as a result of rifting of the Popelogan–Victoria arc (van Staal *et al.* 1998, 2003; Valverde-Vaquero *et al.* 2006). Back-arc spreading responsible for the Tetagouche–Exploits back-arc basin lasted 15–20 Ma and led to a significant oceanic basin, estimated on

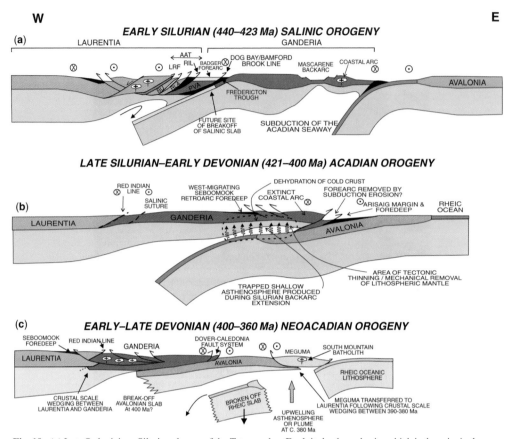

**Fig. 10.** (a) Late Ordovician–Silurian closure of the Tetagouche–Exploits back-arc basin, which is the principal cause of the Salinic orogeny; (**b**) Silurian closure of the Acadian seaway that separated Ganderia and Avalonia, which led to the Acadian orogeny. The early stages of the Acadian (421–417 Ma) were localized preferentially in the back-arc and/or intra-arc basins. Note that most of the fore-arc is assumed to have been subducted to a position beneath the coastal arc. Late stages of the Acadian mainly involve antithetic retro-arc thrusting towards the orogen's hinterland and dextral strike-slip on steep, orogen-parallel faults. (**c**) Accretion of Meguma, which is interpreted to have been accompanied by wedging and breakoff of the downgoing Rheic slab. A new west-dipping subduction zone was probably established outboard of Meguma, necessary to accommodate convergence and Alleghenian collision with Gondwana. Wedging and lower crustal architecture in (b) and (c) is largely based on seismic interpretations (Keen *et al.* 1991; van der Velden *et al.* 2004).

palaeontological and palaeomagnetic evidence to have been in the order of 1000–1500 km wide. The Japan Sea is a close modern analogue (van Staal *et al.* 1991, 2003). Closure of the back-arc basin started at *c.* 450 Ma, immediately after the Popelogan–Victoria arc had accreted to Laurentia at *c.* 455 Ma (van Staal 1994; van Staal *et al.* 2008). Closure was initiated following stepping-back of the west-dipping subduction zone that had rimmed Laurentia during Taconic 3 (Fig. 10a; van Staal *et al.* 1998). Subduction of the backarc's oceanic lithosphere led to sporadic Late Ordovician–Early Silurian (445–435 Ma) arc magmatism in Newfoundland (Fig. 3; 3rd phase of Notre Dame arc, Whalen *et al.* 2006; van Staal

2007) that spatially overlaps the Taconic-related arc and collision-related magmatism (Figs 1 & 10a; van Staal *et al.* 2007). Mafic and felsic plutonic rocks are present in equal proportions and all samples exhibit well-developed negative Nb anomalies on extended element plots (see fig. 6b of Whalen *et al.* 2006). Geochemical plots for these magmatic rocks (Fig. 13) indicate that mafic components exhibit calc-alkaline arc signatures and felsic components exhibit VAG signatures, which are on average more Rb and Y + Nb rich than 2nd stage Notre Dame arc tonalitic rocks (Fig. 7c). In their $\varepsilon_{Nd}(T)$ values these plutons exhibit both contaminated and juvenile signatures (Whalen *et al.* 2006). Combined, these characteristics suggest this magmatism is subduction-related.

Equivalents of this magmatism (mainly volcanic rocks) occur also in northern New Brunswick (*c.* 430 Ma volcanic rocks of the Weir Formation, Wilson *et al.* 2008) and adjacent Quebec (*c.* 433 Ma volcanic rocks of the Pointe aux Trembles and Lac Raymond formations, David & Gariepy 1990). In northern Maine, this phase of arc magmatism is represented by the *c.* 443 Ma Attean pluton in the Chain Lakes massif (Gerbi *et al.* 2006*b*) and the coeval Quimby volcanic rocks (Moench & Aleinikoff 2003). Correlatives also occur further south in the Bronson Hill arc.

## Salinic accretionary phase (445–430 Ma)

Closure of the Tetagouche–Exploits backarc basin led to punctuated latest Ordovician–Early Silurian accretion of seamounts and isolated back-arc crustal ribbons (e.g. van Staal *et al.* 2003, 2008) into a progressively expanding subduction complex (Fig. 10a), which is exceptionally well-preserved in northern New Brunswick (van Staal *et al.* 2001, 2003, 2008), but poorly-preserved and/or largely covered by younger Siluro–Devonian sedimentary rocks in Newfoundland and Maine. Growth of the subduction complex and progressive stepping back of the subduction zone was concomitant with an eastward migration of relatively deep marine basin(s) situated in the arc-trench gap (Matapedia-Badger basin) and its adjacent hinterland during the Late Ordovician–Early Silurian (Elliot *et al.* 1989; Williams *et al.* 1995; van Staal & de Roo 1995; van Staal 2007; Pollock *et al.* 2007). Zeolite, prehnite–pumpellyite, pumpellyite–actinolite, greenschist and blueschist facies metamorphism (Richter & Roy 1974; van der Pluijm 1986; van Staal *et al.* 1990, 2008; O'Brien 2003; Wilson 2003) accompanied deformation of the rocks in the subduction complex and adjacent fore-arc basin.

Deformation during this pre-collisional stage of the Salinic was mainly restricted to the arc-trench gap region and led to predominantly east- or south-directed ductile-brittle thrusting, localized folding and mélange formation (Fig. 11a, b). Tectonic transport direction in Newfoundland locally varied from SE to south or SW (e.g. O'Brien *et al.* 1997; O'Brien 2003), which possibly reflects the buttressing effects of relatively rigid salients in the modified Laurentian margin, such as the Notre Dame Bay flexure. Deformation in the Brunswick subduction complex in New Brunswick was mainly due to underthrusting and underplating, which led to a general, non-coaxial deformation that was accompanied by some of the best developed high-pressure low-temperature metamorphic tectonites in the Appalachians (van Staal *et al.* 1990, 2008).

Deformation in the Brunswick subduction complex was closely monitored by tectonism in the adjacent Matapedia fore-arc basin, which exhibits several stages of localized Llandoverian deformation (Fig. 11d, e) and sedimentary onlap onto exhumed parts of the accretionary wedge (van Staal & de Roo 1995; Dimitrov *et al.* 2004; van Staal *et al.* 2008). As in New Brunswick, ample evidence is preserved for Llandoverian deformation with SE-directed tectonic transport in the same tectonic setting in Newfoundland. Fossil-bearing olistostromes (e.g. Joey's Cove Mélange), which had developed in front of seafloor-breaching thrusts, were partly overridden and deformed into mélange by the same faults (van der Pluijm 1986; P. F. Williams *et al.* 1988). Elsewhere, Llandovery thrust-related deformation is tightly constrained by syn-tectonic intrusions and/or stratigraphy. The intrusives cut the enclosing fault-related rocks such as mélange, but were also deformed by subsequent increments of the responsible progressive deformation (Zagorevski *et al.* 2007*b*). The regional extent of Llandoverian Salinic deformation is not well known, as it was gradually followed by more widespread deformation during the Wenlock (Figs 11c & 12a), which produced, at least in part, similar structures (O'Brien 2003).

## Salinic collision-related orogenesis (430–422 Ma)

Salinic deformation became regional in extent near the Llandovery–Wenlock boundary (*c.* 429 Ma), now also affecting rocks of the Gander margin foreland, mainly underlain by Cambrian–Ordovician rocks of the Gander Group and correlatives elsewhere, and rocks in the Laurentian hinterland. Progressive widening of the Salinic deformation zone probably reflects the onset of full-scale collision between the Gander margin and Laurentia, which sutured along the Dog Bay Line (DBL) in Newfoundland (Williams *et al.* 1993, Valverde-Vaquero *et al.* 2006) and the Bamford Brook-Liberty-Orrington fault system (Figs 1 & 9a) in New Brunswick and adjacent New England (Ludman *et al.* 1993; van Staal 2005, 2007; Hibbard *et al.* 2006). The suture separates two distinct Middle Ordovician–Silurian sequences (Fig. 3), which are particularly obvious in Newfoundland where Wenlock terrestrial rocks (Botwood Group) are separated from coeval marine sediments (Indian Island Group), each with a distinctly different provenance (Pollock *et al.* 2007). The Lower Silurian provenance of the marine foreland sedimentary rocks east of the suture is Ganderian, whereas rocks deposited in the arc-trench gap west of the suture have a predominantly Laurentian provenance and

**Fig. 11.** (**a**) Uppermost Ordovician–Lower Silurian mélange situated structurally beneath the ophiolitic Fournier block, Brunswick subduction complex, northern New Brunswick. Pen for scale. (**b**) Close-up of mélange matrix. (**c**) Nearly isoclinal pre-Devonian, Salinic folds in Early Llandovery calcareous mudstones of the Matapedia Group (part of Matapedia fore-arc basin associated with the Brunswick subduction complex) near New Brunswick–Maine border. Bottom of photo is *c*. 1.5 m wide. Photo courtesy of R. Wilson of the New Brunswick Geological Survey. (**d**) Angular unconformity between folded (Salinic) Lower Silurian (Llandovery–Wenlock boundary) limestones of the La Veille Formation and upper Silurian calcareous conglomerate of the Simpsons Field Formation (Dimitrov *et al.* 2004), Limestone Point, northern New Brunswick. S. McCutcheon of the New Brunswick Geological Survey outlines bedding (S) and cleavage (Sc) in the La Veille Formation. Hammer parallel to trace of bedding in calcareous conglomerate of Simpsons Field Formation. Cleavage overprints unconformity and is Acadian. (**e**) Same outcrop and rocks (broken

also include volcanic rocks. The lithological differences between the Silurian rocks east and west of the suture are more subtle in Maine and parts of New Brunswick, because the fore-arc basin remained largely marine and was probably simple and sloped (Dickinson & Seely 1979) such that its sediments could spill over from the fore-arc basin into the adjacent trench (Fredericton trough) immediately east of the suture. A distinctive feature of the Fredericton trough is that Silurian volcanic rocks are absent, while they are locally present in the fore-arc basin.

The Bamford Brook-Liberty-Orrington fault system that marks the suture in northern New England was originally a major SE-directed thrust that had emplaced Cambrian–Ordovician metamorphic rocks of the Miramichi and Liberty Orrington belts above Lower Silurian rocks of the Fredericton Trough (e.g. Ludman et al. 1993; Tucker et al. 2001; van Staal et al. 2003). Thrust movements and penetrative ductile deformation in the Fredericton trough took place before intrusion of the Upper Silurian, 422 ± 3 Ma Pocomoonshine gabbro (West et al. 1992) and 418 ± 1 Ma Lincoln syenite (West et al. 2003) in Maine. In New Brunswick, the Bamford Brook fault and strongly folded turbidites of the Fredericton trough were cut by the 414–411 Ma Pokiok Batholith (McLeod et al. 2003). Furthermore, the 420 ± 5 Ma Mohannes granite (Fyffe & Bevier 1992), which intruded the Ordovician rocks of the Cookson Group (St. Croix belt) that underlies the Fredericton trough, was either synchronous or postdated the earliest deformation in the poly-deformed Cookson Group. The latter is a correlative of the Gander Group in eastern Newfoundland (van Staal & Fyffe 1995) and hence forms part of the Gander margin (Fig. 3, van Staal 1994). Unpublished U–Pb ages (V. McNicoll, pers. comm.) revealed that Ordovician rocks of the Miramichi highlands in the hanging wall of the Bamford Brook fault were metamorphosed locally up to migmatite grade, during the Silurian. In addition, a detailed structural analysis by Park & Whitehead (2003) documented early east-vergent Silurian thrusting and localized recumbent folding in the Fredericton trough, followed by steepening of the structures prior to- or coeval with Early Devonian dextral transpressive deformation.

The early movement history accommodated by the DBL in Newfoundland is poorly known, although it involved mélange formation prior to

dextral transpressive deformation (Williams et al. 1993). Several lines of evidence (e.g. metamorphism) suggest that the earliest motion on the DBL and mélange formation involved Silurian reverse movement, which was probably synthetic with the SE-directed Silurian thrust faults that deform the adjacent Badger Group and associated olistostromes (Williams et al. 1993; Valverde-Vaquero et al. 2006; Pollock et al. 2007). The DBL and associated transpressive structures were cut by the late Silurian–Early Devonian (424–411 Ma) phases of the Mount Peyton intrusive suite (Dickson et al. 2007). In general, the rocks on both sides of the DBL underwent a significant phase of Silurian deformation and metamorphism prior to the Devonian (e.g. Dunning et al. 1990). Gander Group rocks east of the DBL in eastern Newfoundland were strongly folded, foliated and metamorphosed up to migmatite grade at c. 425 Ma (d'Lemos et al. 1997; Scofield & d'Lemos 2000) during a regional sinistral transpressive deformation (Holdsworth 1994). Silurian deformation and metamorphism, likewise tightly constrained by U–Pb ages of synkinematic or cross-cutting plutons, migmatites and high-grade metamorphic minerals, has also been documented in Cambrian–Ordovician rocks in southern and southwestern Newfoundland (Dunning et al. 1990; van Staal et al. 1994).

Salinic orogenesis led to significant uplift (Figs 11d & 12b) and locally rapid exhumation of rocks that were incorporated in the hinterland of the orogen, both in the northern New England (Maine, New Brunswick and Quebec) and Newfoundland Appalachians (van Staal & de Roo 1995; van Staal et al. 2007, 2008). Unconformable deposition of upper Llandovery–Wenlock terrestrial clastic rocks above Orovician tectonites, originally buried and metamorphosed during Ordovician Taconic shortening of the Notre Dame arc in Newfoundland, took place during and immediately after probably faulting-associated late Llandovery (c. 430 Ma) uplift and unroofing (van Staal et al. 2007). An excellent example is represented by the regionally extensive Lloyd's River-Hungry Mountain-Lobster Cove fault system (Figs 1 & 3), which originally was a major SE-directed Ordovician thrust system (Fig. 6b) that accommodated underthrusting and accretion of ophiolites of the AAT (Lissenberg et al. 2005a; Lissenberg & van Staal 2006; Zagorevski et al. 2006) to Laurentia. This fault system was reactivated during the Silurian as a SE-directed reverse fault (Dean & Strong 1978),

---

**Fig. 11.** (*Continued*) formation) as (f) unconformably overlain with erosional contact by red conglomerate of the Upper Llandovery (C5–C6) Weir Formation. Red beds were folded and cleaved together with the underlying calcareous sandstones during both the Salinic and Acadian orogenies. (**f**) Folded lower Llandovery calcareous sandstones and shales, which were earlier deformed (early Salinic) into broken formation. Bottom of photo is c. 2m wide.

**Fig. 12.** (**a**) Narrow, Salinic SE-directed shear zone (thrust) near Baie Verte, northern Newfoundland. SE is to the right. Hammer for scale. (**b**) Greenschist facies (cold) mylonite with boudinaged quartz veins marking the Lobster Cove fault near Lobster Harbour, Sunday Cove Island, Newfoundland. Blue ballpoint for scale. Shearbands (poorly visible in photograph) and other shear sense indicators indicate that thrust movement was to the SE over the underlying Silurian (Wenlock) red beds. (**c**) Folded Silurian red beds in the immediate footwall of the Lobster Cove fault. A. Zagorevski for scale. (**d**) Pristine, cross-bedded red beds in footwall of Lobster Cove fault. (**e**) Large southerly-overturned $F_2$ recumbent fold in an apophyse of the $418 \pm 2$ Ma Rose Blanche granite. The granite veins probably intruded during $F_2$ (van Staal *et al.* 1992; Valverde-Vaquero *et al.* 2000) because they cut co-planar south-verging tighter $F_2$ isoclines (out of view), $S_1$ (layer-parallel) and the $S_2$ axial plane foliations in the sillimanite schists and gneisses, east of Port aux Basques,

such that unmetamorphosed Wenlock red beds (Fig. 12c, d) and volcanic rocks now locally occur in the footwall of the fault, whereas uplifted and unroofed Cambrian and Ordovician metamorphic rocks occur in the hanging wall (Fig. 12b). Detritus of the hanging wall occurs in the footwall's Silurian clastic rocks. Hence, the red beds date the deformation associated with Silurian reverse faulting (Zagorevski *et al.* 2007*b*).

Widening of the deformation zone into the retro-arc hinterland of the orogen also led to significant reactivation of Taconic structures in the Humber margin of Laurentia and renewed imbrication and metamorphism in a relatively narrow belt near the Baie Vverte Brompton Line (Taconic suture), both in Newfoundland (Cawood *et al.* 1994; Brem *et al.* 2007) and Quebec (Castonguay *et al.* 2007). Salinic recrystallization and metamorphism were apparently so penetrative in this belt that it may have destroyed nearly all evidence for earlier Taconic metamorphism. Strain localization in this belt may be due to a combination of fabric and thermal softening; the latter possibly related to heat generated during slab-breakoff and/or lithospheric thinning prior to this retro-arc deformation.

Temporal-compositional changes in voluminous 433–425 Ma magmatism west of the RIL have been documented by Whalen *et al.* (2006) and interpreted in terms of a slab-breakoff model of the oceanic segment attached to the downgoing Tetagouche-Exploits back-arc lithosphere (Salinic slab). The compositional spectrum includes calc-alkaline to tholeiitic basalts and also non-arc like mafic compositions and felsic magmatism includes both VAG and WPG type suites (Fig. 13). In their $\varepsilon_{Nd}(T)$ values most of these plutons exhibit juvenile (+1 to +6) signatures, though there are some slightly negative values. In general, these geochemical features substantiate that breakoff of the Salinic slab led to sequential tapping of a combination of asthenospheric, lithospheric and crustal sources both proximal to the RIL and in the hinterland of the orogen. Silurian magmatism overlapped with deformation, is locally associated with high-grade metamorphism and migmatites and appears to become younger towards the foreland. Silurian magmatism east of the RIL and west of the DBL in west-central Newfoundland includes parts of the voluminous Mt. Peyton, Fogo (Currie 2003) and Hodges Hill suites (O'Brien 2003). It exhibits a similar spectrum

of felsic compositions to that west of the RIL but its mafic rocks are limited to transitional calc-alkaline to tholeiitic basalt compositions (Fig. 14). This magmatism, like that west of the RIL, is attributed to breakoff of the Salinic slab (Fig. 10a). However, the magmatism occurring east of the RIL and west of the DBL received only input from depleted mantle sources, which contrasts with coeval magmatism occurring west of the RIL. The latter probably was situated immediately above the upwelling asthenosphere that replaced the sinking Salinic slab, whereas the former was situated near the eastern boundary of the area affected by slab-breakoff. In NW New Brunswick and Gaspé, Quebec, most equivalent Silurian magmatism is obscured by younger cover sequences. Our dataset only includes a small number of mafic rock analyses, which exhibit transitional calc-alkaline to continental tholeiitic basalt compositions (Fig. 15a, b).

## Acadian orogeny

The Acadian orogeny is generally attributed to accretion of Avalonia to Laurentia (e.g. Bird & Dewey 1970; Bradley 1983), a causative mechanism retained herein. Points of contention remain the start and duration of Acadian orogenesis, the location of the suture and the polarity of subduction (Bradley 1983; Robinson *et al.* 1998; Tucker *et al.* 2001; van Staal 2007). The discussions particularly hinge on the position of the suture between Avalonia and Laurentia, the leading edge of which at this stage was represented by Salinic-accreted Ganderia. Several lines of evidence suggest that the suture lies along the Hermitage Bay–Dover fault in Newfoundland and Caledonia fault in New Brunswick (Fig. 1) (Samson *et al.* 2000; Barr *et al.* 1998, 2003; van Staal *et al.* 2004; van Staal 2005; Lin *et al.* 2007). The latter fault is sinistrally offset by the Oak Bay fault (Fig. 1) to continue south of Grand Manan Island beneath the Gulf of Maine and to reappear on land as the Bloody Bluff fault in Massachusetts (Hibbard *et al.* 2006). Part of the problem previously identifying with this suture was that Ganderia and Avalonia both have a Neoproterozoic arc-like basement, which could be separated mainly on the basis of detailed isotopic (Kerr *et al.* 1995; Whalen *et al.* 1996*a*; Samson *et al.* 2000; Potter *et al.* 2008) and tectonothermal

---

**Fig. 12.** (*Continued*) south coast of Newfoundland, which form part of the Gander margin. These Acadian folds occur all along Newfoundland's south coast and are interpreted to have formed during the early stages of collision between Avalonia (lower plate) and Laurentia's leading edge (represented by Ganderia at this stage). S. Colman-Sadd of the Newfoundland Geological Survey for scale. (**f**) Silurian mafic dyke complex (sheeted?) in the Mascarene basin, southern New Brunswick. P. Valverde-Vaquero for scale.

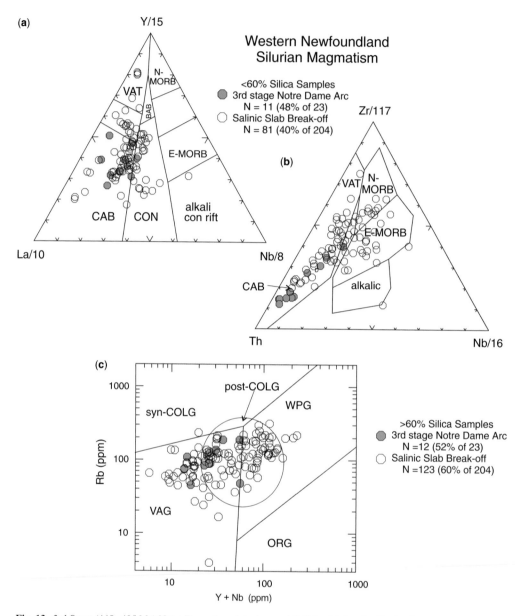

**Fig. 13.** 3rd Stage (445–435 Ma) Notre Dame Arc plutonism and Salinic slab break-off related magmatism (433–423 Ma) in western Newfoundland. Diagrams (**a, b, c**) and abbreviations as in Figure 7. See Appendix 1 for data sources.

studies (Barr & White 1996). More significantly, the Early Palaeozoic tectonic histories of these two terranes are markedly different (Fig. 3) (van Staal *et al.* 2004; van Staal 2007; Lin *et al.* 2007).

The polarity and timing of the Acadian orogeny (Fig. 10b) is best constrained by the following three observations: (1) geometry of Silurian arc and backarc magmatism (442–423 Ma) on the trailing edge of Ganderia facing Avalonia. Backarc is situated west of the arc (van Staal 2007); (2) tectonic slivers, characterized by 420–416 Ma subduction-related fore-arc high pressure–low temperature metamorphism, occur immediately to the east of the coastal arc in southern New Brunswick, very close to the Ganderia–Avalonia suture (White *et al.* 2006); and (3) Silurian shallow water shelf sediments (Arisaig Group) deposited on Avalonia were overlain by a thick sequence of foreland

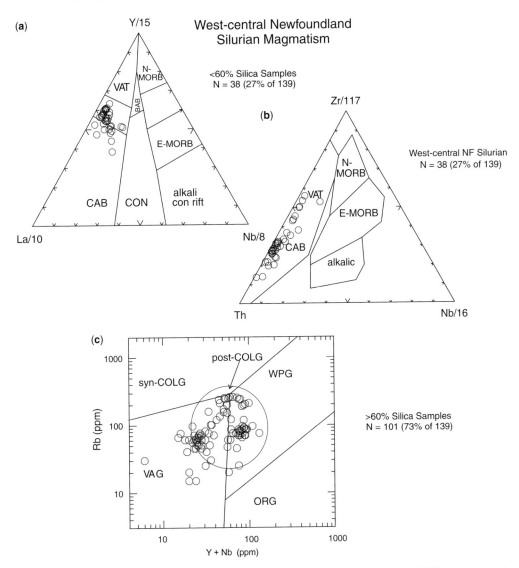

**Fig. 14.** Silurian plutons emplaced between the Red Indian Line (RIL) and the Dog Bay Line (DBL) in west-central Newfoundland. Diagrams (**a, b, c**) and abbreviations as in Figure 7. See Appendix 1 for data sources.

basin deposits during the latest Silurian–Early Devonian, due to tectonic loading of Avalonia (Waldron *et al.* 1996*a*). Avalonia, thus, was the lower plate and subduction was to the west or NW beneath Laurentia (Fig. 10b). The Silurian Acadian arc- and back-arc volcanic rocks (Fig. 12f) can be traced from southern Newfoundland (e.g. La Poile Group and Burgeo granite) through Cape Breton Island (e.g. Money Point Group–Sarach Brook suite and correlatives) to southern New Brunswick (Kingston and Mascarene terranes) (Barr & Jamieson 1991; Price *et al.* 1999;

Barr *et al.* 2002; Miller & Fyffe 2002; Lin *et al.* 2007) and beyond into Massachusetts (Hepburn 2008). They have been recently named the Kingston arc terrane (Barr *et al.* 2002; White *et al.* 2006; van Staal 2007), but were previously referred to as the coastal (volcanic) arc (Bradley 1983), a name we prefer because the arc- and back-arc volcanic rocks define a belt that closely follows the Atlantic coast of eastern North America (Fig. 2).

The timing of tectonic loading of Avalonia's leading edge suggests that the Avalonia–Laurentia collision had started during the Late Silurian at

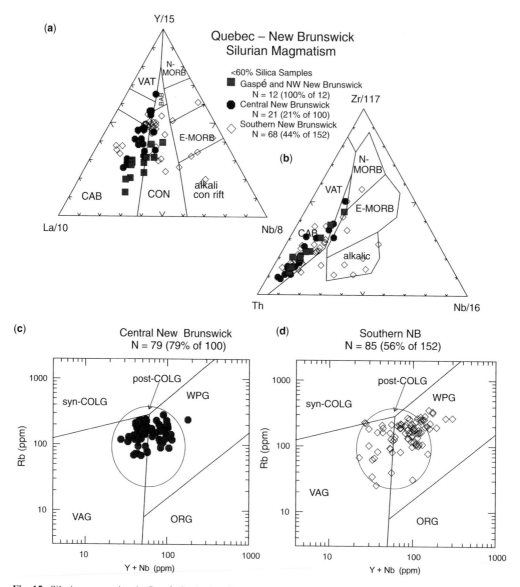

**Fig. 15.** Silurian magmatism in Gaspé (Quebec) and New Brunswick. Diagrams (**a, b, c, d**) and abbreviations as in Figure 7. See Appendix 1 for data sources.

*c.* 421 Ma in maritime Canada, which is virtually coeval with inversion of back-arc and/or intra-arc basins, such as the Mascarene (Fig. 12f) and La Poile basins in New Brunswick and Newfoundland, respectively (O'Brien *et al.* 1991; Fyffe *et al.* 1999) following shut-off of coastal arc volcanism. Deformed and foliated Mascarene back-arc basin rocks in southern New Brunswick, which include volcanic rocks as young as $423 \pm 1$ Ma (van Wagoner *et al.* 2002), were cut by Upper Silurian phases of the St. George batholith (e.g. Utopia and

Bocabec plutons, Fyffe *et al.* 1999). The latter, and consanguineous phases nearby, yielded ages of 423–421 Ma (McLaughlin *et al.* 2003), suggesting a very rapid transition from back-arc and/or intra-arc extension to compression following the onset of collision.

The large collection of new radiometric ages indicates that Salinic- and Acadian-related arc magmatism partly overlapped in time during the Early Silurian (Fig. 10a, b). Hence, while the coastal arc was in the early stages of being constructed upon

Ganderia's trailing edge, the Gander margin's leading edge was in the final stages of being subducted beneath Laurentia (van Staal 2007). It is thus feasible that pre-Acadian transtension-related deformation in the coastal arc (e.g. Doig *et al.* 1990; McLeod *et al.* 2001) induced by oblique subduction of the narrow oceanic seaway that separated Ganderia and Avalonia (Valverde-Vaquero *et al.* 2006) overlaps in time, although not in space (Fig. 2) with the waning stages of Salinic orogenesis in the other parts of Ganderia (e.g. Hibbard 1994).

More problematic is defining the time span of the Acadian orogeny, because the retro-arc deformation front (see below) and associated foreland basin progressively migrated into the Laurentian hinterland during the Devonian (Bradley *et al.* 2000). It reaches its final limit at the Appalachian structural front in the early Middle Devonian at a time when Meguma had already started to collide with Laurentia (Neoacadian, see below).

## Late Silurian (421–416 Ma), early Acadian orogenesis

Structures formed during the early stages of the Acadian occur close to the suture in the region mainly underlain by rocks of the coastal arc-backarc system (Figs 1 & 2), that is, near the Atlantic coast. Metamorphic tectonites related to Acadian subduction have been preserved in a narrow belt underlain by the Pocologan metamorphic suite in southern New Brunswick (White *et al.* 2006). Correlatives may be hidden beneath the seafloor of the Gulf of St. Lawrence, offshore Newfoundland's south coast, but the overall scarcity of identifiable Palaeozoic arc-trench gap rocks between the coastal arc and Avalonia either suggests major subduction erosion of the fore-arc block following establishment of a flat-slab during the Early Devonian (Murphy *et al.* 1999, see below) shortly after the onset of collision (Fig. 10b), or to structural excision related to orogen-parallel transcurrent faulting. Subduction erosion or widening of the subduction channel may have abraded and dragged down most of the fore-arc's crust in the hanging wall of the subduction zone into the mantle to a position below the arc. Such underthrusting of crustal rocks may be important during subsequent Acadian magmatism (e.g. Kay *et al.* 2005, see below). Considering the lack of preservation of arc-trench gap rocks in general, it is not surprising that structures synthetic with west-directed Acadian subduction are relatively rare, although they have been imaged at depth on the seismic profile crossing the Acadian suture in southeastern Newfoundland (van der Velden *et al.* 2004). South- or SE-vergent early Acadian recumbent or overturned folds (Fig. 12e)

and associated thrust or reverse faults have also been mapped along Newfoundland's south coast in Lower Palaeozoic rocks of the Gander margin (Colman-Sadd 1980; Blackwood 1985; van Staal *et al.* 1994). Acadian folds deform an earlier layer-parallel metamorphic foliation, which could be an early Acadian or Salinic structure (van Staal *et al.* 1992). Development of the recumbent or overturned folds and associated shear zones partly overlapped with high-grade metamorphism, dated at *c.* 418–414 Ma (Burgess *et al.* 1995; van Staal *et al.* 1994), and were intruded late syn-kinematically by Upper Silurian (421–418 Ma) biotite ± muscovite granites such as the Gaultois (*c.* 421 Ma, Piasecki 1988; Dunning *et al.* 1990) and Rose Blanche plutons (*c.* 418 Ma, van Staal *et al.* 1992; Valverde-Vaquero *et al.* 2000). Correlative Gander margin rocks in southern New Brunswick (St. Croix belt) also may contain early southerly overturned structures (Fyffe *et al.* 1991), although the dominant Acadian structural transport direction here and in adjacent Maine appears to be towards the NW (Tucker *et al.* 2001; Castonguay *et al.* 2003). Such a movement picture is consistent with the asymmetry of structures developed during the latest Silurian inversion of the La Poile and Mascarene intra-arc and/or back-arc basins in Newfoundland and New Brunswick/Maine, respectively (O'Brien *et al.* 1991; Fyffe *et al.* 1999; Tucker *et al.* 2001). Early Acadian shortening thus appears to have produced bi-vergent structures, recording both SE and NW-directed structural transport and was preferentially localized in the back-arc region.

Early Acadian metamorphism varied regionally, but Gander margin rocks in southern Newfoundland and New Brunswick (St. Croix belt) generally display Barrovian to Buchan style metamorphism (Colman-Sadd 1980; Fyffe *et al.* 1991), although locally more deeply buried rocks have been exhumed (e.g. Burgess *et al.* 1995). Metamorphic isograds generally seem to dip shallowly to the NW or north and high-grade rocks have been juxtaposed locally with low-grade rocks along major faults, suggesting significant and complex post-metamorphic movements along some of the belt-bounding faults.

## Early Devonian (416–400 Ma) late Acadian orogenesis

Acadian orogenesis continued uninterrupted into the Early Devonian. Structural analyses suggest it involved a combination of orogen-normal shortening and dextral movements along orogen-parallel structures (e.g. Hibbard 1994; van Staal & de Roo 1995; Malo & Kirkwood 1995). The structures formed during crustal shortening include upright or steeply inclined folds and high-angle reverse faults. The latter are generally west-vergent, hinterland-directed structures that progressively become younger towards the NW in northern New England and adjacent New Brunswick (Tucker *et al.* 2001; Wilson *et al.* 2004), which is well constrained by migration of a retro-arc foredeep over time (Bradley *et al.* 2000; Bradley & Tucker 2002). Early Devonian reverse faults that overprint Salinic structures have also been observed in central Newfoundland (Rogers *et al.* 2005; Valverde-Vaquero & van Staal 2001; Zagorevski *et al.* 2007*a*, *b*), but the link between migration of the Acadian deformation front and an associated retro-arc foredeep cannot be established or tested in Newfoundland. Devonian sedimentary rocks are rarely preserved and are exposed only in a few isolated areas. Movements on the final Acadian deformation front in westernmost Newfoundland are tightly constrained by deformed foreland basin rocks of the Lochkovian (416–411 Ma) Clam Bank Formation and undeformed Emsian (407–397 Ma) conglomerate of the nearby Red Island Road Formation, which was deposited immediately in front of the Round Head thrust (Stockmal & Waldron 1993; Burden *et al.* 2002; Quinn *et al.* 2004). Acadian deformation lasted longer in Quebec and continued into the Middle/Upper Devonian (380–370 Ma) (Bradley *et al.* 2000). This late stage of Acadian deformation overlapped in age with Neoacadian orogenesis in Meguma (see below).

A large upper amphibolite-facies metamorphic nappe (Meelpaeg allochthon) is exposed in southern and south-central Newfoundland. Its extent and geometry at depth was largely constructed on the basis of seismic profiles (van der Velden *et al.* 2004). Rocks of the allochthon reached peak metamorphism at 418–414 Ma (van Staal *et al.* 1994; Burgess *et al.* 1995; Valverde-Vaquero *et al.* 2000) and must have travelled a fair distance and for a considerable time interval, because they are generally emplaced above low-grade rocks and locally were even transported above unmetamorphosed Lower Devonian sediments (van Staal *et al.* 2005). Attainment of Acadian high-grade metamorphism in these Gander margin rocks so shortly after the onset of the Acadian orogeny suggests that these rocks were already hot, either as a result of being situated in the Acadian backarc (see below) prior to collision and establishment of the flat-slab or were already metamorphosed to high-grade conditions during the immediately preceding Salinic orogeny and remained buried and relatively hot until their involvement in the Acadian. The latter is consistent with the presence of a well-developed pre-418 Ma $S_1$ metamorphic layering or schistosity (van Staal *et al.* 1992, 1994) and age constraints on Salinic metamorphism

(*c.* 425 Ma) from correlative Gander margin rocks along strike (d'Lemos *et al.* 1997; Scofield & d'Lemos 2000). The western tectonic boundary of the Meelpaeg allochthon, the Cape Ray Fault-Victoria Lake shear zone, is generally a major ductile, moderately to shallowly east-dipping, thrust-sense shear zone. Locally, where the fault zone had steepened significantly, it also has accommodated dextral oblique motion (Dubé *et al.* 1996; Valverde-Vaquero & van Staal 2001; van der Velden *et al.* 2004). Mylonites yielded biotite and muscovite cooling ages between 400 and 390 Ma (Burgess *et al.* 1995; Dubé *et al.* 1996), indicating that the high-grade rocks were mostly exhumed by the end of the Early Devonian. These thrust-related structures were cut and stitched by Middle Devonian (*c.* 386 Ma) granite (Dubé *et al.* 1996), supporting other evidence that Acadian orogenesis was mainly confined to the Early Devonian.

Acadian deformation reoriented and modified the earlier established Taconic and Salinic structural architecture significantly. Shallow east-dipping thrusts cut and offset major west-dipping structures at depth, such that it produced a wedge geometry, which was imaged on seismic profiles (van der Velden *et al.* 2004). On the surface, Taconic and/or Salinic folds were refolded (Fig. 11c), whereas the predominantly shallowly west-dipping fault panels were folded into steep or westerly overturned structures and/or cut by new east-dipping low-angle reverse faults (Zagorevski *et al.* 2007*b*). Regional shear associated with west-directed transport locally rotated and overturned older Salinic-generated, west-dipping fault panels into markedly east-dipping overturned panels. Overturning also rotated the regional cleavage into a new orientation such that it now consistently has a shallower dip than the fault panels (Dean & Strong 1978; Lafrance & Williams 1992).

The progressive hinterland migration of the Acadian deformation front was interpreted by Murphy *et al.* (1999) as being due to establishment of a west-dipping 'flat-slab', analogous to the Laramide in the western USA and the present-day Andes in central Chile and Argentina (Kay & Abruzzi 1996) (Fig. 10b). The cause of the inferred Acadian flat-slab setting is likely related in some fashion to prolonged subduction of Avalonia beneath Laurentia, the reason of which is contentious. However, Avalonia's A–subduction and resultant widening of the Acadian deformation zone towards the hinterland during the Early Devonian coincides with rapid southerly palaeolatitudinal motion of Laurentia (van Staal *et al.* 1998, their fig. 8). Hence, construction of the wide Acadian mountain belt may be to some extent analogous to formation of the Andes and North American Cordillera as a result of the Mesozoic and younger westward (trenchward) motion of the America's following the diachronous opening of Atlantic Ocean.

Acadian convergence probably was strongly partitioned, because the reverse hinterland-directed movements were coeval with dextral transcurrent motion on steeply dipping faults (King & Barr 2004) near the suture with Avalonia (Holdsworth 1994; Doig *et al.* 1990; Nance & Dallmeyer 1993; Park *et al.* 1994). Progressive steepening of the Acadian reverse faults over time probably facilitated transcurrent motion in the hinterland as well (e.g. van Staal & de Roo 1995).

## Late Silurian-Middle Devonian (Acadian) magmatism

Late Silurian (423–416 Ma) magmatism is voluminous both in Newfoundland, New Brunswick, Gaspé and Maine (Figs 1 & 2) and represents a high-flux event, although we will argue below that not all rocks of this age are related to the Acadian orogeny. Upper Silurian magmatic rocks occur in two geographically separated belts, which are both represented by volcanic and plutonic rocks. The most westerly belt is spatially associated with the part of Ganderia where Salinic structures and metamorphism are well-developed, whereas the easterly belt is generally not. In Newfoundland, the westerly belt is represented by such igneous rocks as the Late Silurian phases of the 424 ± 2 Ma Mt. Peyton intrusive suite (Dickson *et al.* 2007), 422 ± 2 Ma Fogo Island pluton or batholith (Aydin 1995), the 422 ± 2 Ma Port Albert bimodal dikes (Elliot *et al.* 1991), 423+3/−2 Ma Stony Lake volcanic rocks (Dunning *et al.* 1990) and other related volcanic rocks (420–417 Ma, V. McNicoll, pers. comm. 2006).

Silurian magmatism emplaced west of the DBL in Newfoundland, NW New Brunswick, Maine and adjacent Gaspé has been discussed previously and is illustrated in Figures 13, 14 and 15. The equivalent Late Silurian magmatic event in central New Brunswick (Fig. 15a–c) is mainly exposed in the Miramichi highlands. This magmatic phase includes the 418 ± 1 Ma Mt. Elizabeth and 417 ± 1 Ma North Pole Stream plutonic suites (Bevier & Whalen 1990) and volcanic rocks such as those of the *c.* 423 ± 3 Ma Benjamin and Bryan Point formations (Walker *et al.* 1993). These rocks generally postdate structures formed during the Salinic orogeny, but always predate development of Acadian structures in the areas they occur, that is, they formed NW of the migrating Acadian deformation front (Bradley & Tucker 2002). Mafic rocks, which comprise *c.* 21% of this magmatic suite, exhibit calc-alkalic to continental

tholeiitic and non-arc features (Fig. 15a, b). Felsic samples (Fig. 15c) form two sample clusters, one within the VAG field and the other within the WPG. Most of the latter Y + Nb-enriched samples are from the *c.* 418 Ma Mt. Elizabeth complex, which Whalen (1990) documented as being comprised of contemporaneous I-, A- and mafic-type suites. The spectrum and proportions of mafic and felsic compositions is remarkably similar to both western and west-central Newfoundland Silurian magmatism (Figs 13 & 14) and, for this reason, we interpret these 423–416 Ma magmatic rocks as a final, post-kinematic pulse of magmatism related to breakoff of the Salinic slab (cf. Whalen *et al.* 2006) Hence, if correct, they are not related to the Acadian orogeny.

The Upper Silurian (423–416 Ma) magmatic rocks that occur east of the Fredericton trough in southern New Brunswick, adjacent Maine and east or south of the DBL in Newfoundland are spatially associated with the Acadian suture and are related to the early stages of the Acadian collision between Avalonia and the leading edge of Laurentia (represented at this stage by the accreted Ganderia), since they generally formed syn- to post-kinematically with respect to the associated early Acadian structures and metamorphism in this part of the Canadian Appalachians (see above). The relative abundance of isotopic and geochemical data in Newfoundland (see below) revealed that this magmatic suite can be subdivided into distal and proximal subgroups based on their nearness to the DBL. This subdivision probably correlates with the different crustal evolutions experienced by the inboard (distal group) and outboard (proximal) segments of the Gander margin during the Ordovician and Silurian prior to Acadian orogenesis (see van Staal 1994; Valverde-Vaquero *et al.* 2006). The plutonic rocks of the proximal subgroup in Newfoundland, which were emplaced immediately east of the DBL, are comprised of 27% mafic rocks; the rest are felsic. The mafic rocks exhibit mainly calc-alkaline arc and minor arc tholeiite compositions (Fig. 17a, b), whereas the felsic plutonic rocks plot exclusively within the VAG field. The time equivalent mafic rocks of the distal subgroup in part have similar compositions, but also include a significant proportion of rocks that exhibit continental tholeiite and E-MORB characteristics, suggesting a significant contribution from the lithospheric mantle (Fig. 17a, b). Felsic plutonic rocks from the distal subgroup range to more Rb- and Y + Nb-enriched compositions, such that they straddle the boundaries between the fields for VAG, WPG and the syn-COLG (Fig. 17c). The latter characteristics suggest that these granites received a greater contribution from evolved high-Rb crust than the granites emplaced in proximity to the DBL in Newfoundland (Figs 13c, 14c & 17), which is consistent with less tectonic reworking of the Gander margin crust towards the east during Ordovician-Silurian tectonism. Protoliths of the plutonic rocks of the proximal subgroup probably involved depleted-mantle-derived Ordovician arc and associated sedimentary rocks originally formed near the active leading edge of Ganderia (Penobscot & Popelogan–Victoria arcs and their basement) prior to opening and closing of the Tetagouche–Exploits back-arc basin and accretion of Ganderia to Laurentia (van Staal *et al.* 1998; Zagorevski *et al.* 2007*a*). This interpretation is

supported by the generally positive (juvenile) $\varepsilon_{Nd}(T)$ values of the proximal plutons and negative (old crust contaminated) eNd (T) signatures of the distal plutons (Kerr *et al.* 1995).

In southern New Brunswick, the Acadian-related plutonic rocks comprise an Upper Silurian suite dominated by intermediate to granitic rocks with minor associated gabbro (McLaughlin *et al.* 2003). They also include the *c.* 423 Ma volcanic rocks of the Mascarene back-arc (Van Wagoner *et al.* 2002). Mafic compositions included in this magmatic belt, like their equivalents in Newfoundland east of the DBL (Fig. 17), exhibit a diverse spectrum

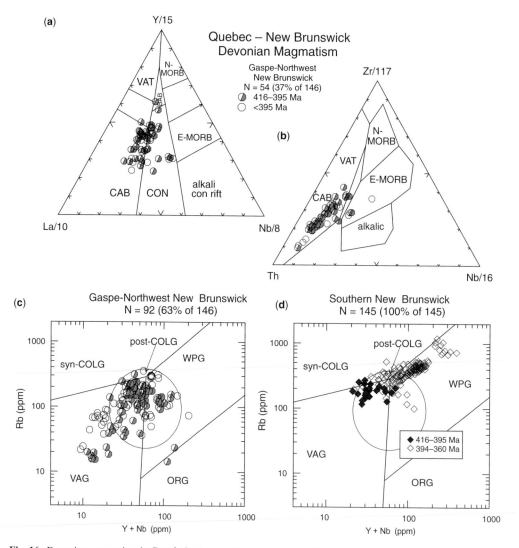

**Fig. 16.** Devonian magmatism in Gaspé (Quebec) and New Brunswick. Diagrams (**a, b, c, d**) and abbreviations as in Figure 7; samples have been subdivided according to age, as indicated in symbol legends, see text for discussion. See Appendix 1 for data sources.

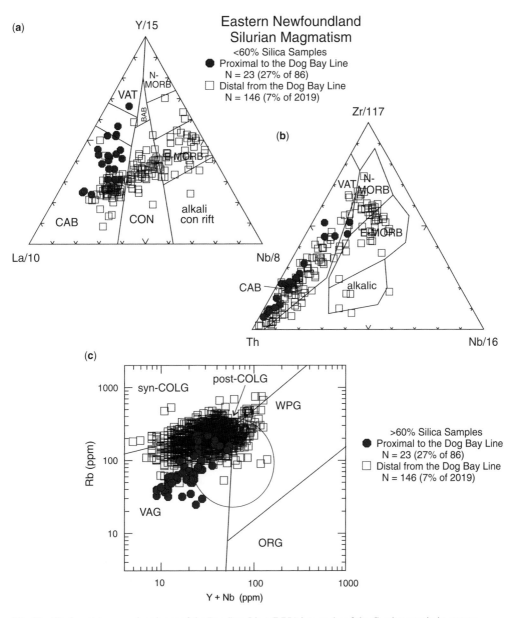

**Fig. 17.** Silurian plutons emplaced east of the Dog Bay Line (DBL) into rocks of the Gander margin in eastern Newfoundland. Subdivision into proximal and distal is based on proximity of the plutons with respect to the DBL (see text). Diagrams (**a, b, c**) and abbreviations as in Figure 7. See Appendix 1 for data sources.

of arc- to non-arc like compositions (Fig. 15a, b), but represent a higher proportion (*c.* 44%) of analyzed samples. Felsic compositions (Fig. 15c) lack the Rb-enrichment observed in Newfoundland (Fig. 17c) and include two distinct sample clusters, one within the VAG field and the other in the WPG field, though both also fall mainly within the overlapping syn-COLG field. The Y + Nb-enriched

data cluster consists of samples from the *c.* 423 Ma Utopia and *c.* 422 Ma Welsford phases of the S. George batholith.

Overall, the early Acadian-related igneous rocks compositionally resemble late Acadian, Early Devonian (416–395 Ma) plutons formed further to the NW in New Brunswick (Fig. 16) and in the equivalent area in Newfoundland (Figs 13, 14 & 15).

The regional distribution of Early Devonian Acadian magmatism suggests the existence of a time-transgressive belt-like pattern (see compilations of Kerr 1997; Whalen *et al.* 1996*a*; Tucker *et al.* 2001; Hibbard *et al.* 2006), which seems to track the progressive migration of the Acadian deformation front, at least locally (Bradley & Tucker 2002). This relationship needs more testing, but if correct, suggests that Early Devonian Acadian magmatism was associated with protracted crustal shortening and related in some way to the progressive shallowing of the subducting slab, which presumably started to happen shortly after the arrival of buoyant Avalonia at the trench at *c.* 421 Ma. Migration of the deformation and magmatic front towards the west suggest that Avalonia and its attached oceanic lithosphere of the Acadian seaway (Acadian slab) remained being thrust beneath composite Laurentia until at least 400 Ma, if not longer. The Acadian magmatic pulse is thus associated with intra-crustal shortening and hence, unlikely related in any fashion to regional extension and/or delamination of the lithospheric mantle (cf. Dostal *et al.* 1989, 1993). This conclusion is consistent with the relatively high ratio of intrusive with respect to extrusive rocks in the Acadian magmatic suite. Magmas intruding in thickened crust during contraction tend to pond and evolve at depth, probably because they have difficulty ascending to crustal levels above the brittle-ductile transition. Critical to understanding the duration and genesis of Early Devonian Acadian magmatism is the boundary condition imposed by the time of separation and sinking of the oceanic slab that was originally attached to Avalonia (= breakoff of Acadian slab). If this process happened, it would signal the end of the Acadian-related convergence. Breakoff may have been facilitated by eclogititation, and hence steepening of the leading oceanic segment attached to the trailing overriding, shallow subducting buoyant segment (including Avalonia) of the Acadian slab (Haschke *et al.* 2002). The area affected by the early Acadian magmatic phase was re-intruded during Middle–Upper Devonian magmatic flare-up represented mainly by plutons, which range in age between 395 and 375 Ma in Newfoundland, locally stitching the suture zone between Laurentia's leading edge (southeastern Ganderia) and Avalonia. This magmatism includes S-type and I-type granites with the latter ranging to HFSE-enriched A-type (or WPG-type) compositions (Fig. 18c). Consanguineous magmatism also took place near Laurentia's leading edge (southeastern Ganderia) in New Brunswick and Maine (Fig. 17d), although no evidence for stitching the Acadian suture with Avalonia.

Middle–Late Devonian magmatic rocks in southern New Brunswick consists exclusively of felsic rocks (Fig. 16d). Contemporaneous plutonic rocks in eastern Newfoundland (Fig. 18) also exhibits a similar paucity (<3% or 26 of 1182 samples) of mafic rocks. The continuity of crustal shortening and lack of evidence for regional extension during this period, combined with the predominance of felsic over mafic compositions and the diverse signatures of granites (VAG, WPG and syn-COLG) included, leads us to relate the Middle–Late Devonian (395–375 Ma) magmatism to breakoff of the Acadian slab, which seems to have taken place immediately after the 400–395 Ma docking of Meguma (see below), suggesting a causative relationship, rather than delamination or subduction of a plume, as was proposed earlier by Murphy *et al.* (1999). Furthermore, the parallelism of this magmatism with the Avalonia–Laurentia plate boundary from Newfoundland to Massachusetts (Hibbard *et al.* 2006) would require a fortuitous linearity of the plume head approximately parallel to the Laurentian margin in the model of Murphy *et al.* (1999). However, comparison of the Middle–Late Devonian magmatism with the better substantiated Silurian slab-breakoff-related magmatism of western Newfoundland (Fig. 13) indicates the Devonian magmatism was dominated by melting of more evolved (higher LILE) crustal materials and that mafic magmas, which likely supplied the thermal flux for the large-scale crustal melting, probably ponded at depth and were unable to move to high crustal levels, because the orogen was still under compression.

If the postulated timing of Acadian slab-breakoff is correct, it implies that the voluminous phase of 415–395 Ma syn-convergence Acadian magmatism was related to shallow or flat-slab subduction and a progressively widening zone of crustal shortening in the hinterland of the orogen. In New Brunswick, where both better age constraints and more REE data are available, mafic rocks formed during this period exhibit calc-alkaline arc, continental tholeiite and non-arc signatures (Fig. 16a, b). Contemporaneous felsic rocks mostly show VAG signatures with many samples plotting where VAG and post-COLG granites overlap (Fig. 16c, d). Relative to immediately preceding and subsequent felsic magmatism, the paucity of WPG compositions is notable, indicating little input from the lithospheric mantle. Another unusual feature of this felsic magmatism is that a high proportion of samples exhibit elevated $(La/Yb)_{CN}$ values (Fig. 19), which are indicative of formation under deep garnet stable $P–T$ conditions, as was suggested for Archaean TTG magmas (cf. Martin 1986). High $(La/Yb)_{CN}$ 'adakitic' magmas have been document by Kay *et al.* (2005) as a characteristic of central Chilean-Argentinian Andean flat subduction and interpreted as resulting from high-pressure

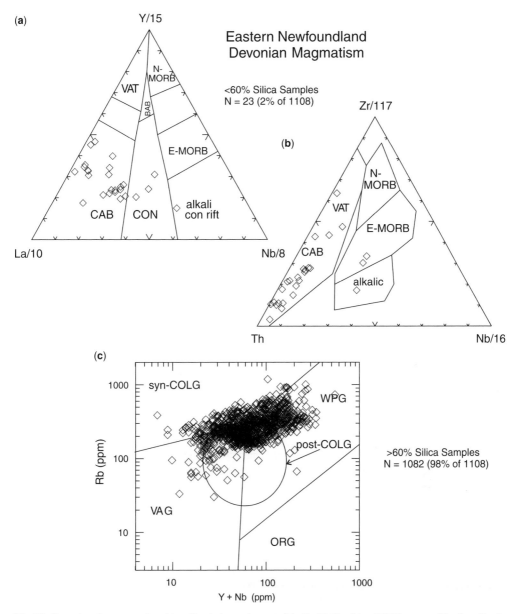

**Fig. 18.** Devonian plutons, emplaced into Ganderian rocks east of the Red Indian Line (RIL) in eastern Newfoundland. Diagrams (**a, b, c**) and abbreviations as in Figure 7. See Appendix 1 for data sources.

metamorphism and partial melting of subducted fore-arc material as it entered the asthenospheric mantle wedge beneath the arc.

We discussed earlier that the fore-arc to the Acadian coastal arc was largely tectonically removed and postulated that it was subducted beneath the leading edge of Laurentia (represented at this stage by Ganderia). This feature together with the apparent migration of the structural front towards the back-arc region suggests that an Andean flat-slab-like setting is an attractive analogue for the Early Devonian Acadian magmatism. The principal difference is that flat subduction in the Andes solely involves oceanic lithosphere, whereas the Acadian flat-slab also comprises progressive underthrusting of Avalonia beneath Laurentia. Shallowing and dehydration of the subducted Acadian slab and progressive thinning and

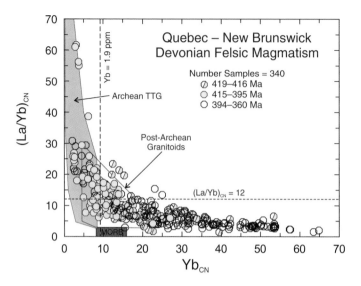

**Fig. 19.** Felsic Devonian magmatism from Gaspé (Quebec) and New Brunswick, subdivided into 2 age groups, with flat subduction of the Acadian slab being postulated for the 416–395 Ma period, plotted on a chondrite normalized La/Yb vs Yb diagram with fields for Archaean TTG (tonalite-trondhjemite-granodiorite), post-Archaean granitoids and MORB from Martin (1986).

cooling of the overlying asthenospheric wedge, would over time have inhibited typical arc magmatism and forced progressive retreat of the potential mantle melt zone towards the rear of the now extinct coastal arc. In New Brunswick, this may help explain why during this period there is an apparent trailing off of magmatism, prior to the Middle to late Devonian 'flare-up', which we related to breakoff of the Acadian slab. However, areas with thin lithosphere formed during the immediately preceding Silurian extension in the back-arc region of the coastal arc and the collapsed Salinic collision zone, beneath which the Salinic slab had broken off (van Staal & de Roo 1995; Whalen et al. 2006) may have trapped pockets or refuges of asthenosphere above the subsequent Acadian flat-slab stage (Fig. 10b). Dehydration of the Acadian flat-slab and fluid-fluxing of the asthenosphere in such trapped pockets may have triggered their partial melting and thus may have been responsible for producing the syn-Acadian upper plate, arc-like magmatism during this phase. We introduced these asthenospheric pockets to explain the mantle-derived component of the syn-Acadian magmatism in a flat-slab tectonic framework, which allows progressive migration of the magmatic front into the orogen's hinterland. Shallow subduction of the cold Acadian slab would have rapidly displaced the fertile mantle wedge and cooled the overlying mantle, and hence would have inhibited formation of any mantle-derived arc to non-arc magmatism (van Hunen et al. 2002)

without such asthenospheric pockets. In general, Silurian back-arc and Salinic slab breakoff-related magmatism, which lasted until the onset of the flat-slab phase (see above) may have played an important role in the subsequent Acadian orogenesis, because it may also have been an important source for the crustal heat necessary for the low to medium pressure-high temperature metamorphism that typifies the back-arc region from the onset of Acadian deformation and preferentially localized deformation by thermal and/or fluid-induced weakening (Hyndman et al. 2005).

## Neoacadian

The Neoacadian orogeny refers to late Early Devonian–Early Carboniferous (400–350 Ma) deformation, metamorphism and magmatism spatially and genetically linked to docking of the Meguma terrane to Laurentia (Hicks et al. 1999; van Staal 2005, 2007). The leading edge of the latter was represented at this stage by Acadian-accreted Avalonia. The name Neoacadian was originally introduced by Robinson et al. (1998), to cover strong Late Devonian–Early Carboniferous deformation and metamorphism in southern New England, which was distinct from Acadian orogenesis in this part of the northern Appalachians. Neoacadian orogenesis, as defined by van Staal (2005, 2007), overlaps in time, but not in space with the tail end of Acadian deformation and magmatism in Laurentia's hinterland near the Appalachian front in New England (see above). Such overlap is absent in Newfoundland, because Newfoundland seems to have escaped most, if not all, effects of Meguma's accretion, which took place further to the south (van Staal 2005, 2007). However, a possible dynamic linkage between the Neoacadian and prolonged continuation of the Acadian into the Middle Devonian in New England cannot be dismissed at present. The name Neoacadian is therefore appropriate, because it permits a possible tectonic link with the Acadian. Such a linkage is indeed supported by the similarities between the tectonic processes responsible for both the Acadian and Neoacadian orogenies (Murphy *et al.* 1999; van Staal 2007), discussed further below.

The provenance and tectonic setting of Meguma during the Early Palaeozoic is contentious and in need of more research. On the basis of available data, we favour it as having been a single Gondwanan-derived microcontinent, but alternative settings such as a promontory on Gondwana itself or forming part of Avalonia to start with (Murphy *et al.* 2004) cannot be overruled at present. Its Palaeozoic stratigraphy is markedly different from that of Avalonia (Fig. 3), although they also share some similarities, such as a large Middle Ordovician hiatus and a palaeontological linkage by at least the Late Silurian (Bouyx *et al.* 1997).

The dominant structural style of Meguma is characterized by a regional set of NW-trending shallowly plunging, upright folds and associated shear zones. The folds are progressively reoriented to a more easterly trend on approaching the dextral Cobequid-Chedabucto fault (CCF in Fig. 1), which represent the boundary with Avalonia on land. The combination of transcurrent shearing and upright folding is generally interpreted to represent a dextral transpressive deformation regime formed in response to dextral oblique convergence between Meguma and Avalonia. How this convergence was accommodated at depth is not well constrained. Geophysical evidence suggests that a wedge of Avalonian crust was thrust beneath Meguma, whereas upper mantle reflectors suggest a NW-dipping subduction zone was present beneath Nova Scotia (Keen *et al.* 1991). Murphy *et al.* (1999) postulated that the dip of the NW-dipping subduction zone was very shallow (flat-slab) due to interaction with a rising plume. Regardless of the cause of the postulated flat-slab, it explains the scarcity or absence of late Early Devonian or younger arc magmatism in Laurentia's leading edge, which at this stage is represented by Acadian accreted Avalonia. The lithospheric configuration of Meguma partly underlain by Avalonian crust suggests wedging of the downgoing Rheic plate by the leading (Avalonian) edge of Laurentia when Meguma entered the

trench of the NW-dipping subduction zone (Fig. 10c). Such a process explains Meguma's accretion and transfer to the upper (Laurentian) plate and the continuation of NW-directed thrusting along the Cobequid–Chedabucto fault system into the Carboniferous (Waldron *et al.* 1989; Pe-Piper & Piper 2002). A temporal link between the break-off of the Avalonian slab (see above) and Meguma's docking and large-scale crustal wedging also suggest a genetic relationship.

Meguma was subjected to extensive Middle–Late Devonian magmatism, which was dominated by peraluminous (S-type) felsic granitoid rocks and included only rare (*c.* 2%) mafic rocks (Fig. 20). Mafic rocks exhibit calc-alkaline to continental tholeiite compositions whereas the voluminous felsic plutonic rocks span the upper portion of the VAG field and the lower to middle portions of the syn-COLG field. Compared to almost all of our other spatial-temporal groups, this felsic plutonism exhibits a total lack of high Y + Nb (WPG type) compositions, suggesting no involvement of enriched lithospheric mantle. This magmatism was interpreted to have occurred after transfer of Meguma to Laurentia and after breakoff of the Rheic slab that was present beneath Meguma (Fig. 10c) (van Staal 2007; Moran *et al.* 2007). However, comparison of its geochemical signatures and the relative high proportion of felsic plutonic rocks with other magmatic events attributed to slab-breakoff (e.g. Figs 13 & 14) suggest a slightly different petrogenesis. Rather the syn- to post-collisional S-type granitoid magmatism may be a product of partial melting of deeply buried fertile Meguma sediments following crustal thickening as a result of Avalonia's wedging, possibly facilitated by thermal input from mantle-derived mafic magmas that ponded at depth. If the latter is correct, mafic magma emplacement is probably related to asthenospheric upwelling following Rheic slab breakoff (Fig. 10c) or alternatively, delamination of Meguma's lithospheric mantle.

## Summary of orogenic events

Four major phases of orogenesis related to accretion of ribbon-shaped crustal blocks to the Laurentian margin have been recognized in the Canadian segment of the northern Appalachians: the Taconic, Salinic, Acadian and Neoacadian orogenies. The Taconic is a composite of three, kinematically different, but temporally and spatially closely related events (Taconic 1, 2 & 3), all involving peri-Laurentian rocks. They have been grouped together under the Taconic banner, because they may not be separable elsewhere in the Appalachians where critical age relationships have been obscured

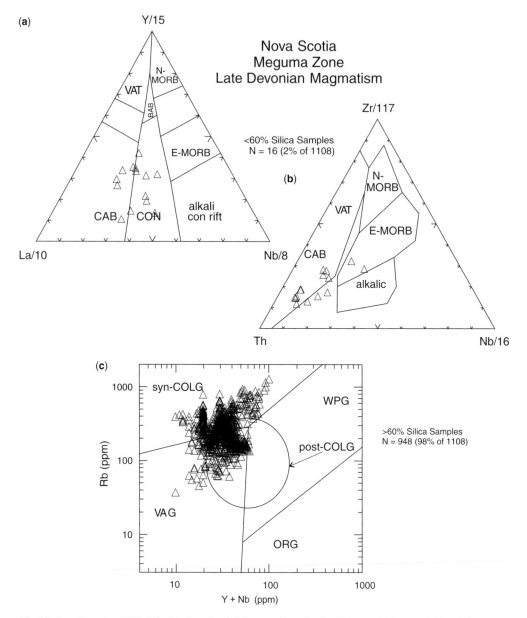

**Fig. 20.** Late Devonian (377–368 Ma) plutonism in Meguma, Nova Scotia. Diagrams (**a, b, c**) and abbreviations as in Figure 7. See Appendix 1 for data sources.

by superimposed orogenesis (van Staal *et al.* 2007). Taconic 1 represents a short-lived, latest Cambrian obduction (500–493 Ma) of an infant oceanic arc onto the peri-Laurentian Dashwoods microcontinent (Fig. 4a) (van Staal *et al.* 2007) and has only been recognized with some degree of certainty in Newfoundland. It mainly produced a regionally extensive mélange in Newfoundland (Fig. 6a). Evidence for the nature and extent of the ductile

structures and metamorphism generated during this event is cryptic and largely based on cross-cutting relationships with younger, Taconic 2-related arc magmatic rocks. Taconic 2 was principally due to dextral oblique collision between Dashwoods and its Notre Dame arc suprastructure with the downgoing Laurentian margin (Figs 4b & 8). Collision had started by at least 475 Ma (Waldron & van Staal 2001), but locally possibly already at

480 Ma (Knight *et al.* 1991) and A–subduction had ended before 450 Ma in most parts of the Canadian Appalachians. Penetrative regional metamorphism appears to have been short-lived (469–457 Ma) and has a temporal and spatial link to a widespread tonalite flare-up in upper plate rocks in Newfoundland (Fig. 5a). These voluminous arc-type tonalites and associated non-arc-like mafic rocks (Fig. 7), are thought to be related to slab-breakoff (van Staal *et al.* 2007). Associated reverse shear zones were predominantly synthetic with the east-directed subduction of the Laurentian margin, although antithetic structures were also generated along Dashwoods' eastern margin (Fig. 7b). The latter structures accommodated underthrusting of supra-subduction zone oceanic elements of the AAT (Fig. 5a) (Lissenberg & van Staal 2006).

Structures and metamorphism generated during Taconic 3 are localized in volcanic and sedimentary rocks that occur in a narrow belt adjacent to the RIL. They formed as a result of collision between the peri-Gondwanan Popelogan–Victoria arc built upon the leading edge of Ganderia and the Laurentian Red Indian Lake arc, which at this time formed the leading edge of the Laurentian plate, between 460 and 450 Ma. The Red Indian Lake arc had earlier re-accreted to Laurentia due to closure of the Lloyds River back-arc basin and incorporated into the AAT during Taconic 2 (Zagorevski *et al.* 2006). Arc-arc collision was shortlived and lasted <10 Ma (van Staal *et al.* 1998; Zagorevski *et al.* 2007*b*, *c*). Tectonic transport was dominantly to the SE and principally affected the upper plate rocks (AAT). Accretion of the Popelogan–Victoria arc thus terminated Taconic orogenesis. Docking of the Popelogan–Victoria arc also implies closure of the main Iapetan oceanic tract, leaving only the Tetagouche–Exploits back-arc basin basins and the oceanic seaway between Ganderia and Avalonia open (van Staal 2005).

The Salinic orogeny was due to closure of the Tetagouche–Exploits back-arc basin (van Staal 1994). Closure was initiated at *c*. 450 Ma, immediately after the Late Ordovician Taconic 3 arc–arc collision, which forced the west-facing subduction zone that dipped beneath Laurentia to step back behind the accreted Popelogan–Victoria arc into the backarc basin, which was floored by oceanic lithosphere. Back-arc closure produced both arc-like mafic and felsic plutonism (Fig. 13). The Gander margin was accreted to Laurentia during the late Llandovery–Wenlock (433–423 Ma) during sinistral oblique convergence. The collision appears to have taken place slightly earlier in Newfoundland (Whalen *et al.* 2006) than New Brunswick (van Staal *et al.* 2003). Silurian syn-collision magmatism (433–423 Ma) (Figs 13 & 14) was due to slab breakoff. Tectonic transport was predominantly to

the SE with early thrust-related structures progressively being steepened, in part due to refolding into upright structures, during terminal collision.

The Acadian orogeny started shortly after or during the waning stages of the Salinic during the Late Silurian at *c*. 421 Ma due to subduction of Avalonia beneath Laurentia following closure of the narrow intervening Acadian seaway (Fig. 10b). The Acadian lasted at least until *c*. 400 Ma in Newfoundland and into the Middle–Late Devonian (*c*. 380 Ma) in Quebec. The dominant orthogonal tectonic transport was to the west, antithetic with respect to a westerly-dipping, progressively shallowing, flat-slab-like subduction zone. The sense of obliquity was dextral. Structures and magmatism (Figs 16 & 18) progressively became younger to the west (Bradley *et al.* 2000). The Acadian structural setting is analogous to the present-day Andes at the latitude of central Chile and Argentina where the Pacific plate subducts very shallowly (flat-slab) to the east, producing high $(La/Yb)_N$ 'garnet-signature' magmatism (Fig. 20). Syn-collision Early Devonian magmatism also shows similarities to that associated with the Andean flat-slab segment and progressively diminished over time until the postulated breakoff of the Acadian slab at 395–390 Ma.

The late Early–Late Devonian/Early Carboniferous (395–350 Ma) Neoacadian orogeny, with its voluminous, spatially restricted syn-collisional, S-type granitoid magmatism (Fig. 18), was due to final docking of Meguma to Laurentia and stepping back of the subduction zone into the Rheic Ocean. The plate tectonic setting was interpreted to be similar to the Acadian, that is, a flat-slab (Murphy *et al.* 1999).

## Conclusions

With the exception of localized deformation associated with major ductile-brittle fault zones, the Canadian Appalachians escaped most of the intense orogenesis associated with Alleghanian collision seen further to the south in the Appalachians. Hence, they provide a window to study the distribution and nature of orogenic events that took place prior to continent–continent collision and assembly of Pangaea. Careful dating and delineating the regional extent of structures and their associated metamorphism and magmatism in the Canadian and immediately adjacent New England Appalachians revealed the following important conclusions: (1) collisional orogenesis was episodic and comprises four different phases. Each phase was generally short-lived and linked to oblique accretion of peri-Laurentian and peri-Gondwanan crustal ribbons to the Laurentian margin, which was progressively expanding eastwards over time; and (2) the locus of initial collision progressively migrated to the east, which after the Middle Ordovician was consistently the direction of the orogen's foreland (Figs 1 & 2). The Late Ordovician and younger foreland-directed migration of the locus of collision-related deformation and orogenesis was concurrent with progressively stepping back of a west-dipping subduction zone behind each of the accreted blocks. Deviations and complications from this overall simple pattern were introduced by local reactivation of older structures, widening of the A–subduction channel and hinterland migration of structures and associated magmatic rocks during the Acadian and possibly also during the Neoacadian. The Acadian hinterland migration, which lasted >20 Ma, is a major deviation to the overall observed pattern. It is interpreted to be due to establishment of a flat-slab (Fig. 10b) with strong coupling between the upper- and downgoing lower plates, analogous to the flat-slab region in the southern Andes. The resultant Acadian deformation produced reverse and transcurrent faults, but most characteristically is the penetrative steeply inclined or upright folding present throughout the northern Appalachians.

Deformation during each accretionary/collisional event appears to start in the arc and/or arc-trench gap and subsequently widens over time. Structures in the arc-trench gap and collision zone commonly show a dominant transport direction. The latter can be deduced locally from the sense of overturning of folds, shear sense indicators in shear zones, and the geometry and stratigraphy of fault-bounded panels (e.g. 'old over young'). The reverse fault zones generally, but not invariably, were synthetic with the polarity of subduction deduced from the position of other tectonic elements (e.g. position of arc, back-arc and foreland basin) during the early stages of collision. Even during the Taconic 3 arc-arc collision, which resulted from closure of the Iapetus main oceanic tract, and involved two subduction zones with opposite polarity (Fig. 5b) (van Staal *et al.* 1998), one sense of overthrusting dominates, which was synthetic with the sense of subduction of the downgoing arc (Zagorevski *et al.* 2007c). To a first-order approximation, our observations are consistent with the generally held belief that the lithospheric shear couple generated during the obliquely sliding and sinking of one plate beneath another during A–subduction controls the overall deformation pattern during at least the early stages of collision, which is a time when there is probably relatively strong coupling between the downgoing and overriding plate. A non-coaxial deformation (simple shear dominated), in general, is a more efficient deformation mechanism than a bulk coaxial deformation (pure shear), because in the latter case penetrative fabric development will progressively

harden the rocks and prohibit development of long-lived movement zones (Lin *et al.* 1998). Relative hardening of the deforming material is probably also the main reason why collisional belts seem to widen over time, because new material at the margins of the belts becomes easier to deform than their interiors (Means 1995). An example is provided by the Taconic 2 collision, where upper plate arc volcanic rocks and sediments were buried and heated when they were incorporated into a progressively widening A–subduction zone during the final stages of collision (Fig. 4b). Here, the upper plate rocks were thermally weakened by a regionally very voluminous, but short-lived (5–7 Ma) tonalite bloom (Fig. 5b) (van Staal *et al.* 2007).

The significance of the oblique component accommodated by the shear zones is a contentious issue in the Appalachian–Caledonian orogen. Some workers (van Staal *et al.* 1998; Brem *et al.* 2007) have speculated that some of the block or terrane bounding faults may have accommodated large orogen-parallel movements during convergence. However, with few exceptions (e.g. Malo & Kirkwood 1995; Zagorevski *et al.* 2006; Brem *et al.* 2007; Brem 2007), evidence for major horizontal translations is commonly ambiguous and there are few examples of structural duplications due to strike-slip (e.g. Lin *et al.* 2007). Nevertheless, a consistent sense of horizontal shear in a set of coeval shear zones is generally thought to reflect the overall obliquity of convergence between the opposing blocks. If correct, the horizontal component of convergence between Laurentia and the various blocks has changed polarity from dextral to sinistral several times between the Ordovician and Late Devonian. For example, most Early Devonian orogen-parallel faults accommodated some component of dextral movements, which likely represents an orogen wide partitioning of the oblique component between Laurtentia and Avalonia. Complications arise where accreting crustal ribbons got caught between two different subduction zones, which were active at the same time. For example, the Dashwoods microcontinent was bounded by two coeval subduction zones with opposite polarity and obliquity of convergence (van Staal *et al.* 2007). Careful dating of the bounding shear zones revealed a dextral sense of shear along its contact with the Laurentian margin (Brem *et al.* 2007), while a sinistral-oblique reverse sense of shear operated coevally along its contact with the AAT on its eastern margin (Lissenberg *et al.* 2005a; Lissenberg & van Staal 2006; Zagorevski *et al.* 2006, 2007b). Dashwoods thus was moving towards the south with respect to both. In general, careful structural analysis combined with age-dating indicate that strike-parallel translations were important in the straight, steeply dipping fault zones, such as the Long Range–Cabot fault system in Newfoundland (Brem 2007; Brem *et al.* 2007), However, most major fault zones are markedly curviplanar. For example, the Lloyds River–Lobster Cove fault system that separates Dashwoods from the AAT (Lissenberg & van Staal 2006; Zagorevski *et al.* 2007*b*) locally bends through angles of >90°, which indicates it is unlikely that it could have accommodated major strike-slip translations along its whole length. Indeed shear sense indicators and old over young relationship suggest it was mainly a major SE-directed reverse fault instead. Nevertheless, some steepened and relatively straight segments of this fault were subsequently reused as strike-parallel movement zones (e.g. Lafrance 1989). Such reactivation of fault zones complicates interpretation of their overall kinematic movement picture, because the younger deformation tends to destroy most, if not all evidence for the earlier movements.

Slab-breakoff-type magmatism was repetitive (Whalen *et al.* 2006; van Staal *et al.* 2007) and appears to have accompanied every accretion event involving a microcontinent, with the possible exception of Meguma. Rather than simply viewing them as arc batholiths formed as a result of subduction, such replication of slab-breakoff magmatism represents a new way of looking at voluminous, short-lived magmatic flare-ups of syn-orogenic granitoid rocks within mountain belts. The latter model can account for the geochemical characteristics of the observed syn-orogenic Appalachian plutonism, whereas the arc model cannot.

This manuscript is a product of the 2005–2010 TGI-3 Appalachian program and we appreciate the continuous financial and logistical support given to us by its manager, S. Hanmer. In addition, we would like to thank our Canadian and New England Appalachian colleagues for discussion and supplying us with many ideas over the years through fieldtrips and intellectual osmosis. Of these, we particularly like to mention S. Barr, S. Castonguay, S. Colman-Sadd, J. Dewey, L. Fyffe, J. Hibbard, P. Karabinos, A. Kerr, D. Lavoie, S. Lin, B. Murphy, B. O'Brien, D. Rankin, D. Reusch, P. Robinson, D. Stewart, A. Tremblay, D. West, C. White, R. Wilson, H. Williams and R. Wintsch. S. Barr, B. Clarke, M. McLeod, D. Lenz and R. Wilson supplied geochemical datasets in digital format. Brendan Murphy invited us to submit this manuscript and showed immense patience waiting for the final version. Constructive and thoughtful reviews by S. Barr and P. Ryan improved the manuscript. Geological Survey of Canada contribution 20080031.

# Appendix 1

In this study, geochemical data compiled from various sources has been split or grouped based on both temporal and spatial criteria. The temporal subdivisions are Early Ordovician (489–475 Ma) and Middle Ordovician (470–455 Ma). Silurian (<445 and >416 Ma) and Devonian (<416 and >360 Ma). In Newfoundland, the spatial subdivisions (see Figs 1 & 2) are: (a) western Newfoundland [area west of the Red Indian Line (RIL)]; (b) west-central Newfoundland [area east of the RIL but west of the Dog Bay Line (DBL)]; (c) eastern Newfoundland east of the DBL. Newfoundland data from west of the RIL includes both coeval plutonic and volcanic rocks and was compiled from Whalen *et al.* (2006), van Staal *et al.* (2007) and Whalen (unpublished data). Data from east of the RIL is exclusively from plutonic rocks and was compiled from Kerr *et al.* (1994) and Kerr (1997). The very large database of Kerr *et al.* (1994) was filtered on the presence of the appropriate trace elements for chosen plots (see below) and on whether trace element concentrations for these elements were near detection limits (i.e. an attempt was made to exclude unreliable or imprecise data).

In the southern Canadian Appalachians, the spatial subdivisions are: (a) Quebec–NW New Brunswick, which consists of the area east of the Baie Verte–Brompton Line in Gaspé to immediately west of the Fredericton trough in New Brunswick; (b) southern New Brunswick (Whalen *et al.* 1996*b*), which consist of the area east of the Fredericton trough and west of the Caledonia fault; and (c) Nova Scotia's Meguma zone, which consist of the area east of the Cobequid–Chedabucto fault. The Gaspé and New Brunswick dataset, which includes both Silurian to Devonian plutonic and volcanic rock analyses, was compiled from Bédard (1986), Dostal *et al.* (1989, 1993), McLeod (1990), Whalen (1993*a*, *b*), van Wagoner *et al.* (2001, 2002), McLaughlin *et al.* (2003), Yang *et al.* (2003) and Wilson *et al.* (2005). The Meguma zone dataset consists exclusively of plutonic rocks and is mainly from the large (1153 samples) compilation by Tate & Merrett (1994), which was filtered for samples that included Rb, Y and Nb data, giving 948 samples. As this dataset only included 2 samples that had <60% silica, an additional 14 mafic samples were compiled from Currie *et al.* (1998) and Tate & Clarke (1995).

It has become well established that mafic (silica <60 wt.%) rock trace element signatures reflect their tectonic settings of formation. For this reason, a large array of mafic rock tectonomagmatic discrimination diagrams employing a wide range of trace elements have been developed for this purpose over the last 30 years. However, a very restricted number of trace elements (e.g. only La and Ce of the REE, and no Ta or Hf) were available in the Kerr *et al.* (1994) Newfoundland compilation, in contrast to the more extensive trace element array available for most New Brunswick samples. This restricted which tectonomagmatic classification diagrams could be employed, because we wished to utilize the same plots for both 'transect' areas. Based in part on past experience (e.g. Whalen *et al.* 2006) and also for the sake of brevity, only two mafic rock plots are employed herein: (a) Wood

(1980) Th–Zr–Nb diagram; and (b) Cabanis & Lecolle (1989) La–Y–Nb plot. Previous studies (e.g. Whalen 1993*a*; Whalen *et al.* 1998, 2001) have demonstrated that mafic end-members of plutonic suites provide both valuable insights into a suite's tectonic setting and an 'independent' cross-check on setting indicated by felsic end-member trace element signatures. As well, knowledge of the proportions of mafic to felsic compositions that comprise a magmatic association represents key information for constraining tectonic setting and helps establish the role played by directly mantle-derived magmas. For example, magmatic suites well-established to have formed in magmatic arc settings, almost invariably include a significant proportion of mafic (basaltic plus andesitic) rocks (Whalen *et al.* 1997*b*). For this reason, both numbers of samples involved and the percentage they represent of the total are shown on geochemical diagrams.

Felsic (silica >60 wt.%) rocks were plotted on both the Pearce (1996) Rb–Nb + Y diagram and the Pearce *et al.* (1984) Nb–Y diagram. As the latter was found to not contribute much additional information, it is not included herein. Previous experience (e.g. Whalen *et al.* 2006) has shown that these granitoid tectonomagmatic discrimination diagrams can be very useful for subdividing granitoid rocks into groups derived from different sources and/or under contrasting tectonic conditions but that these geochemical signatures need to be integrated with geological and other parameters to arrive at an optimum interpretation of granitoid tectonic conditions of formation. Our interpretations of what these diagrams indicate about contributing granitoid source materials in this study follow fairly closely those made by Pearce (1996) in a review of sources and settings of granitic rocks. Some relevant points concerning the Rb–Y + Nb diagram are: (a) VAG reflect derivation from depleted mantle sources modified by a subducted component, that is, low Rb and Y + Nb, but concentrations of both increase with arc maturity within the VAG field; (b) intraplate granites (WPG) show evidence of input from enriched mantle (lithosphere and/or athenosphere) sources (i.e. high Y + Nb); (c) syn-collision granites (syn-COLG) either represent pure crustal melts or strongly crustally contaminated mantle melts and thus exhibit very high Rb; and (d) post-collisional granites reflect derivation from diverse tectonically juxtaposed crustal and mantle sources, including enriched-mantle materials.

# References

AYDIN, N. S. 1995. *Petrology of a composite mafic-felsic rocks of the Fogo Island batholith: a window to mafic magma chamber processes and the role of the mantle in the petrogenesis of granitoid rocks.* Unpublished PhD thesis, Memorial university of Newfoundland, St. John's, Newfoundland.

BARR, S. M. & JAMIESON, R. A. 1991. Tectonic setting and regional correlation of Ordovician-Silurian rocks of the Aspy terrane, Cape Breton Island, Nova Scotia. *Canadian Journal of Earth Sciences*, **28**, 1769–1779.

BARR, S. M. & WHITE, C. E. 1996. Contrasts in Late Precambrian-Early Paleozoic tectonothermal history between Avalon Composite terrane *sensu stricto* and other possible Peri-Gondwanan terranes in southern New Brunswick and Cape Breton. *In*: NANCE, R. D. & THOMPSON, M. D. (eds) *Avalonian and Related Peri-Gondwanan Terranes of the Circum-North Atlantic.* Geological Society of America, Special Paper, **304**, 95–108.

BARR, S. M., WHITE, C. E. & MILLER, B. V. 2002. The Kingston Terrane, southern New Brunswick, Canada: evidence for an Early Silurian volcanic arc. *Geological Society of America Bulletin*, **114**, 964–982.

BARR, S. M., WHITE, C. E. & MILLER, B. V. 2003. Age and geochemistry of Late Neoproterozoic and Early Cambrian igneous rocks in southern New Brunswick: similarities and contrasts. *Atlantic Geology*, **39**, 55–73.

BARR, S. M., RAESIDE, R. P. & WHITE, C. E. 1998. Geological correlations between Cape Breton Island and Newfoundland, northern Appalachian orogen. *Canadian Journal of Earth Science*, **35**, 1252–1270.

BÉDARD, J. H. 1986. Pre-Acadian magmatic suites of the southeastern Gaspe Peninsula. *Geological Society of America Bulletin*, **97**, 1177–1191.

BEVIER, M. L. & WHALEN, J. B. 1990. Tectonic significance of Silurian magmatism in the Canadian Appalachians. *Geology*, **18**, 411–414.

BIRD, J. M. & DEWEY, J. F. 1970. Lithosphere plate-continental margin tectonics and the evolution of the Appalachian Orogen. *Geological Society of America Bulletin*, **81**, 1031–1060.

BLACKWOOD, R. F. 1985. *Geology of the Facheux Bay area (11P/9), Newfoundland.* Mineral Development Division, Department of Mines and Energy, Government of Newfoundland and Labrador, Report **85–4**.

BOSTOCK, H. H. 1988. Geology and petrochemistry of the Ordovician volcano-plutonic Robert's Arm Group, Notre Dame Bay, Newfoundland. *Geological Survey of Canada, Bulletin*, **369**, 1–82.

BOUCOT, A. J. 1962. Appalachian Siluro-Devonian. *In*: COE, K. (ed.) *Some Aspects of the Variscan Fold Belt.* Manchester University Press, UK, 155–177.

BOUYX, E., BLAISE, J. *ET AL.* 1997. Biostratigraphie et paléobiogeographie du Siluro-Dévonien de la Zone de Meguma (Nouvelle-Écosse, Canadian). *Canadian Journal of Earth Science*, **34**, 1295–1309.

BRADLEY, D. C. 1983. Tectonics of the Acadian Orogeny in New England and adjacent Canada. *Journal of Geology*, **91**, 381–400.

BRADLEY, D. C. & TUCKER, R. D. 2002. Emsian synorogenic paleogeography of the Maine Appalachians. *Journal of Geology*, **110**, 483–492.

BRADLEY, D. C., TUCKER, R. D., LUX, D., HARRIS, A. G. & MCGREGOR, D. C. 2000. *Migration of the Acadian orogen and foreland basin across the northern Appalachians of Maine and adjacent areas.* U.S. Geological Survey, Professional Paper, **1624**, 49.

BREM, A. G. 2007. *The late Neoproterozoic to Early Paleozoic evolution of the Long Range Mountains, southwestern Newfoundland.* Unpublished PhD thesis, University of Waterloo, Canada.

BREM, A. G., LIN, S., VAN STAAL, C. R., DAVIS, D. D. & MCNICOLL, V. C. 2007. The Middle Ordovician to Early Silurian voyage of the Dashwoods microcontinent, west Newfoundland; based on new U/Pb and $^{40}Ar/^{39}Ar$ geochronological, and kinematic constraints. *American Journal of Science*, **307**, 311–338.

BURDEN, E. T., QUINN, L., NOWLAN, G. S. & BAILEY-NILL, L. A. 2002. Palynology and micropaleontology of the Clam Bank Formation (Lower Devonian) of western Newfoundland. *Palynology*, **26**, 185–215.

BURGESS, J. L., BROWN, M., DALLMEYER, R. D. & VAN STAAL, C. R. 1995. Microstructure, metamorphism, thermochronology and *P-T-t* deformation history of the Port aux Basques gneisses, south-west Newfoundland, *Canadian Journal of Metamorphic Geology*, **13**, 751–776.

CABANIS, B. & LECOLLE, M. 1989. Le diagramme La/10–Y/15–Nb/8: un outil pour la discrimination des series volcaniques et la mise en evidence des processus de mélange et/ou de contamination crustal. *Comptus Rendus de L'Acadamie des Sciences, Serie 2*, **309**, 2023–2029.

CASTONGUAY, S., RUFFET, G. & TREMBLAY, A. 2007. Dating polyphase deformation across low-grade metamorphic belts: an example based on $^{40}Ar/^{39}Ar$ muscovite age constraints from the southern Quebec Appalachians, Canada. *Geological Society of America Bulletin*, **119**, 978–992.

CASTONGUAY, S., RUFFET, G., TREMBLAY, A. & FERAUD, G. 2001. Tectonometamorphic evolution of the southern Quebec Appalachians: $^{40}Ar/^{39}Ar$ evidence for Middle Ordovician crustal thickening and Silurian-Early Devonian exhumation of the internal Humber zone. *Geological Society of America Bulletin*, **113**, 144–160.

CASTONGUAY, S., WATTERS, S. & RAVENELLE, J. F. 2003. *Preliminary report on the structural geology of the Clarence Stream-Moores Mills area, southwestern New Brunswick: implications for gold exploration.* Geological Survey of Canada, Current Research, **2003-D2**, 1–10.

CAWOOD, P. A., DUNNING, G. A., LUX, D. & VAN GOOL, J. A. M. 1994. Timing of peak metamorphism and deformation along the Appalachian margin of Laurentia in Newfoundland: Silurian, not Ordovician. *Geology*, **22**, 399–402.

CHEW, D. M., DALY, J. S., PAGE, L. M. & KENNEDY, M. J. 2003. Grampian orogenesis and the development of blueschist-facies metamorphism in western Ireland. *Journal of the Geological Society, London*, **160**, 911–924.

COLMAN-SADD, S. D. 1980. Geology of south-central Newfoundland and evolution of the eastern margin of Iapetus. *American Journal of Science*, **280**, 991–1017.

COLMAN-SADD, S. P., DUNNING, G. R. & DEC, T. 1992. Dunnage-Gander relationships and Ordovician orogeny in central Newfoundland: a sediment provenance and U/Pb study. *American Journal of Science*, **292**, 317–355.

CURRIE, K. L. 1995. Plutonic rocks; Chapter 8. *In*: WILLIAMS, H. (ed.) *The Appalachian/Caledonian Orogen: Canada and Greenland*. Geological Survey

of Canada, Geology of Canada, **6**, 631–680 (also Geological Society of America, the Geology of North America, **F-1**).

CURRIE, K. L. 2003. Emplacement of the Fogo Island batholith. *Atlantic Geology*, **39**, 79–96.

CURRIE, K. L., WHALEN, J. B., DAVIS, W. J., LONGSTAFFE, F. J. & COUSENS, B. L. 1998. Geochemical evolution of peraluminous plutons in southern Nova Scotia, Canada – a pegmatite-poor suite. *Lithos*, **44**, 117–140.

DAVID, J. & GARIÉPY, C. 1990. Early Silurian orogenic andesites from the central Quebec Appalachians. *Canadian Journal Earth Sciences*, **27**, 632–643.

DEAN, P. L. & STRONG, D. F. 1978. Folded thrust faults in Notre Dame Bay, central Newfoundland. *American Journal of Science*, **277**, 97–108.

DEWEY, J. F. 2002. Transtension in arcs and orogens. *International Geology Reviews*, **44**, 402–439.

DEWEY, J. F. & SHACKLETON, R. M. 1984. A model for the evolution of the Grampian tract in the early Caledonides and Appalachians. *Nature*, **312**(5990), 115–121.

DICKINSON, W. R. & SEELY, D. R. 1979. Structure and stratigraphy of fore-arc regions. *American Association of Petroleum Geologists Bulletin*, **63**, 2–31.

DICKSON, L., MCNICOLL, V. J., NOWLAN, G. S. & DUNNING, G. R. 2007. The Indian Islands Group and its relationships to adjacent units: recent data. *In*: *Current Research 2007*. Newfoundland and Labrador Department of Natural Resources, Geological Survey Report, **07-1**, 1–9.

DIMITROV, I., MCCUTCHEON, S. R. & WILLIAMS, P. F. 2004. Stratigraphic and structural observations in Silurian rocks between Pointe Rochette and the Southeast Upsalquitch River, northern New Brunswick: A progress report. *In*: MARTIN, G. L. (ed.) *Geological Investigations in New Brunswick for 2003*. New Brunswick Department of Natural Resources; Minerals, Policy and Planning Division: Mineral Resource Report, **2004-4**, 41–74.

D'LEMOS, R. S., SCHOFIELD, D. I., HOLDSWORTH, R. E. & KING, T. R. 1997. Deep crustal and local rheological controls on the siting and reactivation of fault and shear zones, northeastern Newfoundland. *Journal of the Geological Society, London*, **154**, 117–121.

DOIG, R., NANCE, R. D., MURPHY, J. B. & CASSEDAY, R. P. 1990. Evidence for Silurian sinistral accretion of Avalon composite terrane in Canada. *Journal of the Geological Society, London*, **147**, 927–930.

DOSTAL, J., LAURENT, R. & KEPPIE, J. D. 1993. Late Silurian–Early Devonian rifting during dextral transpression in the southern Gaspé Peninsula (Quebec): petrogenesis of volcanic rocks. *Canadian Journal of Earth Sciences*, **30**, 2283–2294.

DOSTAL, J., WILSON, R. A. & KEPPIE, J. D. 1989. Geochemistry of Siluro-Devonian Tobique volcanic belt in northern and central New Brunswick (Canada): tectonic implications. *Canadian Journal of Earth Sciences*, **26**, 1282–1296.

DUBÉ, B., DUNNING, G. R., LAUZIERE, K. & RODDICK, J. C. 1996. New insights into the Appalachian Orogen from geology and geochronology along the Cape Ray fault zone, southwest Newfoundland. *Geological Society of America Bulletin*, **108**, 101–116.

DUNNING, G., O'BRIEN, S. J., COLMAN-SADD, S. P., BLACKWOOD, R., DICKSON, W. L., O'NEILL, P. P. & KROGH, T. E. 1990. Silurian orogeny in the Newfoundland Appalachians. *Journal of Geology*, **98**, 895–913.

ELLIOTT, C. G., BARNES, C. R. & WILLIAMS, P. F. 1989. Southwest New World Island stratigraphy: new fossil data, new implications for the history of the Central Mobile Belt, Newfoundland. *Canadian Journal of Earth Sciences*, **26**, 2062–2074.

ELLIOTT, C. G., DUNNING, G. R. & WILLIAMS, P. F. 1991. New constraints on the timing of deformation in eastern Notre Dame Bay, Newfoundland, from U/Pb zircon ages of felsic intrusions. *Geological Society of America Bulletin*, **103**, 125–135.

FOX, D. & VAN BERKEL, J. T. 1988. Mafic-ultramafic occurrences in metasedimentary rocks of southwestern Newfoundland. *In: Current Research, Part B.* Geological Survey of Canada, Paper **88-1B**, 41–48.

FORTEY, R. A. & COCKS, L. R. 2003. Palaeontological evidence bearing on global Ordovician-Silurian continental reconstructions. *Earth Science Reviews*, **61**, 245–307.

FRIEDRICH, A. M., HODGES, K. V., BOWRING, S. A. & MARTIN, M. W. 1999. Geochronological constraints on the magmatic, metamorphic and thermal evolution of the Connemara Caledonides, western Ireland. *Journal of the Geological Society, London*, **156**, 1217–1230.

FYFFE, L. R. & BEVIER, M. L. 1992. A U–Pb date on the Mohannes pluton of southwestern New Brunswick. *Atlantic Geology*, **28**, 198.

FYFFE, L. R., MCLEOD, M. J. & RUITENBERG, A. A. 1991. A geotraverse across the St. Croix-Avalon terrane boundary, southern New Brunswick; *In:* LUDMAN, A. (ed.) *Geology of the Coastal Lithotectonic Block and Neighbouring Terranes, Eastern Maine and Southern New Brunswick.* New England Intercollegiate Geological Conference, NEIGC 1991, Field Trip **A-2**, 13–54.

FYFFE, L. R., PICKERILL, R. K. & STRINGER, P. 1999. Stratigraphy, sedimentology and structure of the Oak Bay and Waweig formations, Mascarene basins: implications for the Paleotectonic evolution of southwestern New Brunswick. *Atlantic Geology*, **35**, 59–84.

GERBI, C., JOHNSON, S. E. & ALEINIKOFF, J. N. 2006a. Origin and orogenic role of the chain Lakes massif, Maine and Quebec. *Canadian Journal of Earth Sciences*, **43**, 339–366.

GERBI, C., JOHNSON, S. E., ALEINIKOFF, J. N., BEDARD, J. H., DUNNING, G. & FANNING, C. M. 2006b. Early Paleozoic development of the Maine-Quebec Boundary Mountains region. *Canadian Journal of Earth Sciences*, **43**, 367–389.

HALL, L. & VAN STAAL, C. R. 1999. *Geology of Southern End of Long Range Mountains (Dashwoods Subzone), Newfoundland.* Geological Survey of Canada, Open File **3727**, scale 1:50 000.

HAMILTON, M. A. & MURPHY, J. B. 2004. Tectonic significance of a Llanvirn age for the Dunn Point volcanic rocks, Avalon terrane, Nova Scotia, Canada: implications for the evolution of the Iapetus and Rheic Oceans. *Tectonophysics*, **379**, 199–209.

HASCHKE, M. R., SCHEUBER, E., GUNTHER, A. & REUTER, K. J. 2002. Evolutionary cycles during the Andean orogeny: repeated slab breakoff and flat subduction. *Terra Nova*, **14**, 49–55.

HEPBURN, J. C. 2008. Mid-Paleozoic arc accretion on the eastern side of the Appalachian orogen, eastern Massachusetts and adjacent areas. *Abstracts with Programs.* Geological Society of America, 43th Annual Meeting, NE-section, 75.

HIBBARD, J. P. 1994. Kinematics of Acadian deformation in the Northern and Newfoundland Appalachians. *Journal of Geology*, **102**, 215–228.

HIBBARD, J. P. 2000. Docking Carolina: Mid Paleozoic accretion in the southern Appalachians. *Geology*, **28**, 127–130.

HIBBARD, J. P., VAN STAAL, C. R. & RANKIN, D. W. 2007. A comparative analysis of pre-Silurian building blocks of the Northern and Southern Appalachians. *American Journal of Science*, **307**, 23–45.

HIBBARD, J. P., VAN STAAL, C. R., RANKIN, D. & WILLIAMS, H. 2006. *Geology, Lithotectonic Map of the Appalachian Orogen, Canada–United States of America.* Geological Survey of Canada, Map **2096A**, scale 1:1 500 000.

HICKS, R. J., JAMIESON, R. A. & REYNOLDS, P. 1999. Detrital and metamorphic $^{40}$Ar/$^{39}$Ar ages from muscovite and whole-rock samples, Meguma Supergroup, southern Nova Scotia. *Canadian Journal of Earth Sciences*, **36**, 23–32.

HOLDSWORTH, R. E. 1994. Structural evolution of the Gander-Avalon terrane boundary: a reactivated transpression zone in the NE Newfoundland Appalachians. *Journal of the Geological Society, London*, **151**, 629–646.

HUOT, F., HÉBERT, R. & TURCOTTE, B. 2002. A multistage magmatic history for the genesis of the Orford ophiolite (Quebec, Canada): a study of the Mont Chagnon Massif. *Canadian Journal of Earth Sciences*, **39**, 1201–1217.

HUTCHINSON, D. R., KLITGORD, K. D., LEE, M. W. & TREHU, A. M. 1988. U.S. Geological Survey deep seismic reflection profile across the Gulf of Maine. *Geological Society of America Bulletin*, **100**, 172–184.

HYNDMAN, R. D., CURRIE, C. A. & MAZZOTTI, S. P. 2005. Subduction zone backarcs, mobile belts, and orogenic heat. *GSA Today*, **15**, 4–10.

JAMIESON, R. A. 1988. Metamorphic *P-T-t* data from western Newfoundland and Cape Breton-implications for Taconian and Acadian tectonics. *Geological Society of Canada, Program with Abstracts*, **13**, A60.

JOHNSON, R. J. E. & VAN DER VOO, R. 1986. Paleomagnetism of the late Precambrian Fourchu Group, Cape Breton Island, Nova Scotia. *Canadian Journal of Earth Sciences*, **23**, 1673–1685.

KARABINOS, P., SAMSON, S. D., HEPBURN, J. C. & STOLL, H. M. 1998. Taconian orogeny in the New England Appalachians: collision between Laurentia and the Shelburne Falls arc. *Geology*, **26**, 215–218.

KAY, S. M. & ABRUZZI, J. M. 1996. Magmatic evidence for Neogene lithospheric evolution of the central Andean 'flat-slab' between 30°S and 32°S. *Tectonophysics*, **259**, 15–28.

KAY, S. M., GODOY, E. & KURTZ, A. 2005. Episodic arc migration, crustal thickening, subduction erosion, and magmatism in the south-central Andes. *Geological Society of America Bulletin*, **117**, 67–88.

KEEN, C. E., KAY, W. A., KEPPIE, D., MARILLIER, F., PE-PIPER, G. & WALDRON, J. W. F. 1991. Deep seismic reflection data from the Bay of Fundy and Gulf of Maine: tectonic implications for the northern Appalachians. *Canadian Journal of Earth Sciences*, **28**, 1096–1111.

KEPPIE, J. D. & KROGH, T. E. 2000. 440Ma igneous activity in the Meguma terrane, Nova Scotia, Canada: part of the Appalachian overstep sequence? *American Journal of Science*, **300**, 528–538.

KERR, A. 1997. Space-time composition relationships among Appalachian-cycle plutonic suites in Newfoundland. *In*: SINHA, K., WHALEN, J. B. & HOGAN, J. P. (eds) *The Nature of Magmatism in the Appalachian Orogen.* Geological Society of America, Memoir, **191**, 193–220.

KERR, A., HAYES, J. P., COLMAN-SADD, S. P., DICKSON, W. L. & BUTLER, A. J. 1994. *An integrated lithogeochemical database for the granitoid plutonic series of Newfoundland.* Newfoundland Department of Mines and Energy, Geological Survey Branch, Open File Report NFLD/2377, 1–44.

KERR, A., JENNER, G. J. & FRYER, B. J. 1995. Sm–Nd isotopic geochemistry of Precambrian to Paleozoic granitoid suites and the deep-crustal structure of the southeast margin of the Newfoundland Appalachians. *Canadian Journal of Earth Sciences*, **32**, 224–245.

KING, M. S. & BARR, S. M. 2004. Magnetic and gravity models across terrane boundaries in southern New Brunswick, Canada. *Canadian Journal of Earth Sciences*, **41**, 1027–1047.

KNIGHT, I., JAMES, N. P. & LANE, T. E. 1991. The Ordovician St. George Unconformity, northern Appalachians: the relationship of plate convergence at the St. Lawrence promontory to the Sauk/Tippecanoe sequence boundary. *Geological Society of America Bulletin*, **103**, 1200–1225.

KROGH, T. E. & KEPPIE, J. D. 1990. Age of detrital zircon and titanite in the Meguma Group, southern Nova Scotia, Canada: clues to the origin of the Meguma Terrane. *Tectonophysics*, **177**, 307–323.

LAFRANCE, B. 1989. Structural evolution of a transpression zone in north central Newfoundland. *Journal of Structural Geology*, **11**, 705–716.

LAFRANCE, B. & WILLIAMS, P. F. 1992. Silurian deformation in eastern Notre Dame Bay, Newfoundland. *Canadian Journal of Earth Sciences*, **29**, 1899–1914.

LAIRD, J., TRZCIENSKI, W. E., JR. & BOTHNER, W. A. 1993. *High pressure Taconian and subsequent polymetamorphism of southern Quebec and northern Vermont.* Department of Geology and Geography, University of Massachusetts, Contribution, **67–2**, 1–32.

LANDING, E. 1996. Avalon: Insular continent by the latest Precambrian. *In*: NANCE, R. D. & THOMPSON, M. D. (eds) *Avalonian and Related Peri-Gondwanan Terranes of the Circum-North Atlantic.* Geological Society of America, Special Paper, **304**, 29–63.

LIN, S., DAVIS, D. D., BARR, S. M., CHEN, Y., VAN STAAL, C. R. & CONSTANTIN, M. 2007. U–Pb geochronological constraints on the geological evolution and regional correlation of the Aspy terrane, Cape Breton Island, Canadian Appalachians. *American Journal of Science*, **307**, 371–398.

LIN, S., JIANG, D. & WILLIAMS, P. F. 1998. Transpression (or transtension) zones of triclinic symmetry: natural example and theoretical modeling, *In*: HOLDSWORTH, R. E., STACHAN, R. A. & DEWEY, J. F. (eds) *Continental transpressional and transtensional tectonics.* Geological Society, London, Special Publications, **135**, 41–57.

LISS, M. J., VAN DER PLUIJM, B. A. & VAN DER VOO, R. 1994. Avalonian proximity of the Ordovician Miramichi terrane, northern New Brunswick, northern Appalachians: Paleomagnetic evidence for rifting and back-arc basin formation at the southern margin of Iapetus. *Tectonophysics*, **227**, 17–30.

LISSENBERG, C. J. & VAN STAAL, C. R. 2006. Feedback between deformation and magmatism in the Lloyd's River Fault zone, Central Newfoundland: An example of episodic fault reactivation in an accretionary setting. *Tectonics*, **25**, TC4004.

LISSENBERG, C. J., ZAGOREVSKI, A., MCNICOLL, V. J., VAN STAAL, C. R. & WHALEN, J. B. 2005a. Assembly of the Annieopsquotch Accretionary Tract, southwest Newfoundland: Age- and geodynamic constraints from syn-kinematic intrusions. *Journal of Geology*, **113**, 553–570.

LISSENBERG, C. J., VAN STAAL, C. R., BÉDARD, J. H. & ZAGOREVSKI, A. 2005b. Geochemical constraints on the origin of the Annieopsquotch ophiolite belt, southwest Newfoundland. *Geological Society of America Bulletin*, **117**, 1413–1426.

LUDMAN, A., HOPECK, J. T. & BROCK, P. C. 1993. Nature of the Acadian Orogeny in eastern Maine. *In*: ROY, D. C. & SKEHAN, J. W. (eds) *The Acadian Orogeny: Recent studies in New England, Maritime Canada, and the Autochthonous Foreland.* Geological Society of America, Special Paper, **275**, 67–84.

MACDONALD, L. A., BARR, S. M., WHITE, C. E. & KETCHUM, J. W. F. 2002. Petrology, age, and tectonic setting of the White Rock Formation, Meguma terrane, Nova Scotia: Evidence for Silurian continental rifting. *Canadian Journal of Earth Sciences*, **39**, 259–277.

MACNIOCAILL, C. 2000. A new Silurian palaeolatitude for eastern Avalonia and evidence for crustal rotations in the Avalonian margin of southwestern Ireland. *Geophysical Journal International*, **141**, 661–671.

MALO, M. & KIRKWOOD, D. 1995. Faulting and progressive strain history of the Gaspé Peninsula in post-Taconian time: A review. *In*: HIBBARD, J. P., VAN STAAL, C. R. & CAWOOD, P. A. (eds) *Current Perspectives in the Appalachian–Caledonian Orogen.* Geological Association of Canada, Special Paper, **41**, 267–282.

MARTIN, H. 1986. Effect of steeper Archean geothermal gradient on geochemistry of subduction-zone magmas. *Geology*, **14**, 753–756.

MCLAUGHLIN, R. J., BARR, S. M., HILL, M. D., THOMPSON, M. D., RAMEZANI, J. & REYNOLDS, P. H. 2003. The Moosehorn plutonic suite, southeastern Maine and southwestern New Brunswick: Age, petrochemistry, and tectonic setting. *Atlantic Geology*, **39**, 123–146.

MCLEOD, M. J. 1990. *Geology, geochemistry, and related mineral deposits of the Saint George batholith.* New

Brunswick Department of Natural Resources and Energy, Mineral Resource Report, **5**, 1–169.

MCLEOD, M. J., JOHNSON, S. C. & KROGH, T. E. 2003. Archived U–Pb zircon dates from southern New Brunswick. *Atlantic Geology*, **39**, 209–225.

MCLEOD, M. J., PICKERILL, R. K. & LUX, R. D. 2001. Mafic intrusions on Campobello Island: Implications for New Brunswick-Maine correlations. *Atlantic Geology*, **37**, 17–40.

MCNAMARA, A. K., MACNIOCAILL, C., VAN DER PLUIJM, B. A. & VAN DER VOO, R. 2001. West African proximity of the Avalon terrane in the latest Precambrian. *Geological Society America Bulletin*, **113**, 1161–1170.

MCNICOLL, V., VAN STAAL, C. R. & WALDRON, J. W. F. 2001. Accretionary history of the Northern Appalachians: SHRIMP study of Ordovician-Silurian syntectonic sediments in the Canadian Appalachians. Abstract Volume, GAC/MAC Annual meeting, St. John's Nfld., 1–100.

MEANS, W. D. 1995. Shear zones and rock history. *Tectonophysics*, **247**, 157–160.

MILLER, B. V. & FYFFE, L. R. 2002. Geochronology of the Letete and Waweig formations, Mascarene Group, southwestern New Brunswick. *Atlantic Geology*, **38**, 29–36.

MILLER, B. V., FETTER, A. H. & STEWART, K. G. 2006. Plutonism in three orogenic pulses, eastern Blue Ridge province, southern Appalachians. *Geological Society of America Bulletin*, **118**, 171–184.

MOENCH, R. H. & ALEINIKOFF, J. N. 2003. Stratigraphy, geochronology and accretionary terrane setting of two Bronson Hill arc sequences, northern New England. *Physics and Chemistry of the Earth*, **28**, 113–160.

MORAN, P. C., BARR, S. M., WHITE, C. E. & HAMILTON, M. A. 2007. Petrology, age and tectonic setting of the Seal Island Pluton, offshore southwestern Nova Scotia. *Canadian Journal of Earth Sciences*, **44**, 1–12.

MURPHY, J. B., NANCE, R. D. & KEPPIE, J. D. 2002. Discussion and reply: West African proximity of the Avalon terrane in the latest Precambrian. *Geological Society America Bulletin*, **114**, 1049–11052.

MURPHY, J. B., KEPPIE, J. D., DOSTAL, J. & COUSINS, B. l. 1996. Repeated Late Neoproterozoic-Silurian lower crustal melting beneath the Antigonish Highlands, Nova Scotia: Nd-isotopic evidence and tectonic interpretations. *In*: NANCE, R. D. & THOMPSON, M. D. (eds) *Avalonian and Related Peri-Gondwanan Terranes of the Circum-North Atlantic*. Geological Society of America, Special Paper, **304**, 109–120.

MURPHY, J. B., VAN STAAL, C. R. & KEPPIE, J. D. 1999. Middle to Late Paleozoic Acadian Orogeny in the northern Appalachians: A Laramide-style plume-modified orogeny. *Geology*, **27**, 653–656.

MURPHY, J. B., FERNANDEZ-SUAREZ, J., KEPPIE, J. D. & JEFFRIES, T. 2004. Contiguous rather than discrete Paleozoic histories for the Avalon and Meguma terranes based on detrital zircon data. *Geology*, **32**, 585–588.

NANCE, R. D. & DALLMEYER, R. D. 1993. $^{40}$Ar/$^{39}$Ar amphibole ages from the Kingston Complex, New Brunswick: Evidence for Silurian-Devonian tectonothermal activity and implications for the accretion of the Avalon composite terrane. *Journal of Geology*, **101**, 375–388.

O'BRIEN, B. 2003. *Geology of the central Notre Dame Bay region (parts of NTS areas 2E/3,6,11), northeastern Newfoundland*. Government of Newfoundland and Labrador Department of Mines and Energy, Report, **03–03**, 1–147.

O'BRIEN, B., O'BRIEN, S. & DUNNING, G. R. 1991. Silurian cover, Late Precambrian-Early Ordovician basement, and the chronology of Silurian orogenesis in the Hermitage Flexure (Newfoundland Appalachians). *American Journal of Science*, **291**, 760–799.

O'BRIEN, S. J., O'BRIEN, B. H., DUNNING, G. R. & TUCKER, R. D. 1996. Late Neoproterozoic Avalonian and related peri-Gondwanan rocks of the Newfoundland Appalachians. *In*: NANCE, R. D. & THOMPSON, M. D. (eds) *Avalonian and Related Peri-Gondwanan Terranes of the Circum-North Atlantic*. Geological Society of America, Special Paper, **304**, 9–28.

O'BRIEN, B., SWINDEN, H. S., DUNNING, G. R., WILLIAMS, S. H. & O'BRIEN, F. 1997. A peri-Gondwanan arc-back arc complex in Iapetus: Early-Mid Ordovician evolution of the Exploits Group, Newfoundland. *American Journal of Science*, **297**, 220–272.

PARK, A. F. & WHITEHEAD, J. 2003. Structural transect through Silurian turbidites of the fredericton Belt southwest of Fredericton, New Brunswick: The role of the Fredericton Fault in late Iapetus convergence. *Atlantic Geology*, **39**, 227–237.

PARK, A. F., WILLIAMS, P. F., RALSER, S. & LEGER, A. 1994. Geometry and kinematics of a major crustal shear zone segment in the Appalachians of southern New Brunswick. *Canadian Journal of Earth Sciences*, **31**, 1523–1535.

PEARCE, J. A. 1996. Sources and settings of granitic rocks. *Episodes*, **19**, 120–125.

PEARCE, J. A., HARRIS, N. B. W. & TINDLE, A. G. 1984. Trace element discrimination diagrams for the tectonic interpretation of granitic rocks. *Journal of Petrology*, **25**, 956–983.

PEHRSSON, S., VAN STAAL, C. R., HERD, R. K. & MCNICOLL, V. 2003. The Cormacks Lake complex, Dashwoods Subzone: a window into the deeper levels of the Notre Dame Arc. *In*: PEREIRA, C. P. G., WALSH, D. G. & KEAN, B. F. (eds) *Current Research 2002*. Newfoundland Department of Mines and Energy, Geological Survey, Report, **2002–1**, 115–125.

PE-PIPER, G. & JANSA, L. F. 1999. Pre-mesozoic basement rocks offshore Nova Scotia, Canada: New constraints on the accretion history of the Meguma terrane. *Geological Society America Bulletin*, **111**, 1773–1791.

PE-PIPER, G. & PIPER, D. J. W. 2002. A synopsis of the geology of the Cobequid Highlands, Nova Scotia. *Atlantic Geology*, **38**, 145–160.

PIASECKI, M. A. J. 1988. *A major ductile shear zone in the Bay d'Espoir area, Gander terrane, southeastern Newfoundland*. Newfoundland Department of Mines and Energy, Mineral Development Division, Report, **88–1**, 135–144.

PINCIVY, A., MALO, M., RUFFET, G., TREMBLAY, A. & SACKS, P. E. 2003. Regional metamorphism of the

Appalachian Humber zone of Gaspe Peninsula: $^{40}Ar/^{39}Ar$ evidence for crustal thickening during the Taconian orogeny. *Canadian Journal of Earth Sciences*, **40**, 301–315.

POLLOCK, J. C., WILTON, D. H. C., VAN STAAL, C. R. & MORRISSEY, K. D. 2007. U–Pb detrital zircon geochronological constraints on the early Silurian collision of Ganderia and Laurentia along the Dog Bay Line: The terminal Iapetan suture in the Newfoundland Appalachians. *American Journal of Science*, **307**, 399–433.

POTTER, J., LONGSTAFFE, F. J. & BARR, S. M. 2008. Regional $^{18}O$ depletion of Neoproterozoic igneous rocks from Avalonia, Cape Breton Island and southern New Brunswick, Canada. *Geological Society of America Bulletin*, **120**, 347–367.

PRICE, J. R., BARR, S. M., RAESIDE, R. P. & REYNOLDS, P. H. 1999. Petrology, tectonic setting, and $^{40}Ar/^{39}Ar$ (hornblende) dating of the Late Ordovician–Early Silurian Belle Cote Road orthogneiss, western Cape Breton Highlands, Nova Scotia. *Atlantic Geology*, **35**, 1–17.

PRIGMORE, J. K., BULLER, A. J. & WOODCOCK, N. H. 1997. Rifting during separation of eastern Avalonia from Gondwana: evidence from subsidence analysis. *Geology*, **25**, 203–206.

QUINN, L., BASHFORTH, A. R., BURDEN, E. T., GILLESPIE, H., SPRINGER, R. K. & WILLIAMS, S. H. 2004. The Red Island Road Formation: Early Devonian terrestrial fill in the Anticosti foreland basin, western Newfoundland. *Canadian Journal of Earth Sciences*, **41**, 587–602.

RATCLIFFE, N. M., HAMES, W. E. & STANLEY, R. S. 1998. Interpretation of ages of arc magmatism and collisional tectonics in the Taconian Orogen of western New England. *American Journal of Earth Sciences*, **298**, 791–797.

RATCLIFFE, N. M., HARRIS, A. G. & WALSH, G. J. 1999. Tectonic and regional metamorphic implications of the discovery of Middle Ordovician conodonts in cover rocks east of Green Mountain massif, Vermont. *Canadian Journal of Earth Sciences*, **36**, 371–382.

RICHTER, D. A. & ROY, D. C. 1974. Sub-greenschist metamorphic assemblages in northern Maine. *Canadian Mineralogist*, **12**, 469–474.

ROBINSON, P., TUCKER, R. D., BRADLEY, D., BERRY, H. N. & OSBERG, P. H. 1998. Paleozoic orogens in New England, USA. *Geologiska Fóreningens Stockholm Forhandlingar*, **120**, 119–148.

ROGERS, N. & VAN STAAL, C. R. 2003. Volcanology and tectonic setting of the northern Bathurst Mining Camp: Part II – Mafic volcanic constraints on back-arc opening. *Economic Geology, Monograph*, **11**, 181–202.

ROGERS, N., VAN STAAL, C. R. & VALVERDE-VAQUERO, P. 2005. *Geology, Noel Paul's Brook, Newfoundland (NTS 12-A/9)*. Geological Survey of Canada Open File OF**4547**, scale 1:50 000, 1 sheet.

ROGERS, N., VAN STAAL, C. R., McNICOLL, V., POLLOCK, J., ZAGOREVSKI, A. & WHALEN, J. 2006. Neoproterozoic and Cambrian arc magmatism along the eastern margin of the Victoria Lake Supergroup: A remnant of Ganderian basement in central

Newfoundland? *Precambrian Research*, **148**, 320–341.

SACKS, P. E., MALO, M., TRZCIENSKI, W. E., PINCIVY, A. & GOSSELIN, P. 2004. Taconian and Acadian transpression between the internal Humber Zone and the Gaspe belt in the Gaspe Peninsula: tectonic history of the Shickshock Sud fault zone. *Canadian Journal of Earth Sciences*, **41**, 635–653.

SAMSON, S. D., BARR, S. M. & WHITE, C. E. 2000. Nd isotopic characteristics of terranes within the Avalon Zone, southern New Brunswick. *Canadian Journal of Earth Sciences*, **37**, 1039–1052.

SCHENK, P. E. 1997. Sequence stratigraphy and provenance on Gondwana's margin: The Meguma zone (Cambrian to Devonian) of Nova Scotia, Canada. *Geological Society America Bulletin*, **109**, 395–409.

SCHOFIELD, D. I. & D'LEMOS, R. S. 2000. Granite petrogenesis in the Gander Zone, NE Newfoundland: mixing of melts from multiple sources and the role of lithospheric delamination. *Canadian Journal of Earth Sciences*, **37**, 535–547.

SCHULTZ, K. J., STEWART, D. B., TUCKER, R. D., POLLOCK, J. & AYUSO, R. A. 2008. The Ellsworth terrane coastal Maine: Geochronology, geochemistry, and Nd–Pb isotopic composition – Implications for the rifting of Ganderia. *Geological Society of America Bulletin*, **120**, 1134–1158.

STOCKMAL, G. S. & WALDRON, J. W. F. 1993. Structural and tectonic evolution of the Humber Zone, western Newfoundland: Implications of cross sections through the Appalachian structural front, Port au Port Peninsula. *Tectonics*, **12**, 1056–1075.

SWINDEN, H. S., JENNER, G. A. & SZYBINSKI, Z. A. 1997. Magmatic and tectonic evolution of the Cambrian-Ordovician Laurentian margin of Impetus: Geochemical and isotopic constraints from the Notre Dame subzone, Newfoundland. *In*: SINHA, K., WHALEN, J. B. & HOGAN, J. P. (eds) *The Nature of Magmatism in the Appalachian Orogen*. Geological Society of America, Memoir, **191**, 367–395.

SZYBINSKI, Z. A. 1995. *Paleotectonic and structural setting of thye western Notre Dame Bay area, Newfoundland Appalachians*. Unpublished PhD thesis, Memorial University of Newfoundland, St. John's, Newfoundland.

TATE, M. C. & CLARKE, D. B. 1995. Petrogenesis and regional tectonic significance of Late Devonian mafic intrusions in the Meguma Zone, Nova Scotia. *Canadian Journal of Earth Sciences*, **32**, 1883–1898.

TATE, M. C. & MERRETT, D. 1994. *Compilation of major oxide, trace element and rare-earth element analyses for Late Devonian peraluminous granitoids in the Meguma Zone, Nova Scotia*. Nova Scotia Department of Natural Resources, Open File Report, **94–14**, 71.

THURLOW, J. G., SPENCER, C. P., BOERNER, D. E., REED, L. E. & WRIGHT, J. A. 1992. Geophysical interpretation of a high resolution reflection seismic survey at the Buchans mine, Newfoundland. *Canadian Journal of Earth Sciences*, **29**, 2022–2037.

TREMBLAY, A. & PINET, N. 1994. Distribution and characteristics of Taconian and Acadian deformation, southern Quebec Appalachians. *Geological Society of America Bulletin*, **106**, 1172–1181.

TUCKER, R. D., OSBERG, P. H. & BERRY, H. N. 2001. The geology of a part of Acadia and the nature of the Acadian orogeny across central and eastern Maine. *American Journal of Science*, **301**, 205–260.

VALVERDE-VAQUERO, P., DUNNING, G. & VAN STAAL, C. R. 2000. The Margaree orthogneiss (Port aux Basques Complex, SW Newfoundland) evolution of a peri-Gondwanan, Mid Arenig–Early Llanvirn, mafic-felsic igneous complex. *Canadian Journal of Earth Sciences*, **37**, 1691–1710.

VALVERDE-VAQUERO, P. & VAN STAAL, C. R. 2001. Relationships between the Dunnage–Gander zones in the Victoria Lake–Peter Strides Pond area. *In*: PEREIRA, C. P. G., WALSH, D. G. & KEAN, B. F. (eds) *Current Research 2001*. Newfoundland Department of Mines and Energy, Geological Survey Report, **2000–1**, 1–9.

VALVERDE-VAQUERO, P., VAN STAAL, C. R., MCNICOLL, V. & DUNNING, G. 2006. Middle Ordovician magmatism and metamorphism along the Gander margin in Central Newfoundland. *Journal of the Geological Society, London*, **163**, 347–362.

VAN DER PLUIJM, B. A. 1986. Geology of eastern New World Island, Newfoundland: An accretionary terrane in the northeastern Appalachians. *Geological Society of America Bulletin*, **97**, 932–945.

VAN DER VELDEN, A. J., VAN STAAL, C. R. & COOK, F. A. 2004. Crustal structure, fossil subduction and the tectonic evolution of the Newfoundland Appalachians: Evidence from a reprocessed seismic reflection survey. *Geological Society America Bulletin*, **116**, 1485–1498.

VAN DER VOO, R. & JOHNSON, R. J. E. 1985. Paleomagnetism of the Dunn Point Formation (Nova Scotia): High paleolatitudes for the Avalon terrane in the Late Ordovician. *Geophysical Research Letters*, **12**, 337–340.

VAN HUNEN, J., VAN DER BERG, A. P. & VLAAR, N. J. 2002. On the role of subducting oceanic plateaus in the development of shallow flat subduction. *Tectonophysics*, **352**(3–4), 317–333.

VAN STAAL, C. R. 1994. The Brunswick subduction complex in the Canadian Appalachians: Record of the Late Ordovician to Late Silurian collision between Laurentia and the Gander margin of Avalon. *Tectonics*, **13**, 946–962.

VAN STAAL, C. R. 2005. The Northern Appalachians. *In*: SELLEY, R. C., COCKS, L. R. & PLIMER, I. R. (eds) *Encyclopedia of Geology*. Elsevier, Oxford, **4**, 81–91.

VAN STAAL, C. R. 2007. Pre-Carboniferous tectonic evolution and metallogeny of the Canadian Appalachians, *In*: GOODFELLOW, W. D. (ed.) *Mineral Deposits of Canada: A Synthesis of Major Deposit-types, District Metallogeny, the Evolution of Geological Provinces, and Exploration Methods*. Geological Association of Canada, Mineral Deposit Division, Special Publication, **5**, 793–818.

VAN STAAL, C. R. & DE ROO, J. A. 1995. Mid-Paleozoic tectonic evolution of the Appalachian Central Mobile Belt in northern New Brunswick, Canada: Collision, extensional collapse, and dextral transpression. *In*: HIBBARD, J., VAN STAAL, C. R. & CAWOOD, P. (eds) *Current Perspectives in the Appalachian-Caledonian*

*Orogen*. Geological Association of Canada, Special Paper, **41**, 367–389.

VAN STAAL, C. R. & HATCHER, R. D., JR. 2009. Global setting of Ordovician orogenesis. *In*: FINNEY, S. C. & BERRY, W. B. N. (eds) *Global Ordovician Earth System*. Geological Society of America, Special Paper (in press).

VAN STAAL, C. R., RAVENHURST, C., WINCHESTER, J. A., RODDICK, J. C. & LANGTON, J. P. 1990. Post-Taconic blueschist suture in the northern Appalachians of northern New Brunswick, Canada. *Geology*, **18**, 1073–1077.

VAN STAAL, C. R., WINCHESTER, J. A. & BEDARD, J. H. 1991. Geochemical variations in Ordovician volcanic rocks of the northern Miramichi Highlands and their tectonic significance. *Canadian Journal of Earth Sciences*, **28**, 1031–1049.

VAN STAAL, C. R., WINCHESTER, J. A., BROWN, M. & BURGESS, J. L. 1992. A reconnaissance geotraverse through southwestern Newfoundland. Current Research, Part D, Geological Survey of Canada, Paper **92-1D**, 133–144.

VAN STAAL, C. R., DUNNING, G., VALVERDE, P., BURGESS, J. & BROWN, M. 1994. Arenig and younger evolution of the Gander margin: A comparison of the New Brunswick and Newfoundland segments. Geological Association of Canada, NUNA conference: New perspectives in the Appalachian–Caledonian Orogen, Program and Abstracts. *Atlantic Geology*, **30**, 178–179.

VAN STAAL, C. R., SULLIVAN, R. W. & WHALEN, J. B. 1996. Provenance and tectonic history of the Gander Margin in the Caledonian/Appalachian Orogen: Implications for the origin and assembly of Avalonia. *In*: NANCE, R. D. & THOMPSON, M. D. (eds) *Avalonian and Related Peri-Gondwanan Terranes of the Circum-North Atlantic*. Geological Society of America, Special Paper, **304**, 347–367.

VAN STAAL, C. R., DEWEY, J. F., MACNIOCAILL, C. & MCKERROW, S. 1998. The Cambrian-Silurian tectonic evolution of the northern Appalachians: History of a complex, southwest Pacific-type segment of Iapetus. *In*: BLUNDELL, D. J. & SCOTT, A. C. (eds) *Lyell: the Past is the Key to the Present*. Geological Society, London, Special Publications, **143**, 199–242.

VAN STAAL, C. R., ROGERS, N. & TAYLOR, B. E. 2001. Formation of low temperature mylonites and phyllonites by alkali-metasomatic weakening of felsic volcanic rocks during a progressive, subduction-related deformation. *Journal of Structural Geology*, **23**, 903–921.

VAN STAAL, C. R., WILSON, R. A. *ET AL*. 2003. Geology and tectonic history of the Bathurst Supergroup and its relationships to coeval rocks in southwestern New Brunswick and adjacent Maine – a synthesis. *In*: GOODFELLOW, W. D., MCCUTCHEON, S. R. & PETER, J. M. (eds) *Massive Sulfide Deposits of the Bathurst Mining Camp, New Brunswick, and Northern Maine*. Economic Geology, Monograph, **11**, 37–60.

VAN STAAL, C. R., MCNICOLL, V., VALVERDE-VAQUERO, P., BARR, S., FYFFE, L. R. & REUSCH, D. N. 2004. Ganderia, Avalonia, and the Salinic and Acadian orogenies. *Abstracts with Programs*,

*Geological Society of America, 39th Annual Meeting, NE-section,* 128–129.

VAN STAAL, C. R., VALVERDE-VAQUERO, P., ZAGOR-EVSKI, A., BOUTSMA, S., PEHRSSON, S., VAN NOORDEN, M. & McNICOLL, V. 2005. *Geology, King George IV Lake, Newfoundland (NTS 12-A/04).* Geological Survey of Canada, Open File **1665**, scale 1:50 000, 1 sheet.

VAN STAAL, C. R., WHALEN, J. B. *ET AL*. 2007. The Notre Dame arc and the Taconic Orogeny in Newfoundland. *In*: HATCHER, R. D., JR., CARLSON, M. P., McBRIDE, J. H. & MARTÍNEZ CATALÁN, J. R. (eds) *4-D Framework of Continental Crust.* Geological Society of America, Memoir, **200**, 511–552.

VAN STAAL, C. R., CURRIE, K. L., ROWBOTHAM, G., GOODFELLOW, W. & ROGERS, N. 2008. Pressure–temperature paths and exhumation of Late Ordovician–Early Silurian blueschists and associated metamorphic nappes of the Salinic Brunswick subduction complex, northern Appalachians. *Geological Society of America Bulletin*, **120**, 1455–1477.

VAN STAAL, C. R., CASTONGUAY, S., McNICOLL, V., BREM, A., HIBBARD, J., SKULSKI, T. & JOYCE, N. 2009. Taconic arc–continent collision confirmed in the Newfoundland Appalachians. *Geological Society of America, 44th Annual Meeting, NE–section,* p. 4.

VAN WAGONER, N. A., LEYBOURNE, M. I., DADD, K. A. & HUSKINS, M. L. A. 2001. The Silurian(?) Passamaquoddy Bay mafic dyke swarm, New Brunswick: Petrogenesis and tectonic implications. *Canadian Journal of Earth Sciences*, **38**, 1565–1578.

VAN WAGONER, N. A., LEYBOURNE, M. I., DADD, K. A., BALDWIN, D. K. & McNEIL, W. 2002. Late Silurian bimodal volcanism of southwestern New Brunswick, Canada: Products of continental extension. *Geological Society of America Bulletin*, **114**, 400–418.

WALDRON, J. W. F. & VAN STAAL, C. R. 2001. Taconic orogeny and the accretion of the Dashwoods block: A peri-Laurentian microcontinent in the Iapetus Ocean. *Geology*, **29**, 811–814.

WALDRON, J. W. F., PIPER, D. J. W. & PE-PIPER, G. 1989. Deformation of the Cape Chignecto Pluton, Cobequid Highlands, Nova Scotia: Thrusting at the Meguma-Avalon boundary. *Atlantic Geology*, **25**, 51–62.

WALDRON, J. W. F., ANDERSON, S. D., CAWOOD, P. A., GOODWIN, L. B., HALL, J., JAMIESON, R. A., PALMER, S. E., STOCKMAL, G. S. & WILLIAMS, P. F. 1998. Evolution of the Appalachian Laurentian margin; Lithoprobe results in western Newfoundland. Lithoprobe East transect–Le transect est du projet Lithoprobe: Ottawa, ON, Canada. *Canadian Journal of Earth Sciences*, **35**, 1271–1287.

WALDRON, J. W. F., MURPHY, J. B., MELCHIN, M. J. & DAVIS, G. 1996. Silurian tectonics of Western Avalonia: Strain-corrected subsidence history of the Arisaig Group, Nova Scotia. *Journal of Geology*, **104**, 677–694.

WALDRON, J. W. F., HENRY, A. D., BRADLEY, J. C. & PALMER, S. E. 2003. Development of a folded thrust stack: Humber Arm Allochthon, Bay of Islands, Newfoundland Appalachians. *Canadian Journal of Earth Sciences*, **40**, 237–253.

WALDRON, J. W. F., WHITE, C. E., BARR, S. M., SIMONETTI, A. & HEAMAN, L. M. 2009. Provenance of the meguma terrane, Nova Scotia: rifted margin of early Paleozoic Gondwana. *Canadian Journal of Earth Sciences*, **46**, 1–8.

WALKER, J. A., GOWER, S. & McCUTCHEON, S. R. 1993. Antinouri Lake-Nicolas Denys project, Gloucester and Restigouche counties, New Brunswick. *In*: ABBOTT, S. A. (ed.) *Current research 1993.* New Brunswick Department of Natural Resources and Energy, Minerals and Energy Division, Information circular **93-1**, 58–70.

WEST, D. P., LUDMAN, A. & LUX, D. R. 1992. Silurian age for the Pocomoonshine Gabbro-Diorite, southeastern Maine and its regional tectonic implications. *American Journal of Earth Sciences*, **292**, 253–273.

WEST, D. P., BEAL, H. M. & GROVER, T. 2003. Silurian deformation and metamorphism of Ordovician arc rocks of the Casco Bay Group, south-central Maine. *Canadian Journal of Earth Sciences*, **40**, 887–905.

WHALEN, J. B. 1990. *Geology of a northern portion of the Central Plutonic Belt, New Brunswick.* Geological Survey of Canada, Map **1751A**, scale 1:100 000.

WHALEN, J. B. 1993a. Geology, petrography and geochemistry of Appalachian granites in New Brunswick and Gaspesie, Quebec. *Geological Survey of Canada, Bulletin*, **436**, 124.

WHALEN, J. B. 1993b. *Petrographic, geochemical, and isotopic data on granites in New Brunswick and Gaspesie, Quebec.* In digital format and sample location maps. Geological Survey of Canada, Open File **2570**.

WHALEN, J. B., FYFFE, L. R., LONGSTAFFE, F. & JENNER, G. A. 1996a. The position and nature of the Gander–Avalon boundary, southern New Brunswick, based on geochemical and isotopic data from granitoid rocks. *Canadian Journal of Earth Sciences*, **33**, 129–139.

WHALEN, J. B., JENNER, G. A., LONGSTAFFE, F. J. & HEGNER, E. 1996b. Nature and evolution of the eastern margin of Iapetus: Geochemical and isotopic constraints from Siluro-Devonian granitoid plutons in the New Brunswick Appalachians. *Canadian Journal of Earth Sciences*, **33**, 140–155.

WHALEN, J. B., JENNER, G. A., LONGSTAFFE, F. J., GARIÉPY, C. & FRYER, B. J. 1997a. Implications of granitoid geochemical and isotopic (Nd, O, Pb) data from the Cambrian-Ordovician Notre Dame arc for the evolution of the Central Mobile belt, Newfoundland Appalachians. *In*: SINHA, K., WHALEN, J. B. & HOGAN, J. P. (eds) *The Nature of Magmatism in the Appalachian Orogen.* Geological Society of America, Memoir, **191**, 367–395.

WHALEN, J. B., VAN STAAL, C. R., LONGSTAFFE, F. J., GARIEPY, C. & JENNER, G. A. 1997b. Insights into tectonostratigraphic zone identification in southwestern Newfoundland based on isotopic (Nd, O, Pb) and geochemical data. *Atlantic Geology*, **33**, 231–241.

WHALEN, J. B., SYME, E. C. & STERN, R. A. 1998. Geochemical and Nd isotopic evolution of Paleoproterozoic arc-type magmatism in the Flin Flon Belt, Trans-Hudson Orogen, Canada. *Canadian Journal of Earth Sciences*, **35**, 227–250.

WHALEN, J. B., MCNICOLL, V. J., VAN STAAL, C. R., LISSENBERG, C. J., LONGSTAFFE, F. J., JENNER, G. A. & VAN BREEMEN, O. 2006. Spatial, temporal and geochemical characteristics of Silurian collision-zone magmatism: An example of a rapidly evolving magmatic system related to slab break-off. *Lithos*, **89**, 377–404.

WHITE, C. E., BARR, S. M., BEVIER, M. L. & KAMO, S. 1994. A revised interpretation of the Cambrian and Ordovician rocks in the Bourinot belt of central Cape Breton Island, Nova Scotia. *Atlantic Geology*, **30**, 123–142.

WHITE, C. E., BARR, S. M., REYNOLDS, P. H., GRACE, E. & MCMULLIN, D. W. A. 2006. The Pocologan metamorphic suite: High-pressure metamorphism in a Silurian fore-arc complex, Kingston terrane, southern New Brunswick. *Canadian Mineralogist*, **44**, 905–927.

WILLIAMS, H. 1979. Appalachian Orogen in Canada. *Canadian Journal of Earth Sciences*, **16**, 792–807.

WILLIAMS, H. & ST. JULIEN, P. 1982. The Baie Verte Brompton Line: Early Paleozoic continent ocean interface in the Canadian Appalachians. *In*: ST. JULIEN, P. & BELAND, J. (eds) *Major Structural Zones and Faults of the Northern Appalachians*. Geological Association of Canada, Special Paper, **24**, 177–208.

WILLIAMS, H., DALLMEYER, R. D. & WANLESS, R. K. 1976. Geochronology of the Twillingate Granite and Herring Neck Group, Notre Dame Bay, Newfoundland. *Canadian Journal of Earth Sciences*, **13**, 1591–1601.

WILLIAMS, H., COLMAN-SADD, S. P. & SWINDEN, H. S. 1988. Tectonicstratigraphic subdivisions of central Newfoundland. *Current Research, Part B*. Geological Survey of Canada, Paper, **88-1B**, 91–98.

WILLIAMS, H., CURRIE, K. L. & PIASECKI, M. A. J. 1993. The Dog Bay Line: A major Silurian tectonic boundary in northeast Newfoundland. *Canadian Journal of Earth Sciences*, **30**, 2481–2494.

WILLIAMS, H., LAFRANCE, B., DEAN, P. L., WILLIAMS, P. F., PICKERING, K. T. & VAN DER PLUIJM, B. A. 1995. Badger Belt. *In*: WILLIAMS, H. (ed.) *The Appalachian/Caledonian Orogen: Canada and Greenland*. Geological Survey of Canada, Geology of Canada, **6**, 403–413 (also Geological Society of America, the Geology of North America, **F-1**).

WILLIAMS, P. F., ELLIOTT, C. G. & LAFRANCE, B. D. 1988. Structural geology and melanges of eastern Notre Dame Bay, Newfoundland. *Geological Association of Canada, Annual Meeting, St. John's, Newfoundland Field Trip Guide Book, Trip* **B2**.

WILSON, R. 2003. Geochemistry and petrogenesis of Ordovician arc-related mafic volcanic rocks in the Popelogan Inlier, northern New Brunswick. *Canadian Journal of Earth Sciences*, **40**, 1171–1189.

WILSON, R., BURDEN, E. T., BERTRAND, R., ASSELIN, E. & MCRACKEN, A. D. 2004. Stratigraphy and tectono-sedimentary evolution of the Late Ordovician to Middle Devonian Gaspe belt in northern New Brunswick: evidence from the Restigouche area. *Canadian Journal of Earth Sciences*, **41**, 527–551.

WILSON, R. A., KAMO, S. & BURDEN, E. T. 2005. Geology of the Val d'Amour Formation: revisiting the type area of the Dalhousie Group, northern New Brunswick. *In*: MARTIN, G. L. (ed.) *Geological Investigations in New Brunswick for 2004*. New Brunswick Department of Natural Resources; Minerals, Policy and Planning Division, Mineral Resource Report, **2005-1**, 167–212.

WILSON, R. A., VAN STAAL, C. R. & KAMO, S. 2008. Lower Silurian subduction-related volcanic rocks in the Chaleurs Group, northern New Brunswick, Canada. *Canadian Journal of Earth Sciences*, **45**, 981–998.

WOOD, D. A. 1980. The application of a Th-Hf-Ta diagram to problems of tectonomagmatic classification and to establishing the nature of crustal contamination of basaltic lavas of the British Tertiary volcanic province. *Earth and Planetary Science Letters*, **50**, 11–30.

YANG, X. M., LENTZ, D. R. & MCCUTCHEON, S. R. 2003. Petrochemical evolution of subvolcanic granitoid intrusions within the Late Devonian Mount Pleasant Caldera, southwestern New Brunswick, Canada: Comparison of Au versus Sn-W-Mo-polymetallic mineralization systems. *Atlantic Geology*, **39**, 97–121.

ZAGOREVSKI, A., ROGERS, N., VAN STAAL, C. R., MCNICOLL, V., LISSENBERG, C. J. & VALVERDE-VAQUERO, P. 2006. Lower to Middle Ordovician evolution of peri-Laurentian arc and back-arc complexes in the Iapetus: constraints from the Annieopsquotch Accretionary Tract, Central Newfoundland. *Geological Society of America Bulletin*, **118**, 324–342.

ZAGOREVSKI, A., VAN STAAL, C. R., MCNICOLL, V. C. & ROGERS, N. 2007a. Upper Cambrian to Upper Ordovician peri-Gondwanan island arc activity in the Victoria Lake Supergroup, Central Newfoundland: Tectonic development of the Ganderian margin. *American Journal of Science*, **307**, 339–370.

ZAGOREVSKI, A., VAN STAAL, C. R. & MCNICOLL, V. 2007b. Distinct Taconic, Salinic and Acadian deformation along the Iapetus suture zone, Newfoundland Appalachians. *Canadian Journal of Earth Sciences*, **44**, 1567–1585.

ZAGOREVSKI, A., VAN STAAL, C. R., MCNICOLL, V., ROGERS, N. & VALVERDE-VAQUERO, P. 2007c. Tectonic architecture of an arc-arc collision zone, Newfoundland Appalachians. *In*: CLIFT, P. & DRAUT, A. (eds) *Formation and Applications of the Sedimentary Record in Arc-Collision Zones*. Geological Society of America, Special Paper, **436**, 309–333.

ZAGOREVSKI, A., ROGERS, N., VAN STAAL, C. R., MCCLENNAGHAN, S. & HASLAM, R. 2007d. Tectonostratigraphic relationships in the Buchans area: A composite of Ordovician and Silurian terranes. *In*: PEREIRA, C. P. G., WALSH, D. G. & KEAN, B. F. (eds) *Current Research 2007*. Newfoundland Department of Mines and Energy, Geological Survey Report, **07–1**, 103–116.

# From Rodinia to Pangaea: ophiolites from NW Iberia as witness for a long-lived continental margin

SONIA SÁNCHEZ MARTÍNEZ[1], RICARDO ARENAS[1]*, JAVIER FERNÁNDEZ-SUÁREZ[1]
& TERESA E. JEFFRIES[2]

[1]*Departamento de Petrología y Geoquímica e Instituto de Geología Económica (CSIC), Universidad Complutense, 28040 Madrid, Spain*

[2]*Department of Mineralogy, The Natural History Museum, London SW7 5BD, UK*

*Corresponding author (e-mail: arenas@geo.ucm.es)*

**Abstract:** The ophiolites preserved in the Variscan suture of NW Iberia (Galicia) show a broad variability in lithology, geochemistry and chronology. This wide variety rules out the simplest plate tectonic scenario in which these ophiolites would have been exclusively related to the oceanic domain closed during the final Pangaea assembly, that is the Rheic Ocean. The ophiolitic units from Galicia also provide important data about the palaeogeography immediately preceding the opening of this ocean, and some information about pre-Gondwanan supercontinent cycles.

Six different ophiolites can be distinguished in the allochthonous complexes of Galicia: the Purrido, Somozas, Bazar, Vila de Cruces, Moeche and Careón units. The Purrido Ophiolite is constituted by metagabbroic amphibolites with igneous protoliths dated at $1159 \pm 39$ Ma (Mesoproterozoic), and geochemical affinities typical of island-arc tholeiites. These mafic rocks can be interpreted as one of the scarce members of the pre-Rodinian ophiolites, and they were probably generated in a back-arc setting in the periphery of the West African Craton. The Somozas Ophiolitic Mélange consists of a mixing of submarine volcanic rocks (pillow-lavas, submarine breccias, pillow-breccias, hyaloclastites), diabases, gabbros, microgabbros, diorites and granitoids, surrounded by a matrix of serpentinites or, less frequently, phyllites. Two granitic samples from this mélange yield U–Pb ages ranging between c. 527 and 503 Ma (Cambrian), which together with the characteristic arc signatures obtained in all the studied igneous rocks suggest that this ophiolite was generated in a peri-Gondwanan volcanic arc. The Bazar Ophiolite is formed by different tectonic slices with high temperature amphibolites, granulites, metagabbros and ultramafic rocks. The amphibolites are the most abundant rock type and show typical N–MORB compositions with igneous protoliths dated at $498 \pm 2$ Ma (Cambrian). The high-temperature metamorphism affecting some parts of the unit has been dated at c. 480 Ma (lower Ordovician), and it is considered to be related to the development of an oceanic accretionary complex under the volcanic arc represented by the upper units of the allochthonous complexes of Galicia. Considering the most common palaeogeographic reconstructions for the Cambrian period, it is suggested that the oceanic lithosphere represented by the Bazar Ophiolite was formed into the peri-Gondwanan oceanic domain prior to the rifting of the Avalonian microcontinent, that is the Iapetus–Tornquist Ocean. According to current data about the Vila de Cruces Unit, it can be interpreted as a composite terrane, whose lithologies have U–Pb ages ranging from 1176–497 Ma, but constituted by metaigneous rocks with arc signatures. This dataset has been interpreted in relation to the development of a back-arc basin around the Cambrian–Ordovician limit, involving a Mesoproterozoic basement and the reactivation of a former suture. The opening of this back-arc basin can also be identified as the birth of the Rheic Ocean, and probably it would also include the lithological succession belonging to the Moeche Unit, although its basic rocks exhibit compositions with more oceanic character. Finally, the Careón Ophiolite includes remnants of an oceanic lithosphere generated in a supra-subduction zone setting at $395 \pm 2$ Ma (middle Devonian). This ophiolite was formed in a contractive Rheic Ocean, shortly preceding the closure of this ocean. This is the only ophiolite in Galicia that can be related to mature stages of the Rheic Ocean, although as it is commonly observed in other regions the N–MORB crust is not preserved. This common oceanic crust has disappeared during subduction, probably in an intra-oceanic setting and during the generation of the igneous section preserved in the Careón Ophiolite.

The axial zone of the Variscan belt includes different exotic terranes extending from the western Iberian Peninsula through the French Armorican and Central Massifs to the Bohemian Massif (Fig. 1). These terranes include several ophiolitic units of different age with a complex tectonothermal evolution involving high-pressure events of pre-Variscan and Variscan ages. Considering that the

*From*: MURPHY, J. B., KEPPIE, J. D. & HYNES, A. J. (eds) *Ancient Orogens and Modern Analogues.*
Geological Society, London, Special Publications, **327**, 317–341.
DOI: 10.1144/SP327.14   0305-8719/09/$15.00 © The Geological Society of London 2009.

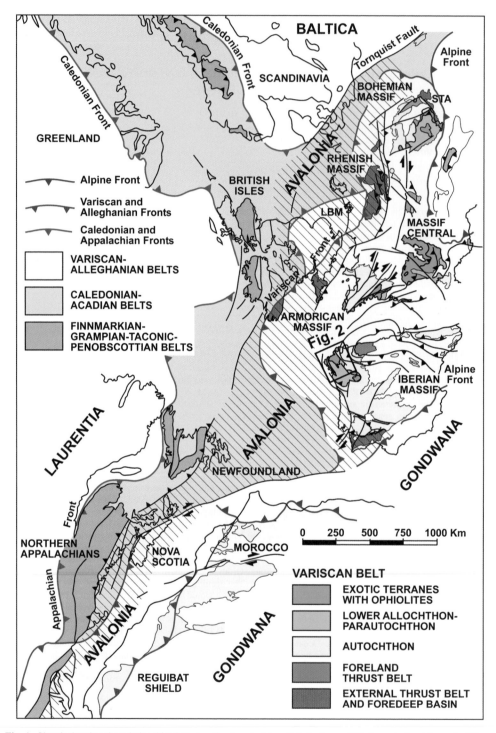

**Fig. 1.** Sketch showing the relationships between the Appalachian, Caledonian and Variscan belts at the end of the Variscan convergence. The striped area represents the approximate extent of Avalonia. LBM, London–Brabant Massif; STA, Silesian Terrane Assemblage. The approximate location of the NW Iberia map of Figure 2 is also shown. From Martínez Catalán *et al.* (2007*a*).

Variscan orogenic belt was generated as a result of the collision of Laurussia and Gondwana during the final assembly of Pangaea, the ophiolites preserved along the Variscan suture zone should be related to the oceanic domain that was closed in the course of the collision, that is the Rheic Ocean (Matte 2001; Stampfli & Borel 2002; Winchester et al. 2002; Murphy & Nance 2003). This interpretation, although consistent with the overall plate tectonic scenario, is simplistic in light of recent data on the NW Iberia ophiolites. These ophiolites show a remarkable geological and geochemical diversity and offer an outstanding case study to explore not only the nature of the evolved Rheic Ocean (Sánchez Martínez et al. 2007a) but also the initial stages of its development (Arenas et al. 2007a), some features of the peri-Gondwanan pre-Rheic oceanic realms and even palaeogeographic scenarios related to the assembly and dispersal of the Rodinia supercontinent (Sánchez Martínez et al. 2006).

This work is aimed at providing a synthetic view of the geological, geochronological and geochemical features of the NW Iberian ophiolites and their interpreted palaeogeographic and palaeotectonic implications. It is focused on the ophiolitic units from Galicia, NW Spain, which are included in the allochthonous complexes of Cabo Ortegal and Órdenes (Figs 1 & 2). The striking diversity of the ophiolites from NW Iberia is an excellent reminder of the complexity of the terranes with oceanic affinity involved in orogenic sutures. At the same time, such diversity offers a natural example to study processes that occurred at the palaeocontinental margins and oceanic realms involved in the Variscan collision. If we assume that the features of the Variscan suture preserved in NW Iberia are not unique at global scale, it is logical to consider that the complexity of oceanic units preserved in such zones of different orogenic belts may be greater than it is commonly assumed and reported. This may spur further studies that should lead to a better and more complete understanding of ancient oceanic realms.

## The Variscan suture of NW Iberia

The Variscan suture in NW Iberia is preserved in several allochthonous complexes of synformal structure and is highlighted by an ensemble of mafic rock units of ophiolitic affinity (Arenas et al. 1986). In Galicia (NW Spain), six ophiolitic units have been described so far, these are named the Purrido, Somozas (ophiolitic mélange), Bazar, Vila de Cruces, Moeche and Careón (Fig. 2). Based on their structural position, five of these six ophiolitic units have been traditionally grouped in two ensembles: the lower and upper ophiolitic units (Arenas et al. 2007b) with the Somozas Ophiolitic Mélange being considered as a structurally different case (Fig. 2). Other ophiolitic units exist in the Trás-os-Montes region, in northern Portugal, in the allochthonous complexes of Morais and Bragança. These ophiolites have not been included in this review, but recent descriptions of these mafic ensembles and new U–Pb data are reported in Pin et al. (2006; and references therein). The mafic units involved in the Variscan suture in NW Iberia appear between two allochthonous terranes of contrasting geological features, the so-called basal and upper units (Fig. 2).

The upper units are constituted by metasedimentary rocks intruded by igneous rocks with a wide compositional range (gabbros to granitoids). The thickness of these units may reach tens of kilometres. The upper units have been interpreted as remnants of a magmatic arc whose main activity took place at the Cambrian–Ordovician boundary (c. 500 Ma). The signature of detrital zircons in greywackes from turbidite beds located towards the top of these units suggests that the arc was generated in the periphery of the northern margin of west Gondwana (Fernández Suárez et al. 2003). This arc would have rifted from the Gondwanan margin during the initial stages of the opening of the Rheic Ocean and drifted northwards as an island arc that eventually collided with Laurussia (Gómez Barreiro et al. 2007). The tectonothermal evolution of the upper units is complex and of a polymetamorphic nature (Fernández-Suárez et al. 2002). The arc records a metamorphic episode of intermediate pressure accompanied by intense deformation that took place between 498 and 480 Ma (Abati et al. 1999, 2007). This event is interpreted as generated as a consequence of the accretionary dynamics of the arc. A later HP–HT event is recorded in the lower part of the upper units and has been dated at 420–400 Ma (Gómez Barreiro et al. 2006; Fernández-Suárez et al. 2007) and it has been interpreted as a result of the collision (and understacking) of the arc with/under Laurussia, broadly coeval with the collision of Avalonia. The upper units also record Variscan tectonothermal events related to the final stages of Pangaea assembly.

The basal units consist of a rock ensemble of continental affinity constituted by metasedimentary rocks, orthogneisses and mafic rocks that have been interpreted to represent a section of the most external part of the Gondwanan margin (Martínez Catalán et al. 1996). The granitic orthogneisses are abundant, mostly with early Ordovician ages and the igneous protoliths show both calc-alkaline (older intrusions) and peralkaline (younger intrusions) geochemical affinity. The presence of these

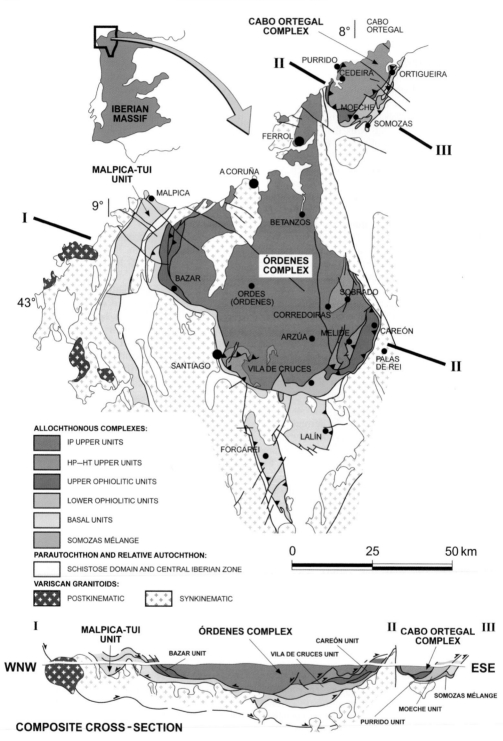

**Fig. 2.** Schematic geological map and cross-section of the allochthonous complexes of Galicia, showing the distribution and structural relationships of the most important terranes. Six different ophiolitic units can be distinguished in the Cabo Ortegal and Órdenes complexes, the Purrido, Somozas, Moeche, Vila de Cruces, Careón and Bazar ophiolites.

granitoids and coeval mafic rocks are evidence for magmatic activity at the Gondwanan margin after the separation of the volcanic arc represented by the upper units. The tectonothermal evolution of the basal units is characterized by the ubiquitous presence of high-pressure and low- to intermediate-temperature metamorphism reaching the (glaucophane bearing) eclogite and blueschist facies (Arenas *et al.* 1995, 1997). The age of this metamorphic episode is dated at *c.* 370 Ma (Rodríguez *et al.* 2003). This event is considered to be generated by the northward directed subduction of the external margin of Gondwana under Laurussia, heralding the Variscan collision *s.s.* The basal units record the earliest Variscan deformation event recognized in the basement of western Europe, this event being coeval with the closure of the Rheic Ocean. This ocean possible attained its maximum width at the time of the collision of the upper units with Laurussia and from that point it contracted until it was closed with the subduction of the Gondwanan margin under Laurussia.

## The ophiolitic units of NW Iberia

### Purrido Ophiolite

The Purrido ophiolitic unit, *c.* 300 m thick, is exposed in the western part of the Cabo Ortegal Complex, under the lower part of the upper units affected by high-P and high-T metamorphism (Fig. 2). It is a monotonous mafic sequence constituted by common and garnet-bearing amphibolites showing an intense planolinear fabric and mineral associations typical of medium-pressure amphibolite facies. No relics of pre-metamorphic lithologies have been identified so far but their textural aspect (often resembling that of flasergabbros) suggests a possible gabbroic protolith. The massive amphibolite unit is bound at its top by garnet–staurolite–kyanite medium pressure paragneisses that crop out within the Carreiro shear zone (Fig. 3a). Detrital zircon age data indicate a maximum sedimentation age of *c.* 470 Ma for the sedimentary precursor of the paragneisses (Sánchez Martínez 2009). The nature of the contact between amphibolites and paragneisses is uncertain although these gneisses are different from those within the overlying units affected by high-P and high-T metamorphism and, like the amphibolites themselves, cannot be considered part of the upper units.

U–Pb dating of zircons from these amphibolites by LA–ICP–MS (Sánchez Martínez *et al.* 2006) yielded a poorly constrained Mesoproterozoic crystallization age of 1159 ± 39 Ma and older zircons whose ages range from 1265–1658 Ma that have been interpreted as xenocrysts. The amphibolite sample studied also contained younger zircons

(816 ± 15 Ma and 428 ± 5 Ma). The latter overlaps with the LA–ICP–MS U–Pb age obtained in two rutile crystals extracted from the overlying paragneisses (412 ± 19 Ma and 428 ± 11 Ma; Sánchez Martínez 2009). This age is interpreted to represent a metamorphic event that affected both the amphibolites and the paragneisses.

The common amphibolites of the Purrido Ophiolite have geochemical features that fit those of island-arc tholeiites, as shown in the Th–Hf–Ta diagram of Figure 3b (Wood 1980). This interpretation is consistent with normalized abundance patterns of immobile trace elements (Pearce 1996) (Fig. 3c). Although this pattern is flat and close to unity (Th and Ce concentrations are similar to those of N–MORB, whereas Nb, Zr, Ti and Y are variably depleted), there is a larger negative Nb anomaly suggesting a subduction-related origin component (Pearce 1996).

### Somozas Ophiolitic Mélange

In the eastern part of the Cabo Ortegal Complex, at the thrust front of the allochthonous complexes of Galicia, there is a tectonic mélange of considerable size (Fig. 2). In detail, the mélange is formed by complex tectonic imbrications involving an ophiolitic unit and slices of the basal units and the Parauthochton. The mélange contains metric to hectometric lens-shaped tectonic blocks and slices constituted by ophiolites, metasediments and high-T felsic orthogneisses and amphibolites wrapped around by a low-T, highly sheared matrix of serpentinites or phyllites. The upper main slice of the mélange reaches 500 m in thickness and represents a typical ophiolitic mélange with minor metasedimentary rocks and few high-T tectonic blocks. The ophiolitic lithologies of the Somozas Mélange are submarine metavolcanic rocks (pillow-lavas, submarine breccias, pillow-breccias and hyaloclastites), diabases, gabbros, micro-gabbros, diorites, granitoids and highly serpentinized spinel-bearing ultramafic rocks (Fig. 4a). The serpentinites are the most abundant rock type in the ophiolitic mélange. The igneous rocks are always affected by a hydrothermal metamorphism that did not obliterate the igneous textures but caused generalized greenschist to amphibolite facies recrystallization.

A sample taken from a hectometric tectonic block of granitoid was dated using U–Pb–LA–ICP–MS geochronology (Arenas *et al.* 2007c) giving an age of 527 ± 5 Ma (Cambrian). Another sample of granite from a different hectometric tectonic block that also includes gabbros and diorites yielded an age of 503 ± 5 Ma age (Cambrian) using U–Pb (SHRIMP–RG). A sample of conglomerates taken from a large tectonic block included in serpentinites yielded several age

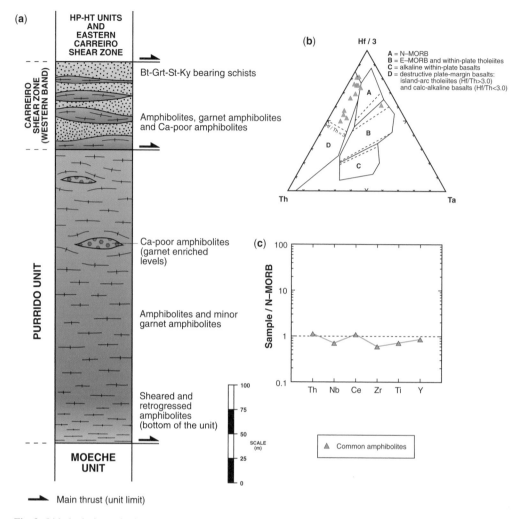

**Fig. 3.** Lithological constitution and geochemistry of the Purrido Ophiolite. (**a**) Characteristic section of the ophiolite. (**b**) Th–Hf–Ta diagram (Wood 1980) with the projection of the most representative amphibolites. (**c**) Normal mid-ocean-ridge basalt (N–MORB)-normalized trace-element pattern of the common amphibolites (average composition); selected elements and normalizing values according to the criteria of Pearce (1996).

groups suggesting that sedimentation of this rock occurred in the periphery of the West-African Craton (LA–ICP–MS; 59 concordant or subconcordant detrital zircons). This conglomerate exhibits an age group with a large number of zircons (19 zircons) with ages between 630 and 497 Ma, probably representing Pan-African events and also the activity in the volcanic arc where the ophiolite was generated. The maximum age of sedimentation of this conglomerate as inferred from the two youngest concordant zircons is $465 \pm 5$ Ma. This age can be considered as a reference age for the end of the magmatic activity in the volcanic-arc located in the periphery of Gondwana.

The ophiolite involved in the Somozas Mélange was the first to be studied with some detail in the NW Iberian Massif. Arenas (1988) described its distribution, lithological constitution and geochemical characteristics, suggesting that it could be considered as a typical mid-ocean ridge ophiolite (harzburgite ophiolite type, HOT). However, some of the studied ophiolitic lithologies showed geochemical features that are not entirely compatible with this origin. New geochronological and geochemical data of the igneous and sedimentary lithologies involved in the Somozas Mélange allow this interpretation to be refined. According to their chemical characteristics two different groups of igneous rocks can be

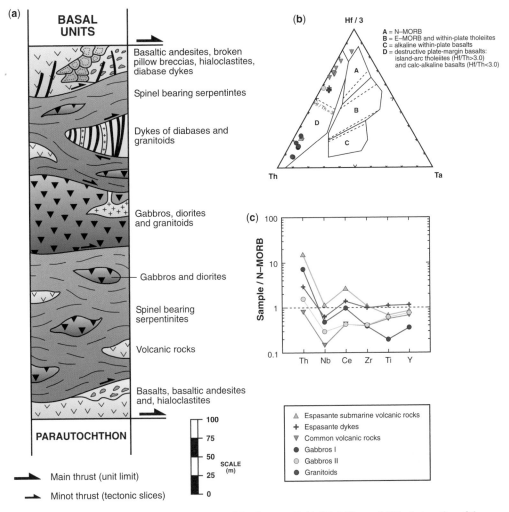

**Fig. 4.** Lithological constitution and geochemistry of the Somozas Ophiolitic Mélange (**a**) Typical section of the ophiolitic mélange. (**b**) Th–Hf–Ta diagram (Wood 1980) with the projection of the most representative lithologies of the ophiolite; they include basalts, basaltic andesites, diabases and two main types of gabbros and diorites. (**c**) Normal mid-ocean-ridge basalt (N–MORB) normalized trace-element patterns (average composition); selected elements and normalizing values after Pearce (1996).

distinguished. A first group is formed by gabbros, diorites, granitoids and basalts–basaltic andesites with calc-alkaline affinities and close similarity with igneous suites generated in volcanic arcs (Fig. 4b). The second group is constituted by gabbros, diabase dikes and common basaltic rocks with chemical compositions typical of island–arc tholeiites, and therefore also interpreted as generated in supra-subduction zone settings (Fig. 4b). The general patterns exhibited by selected trace elements normalized to the N–MORB composition are consistent with this interpretation (Fig. 4c). The recognition of the original intrusive relationships

between these igneous rocks is difficult since they occur as large tectonic blocks in a mélange. However, the common basaltic rocks have never been found intruded by calc-alkaline dykes or pluto-nic bodies. Moreover, a key outcrop on the coast shows calc-alkaline submarine volcanic rocks intruded by a network of basaltic dykes belonging to the second compositional group. Therefore, it is possible to interpret that the island–arc tholeiites are younger than the calc-alkaline igneous rocks, and they were probably generated after a mature volcanic arc located in the periphery of Gondwana, possibly at its most external margin.

## Bazar Ophiolite

The Bazar Ophiolite is located in the westernmost exposures of the Ordenes Complex. It is constituted by several imbricate slices that contain gabbroic rocks, with minor ultramafic rocks at the base of the unit (Díaz García 1990). The main tectonic slice (the Carballo–Bazar slice) reaches a thickness of 4000 m and is composed of amphibolites and flasergabbros displaying a high-T foliation that evolved from an initial granulite-facies stage. Metre-scale boudin of well-preserved mafic granulites are found within the metagabbros (Fig. 5a). These boudin are wrapped by the high-T foliation

and contain mineral associations indicative of low- to intermediate-pressure conditions (plagioclase + clinopyroxene + orthopyroxene + hornblende + ilmenite ± garnet ± olivine). A different lithological section exists at the base of the Carballo–Bazar slice; it is constituted by relatively well preserved gabbros, leucogabbros, tonalites and ultramafic rocks (Fig. 5a). This section also shows a lower-T metamorphism belonging to the low-T part of the amphibolite facies, and it is therefore interpreted as a different slice.

U–Pb–LA–ICP–MS dating has been performed on a sample of amphibolite from the Carballo–Bazar slice (Sánchez Martínez 2009). A

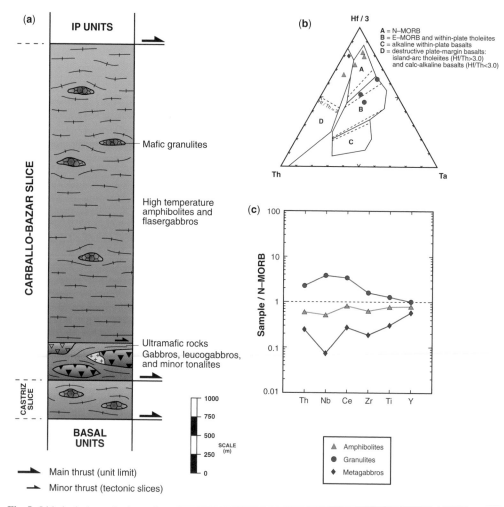

**Fig. 5.** Lithological constitution and geochemistry of the Bazar Ophiolite (**a**) Typical section of the ophiolite; it shows three main different slices although more cryptic imbrications probably do exist into the Carballo-Bazar slice. (**b**) Th– Hf–Ta diagram (Wood 1980) with the projection of the most representative lithologies of the ophiolite; they include amphibolites, mafic granulites and low-T metagabbros. (**c**) Normal mid-ocean-ridge basalt (N–MORB) normalized trace-element patterns (average composition); selected elements and normalizing values after Pearce (1996).

population of 36 zircons gave ages mostly concordant that allowed the definition of two clusters at $498 \pm 2$ Ma (Cambrian) and $483 \pm 2$ Ma (lower Ordovician). These ages are interpreted as the crystallization age of the mafic protolith and the age of the high-T metamorphism, respectively. Granulites preserved as boudin also contain zircon, although it is less abundant than in the studied amphibolite sample. Seven zircons separated from a granulite boudin in the Castriz slice (Fig. 5a) were analysed. Five of these zircons yielded concordant to subconcordant ages between $458 \pm 8$ and $489 \pm 6$ Ma and are considered to be related to the granulite metamorphic event. Two zircons yielded Neoproterozoic ages around 620–625 Ma. Although more data are needed to clarify the meaning of these older ages, they are an indication that the protolith ages of the granulite boudin may be significantly older than those of the host mafic rocks (Sánchez Martínez 2009).

The chemical composition of the common amphibolites in the Bazar Ophiolite indicates an oceanic affinity with N–MORB (i.e. basalts generated in typical mid-ocean ridges or in broad and evolved back-arc basins; Fig. 5b, c), whereas the mafic granulites are transitional between MOR and WP basalts with normalized trace element patterns comparable to those of T–MORB generated in plume–ridge interactions (Pearce 1996). Metagabbros associated with ultramafic rocks from the lower slice show a clearly different composition, characteristic of island–arc tholeiites (Fig. 5b, c). Although, as expressed above, further chronological and geochemical work is needed, the data obtained so far suggest that the Bazar Ophiolite may represent a composite terrane with a complex evolution. The chronological evidence is consistent with the geochemical diversity within the mafic units.

## Vila de Cruces Ophiolite

The Vila de Cruces Ophiolitic unit is located to the S–SE of the Ordenes Complex, lying under the Careón Ophiolite and above the basal units of the allochthonous complexes (Fig. 2). Together with the Moeche Ophiolite, it has been included in the group of the lower ophiolites (Fig. 2; Arenas *et al.* 2007*b*). This ophiolite is *c.* 4000 m in thickness and is a complex structure characterized by the existence of several tectonic slices, including the Sampayo slice with internal imbrications (Fig. 6a). The unit is also characterised by the existence of two generations of often large recumbent folds (Martínez Catalán *et al.* 2002; Arenas *et al.* 2007*a*).

The Vila de Cruces Ophiolite is dominated by mylonitized greenschists, with intercalations of metapelites and metacherts. These alternations of mafic rocks and metasedimentary rocks are consistent with a metabasaltic origin for the greenschists. There are scarce intercalations of mylonitic orthogneisses of tonalitic composition and metre- to decametre-scale lenses of medium- to fine-grained metagabbros. The lower tectonic slice is of ultramafic nature and is in contact with the rest of the unit by an extensional detachment (Fig. 6a). The unit shows variable metamorphic features, from low-grade rocks in the Vila de Cruces slice to medium grade in the Sampayo slice. In addition, a high-P metamorphic gradient has been identified in the upper slice, where the schists (although very retrograded) contain mineral associations with garnet and without biotite.

Geochronological data suggest a complex scenario that may involve a composite terrane rather than a single ophiolitic unit. A tonalitic orthogneiss yielded a U–Pb age of $497 \pm 4$ Ma (Arenas *et al.* 2007*a*). Two gabbro lenses yielded overlapping albeit imprecise U–Pb ages of $1176 \pm 85$ and $1168 + 14/-50$ Ma (Sánchez Martínez 2009). Unfortunately, the greenschists (volumetrically dominant lithology of the unit), have not been dated yet, nor the maximum sedimentation age of the clastic metasedimentary intercalations (zircons were not found in several studied samples of micaschists). However, Sm–Nd data on the greenschists are compatible with a Palaeozoic age for these rocks (Sánchez Martínez 2009). Even though it is evident that further isotopic and radiometric data are needed to clarify the origin and meaning of this unit, at this stage, it is possible to assume with confidence that Vila de Cruces likely represents a composite unit in which at least a Mesoproterozoic element (coeval with that of the Purrido ophiolite, see above) and a Palaeozoic mostly mafic element are present. The regional foliation of the schists, acquired during the Variscan regional deformation has been dated at 363–367 Ma ($^{40}Ar/^{39}Ar$ in two phyllite samples Dallmeyer *et al.* 1997).

The geochemical features of the main lithologies of the Vila de Cruces Unit are shown in Figure 6b, c. The mafic rocks, both the greenschists and the Mesoproterozoic metagabbros, have compositions compatible with those of island arc tholeiites, following the discriminant diagram of Wood (1980) (Fig. 6b). This signature is also highlighted by their marked Nb-negative anomaly (Fig. 6c). The supra-subduction origin of these mafic rocks is also consistent with the volcanic arc signature of the tonalitic orthogneisses (Fig. 6b; Sánchez Martínez 2009).

## Moeche Ophiolite

This unit has traditionally been correlated with the Vila de Cruces Unit, given the similar

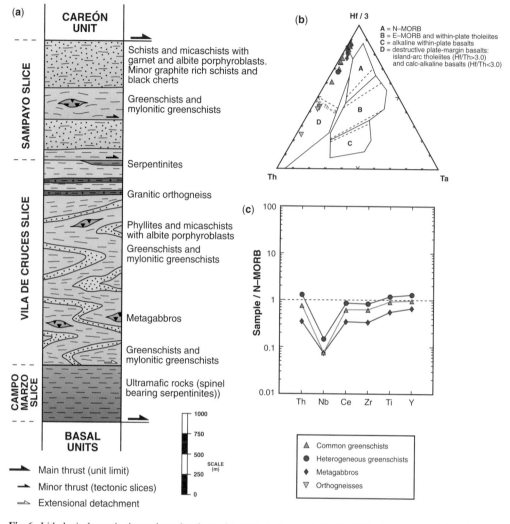

**Fig. 6.** Lithological constitution and geochemistry of the Vila de Cruces Ophiolite (**a**) Typical section of the ophiolite, with three main tectonic slices, which also show internal imbrication and recumbent folding. (**b**) Th–Hf–Ta diagram (Wood 1980) with the projection of the most representative lithologies of the ophiolite; including common and heterogeneous greenschists, metagabbros and orthogneisses. (**c**) Normal mid-ocean-ridge basalt (N–MORB) normalized trace-element patterns (average composition) of the mafic lithologies; selected elements and normalizing values after Pearce (1996).

lithological constitution of both units (Arenas *et al.* 2007*b*), although Vila the Cruces is considerably thicker than Moeche (Fig. 7a). The Moeche Unit is constituted by greenschists of possible metabasaltic origin with abundant intercalations of phyllites and micaschists, some metagabbro bodies and some serpentinite intercalations towards the upper part of the unit. Its internal structure is poorly known, and therefore the possible existence of tectonic imbrications and/or recumbent folds cannot be ruled out (by analogy with the Vila de Cruces Unit).

Sánchez Martínez (2009) has obtained U–Pb detrital zircon ages from micaschists that indicate maximum sedimentation ages between *c.* 469 and 455 Ma (Middle–Upper Ordovician). Older detrital zircon age clusters are compatible with a provenance in the West African Craton and the Cadomian magmatic arc rocks, that is the original sediments were deposited in the periphery of northern Gondwana. Although no radiometric dates are available for the metaigneous rocks of the unit, preliminary Sm–Nd data obtained on the mafic schists (Sánchez Martínez 2009) are compatible with a

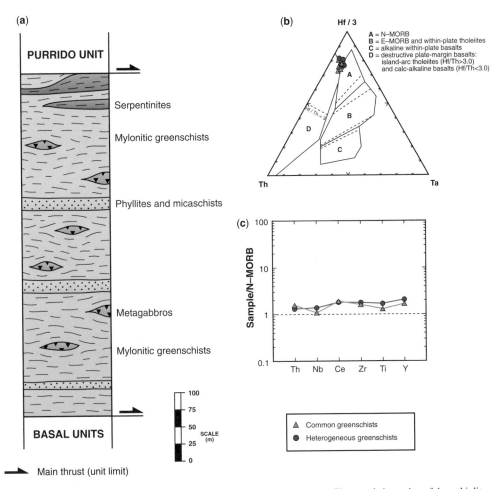

**Fig. 7.** Lithological constitution and geochemistry of the Moeche Ophiolite (**a**) Characteristic section of the ophiolite. (**b**) Th–Hf–Ta diagram (Wood 1980) with the projection of the greenschists of this unit. (**c**) Normal mid-ocean-ridge basalt (N–MORB) normalized trace-element patterns (average composition) of the greenschists; selected elements and normalizing values after Pearce (1996).

Palaeozoic age. The regional foliation has been dated at 364 Ma (Dallmeyer *et al.* 1997, coeval with that of the Vila de Cruces Unit).

The main compositional features of the greens-chists of this unit are shown in the diagrams of Figure 7b, c. They have a composition transitional between N–MORB and island-arc tholeiites. In the Hf–Th–Ta diagram of Wood (1980) they plot in the area straddling both compositional fields (Fig. 7b). They show a slight enrichment in trace elements with respect to N–MORB (Fig. 7c), with a slight Nb–negative anomaly suggesting a supra-subduction environment. Therefore, the mafic schists of the Moeche Unit are compositionally different from their putative analogues in the Vila de Cruces Unit, given that they show an attenuated arc signature and a drift towards N–MORB compositions.

## Careón Ophiolite

The Careón Ophiolite is exposed in the SE part of the Ordenes Complex and exhibits a varied lithological sequence. It is made up of three main tectonic slices, of which the intermediate (Careón) slice is the thickest and contains the most varied lithological ensemble with ultramafic rocks at the base and a sequence of isotropic gabbros towards the top (Fig. 8a). Both the ultramafic rocks and the gabbros are intruded by a swarm of diabase dykes and pegma-toid gabbros. This ultramafic–mafic rocks assem-blage with no record of volcanic or sedimentary rocks is remarkably different from that of typical mid-ocean ridge ophiolites. Díaz García *et al.* (1999) interpreted the Careón Unit as an ophiolite generated in a supra-subduction zone (SSZ) setting.

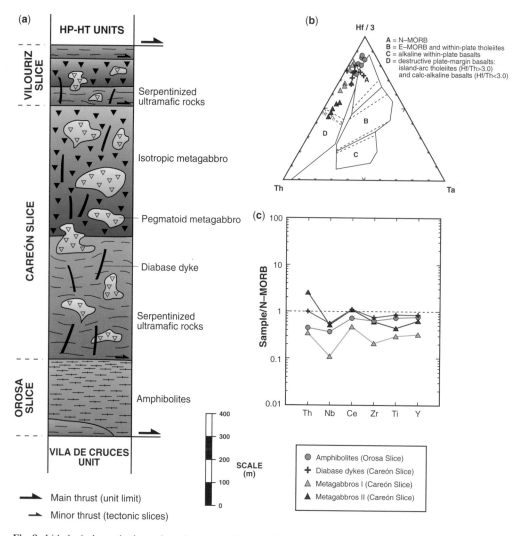

**Fig. 8.** Lithological constitution and geochemistry of the Careón Ophiolite. (**a**) Characteristic section of the ophiolite. (**b**) Th–Hf–Ta diagram (Wood 1980) with the projection of two types of metagabbros of the Careón slice, the diabase dykes and the amphibolites of the Orosa slice. (**c**) Normal mid-ocean-ridge basalt (N–MORB) normalized trace-element patterns (average composition); selected elements and normalizing values after Pearce (1996).

At the contact zones between the different tectonic slices, there is evidence of thermal effects, even the development of metamorphic soles with high-T recrystallization. This is particularly well developed and preserved at the contact between the Careón and Vilouriz slices, where a 2 m-thick level with corundum crystals up to 2 cm was developed. These features of the contacts are interpreted as an indication that the tectonic slices were stacked when the whole section of oceanic lithosphere was still very hot, that is shortly after its generation.

U–Pb dating of zircons from a sample of leucogabbro from the upper part of the Careón slice

yielded a crystallization age of 395 ± 2 Ma (middle Devonian; Díaz García *et al.* 1999), while another sample of leucogabbro from the same slice yielded an identical age of 395 ± 3 Ma (Pin *et al.* 2002). These ages indicate that this ophiolite records the final stages of the Rheic Ocean when it was well into its contractional phase (e.g. Sánchez Martínez *et al.* 2007*a*). On the other hand $^{40}Ar/^{39}Ar$ dating of the amphiboles defining the amphibolite facies foliation yielded an age of 376.8 ± 0.4 Ma (Upper Devonian; Dallmeyer *et al.* 1997) that is interpreted as indicative of the chronology of the accretion of the ophiolite slices to the southern margin of Laurussia. It must be noted that the accretion ages of 363–367 Ma obtained for the Vila de Cruces and Moeche units fit the structural position of the latter under the Careón Unit, indicating a later (younger) accretion.

Figure 8b, c show the geochemical features of the Careón Ophiolite. All the lithologies have compositions typical of island-arc tholeiites. The metagabbros show pronounced Nb-negative anomalies, and slightly more transitional compositions towards N–MORB in the diabase dykes and the amphibolites from the Orosa slice. In the case of the Careón slice (where the oceanic lithosphere section is more complete) the transition from island-arc compositions in the gabbros towards N–MORB in the diabases is interpreted as the geochemical signature of the lithospheric extension above a subduction zone.

## Discussion: origin of the ophiolites of NW Iberia

### A Mesoproterozoic ophiolite in NW Iberia

According to the data presented in previous works, the gabbroic protoliths of the Purrido Ophiolite have yielded a crystallization age of 1159 ± 39 Ma. Taking into account that this is a mafic unit built up by 300 m of massive amphibolites, this assemblage can be interpreted as a pre-Rodinian ophiolite. This group of ophiolites older than 1100–1000 Ma, the estimated assembly age for the Rodinia supercontinent (Dalziel 1991; Hoffman 1991), is relatively small, as only around 35 cases have been reported so far (Moores 2002). Considering the geochemical affinity of its mafic rocks, the Purrido Ophiolite can be assigned to the supra-subduction zone type, suggesting that it was generated in relation to the activity of a volcanic arc, probably in a back-arc setting or a marginal basin. This tectonic setting is also compatible with the presence of inherited zircons with ages ranging between 1658 and 1265 Ma. Thus, this unit would represent a fragment of the plutonic section of a SSZ ophiolite.

The Purrido Ophiolite does not show evidence of having been affected by Mesoproterozoic high-temperature metamorphic events. It is not related to other rocks with continental-affinity of the same age either. Therefore it is unlikely that the Purrido Unit was involved in the Grenville orogenic belt (developed between 1300 and 1000 Ma), and it can be thus interpreted as a section of peri-arc oceanic lithosphere located away from the realm of this mobile belt. Considering the currently accepted palaeogeographic reconstructions of the continents immediately before the assembly of Rodinia, it seems that at c. 1100 Ma Amazonia and West Africa defined a single continental domain located between

the equator and latitude 40°S (Fig. 9a; Kröner & Cordani 2003). Data provided here are consistent with the idea that the Purrido Ophiolite could have been generated in relation to an active arc system in the eastern margin of Amazonia-West Africa (Fig. 9a).

An enigmatic aspect in the evolution of the Purrido Ophiolite is related to its preservation, as its mafic character and probable generation in an oceanic setting should have made its survival difficult. However, part of the ophiolite has been preserved to the present day, and it is involved in the Variscan suture of NW Iberia without apparent connection with other Mesoproterozoic rocks of continental nature. In principle there are two main hypotheses that may account for the preservation of the ophiolite: a fast obduction over the continental margin of the West African Craton, or remaining stable far away from the subduction zone in some kind of marginal basin located between a volcanic arc system and the West African Craton margin itself. It is to be hoped that future investigations focused on the study of the structural history of the Purrido Ophiolite will help to clarify this issue. It can be expected that the structural history of the Purrido amphibolites is not as simple as it may seem, and their evolution could have been polydeformational. These amphibolites show a rather simple amphibolite-facies plano-linear fabric that is inconsistent with the complete absence of igneous relics within the unit. There is no other metaplutonic unit in Galicia with the same thickness, where igneous bodies with primary textures and/or mineralogy have not been preserved. This is recorded in units affected even by granulite-facies high-temperature metamorphism, and it is to be expected to be more common when rocks only reach medium-grade metamorphic conditions, as is the case with the Purrido Ophiolite. Thus, the apparent simplicity of the deformational history of the Purrido Ophiolite may not be real, and we cannot rule out its obduction over the leading margin of a continent in motion soon after its generation in a marginal basin. On the other hand, we can neither discount the possibility that it was preserved roughly intact within the back-arc. The alternative hypothesis, of the obduction of the ophiolite during the continental dispersal that followed the Rodinia fragmentation, after a long stability period in a marginal basin (Fig. 9b), cannot be completely ruled out before having more information about the tectonothermal history of the unit. It can be pointed out that the fragmentation of Rodinia started at 750 Ma (e.g. Torsvik 2003) and that the dated Purrido amphibolite sample had a single zircon that yielded a concordant age of 816 ± 15 Ma.

The presence of Palaeo-Proterozoic or even Archaean continental basement in the Variscan

(a) **1100 Ma**

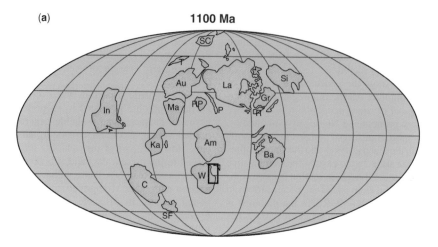

(b) **750 Ma**

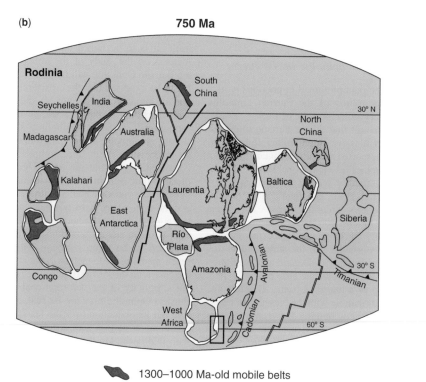

1300–1000 Ma-old mobile belts

**Fig. 9.** (a) Distribution of the main continental blocks at 1100 Ma, during the stages previous to the development of the Grenvillian orogen and the resulting assembly of the Rodinia supercontinent (Kröner & Cordani 2003; based in Pisarevski *et al.* 2003). The frame indicates the most probable region considered for the generation of the mafic rocks constituting the Purrido Ophiolite. Abbreviations: Am, Amazonia; Au, Australia; Ba, Baltica; C, Congo; Gr, Greenland; In, India; Ka, Kalahari; La, Laurentia; Ma, Mawson craton; W, West Africa; P, Pampean terrane; R, Rockall; RP, Río de la Plata; SC, South China; SF, São Francisco; Si, Siberia; T, Tarim. (b) Probable location of the Purrido Ophiolite (framed region) in a reconstruction of the Rodinia supercontinent prior to its fragmentation (*c.* 750 Ma). Rodinia reconstruction by Torsvik (2003).

Belt of western Europe was described a long time ago. This is the case of the Icart gneisses (located on the Channel Islands and Brittany; Samson & D'Lemos 1998) and the submerged granulites of the Bay of Biscay (Guerrot *et al.* 1989), which in both cases are interpreted to be lithological assemblages related to the West African Craton. Considering the information provided in this work, it is clear that the basement of the Variscan Belt also includes – at least – Mesoproterozoic mafic sections with oceanic affinity, as the Purrido Ophiolite, which probably obducted over a continental margin, or remained stable next to it during the Rodinia assembly. We hypothesize that during the stages immediately preceding the fragmentation and dispersal of Rodinia, the Purrido Ophiolite was located next to the eastern margin of the West African Craton, in a position equivalent to that indicated in Figure 9b (Sánchez Martínez *et al.* 2006). In other words, we favour the simplest possible interpretation and we situate this Mesoproterozoic terrane in a position analogous to that currently occupied by its remnants.

Reviewing the palaeographic reconstruction for the times following the Rodinia fragmentation, some of the most outstanding events are connected with the amalgamation of continents to constitute the present central and southern Africa. These different collisions are Neoproterozoic in age and are usually grouped under the term Pan-African orogeny. Dalziel (1997) suggested the existence of a supercontinent of ephemeral life during the late Neoproterozoic, that he named Pannotia (see also Murphy & Nance 2004). During its assembly, the eastern margin of Amazonia–West Africa remained essentially stable and without significant orogenic activity, although it seems clear that the development of an important magmatic arc, the so-called Cadomian–Avalonian Arc, took place along its northern margin (Figs 9 & 10a). Pannotia broke up into four large continents, Laurentia, Baltica, Siberia and Gondwana at 550 Ma ago, but this partition happened without the margin of Amazonia–West Africa being significantly affected (Fig. 10b). In other words, the data available seem to indicate that this large continental margin behaved coherently and was rather stable, at least over a period between 1100 Ma (Fig. 9a) and 550–520 Ma (Fig. 10b). A little modified large continental margin, the Amazonia–West African margin, may have participated in the amalgamation and fragmentation of two supercontinents (Rodinia and Pannotia), to finally constitute most of the northern margin of Gondwana at *c.* 550 Ma. Thus there are reasons to surmise that, on the basis of the apparent stability of the Amazonia–West African margin, the Purrido Ophiolite could have remained next to this margin until the early

Palaeozoic, possibly without being affected by significant deformation.

The Purrido Ophiolite is in contact with detrital metasedimentary rocks from the Carreiro shear zone, which have a maximum sedimentation age of 470 ± 7 Ma (Lower–Middle Ordovician boundary) and were likely deposited in the periphery of the West African Craton as suggested by detrital zircon U–Pb age clusters (Sánchez Martínez 2009). It is not possible to determine if these sediments were initially deposited over the mafic rocks, although the geological link existing between both lithological assemblages and their similar tectonothermal evolution could be interpreted in this way. If this interpretation is correct, and taking into account the data provided by the other ophiolitic units of Galicia that will be discussed below (see also Scotese 2001; Stampfli & Borel 2002), the expected scenario at the early Palaeozoic in the northern Gondwanan margin would be determined by the development of marginal basins behind an active volcanic arc. This volcanic arc has been traditionally situated at the inception of its development at the external margin of Gondwana (see Arenas *et al.* 2007a, b), although an origin from an intra-oceanic subduction zone in the proximity of Gondwana cannot be ruled out. The drift of the arc and its retreat from Gondwana, together with the rifting of other microcontinents from the main continent, as was the case of Avalonia (Stampfli & Borel 2002; Murphy *et al.* 2006), led to the development of a new major oceanic domain, the Rheic Ocean, that determined to a large extent the Palaeozoic paleogeography of the domain presently occupied by western Europe and eastern North America. This situation corresponding to the period between the Cambro–Ordovician boundary and the middle Ordovician is shown in Figure 10c, d.

Figure 11 represents an interpretive model for the development of a marginal basin system that occurred at the early stages of the opening of the Rheic Ocean. This sketch does not reflect a specific moment, but a period of time characterized by the development of the volcanic arc and the beginning of its rift from Gondwana, ranging from *c.* 500–480 Ma. One of the most uncertain aspects of this model is the width of the marginal basin, which is at present impossible to constrain. However, it is possible that it was not too wide at this age (Fig. 10), especially taking into account that the Mesoproterozoic terrane was probably covered by sediments at *c.* 470 Ma, just postdating the moment represented in this diagram, and whose probable source area was the West African Craton or sediments derived from it. Accordingly, it is conceivable that the back-arc was covered by sediments.

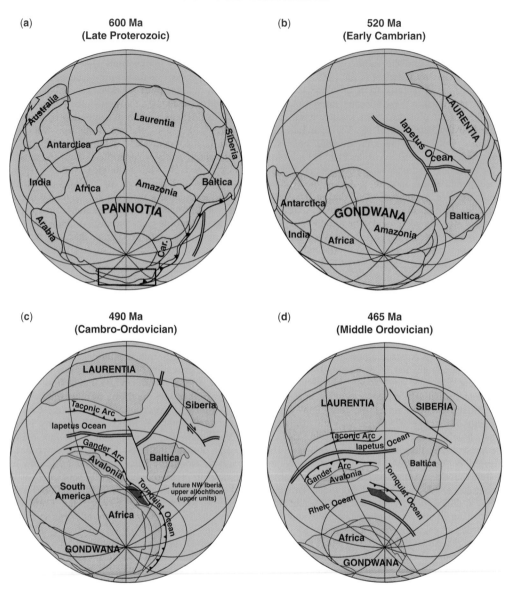

**Fig. 10.** Schematic evolution of continents and oceanic domains between the late Proterozoic (*c.* 600 Ma) and the Middle Ordovician (*c.* 465 Ma) (**a**) Reconstruction of the Pannotia supercontinent; the frame shows the most probable region for the location of the NW Iberia Proterozoic terranes. (**b**) Fragmentation of Pannotia in four main continents (Gondwana, Laurentia, Baltica and Siberia) during the early Cambrian. (**c**) Reconstruction of the continental blocks during the stages preceding the opening of the Rheic Ocean (*c.* 490 Ma). It shows the beginning of the generation of the marginal basin system that gave rise to this ocean. The Rheic Ocean opened in the Northern margin of Gondwana as a consequence of the separation and drift of Avalonia and other peri-Gondwanan terranes. The section of the arc-system probably originating the upper units of the allochthonous complexes of Galicia appears coloured in red. (**d**) The Rheic Ocean domain prior to the closure of the Iapetus and Tornquist oceans and the final assembly of Laurussia. Modified and adapted after Arenas *et al.* (2007*a*), Gómez Barreiro (2007) and Martínez Catalán *et al.* (2007*a*); based on the palaeogeographic reconstruction by Winchester *et al.* (2002).

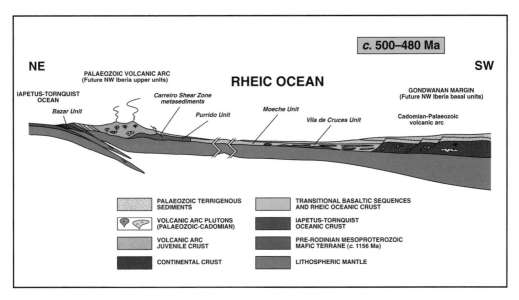

**Fig. 11.** Diagram illustrating the general model suggested in this paper for the beginning of the opening of the Rheic Ocean. The figure shows the position and most probable meaning of four of the ophiolitic units of Galicia, the Purrido, Bazar, Vila de Cruces and Moeche units. Explanation in the text.

In Figure 11, we have considered that the Rheic Ocean opened in a region where the Mesoproterozoic terrane represented by the Purrido Ophiolite was already integrated, and that this opening split this terrane, so that one part of it was separated from Gondwana becoming part of the basement of the drifting volcanic arc, and another part remained attached to the continental margin, thus retaining its original position. The clues for this tentative interpretation lie, on the one hand, in the structural position occupied within the Variscan suture by the units with Mesoproterozoic remnants, and on the other hand, in the tectonothermal evolution recorded by the terrane itself, arguments that will be addressed below.

In principle, the most straightforward way to know the concrete position of a terrane of the orogenic wedge in relation to the suture, that is to the Rheic Ocean, is to determine its structural relationship with the ophiolites derived from the true and characteristic Rheic domain, assuming that the polarity of the accretion in the orogenic wedge is known. The only ophiolite in Galicia that can be assigned to an evolved Rheic Ocean is that represented by the Careón Unit, since the other ophiolitic units were developed in early stages of the history of this ocean or they are unrelated to it. Taking into account that the Purrido and Careón units are not in contact with each other, it is impossible to know if the Purrido Unit remained attached to Gondwana during the opening of the Rheic Ocean or if it drifted towards Laurussia (i.e. it was situated to

the north of the mid-Rheic ridge). However, data relating to the tectonothermal evolution of the Purrido Ophiolite and the metasedimentary rocks involved in the Carreiro shear zone seem to indicate that the second interpretation is more plausible, and therefore it is depicted in Figure 11. The sample of metasedimentary rock from the Carreiro shear zone contains rutile dated by LA-ICP-MS, which yielded U–Pb ages of $412 \pm 19$ and $428 \pm 11$ Ma; moreover, the sample of Purrido amphibolite also dated by LA-ICP-MS contains a zircon of c. $428 \pm 5$ Ma. Regardless of the existence of other younger zircons in both lithologies (Variscan reworking), these U–Pb data, though scarce, suggest that both ensembles underwent a metamorphic event during the Middle–Late Silurian that happened when the Rheic Ocean was still at the height of its spreading and probably at the time when its width was maximum. This metamorphic event clearly links the tectonothermal evolution of the Purrido Ophiolite and metasedimentary rock located on top, with that of the upper units of the allochthonous complexes of the NW Iberian Massif. More precisely, the ages are analogous to those obtained in the basal slice of the upper units, affected by high-P and high-T metamorphism, that have been interpreted in connection with the accretion of an arc to the southern margin of Laurussia during the stages preceding the onset of the contraction of the Rheic Ocean (Gómez-Barreiro *et al.* 2006; Sánchez Martínez *et al.* 2007a; Arenas *et al.* 2007b; Fernández-Suárez *et al.* 2007). Accordingly, the data presented in this paper seem to indicate that a part of the Mesoproterozoic terrane, represented by the Purrido Unit, should be part of the basement of the volcanic arc that split up from Gondwana at the Cambrian–Ordovician boundary. Together with this arc, it drifted to the north, until its final accretion to the southern margin of Laurussia in the Silurian, the time when the Rheic Ocean probably started to contract.

## Remnants of a Cambrian mature peri-Gondwanan volcanic arc

In the Somozas Ophiolitic Mélange, the ophiolitic lithologies appear imbricated with continental rocks probably related to the palaeomargin of Gondwana (basal units of the allochthonous complexes and Schistose Domain). These ophiolites are those occupying the lowest structural position in NW Iberia (Arenas *et al.* 2007*b*), as they appear below other ophiolitic units developed in the southern domain of the Rheic Ocean. Based on its geological context, the original paleogeographic position of the ophiolites involved in the Somozas Mélange was likely located in the oceanic domain attached to the Gondwana margin or even in the more external margin of this supercontinent, depending on the meaning of this ophiolite. According to the compositional features of the rocks that constitute this ophiolitic mélange, we suggest an origin in a peri-Gondwanan volcanic arc. As discussed in the previous section, this arc had a long period of activity and reached a mature stage, which in the Somozas mélange is manifest in the presence of granitoid rocks. Based on the U–Pb zircon ages obtained from these granitoids within the mélange, the arc was probably active during most of the Cambrian period (*c.* 530–503 Ma; Arenas *et al.* 2007*c*).

The metaigneous lithologies of the Somozas Ophiolitic Mélange belong to two compositional groups linked by their relationship with a volcanic arc. The older group has calc-alkaline compositions whereas the younger igneous rocks display compositional features of island-arc tholeiites. The mutual relationship between both groups can be established by the presence of a network of diabases belonging to the second group intruding the calc-alkaline submarine volcanics. Furthermore, the evolution from a calc-alkaline magmatism typical of mature arcs to a younger one which shows affinities with island-arc tholeiites, can be related to the opening of an intra-arc basin. The progressive widening of this basin could explain the rifting of the external part of the arc, and its subsequent drift leaving behind a new oceanic domain, the Rheic Ocean. According to the materials involved in the Somozas Mélange, which were originally located in the periphery of

the Gondwanan edge, the birth date of this ocean can be constrained to a relatively narrow period. This is the time period ranging from the peak activity in a mature volcanic arc (*c.* 527–503 Ma) to the development of a basin which includes detritus derived from rocks representing the final magmatic activity in the arc (*c.* 465 $\pm$ 5 Ma; Arenas *et al.* 2007*c*).

## Remnants of the Cambrian peri-arc ocean?

The results obtained in the Bazar Ophiolite have been quite surprising because of their complexity (Sánchez Martínez 2009; Sánchez Martínez *et al.* 2007*b*). When it comes to analysing the significance of the new data, it has to be considered that the Bazar Unit has traditionally been correlated with the Careón Unit, which was generated during middle Devonian times (Arenas *et al.* 2007*b*) and considered a typical representative of the lithosphere of the Rheic Ocean. The new U–Pb geochronological data are preliminary in some of the lithologies belonging to the Bazar Unit. However, the new U–Pb data for the mafic rocks seem to indicate that the Bazar Unit represents a composite mafic unit, constituted by materials of Neoproterozoic (*c.* 624 Ma) to late Cambrian (*c.* 498 Ma) age. This diversity of ages is also consistent with the compositional variability between the different lithological types (from N–MORB to island-arc tholeiites generated at supra-subduction zone settings), as well as with their different metamorphic evolution.

The most common rock type in the Bazar Unit is high-temperature amphibolite, which contains a large amount of zircon with U–Pb ages grouped into two populations. The first, with a mean age of 498 $\pm$ 2 Ma (Upper Cambrian), is interpreted to reflect the crystallization age of the gabbroic protoliths, while the second, with a mean age of 483 $\pm$ 2 Ma (Lower Ordovician), has been interpreted to date the granulitic metamorphism undergone by the unit. If we consider that the Bazar amphibolites have a chemical composition typical of N–MORB, and that hence they were probably generated in a mid-ocean ridge, it is necessary to consider a scenario where the common oceanic lithosphere was accreted and underwent high-temperature metamorphism. Considering the nature of metamorphism, the most probable tectonic setting would be an active and maybe relatively juvenile volcanic arc, below which the accretion of the oceanic lithosphere that fed the magmatism of this arc (see e.g. the work of Peacock 1990) took place. Considering the palaeogeographic models accepted for the peri-Gondwanan realm during Cambrian and Ordovician times, as well as the data obtained in the other units of the allochthonous complexes of the NW Iberian Massif, it is likely that the Bazar amphibolites derive from the oceanic lithosphere exterior to the peri-Gondwanan arc system (Fig. 11). This lithosphere was accreted under this arc system at a subduction zone dipping towards the arc and Gondwana, and that would be the cause generating the arcs themselves, the peri-Gondwanan marginal basin system and the supra-subduction geochemical signatures identified in the Palaeozoic metabasites of Vila de Cruces and Moeche units. In other words, the Bazar amphibolites probably represent the crust of the Iapetus-Tornquist Ocean, being consumed and reducing its width while it was giving ground to the new ocean floor that was generated at the spreading Rheic domain (Fig. 11). The structural position of the Bazar amphibolites in NW Iberian Massif reflects the described dynamic scenario, since they are located below the arc-derived terrane (the upper units) and on the side that should be facing the exterior ocean (the western part of these units).

The meaning of the Neoproterozoic rocks is more uncertain. However, if confirmed with further data, their presence would add evidence for the protracted and complex evolution of the Proterozoic margins of Amazonia–West Africa, as already highlighted by previous studies (e.g. Sánchez Martínez et al. 2006).

## Two different elements of an evolving back-arc basin

The model favoured here considers that the Vila de Cruces Unit remained attached to the margin of Gondwana until the formation of Pangaea (Fig. 11). This interpretation is based on the structural position occupied by the unit within the Variscan orogenic wedge, where it appears below the ophiolites of the Rheic Ocean represented by the Careón Unit (Fig. 2). Moreover, evidence for the metamorphic event developed at 410–430 Ma has not been found in this unit, which further supports the idea that it did not accrete to the southern margin of Laurussia, as the Purrido Unit, and that its Palaeozoic tectonothermal evolution is exclusively Variscan and related to the main stages of the development of Pangaea.

As discussed in previous sections, the Vila de Cruces Unit shows a great complexity, since it is constituted by lithologies very different in age. Thus, it represents a composite terrane formed during a long period of time. This unit contains remnants of a Mesoproterozoic terrane which is similar to (or the same as) that represented by the Purrido Ophiolite: two metagabbros of the Vila de Cruces Unit have been dated at c. 1170 Ma (Sánchez Martínez 2009). These metagabbros appear as lenticular inclusions of metric to decametric size within the strongly deformed greenschists that represent the predominant lithology in this unit. Unfortunately, the age of these greenschists is not well constrained, hindering their interpretation. However, available Sm–Nd data indicate that at least part of these greenschists could be Palaeozoic. Meanwhile, their geochemical characteristics are homogeneous and indicative of an origin in a supra-subduction zone setting. The Vila de Cruces Unit also contains orthogneisses dated at 497 ± 4 Ma (Cambrian) with typical volcanic arc geochemical features, as well as intercalations of metasedimentary rocks (in which no zircon was found), but considering their lithological resemblance and analogous tectonothermal evolution, it is reasonable to link them with those present in the Moeche Unit (that will be discussed below), that yield Palaeozoic maximum sedimentation ages and were deposited around the periphery of the West African Craton.

In short, the Vila de Cruces Unit represents a composite terrane in which a Mesoproterozoic mafic basement similar to that forming the Purrido Ophiolite was involved, and new Palaeozoic crust possibly developed around the Cambrian–Ordovician boundary, as the result of oceanic magmatism associated to a supra-subduction zone. The Mesoproterozoic mafic basement should be intensely intruded by magmas generated in the supra-subduction domain during the development of this new Palaeozoic crust, until being reduced to a few remnants within the Palaeozoic rocks. These remnants would be represented by the metagabbros dated at c. 1170 Ma. Therefore, the Vila de Cruces Unit contains the part of the Mesoproterozoic basement that remained attached to Gondwana when the Rheic Ocean opened (Fig. 11). It has to be highlighted that according to the data provided in this work it seems clear that this ocean opened in the proximity of an ancient suture, and possibly along it. This possibility was already proposed by Murphy et al. (2006), albeit these authors developed this idea to suggest that Avalonia was initially accreted to the margin of Amazonia–West African Craton, generating the suture from which the subsequent separation and drifting of Avalonia from Gondwana, opening the Rheic Ocean (Fig. 10), took place. Also, according to our data, it is possible that the opening of this ocean took place in a region from the northern margin of Gondwana characterized by the existence of more than one Proterozoic suture. It is pertinent to indicate here that different data suggest the possible existence of a Mesoproterozoic basement in the lower crust of NW Iberia. Mesoproterozoic ages are common in detrital zircons and micas from the core of the Ibero–Armorican arc (Gutiérrez-Alonso et al. 2005, and references therein). Moreover, most of the

Ordovician volcanic rocks from NW Iberia yield Mesoproterozoic Nd model ages (Murphy *et al.* 2008). The nature and extent of such a basement are unknown, but it is conceivable that some correlation with the Mesoproterozoic rocks of the Vila de Cruces Unit may exist. In this way, this unit can include the thinned Mesoproterozoic basement located in the transitional crust of the more external margin of Gondwana (Fig. 11). This basement, as part of the more complex Vila de Cruces Unit, would have been thrust over the continental margin (basal units of the NW Iberia allochthonous complexes) during the early stages of the Variscan collision.

The Moeche Unit is made up of strongly deformed greenschists alternating with phyllite levels. It thus bears a close lithological similarity to the Vila de Cruces Unit, with which it is usually correlated (Arenas *et al.* 2007*b*). The correlation between these units is also supported by an analogous tectonothermal evolution, with identical $^{40}Ar/^{39}Ar$ ages for the main regional foliation (Dallmeyer *et al.* 1997) and the absence of the metamorphic event dated around 410–430 Ma in the Purrido Ophiolite. However, there are no Palaeozoic orthogneisses in the Moeche Unit like those forming part of the Vila de Cruces Unit, nor metagabbros of Mesoproterozoic age like those appearing in that unit. Nevertheless, taking into account the limited extent of the outcrops of these metagabbros and our limited sampling, we cannot rule out that not-yet-found remnants of this Mesoproterozoic mafic terrane are present in the Moeche Unit.

As in the case of the Vila de Cruces Unit, there is an uncertainty about the age of the mylonitic greenschists, that represent the main lithological type of this terrane, to interpret the meaning of the Moeche Unit. These mafic greenschists appear interbedded with levels of phyllite whose detrital zircons indicate maximum sedimentation ages ranging between 470 and 455 Ma (Middle–Upper Ordovician), and a provenance from source areas located in the proximity of the West African Craton. In other words, these phyllites represent detrital sediments probably deposited during middle Ordovician times. Regarding the mafic greenschists, their textural features indicate that they probably derive from basic volcanic rocks, and their distribution in the field alternating with levels of Ordovician sediments suggests similar ages. The Sm–Nd isotopic composition of these greenschists is also more compatible with a Palaeozoic age, although our data are very scarce and should be completed in the future.

Based on the above, the data presented in this paper allow us to interpret the Moeche Unit as an ensemble of rocks mostly of Palaeozoic age, although it cannot be completely ruled out that remnants of the Mesoproterozoic mafic terrane may be present. Taking into account their geochemical features, the greenschists of this unit appear as basaltic types similar to N–MORB but with compositional features indicating that their generation took place close to a subduction zone. As a result, we interpret that the origin of the Moeche Unit should be located at a more outboard position within the peri-Gondwanan marginal basin system (Fig. 11). In other words, this unit can be considered as transitional between the Vila de Cruces Unit and the typical lithosphere of the Rheic Ocean. It must be clarified that this kind of oceanic lithosphere that would be characteristic of the Rheic realm has neither been found in the allochthonous complexes of Galicia nor is it likely to have been generated during the Middle Ordovician. In any case, it should be constituted by typical N–MORB basalts lacking alternating levels of detrital rocks.

## A Middle Devonian supra-subduction ophiolite from a contractive Rheic Ocean

The age of the Careón Ophiolite has been known since a U–Pb dating study by TIMS revealed that the metagabbros of this unit were generated $395 \pm 2$ Ma ago, that is, during the Middle Devonian (Díaz García *et al.* 1999). If the Careón Ophiolite is the sole section of lithosphere of the Rheic Ocean preserved in the Variscan suture of NW Iberian Massif, except for the infant stages of the generation of the peri-Gondwanan marginal basins, it is too young to be representative of the typical lithosphere of this ocean. Moreover, the lithological constitution of the Careón Ophiolite cannot be considered characteristic of the most common oceanic lithospheres. This led Díaz García *et al.* (1999) to interpret this ophiolite as generated in a supra-subduction zone setting. This interpretation was also favoured by Pin *et al.* (2002) who studied the chemical composition of the most representative metabasites of the Careón Unit. This idea is also consistent with the new data based on the study and geochemical interpretation of a group of metabasite samples (Sánchez Martínez *et al.* 2007*a*).

According to the palaeogeographic models accepted for the Middle Ordovician, the width of the Rheic Ocean should increase quickly as Avalonia and other related peri-Gondwanan terranes (as that represented by the upper units of the allochthonous complexes of Galicia) drifted to the north away from Gondwana (Fig. 10c, d). This widening of the Rheic Ocean was taking place while the Iapetus and Tornquist oceans were undergoing a fast contraction. In general, it is accepted that the Rheic Ocean started to close during Ludlow times (*c.* 420 Ma), right after the accretion of Avalonia to Laurussia and the resulting closure of the Iapetus Ocean (Stampfli & Borel 2002). At the same time, it seems that another peri-Gondwanan terrane derived from a volcanic arc was accreted to the southern margin of Baltica–Avalonia (Gómez-Barreiro *et al.* 2007). That terrane constitutes the upper units of the allochthonous complexes of NW Iberian Massif, but also

other equivalent allochthonous terranes in the French Massif Central and in the Bohemian Massif (see Sánchez Martínez *et al.* 2007*a*). It is characterized by the presence of a high-pressure high-temperature metamorphic event dated by U–Pb and $^{40}Ar/^{39}Ar$ at 400–425 Ma (Gómez-Barreiro *et al.* 2006; Fernández-Suárez *et al.* 2007), probably

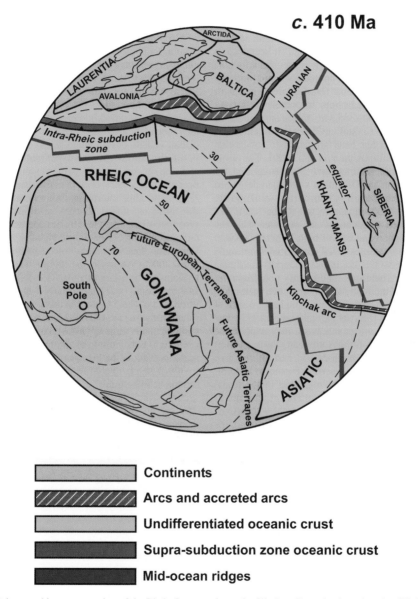

**Fig. 12.** Palaeographic reconstruction of the Rheic Ocean realm at the Silurian–Devonian boundary (modified after Sánchez Martínez *et al.* 2007*a*). It shows the generation of new oceanic lithosphere associated with intra-oceanic subduction directed to the NE. It also shows the situation in the southern margin of Laurussia of the accreted arc from which the upper units of the allochthonous complexes of NW Iberia are considered to derive. The general position of the main continents is a modification from the original palaeogeographical reconstruction by Stampfli & Borel (2002).

recording its accretion to the southern margin of Laurussia. As discussed in previous sections, the detection of this metamorphic event in the Purrido Ophiolite and in the paragneisses of the Carreiro shear zone allowed us to interpret that both lithological ensembles moved together, being part of this terrane, therefore including remnants of a mafic Mesoproterozoic terrane.

It is a matter of controversy whether a single (the Rheic Ocean; Linnemann *et al.* 2004; Murphy *et al.* 2006) or multiple oceanic domains (Franke 2000; Winchester *et al.* 2002) existed to the south of Avalonia in the Silurian and Devonian. It seems clear that the southern continental margin originally bordering the Rheic Ocean is presently represented by the Saxo–Thuringia and Ossa–Morena zones of the European Variscan belt. Important arc-related magmatism *c.* 360–335 Ma has been studied in the Mid-German Crystalline Rise (Saxo–Thuringia Zone), where it has been attributed to subduction toward the south (Altherr *et al.* 1999). The same age and tectonic setting have been suggested for the Late Devonian–Dinantian volcanism described in the French Massif Central (Pin & Paquette 2002). This subduction directed to the south and the associated magmatism are younger than the first deformation and coeval high-P metamorphism affecting the most external margin of Gondwana (dated at 365–370 Ma in NW Iberia; Rodríguez *et al.* 2003). Therefore, they mainly occurred after the closure of the Rheic Ocean and have been interpreted in relation to the opening and later closure of a foredeep basin (Martínez Catalán *et al.* 1997). The opening and closure of this basin may explain the double vergence of the European Variscan Belt (Matte 1991). The south-facing part of the belt shows the oldest tectonothermal evolution and preserves information about the closure of the Rheic Ocean, which would have been coeval with north-directed subduction (Matte 1991; Martínez Catalán *et al.* 1997, 2007*b*). On the other hand, the north-facing part of the belt is younger, and its development was probably preceded by subduction toward the south and probably also by an important extensional event. The general absence of large Silurian–Devonian volcanic arcs associated with the closure of the Rheic Ocean is also a characteristic of the European margin of Laurussia–Avalon. In this way, the Devonian volcanic rocks described in southern Avalonia in England have been interpreted as originating in an extensional setting (Floyd 1982), as has been voluminous latest Devonian–Visean volcanism described in the continuation of Avalonia in the South Portuguese zone of SW Iberia. So, it can be concluded that neither in the southern margin of Avalonia, nor in the related peri-Gondwanan terranes accreted to the south of Laurussia, as that represented in the

upper units of the allochthonous complexes of NW Iberia, is there evidence for large late Silurian or Devonian volcanic arcs developed from subduction to the north. The same general absence of volcanic arcs of this age is typical for the terranes located in the northern margin of Gondwana, which were finally involved in the Variscan orogeny in Europe (Fig. 12). However, our data require significant generation of oceanic lithosphere in the early to middle Devonian, while the Rheic Ocean was evidently contracting.

Considering the lithological section of the Careón Ophiolite, its suprasubduction zone geochemical affinity, and the general characteristics of the European Variscan belt, we suggest that the Rheic Ocean was closed mainly by intraoceanic subduction directed to the north (Fig. 12; Sánchez Martínez *et al.* 2007*a*). This subduction was probably located near the northern margin of the ocean, and its development involved consumption of old and cold N–MORB-type oceanic lithosphere and the generation of limited volumes of new oceanic lithosphere of suprasubduction zone type (Fig. 12). This interpretation is compatible with the rarity or virtual absence of common MOR–type ophiolites, like those associated with divergent tectonic settings (Boudier & Nicolas 1985) in the European Variscan belt (see Sánchez Martínez *et al.* 2007*a*; Arenas *et al.* 2007*a*). The model also explains the scarcity of older (pre-Silurian) ophiolites that could be related to early stages of the evolution of the Rheic Ocean. In this respect, it is worth mentioning that the old ophiolitic units preserved in Galicia, that is those of Ordovician or older age, represent an outstanding example on a continental scale, and allow the history of the oceanic domains involved in the evolution of the western European basement to be traced.

This research has been funded by projects CGL2004-04306-CO2-02/BTE and CGL2007-65338-CO2-01/BTE of the Spanish Agency Dirección General de Investigación (Ministerio de Educación y Ciencia). The authors thank G. G. Alonso and F. Pereira for insightful reviews.

# References

ABATI, J., DUNNING, G. R., ARENAS, R., DÍAZ GARCÍA, F., GONZÁLEZ CUADRA, P., MARTÍNEZ CATALÁN, J. R. & ANDONAEGUI, P. 1999. Early Ordovician orogenic event in Galicia (NW Spain): evidence from U–Pb ages in the uppermost unit of the Órdenes Complex. *Earth and Planetary Science Letters*, **165**, 213–228.

ABATI, J., CASTIÑEIRAS, P., ARENAS, R., FERNÁNDEZ-SUÁREZ, J., GÓMEZ-BARREIRO, J. & WOODEN, J. 2007. Using SHRIMP zircon dating to unravel tectonothermal events in arc environments. The early Palaeozoic arc of NW Iberia revisited. *Terra Nova*, **19**, 1–8.

ALTHERR, R., HENES-KLAIBER, U., HEGNES, E. &
SATIR, M. 1999. Plutonism in the Variscan Odenwald
(Germany): from subduction to collision. *International
Journal of Earth Sciences*, **88**, 422–443.

ARENAS, R. 1988. Evolución petrológica y geoquímica de
la unidad alóctona inferior del complejo metamórfico
básico-ultrabásico de Cabo Ortegal (Unidad de
Moeche) y del Silúrico paraautóctono, Cadena Hercínica Ibérica (NW de España). *Corpus Geologicum
Gallaeciae*, **4**, 1–543.

ARENAS, R., GIL IBARGUCHI, J. I. *ET AL*. 1986. Tectonostratigraphic units in the complexes with mafic and
related rocks of the NW of the Iberian Massif. *Hercynica*, **II**, 87–110.

ARENAS, R., RUBIO PASCUAL, F. J., DÍAZ GARCÍA, F. &
MARTÍNEZ CATALÁN, J. R. 1995. High-pressure
micro-inclusions and development of an inverted
metamorphic gradient in the Santiago Schists
(Órdenes Complex, NW Iberian Massif, Spain): evidence of subduction and syn-collisional decompression. *Journal of Metamorphic Geology*, **13**, 141–164.

ARENAS, R., ABATI, J., MARTÍNEZ CATALÁN, J. R.,
DÍAZ GARCÍA, F. & RUBIO PASCUAL, F. 1997.
*P–T* evolution of eclogites from the Agualada Unit
(Órdenes Complex, NW Iberian Massif, Spain):
Implications for crustal subduction. *Lithos*, **40**,
221–242.

ARENAS, R., MARTÍNEZ CATALÁN, J. R., SÁNCHEZ
MARTÍNEZ, S., FERNÁNDEZ-SUÁREZ, J., ANDONAE-
GUI, P. & PEARCE, J. A. 2007a. The Vila de Cruces
Ophiolite: a remnant of the early Rheic Ocean in the
Variscan suture of Galicia (NW Iberian Massif).
*Journal of Geology*, **115**, 129–148.

ARENAS, R., MARTÍNEZ CATALÁN, J. R. *ET AL*. 2007b.
Paleozoic ophiolites in the Variscan suture of Galicia
(northwest Spain): distribution, characteristics and
meaning. *In*: HATCHER, R. D., JR., CARLSON, M. P.,
McBRIDE, J. H. & MARTÍNEZ CATALÁN, J. R.
(eds) *4-D Framework of Continental Crust*. Geological
Society of America Memoirs, **200**, 425–444.

ARENAS, R., SÁNCHEZ MARTÍNEZ, S., CASTIÑEIRAS, P.,
FERNÁNDEZ-SUÁREZ, J. & JEFFRIES, T. 2007c.
Geochemistry and geochronology of the ophiolite
involved in the Somozas Mélange: new insights
on the birth of the Rheic Ocean. *In*: ARENAS, R.,
MARTÍNEZ CATALÁN, J. R., ABATI, J. & SÁNCHEZ
MARTÍNEZ, S. (eds) *The International Geoscience
Programme IGCP 497, 'The Rheic Ocean: Its Origin,
Evolution and Correlatives', Galicia Meeting 2007:
Field Trip Guide & Conference Abstracts Volume*,
151–153. Publicaciones del Instituto Geológico y
Minero de España.

BOUDIER, F. & NICOLAS, A. 1985. Harzburgite and lherzolite subtypes in ophiolitic and oceanic environments.
*Earth and Planetary Science Letters*, **76**, 84–92.

DALLMEYER, R. D., MARTÍNEZ CATALÁN, J. R. *ET AL*.
1997. Diachronous Variscan tectonothermal activity
in the NW Iberian Massif: evidence from $^{40}Ar/^{39}Ar$
dating of regional fabrics. *Tectonophysics*, **277**,
307–337.

DALZIEL, I. W. D. 1991. Pacific margins of Laurentia and
East Antarctica-Australia as a conjugate rift pair: evidence and implications for an Eocambrian supercontinent. *Geology*, **19**, 598–601.

DALZIEL, I. W. D. 1997. Neoproterozoic-Paleozoic
geography and tectonics: review, hypothesis, environmental speculation. *Geological Society of America
Bulletin*, **109**, 16–42.

DÍAZ GARCÍA, F. 1990. La geología del sector occidental
del Complejo de Órdenes (Cordillera Hercínica, NW
de España). *Nova Terra*, **3**, 1–230.

DÍAZ GARCÍA, F., ARENAS, R., MARTÍNEZ CATALÁN,
J. R., GONZÁLEZ DEL TÁNAGO, J. & DUNNING, G.
1999. Tectonic evolution of the Careón ophiolite
(Northwest Spain): a remnant of oceanic lithosphere
in the Variscan belt. *Journal of Geology*, **107**,
587–605.

FERNÁNDEZ-SUÁREZ, J., CORFU, F. *ET AL*. 2002. U–Pb
evidence for a polyorogenic evolution of the HP–HT
units of the NW Iberian Massif. *Contributions to
Mineralogy and Petrology*, **143**, 236–253.

FERNÁNDEZ-SUÁREZ, J., DÍAZ GARCÍA, F., JEFFRIES,
T. E., ARENAS, R. & ABATI, J. 2003. Constraints on
the provenance of the uppermost allochthonous
terrane of the NW Iberian Massif: inferences from detrital zircon U–Pb ages. *Terra Nova*, **15**, 138–144.

FERNÁNDEZ-SUÁREZ, J., ARENAS, R., ABATI, J.,
MARTÍNEZ-CATALÁN, J. R., WHITEHOUSE, M. J. &
JEFFRIES, T. E. 2007. U–Pb chronometry of polymetamorphic high-pressure granulites: an example from
the allochthonous terranes of the NW Iberian Variscan
belt. *In*: HATCHER, R. D., JR., CARLSON, M. P.,
McBRIDE, J. H. & MARTÍNEZ CATALÁN, J. R.
(eds) *4-D Framework of Continental Crust*. Geological
Society of America Memoirs, **200**, 469–488.

FLOYD, P. A. 1982. Chemical variation in Hercynian
basalts relative to plate tectonics. *Journal of the Geological Society, London*, **139**, 507–520.

FRANKE, W. 2000. The mid-European segment of the Variscides: tectonostratigraphic units, terrane boundaries
and plate tectonic evolution. *In*: FRANKE, W.,
HAAK, V., ONCKEN, O. & TANNER, D. (eds) *Orogenic Processes: Quantification and Modelling in the
Variscan Belt*. Geological Society, London, Special
Publications, **179**, 35–61.

GÓMEZ-BARREIRO, J., WIJBRANS, J. R. *ET AL*. 2006.
$^{40}Ar/^{39}Ar$ laserprobe dating of mylonitic fabrics in a
polyorogenic terrane of NW Iberia. *Journal of the Geological Society, London*, **163**, 61–73.

GÓMEZ-BARREIRO, J., MARTÍNEZ CATALÁN, J. R. *ET AL*.
2007. Tectonic evolution of the upper allochthon
of the Órdenes Complex (northwestern Iberian
Massif): structural constraints to a polyorogenic peri-
Gondwanan terrane. *In*: LINNEMANN, U., NANCE,
R. D., KRAFT, P. & ZULAUF, G. (eds) *The Evolution
of the Rheic Ocean: From Avalonian–Cadomian Active
Margin to Alleghenian–Variscan collision*. Geological
Society of America Special Papers, **423**, 315–332.

GUERROT, C., PEUCAT, J. J., CAPDEVILA, R. & DOSSO,
L. 1989. Archean protoliths within Early Proterozoic
granulitic crust of the west European Hercynian
belt: possible relics of the west African craton.
*Geology*, **17**, 241–244.

GUTIÉRREZ-ALONSO, G., FERNÁNDEZ-SUÁREZ, J.,
COLLINS, A. S., ABAD, I. & NIETO, F. 2005. Amazonian Mesoproterozoic basement in the core of the
Ibero-Armorican arc: $^{40}Ar/^{39}Ar$ detrital mica ages
complement the zircoñs tale. *Geology*, **33**, 637–640.

HOFFMAN, P. F. 1991. Did the breakout of Laurentia turn Gondwana inside-out? *Science*, **252**, 1409–1412.

KRÖNER, A. & CORDANI, U. 2003. African, southern Indian and South American cratons were not part of the Rodinia supercontinent: evidence from field relationships and geochronology. *Tectonophysics*, **375**, 325–352.

LINNEMANN, U., MCNAUGHTON, N. J., ROMER, R. L., GEHMLICH, M., DROST, K. & TONK, C. 2004. West African provenance for Saxo-Thuringia (Bohemian Massif): Did Armorica ever leave pre-Pangean Gondwana? – U/Pb-SHRIMP zircon evidence and the Nd-isotopic record. *International Journal of Earth Sciences*, **93**, 683–705.

MARTÍNEZ CATALÁN, J. R., ARENAS, R., DÍAZ GARCÍA, F., RUBIO PASCUAL, F. J., ABATI, J. & MARQUÍNEZ, J. 1996. Variscan exhumation of a subducted Paleozoic continental margin: the basal units of the Órdenes Complex, Galicia, NW Spain. *Tectonics*, **15**, 106–121.

MARTÍNEZ CATALÁN, J. R., ARENAS, R., DÍAZ GARCÍA, F. & ABATI, J. 1997. Variscan accretionary complex of northwest Iberia: terrane correlation and succession of tectonothermal events. *Geology*, **25**, 1103–1106.

MARTÍNEZ CATALÁN, J. R., DÍAZ GARCÍA, F. *ET AL.* 2002. Thrust and detachment systems in the Órdenes Complex (northwestern Spain): implications for the Variscan-Appalachian Dynamics. *In*: MARTÍNEZ CATALÁN, J. R., HATCHER, R. D., ARENAS, R., DÍAZ GARCÍA, F. (eds) *Variscan-Appalachians Dynamics: the Building of the Late Paleozoic Basement.* Geological Society of America Special Papers, **364**, 163–182.

MARTÍNEZ CATALÁN, J. R., ARENAS, R. *ET AL.* 2007*a*. The rootless Variscan suture of NW Iberia (Galicia, Spain). Field trip guide. *In*: ARENAS, R., MARTÍNEZ CATALÁN, J. R., ABATI, J. & SÁNCHEZ MARTÍNEZ, S. (eds) *The International Geoscience Programme IGCP 497, 'The Rheic Ocean: Its Origin, Evolution and Correlatives', Galicia Meeting 2007: field Trip Guide & Conference Abstracts Volume, 3–117.* Publicaciones del Instituto Geológico y Minero de España.

MARTÍNEZ CATALÁN, J. R., ARENAS, R. *ET AL.* 2007*b*. Space and time in the tectonic evolution of the northwestern Iberian Massif. Implications for the comprehension of the Variscan belt. *In*: HATCHER, R. D., JR., CARLSON, M. P., MCBRIDE, J. H. & MARTÍNEZ CATALÁN, J. R. (eds) *4-D framework of Continental Crust.* Geological Society of America Memoirs, **200**, 403–423.

MATTE, P. 1991. Accretionary history and crustal evolution of the Variscan belt in Western Europe. *Tectonophysics*, **196**, 309–337.

MATTE, P. 2001. The Variscan collage and orogeny (480–290 Ma) and the tectonic definition of the Armorica microplate: a review. *Terra Nova*, **13**, 122–128.

MOORES, E. M. 2002. Pre-1 Ga (pre-Rodinian) ophiolites: their tectonic and environmental implications. *Geological Society of America Bulletin*, **114**, 80–95.

MURPHY, J. B. & NANCE, R. D. 2003. Do supercontinents introvert or extrovert?: Sm–Nd isotopic evidence. *Geology*, **31**, 873–876.

MURPHY, J. B. & NANCE, R. D. 2004. How do supercontinents assemble? *American Scientist*, **92**, 324–333.

MURPHY, J. B., GUTIÉRREZ-ALONSO, G. *ET AL.* 2006. Origin of the Rheic Ocean: rifting along a Neoproterozoic suture? *Geology*, **34**, 325–328.

MURPHY, J. B., GUTIÉRREZ-ALONSO, G., FERNÁNDEZ-SUÁREZ, J. & BRAID, J. A. 2008. Probing crustal and mantle lithosphere origin through Ordovician volcanic rocks along the Iberian passive margin of Gondwana. *Tectonophysics*, **461**, 166–180.

PEACOCK, S. M. 1990. Numerical-simulation of metamorphic pressure-temperature-time paths and fluid production in subducting slabs. *Tectonics*, **9**, 1197–1211.

PEARCE, J. A. 1996. A users guide to basalt discrimination diagrams. *In*: WYMAN, D. A. (ed.) *Trace Element Geochemistry of Volcanic Rocks: Application for Massive Sulphide Exploration.* Short Course Notes, Geological Association of Canada, **12**, 79–113.

PIN, C. & PAQUETTE, J. L. 2002. Le magmatisme basique calcoalcalin d'âge dévono-dinantien du nord du Massif Central, témoin d'une marge active hercynienen: arguments géochimiques et isotopiques Sr/Nd. *Geodinamica Acta*, **15**, 63–77.

PIN, C., PAQUETT, J. L., SANTOS ZALDUEGUI, J. F. & GIL IBARGUCHI, J. I. 2002. Early Devonian suprasubduction zone ophiolite related to incipient collisional processes in the Western Variscan Belt: the Sierra de Careón unit, Órdenes Complex, Galicia. *In*: MARTÍNEZ CATALÁN, J. R., HATCHER, R. D., ARENAS, R. & DÍAZ GARCÍA, F. (eds) *Variscan-Appalachians Dynamics: the Building of the Late Paleozoic Basement.* Geological Society of America Special Papers, **364**, 54–71.

PIN, C., PAQUETTE, J. L., ÁBALOS, B., SANTOS, F. J. & IBARGUCHI, J. I. 2006. Composite origin of an early Variscan transported suture: ophiolitic units of the Morais Nappe Complex (north Portugal). *Tectonics*, **25**, 1–19.

PISAREVSKY, S. A., WINGATE, M. T. D., POWELL, C. MCA., JOHNSON, S. P. & EVANS, D. A. D. 2003. Models of Rodinia assembly and fragmentation. *In*: YOSHIDA, M., WINDLEY, B. F. & DASGUPTA, S. (eds) *Proterozoic East Gondwana: Supercontinent Assembly and Break-up.* Geological Society, London, Special Publications, **206**, 35–55.

RODRÍGUEZ, J., COSCA, M. A., GIL IBARGUCHI, J. I. & DALLMEYER, R. D. 2003. Strain partitioning and preservation of $^{40}Ar/^{39}Ar$ ages during Variscan exhumation of a subducted crust (Malpica-Tui complex, NW Spain). *Lithos*, **70**, 111–139.

SAMSON, S. D. & D'LEMOS, R. 1998. U–Pb geochronology and Sm-Nd isotopic composition of Proterozoic gneisses, Channel Island, UK. *Journal of the Geological Society, London*, **155**, 609–618.

SÁNCHEZ MARTÍNEZ, S. 2009. Geoquímica y Geocronología de las ofiolitas de Galicia. *Nova Terra*, **37**, 1–351.

SÁNCHEZ MARTÍNEZ, S., JEFFRIES, T., ARENAS, R., FERNÁNDEZ-SUÁREZ, J. & GARCÍA-SÁNCHEZ, R. 2006. A pre-Rodinian ophiolite involved in the Variscan suture of Galicia (Cabo Ortegal Complex, NW Spain). *Journal of the Geological Society, London*, **163**, 737–740.

SÁNCHEZ MARTÍNEZ, S., ARENAS, R., DÍAZ GARCÍA, F., MARTÍNEZ CATALÁN, J. R., GÓMEZ-BARREIRO,

J. & PEARCE, J. A. 2007a. Careón Ophiolite, NW Spain: suprasubduction zone setting for the youngest Rheic Ocean floor. *Geology*, **35**, 53–56.

SÁNCHEZ MARTÍNEZ, S., ARENAS, R., FERNÁNDEZ-SUÁREZ, J., JEFFRIES, T., GARCÍA-SÁNCHEZ, R. & ABATI, J. 2007b. The Bazar Ophiolite (NW Iberia): a remnant of the Iapetus-Tornquist Ocean in the Variscan suture?. *In*: ARENAS, R., MARTÍNEZ CATALÁN, J. R., ABATI, J. & SÁNCHEZ MARTÍNEZ, S. (eds) *The International Geoscience Programme IGCP 497, 'The Rheic Ocean: its Origin, Evolution and Correlatives', Galicia Meeting 2007: Field Trip Guide & Conference Abstracts Volume*, 150–151. Publicaciones del Instituto Geológico y Minero de España.

SCOTESE, C. R. 2001. *Atlas of Earth History: PALEOMAP Project*. Arlington, Texas, 1–52.

STAMPFLI, G. M. & BOREL, G. D. 2002. A plate tectonic model for the Paleozoic and Mesozoic constrained by dynamic plate boundaries and restored synthetic oceanic isochrons. *Earth and Planetary Science Letters*, **196**, 17–33.

TORSVIK, T. H. 2003. The Rodinia jigsaw puzzle. *Science*, **300**, 1379–1381.

WINCHESTER, J. A., PHARAOH, T. C. & VERNIERS, J. 2002. Paleozoic amalgamation of Central Europe: an introduction & synthesis of new results from recent geological and geophysical investigations. *In*: WINCHESTER, J. A., PHARAOH, T. C. & VERNIERS, J. (eds) *Palaeozoic Amalgamation of Central Europe*. Geological Society, London, Special Publications, **201**, 1–18.

WOOD, D. A. 1980. The application of a Th–Hf–Ta diagram to problems of tectomagmatic classification and to establishing the nature of crustal contamination of basaltic lavas of the British Tertiary Volcanic Province. *Earth and Planetary Science Letters*, **50**, 11–30.

# Rheic Ocean mafic complexes: overview and synthesis

J. BRENDAN MURPHY[1]*, GABRIEL GUTIÉRREZ-ALONSO[2], R. DAMIAN NANCE[3],
JAVIER FERNÁNDEZ-SUÁREZ[4], J. DUNCAN KEPPIE[5], CECILIO QUESADA[6],
JAROSLAV DOSTAL[7] & JAMES A. BRAID[1]

[1]*Department of Earth Sciences, St. Francis Xavier University, Antigonish, Nova Scotia,
B2G 2W5 Canada*

[2]*Departamento de Geología, Universidad de Salamanca, 33708 Salamanca, Spain*

[3]*Department of Geological Sciences, Ohio University, Athens, Ohio 45701, USA*

[4]*Departamento de Petrología y Geoquímica, Universidad Complutense,
28040, Madrid, Spain*

[5]*Departamento de Geología Regional, Instituto de Geologia, Universidad Nacional
Autonoma de Mexico, 04510 Mexico D.F. México*

[6]*IGME, c/Ríos Rosas 23, 28003, Madrid, Spain*

[7]*Department of Geology, St. Mary's University, Halifax, Nova Scotia, B3H 3C3 Canada*

*\*Corresponding author (e-mail: bmurphy@stfx.ca)*

**Abstract:** The Rheic Ocean formed during the Late Cambrian–Early Ordovician when peri-Gondwanan terranes (e.g. Avalonia) drifted from the northern margin of Gondwana, and was consumed during the collision between Laurussia and Gondwana and the amalgamation of Pangaea. Several mafic complexes, from the Acatlán Complex in Mexico to the Bohemian Massif in eastern Europe, have been interpreted to represent vestiges of the Rheic Ocean. Most of these complexes are either Late Cambrian–Early Ordovician or Late Palaeozoic in age. Late Cambrian–Early Ordovician complexes are predominantly rift-related continental tholeiites, derived from an enriched *c.* 1.0 Ga subcontinental lithospheric mantle, and are associated with crustally-derived felsic volcanic rocks. These complexes are widespread and virtually coeval along the length of the Gondwanan margin. They reflect magmatism that accompanied the early stages of rifting and the formation of the Rheic Ocean, and they remained along the Gondwanan margin to form part of a passive margin succession as Avalonia and other peri-Gondwanan terranes drifted northward. True ophiolitic complexes of this age are rare, a notable exception occurring in NW Iberia where they display ensimatic arc geochemical affinities. These complexes were thrust over, or extruded into, the Gondwanan margin during the Late Devonian–Carboniferous collision between Gondwana and Laurussia (Variscan orogeny). The Late Palaeozoic mafic complexes (Devonian and Carboniferous) preserve many of the lithotectonic and/or chemical characteristics of ophiolites. They are characterized by derivation from an anomalous mantle which displays time-integrated depletion in Nd relative to Sm. Devonian ophiolites pre-date closure of the Rheic Ocean. Although their tectonic setting is controversial, there is a consensus that most of them reflect narrow tracts of oceanic crust that originated along the Laurussian margin, but were thrust over Gondwana during Variscan orogenesis. The relationship of the Carboniferous ophiolites to the Rheic Ocean *sensu stricto* is unclear, but some of them apparently formed in a strike-slip regimes within a collisional setting directly related to the final stages of the closure of the Rheic Ocean.

## Introduction

Since the publication of Tuzo Wilson's provocative (1966) paper 'Did the Atlantic close and then re-open?' the Palaeozoic evolution of the Appalachian–Caledonide–Variscan orogen has been key to understanding the development of Pangaea. The Palaeozoic 'Proto-Atlantic' ocean he envisaged between the rocks with 'Pacific fauna' of Laurentia and those with 'Atlantic fauna' of Baltica–Gondwana became known as the Iapetus Ocean (Harland & Geyer 1972) and represents the first application of modern plate tectonic principles to the pre-Mesozoic world. Since that time, volumes of research have been published on vestiges of the Iapetus Ocean preserved as ophiolitic complexes, primarily along the Laurentian margin.

More recent palaeogeographic reconstructions, primarily based on faunal and palaeomagnetic data, have seen the orthogonal opening and closing

*From*: MURPHY, J. B., KEPPIE, J. D. & HYNES, A. J. (eds) *Ancient Orogens and Modern Analogues.*
Geological Society, London, Special Publications, **327**, 343–369.
DOI: 10.1144/SP327.15   0305-8719/09/$15.00 © The Geological Society of London 2009.

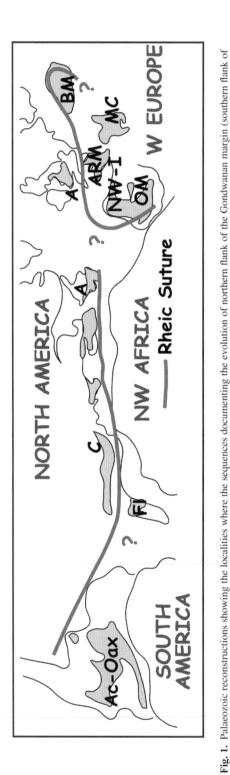

**Fig. 1.** Palaeozoic reconstructions showing the localities where the sequences documenting the evolution of northern flank of the Gondwanan margin (southern flank of the Rheic Ocean) are preserved shown in a Pangaean reconstruction. Ac–Oax, Acatlán–Oaxaquia; Fl, Florida; C, Carolinia; A, Avalonia; O-M, Ossa Morena; NW-I, NW Iberia; Arm, Armorica; MC, Massif Central; BM, Bohemian massif.

model of Wilson (1966) superceded by more actualistic models (Figs 1 & 2) involving terrane transfer from Gondwana to Baltica and Laurentia prior to the terminal collision between Gondwana and Laurussia (Keppie 1985; McKerrow & Scotese 1990; Cocks & Fortey 1988; Cocks & Torsvik 2002; Murphy *et al.* 2006*a*; Gómez-Barreiro *et al.* 2007). These studies show: (i) that the ophiolitic complexes were formed and obducted in the early stages (by the Early Ordovician) of development of the Iapetus Ocean (e.g. van Staal *et al.* 1998); (ii) that the ophiolites are generally supra-subduction zone (back-arc and fore-arc) bodies (Jenner & Swinden 1993; MacLachlan & Dunning 1998; Bédard *et al.* 1998; Bédard & Stevenson 1999; Pin *et al.* 2006; Sánchez Martínez *et al.* 2007); and (iii) that closure of the Iapetus Ocean had occurred by the mid-Silurian and so preceded the formation of Pangaea by >100 Ma (e.g. Williams 1979; Keppie 1985, 1993). The northern realm of the Iapetus Ocean was closed by the Early–mid-Silurian collision between Baltica and Laurentia to form Laurussia (Scandian orogeny: Roberts & Gee 1985; Roberts & Stephens 2000; Winchester *et al.* 2002; Brueckner & Van Roermund 2004). In the central and southern realms of the Iapetus Ocean, however, closure is attributed to the accretion of smaller terranes (such as Avalonia, Ganderia and Carolinia) that had rifted away from their former positions along the margin of Gondwana (e.g. Murphy *et al.* 1995, 2005*b*; Hibbard 2000). In fact, we now know that much of the Appalachian-Caledonide orogen underlain by rocks containing the 'Atlantic fauna' of Wilson (1966) belongs to Avalonia, a terrane that separated from Gondwana in the Late Cambrian–Early Ordovician, giving rise to the Rheic Ocean between Avalonia and Gondwana. In Ordovician times, Avalonia consequently separated a closing Iapetus Ocean to the north from an opening Rheic Ocean to the south (Fig. 2). As a result, it was not the demise of the Iapetus Ocean that gave rise to Pangaea, but the closure of its successor, the Rheic Ocean.

Despite its obvious importance to the assembly of Pangaea, studies of the Rheic Ocean have lagged behind those of the Iapetus Ocean. A variety of mafic complexes, interpreted by most authors as ophiolites, are potentially the remnants of this ocean and record important tectonothermal events during its evolution. However, a modern synthesis of these complexes and their tectonic significance is lacking. In this paper, we first provide the regional context in which Rheic Ocean ophiolites and other mafic complexes can be evaluated. We then provide an overview of these complexes, assess whether they represent ophiolites, and interpret their significance in terms of understanding Pangaean geology. Although more data are available for some mafic

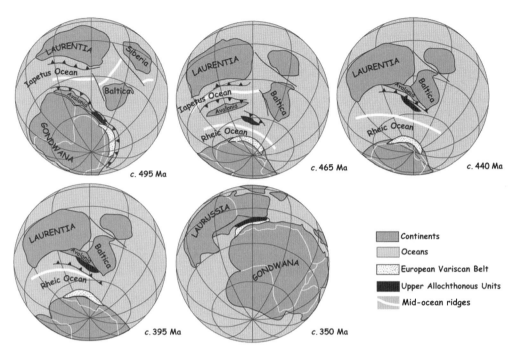

**Fig. 2.** Palaeozoic reconstructions (modified from Stampfli *et al.* 2002) showing the location of Avalonian-type terranes along the Gondwanan margin at 495 Ma, their separation from Gondwana and northward drift of *c.* 2000 km by 465 Ma, and their accretion to Baltica by 440 Ma, followed by Laurentia, leading to the closure of the Iapetus Ocean. Closure of the Rheic ocean was accomplished by northwesterly-directed subduction beneath the Laurussian margin (e.g. 395 Ma) leading to the amalgamation of Pangaea (e.g. 350 Ma).

complexes than others, a first-order similarity exists within and between them that allows their regional tectonic significance to be evaluated.

## Evolution of the Rheic Ocean: global context

The tectonic evolution of the Palaeozoic Era is dominated by early Palaeozoic ocean development and continental dispersal, followed by convergence in the mid- to Late Palaeozoic, which culminated in continental collisions that led to the Late Carboniferous–Early Permian amalgamation of the supercontinent Pangaea (Fig. 2). The Appalachian–Caledonide–Variscan orogen of North America and Europe formed within the interior of Pangaea as the result of a series of orogenic pulses related to the closure of two key oceans, the Iapetus Ocean by mid-Silurian, and the Rheic Ocean in the Carboniferous. Subduction of the Iapetus Ocean initiated along its northern flank in the Late Cambrian and led to arc–continent collisions and ophiolite obduction (events assigned to the Taconic orogeny, Williams 1979; Keppie 1985; Cawood *et al.* 1996; van Staal *et al.* 1998;

Winchester *et al.* 2002; van Staal *et al.* 2008), and to collision between Laurentia and Baltica to form Laurussia (events assigned to the Scandian orogeny, Roberts & Gee 1985).

Subduction initiation within the Iapetus Ocean was broadly coeval with the *c.* 500 Ma formation of the Rheic Ocean. The Rheic Ocean originated when ribbon terranes (collectively called peri-Gondwanan) drifted from the Gondwanan margin. By 460 Ma, for example, one of these terranes (Avalonia) had drifted 2000 km north of the Gondwanan margin (Hamilton & Murphy 2004). As these terranes formed during the inception of the Rheic Ocean, their locations broadly coincided with the boundary between the contracting Iapetus Ocean to the north and the expanding Rheic Ocean to the south (Fig. 2, Cocks & Torsvik 2002; Stampfli & Borel 2002; Murphy *et al.* 2006a). By the end of the Ordovician, most studies suggest that these peri-Gondwanan terranes had collided with Baltica; however, the timing of their accretion to Laurentia is controversial. Some authors favour accretion at various times between the Early Silurian and Late Devonian (e.g. van Staal *et al.* 1998; Hibbard *et al.* 2002; van Staal *et al.* 2008), some favour a major Silurian accretionary event, the Salinic orogeny (e.g. Keppie *et al.* 2003*b*; Murphy *et al.* 2002*b*), whereas other favour accretion in the Devonian, the Acadian orogeny (Hatcher 2002). Irrespective of these important details, there is a general consensus that: (i) from the Late Silurian–Late Carboniferous, the Rheic Ocean separated Gondwana from Laurussia; (ii) the closure of this ocean led to the assembly of Pangaea through the collision of these two continental landmasses; and (iii) this collision gave rise to the Variscan orogeny in Europe and the Alleghanian orogeny in North America. It is also widely accepted that the Rheic Ocean did not close completely during Variscan orogenesis, but remained open to the east, where it is known as the Paleotethys Ocean (e.g. Stampfli *et al.* 2002).

Mafic complexes that potentially represent vestiges of Rheic Ocean geology occur from the Acatlán Complex in Mexico (e.g. Nance *et al.* 2006, 2007; Keppie *et al.* 2008*a*) to the Bohemian Massif in eastern Europe (e.g. Oliver *et al.* 1993; Zelazniewicz *et al.* 1998) to the NE (Fig. 1; current co-ordinates). The Acatlán Complex is the largest inlier of Palaeozoic rocks in Mexico, outcropping over an area equivalent in size to Massachusetts (Fig. 3). The origin of the Acatlán Complex is controversial. According to some authors, its mafic assemblages represent vestiges of the Iapetus Ocean (e.g. Yàñez *et al.* 1991; Ortega-Gutiérrez *et al.* 1999; Talavera-Mendoza *et al.* 2005; Vega-Granillo *et al.* 2007). However, recent age data suggest that much of its tectonothermal evolution postdates closure of the Iapetus Ocean, and that its history is more compatible with an evolution within the Rheic Ocean (Murphy *et al.* 2006*b*; Nance *et al.* 2006, 2007; Middleton *et al.* 2007; Keppie *et al.* 2008*a*).

Rheic Ocean mafic complexes are not exposed in northern South America or in North America. Geophysical data from the southeastern United States, however, has led to the identification of deep seismic reflection and magnetic anomalies (known as the Brunswick and East Coast Anomalies), which are interpreted as mafic bodies that reflect the subsurface expression of the Rheic Ocean suture in the southern Appalachian orogen (McBride & Nelson 1988, 1991).

In Europe, mafic complexes are found within the Variscan orogen (Fig. 4), which is a *c.* 1000 km-wide curvilinear belt with several fault-bounded tectonostratigraphic zones that can be traced the length of the orogen (e.g. Franke 1989). Mafic complexes occur predominantly in terranes adjacent to the northern (e.g. southern Iberia, Lizard Complex) and southern (NW Iberia, Massif Central) margins of the Rheic Ocean. Curvature of the tectonostratigraphic zones is interpreted either as an orocline (Weil *et al.* 2001; Gutiérrez-Alonso *et al.* 2003) or as the result of a pre-orogenic promontory

of Gondwana that acted as an indenter during Late Palaeozoic collisional orogenesis (Quesada *et al.* 1991; Quesada & Dallmeyer 1994). In both interpretations, the Palaeozoic rocks are genetically related to one another and to the evolution of the Rheic Ocean. However, the contrasting views affect the interpretation of the palaeogeography of the northern Gondwanan margin during Variscan orogenesis and the tectonic setting associated with emplacement of these mafic complexes.

## Acatlán Complex, Mexico

### General Geology

The Acatlán Complex forms the Mixteca terrane of southern Mexico (Fig. 3a) and is dominated by eclogitic sedimentary and meta-igneous rocks (Piaxtla Suite) and tectonic slices of phyllites, psammites and basalt. The complex is faulted against Mesoproterozoic (*c.* 1 Ga) granulite facies gneisses of the Oaxacan Complex, which forms the basement of southern Oaxaquia (Ortega-Gutiérrez *et al.* 1995, 1999), a terrane that was likely attached to the northern portion of the Amazonian craton throughout the Palaeozoic (Keppie & Ramos 1999; Keppie 2004). Oaxaquia extends northwards along the backbone of Mexico to the suture zone with Laurentia near the US–Mexican border (Fig. 3a). The Oaxacan Complex is unconformably overlain by latest Cambrian–earliest Ordovician platformal strata (Tiñu Formation) that contain Gondwanan fauna (Landing *et al.* 2007). These, in turn, are overlain by Carboniferous–Permian carbonates and clastic rocks (Centeno-Garcia & Keppie 1999). The boundary between the Mixteca and Oaxaquia terranes is a Permian dextral shear zone (Fig. 3b; Elías-Herrera & Ortega-Gutiérrez 2002).

The evolution of the Acatlán Complex spans the Ordovician–Jurassic with tectonothermal activity reflecting: (i) the development of an Ordovician rift-passive margin on an Oaxacan-aged basement, which corresponds with the southern flank of the Rheic Ocean (Keppie 2004; Miller *et al.* 2007; Keppie *et al.* 2008*a*; Morales-Gámez *et al.* 2008; Ramos-Arias *et al.* 2008); (ii) subduction-related, eclogite facies metamorphism and exhumation in the Late Devonian–Mississippian (Middleton *et al.* 2007) during and after the amalgamation of Pangaea; (iii) subduction along the paleo-Pacific margin following the assembly of Pangaea (Nance *et al.* 2007; Keppie *et al.* 2008*a*); and (iv) Jurassic plume-related activity coeval with Pangaea breakup and the opening of the Gulf of Mexico (Keppie *et al.* 2004*a*; Nance *et al.* 2006).

The Acatlán Complex contains remnants of oceanic lithosphere preserved as high-pressure

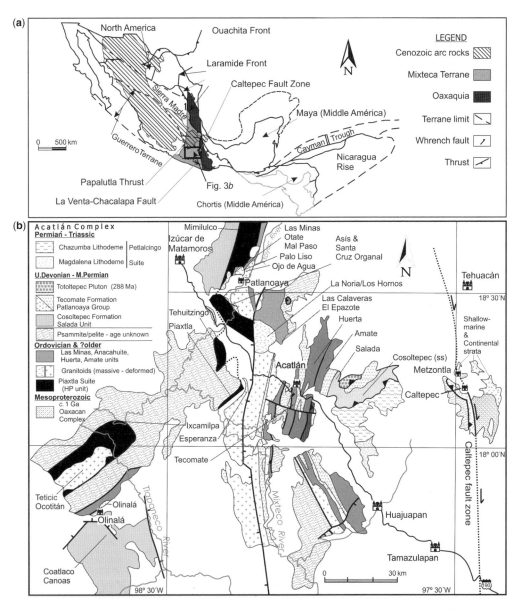

**Fig. 3.** Simplified geological map of the Acatlán Complex, southern Mexico (modified from Ortega-Gutiérrez *et al.* 1999; Keppie *et al.* 2004*b*, 2008*a*; Nance *et al.* 2006).

rocks (serpentinites, blueschists and eclogites) that have been emplaced into low grade psammites and pelites containing mid-ocean ridge (MORB) and ocean island (OIB) pillow basalts interbedded with cherts. These rocks have been interpreted as vestiges of the Iapetus Ocean (Ortega-Gutiérrez *et al.* 1999; Talavera-Mendoza *et al.* 2005; Vega-Granillo *et al.* 2007), but other studies suggest that they are more likely to have originated in the Rheic Ocean (Keppie & Ramos 1999; Keppie 2004; Keppie

*et al.* 2004*a*, *b*, 2006, 2007, 2008*a*; Murphy *et al.* 2006*b*; Nance *et al.* 2006, 2007; Middleton *et al.* 2007; Miller *et al.* 2007; Ramos-Arias *et al.* 2008; Morales-Gámez *et al.* 2008).

The Carboniferous protolith ages of some of these mafic-ultramafic rocks clearly post-date the Iapetus Ocean. However, the occurrence of Mid-continent (USA) fauna in the Mississippian rocks overlying the *c.* 1.0 Ga Oaxacan Complex indicates that Pangaea had already amalgamated by the

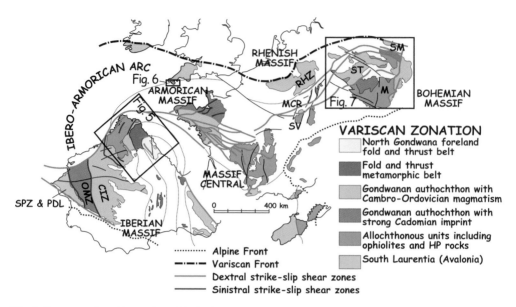

**Fig. 4.** Tectonostratigraphic zonation of the Variscan orogen in Europe (after Franke 2000; Martínez Catalán *et al.* 2007).

beginning of the Carboniferous (Navarro-Santillán *et al.* 2002). The Acatlán Complex lies *c.* 1000 km to the south of the suture between Oaxaquia–Amazonia and Laurentia (Keppie 2004), so whereas some of these mafic–ultramafic rocks may have originated within the Rheic Ocean, their metamorphism and exhumation probably occurred on the western margin of Pangaea within the Panthalassa Ocean.

The oldest dated igneous rocks in the Acatlán Complex are *c.* 440–480 Ma granitoid rocks and amphibolites (Keppie *et al.* 2004*a*, 2008*b*; Talavera-Mendoza *et al.* 2005; Miller *et al.* 2007). Mafic units within the Acatlán Complex occur in several fault slices and may preserve oceanic vestiges of different ages and tectonic settings (Vega-Granillo *et al.* 2007). Hence, the complex has been subdivided into lithodemes (Keppie *et al.* 2006, 2008*a*; Middleton *et al.* 2007; Ramos-Arias *et al.* 2008; Morales-Gámez *et al.* 2008; Grodzicki *et al.* 2008) and tectonic synthesis of these units is a long-term goal of current research. Here, we summarize recent advances in our understanding of some of these complexes in order to illustrate the variability.

### Ordovician–Early Silurian complexes

The largest ultramafic body, the Tehuitzingo serpentinite, occurs in the middle of the Acatlán Complex (Fig. 3), and contains small eclogitic lenses indicating high-pressure metamorphism probably associated with a subduction zone (Proenza *et al.*

2004). Analysis of chromites within the serpentinite suggests that the protoliths were formed in a supra-subduction zone setting. Unfortunately, no protolith or metamorphic age data are available. However, along strike to the north, in the San Francisco de Asís area, retrograde eclogite facies rocks have yielded a concordant U–Pb TIMS zircon age of $346 \pm 2$ Ma, 350–330 Ma U–Pb SHRIMP ages for decompression migmatites, and a *c.* 350 Ma $^{40}Ar/^{39}Ar$ muscovite plateau age (Middleton *et al.* 2007) that are inferred to date peak metamorphism and rapid exhumation. Here, the protoliths include rift-passive margin sedimentary rocks intruded by a bimodal igneous assemblage dated at 470–420 Ma (U–Pb SHRIMP ages for megacrystic granitoids, Murphy *et al.* 2006*b*) and $442 \pm 2$ Ma (U–Pb SHRIMP ages for zircon cores in amphibolite dikes: Elías-Herrera *et al.* 2007). Further north, in the Patlanoaya area, retrogressed eclogite gave an U–Pb zircon age of $353 \pm 1$ Ma, and blueschists have yielded *c.* 342 Ma $^{40}Ar/^{39}Ar$ ages for both glaucophane and phengite (Elías-Herrera *et al.* 2007). Similarly, Vega-Granillo *et al.* (2007) reported $^{40}Ar/^{39}Ar$ plateau ages for phengite of *c.* 347–333 Ma from several other high-pressure suites. Thus, the subduction-related metamorphism is Mississippian in age. The rift-passive margin tectonic setting for the protoliths of most of the high-pressure rocks is almost identical to that of the latest Cambrian–earliest Ordovician Tiñu Formation, which lies unconformably upon the *c.* 1 Ga Oaxacan Complex (Murphy *et al.* 2005*b*).

These relationships suggest that the leading edge of the continental margin was first subducted and then obducted over the inner continental margin (Nance *et al.* 2006, 2007). In this model, the high-pressure blueschist–eclogitic rocks mark an oceanic suture. However, the geological record of the Acatlán Complex on either side of the high-pressure rocks is remarkably similar. This has led Keppie *et al.* (2008*a*) to conclude that high pressure rocks represent material that was removed by subduction erosion from the upper plate, carried down the subduction channel and then extruded into the upper plate.

The Asís Lithodeme of the Piaxtla Suite (Fig. 3) is composed mainly of medium-to high-grade metapsammitic and metapelitic rocks, and thin bands of amphibolite (interpreted as dykes, Murphy *et al.* 2006*b*; Middleton *et al.* 2007) that are intruded by megacrystic granite, the margins of which are mylonitized. Geochemical and Sm–Nd and U–Pb isotopic data for the metasedimentary rocks closely match those of the Tiñu Formation overlying the neighbouring Oaxacan Complex. Both contain a significant mafic component (suggesting proximal sources, e.g. Nesbitt & Young 1996), Mesoproterozoic $T_{DM}$ ages (1.5–1.83 Ga) (Murphy *et al.* 2006*a, b*), and *c.* 990–1200 Ma detrital zircons (Gillis *et al.* 2005; Middleton *et al.* 2007). Since the Tiñu Formation data are interpreted to reflect derivation mainly from the underlying Oaxacan Complex, the protoliths of the Asís Lithodeme metasedimentary rocks have been attributed to a similar rift-passive margin setting (Keppie *et al.* 2001, 2003*a*; Solari *et al.* 2003; Ortega-Obregón *et al.* 2003; Cameron *et al.* 2004; Murphy *et al.* 2006*b*). Although poorly constrained, the *c.* 700–470 Ma age range for deposition of the Asís Lithodeme, together with its geochemical, isotopic and detrital zircon population characteristics, are consistent with the correlation of its protoliths with the Tiñu Formation, and consequently an origin along the Gondwanan margin of the Rheic Ocean.

The amphibolites display geochemical trends typical of differentiated continental tholeiites (Murphy *et al.* 2006*b*). They contain high $FeO_t/MgO$ and $TiO_2$, slight enrichment in LREE and trace element patterns typical of derivation from an enriched mantle source (Murphy *et al.* 2006*b*) with no obvious evidence of crustal contamination (such as negative Nb, Ta or Zr anomalies). Sm–Nd isotopic analyses yield $\varepsilon_{Nd(t)}$ ranging from $+2.8$ to $+4.6$ (t = 475 Ma) and $T_{DM}$ from 0.75–1.27 Ga (Murphy *et al.* 2006*b*). The $\varepsilon_{Nd(t)}$ values are considerably lower than values expected for juvenile magmas from a depleted mantle source and are interpreted to reflect derivation from the sub-continental lithospheric mantle. Taken together,

the mafic rocks are interpreted to be differentiated continental tholeiites derived from an enriched mantle source. Zircon cores have yielded an age of $442 \pm 2$ Ma (Elías-Herrera *et al.* 2004), which is interpreted as the time of intrusion.

The megacrystic granites have crustal signatures and 1.5–1.8 Ga $T_{DM}$ ages that are very similar to those of the Oaxacan Complex (Patchett & Ruiz 1987; Ruiz *et al.* 1988; Weber & Köhler 1999) and to Ordovician granitoid plutons in the northern Acatlán Complex (Talavera-Mendoza *et al.* 2005; Miller *et al.* 2007). Collectively, these data suggest that the Asís granitic magma was derived by crustal anatexis of the Oaxaquia basement, which is inferred to underlie the Acatlán Complex (Keppie 2004; Keppie *et al.* 2008*a*). Although the available geochronological data are not sufficient to resolve the relative timing of basaltic and granitoid magmatism in the Asís Lithodeme, regional considerations are consistent with the interpretation that the amphibolites and granitoids represent part of a bimodal suite.

The continental affinity of some of the rocks in the Asís Lithodeme (Murphy *et al.* 2006*b*; Middleton *et al.* 2007) suggests that these rocks are not ophiolites. Instead, the Asís Lithodeme igneous rocks are interpreted as a bimodal suite formed during continental rifting and crustal extension associated with the opening of the Rheic Ocean (Murphy *et al.* 2006*b*; Keppie *et al.* 2008*a, b*).

Recent geochronological data shows that these rocks underwent eclogite facies metamorphism at *c.* 346 Ma, followed by amphibolite facies metamorphism and migmatization associated with rapid uplift in the Visean (Middle Mississippian, Elías-Herrera *et al.* 2004; Middleton *et al.* 2007; Vega-Granillo *et al.* 2007). These data are attributed to either subduction and obduction of the leading edge of Gondwana during closure of the Rheic Ocean and the amalgamation of Pangaea (Nance *et al.* 2007), or to subduction erosion followed by high-pressure metamorphism and extrusion into the upper plate (Keppie *et al.* 2008*a*).

### Cosoltepec Formation

The Cosoltepec Formation (Fig. 3b) forms a major part of the Acatlán Complex and consists of unfossiliferous phyllites and psammites with tectonic slices of oceanic basalt and intercalated red chert (Keppie *et al.* 2007). The formation, as currently defined, is a composite unit (Keppie *et al.* 2008*b*; Morales-Gámez *et al.* 2008; Ramos-Arias *et al.* 2008). Some metasedimentary rocks assigned to the formation are intruded by mid-Upper Ordovician granitoids (Miller *et al.* 2007; Keppie *et al.* 2008*b*), whereas metasedimentary rocks from the type area have yielded detrital zircons as young as *c.* 410 Ma (Talavera-Mendoza *et al.* 2005). In contrast to the Asís lithodeme, an oceanic affinity has been proposed for this formation, based primarily upon the occurrence of pillow basalts in the formation (e.g. Ramírez Espinoza 2001). Recently published geochemical data indicate that the basalts are predominantly MORB and continental tholeiites with flat or depleted LREE patterns, and basalts and andesites with OIB affinities, which have distinctly fractionated LREE-enriched patterns (Keppie *et al.* 2007, 2008*b*). Current age data indicate deposition during the Ordovician and Carboniferous (Keppie *et al.* 2008*b*; Morales-Gámez *et al.* 2008; Grodzicki *et al.* 2008). In places, these units are inferred to be unconformably overlain by either uppermost Devonian (Fammenian) strata (Vachard & Flores de Dios 2002) or the Permian Tecomate Formation (Keppie *et al.* 2008*a*).

The widespread tectonic interleaving with the clastic metasediments was interpreted by Keppie *et al.* (2007) to reflect either deposition of the sedimentary rocks directly upon the ocean-floor basalts or extrusion of the basalts in areas removed from continentally-derived sediments. However, some mafic dykes intrude the clastic rocks (Keppie *et al.* 2008*b*; Ramos-Arias *et al.* 2008; Morales-Gámez *et al.* 2008). Deformation and greenschist to sub-greenschist facies metamorphism commenced immediately prior to deposition of the late Fammenian strata, but continued into the Mississippian (Ramos-Arias *et al.* 2008).

In the western portion of the Acatlán Complex, basalts occur in an unnamed sequence that was formerly assigned to the Cosoltepec Formation near Olinalá (Fig. 3b). Here, the rocks consist of two tectonically interleaved units (Grodzinski *et al.* 2008). The Canoas unit consists of interbedded psammites and pelites, the youngest concordant detrital zircon from which yields an age of 462 ± 15 Ma (Middle Ordovician). The Coatlaco unit consists of interbedded quartzite and pillow basalts. An average of the 8 youngest detrital zircons in the Coatlaco quartzite yields an age of 357 ± 8 Ma (*c.* Devono–Carboniferous boundary), which is interpreted as the present best estimate for the depositional age of these basalts. The basalts are tholeiitic with within-plate and MORB-like affinities that together with the interbedded cherts suggest an oceanic tectonic setting. The basaltic magmatism was coeval with rapid exhumation of eclogitic rocks of the Piaxtla Suite, and Grodinski *et al.* (2008) consequently suggest a link between their eruption and the exhumation of the Piaxtla Suite.

Tectonic juxtaposition of the low-grade clastic rocks with the eclogite-bearing rocks of the Asís Lithodeme suggests that the Late Devonian–Mississippian deformation was related to exhumation following subduction, with the Cosoltepec Formation and its mafic lenses derived from the overriding plate and the Asís lithodeme eclogitic rocks forming a subducted part of the upper plate. The occurrence of Mid-continent (USA) fauna in Mississippian rocks resting above the Tiñu Formation (Navarro-Santillán *et al.* 2002) suggests that Pangaea had amalgamated by this time, implying that these events took place on the western margin of Pangaea.

## Iberia

### General geology

Palaeozoic rocks of Iberia are divided into tectono-stratigraphic zones (e.g. Lotze 1945; Julivert *et al.* 1972; Farias *et al.* 1987; Quesada 1991; Fig. 4) based on their Lower Palaeozoic sedimentary differences, which are interpreted to reflect their relative proximity to the Gondwanan margin. Boundaries between these zones are major Variscan thrusts and reverse faults that were in some cases reactivated extensionally in the aftermath of the Variscan orogeny. The Cantabrian Zone (CZ) preserves a coastal environment, whereas the West Asturian–Leonese (WALZ), Central Iberian (CIZ), Schistose Domain (also known as the Galicia-Tras–os-Montes Domain, SGTM) and Ossa–Morena (OMZ) zones preserve the more outboard tectonostratigraphy (Julivert *et al.* 1972; Quesada 1991; Ribeiro *et al.* 1990; Pérez Estaún *et al.* 1990; Quesada *et al.* 1991; Aramburu *et al.* 2002; Martínez Catalán *et al.* 1997, 1999; Marcos *et al.* 2004; Gutiérrez-Marco *et al.* 1999; Robardet 2002, 2003; Robardet & Gutiérrez-Marco 2004). The SGTM is considered to be parauthochthonous (Farias *et al.* 1987; Ribeiro *et al.* 1990) and structurally overlies the authochthonous CIZ with which it shares igneous and stratigraphic affinities and is interpreted to consist of the most internal parts of the Gondwanan passive margin. Structurally overlying the Gondwana northern passive margin as klippen, Variscan allochthonous complexes in NW Iberia include ophiolitic mafic complexes whose correlatives extend into the Massif Central of France. In southern Iberia, ophiolitic complexes (e.g. the Beja–Acebuches ophiolite (BAO), Aracena massif) occur as discontinuous, dismembered bodies along the boundary between the Pulo do Lobo Zone (PDLZ) and the OMZ (e.g. Silva *et al.* 1990). The PDLZ is also characterized by an accretionary prism of Middle–Late Devonian age (Robardet 2003; Crowley *et al.* 2000) and together with the BAO is thought to represent a suture reflecting the closure of the Rheic Ocean which spatially juxtaposed the Iberian para-autochthon (OMZ) and the South Portuguese Zone (SPZ), a suspect exotic terrane. Although the oldest exposed rocks in the

SPZ are Devonian in age, several studies suggest the terrane is underlain by Avalonian basement, which was attached to Laurussia at the time of Variscan orogenesis (Oliveira & Quesada 1998; de la Rosa *et al.* 2002). Although highly controversial, some authors also propose the presence of another Variscan suture in the intensely deformed Unidad Central that bounds the CIZ and the OMZ, locating a narrow ocean subsidiary to the Rheic, between both zones (see Simancas *et al.* 2002, and references therein).

## Palaeozoic stratigraphy along the Northern Gondwana Margin of Iberia

Palaeozoic mafic complexes in Iberia occur in different tectonostratigraphic zones that record different aspects of Rheic Ocean evolution. Mafic complexes are predominantly either Late Cambrian–Early Ordovician or Devonian in age; although some Mesoproterozoic mafic rocks have been recently discovered (Sánchez Martínez *et al.* 2006).

Early Cambrian rocks are dominated by siliciclastic rocks that were deposited on top of Ediacaran strata. In most locations, the contact between Early Cambrian and Ediacaran successions is an unconformity. The Early and Middle Cambrian successions comprise the base of a passive margin sequence that continued throughout most of the Palaeozoic. The Late Cambrian–Middle Ordovician stratigraphy is preserved in the CIZ, WALZ and OMZ and is dominated by a rift-to-drift transition, most notably by the Lower Ordovician laterally extensive, mostly detrital sequence, known as the Armorican Quartzite, overlying Llanvirn black shales (e.g. Young 1990; Gutiérrez-Alonso *et al.* 2007) and accompanying igneous suites. Although heterogeneously deformed by Variscan orogenesis, NW Iberia preserves the most complete Late Cambrian–Devonian passive margin sequence, which sits on top of Mesoproterozoic (*c.* 1.1–1.4 Ga) and Late Neoproterozoic (*c.* 750–550 Ma) basement rocks (Fernández-Suárez *et al.* 2000, 2002*a*, *b*; Gutiérrez-Alonso *et al.* 2003, 2005). The stratigraphy records the rift-to-drift transition represented by the Late Cambrian–Ordovician Armorican quartzite and overlying black shales. Silurian and Devonian rocks outcrop extensively in all zones except the WALZ, predominantly in the core of some synclinal structures in the CIZ and OMZ (Arenas *et al.* 2007*a*, *b*; Martínez Catalán *et al.* 1997, 1999, 2007). Silurian strata include a wide variety of lithologies that are mostly pelagic, including black shales (Robardet & Gutiérrez-Marco 2004). Devonian strata have nearshore and offshore intercalations and, in the CZ, include abundant reef deposits. Early–Middle Devonian

rocks preserve either an Eifelian–Givetian unconformity or a hiatus with some volcanic rocks present (Puschmann 1967; Gutiérrez-Alonso *et al.* 2008). The Upper Devonian sedimentary sequences extend from the CZ (coastal environment) to the OMZ (outermost shelf) and record a significant diachronous increase in subsidence towards the coast, and herald the loading of the margin during the progressive collision between Gondwana and Laurussia (Dallmeyer *et al.* 1997; Martínez Catalán *et al.* 2007).

Much of the passive margin sequence was subducted and exhumed during Variscan orogenesis and occurs as high- to low-pressure (eclogite, amphibolite and greenschist facies) units in allochthonous complexes that are preserved as klippen in NW Iberia (Martínez Catalán *et al.* 1997; Arenas *et al.* 2007*a*, *b*).

## Palaeozoic igneous activity along the northern Gondwanan margin

The CZ includes several Lower Palaeozoic volcanic events, the most common of which are Lower Ordovician in age (Loeschke & Zeidler 1982; Heinz *et al.* 1985; Gallastegui *et al.* 1992; Barrero & Corretgé 2002; Gutiérrez-Alonso *et al.* 2003). The most voluminous volcanics of this age, however, occur in the northern CIZ, including the 'Ollo de Sapo' belt, which are dominated by felsic volcanics interpreted as rift-related intra-crustal melts (Valverde-Vaquero & Dunning 2000; Castro *et al.* 1999, 2003; Díez Montes 2006; Bea *et al.* 2006; Zeck *et al.* 2007). Late Ordovician mafic volcanic activity has a continental, within-plate, alkalic signature. Sm–Nd isotopic data yield $\varepsilon_{Nd}$ values ranging from +1.0 to +1.1, which are well below the values for contemporary depleted mantle and are interpreted to reflect derivation from an old (*c.* 0.9–1.1 Ga) sub-continental mantle lithosphere, whereas the coeval felsic rocks were generated by partial melting of Mesoproterozoic–Palaeoproterozoic crust (Murphy *et al.* 2008).

Five allochthonous units described in NW Iberia (Ortegal, Ordenes, Malpica–Tui, Morais and Bragança, Fig. 5) share a common organization that indicate their collective origin as a single allochthonous slice obducted on top of the Gondwana margin. The five allochthonous units represent the product of the subsequent dismemberment of the obducted slice and are preserved in the cores of late Variscan synforms. The SGTM in this slice consists of a *c.* 475 Ma thick, interbedded siliciclastic volcanic sequence (Valverde-Vaquero *et al.* 2005), and represents the outermost portion of the Gondwanan passive margin (Farias *et al.* 1987; Martínez-Catalán *et al.* 1996; Marcos *et al.* 2002; Pereira *et al.* 2006). Above this unit, and present in all the complexes, the Basal Unit represents the basement of the Gondwanan margin and is composed of metasediments of unknown age intruded by a 490–460 Ma bimodal igneous suite with alkaline affinities (Pin *et al.* 1992; Montero 1993; Santos Zalduegui *et al.* 1995; Abati *et al.* 2003) that are interpreted to be related to the rifting episode that caused the opening of the Rheic Ocean (Ribeiro & Floor 1987; Pin *et al.* 1992). Sm/Nd isotopic values indicate that the mafic rocks were derived from partial melting of a *c.* 1.0 Ga sub-continental lithospheric mantle (Murphy *et al.* 2008). The Basal Unit is thought to represent crust that was rifted from Gondwana during the formation of the Rheic Ocean, and returned to Gondwana during the collision between Gondwana and Laurussia. The Basal Unit underwent HP conditions around 370 Ma (e.g. Arenas *et al.* 1995; Gil Ibarguchi 1995; Santos Zalduegui *et al.* 1995; Martínez Catalán *et al.* 1996; Rubio Pascual *et al.* 2002; Rodríguez *et al.* 2003; Rodríguez Aller 2005), an age that is considered as the collision age between Gondwana and Laurussia in this sector of the Variscan belt.

Two ophiolitic units of Early Ordovician and Devonian age structurally overlie the Basal Unit (Martínez Catalán *et al.* 1997). The Lower Ophiolite is variously known as the Moeche (Ortegal Complex), Vila de Cruces (Ordenes Complex, where it has been dated at 497 ± 4 Ma, Arenas *et al.* 2007*a*) or the Izeda-Remondes (Morais Complex) and also occurs in the Bragança Complex. These complexes preserve a complete ophiolitic suite that was metamorphosed under greenschist conditions. Available geochemical data indicate that the ophiolites of the Ortegal and the Ordenes complexes have a supra-subduction zone affinity (Sánchez Martínez *et al.* 2007; Arenas *et al.* 2007*a*), whereas the ophiolites in the Morais Complex have a MORB signature (Pin *et al.* 2006). Sm–Nd isotopic values range between +8.0 and +9.0, suggesting a MORB source reservoir that was depleted in Nd relative to Sm on a time-integrated basis (Pin *et al.* 2006). High-pressure metamorphism has been recognized only in the Vila de Cruces unit (Arenas *et al.* 2007*a*). The main deformation event has been dated at *c.* 365 Ma (Dallmeyer & Gil Ibarguchi 1990; Dallmeyer *et al.* 1997).

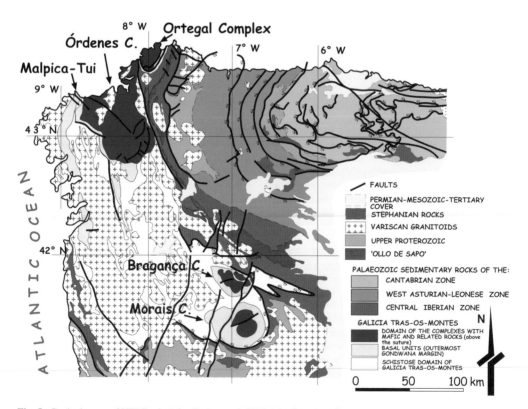

**Fig. 5.** Geologic map of NW Iberia (after Farias *et al.* 1987; Martínez Catalán *et al.* 1997, 2007) emphasizing the allochthonous complexes that contain mafic complexes.

The Upper Ophiolite includes the Careón and Bazar units (Ordenes Complex), the Morais–Talhinas Unit (Morais Complex), the Soeira Unit (Bragança Complex) and Purrido Unit (Ortegal Complex). The Careón Unit has been dated at c. 395 Ma (Díaz-García et al. 1999; Pin et al. 2002) and the Morais–Talhinas Unit between c. 396 and 405 Ma (Pin et al. 2006). Geochemical data indicate that these ophiolites were formed in a suprasubduction setting (Pin et al. 2006; Sánchez-Martinez et al. 2007). Sm–Nd isotopic data yield $\varepsilon_{Nd}$ values that vary from +7.5 to +8.9 and are higher than the contemporary depleted mantle, indicating derivation from an anomalously depleted source (Pin et al. 2006). Amphibolite facies metamorphism in the Careón unit yielded an age of c. 375 Ma (Dallmeyer et al. 1997), but older ages of c. 385 Ma were found in Ortegal (Peucat et al. 1990) and Bragança and Morais (Dallmeyer et al. 1991).

Until recently, the aforementioned units were correlated with the Purrido Unit (Vogel 1967; Azcárraga 2000) located in the Ortegal Complex. However, recent U–Pb (zircon) data (Sánchez Martínez et al. 2006) suggest a 1.1 Ga age for at least part of this unit. Like the other ophiolites, this unit underwent amphibolite facies metamorphism at c. 390 Ma (Peucat et al. 1990). However, the Purrido ophiolite may be a composite unit. The $\varepsilon_{Nd}$ calculated for a crystallization age of 1.1 Ga is +10.1, which is well above the depleted mantle value for that age (DePaolo 1988), suggesting that the protolith age of the sample is much younger than 1.1 Ga (Murphy & Gutiérrez-Alonso 2008). The high $\varepsilon_{Nd}$, together with the negative Nb–Ta anomalies suggest that the juvenile mantle source was contaminated by a subduction component that was itself derived from juvenile crust (Murphy & Gutiérrez-Alonso 2008). Such contamination is characteristic of the incipient stages of oceanic arc development (e.g. Stern 2004). The high $\varepsilon_{Nd}$ values are also characteristic of several Devonian and Carboniferous ophiolites in the Variscan orogen, including the Morais Complex (Pin et al. 2006), the Lizard Complex (Davies et al. 1984), Aracena (Castro et al. 1996) and Massif Central (Pin & Paquette 2002). These complexes have εNd values that are all characterized by $\varepsilon_{Nd}$ values that are equivalent to or slightly higher than the isotopic composition of the contemporary depleted mantle (DePaolo 1981, 1988) suggesting derivation from a mantle source that was depleted in Nd relative to Sm on a time-integrated basis.

Structurally overlying the ophiolites are highly metamorphosed continental rocks, divided into a high pressure-high temperature (HP–HT) unit and an intermediate pressure (IP) unit. The protoliths and geochemical affinities of the HP–HT unit are broadly consistent with an ensialic arc environment (Arenas et al. 1986; Andonaegui et al. 2002; Santos Zalduegui et al. 2002; Abati et al. 2003; Castiñeiras 2005), possibly adjacent to a continental margin (Peucat et al. 1990). They may reflect either residues of anatexis (Drury 1980), crystal fractionates from melts derived from a primitive mantle source (Gravestock 1992) or crystallization of a stratiform gabbroic complex at the base of the continental crust (Galán & Marcos 1997). The IP unit is composed of terrigenous Ordovician sediments, younger than 480 Ma (Fernández-Suárez et al. 2003) and c. 500 Ma (Abati et al. 1999) calcalkaline igneous rocks.

The HT–LP protoliths are c. 490 Ma (conventional and SHRIMP U–Pb on zircon; Peucat et al. 1990; Ordóñez-Casado et al. 2001; Fernández-Suárez et al. 2007; and a Sm–Nd isochron, Santos Zalduegui et al. 2002), and a similar age has been proposed for the magmatic activity recorded in the IP units. This age is overprinted by a strong subduction-related HP metamorphism before c. 400 Ma (Schäfer et al. 1993; Santos Zalduegui et al. 1996; Ordóñez-Casado et al. 2001; Fernández-Suárez et al. 2002a, b, 2007; Roger & Matte 2005; Gómez Barreiro et al. 2006) and the subsequent exhumation-related HT event at c. 390 Ma (Dallmeyer et al. 1991, 1997; Valverde Vaquero & Fernández 1996; Gómez Barreiro et al. 2006; Fernández-Suárez et al. 2007). The ultramafic lithologies present in the HP–HT unit have been variously interpreted as the vestiges of subducted, imbricated oceanic slabs (Gil Ibarguchi et al. 1990), residual harzburgites related to ophiolitic complexes (Santos Zalduegui et al. 1995), exhumed subcontinental mantle (Peucat et al. 1990; Ábalos et al. 1996), subducted back-arc ophiolites (Peucat et al. 1990) or the crust–mantle interface beneath a magmatic arc (Moreno et al. 2001).

Mafic to intermediate samples from the Upper Unit have arc-related major and trace element patterns, and are characterized by relatively low $\varepsilon_{Nd}$ (−1.2 to +2.0, t = 395 Ma) and high $T_{DM}$ values which suggest derivation from the subcontinental lithospheric mantle (SCLM) (Murphy & Gutiérrez-Alonso 2008). Ultramafic HP–HT rocks include eclogite and peridotite and Sm–Nd isotopic systematics suggest contrasting sources. The eclogite has high $\varepsilon_{Nd}$ (+9.9, t = 395), suggesting derivation from a juvenile mantle source. The peridotite has much lower $\varepsilon_{Nd}$ and $^{147}Sm/^{144}Nd$, values that are similar to the SCLM source.

Both HP and IP units probably represent part of the arc that separated the closing Iapetus ocean from the spreading Rheic ocean and was accreted to, and partially subducted beneath, the Laurussian margin in early Devonian times (Fernández-Suárez et al. 2007; Gómez Barreiro et al. 2007) reaching a pressure of at least 1.8 GPa (Gil Ibarguchi et al.

1990; Mendía Aranguren 2000). The emplacement of the allochthonous complexes, including the ophiolitic mafic units, occurred after the 375–365 Ma collision of the Gondwana margin with the trench (Santos Zalduegui *et al.* 1995; Rodríguez *et al.* 2003).

## SW Iberia

In southwestern Iberia, mafic complexes occur along the contact between the OMZ and the SPZ as a sequence of dismembered amphibolites (Beja Acebuches Ophiolite, Munhá *et al.* 1986). The occurrence of these potential oceanic remnants together with the bounding PDLZ oceanic domain (proposed accretionary prism; Eden 1991) to the south, suggests the presence of a potential Variscan suture zone linking the OMZ (Gondwanan para-autochthon) to the SPZ (Laurussia, Avalonia?).

The OMZ consists of a Neoproterozoic continental arc basement that was accreted to the CIZ (Gondwana) during the Cadomian orogeny. This basement is overlain by Cambrian–Ordovician rift-related bimodal volcanic and sedimentary rocks (e.g. Quesada 1990, 1998). Trace element data suggest that the mafic rocks were derived from an asthenospheric source variably enriched in LILE and LREE similar to that of basalts found in modern ocean island and continental to oceanic rift settings (Sanchez-Garcia *et al.* 2003). Felsic magmas are attributed to crustal anatexis. The magmatism is attributed to a mantle plume that is genetically related to the opening of the Rheic Ocean (Sánchez-García *et al.* 2003).

The SPZ consists of pelites and quartzites of upper Devonian age, bimodal volcanics of Tournaisian–lower Visean age (forming the 'Iberian Pyrite Belt' with its associated important massive sulphide ore-bodies) and flysch deposits of upper Visean and Namurian–Westphalian age. During the Variscan Orogeny it has been suggested that the SPZ underwent a transition from a passive margin shelf-type environment to a syn-orogenic flysch type setting (Silva 1990; Quesada *et al.* 1991). This strong sedimentary and tectonic polarity is dominated by SW vergent folding, thrust displacement and a general decrease of deformation intensity in the same direction, suggesting that this syntectonic sedimentation propagated toward the SW (Silva 1990; Quesada *et al.* 1991) during the Variscan Orogeny.

The PDLZ, which occurs along the northern margin of the SPZ, has been interpreted as an accretionary complex (Eden 1991). The PDLZ is tectonically overlain by the potential ophiolitic units (BAO), suggesting a change from a passive to an active margin along the OMZ during the mid-late Palaeozoic. Despite these plate scale

implications, the timing of deposition and deformation of units within the PDLZ remain poorly constrained. Limited dating of spores in the upper flysch sequences yielded Late Devonian–early Carboniferous ages (Eden 1991; Oliveira & Quesada 1998), suggesting an Early–Middle Devonian age for the basal formation of the accretionary complex.

Mafic complexes are preserved along the northern margin of the PDLZ in the southern part of the Aracena Metamorphic Belt (Munhá *et al.* 1986; Quesada & Dallmeyer 1994; Castro *et al.* 1996). Amphibolites (BAO) in the Aracena metamorphic belt have MORB-like geochemical characteristics and $\varepsilon_{Nd}$ values (t = 350 Ma) close to or above the depleted mantle curve at +7.9 to +9.2. They are thought to have originated in an ocean ridge setting (Castro *et al.* 1996) located adjacent to the Gondwanan margin (Quesada *et al.* 2006). The Beja mafic complex has also been interpreted as a potential ophiolite (e.g. Andrade *et al.* 1976) or a remnant of an ensimatic arc. U–Pb (zircon) data yield a 350 Ma protolith age (Pin & Paquette 1997; Pin *et al.* 2008), and cooling below 500 °C at 340 Ma (Ar/Ar hornblende, Dallmeyer & Tucker 1993; Pin *et al.* 2008) reflects exhumation relative to the colliding SPZ during Variscan orogenesis. $\varepsilon_{Nd}$ values for the gabbros (t = 350 Ma) show a wide range from +4.0 to −6.1 (Pin & Paquette 1997). The higher values are from mafic cumulates, whereas the lower values, which correspond to higher Sr initial values, are attributed to crustal contamination. $\varepsilon_{Nd}$ values from metabasalts (+8 to +9), are higher than that of the model depleted mantle curve of DePaolo (1981). These isotope data, together with trace element and REE analyses, are consistent with a MORB environment (Pin & Paquette 1997). Recent Shrimp U/Pb (zircon) age dates from the MORB amphibolites of the BAO and Beja Gabbro suggest a crystallization age for the mafic protoliths of 332 ± 3–340 ± 4 Ma (Azor *et al.* 2008).

To date, most researchers contend that the SPZ is underlain by Avalonian basement. As Avalonia had accreted to Laurussia by the mid-Silurian (Quesada 1990), the BAO and PDLZ potentially preserve a Rheic Ocean suture and record the accretionary processes that reflect the amalgamation of Pangaea. A plethora of tectonic models have been ascribed to accommodate the current spatial juxtaposition of the SPZ and OMZ based on this Avalonian affinity (Quesada & Dallmeyer 1994; Castro *et al.* 1996; Onézime *et al.* 2003). However, the recent discovery of relatively young amphibolite protolith ages (Azor 2008) suggests the BAO is not genetically linked to other oceanic units around the Rheno-Hercynian belt (i.e. Lizard). Although this does not preclude the possibility that the SPZ is underlain by Avalonian basement, it does suggest the need for a re-evaluation of the genesis and tectonothermal history of the BAO oceanic units.

## Lizard Complex, U.K.

The Lizard Complex of SW England (Fig. 6) has long been interpreted as an Early Devonian ophiolite (e.g. Bromley 1979; Barnes & Andrews 1984; Davies 1984; Floyd 1984; Gibbons & Thompson 1991; Cook *et al.* 2000, 2002; Sandeman *et al.* 1997, 2000; Nutman *et al.* 2001) obducted onto the southern continental margin of Avalonia. According to the tectonostratigraphic zonation of the Variscan orogen (Franke 1989), the Lizard Complex is potentially correlative with the

ophiolitic complexes of the Pulo de Lobo shear zone of Iberia and the Rhenohercynian zone of northern Germany (Figs 4 & 7).

Geophysical evidence suggests that the Lizard Complex is a thin-skinned, gently southward-dipping structural slice <1 km thick (e.g. Doody & Brooks, 1986; BIRPS & ECORS 1986). The complex predominantly consists of peridotite, amphibolite and gabbro with local sheeted dykes, subordinate granite and metasediment. The lowest structural unit, known as the Basal Unit, contains hornblende schists that are interpreted to represent pillow lavas, dykes and interflow sediments. The Basal Unit structurally overlies the early Ordovician Man O'War Gneiss, interpreted as part of the Avalonian continental margin.

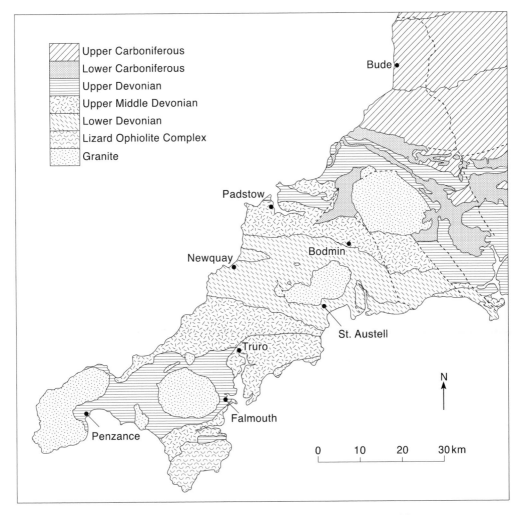

**Fig. 6.** Simplified geological map of southwestern Britain modified from British Geological Survey.

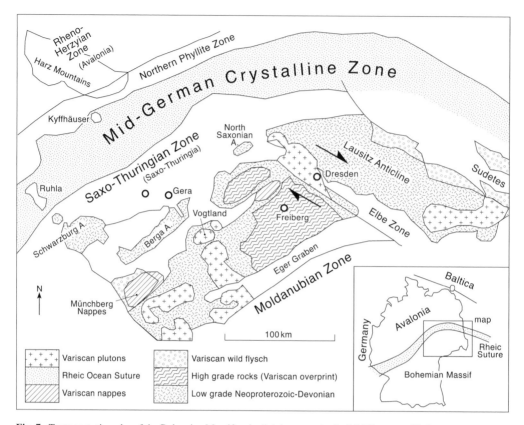

**Fig. 7.** Tectonostratigraphy of the Bohemian Massif and adjoining areas in the Mid-European Variscan orogen (after Linnemann *et al.* 2004). *Inset*: Simplified map showing the large suture zones between Bohemia, Avalonia and Baltica.

Although the age of the ophiolite is widely believed to be 397 ± 2 Ma (U–Pb, zircon; Clark *et al.* 1998), the age of obduction is controversial. A gneiss, thought to represent basement melted during obduction of the ophiolite, has yielded an age of 376 Ma (Sandeman *et al.* 2000). However, on the basis of SHRIMP data, Nutman *et al.* (2001) argued that obduction occurred between 400 and 380 Ma, that is soon after crystallization of the oceanic crust. This was followed by a later episode of thrusting that post-dates deposition of underlying Late Devonian clastic rocks.

The Lizard Complex has MORB-like geochemistry (e.g. Cooke *et al.* 2000) and $\varepsilon_{Nd}$ values (t = 397 Ma) for the mafic dykes and plagiogranite ranging from +9.0 to +11.8 (Davies 1984) which are higher than the model depleted mantle of equivalent (i.e. *c.* 397 Ma) age (DePaolo 1981, 1988). The high $\varepsilon_{Nd}$ values implies derivation from a highly depleted mantle at *c.* 397 Ma which, together with regional constraints, suggest an oceanic lithospheric source generated during the

later stages of Rheic Ocean evolution. Recently, Cook *et al.* (2000) proposed that the Lizard peridotites formed part of a non-volcanic rifted margin that was exhumed from depths of 15.7–7.5 kbar along extensional shear zones before intrusion of the gabbro and the sheeted dykes. If so, the peridotite would not be a cogenetic member of the ophiolitic complex.

The ophiolite preserves pre-obduction fabrics that record high-temperature deformation within an oceanic environment (Gibbons & Thompson 1991; Roberts *et al.* 1993). For example, the peridotite contains steep mylonitic fabrics and stretching lineations that are cut by the gabbros and sheeted dykes, which are interpreted to have formed during thinning of the upper mantle prior to the formation of oceanic crust. The Lizard Complex is widely believed to represent a vestige of a narrow (*c.* 400 km) Red Sea-type oceanic tract, which may (Linnemann *et al.* 2004) or may not (Tait *et al.* 2000) have been attached to Gondwana at the time. The width of the putative ocean tract is apparently too small to be resolved palaeomagnetically. Kinematic data along intra-oceanic shear zones together with and the obliquity of the NW–SE orientation of the sheeted dykes to the E–W faults in Armorica is interpreted to reflect dextral transtension along an intra-continental strike–slip zone between Avalonia and Armorica (Badham 1992; Cook et al. 2000, 2002). If so, the Lizard Complex may not be a vestige of the Rheic Ocean proper, but may instead represent a narrow oceanic tract, similar to the modern Gulf of California, formed adjacent to the northern margin of the Rheic Ocean as suggested by Davies (1984).

## Eastern Europe

The Bohemian Massif (Fig. 7) is divided into four tectonostratigraphic zones; from north to south, these are the Rhenohercynian zone, the Mid-German Crystalline Zone, and the Saxothuringian and Moldanubian zones (e.g. Franke 2000). The Rhenohercynian zone is correlated with East Avalonia (Figs 4 & 7), and therefore was part of Laurussia in the Late Palaeozoic (Fig. 2). The major crustal units south of the Mid-German Crystalline Zone belong to the Bohemian Massif, which is interpreted to be the northern margin of the peri-Gondwanan Armorican Microplate that was tectonically dismembered during Late Devonian–Carboniferous orogenesis (Matte 2001). The Mid-German Crystalline Zone is therefore interpreted to preserve part of the suture zone of the Rheic Ocean (Fig. 2). Although potential source rocks are not exposed, detrital chrome-spinel in Frasnian to Visean greywackes in the Harz Mountains of the Rheno-Hercynian zone is interpreted to have been derived from pre-Givetian Alpine-type peridotites in the German Mid-Crystalline Rise (Ganssloser 1999).

Saxo–Thuringia is considered to reflect the northern margin of the Armorican plate and is underlain by Late Neoproterozoic arc and back-arc complexes, and 540 Ma post-tectonic granitoids developed during the Cadomian Orogeny (Linnemann et al. 2000; Linnemann & Romer 2002). These rocks are unconformably overlain by a Lower–Middle Cambrian platformal succession, followed disconformably in the Upper Cambrian and Lower Ordovician by a thick (c. 3 km) transgressive sequence thought to reflect the drift of Avalonia from the Gondwanan margin (Linnemann et al. 2004). A West African affinity for Saxo–Thuringia is suggested by Cambro–Ordovician fauna and the presence of latest Ordovician glaciomarine diamictites of the Saharan glaciation (Linnemann et al. 2004). The Lower Silurian–Lower Carboniferous stratigraphy is dominated by deep-water clastic and carbonate sediments deposited along the southern flank of the Rheic Ocean.

The relationship between the Saxo–Thuringian and Moldanubian zones is unclear because the contact is obscured by Late Mesozoic and Early Cenozoic strata. Matte (2001) interpreted these zones as a coherent microplate (Armorica), whereas Franke (2000) proposed that an oceanic tract existed between them. Palaeomagnetic data have been used to support (Tait et al. 2000) and refute (McKerrow et al. 2000) the existence of such an oceanic tract. Amphibolites are abundant in the Moldanubian zone of the eastern Bohemian Massif, and are traditionally divided into three units: the Rehberg ophiolite, the Buschandlwand amphibolite and the Raabs Group. The Rehberg ophiolite consists of metamorphosed (amphibolite facies) ultramafic to mafic plutons, a gabbro/dyke complex, basaltic, andesitic and rhyolitic volcanics, and siliciclastic sediments. Geochemical data are consistent with a supra-subduction zone environment (Höck et al. 1997). These traits are similar to the Letovice ophiolite in Moravia, which contains both MORB and tholeiitic island arc basalts and is interpreted to have formed in a narrow basin floored by oceanic crust between the Moldanubian block to the west and the Brunovistulian–Moravian block to the east (Höck et al. 1997).

$^{40}Ar/^{39}Ar$ data from the Letovice ophiolite yield ages of $328 \pm 7$ and $332 \pm 5$ Ma, which are interpreted to date the age of metamorphism (MacIntyre et al. 1992). They are consistent with cooling ages documented by Dallmeyer et al. (1992) from the Moldanubian and Moravian zones, respectively (Höck et al. 1997). The protolith ages of the ophiolites are unknown. However, in the Sudetes Mountains to the NE, ophiolites of the Central Sudetic terrane (Cymerman et al. 1997) occur between the Saxo–Thuringian and Mondanubian zones and have yielded c. 420 Ma crystallization ages (Ślęża ophiolite, U–Pb zircon, Oliver et al. 1993) and c. 350 Ma Sm–Nd whole rock ages (Pin et al. 1988) interpreted to reflect the age of metamorphism (Cymerman et al. 1997). Elsewhere within the Central Sudetic terrane, the Klodzko Metamorphic Complex (KMC) contains Neoproterozoic–Devonian metasedimentary and meta-igneous rocks exposed in a sequence of thrusts. This complex is interpreted to preserve remnants of the northern Gondwanan margin that was telescoped during the Variscan orogeny (Cymerman et al. 1997). Kryza et al. (2003) identified two dominant Palaeozoic meta-igneous units within the KMC: (1) a Cambro–Ordovician bimodal, rift-related suite similar to several suites developed along the northern periphery of Gondwana (most notably in the Ossa Morena Zone of Iberia; Sánchez-García et al. 2003); and (2) a Devonian, predominantly mafic unit in the northern part of the complex, which has a $\varepsilon_{Nd}$ value of +6.8 (assuming a youngest possible crystallization age of 400 Ma), typical of magmas derived from a depleted mantle source.

According to Franke (2000), the oceanic rocks between the Saxo–Thuringian and Moldanubian zones represent an oceanic tract of unknown width. Alternatively, they have been interpreted as nappes transported southward from the Rheic suture zone (Kroner & Hahn 2003).

## Massif Central

The Massif Central is a portion of the Gondwanan margin that underwent polyphase orogenic activity between the Late Silurian and Late Carboniferous. It is dominated by allochthonous complexes including ophiolites and high-pressure metamorphic units. Early orogenic activity produced coesite-bearing eclogitic rocks formed at a depth of *c.* 90 km between 420 and 400 Ma (Lardeaux *et al.* 2001). Unroofing to a depth of *c.* 30 km by 380–360 Ma was accompanied by crustal extension and gravitational collapse (Burg *et al.* 1994; Faure 1995). Middle–Late Devonian nappes contain a gneissic series consisting of Late Neoproterozoic–Early Palaeozoic metasediments, intruded by Early Ordovician plutons (e.g. Ledru *et al.* 1994). Early Ordovician bimodal igneous rocks are interpreted by Pin & Marini (1993) to be a local representative of a regional rifting event. The mafic magmas have a plume-related chemistry and the felsic magmas are attributed to crustal melting.

Voluminous Devonian–Carboniferous magmatism occurs throughout the Massif Central. In the NE, the massif is dominated by mid- to Late Devonian calc-alkaline to tholeiitic volcanic and plutonic rocks (Thiéblemont & Cabanis 1986). Low-grade Late Devonian–Early Carboniferous (*c.* 366–358 Ma: U–Pb zircon, Pin & Paquette 1997) bimodal basalts, rhyolites and genetically related trondhjemite intrusions occur in the allochthonous Brévenne Series, which was tectonically juxtaposed against high-grade rocks to the north and south by Carboniferous dextral strike-slip motion (Costa *et al.* 1993). The metabasalts are characterized by an enrichment in incompatible elements (e.g. Th and LREE), depletion in high field strength elements, and positive $\varepsilon_{Nd_t}$ values (from $+5$ to $+8$ at $t = 360$ Ma). Taken together, these data are consistent with a MORB mantle source that was contaminated by subduction-derived fluids. The metarhyolites are enriched in LILE, have high $\varepsilon_{Nd_t}$ values (from $+4.7$ to $+6.8$) and are interpreted to have been derived from the basalts by fractional crystallization (Pin & Paquette 1997). The trondhjemites, on the other hand, have much lower $\varepsilon_{ND_t}$ values (from $-1.0$ to $+2.2$) and their genesis is thought to involve an important contribution of continental crust. The magmatism

is interpreted to reflect incipient rifting of a volcanic arc along a continental margin associated with southward-dipping subduction (Pin & Paquette 2002). In the western massif, *c.* 370–380 Ma arc plutons (Pin & Paquette 2002) are interpreted to reflect either northward- (Shaw *et al.* 1993) or southward-directed subduction (Faure *et al.* 1997). In the South Vosges, a series of thrusts expose Late Devonian tholeiitic basalts along with siliceous shales, radiolarian cherts and rare limestone, a sequence that is thought to reflect deposition in an ocean basin (Schneider *et al.* 1990).

## Discussion

Although the study of mafic complexes within the Rheic Ocean has lagged behind that of their counterparts in the Iapetus Ocean, the available data are sufficient to identify some first-order characteristics: (1) those of Late Cambrian–Early Ordovician age are related to the initial rifting of the Rheic Ocean along the northern Gondwanan margin; (2) those of Devonian and Carboniferous age are ophiolitic and represent fragments of Rheic or Rheic-related ocean floor, probably generated along the Avalonian margin of Laurussia; and (3) those of Carboniferous age appear to be oceanic in origin, but their relationship to the Rheic Ocean, *sensu stricto*, is unclear.

Late Cambrian–Early Ordovician mafic complexes occur from Mexico to Bohemia, indicating that rifting was coeval along much of the northern Gondwanan margin. In most of these complexes, the basalts are differentiated continental tholeiites and are part of a bimodal suite. Where isotopic data are available (Mexico, NW Iberia) the basalts appear to have been derived from either the sub-continental lithospheric mantle (SCLM) that was metasomatically altered at *c.* 1.0 Ga, or from juvenile sources.

Basalts derived from SCLM are typically part of a bimodal complex, with varying proportions of felsic magmas that were derived by crustal melting. Passive margin sedimentation accompanied this rift-related magmatism, as documented, from west to east, by siliciclastic rocks at the base of the Tiñu Formation in Mexico and by the Armorican Quartzite in mainland Europe. The genetic relationship of the basalts with crustally derived melts and their association with passive margin sediments indicates that they are not ophiolites. Instead, these complexes developed along the passive margin of northern Gondwana and reflect the rift and drift of Avalonia and related peri-Gondwanan terranes during the formation of the Rheic Ocean. Passive margin sedimentation along this margin appears to have continued virtually uninterrupted until the Devonian.

Late Cambrian–Early Ordovician ophiolites are rare. The best documented examples occur in NW Iberia and are interpreted to be remnants of the Rheic Ocean (or subsidiary oceans) that were obducted during the Variscan Orogeny. These ophiolite bodies typically have an ensimatic island arc tholeiitic chemistry and are interpreted to have been generated in a back-arc environment during the first stages of the opening of the Rheic Ocean (Arenas *et al.* 2007*a*). In Mexico, mafic complexes between *c.* 480 and 400 Ma are not well documented. However, unit formerly assigned to the Cosoltopec Formation clearly contain Ordovician and Late Devonian–Carboniferous volcanic rocks with MORB and OIB chemical affinities and oceanic sediments that are candidates for ophiolites.

The Devonian (e.g. Lizard, southern Britain) and Carboniferous (Aracena, Beja-Acebuches; SW Iberia) mafic complexes are ophiolites with geochemical affinities and juvenile Sm–Nd isotopic characteristics that are typical of MORB compositions (e.g. Castro *et al.* 1996; Davies 1984; Cook *et al.* 2000). Unlike the Ordovician ophiolites, there is nothing in the geochemistry or isotopic systematics of these complexes that suggests an arc-related environment. Indeed these ophiolites typically have higher $\varepsilon_{Nd}$ than the theoretical contemporary depleted mantle value, indicating derivation from a similar, but isotopically unusual mantle, with time-integrated depletion of Nd relative to Sm. A mantle composition as unusually depleted as this could form if a significant fraction of basalt was removed from it during an earlier (unspecified) tectonothermal episode. However, the relationship of the Devonian and Carboniferous ophiolites to Rheic ocean evolution may be different. The Devonian Lizard Complex, for example, is widely believed to represent a narrow oceanic tract formed by dextral strike-slip motion along the southern margin of Avalonia. Its Devonian age indicates that its origin pre-dates the closure of the Rheic Ocean. The relationship of the Carboniferous Aracena and Beja ophiolites to the Rheic ocean *sensu stricto* is enigmatic, because they potentially post-date the collision between the SPZ and the OMZ, which represents closure of the Rheic Ocean in SW Iberia (Quesada 1990; Azor *et al.* 2008). These ophiolites may have formed in a sinistral strike slip environment associated with oblique subduction and collision between these zones (Quesada 1990) that, according to Pin *et al.* (2008), may also have involved slab breakoff, or a post-collisional transtensional event (Azor *et al.* 2008).

Unfortunately, Sm–Nd isotopic data from other mafic complexes, required to test either the regional extent of this unusual mantle, or the processes that gave rise to it, are lacking. However, several ophiolite complexes, such as those of NW Iberia and the Modanubian Zone of Bohemia have chemistries consistent with the settings of these Late Palaeozoic ophiolites, although whether these ophiolites formed before or during collisional processes that accompanied Rheic Ocean closure is unclear.

Devonian and Carboniferous mafic complexes appear to be isolated, individually enigmatic bodies, although some common traits are shared. Interpretations of these complexes depend to some degree on resolution of the controversy about the existence of a separate Armorican microplate. If this microplate existed, then some of the mafic complexes may reflect the formation of that plate, or the evolution of the ocean (known as the Palaeotethys Ocean), which is thought to have formed when the Armorican plate separated from Gondwana.

Deformation and metamorphism of Rheic Ocean mafic complexes occurred in the Carboniferous associated with the collision of Gondwana with Laurasia. Late Cambrian–Early Ordovician sequences lack regional-scale tectonothermal activity until the Carboniferous deformation and metamorphism. From the Acatlán Complex of Mexico to the Bohemian Massif, metamorphism up to eclogite facies followed by partial melting collectively reflect subduction and/or subduction erosion followed by rapid exhumation of the leading edge of the Gondwanan margin. These relationships suggest that this sequence was on the lower plate during continental collision (e.g. Martínez Catalán *et al.* 2007; Middleton *et al.* 2007), or the upper plate removed by subduction erosion (Keppie *et al.* 2008*a*).

The classic exposures in NW Iberia indicate that the Gondwanan passive margin succession was overthrust by nappes, which include ophiolite complexes derived from the remnants of the Rheic Ocean and possibly from Laurussia (Martínez-Catalán *et al.* 1996, 2007). The nappes also contain ophiolitic complexes that are interpreted to be remnants of the Rheic Ocean that were obducted during the Variscan orogeny.

In conclusion, the expression of the Rheic Ocean suture zone can be traced from the Acatlán Complex in Mexico, through Iberia and the Mid-German Crystalline Rise, to the northern margin of the Bohemian massif. Mafic complexes document important aspects of the evolution of this ocean from its origin to the genesis of ophiolites emplaced during its closure. Although much work remains to be done, the available data demonstrate that a more precise knowledge of the evolution of these mafic complexes will help to constrain the geodynamic setting that attended the formation of Pangaea.

JBM acknowledges the continuing support of NSERC. (Canada) through Discovery and Research Capacity

Development grants. G.G.-A. funding comes from Spanish Education and Science Ministry Project Grant CGL2006-00902 (ODRE.) and the Mobility Program Grant PR2007-0475 and the hospitality of St. Francis Xavier University. RDN acknowledges funding from NSF (EAR-0308105) and an Ohio University Baker Award. JDK would like to acknowledge a Papiit grant IN100108 and a CONACyT grant (CB–2005–1: 24894). We are grateful to D. Thorkelson and A. Hynes for constructive reviews.

# References

ÁBALOS, B., AZCÁRRAGA, J., GIL IBARGUCHI, J. I., MENDIA, M. & SANTOS ZALDUEGUI, J. F. 1996. Flow stress, strain rate and effective viscosity evaluation in a high-pressure metamorphic nappe (Cabo Ortegal, Spain). *Journal of Metamorphic Geology*, **14**, 227–248.

ABATI, J., DUNNING, G. R., ARENAS, R., DÍAZ GARCÍA, F., GONZÁLEZ CUADRA, P., MARTÍNEZ CATALÁN, J. R. & ANDONAEGUI, P. 1999. Early Ordovician orogenic event in Galicia (NW Spain): evidence from U–Pb ages in the uppermost unit of the Ordenes Complex. *Earth and Planetary Science Letters*, **165**, 213–228.

ABATI, J., ARENAS, R., MARTÍNEZ CATALÁN, J. R. & DÍAZ GARCÍA, F. 2003. Anticlockwise P–T Patch of Granulites from the Monte Castelo Gabbro (Órdenes Complex, NW Spain). *Journal of Petrology*, **44**, 305–327.

ANDONAEGUI, P., GONZÁLEZ DEL TANAGO, J., ARENAS, R., ABATI, J., MARTÍNEZ CATALÁN, J. R., PEINADO, M. & DÍAZ GARCÍA, F. 2002. Tectonic setting of the Monte Castelo gabbro (Ordenes Complex, northwestern Iberian Massif): evidence for an arc-related terrane in the hangingwall to the Variscan suture. *In*: MARTÍNEZ CATALÁN, J. R., HATCHER, R. D., JR., ARENAS, R. & DÍAZ GARCÍA, F. (eds) *Variscan-Appalachian Dynamics: The Building of the Late Paleozoic Basement*. Geological Society of America, Special Paper, **364**, 37–56.

ARAMBURU, C., TRUYOLS, J. *ET AL.* 2002. El Paleozoico Inferior de la Zona Cantábrica. *In*: RABANO, I., GUTIÉRREZ-MARCO, J. C. & SAAVEDRA, J. (eds) *El Paleozoico Inferior de Ibero-América*. Universidad de Extremadura, 397–421.

ARENAS, R., GIL IBARGUCHI, J. I. *ET AL.* 1986. Tectonostratigraphic units in the complexes with mafic and related rocks of the NW of the Iberian Massif. *Hercynica*, **2**, 87–110.

ARENAS, R., RUBIO PASCUAL, F. J., DÍAZ GARCÍA, F. & MARTÍNEZ CATALÁN, J. R. 1995. High-pressure micro-inclusions and development of an inverted metamorphic gradient in the Santiago Schists (Ordenes Complex, NW Iberian Massif, Spain): evidence of subduction and syn-collisional decompression. *Journal of Metamorphic Geology*, **13**, 141–164.

ARENAS, R., MARTÍNEZ CATALÁN, J. R., SÁNCHEZ MARTÍNEZ, S., FERNÁNDEZ-SUÁREZ, J., ANDONAEGUI, P., PEARCE, J. A. & CORFU, F. 2007a. The Vila de Cruces Ophiolite: a Remnant of the Early Rheic Ocean in the Variscan Suture of Galicia

(Northwest Iberian Massif). *Journal of Geology*, **115**, 129–148.

ARENAS, R., MARTÍNEZ CATALÁN, J. R. *ET AL.* 2007b. Paleozoic ophiolites in the Variscan suture of Galicia (northwest Spain). *In*: HATCHER, R. D., JR., CARLSON, M. P., MCBRIDE, J. H. & MARTÍNEZ CATALÁN, J. R. (eds) *4-D Framework of Continental Crust*. Geological Society, London, Memoirs, **200**, 403–423.

AZCÁRRAGA, J. 2000. Evolución tectónica y metamórfica de los mantos inferiores de grado alto y alta presión del Complejo de Cabo Ortegal. *Serie Nova Terra*, **17**, 1–306.

AZOR, A., RUBATTO, D., SIMANCAS, J. F., GONZÁLEZ LODEIRO, F., MARTÍNEZ POYATOS, D., MARTÍN PARRA, L. M. & MATAS, J. 2008. Rheic Ocean ophiolitic remnants in southern Iberia questioned by SHRIMP U–Pb zircon ages on the Beja-Acebuches amphibolites. *Tectonics*, **27**, TC5006, doi: 10.1029/2008TC002306, 1–11.

BARNES, R. P. & ANDREWS, J. R. 1984. Hot or cold emplacement of the Lizard Complex? *Journal of the Geological Society, London*, **141**, 37–39.

BARRERO, M. & CORRETGÉ, L. G. 2002. La diferenciación petrográfica y geoquímica de las rocas ígneas del Arroyo Farandón (Asturias). *Geogaceta*, **32**, 135–138.

BADHAM, J. P. N. 1982. Strike-slip orogens–an explanation for the Hercynides. *Journal of the Geological Society, London*, **139**, 493–504.

BEA, F., MONTERO, P., TALAVERA, C. & ZINGER, T. 2006. A revised Ordovician age for the oldest magmatism of Central Iberia: U–Pb ion-microprobe and LA-ICPMS dating of the Miranda do Douro orthogneiss. *Geologica Acta*, **4**, 395–401.

BÉDARD, J. H. & STEVENSON, R. 1999. The Caldwell Group lavas of southern Quebec: MORB-like tholeiites associated with the opening of the Iapetus Ocean. *Canadian Journal of Earth Sciences*, **36**, 999–1019.

BÉDARD, J. H., LAUZIÈRE, K., TREMBLAY, A. & SANGSTER, A. 1998. Evidence for forearc seafloor-spreading from the Betts Cove ophiolite, Newfoundland: oceanic crust of boninitic affinity. *Tectonophysics*, **284**, 233–245.

BIRPS & ECORS 1986. Deep seismic reflection profiling between England, France and Ireland. *Journal of the Geological Society, London*, **143**, 45–52.

BROMLEY, A. V. 1979. Ophiolitic origin of the Lizard Complex. *Cambourne School of Mines Journal*, **79**, 25–38.

BRUECKNER, H. K. & VAN ROERMUND, H. L. M. 2004. Dunk tectonics: a multiple subduction/eduction model for the evolution of the Scandinavian Caledonides. *Tectonics*, **23**(TC2004), 1–20, doi: 10.1029/2003TC001502.

BURG, J. P., VAN DEN DRIESSCHE, J. & BRUN, J. P. 1994. Syn- to post-thickening extension in the Variscan Belt of Western Europe: mode and structural consequences. *Géologie de la France*, **3**, 33–51.

CAMERON, K. L., LÓPEZ, R., ORTEGA-GUTIÉRREZ, F., SOLARI, L. A., KEPPIE, J. D. & SCHULZE, C. 2004. U–Pb constraints and Pb isotopic compositions of leached feldspars: constraints on the origin and evolution of Grenvillian rocks from eastern and southern Mexico. *In*: TOLLO, R. P., CORRIVEAU, L.,

MCLELLAND, J. & BARTHOLOMEW, M. J. (eds) *Proterozoic Tectonic Evolution of the Grenville Orogen in North America.* Geological Society of America Memoir, **197**, 755–770.

CASTIÑEIRAS, P. 2005. Origen y evolución tectonotermal de las unidades de O Pino y Cariño (Complejos alóctonos de Galicia). *Laboratorio Xeolóxico de Laxe, Serie Nova Terra*, **28**, 1–279.

CASTRO, A., FERNÁNDEZ, C., DE LA ROSA, J. D., MORENO VENTAS, I. & ROGERS, G. 1996. Significance of MORB-derived amphibolites from the Aracena metamorphic belt, southwest Spain. *Journal of Petrology*, **37**, 235–260.

CASTRO, A., PATIÑO DOUCE, A. E., CORRETGÉ, L. G., DE LA ROSA, J. D., EL-BIAD, M. & EL-HMIDI, H. 1999. Origin of peraluminous granites and granodiorites, Iberian massif, Spain: an experimental test of granite petrogénesis. *Contributions to Mineralogy and Petrology*, **135**, 255–276.

CASTRO, A., CORRETGÉ, L. G., DE LA ROSA, J. D., FERNÁNDEZ, C., LÓPEZ, S., GARCÍA-MORENO, O. & CHACÓN, H. 2003. The Appinite-Migmatite Complex of Sanabria, NW·Iberian Massif, Spain. *Journal of Petrology*, **44**, 1309–1344.

CAWOOD, P. A., VAN GOOL, J. A. M. & DUNNING, G. R. 1996. Geological development of eastern Humber and western Dunnage zone: corner Brook–Glover Island region. *Canadian Journal of Earth Sciences*, **33**, 182–198.

CENTENO-GARCÍA, E. & KEPPIE, J. D. 1999. Latest Paleozoic-early Mesozoic structures in the central Oaxaca terrane of southern Mexico: deformation near a triple junction. *Tectonophysics*, **301**, 231–242.

CLARK, A. H., SCOTT, D. J., SANDEMAN, H. A. & BROMLEY, A. V. 1998. Siegenian generation of the Lizard ophiolite: U–Pb zircon age data for plagiogranite, Porthkerris, Cornwall. *Journal of the Geological Society, London*, **155**, 595–598.

COCKS, L. R. M. & FORTEY, R. A. 1988. Lower Paleozoic facies and faunas around Gondwana. *In*: AUDLEY-CHARLES, M. G. & HALLAM, A. (eds) *Gondwana and Tethys.* Geological Society, London, Special Publications, **37**, 183–200.

COCKS, L. R. M. & TORSVIK, T. H. 2002. Earth geography from 500 to 400 million years ago: a faunal and palaeomagnetic review. *Journal of the Geological Society, London*, **159**, 631–644.

COOK, C. A., HOLDSWORTH, R. E., STYLES, M. T. & PEARCE, J. A. 2000. Pre-emplacement structural history recorded by mantle peridotites: an example from the Lizard Complex, SW England. *Journal of the Geological Society, London*, **157**, 1049–1064.

COOK, C. A., HOLDSWORTH, R. E. & STYLES, M. T. 2002. The emplacement of peridotites and associated oceanic rocks from the Lizard Complex, southwest England. *Geological Magazine*, **139**, 2745.

COSTA, S., MALUSKI, H. & LARDEAUX, J. M. 1993. $^{40}Ar–^{39}Ar$ chronology of Variscan tectono-metamorphic events in an exhumed crustal nappe: the Monts du Lyonnais complex (Massif Central, France). *Chemical Geology*, **105**, 339–359.

CROWLEY, Q. G., FLOYD, P. A., WINCHESTER, J. A., FRANKE, W. & HOLLAND, J. G. 2000. Early Palaeozoic rift-related magmatism in Variscan Europe:

fragmentation of the Armorican Terrane Assemblage. *Terra Nova*, **12**, 171–180.

CYMERMAN, Z., PIASECKI, M. A. J. & SESTON, R. 1997. Terranes and terrane boundaries in the Sudetes. *Geological Magazine*, **134**, 717–725.

DALLMEYER, R. D. & GIL IBARGUCHI, J. I. 1990. Age of amphibolitic metamorphism in the ophiolitic unit of the Morais allochthon (Portugal): implications for early Hercynian orogenesis in the Iberian Massif. *Journal of the Geological Society, London*, **147**, 873–878.

DALLMEYER, R. D. & TUCKER, R. D. 1993. U–Pb zircon age for the Lagoa augen gneiss, Morais complex, Portugal: tectonic implications. *Journal of the Geological Society, London*, **150**, 405–410.

DALLMEYER, R. D., RIBEIRO, A. & MARQUES, F. 1991. Polyphase Variscan emplacement of exotic terranes (Morais and Bragança Massifs) onto Iberian successions: evidence from $^{40}Ar/^{39}Ar$ mineral ages. *Lithos*, **27**, 133–144.

DALLMEYER, R. D., NEUBAUER, F. & HÖCK, V. 1992. Chronology of late Paleozoic tectonothermal activity in the southeastern Bohemian massif, Austria (Moldanubian and Moravo-Silesian zones) – $^{40}Ar/^{39}Ar$ mineral age controls. *Tectonophysics*, **210**, 135–153.

DALLMEYER, R. D. & MARTÍNEZ CATALÁN, J. R. 1997. Diachronous Variscan tectonothermal activity in the NW Iberian Massif: evidence from $^{40}Ar/^{39}Ar$ dating of regional fabrics. *Tectonophysics*, **227**, 307–337.

DAVIES, G. R. 1984. Isotopic evolution of the Lizard Complex. *Journal of the Geological Society, London*, **141**, 3–14.

DE LA ROSA, J. D., JENNER, G. A. & CASTRO, A. 2002. A study of inherited zircons in granitoid rocks from the South Portuguese and Ossa-Morena Zones, Iberian Massif: support for the exotic origin of the South Portuguese Zone. *Tectonophysics*, **352**, 245–256.

DEPAOLO, D. J. 1981. Neodymium isotopes in the Colorado Front Range and crust-mantle evolution in the Proterozoic. *Nature*, **29**, 193–196.

DEPAOLO, D. J. 1988. *Neodymium Isotope Geochemistry: An Introduction.* Springer Verlag, New York.

DÍAZ GARCÍA, F., ARENAS, R., MARTÍNEZ CATALÁN, J. R., GONZÁLEZ DEL TÁNAGO, J. & DUNNING, G. R. 1999. Tectonic evolution of the Careón ophiolite (Northwest Spain): a remnant of oceanic lithosphere in the Variscan belt. *Journal of Geology*, **107**, 587–605.

DÍEZ MONTES, A. 2006. *La geología del Dominio 'Ollo de Sapo' en las comarcas de Sanabria y Terra do Bolo.* PhD thesis, University of Salamanca, Spain.

DOODY, J. J. & BROOKS, M. 1986. Seismic refraction investigation of the structural setting of the Lizard and Start complexes, SW England. *Journal of the Geological Society, London*, **143**, 135–140.

DRURY, S. A. 1980. The geochemistry of high-pressure gneisses from Cabo Ortegal (NW Spain): residues of deep anatexis. *Geol Mijbouw*, **59**, 61–64.

EDEN, C. 1991. *Tectonostratigraphic analysis of the northern extend of the oceanic exotic terrane, northwestern Huelva Province, Spain.* PhD thesis, Earth Sciences, Southampton.

ELÍAS-HERRERA, M. & ORTEGA-GUTIÉRREZ, F. 2002. Caltepec fault zone: an Early Permian dextral transpressional boundary between the Proterozoic Oaxacan and Paleozoic Acatlán complexes, southern Mexico, and regional implications. *Tectonics*, **21**(3), 4–1 to 4–18, doi: 10.1029/200TC001278.

ELÍAS-HERRERA, M., ORTEGA-GUTIÉRREZ, F., SÁNCHEZ-ZAVALA, J. L., REYES-SALAS, A. M., MACÍAS-ROMO, C. & IRIONDO, A. 2004. *New geochronological and stratigraphic data related to the Paleozoic evolution of the high-pressure Piaxtla Grop, Acatlán Complex, southern Mexico.* IV reunion nacional de ciencias de la tierra, Libro de Resumenes, 1–150.

ELÍAS-HERRERA, M., MACÍAS-ROMO, C., ORTEGA-GUTIÉRREZ, F., SÁNCHEZ-ZAVALA, J. L., IRIONDO, A. & ORTEGA-RIVERA, A. 2007. Conflicting Stratigraphic and Geochronologic Data From the Acatlán Complex: 'Ordovician' Granites Intrude Metamorphic and Sedimentary Rocks of Devonian-Permian age. *Eos Transactions of the American Geophysical Union*, **88**(23), Joint Assembly Supplement, Abstract T41A-12.

FARIAS, P., GALLASTEGUI, G. *ET AL.* 1987. Aportaciones al conocimiento de la litoestratigrafía y estructura de Galicia Central. Faculdade de Ciências do Porto. *Memorias do Museu e Laboratório Mineralógico e Geológico*, **1**, 411–431.

FAURE, M. 1995. Late orogenic Carboniferous extensions in the Variscan French Massif Central. *Tectonics*, **14**, 132–153.

FAURE, M., LELOIX, C. & ROIG, J. Y. 1997. L'évolution polycyclique de la chaîne hercynienne. *Bulletin de la Societé.Géologique de France*, **168**, 695–705.

FERNÁNDEZ-SUÁREZ, J., GUTIÉRREZ ALONSO, G., JENNER, G. A. & TUBRETT, M. N. 2000. New ideas on the Proterozoic–Early Paleozoic evolution of NW Iberia: insights from U–Pb detrital zircon ages. *Precambrian Research*, **102**, 185–206, doi: 10.1016/S0301-9268(00)00065-6.

FERNÁNDEZ-SUÁREZ, J., GUTIÉRREZ ALONSO, G., COX, R. & JENNER, G. A. 2002*a*. Assembly of the Armorica microplate: a strike-slip terrane delivery? Evidence from U–Pb ages of detrital zircons. *Journal of Geology*, **110**, 619–626, doi: 10.1086/341760.

FERNÁNDEZ-SUÁREZ, J., CORFU, F. *ET AL.* 2002*b*. U–Pb evidence for a polyorogenic evolution of the HP–HT units of the NW Iberian Massif. *Contributions to Mineralogy and Petrology*, **143**, 236–253.

FERNÁNDEZ-SUÁREZ, J., DÍAZ-GARCIÁ, F., JEFFRIES, T. E., ARENAS, R. & ABATI, J. 2003. Constraints on the peovenance of the uppermost allochthonous terrane of the NW Iberian Massif. Inferences from detrital zircon U–Pb age. *Terra Nova*, **15**, 138–144.

FERNÁNDEZ-SUÁREZ, J., ARENAS, R., ABATI, J., MARTÍNEZ CATALÁN, J. R., WHITEHOUSE, M. J. & JEFFRIES, T. 2007. U–Pb chronometry of polymetamorphic high-pressure granulites: an example from the allochthonous terranes of the NW Iberian Variscan belt. *In*: HATCHER, R. D., JR., CARLSON, M. P., MCBRIDE, J. H. & MARTÍNEZ CATALÁN, J. R. (eds) *Four-D evolution of continental crust*. Geological Society of America Memoir, **200**, 469–488.

FLOYD, P. A. 1984. Geochemical characteristics and comparison of the basic rocks of the Lizard Complex and the basaltic lavas within the Hercynian troughs of SW England. *Journal of the Geological Society, London*, **141**, 61–70.

FRANKE, W. 1989. Variscan plate-tectonics in Central-Europe – current ideas and open questions. *Tectonophysics*, **169**, 221–228.

FRANKE, W. 2000. The mid-European segment of the Variscides: tectonometamorphic units, terrane boundaries and plate tectonic evolution. *Geological Society, London, Special Publications*, **179**, 35–61.

GALÁN, G. & MARCOS, A. 1997. Geochemical evolution of high-pressure mafic granulites from the Bacariza formation (Cabo Ortegal Complex, NW Spain): an example of a heterogeneous lower crust. *Geologische Rundschau*, **86**, 539–555.

GALLASTEGUI, G., MARTÍN PARRA, L. M., DE PABLO MACIÁ, J. G. & RODRÍGUEZ FERNÁNDEZ, L. R. 1987. Las metavulcanitas del Dominio Esquistoso de Galicia tras os Montes: petrografía, geoquímica y ambiente geotectónico (Galicia, NO de España). *Cuadernos do Laboratorio Xeoloxico de Laxe, Coruña*, **15**, 127–139.

GANSSLOSER, M. 1999. Detrital chromian spinels in Rhenohercynian greywackes and sandstones (Givetian-Visean, Variscides, Germany) as indicators of ultramafic source rocks. *Geological Magazine*, **136**, 437–451.

GIBBONS, W. & THOMPSON, L. 1991. Ophiolitic mylonites in the Lizard complex: ductile extension in the lower crust. *Geology*, **19**, 1009–1012.

GILLIS, R. J., GEHRELS, G. E., RUIZ, J. & FLORES DE DIOS GONZÁLEZ, A. 2005, Detrital zircon provenance of Cambrian-Ordovician and Carboniferous strata of the Oaxaca terrane, southern Mexico. *Sedimentary Geology*, **182**, 87–100.

GIL IBARGUCHI, J. I. 1995. Petrology of jadeite metagranite and associated orthogneiss from the Malpica-Tuy Allochton (Northwest Spain). *European Journal of Mineralogy*, **7**, 403–415.

GIL IBARGUCHI, J. I. & DALLMEYER, R. D. 1991. Hercynian blueschist metamorphism in North Portugal: tectonothermal implications. *Journal of Metamorphic Geology*, **9**, 539–549.

GÓMEZ BARREIRO, J., WIJBRANS, J. R., CASTINEIRAS, P., MARTÍNEZ CATÁLAN, J. R., ARENAS, R., DÍAZ GARCÍA, F. & ABATI, J. 2006. Ar-40/Ar-39 laserprobe dating of mylonitic fabrics in a polyorogenic terrane of NW Iberia. *Journal of the Geological Society, London*, **163**, 61–73.

GÓMEZ-BARREIRO, J., MARTÍNEZ CATALÁN, J. R., ARENAS, R., CASTIÑEIRAS, P., ABATI, J., DÍAZ GARCÍA, F. & WIJBRANS, J. R. 2007. Tectonic evolution of the upper allochthon of the Ordenes Complex (northwestern Iberian Massif): structural constraints to a polyorogenic peri-Gondwanan terrane. *In*: LINNEMANN, U., NANCE, R. D., KRAFT, P. & ZULAUF, G. (eds) *The Evolution of the Rheic Ocean: From Avalonian–Cadomian Active Margin to Alleghenian–Variscan Collision*. Geological Society of America Special Paper, **423**, 315–332.

GRAVESTOCK, P. J. 1992. *The chemical causes of uppermost mantle heterogenities.* PhD thesis, Open University.

GRODZICKI, K. R., NANCE, R. D., KEPPIE, J. D., DOSTAL, J. & MURPHY, J. B. 2008. Structural, geochemical and geochronological analysis of metasedimentary and metavolcanic rocks of the Coatlaco area, Acatlán Complex, southern Mexico. *Tectonophysics*, **461**, 311–323.

GUTIÉRREZ-ALONSO, G. & FERNÁNDEZ-SUÁREZ, J. 1996. Geología y Geoquímica del granitoide pre-Varisco de Puente de Selce (Antiforme del Narcea, Asturias). *Revista de la Sociedad Geológica de España*, **9**, 227–239.

GUTIÉRREZ-MARCO, J. C., ARAMBURU, C. *ET AL.* 1999. Revisión bioestratigráfica de las pizarras del Ordovícico Medio en el noroeste de España zonas Cantábrica, Asturoccidental-leonesa y Centroibérica septentrional. *Acta Geologica Hispanica*, **34**, 3–87.

GUTIÉRREZ-ALONSO, G., FERNÁNDEZ-SUÁREZ, J., JEFFRIES, T. E., JENNER, J. E., TUBRETT, M. N., COX, R. & JACKSON, S. E. 2003. Terrane accretion and dispersal in the northern Gondwana margin: an Early Paleozoic analogue of a long lived active margin. *Tectonophysics*, **65**, 221–232, doi: 10.1016/S0040-1951(03)00023-4.

GUTIÉRREZ-ALONSO, G., FERNÁNDEZ-SUÁREZ, J., COLLINS, A. S., ABAD, I. & NIETO, F. 2005. Amazonian Mesoproterozoic basement in the core of the Ibero-Armorican Arc: $^{40}Ar/^{39}Ar$ detrital mica ages complement the zircon's tale. *Geology*, **33**, 637–640.

GUTIÉRREZ-ALONSO, G., FERNÁNDEZ-SUÁREZ, J., GUTIERREZ-MARCO, J. C., CORFU, F., MURPHY, J. B. & SUÁREZ, M. 2007. U–Pb depositional age for the upper Barrios Formation (Armorican Quartzite facies) in the Cantabrian zone of Iberia: Implications for stratigraphic correlation and paleogeography. *In*: LINNEMANN, U., NANCE, R. D., KRAFT, P. & ZULAUF, G. (eds) *The Evolution of the Rheic Ocean: From Avalonian–Cadomian Active Margin to Alleghenian–Variscan Collision*. Geological Society of America Special Paper, **423**, 287–296, doi: 10.1130/2007.2423(13).

GUTIÉRREZ-ALONSO, G., MURPHY, J. B., FERNÁNDEZ-SUÁREZ, J. & HAMILTON, M. 2008. A Devonian age for the El Castillo Volcanic rocks (Salamanca, Central Iberian Zone): implications for rifting along the northern Gondwana margin and the evolution of the Rheic ocean. *Tectonophysics*, **461**, 157–165.

HAMILTON, M. A. & MURPHY, J. B. 2004. Tectonic significance of a Llanvirn age for the Dunn Point volcanic rocks, Avalon terrane, Nova Scotia, Canada: implications for the evolution of the Iapetus and Rheic oceans. *Tectonophysics*, **379**, 199–209.

HARLAND, W. B. & GAYER, R. A. 1972, The Arctic Caledonides and earlier oceans. *Geological Magazine*, **109**, 289–314.

HATCHER, R. D. 2002. Alleghenian (Appalachian) orogeny, a product of zipper tectonics: rotational, trans pressive continent-continent collision and closing of ancient oceans along irregular margins. *In*: MARTÍNEZ CATALÁN, J. R., HATCHER, R. D., JR., ARENAS, R. & DÍAZ GARCÍA, F. (eds) *Variscan–Appalachian Dynamics: The Building of the Late Paleozoic Basement*. Geological Society of America, Special Paper, **364**, 199–208.

HEINZ, W., LOESCHKE, J. & VAVRA, G. 1985. Phreatomagmatic volcanism during the Ordovician of the Cantabrian Mountains. *Geologische Rundschau*, **74**, 623–639.

HIBBARD, J. 2000. Docking Carolina: Mid-Paleozoic accretion in the southern Appalachians. *Geology*, **28**, 127–130.

HIBBARD, J. P., STODDARD, E. F., STODDARD, E. F., SECOR, D. T. & DENNIS, A. J. 2002. The Carolina Zone: overview of Neoproterozoic to Early Paleozoic peri-Gondwanan terranes along the eastern flank of the southern Appalachians. *Earth Science Review*, **57**, 299–339.

HÖCK, V., MONTAG, O. & LEICHMANN, J. 1997. Ophiolite remnants at the eastern margin on the Bohemian Massif, and their bearing on the tectonic evolution. *Mineralogy and Petrology*, **60**, 267–287.

JENNER, G. A. & SWINDEN, H. S. 1993. The Pipestone Pond Complex, central Newfoundland: complex magmatism in an eastern Dunnage ophiolite. *Canadian Journal of Earth Sciences*, **30**, 434–448.

JULIVERT, M., MARCOS, A. & TRUYOLS, J. 1972. L'evolutión paléogéographique du nord-ouest de l'Espagne pendant l'Ordovicien-Silurien. *Bulletin de la Societé Géologique et Minéralogique de Bretagne*, **4**, 1–7.

KEPPIE, J. D. 1985. The Appalachian Collage. *In*: GEE, D. G. & STURT, B. (eds) *The Caledonide Orogen, Scandinavia, and Related Areas*. J. Wiley & Sons, New York, 1217–1226.

KEPPIE, J. D. 1993. Synthesis of Paleozoic deformational events and terrane accretion in the Canadian Appalachians. *Geologische Rundschau*, **82**, 381–431.

KEPPIE, J. D. 2004. Terranes of Mexico revisited: A 1.3 billion year odyssey. *International Geology Review*, **46**, 765–794.

KEPPIE, J. D. & RAMOS, V. S. 1999. Odyssey of terranes in the Iapetus and Rheic Oceans during the Paleozoic. *In*: KEPPIE, J. D. & RAMOS, V. A. (eds) *Laurentia-Gondwana Connections Before Pangea*. Geological Society of America, Special Paper, **336**, 267–276.

KEPPIE, J. D., DOSTAL, J., ORTEGA-GUTIÉRREZ, F. & LÓPEZ, R. 2001. A Grenvillian arc on the margin of Amazonia: evidence from the southern Oaxacan Complex, southern Mexico. *Precambrian Research*, **112**, 165–181.

KEPPIE, J. D., DOSTAL, J., CAMERON, K. L., SOLARI, L. A., ORTEGA-GUTIÉRREZ, F. & LÓPEZ, R. 2003a. Geochronology and geochemistry of Grenvillian igneous suites in the northern Oaxacan Complex, southern Mexico: tectonic implications. *Precambrian Research*, **120**, 365–389.

KEPPIE, J. D., NANCE, R. D., MURPHY, J. B. & DOSTAL, J. 2003b. Tethyan, Mediterranean, and Pacific analogues for the Neoproterozoic–Paleozoic birth and development of peri-Gondwanan terranes and their transfer to Laurentia and Laurussia. *Tectonophysics*, **365**, 195–220.

KEPPIE, J. D., SANDBERG, C. A., MILLER, B. V., SÁNCHEZ-ZAVALA, J. L., NANCE, R. D. & POOLE, F. G. 2004a. Implications of latest Pennsylvanian to Middle Permian paleontological and U–Pb SHRIMP data from the Tecomate Formation to re-dating tectonothermal events in the Acatlán

Complex, southern Mexico. *International Geology Review*, **46**, 745–753.

KEPPIE, J. D., NANCE, R. D. *ET AL.* 2004*b*. Mid-Jurassic tectonothermal event superposed on a Paleozoic geological record in the Acatlán Complex of southern Mexico: hotspot activity during the breakup of Pangea. *Gondwana Research*, **7**, 239–260.

KEPPIE, J. D., NANCE, R. D., FERNÁNDEZ-SUÁREZ, J., STOREY, C. D., JEFFRIES, T. E. & MURPHY, J. B. 2006. Detrital zircon data from the eastern Mixteca terrane, southern Mexico: evidence for an Ordovician-Mississippian continental rise and a Permo-Triassic clastic wedge adjacent to Oaxaquia. *International Geology Review*, **48**, 97–111.

KEPPIE, J. D., DOSTAL, J. & ELIÁS-HERRERA, M. 2007. Ordovician-Devonian oceanic basalts in the Cosoltepec Formation, Acatlán Complex, southern Mexico: vestiges of the Rheic Ocean? *In*: LINNEMANN, U., NANCE, R. D., KRAFT, P. & ZULAUF, G. (eds) *The Evolution of the Rheic Ocean: From Avalonian–Cadomian active margin to Alleghenian–Variscan collision*. Geological Society of America, Special Paper, **423**, 477–488.

KEPPIE, J. D., DOSTAL, J., MURPHY, J. B. & NANCE, R. D. 2008*a*. Synthesis and tectonic interpretation of the westernmost Paleozoic Variscan orogen in southern Mexico: from rifted Rheic margin to active Pacific margin. *Tectonophysics*, doi: http://dx.doi.org/10.1016/j.tecto.2008.01.012.

KEPPIE, J. D., DOSTAL, J. *ET AL.* 2008*b*. Ordovician-earliest Silurian rift tholeiites in the Acatlán Complex, southern Mexico: evidence of rifting on the southern margin of the Rheic Ocean. *Tectonophysics*, **461**, 130–156; doi: 10.1016/j.tecto.2008.01.010.

KRONER, U. & HAHN, T. 2003. Sedimentation, Deformation und Metamorphose im Saxothuringikum während der variszsichen Orogenese: Die komplexe Entwicklung von Nord-Gondwana während kontinentaler Subduktion und schiefer Kollision. *In*: LINNEMANN, U. (ed.) *Das Saxothuringikum*. Staatliche Naturhistorische Sammlungen Dresden, 133–146.

KRYZA, R., MAZUR, S. & PIN, C. 2003. Subduction- and non-subduction-related igneous rocks in the Central-European Variscides: geochemical and Nd isotope evidence for a composite origin of the Kodzko-Metamorphic Complex, Polish Sudetes. *Geodinamica Acta*, **16**, 39–57.

LANDING, E., KEPPIE, J. D. & WESTROP, S. 2007. Terminal Cambrian and lowest Ordovician of Mexican West Gondwana: biotas and sequence stratigraphy of the Tiñu Formation. *Geological Magazine*, **144**, 900–936.

LARDEAUX, J. M., LEDRU, P., DANIEL, I. & DUCHENE, S. 2001. The Variscan French Massif Central – a new addition to the ultra-high pressure metamorphic 'club': exhumation processes and geodynamic consequences. *Tectonophysics*, **332**, 143–167.

LEDRU, P., AUTRAN, A. & SANTALLIER, D. 1994. Lithostratigraphy of Variscan terranes in the French Massif Central. A basic for paleogeographical reconstruction. *In*: KEPPIE, J. D. (ed.) *Pre-Mesozoic geology in France and related areas*. Springer Verlag, Berlin, 276–288.

LINNEMANN, U. & ROMER, R. L. 2002. The Cadomian Orogeny in Saxo-Thuringia, Germany: geochemical and Nd–Sr–Pb isotopic characterisation of marginal basins with constraints to geotectonic setting and provenance. *Tectonophysics*, **352**, 33–64.

LINNEMANN, U., GEHMLICH, M. *ET AL.* 2000. Cadomian subduction to Early Palaeozoic rifting: the evolution of Saxo-Thuringia at the margin of Gondwana in the light of single zircon geochronology and basin development (Central European Variscides, Germany). *In*: FRANKE, W., HAAK, V., ONCKEN, O. & TANNER, D. (eds) *Orogenic Processes: Quantification and Modelling in the Variscan Belt*. Geological Society, London, Special Publications, **179**, 131–153.

LINNEMANN, U., MCNAUGHTON, N. J., ROMER, R. L., GEHMLICH, M., DROST, K. & TONK, C. 2004. West African provenance for Saxo-Thuringia (Bohemian Massif): did Armorica ever leave pre-Pangean Gondwana? – U/Pb-SHRIMP zircon evidence and the Nd-isotopic record. *International Journal of Earth Sciences*, **93**, 683–705.

LOESCHKE, W. H. J. & ZEIDLER, N. 1982. Early Paleozoic sills in the Cantabrian Mountains (Spain) and their geotectonic environment. *Jahrbuch für Geologie und Paläontologie, Abhandlungen*, **7**, 419–439.

LOTZE, F. 1945. Zur Gliederung der Varisziden der Iberischen Meseta. *Geotektonische Forschungen*, **6**, 78–92.

MACINTYRE, R. M., BOWES, D. R., HAMIDULLAH, S. & ONSTOTT, T. C. 1992. K–Ar and Ar–Ar isotopic study on meta-amphiboles from meta-ophiolite complexes, Eastern Bohemian Massif. *Proceedings, 1st International Conference Bohemian Massif, Prague 1988*, 195–199.

MACLACHLAN, K. & DUNNING, G. 1998. U–Pb ages and tectono-magmatic evolution of Middle Ordovician volcanic rocks of the Wild Bight Group, Newfoundland Appalachians. *Canadian Journal of Earth Sciences*, **35**, 998–1017.

MARCOS, A., FARIAS, P., GALÁN, G., FERNÁNDEZ, F. J. & LLANA-FÚNEZ, S. 2002. The tectonic framework of the Cabo Ortegal Complex: a slab of lower crust exhumed in the Variscan Orogen (NW Iberian Peninsula). *In*: MARTÍNEZ CATALÁN, J. R., HATCHER, R. D., JR., ARENAS, R. & DÍAZ GARCÍA, F. (eds) *Variscan–Appalachian Dynamics: The Building of the Late Paleozoic Basement*. Geological Society of America, Special Paper, **364**, 143–162.

MARCOS, A., MARTÍNEZ CATALÁN, J. R., GUTIÉRREZ-MARCO, J. C. & PÉREZ-ESTAÚN, A. 2004. Estratigrafía y paleogeografía. *In*: VERA, J. A. (ed.) *Geología de España*. S.G.E.-I.G.M.E., Madrid, 49–52.

MARTÍNEZ CATALÁN, J. R., HACAR RODRÍGUEZ, M. P., VILLAR ALONSO, M. P., PÉREZ-ESTAÚN, A. & GONZÁLEZ LODEIRO, F. 1992. Paleozoic extensional tectonics in the limit between the West Asturias-Leonese and the Central Iberian Zones of the Bariscan Fold-Belt in NW Spain. *Geologische Rundschau*, **81**, 545–560.

MARTÍNEZ CATALÁN, J. R., ARENAS, R., DÍAZ GARCÍA, F., RUBIO PASCUAL, F. J., ABATI, J. & MARQUÍNEZ, J. 1996. Variscan exhumation of a subducted Paleozoic continental margin: the basal units of the Órdenes Complex, Galicia, NW Spain. *Tectonics*, **15**, 106–121.

MARTÍNEZ CATALÁN, J. R., ARENAS, R., DÍAZ GARCÍA, F. & ABATI, J. 1997. Variscan accretionary complex of northwest Iberia: terrane correlation and succession of techtonothermal events. *Geology*, **25**, 1103–1106.

MARTÍNEZ CATALÁN, J. R., ARENAS, R., DÍAZ GARCÍA, F. & ABATI, J. 1999. Allochthonous units in the Variscan belt of NW Iberia. Terranes and accretionary history. *In*: SINHA, A. K. (ed.) *Basement Tectonics*. Kluwer Academic Publishers, **13**, 65–84.

MARTÍNEZ CATALÁN, J. R., ARENAS, R. *ET AL.* 2007. Space and time in the tectonic evolution of the northwestern Iberian Massif. Implications for the comprehension of the Variscan belt. *In*: HATCHER, R. D., JR., CARLSON, M. P., MCBRIDE, J. H. & MARTÍNEZ CATALÁN, J. R. (eds) *4-D framework of Continental Crust*. Geological Society, London, Memoirs, **200**, 1–435.

MATTE, P. 2001. The Variscan collage and orogeny (480–290 Ma) and the tectonic definition of the Armorica microplate: a review. *Terra Nova*, **13**, 122–128.

MCBRIDE, J. & NELSON, D. 1991. Deep seismic reflection constraints on Palaeozoic crustal structure and definition of the Moho in the buried southern Appalachian orogen. Continental lithosphere: deep Seismic Reflections. *Geodynamics*, **22**, 1–20.

MCBRIDE, J. & NELSON, D. 1988. Intergration of COCORP deep reflection and magnetic anomaly analysis in the southeastern US: implications for the origin of the Brunswick and East Coast magnetic anomalies. *Geological Society of America Bulletin*, **100**, 436–455.

MCKERROW, W. S. & SCOTESE, C. R. (eds) 1990. *Palaeozoic palaeogeography and biogeography*. Geological Society of London, London, Memoirs, 12.

MCKERROW, W. S., MACNIOCAIL, C., AHLBERG, P. E., CLAYTON, G., CLEAL, C. J. & EAGAR, M. C. 2000. The Late Palaeozoic relations between Gondwana and Laurussia. *Geological Society, London, Special Publications*, **179**, 9–20.

MIDDLETON, M., KEPPIE, J. D., MURPHY, J. B., MILLER, B. V., NANCE, R. D., ORTEGA-RIVERA, A. & LEE, J. K. W. 2007. *P–T–t* constraints on exumation following subduction in the Rheic Ocean from eclogitic rocks in the Acatlán Complex of southern Mexico. *In*: LINNEMANN, U., NANCE, R. D., KRAFT, P. & ZULAUF, G. (eds) *The Evolution of the Rheic Ocean: From Avalonian–Cadomian Active Margin to Alleghenian–Variscan Collision*. Geological Society of America, Special Paper, **423**, 489–509.

MONTERO, M. P. 1993. *Geoquímica y petrogénesis del complejo peralcalino de la Sierra del Galiñeiro (Pontevedra, España)*. PhD thesis, University of Oviedo, Spain.

MORALES-GÁMEZ, M., KEPPIE, J. D. & NORMAN, M. 2008. Ordovician-Silurian rift-passive margin on the Mexican margin of the Rheic Ocean overlain by Permian periarc rocks: evidence from the Acatlán Complex, southern Mexico. *In*: PEREIRA, M. F., BOZKURT, E., QUESADA, C. & STRACHAN, R. (eds) *The Foundations and Birth of the Rheic Ocean: Avalonian–Cadomian Orogenic Processes and Early Palaeozoic Rifting at the North-Gondwana Margin. Tectonophysics*, **461**, 291–310; http://dx.doi.org/10.1016/j.tecto.2008.01.014

MORENO, T., GIBBONS, W., PRICHARD, H. & LUNAR, R. 2001. Platiniferous chromitite and the tectonic setting of ultramafic rocks in Cabo Ortegal (north-west Spain). *Journal of the Geological Society, London*, **158**, 601–614.

MILLER, B. V., DOSTAL, J., KEPPIE, J. D., NANCE, R. D., ORTEGA-RIVERA, A. & LEE, J. W. K. 2007. Ordovician calc-alkaline granitoids in the Acatlán Complex, southern Mexico: geochemical and geochronological data and implications for tectonic of the Gondwanan margin of the Rheic Ocean. *In*: LINNEMANN, U., NANCE, R. D., ZULAF, G. & KRAFT, P. (eds) *The Evolution of the Rheic Ocean: From Avalonian–Cadomian Active Margin to Alleghenian–Variscan Collision*. Geological Society of America, Special Paper, **423**, 465–475.

MUNHÁ, J., OLIVEIRA, J. T., RIBEIRO, A., OLIVEIRA, V., QUESADA, C. & KERRICH, R. 1986. Beja-Acebuches ophiolite: characterization and geodynamic significance. *Maleo*, **2**(2), 13, 30.

MURPHY, J. B. & GUTIÉRREZ-ALONSO, G. 2008. The origin of the Variscan NW Iberian allochthonous complexes in NW Iberia: Sm/Nd isotopic constraints. *Canadian Journal of Earth Sciences*, **45**, 651–668.

MURPHY, J. B., RICE, R. J., STOKES, T. R. & KEPPIE, D. F. 1995. The St. Mary's Basin, central mainland Nova Scotia: late Paleozoic basin formation and deformation along the Avalon-Meguma Terrane boundary, Canadian Appalachians. *In*: HIBBARD, J., VAN STAAL, C. & CAWOOD, P. (eds) *New perspectives in the Caledonian–Appalachian orogen*. Geological Association of Canada, Special Paper, **41**, 409–420.

MURPHY, J. B., KEPPIE, J. D., BRAID, J. & NANCE, R. D. 2005*b*. Geochemistry of the Tremadocian Tiñu Formation (southern Mexico): provenance in the underlying ∼1 Ga Oaxacan Complex on the southern margin of the Rheic Ocean. *International Geology Review*, **47**, 887–900.

MURPHY, J. B., GUTIÉRREZ-ALONSO, G. *ET AL.* 2006*a*. Origin of the Rheic ocean: rifting along a Neoproterozoic Suture? *Geology*, **34**, 325–328.

MURPHY, J. B., KEPPIE, J. D., NANCE, R. D., MILLER, B. V., MIDDLETON, M., FERNÁNDEZ-SUÁREZ, J. & JEFFRIES, T. E. 2006*b*. Geochemistry and U–Pb protolith ages of eclogitic rocks of the Asís Lithodeme, Piaxtla Suite, Acatlán Complex, southern Mexico: tectonothermal activity along the southern margin of the Rheic Ocean. *Journal of the Geological Society, London*, **163**, 683–695.

NANCE, R. D., MILLER, B. V., KEPPIE, J. D., MURPHY, J. B. & DOSTAL, J. 2006. The Acatlán Complex, southern Mexico: record of Pangea assembly to breakup. *Geology*, **34**, 857–860.

NANCE, R. D., MILLER, B. V., KEPPIE, J. D., MURPHY, J. B. & DOSTAL, J. 2007. Vestige of the Rheic Ocean in North America: the Acatlán Complex of southern México. *In*: LINNEMANN, U., NANCE, R. D., ZULAF, G. & KRAFT, P. (eds) *The Evolution of the Rheic Ocean: From Avalonian–Cadomian active margin to Alleghenian–Variscan collision*. Geological Society of America, Special Paper, **423**, 437–452.

NAVARRO-SANTILLÁN, D., SOUR-TOVAR, F. & CENTENO-GARCÍA, E. 2002. Lower Mississippian

(Osagean) brachiopods from the Santiago Formation, Oaxaca, Mexico: stratigraphic and tectonic implications. *Journal of South American Earth Sciences*, **15**, 327–336.

NESBITT, H. W. & YOUNG, G. M. 1996. Petrogenesis of sediments in the absence of chemical weathering: effects of abrasion and sorting on bulk composition and mineralogy. *Sedimentology*, **43**, 341–358.

NUTMAN, A. P., GREEN, D. H., COOK, C. A., STYLES, M. T. & HOLDSWORTH, R. E. 2001. SHRIMP U–Pb zircon dating of the exhumation of the Lizard Peridotite and its emplacement over crustal rocks: constraints for tectonic models. *Journal of the Geological Society, London*, **158**, 809–820.

OLIVEIRA, J. T. & QUESADA, C. 1998. A comparison of stratigraphy, structure, and palaeogeography of the South Portuguese zone and south-west England, European Variscides. *Geoscience in South-west England*, **9**, 141–150.

OLÍVER, G., CORFU, F. & KROGH, T. 1993. U–Pb ages from southwest Poland: evidence for a Caledonian suture zone between Baltica and Gondwana. *Journal of the Geological Society, London*, **150**, 355–369.

ONÉZIME, J., CHARVET, J., FAURE, M., BOURDIER, J. L. & CHAUVET, A. 2003. A new geodynamic interpretation for the South Portuguese Zone (SW Iberia) and the Iberian Pyrite Belt genesis. *Tectonics*, **22**, 4.

ORDÓÑEZ-CASADO, B., GEBAUER, D., SCHÄFER, H.-J., GIL IBARGUCHI, J. I. & PEUCAT, J. J. 2001. A single Devonian subduction event for the HP/HT metamorphism of the Cabo Ortegal Complex within the Iberian Massif. *Tectonophysics*, **332**, 359–385.

ORTEGA-GUTIÉRREZ, F., RUIZ, J. E. & CENTENO-GARCÍA, E. 1995. Oaxaquia, a Proterozoic microcontinent accreted to North America during the late Paleozoic. *Geology*, **23**, 1127–1130.

ORTEGA-GUTIÉRREZ, F., ELÍAS-HERRERA, M., REYES-SALAS, M., MACÍAS-ROMO, C. & LÓPEZ, R. 1999. Late Ordovician-Early Silurian continental collision orogeny in southern Mexico and its bearing on Gondwana-Laurentia connections. *Geology*, **27**, 719–722.

ORTEGA-OBREGÓN, C. & KEPPIE, J. D. *ET AL.* 2003. Geochronology and geochemistry of the 917 Ma, calc-alkaline Etla granitoid pluton (Oaxaca, southern México): evidence of post-Grenvillian subduction along the northern margin of Amazonia. *International Geology Review*, **45**, 596–610.

PATCHETT, P. J. & RUIZ, J. 1987. Nd isotopic ages of crustal formation and metamorphism in the Precambrian of eastern and southern Mexico. *Contributions to Mineralogy and Petrology*, **96**, 523–528.

PÉREZ-ESTAÚN, A., BASTIDA, F., MARTÍNEZ-CATALÁN, J. R., GUTIÉRREZ-MARCO, J. C., MARCOS, A. & PULGAR, J. A. 1990. West Asturias-Leonese Zone: Stratigraphy. *In*: DALLMEYER, R. D. & MARTÍNEZ-GARCÍA, E. (eds) *Pre-Mesozoic Geology of Iberia*. Springer Verlag, Berlin, 92–102.

PEUCAT, J. J., BERNARD-GRIFFITHS, J., GIL IBARGUCHI, J. I., DALLMEYER, R. D., MENOT, R. P., CORNICHET, J. & IGLESIAS PONCE DE LEÓN, M. 1990. Geochemical and geochronological cross section of the deep Variscan crust: the Cabo Ortegal

high-pressure nappe (northwestern Spain). *Tectonophysics*, **177**, 263–292.

PEREIRA, M. F., CHICHORRO, M., LINNEMANN, U., EGUILUZ, L. & SILVA, J. B. 2006. Inherited arc signature in the Ediacaran and Early Cambrian basins of the Ossa-Morena Zone (Iberian Massif, Portugal): paleogeographic link with West-Central European and North African Cadomian correlatives. *Precambrian Research*, **144**, 297–315.

PIN, C. & MARINI, F. 1993. Early Ordovician continental break-up in Variscan Europe: Nd–Sr isotope and trace element evidence from bimodal igneous associations of the Southern Massif Central, France. *Lithos*, **29**, 177–196.

PIN, C. & PAQUETTE, J.-L. 1997. A mantle-derived bimodal suite in the Hercynian belt: Nd isotope and trace element evidence for a subduction-related rift origin of the Late Devonian Brevenne metavolcanics, Massif Central (France). *Contributions to Mineralogy & Petrology*, **129**, 222–238.

PIN, C. & PAQUETTE, J.-L. 2002. Sr–Nd isotope and trace element evidence for a Late Devonian active margin in northern Massif Central (France). *Geodynamica Acta*, **15**, 63–77.

PIN, C., MAJEROWICZ, A. & WOJCIECHOWSKA, I. 1988. Upper Palaeozoic oceanic crust in the Polish Sudetes: Nd–Sr isotope and trace element evidence. *Lithos*, **21**, 195–209.

PIN, C., ORTEGA CUESTA, L. A. & GIL IBARGUCHI, J. I. 1992. Mantle derived, early Paleozoic A-type metagranitoids from the NW Iberian massif: Nd isotope and trace-element constraints. *Bulletin de la Societe Geologique de la France*, **163**, 483–494.

PIN, C., PAQUETTE, J. L., SANTOS ZALDUEGUI, J. F. & GIL IBARGUCHI, J. I. 2002. Devonian suprasubduction zone ophiolite related to incipient collisional processes in the Western Variscan Belt: the Sierra de Careón Unit, Órdenes Complex, Galicia. *In*: MARTÍNEZ CATALÁN, J. R., HATCHER, R. D., ARENAS, R. & DÍAZ GARCÍA, F. (eds) *Variscan-Appalachian Dynamics: The Building of the Late Paleozoic Basement*. Geological Society of America, Special Paper, **364**, 57–71.

PIN, C., PAQUETTE, J. L., ÁBALOS, B., SANTOS ZALDUEGUI, J. F. & GIL IBARGUCHI, J. I. 2006. Composite origin of an early Variscan transported suture: ophiolitic units of the Morais Nappe Complex (north Portugal). *Tectonics*, **25**, TC5001, doi: 10.1029/2006TC001971.

PIN, C., FONSECA, P. E., PAQUETTE, J.-L., CASTRO, P. & MATTE, P. 2008. The *c.* 350 Ma Beja Igneous Complex: a record of transcurrent slab break-off in the Southern Iberia Variscan Belt? *Tectonophysics*, **461**, 356–377; doi: 10.1016/j.tecto.2008.06.001.

PROENZA, J. A., ORTEGA-GUTIÉRREZ, F., CAMPRUBÍ, A., TRITLLA, J., ELÍAS-HERRERA, M. & REYES-SALAS, M. 2004. Paleozoic serpentinite-enclosed chromitites from Tehuitzingo (Acatlán Complex, southern Mexico): a petrological and mineralogical study. *Journal of South America Earth Sciences*, **16**, 649–666.

PUSCHMANN, H. 1967. Zum Problem derSchichtlücken im Devon der Sierra Morena (Spanien). *Geologische Rundschau*, **56**, 528–542.

QUESADA, C. 1990. Precambrian terranes in the Iberian Variscan foldbelt. *In*: STRACHAN, R. A. & TAYLOR, G. K. (eds) *Avalonian and Cadomian Geology of the North Atlantic*. Blackie and Son, Glasgow, 109–133.

QUESADA, C. 1991. Geological constraints on the Paleozoic tectonic evolution of tectonostratigraphic terranes in Iberia. *Tectonophysics*, **185**, 225–245.

QUESADA, C. 1998. A reappraisal of the structure of the Spanish segment of the Iberian pyrite belt. *Mineralium Deposita*, **33**, 31–44.

QUESADA, C. & DALLMEYER, R. D. 1994. Tectonothermal evolution of the Badajoz–Córdoba shear zone (SW Iberia): characteristics and $^{40}Ar/^{39}Ar$ mineral age constraints. *Tectonophysics*, **231**, 195–213.

QUESADA, C., BELLIDO, F. *ET AL.* 1991. Terranes within the Iberian Massif: correlations with West African sequences. *In*: DALLMEYER, R. D. & LECORCHÉ, J. P. (eds) *The West African Orogens and Circum–Atlantic Correlations*. Springer-Verlag, Berlin, 267–294.

QUESADA, C., SÁNCHEZ-GARCÍA, T., BELLIDO, F., LÓPEZ-GUIJARRO, R., ARMENDÁRIZ, M. & BRAID, J. 2006. Field trip guide (Spain). Introduction: the Ossa–Morena Zone – from Neoproterozoic arc trough Early Paleozoic rifting to late Paleozoic orogeny. *In*: PEREIRA, M. F. & QUESADA, C. (eds) *Ediacaran to Visean crustal growth processes in the Ossa–Morena Zone (SW Iberia)*. Évora Meeting 2006, Conference abstracts and field trip guide. Instituto Geológico Minero de España, 51–73.

RAMÍREZ ESPINOZA, J. 2001. *Tectono-magmatic evolution of the Paleozoic Acatlán Complex, southern Mexico and its correlation with the Appalachian system*. Unpublished PhD thesis, University of Arizona, 1–170.

RAMOS-ARIAS, M. A., KEPPIE, J. D., ORTEGA-RIVERA, A. & LEE, J. W. K. 2008. Extensional deformation on the western margin of Pangea, Patlanoaya area, Acatlán Complex, southern Mexico. *Tectonophysics*, **448**(1–4), 60–76.

RIBEIRO, M. L. & FLOOR, P. 1987. Magmatismo peralcalino no Maciço Hespérico: sua distribuçao e significado geodinámico. *In*: BEA, F., CARNICERO, A., GONZALO, J. C., LÓPEZ-PLAZA, M. & RODRÍGUEZ ALONSO, M. D. (eds), *Geología de los granitoides y rocas asociadas del Macizo Hespérico*. Rueda, Madrid, 211–221.

RIBEIRO, A., PEREIRA, E. & DIAS, R. 1990. Structure in the NW of the Iberian Peninsula. *In*: DALLMEYER, R. D. & MARTÍNEZ-GARCÍA, E. (eds) *Pre-Mesozoic Geology of Iberia*. Springer Verlag, Berlin, 221–236.

ROBARDET, M. 2002. Alternative approach to the Variscan Belt in Southwestern Europe: preorogenic paleobiogeographical constraints. *In*: MARTÍNEZ-CATALÁN, J. R., HATCHER, R. D., ARENAS, R. & DÍAZ GARCÍA, F. (eds) *Variscan–Appalachian Dynamics: The Building of the Late Paleozoic Basement*. Geological Society of America, Special Paper, **364**, 1–15.

ROBARDET, M. 2003. The Armorica microplate: fact or fiction? Critical review of the concept and contradictory paleobiogeographical data. *Palaeogeography, Palaeoclimatology & Palaeoecology*, **195**, 125–148.

ROBARDET, M. & GUTIÉRREZ-MARCO, J. C. 2004. The Ordovician, Silurian and Devonian sedimentary rocks of the Ossa-Morena Zone (SW Iberian Peninsula, Spain). *Journal of Iberian Geology*, **30**, 73–92.

ROBERTS, D. & GEE, D. G. 1985. An introduction to the structure of the Scandinavian Caledonides. *In*: GEE, D. G. & STURT, B. A. (eds) *The Caledonide Orogen–Scandinavia and Related Areas*. Wiley, Chichester, 55–68.

ROBERTS, D. & STEPHENS, M. G. 2000. Caledonian Orogenic Belt. *In*: LUNDQVIST, T. & AUTIO, S. (eds) *Description to the Bedrock Map of Central Fennoscandia (Mid-Norden)*. Geological Survey of Finland, Special Papers, **28**, 79–108.

ROBERTS, S., ANDREWS, J. R., BULL, J. M. & SANDERSON, D. J. 1993. Slow-spreading ridge-axis tectonics: evidence from the Lizard complex, UK. *Earth and Planetary Science Letters*, **116**, 101–112.

RODRÍGUEZ ALLER, J. 2005. Recristalización y deformación de litologías supracorticales sometidas a matamorfismo de alta presión (Complejo de Malpica-Tuy, NO del Macizo Ibérico). *Laboratorio Xeolóxico de Laxe, Serie Nova Terra*, **29**, 1–572.

RODRÍGUEZ, J., COSCA, M. A., GIL IBARGUCHI, J. I. & DALLMEYER, R. D. 2003. Strain partitioning and preservation of $^{40}Ar/^{39}Ar$ ages during Variscan exhumation of a subducted crust (Malpica-Tui complex, NW Spain). *Lithos*, **70**, 111–139.

ROGER, F. & MATTE, P. 2005. Early Variscan HP metamorphism in the western Iberian Allochthon-A 390 Ma U–Pb age for the Braganca eclogite (NW Portugal). *International Journal of Earth Sciences*, **94**, 173–179.

RUBIO PASCUAL, F. J., ARENAS, R., DÍAZ GARCÍA, F., MARTÍNEZ CATALÁN, J. R. & ABATI, J. 2002. Eclogites and eclogite–amphibolites from the Santiago Unit Ordenes Complex, NW Iberian Massif, Spain): a case study of contrasting high-pressure metabasites in a context of crustal subduction. *In*: MARTÍNEZ CATALÁN, J. R., HATCHER, R. D., ARENAS, R. & DÍAZ GARCÍA, F. (eds) *Variscan–Appalachian Dynamics: The Building of the Late Paleozoic Basement*. Geological Society of America, Special Papers, **364**, 105–124.

RUIZ, J., PATCHETT, P. J. & ORTEGA-GUTIÉRREZ, F. 1988. Proterozoic and Phanerozoic basement terranes of Mexico from Nd isotopic studies, *Geological Society of America Bulletin*, **100**, 274–281.

SÁNCHEZ MARTÍNEZ, S., ARENAS, R., DÍAZ GARCÍA, F., MARTÍNEZ CATALÁN, J. R. & GÓMEZ-BARREIRO, J. 2007. Careón ophiolite, NW Spain: suprasubduction zone setting for the youngest Rheic Ocean floor. *Geology*, **7**, 53–56; doi: 10.1130/G23024A.

SÁNCHEZ-GARCÍA, T., BELLIDO, F. & QUESADA, C. 2003. Geodynamic setting and geochemical signatures of Cambrian-Ordovician rift-related igneous rocks (Ossa-Morena Zone, SW Iberia). *Tectonophysics*, **365**, 233–255.

SÁNCHEZ-MARTÍNEZ, S, JEFFRIES, T., ARENAS, R., FERNÁNDEZ-SUÁREZ, J. & GARCÍA-SÁNCHEZ, R. 2006. A pre-Rodinian ophiolite involved in the Variscan suture of Galicia (Cabo Ortegal Complex, NW

Spain). *Journal of the Geological Society, London*, **163**, 737–740.

SÁNCHEZ-MARTÍNEZ, S., ARENAS, R., DÍAZ GARCÍA, F., MARTÍNEZ CATALÁN, J. R. & GÓMEZ BARREIRO, J. 2007. Careón ophiolite, NW Spain: Suprasubduction zone setting for the youngest Rheic Ocean floor. *Geology*, **35**, 53–56.

SANDEMAN, H. A., CLARK, A. H., SCOTT, D. J. & MALPAS, J. G. 2000. The Kennack Gneiss of the Lizard Peninsula, Cornwall, S. W. England: comingling and mixing of mafic and felsic magmas during Givetian continental incorporation of the Lizard ophiolite. *Journal of the Geological Society, London*, **157**, 1227–1242.

SANDEMAN, H. A., CLARK, A. H., STYLES, M. T., SCOTT, D. J., MALPAS, J. G. & FARRAR, E. 1997. Geochemistry and U–Pb and $^{40}$Ar/$^{39}$Ar geochronology of the Man of War Gneiss, Lizard Complex, SW England: pre-Hercynian arc-type crust with a Sudetan-Iberian connection. *Journal of the Geological Society, London*, **154**, 403–411.

SANTOS ZALDUEGUI, J. F., SCHÄRER, U. & GIL IBARGUCHI, J. I. 1995. Isotope constraints on the age and origin of magmatism and metamorphism in the Malpica-Tuy allochthon, Galicia, NW-Spain. *Chemical Geology*, **121**, 91–103.

SANTOS ZALDUEGUI, F. J., SCHARER, U., GIL IBARGUCHI, J. I. & GIRARDEAU, J. 2002. Genesis of pyroxenite-rich peridotite at Cabo Ortegal (NW Spain): geochemical and Pb–Sr–Nd isotope data. *Journal of Petrology*, **43**, 17–43.

SCHÄFER, H. J., GEBAUER, D., GIL IBARGUCHI, J. I. & PEUCAT, J. J. 1993. Ion-microprobe U–Pb zircon dating on the HP/HT Cabo Ortegal Complex (Galicia, NW Spain): preliminary results. *Terra Abstracts*, **5**, 22.

SCHNEIDER, J. L., HASSENFORDER, B. & PAICHELER, J.-C. 1990. Une ou plusieurs 'lignes des klippes' dans les Vosges du Sud (France)? Nouvelles données sur la nature des 'klippes' et leur signification dans la dynamique varisque. *Comptes Rendus Academie de Sciences, Paris*, **311**(II), 1221–1226.

SHAW, A., DOWNES, H. & THIRLWALL, M. F. 1993. The quartz-diorites of Limousin: elemental and isotopic evidence for Devono-Carboniferous subduction in the Hercynian belt of the French Massif Central. *Chemical Geology*, **107**, 1–18.

SILVA, J. B., OLIVEIRA, J. T. & RIBEIRO, A. 1990. Structural outline of the South Portuguese Zone. *In*: DALLMEYER, R. D. & MARTÍNEZ GARCÍA, E. (eds) *Pre-Mesozoic Geology of Iberia*. Springer Verlag, 348–362.

SIMANCAS, J. F., GONZÁLEZ LODEIRO, F., EXPÓSITO, I., AZOR, A. & MARTÍNEZ POYATOS, D. 2002. Opposite subduction polarities connected by transform. faults in the Iberian Massif and western European Variscides. *In*: MARTÍNEZ CATALÁN, J. R., HATCHER, R. D., ARENAS, R. & DÍAZ GARCÍA, F. *Variscan–Appalachian dynamics: the building of the late Paleozoic basement*. Geological Society of America, Special Paper, **364**, 253–262.

SOLARI, L. A., KEPPIE, J. D., ORTEGA-GUTIÉRREZ, F., CAMERON, K. L., LÓPEZ, R. & HAMES, W. E. 2003. 990 and 1100 Ma Grenvillian tectonothermal events in the northern Oaxacan Complex, southern Mexico: roots of an orogen. *Tectonophysics*, **365**, 257–282.

STAMPFLI, G. M., VON RAUMER, J. F. & BOREL, G. D. 2002. Paleozoic evolution of pre-Variscan terranes: from Gondwana to the Variscan collision. *Geological Society of America, Special Paper*, **364**, 263–280.

STAMPFLI, G. M. & BOREL, G. D. 2002. A plate tectonic model for the Paleozoic and Mesozoic constrained by dynamic plate boundaries and restored synthetic oceanic isochrones. *Earth and Planetary Science Letters*, **196**, 17–33.

STERN, R. J. 2004. Subduction initiation: spontaneous and induced. *Earth and Planetary Science Letters*, **226**, 275–292.

TAIT, J., SCHÄTZ, M., BACHTADSE, V. & SOFFEL, H. 2000. Palaeomagnetism and Palaeozoic palaeogeography of Gondwana and European terranes. *Geological Society, London, Special Publications*, **179**, 21–34.

TALAVERA-MENDOZA, O., RUÍZ, J., GEHRELS, G. E., MEZA-FIGUEROA, D. M., VEGA-GRANILLO, R. & CAMPA-URANGA, M. F. 2005. U–Pb geochronology of the Acatlán Complex and implications for the Paleozoic paleogeography and tectonic evolution of southern Mexico. *Earth and Planetary Science Letters*, **235**, 682–699.

THIÉBLEMONT, D. & CABANIS, B. 1986. Découverte d'une association de volcanites d'arc et de basaltes de type 'MORB' dans la formation paléo-volcanique silurienne de la Meilleraie, Vendée, France. *Comptes Rendus Academie de Sciences, Paris*, **311**(II), 1221–1226.

VACHARD, D. & FLORES DE DIOS, A. 2002. Discovery of latest Devonian/earliest Mississippian microfossils in San Salvador Patlanoaya (Puebla, Mexico): biogeographic and geodynamic consequences. *Compte Rendu Geoscience*, **334**, 1095–1101.

VALVERDE VAQUERO, P. & FERNÁNDEZ, F. J. 1996. Edad de enfriamiento U–Pb en rutilos del Gneiss de Chimparra (Cabo Ortegal, NO de España). *Geogaceta*, **20**(2), 475–478.

VALVERDE-VAQUERO, P. & DUNNING, G. R. 2000. New U–Pb ages for Early Ordovician magmatism in Central Spain. *Journal of the Geological Society, London*, **157**, 15–26.

VALVERDE-VAQUERO, P., MARCOS, A., FARIAS, P. & GALLASTEGUI, G. 2005. U–Pb dating of Ordovician felsic volcanism in the Schistose Domain of the Galicia-Trás-os-Montes Zone near Cabo Ortegal (NW Spain). *Geologica Acta*, **3**(1), 27–37.

VAN STAAL, C. R., DEWEY, J. F., MAC NIOCAILL, C. & MCKERROW, W. S. 1998. The Cambrian-Silurian tectonic evolution of the Northern Appalachians and British Caledonides: history of a complex, west and southwest Pacific-type segment of Iapetus. *In*: BLUNDELL, D. & SCOTT, A. C. (eds) *Lyell: The Past is the Key to the Present*. Geological Society, London, Special Publications, **143**, 199–242.

VAN STAAL, C. R., CURRIE, K. L., ROWBOTHAN, G., ROGERS, N. & GOODFELLOW, W. 2008. *P–T* paths and exhumation of Late Ordovician-Early Silurian blueschists and associated metamorphic nappes of the Salinic Brunswick subduction complex, northern Appalachians. *Geological Society of America Bulletin*, **120**, 1455–1477; doi: 10.1130/B26324.1.

VEGA-GRANILLO, R., TALAVERA-MENDOZA, O., MEZA-FIGUEROA, D., RUIZ, J., GEHRELS, G. E., LÓPEZ-MARTÍNEZ, M. & DE LA CRUZ-VARGAS, J. C. 2007. Pressure–temperature–time evolution of Paleozoic high-pressure rocks of the Acatlán Complex (southern Mexico): implications for the evolution of the Iapetus and Rheic Oceans. *Geological Society of America, Bulletin*, **119**, 1249–1264.

VOGEL, D. E. 1967. Petrology of an eclogite- and pyrigarnite-bearing polymetamorphic rock complex at Cabo Ortegal, NW Spain. *Leidse Geologische Mededelingen*, **40**, 121–213.

WEBER, B. & KÖHLER, H. 1999. Sm/Nd, Rb/Sr and U–Pb geochronology of a Grenville terrane in southern Mexico: origin and geologic history of the Guichicovi complex. *Precambrian Research*, **96**, 245–262.

WEIL, A. B., VAN DER VOO, R. & VAN DER PLUIJM, B. 2001. New paleomagnetic data from the southern Cantabria-Asturias Arc, northern Spain: implications for true oroclinal rotation and the final amalgamation of Pangea. *Geology*, **29**, 991–994.

WILLIAMS, H. 1979. Appalachian orogen in Canada. *Canadian Journal of Earth Sciences*, **16**, 792–798.

WILSON, J. T. 1966. Did the Atlantic Ocean close and then re-open? *Nature*, **211**, 676–681.

WINCHESTER, J. A. & FLOYD, P. A. 1977. Geochemical discrimination of different magma series and their differentiation products using immobile elements. *Chemical Geology*, **20**, 325–343.

YÁÑEZ, P., RUIZ, J., PATCHETT, P. J., ORTEGA-GUITÉRREZ, F. & GEHRELS, G. 1991. Isotopic studies of the Acatlan complex, southern Mexico: implications for Paleozoic North America tectonics. *Geological Society of America Bulletin*, **103**, 817–828.

YOUNG, T. 1990. Ordovician sedimentary facies and faunas of Southwest Europe: palaeogeographic and tectonic implications. *In*: MCKERROW, W. S. & SCOTESE, C. R. (eds) *Palaeozoic Palaeogeography and Biogeography*. Geological Society, London, Memoirs, **12**, 421–430.

ZECK, H. P., WHITEHOUSE, M. J. & UGIDOS, J. M. 2007. 496 ± 3 Ma zircon ion microprobe age for pre-Hercynian granite, Central Iberian Zone, NE Portugal (earlier claimed 618 ± 9). *Geological Magazine*, **144**, 21–31.

ZELAZNIEWICZ, A., DORR, W. & DUBINSKA, E. 1998. Lower Devonian oceanic crust from U–Pb zircon evidence and Eo-Variscan event in the Sudetes. *In*: FRANKE, W. & HERZBERG, S. (eds) *SPP Orogene Processe 7. Kolloquium, Giessen*. Terra Nostra, **98**(2), 173–176.

# The palaeomagnetically viable, long-lived and all-inclusive Rodinia supercontinent reconstruction

DAVID A. D. EVANS

*Department of Geology & Geophysics, Yale University, New Haven, CT 06520-8109, USA*

*(e-mail: dai.evans@yale.edu)*

**Abstract:** Palaeomagnetic apparent polar wander (APW) paths from the world's cratons at 1300–700 Ma can constrain the palaeogeographic possibilities for a long-lived and all-inclusive Rodinia supercontinent. Laurentia's APW path is the most complete and forms the basis for super-position by other cratons' APW paths to identify possible durations of those cratons' inclusion in Rodinia, and also to generate reconstructions that are constrained both in latitude and longitude relative to Laurentia. Baltica reconstructs adjacent to the SE margin of Greenland, in a standard and geographically 'upright' position, between c. 1050 and 600 Ma. Australia reconstructs adjacent to the pre-Caspian margin of Baltica, geographically 'inverted' such that cratonic portions of Queensland are juxtaposed with that margin via collision at c. 1100 Ma. Arctic North America reconstructs opposite to the CONgo + São Francisco craton at its DAmaride–Lufilian margin (the 'ANACONDA' fit) throughout the interval 1235–755 Ma according to palaeomag-netic poles of those ages from both cratons, and the reconstruction was probably established during the c. 1600–1500 Ma collision. Kalahari lies adjacent to Mawsonland following collision at c. 1200 Ma; the Albany–Fraser orogen continues along-strike to the Sinclair-Kwando-Choma-Kaloma belt of south-central Africa. India, South China and Tarim are in proximity to Western Australia as previously proposed; some of these connections are as old as Palaeoproterozoic whereas others were established at c. 1000 Ma. Siberia contains a succession of mainly sedimentary-derived palaeomagnetic poles with poor age constraints; superposition with the Keweenawan track of the Laurentian APW path produces a position adjacent to western India that could have persisted from Palaeoproterozoic time, along with North China according to its even more poorly dated palaeomagnetic poles. The Amazonia, West Africa and Rio de la Plata cratons are not well constrained by palaeomagnetic data, but they are placed in proximity to western Laurentia. Rift successions of c. 700 Ma in the North American COrdillera and BRAsiliano-Pharuside orogens indicate breakup of these 'COBRA' connections that existed for more than one billion years, following Palaeoproterozoic accretionary assembly. The late Neopro-terozoic transition from Rodinia to Gondwanaland involved rifting events that are recorded on many cratons through the interval c. 800–700 Ma and collisions from c. 650–500 Ma. The pattern of supercontinental transition involved large-scale dextral motion by West Africa and Amazonia, and sinistral motion plus rotation by Kalahari, Australia, India and South China, in a combination of introverted and extroverted styles of motion. The Rodinia model presented here is a marked departure from standard models, which have accommodated recent discordant palaeomagnetic data either by excluding cratons from Rodinia altogether, or by decreasing duration of the supercontinental assembly. I propose that the revised model herein is the only poss-ible long-lived solution to an all-encompassing Rodinia that viably accords with existing palaeomagnetic data.

## Motivation

A full understanding of ancient orogens requires an accurate palaeogeographic framework. By their nature, orogens destroy prior geologic information (via metamorphism, erosion and subduction) and thus challenge our efforts to reconstruct their his-tories. Reconstructing supercontinents is the great-est palaeogeographic challenge of all, combining patchworks of this partially destroyed information from a series of orogens with a complementary record of rifting and passive margin development, and with quantitative kinematic data. The latter record is best constrained by palaeomagnetic information: for Pangaea by seafloor magnetic ano-malies with a supportive record from continent-based palaeomagnetic studies (e.g. Torsvik *et al.* 2008), and for times before Pangaea thus far only by the labourious continent-based method.

Most reconstructions of the early Neoprotero-zoic supercontinent Rodinia (Fig. 1), involving connections between western North America and Australia + Antarctica (Moores 1991; Karlstrom *et al.* 1999; Burrett & Berry 2000; Wingate *et al.* 2002) and eastern North America adjacent to Amazonia (Dalziel 1991; Hoffman 1991; Sadowski & Bettencourt 1996), have in the former case been negated or superseded by subsequent

*From*: MURPHY, J. B., KEPPIE, J. D. & HYNES, A. J. (eds) *Ancient Orogens and Modern Analogues.*
Geological Society, London, Special Publications, **327**, 371–404.
DOI: 10.1144/SP327.16   0305-8719/09/$15.00 © The Geological Society of London 2009.

geochronological and palaeomagnetic results (Pisarevsky *et al.* 2003*a*, *b*), and in the latter instance suffered from minimal palaeomagnetic support (Tohver *et al.* 2002, 2006). Configurations that directly adjoin Laurentia and Siberia (e.g.

Rainbird *et al.* 1998; Sears & Price 2003) are incompatible with recent palaeomagnetic data (reviewed by Pisarevsky & Natapov 2003), as are reconstructions that directly juxtapose Laurentia and Kalahari (Hanson *et al.* 2004). Kalahari and the composite

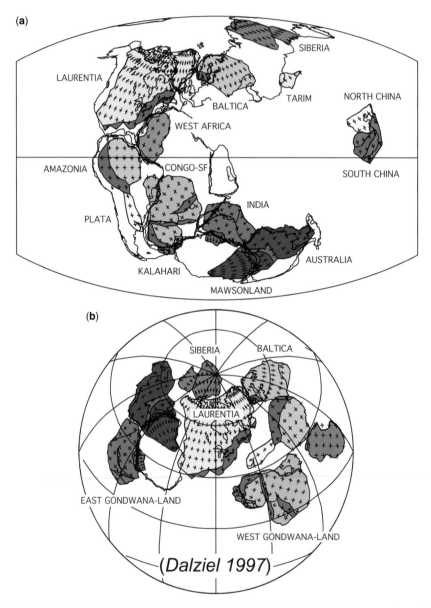

**Fig. 1.** (**a**) Base maps for Rodinian cratons, reconstructed to their relative early Jurassic Pangaea configuration in present South American co-ordinates. Gondwanaland fragments are reconstructed according to McElhinny *et al.* (2003); these and other Euler parameters are listed in Table 1. North and South China, although distinct cratons in Rodinia, are united in this Early Jurassic configuration as indicated by Yang & Besse (2001). Late Mesoproterozoic ('Grenvillian') orogens are shaded grey. Truncated Mercator projection. Panels (**b**–**d**) show previous Rodinia models in the present North American (i.e. Laurentian) reference frame. References are Dalziel (1997), Pisarevsky *et al.* (2003*a*) and Li *et al.* (2008).

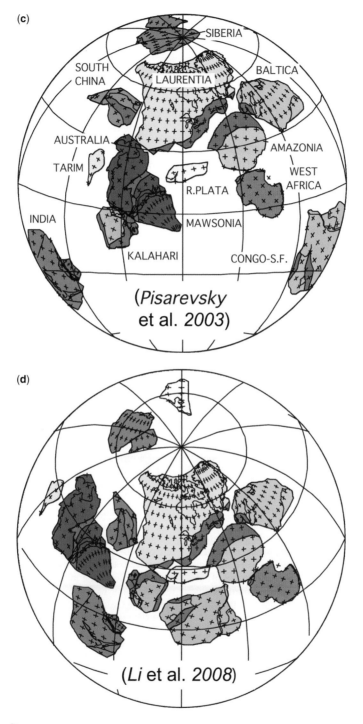

(c)

(*Pisarevsky* et al. *2003*)

(*Li* et al. *2008*)

**Fig. 1.** (*Continued*).

Congo + São Francisco craton have been suggested to be excluded from Rodinia altogether (Kröner & Cordani 2003; Cordani *et al.* 2003).

Four general classes of options exist for how to deal with discrepant palaeomagnetic data: (1) consider potential shortcomings in those data, regarding either palaeomagnetic reliability or age constraints; (2) add loops in the APW path to accommodate the outlying poles; (3) exclude cratons from membership in Rodinia entirely; or (4) restrict the duration of that membership so that it falls between the discrepant palaeomagnetic pole ages. The standard Rodinia model (Hoffman 1991) and its relatively minor variations have been strained to the limits of temporal and spatial constraints, so that some have questioned whether Rodinia even existed at all (Meert & Torsvik 2003); and yet there is still the persistent global tectonostratigraphic evidence for late Mesoproterozoic convergence of cratons, followed by mid-late Neoproterozoic rifting and passive margin development. Given the abundance of focused studies yielding a wealth of new tectonostratigraphic, geochronological and palaeomagnetic data during the last two decades (summarized by Pisarevsky *et al.* 2003*a*; Meert & Torsvik 2003; Li *et al.* 2008), the search for Rodinia may benefit from an entirely fresh perspective. Here I introduce a novel, long-lived Rodinia that is compatible with the most reliable palaeomagnetic data from all of the dozen or so largest cratons during the interval 1300–700 Ma, with minor allowances on the ages of a single set of poles from the São Francisco craton. I show that given these palaeomagnetic data, the new reconstruction is the *only* general model of Rodinia that could have existed for this length of time with all of the largest cratons included in its assembly. The model serves as a palaeomagnetic end-member starting point for further testing and, if desired, relaxation on the assumptions of longevity or inclusion of all the largest cratons. The present analysis is thus most similar to that of Weil *et al.* (1998) in seeking a unified Rodinia model that conforms to the original concept of late Mesoproterozoic assembly and mid-Neoproterozoic dispersal, while incorporating all of the most reliable palaeomagnetic data.

## Methods

Many Rodinia reconstructions have been based primarily on comparisons of the geological records among Meso–Neoproterozoic cratons. For this time interval, we can identify 13 large cratons (Fig. 1), plus many smaller terranes (e.g. Kolyma, Barentsia, Oaxaquia, Yemen). With only a dozen or so large pieces and an abundant well-preserved Meso-Neoproterozoic rock record, the Rodinia

puzzle is tantalizingly solvable. Most of the geological comparisons are purely of regional extent, considering only two or three cratons (examples cited above), generally within the context of Hoffman's (1991) global model. In these geologically-based juxtapositions, palaeomagnetic data have been used in a subsidiary capacity, or ignored altogether, despite the fact that palaeomagnetism is currently the only strictly quantitative method available for reconstructing Rodinia and earlier supercontinents.

Although in a global sense Rodinia assembled in the late Mesoproterozoic and fragmented in the mid-Neoproterozoic (Hoffman 1991; Dalziel 1997; Condie 2002), thereby existing through the interval 1000–800 Ma, there are numerous indications of locally earlier assembly or later breakup. For example, only one side of Laurentia was deformed by orogeny in the late Mesoproterozoic: the Grenvillian (= proto-Appalachian) margin, and this belt did not evolve to a rifted passive margin until after 600 Ma (Cawood *et al.* 2001). Northern Laurentia experienced the *c.* 1600 Ma Forward Orogeny (Maclean & Cook 2004) followed by extensional events with associated large igneous provinces at 1270 Ma (LeCheminant & Heaman 1989) and 720 Ma (Heaman *et al.* 1992). The western margin assembled in the Palaeoproterozoic (Karlstrom *et al.* 2001; Ross 2002) and did not rift until the middle or latest Neoproterozoic (Link *et al.* 1993; Colpron *et al.* 2002). Rifting of Rodinia along these northern and western Laurentian margins, then, split the proto-Laurentian continent through terrains that had been joined for about a billion years. In these instances, if we can find the correct Rodinia juxtapositions, we have also solved part of the configuration of Nuna, which is Rodinia's Palaeoproterozoic supercontinental predecessor (Hoffman 1996). Many other examples of this type exist around the world, essentially wherever a Neoproterozoic rifted margin does not coincide with a 'Grenvillian' orogen (Fig. 1). When we test Rodinia models with palaeomagnetic data, therefore, we must in some cases consider results from rocks as old as *c.* 1800 Ma (e.g. Idnurm & Giddings 1995).

Axisymmetry of the Earth's time-averaged geomagnetic field implies that when individual palaeomagnetic poles from two continents are compared, their relative palaeolongitude remains unconstrained. This shortcoming to palaeomagnetically-based palaeogeographic reconstructions has led to illustrations of Rodinia and older supercontinents that show only a set of latitude-constrained options, further unconstrained by the unknown geomagnetic polarity states of the compared palaeomagnetic data, a degree of freedom for nearly every Precambrian reconstruction (Hanson *et al.* 2004). Among these degrees of freedom in palaeolongitude and hemispheric ambiguity, two or more cratons are juxtaposed in several allowable positions of direct contact for the specific age of pole comparison. If a similar reconstruction emerges from several adjacent time slices, then a long-lived direct connection between the cratons can be considered viable. Examples of this method, called the 'closest approach' technique, are found in Buchan et al. (2000, 2001), Meert & Stuckey (2002) and Pesonen et al. (2003).

A more powerful method of reconstructing ancient supercontinents relies on the coherent motion of all component cratons as part of that supercontinent, for the duration of their conjunction within a single lithospheric plate. Throughout the time interval when constituent cratons are assembled into a supercontinent, and if that assemblage is in motion relative to the Earth's magnetic field reference frame (due to plate tectonics or true polar wander, or both), then all elements of the landmass will share the same palaeomagnetic APW path. After the supercontinent disaggregates, the APW paths diverge (Powell et al. 1993), but their older segments carry a record of the earlier supercontinental motion. As we approach the problem from the present, we see that each craton's APW path contains segments alternating between times of individual plate motion and membership in successive supercontinents. When the cratons are reconstructed to their correct positions in a supercontinent, the APW paths superimpose atop one another (Evans & Pisarevsky 2008). Examples of this type of analysis are found in Weil et al. (1998) and Piper (2000), although both of those studies preceded important new palaeomagnetic data that disallow some of their cratonic juxtapositions. The modified Palaeopangaea reconstruction of Piper (2007) achieves broad-brush palaeomagnetic APW concordance among several cratons, merely as a result of pole averaging (e.g. Siberia), misquoted ages (e.g. Bangemall sills of Australia), or rotation parameters yielding somewhat acceptable pole matches but differing dramatically from the simple cartoon depiction of the reconstruction (e.g. Amazonia, São Francisco + Plata, West Africa and Tanzania + Kalahari) or even producing unacceptable geometric overlaps (northern Australia directly atop Kalahari, and portions of North China directly atop eastern India, in the 'primitive' or pre-1100 Ma reconstruction).

As discussed below, Laurentia has the most complete palaeomagnetic APW path for the interval of c. 1300–750 Ma that is most relevant for testing Rodinia reconstructions. In this paper I use the most reliable palaeomagnetic poles from non-Laurentian cratons to compare with the Laurentian reference APW path and thereby to constrain the possible configurations of a long-lived Rodinia.

A quantitatively viable Rodinia may be found by investigating possible APW superpositions and determining whether the resulting juxtapositions are geologically reasonable for the time intervals under consideration. This method requires equal APW track lengths between coeval poles on any two given cratons; thus it is conceivable that no APW comparisons will be possible between those blocks and that they must have been in relative motion throughout the interval under consideration. Likewise, there is no guarantee that direct juxtapositions of cratons will emerge: some pole comparisons may result in substantial or complete geographic overlap between two or more cratons, which are unallowable, and others may indicate wide separations between blocks, requiring the presence of intervening blocks (or occurrence of rapid true polar wander; see Evans (2003) to legitimize the initial hypothesis of common APW).

Accurate Neoproterozoic craton outlines are important not only for correct geometric fits in Rodinia reconstructions, but they also indicate whether certain palaeomagnetic results from marginal foldbelts apply to a craton or to its allochthonous terranes. Cratonic outlines, drawn in accordance with a broad range of tectonic and stratigraphic studies that are too numerous to cite here, are generally chosen to lie within craton-marginal orogens at the most distal extent of recognizable stratigraphic connections to each adjacent block. Cratons that have split into fragments during the breakup of Pangaea (e.g. Laurentia + Greenland + Rockall, or Kalahari + Falkland + Grunehogna + Ellsworth) must first be reassembled according to seafloor-spreading data combined with geological 'piercing points.' Post-Pangaean fragments are restored to each other according to standard reconstructions (Table 1), with the exception of Kalahari: following restoration of the Falkland Islands (Grunow et al. 1991), Grunehogna is reconstructed to align the Natal and Maud orogenic fronts in the manner suggested by Jacobs & Thomas (2004), and the Ellsworth + Haag province is then rotated to fit into the Natal embayment. The Siberian craton shows restoration of a 25° internal rotation between its northwestern and southeastern (Aldan) portions, associated with development of the Devonian Vilyuy aulacogen, to resolve discrepancies in older palaeomagnetic data (Smethurst et al. 1998; Gallet et al. 2000). Craton boundaries in Antarctica are particularly uncertain, and the present analysis uses conservative estimates of minimal areas attached to each block. Smaller blocks with limited to no palaeomagnetic data, such as Precordillera–Cuyania, Oaxaquia, Barentsia, Azania and various poorly exposed blocks in South America (Dalziel 1997; Collins & Pisarevsky 2005; Fuck et al. 2008), are not described in

**Table 1.** *Pre-Mesozoic reassemblies of Rodinian cratons*

| Craton fragment | Euler rotn. | | | Reference |
|---|---|---|---|---|
| | °N | °E | °CCW | |
| *Rotations to Laurentia* | | | | |
| Greenland | 67.5 | 241.5 | − 13.8 | Roest & Srivastava 1989 |
| Rockall Plateau | 75.3 | 159.6 | − 23.5 | Srivastava & Roest 1989; Royer *et al.* 1992 |
| *Rotations to Kalahari* | | | | |
| Falkland Islands | − 45.3 | 349.2 | 156.3 | Grunow *et al.* 1991 |
| Grunehogna | − 05.3 | 324.5 | 58.6 | After Jacobs & Thomas 2004 |
| Ellsworth-Haag | − 48.9 | 102.8 | 82.8 | Geometric fit |
| *Rotation to Congo* | | | | |
| São Francisco | 46.8 | 329.4 | 55.9 | McElhinny *et al.* 2003 |
| *Rotation to West Africa* | | | | |
| São Luis | 53.0 | 325.0 | 51.0 | McElhinny *et al.* 2003 |
| *Rotations to India* | | | | |
| Enderby Land | − 04.8 | 016.6 | 93.2 | McElhinny *et al.* 2003 |
| Eastern Madagascar | 18.8 | 026.3 | 62.1 | McElhinny *et al.* 2003 |
| Southern Somalia | 28.9 | 040.9 | 64.8 | McElhinny *et al.* 2003 |
| *Rotation to Australia* | | | | |
| Terre Adélie | 01.3 | 037.7 | 30.3 | McElhinny *et al.* 2003 |
| *Reconstruction of Siberia* | | | | |
| NW Siberia to Aldan shield | 60.0 | 115.0 | − 25.0 | Fit pre-Devonian poles from Smethurst *et al.* 1998; Gallet *et al.* 2000 |

detail but are mentioned below where appropriate. Cratons and palaeomagnetic poles are rotated to geometric accuracy via the software created by Cogné (2003). All calculations assume a geocentric axial-dipolar magnetic field, recently verified for the Proterozoic using a compilation of evaporite palaeolatitudes that gave subtropical values as expected (Evans 2006) and a planetary sphere of constant radius.

## Laurentia

Reliable Precambrian palaeomagnetic data are currently so sparse that in only a few instances can we assemble poles into coherent APW paths. In the Rodinia interval, only Laurentia has a well-established path, with ages of *c.* 1270–1000 Ma and tracking younger, with imprecise cooling ages from the Grenville Province (Fig. 2; Weil *et al.* 1998; Pisarevsky *et al.* 2003*a*). The APW path shown in Figure 2 also includes the 1750 Ma grand mean of Irving *et al.* (2004) and representative 'key' poles from *c.* 1450 Ma, as listed by Buchan *et al.* (2000). Although this is not a complete set of data from the 1750–1270 Ma interval,

it adequately represents the general trend with which the most important poles from other cratons may be compared. The sense of vorticity of the Grenville APW loop, at *c.* 1000 Ma, has been debated (Weil *et al.* 1998). The present compilation follows Pisarevsky *et al.* (2003*a*) in selecting the most reliable (Q > 3 in the scheme of Van der Voo 1990) results from late Keweenawan sedimentary rocks, in stratigraphic order, which generates a southward leg of the loop at *c.* 180° longitude, followed by the well-dated Haliburton 'A' pole at 1015 ± 15 Ma (Warnock *et al.* 2000), and then by a northward leg at <180°E longitude. This clockwise sense of the Grenville loop is compatible with the earlier interpretation of Hyodo & Dunlop (1993) but not that of Weil *et al.* (1998). Another set of reliable Laurentian poles is determined for the interval 780–720 Ma, summarized by Buchan *et al.* (2000) and Pisarevsky *et al.* (2003*a*), plus more recent data from stratified successions in western United States (e.g. Weil *et al.* 2004, 2006). Within the intervening 200 Ma gap of no 'key' poles, a recent result from poorly dated but palaeomagnetically stable sedimentary rocks in Svalbard (including a positive soft-sediment slump fold test guaranteeing primary remanence)

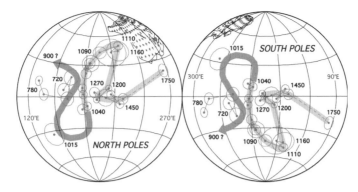

**Fig. 2.** Apparent polar wander path for Laurentia between 1270 and 720 Ma (poles listed in Table 2), which is used as a reference curve for superimposing 'key' poles from other cratons during the Rodinian interval (Buchan *et al.* 2001). Note the large age gap between 1015 and 780 Ma, which is hypothesized here to include a large APW loop at 800 Ma as observed from other cratons (see text and Figs 8 & 10) and from recent high-quality data from Svalbard (Maloof *et al.* 2006).

suggests a large APW loop, hitherto unrecognized for early Neoproterozoic Laurentia (Maloof *et al.* 2006). Regarding these results, it is important to note that this loop is underpinned by data from continuous stratigraphic sections in Svalbard, thus eliminating uncertainties in the reconstruction of Svalbard to Laurentia, or local rotations, as trivial explanations for the divergent pole positions. Global correlations of this new, *c.* 800 Ma, APW loop are discussed in various sections below with the relevant data from other cratons.

Lack of data from the *c.* 1000–800 Ma interval of the Laurentian APW path renders many Rodinian cratonic juxtapositions currently untested; for example, the various reconstructions of Australia + Mawsonland against particular segments of the Laurentian Cordilleran margin (SWEAT, AUSWUS, AUSMEX) all fail palaeomagnetic comparisons at *c.* 750 Ma (Wingate & Giddings 2000) and *c.* 1200 Ma (Pisarevsky *et al.* 2003b), but any one of those reconstructions could be salvaged if it assembled after *c.* 1000 Ma and fragmented by *c.* 800 Ma. In this option for dealing with the discrepant palaeomagnetic data as described above, only Li *et al.* (1995, 2002, 2008) have developed a tectonically reasonable hypothesis by inserting the South China block between Laurentia and Australia. In that model, the Sibao orogen represents the suture between the Australia + Yangtze craton and Cathaysia + Laurentia in a collision at *c.* 900 Ma (Li *et al.* 2008).

## Baltica

The least controversial component of the revised Rodinia reconstruction proposed herein is the placement of Baltica adjacent to eastern Laurentia, as has

been suggested with minor variations throughout the last three decades (Patchett *et al.* 1978; Piper 1980; Bond *et al.* 1984; Gower *et al.* 1990; Hoffman 1991; Dalziel 1997; Weil *et al.* 1998; Hartz & Torsvik 2002; Pisarevsky *et al.* 2003a; Cawood & Pisarevsky 2006). The principal variations among these reconstructions are the latitude of juxtaposition along the Greenland margin and the orientation of Balitca such that various margins (e.g. Caledonide, Timanian, Uralian) are proposed to participate in the direct conjunction with Greenland (Buchan *et al.* 2000; Cawood & Pisarevsky 2006). The reconstruction favoured here is nearly identical to that proposed by Pisarevsky *et al.* (2003a), but with a tighter fit. Pisarevsky *et al.* (2003a) opted for a several hundred-kilometre gap between the present-day margins of SE Greenland and the Norwegian Caledonides, in order to account for palinspastic restoration of Caledonide shortening. These same margins, however, experienced a counteracting amount of extension during Cenozoic initiation of Atlantic Ocean opening as well as Eocambrian opening of Iapetus; the three post-Rodinian alterations to the Greenland and Baltic margins may well have nullified one another, so a tight fit is preferred here (Fig. 3).

Palaeomagnetic poles within the 1100–850 Ma interval from Baltica broadly superimpose onto the Laurentian Grenville APW loop when the two cratons are restored to their proposed Rodinian reconstruction (Fig. 3). The distribution of Baltic poles from this interval has been described in terms of a so-called Sveconorwegian APW loop, with discussions on the ages of individual poles and whether a complete loop is actually circumscribed by the data (Walderhaug *et al.* 1999; Brown & McEnroe 2004; Pisarevsky & Bylund

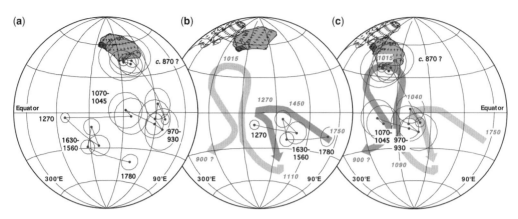

**Fig. 3.** Palaeomagnetic poles from Baltica (grey), compared with the Laurentian reference APW path. As with Figures 4–11, dark-grey shaded cratonic areas are late Mesoproterozoic ('Grenvillian') orogens. (**a**) Poles with their 95% confidence ellipses and ages in Ma, as listed in Table 2, plus Evans & Pisarevsky (2008) for the 1780–1270 Ma data. (**b**) Superposition of the 1780–1270 Ma poles with the pre-Rodinian APW path from Laurentia, rotating Baltica and its poles by (47.5°N, 001.5°E, +49°CCW) after Evans & Pisarevsky (2008). This pre-Rodinia configuration is essentially the same as 'NENA' by Gower *et al.* (1990). (**c**) Rodinian reconstruction (1070–615 Ma) that superimposes the Sveconorwegian APW loop atop the Grenville APW loop of Laurentia (despite recently recognized age mismatches), using the Euler rotation parameters given in Table 3.

2006). Here I tentatively adopt the simple explanation postulated by Pisarevsky & Bylund (2006) that the high-latitude Sveconorwegian poles represent a single, post-900 Ma overprint affecting the southernmost regions of Norway and Sweden, despite lack of independent supporting evidence for such an event (Brown & McEnroe 2004). As a more complex alternative, there might be several oscillatory 'Sveconorwegian' loops in the Baltic APW path, which would be geodynamically explained best by multiple episodes of true polar wander (TPW; Evans 2003). More detailed palaeomagnetic and thermochronological studies of the two cratons from this time interval are needed to resolve these questions.

The proposed reconstruction of Baltica adjacent to SE Greenland at *c.* 1100–600 Ma, like that of Pisarevsky *et al.* (2003*a*), brings the Sveconorwegian orogen in southern Scandinavia close to the Grenville orogen in Labrador with minor right-stepping offset (Gower *et al.* 2008). It also unites the loci of precisely coeval 615 Ma Long Range dykes in Labrador (Kamo *et al.* 1989; Kamo & Gower 1994) and Egersund dykes in southernmost Norway (Bingen *et al.* 1998). Palaeomagnetic results from both of these dyke swarms have yielded scattered results spanning a wide range of inclinations, rendering palaeolatitude comparisons difficult (Murthy *et al.* 1992; Walderhaug *et al.* 2007); however, they are as consistent with the reconstruction introduced here as they are for that of Pisarevsky *et al.* (2003*a*) and Li *et al.* (2008), and this general class of reconstructions is superior

to all other proposed Rodinian juxtapositions of Laurentia and Baltica (Cawood & Pisarevsky 2006).

Palaeomagnetic poles from Baltica and Laurentia during the preceding interval *c.* 1750–1270 Ma are incompatible with the preferred *c.* 1100–600 Ma reconstruction, and suggest instead a modified fit with Baltica's Kola–Timanian margin adjacent to East Greenland (Fig. 3). This fit is essentially the geologically-based Northern Europe + North America (NENA) reconstruction of Gower *et al.* (1990), confirmed palaeomagnetically by Buchan *et al.* (2000) and Evans & Pisarevsky (2008) for the pre-Rodinian interval. Relative to Laurentia, Baltica rotated clockwise *c.* 70° about an Euler pole near Scoresby Sund, some time between 1270 and 1050 Ma, in approximately the same sense as was first proposed by Patchett *et al.* (1978) and Piper (1980). New palaeomagnetic results from the 1122 Ma Salla Dyke in northern Finland are more compatible with a pre-rotation reconstruction than a post-rotation reconstruction, suggesting that the rotation occurred after, or even coincident with, dyke emplacement at *c.* 1120 Ma (Salminen *et al.* 2009). The proposed rotation is consistent with the broad-scale tectonic asymmetry of Baltica (orogeny in west, rifting in east) through the Mesoproterozoic interval (Bogdanova *et al.* 2008). Below it will be shown how this rotation created a broad gulf along the edge of the Rodinia-encircling ocean, Mirovia (McMenamin & McMenamin 1990), which became an isolated sea following further Rodinian amalgamation.

## Australia + Mawsonland

The semi-contiguous Albany–Fraser and Musgrave belts are commonly considered as part of a late Mesoproterozoic suture zone among three constituent Australian cratons (western, northern and southern; Myers et al. 1996), or between a previously united western + northern craton (Li 2000) and the southern, 'Mawson Continent' (Cawood & Korsch 2008). The latter entity extends from the Australian Gawler craton s.s. into Terre Adélie in Antarctica, and possibly as far south as the Transantarctic Mountains near the Miller Range (Goodge et al. 2001; Fitzsimons 2003; Payne et al. 2009). Here the term 'Mawsonland' is formally introduced as a more succinct synonym to 'Mawson Continent.' The Albany–Fraser belt is truncated on its western end by the late Neoproterozoic Pinjarra orogen (Fitzsimons 2003) and on its eastern end by the Palaeozoic Lachlan–Thompson accretionary orogen (Li & Powell 2001), although some local vestiges of Mesoproterozoic orogeny or magmatism are documented in northern Queensland (Blewett & Black 1998) and around Tasmania (Berry et al. 2005; Fioretti et al. 2005). These truncations can provide important piercing points for Rodinia reconstructions, because the Albany–Fraser–Musgrave orogen contains, along its entire length, two episodes of tectonomagmatic activity dated at c. 1320 and c. 1200–1150 Ma (White et al. 1999; Clark et al. 2000). Early consolidation of the Australia + Mawsonland continent allows it to be considered as a single entity in post-1200 Ma Rodinia reconstructions.

Two important ('key') palaeomagnetic poles are available for the c. 1100–750 Ma Rodinian interval: the Bangemall basin sills at 1070 Ma (Wingate et al.

2002), and the Mundine Well Dykes at 755 Ma (Wingate & Giddings 2000). The latter result is supplemented by a pole from oriented borehole core of the Browne Formation (estimated age c. 830–800 Ma), which is the only result among several reported by Pisarevsky et al. (2007) with adequate statistics on the mean direction. Other palaeomagnetic poles from Australia during the Meso–Neoproterozoic interval are problematic, as discussed by Wingate & Evans (2003): they suffer from any combination of poor geochronology, lack of tilt control, and unknown timing of the magnetic remanence acquisition. Similarly, a more recent result from the Alcurra dykes in the Musgrave belt (Schmidt et al. 2006) also suffer from lack of tectonic control, either relative to the palaeohorizontal or in the sense of vertical-axis rotation of the Musgrave region. The principal conclusion of the latter study, that Australia did not assemble until after 1070 Ma, should be treated with caution until further palaeomagnetic studies of Australia's constituent cratons are undertaken.

The great-circle angular distance between the two key poles (32.5°) is identical within error of the angular distance between the two age-correlative interpolated positions on the Laurentian APW path, and this permits the working hypothesis that both cratons could have been part of a single Rodinia plate throughout the intervening time interval. Under this assumption, these two poles can be superimposed on the Laurentian APW path in two options, depending on choice of geomagnetic polarity (Fig. 4). One option points the Albany-Fraser orogen directly into the centre of the northern margin of Laurentia (Fig. 4a), which appears incompatible with the lack of an equivalently aged orogen. Although Hoffman (1991) depicted the Racklan

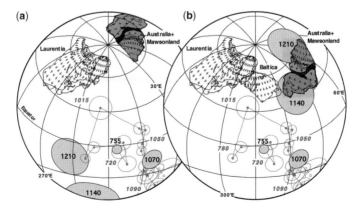

**Fig. 4.** Superposition of 'key' poles from Australia, at 1070 and 755 Ma (Table 2), and the Laurentian APW path (in the North American reference frame) according to one polarity option (**a**) that produces pronounced geologic mismatches between Western Australia and northern Canada, and the alternative, allowable polarity option (**b**) indicating the preferred position adjacent to the Uralian margin of Baltica. Euler parameters for the rotation of Australia and its poles to Laurentia are ($-21.8°$N, $318.4°$E, $+155.3°$CCW) in panel (A), and ($-12.5°$N, $064.0°$E, $+134.5°$CCW) in panel (B).

orogeny in that region as a Mesoproterozoic event, subsequent work indicates a Palaeoproterozoic age for that and related events (Thorkelson *et al.* 2001; Maclean & Cook 2004). There is a poorly understood post-Racklan orogenic event in Yukon (Corn Creek orogeny; Thorkelson *et al.* 2005), but its precise timing and regional extent are unknown. Similarly, although Hoffman (1991) and Dalziel (1997) extended the Grenville orogen northward along the margin of East Greenland, more recent work in that area – plus the once-contiguous eastern Svalbard – dates the 'Grenvillian' tectonomagmatic activity at *c.* 950 Ma (Watt & Thrane 2001; Johansson *et al.* 2005), far younger than the Albany-Fraser belt and negating that potential piercing point. The reconstruction of Australia relative to Laurentia as shown in Figure 4a is also incompatible with the only viable option for the Congo + São Francisco craton, as will be shown below.

The second option for a continuously connected Laurentia and Australia + Mawsonland in 1100–750 Ma Rodinia (Fig. 4b) requires a gap between the two cratons that is neatly filled by Baltica in the reconstruction presented above. In this fit, which juxtaposes the southern Urals and pre-Caspian depression against northern Queensland, poles from earlier ages of 1200–1140 Ma do not superimpose (Fig. 4b), requiring that the postulated connection was not established until *c.* 1100 Ma. Blewett & Black (1998) documented evidence of *c.* 1100 Ma tectonomagmatism in the Cape River province of northern Queensland, which could testify to the inferred collision with Baltica at that time – almost synchronously with Baltica's rotation as described above. Although Proterozoic geology of the pre-Caspian depression is entombed by *c.* 15 km of overlying Phanerozoic sedimentary cover (Volozh *et al.* 2003), the para-autochthonous Bashkirian anticlinorium of the southern Urals exposes the Riphean stratotype succession of the Baltic craton that can be subdivided into three unconformity-bounded successions. Angular unconformity between the Middle and Upper Riphean successions has commonly been attributed to a rift event (Maslov *et al.* 1997) but it could also be the distal expression of collisional tectonism at *c.* 1100 Ma between Baltica and Australia as proposed herein. The Beloretzk terrane, with two stages of deformation bracketing eclogitization, all between 1350 and 970 Ma (Glasmacher *et al.* 2001), could be a sliver of the proposed collision zone.

## Congo + São Francisco

Sharing many tectonic similarities since the Archaean–Palaeoproterozoic, the Congo craton in central Africa and the São Francisco craton in eastern Brazil are almost universally considered to represent a single tectonic entity in Rodinian times (e.g. Brito Neves *et al.* 1999; Alkmim *et al.* 2001). The Congo craton itself is transected by a Mesoproterozoic orogen, the Kibaran belt, which divides the poorly-known western two-thirds of the craton that is largely covered by the Phanerozoic Congo basin (Daly *et al.* 1992) from the relatively well-exposed Tanzania and Bangweulu massifs (and bounding Palaeoproterozoic belts) in the east (De Waele *et al.* 2008). The Kibaran belt has been viewed as either an ensialic orogen (e.g. Klerkx *et al.* 1987) or a subduction-accretionary margin followed by *c.* 1080 Ma continental collision (e.g. Kokonyangi *et al.* 2006). At the southeastern extremity of the craton, the Irumide belt records *c.* 1100–1000 Ma deformation and magmatism (Johnson *et al.* 2005; De Waele *et al.* 2008).

Reliable palaeomagnetic data from the aggregate Congo + São Francisco craton are sparse. In this paper, two poles from Congo are used as the key tie points to the Laurentian master APW curve: the post-Kibaran (e.g. Kabanga–Musongati) layered mafic–ultramafic intrusions (Meert *et al.* 1994) and the Mbozi gabbro (Meert *et al.* 1995). The former pole is constrained by Ar/Ar dating at *c.* 1235 Ma, despite crystallization ages of the complexes as early as *c.* 1400–1350 Ma (Maier *et al.* 2007). The intrusions lie along the boundary between para-autochthonous Tanzania craton to the east, and an orogenic internal zone to the west (Tack *et al.* 1994). Using the Kabanga–Musongati pole to represent the entire Congo craton requires that the subsequent *c.* 1080 Ma deformation was ensialic rather than collisional. Despite the fact that the palaeomagnetically studied Mbozi gabbro is not directly dated, the later-stage syenites in the complex are now constrained by a $748 \pm 6$ Ma zircon U–Pb age (Mbede *et al.* 2004), and this may serve as an approximation of the age of palaeomagnetic remanence. In addition to these poles, the $795 \pm 7$ Ma (Ar/Ar; Deblond *et al.* 2001) Gagwe–Kabuye lavas have yielded a result that appears reliable yet is widely separated from the slightly younger Mbozi pole (Meert *et al.* 1995). Two groups of poles from dykes in Bahia, Brazil (D'Agrella-Filho *et al.* 2004) are also included in the aggregate Congo + São Francisco APW path. These groups of poles, with Ar/Ar ages of *c.* 1080 and 1020 Ma, suggest high-latitude positions for the Congo + São Francisco craton that appeared to negate any direct long-lived Rodinian connections with Laurentia (Weil *et al.* 1998; Pisarevsky *et al.* 2003a; Cordani *et al.* 2003), although collision between the two blocks at *c.* 1000 Ma was considered possible (D'Agrella-Filho *et al.* 1998). As discussed below, these poles are of crucial importance for testing the radical Rodinia revisions proposed in this paper.

Given the 1235 Ma Kabanga–Musongati and c. 750 Ma Mbozi poles superimposed atop the coeval Sudbury dykes and c. 750 Ma poles from Laurentia (Table 2), the two polarity options for this long-lived reconstruction of the two blocks are shown in Figure 5. In the first option (Fig. 5a), there is substantial overlap between the two cratons that cannot be avoided by minor adjustments to the rotations within the uncertainty limits of the poles. This implies that the reconstruction, although palaeomagnetically accurate, is not geologically possible. Other Congo + São Francisco poles are also shown in Figure 5a, to illustrate that they too fall off the Laurentian APW path in the reconstruction of this first polarity option.

The second polarity option for superimposing the 1235 Ma and c. 750 Ma poles between Congo + São Francisco and Laurentia produces the juxtaposition of Arctic North America with CONgo at its DAmaride margin ('ANACONDA'). This reconstruction places the two groups of poles from Bahia dykes (D'Agrella-Filho et al. 2004) atop the c. 1100 Ma Keweenawan poles from Laurentia. This would imply that the c. 1080–1010 Ma Ar/Ar ages from these dykes and baked country rocks (Renne et al. 1990; D'Agrella-Filho et al. 2004) are inaccurately low, a reflection of large scatter in the raw Ar datasets and thus potentially ubiquitous Ar-loss in those rocks. Interestingly, palaeomagnetic polarity reversal asymmetries that are documented in two studies of São Francisco dykes in Bahia, Brazil (D'Agrella-Filho et al. 1990, 2004), precisely superimpose on reversal asymmetries among Keweenawan rocks in Laurentia (Halls & Pesonen 1982) in the ANACONDA reconstruction; this suggests a geomagnetic origin

**Table 2.** *Palaeomagnetic poles used in this study*

| Craton/Rock unit* | Age (Ma)[†] | Unrot. | | A95° | Pole reference |
|---|---|---|---|---|---|
| | | °N | °E | | |
| *Laurentia* | | | | | |
| Franklin–Natkusiak (FN) | c. 720 | 08 | 163 | 4.0 | Buchan et al. 2000 |
| Kwagunt Fm (Kw) | 742 ± 6 | 18 | 166 | 7.0 | Weil et al. 2004 |
| Galeros Fm (G) | >Kw | −02 | 163 | 6.0 | Weil et al. 2004 |
| Tsezotene sills (Ts) | 779 ± 2 | 02 | 138 | 7.0 | Park et al. 1989 |
| Wyoming dykes (Wy) | c. 784 | 13 | 131 | 4.0 | Harlan et al. 1997 |
| Haliburton 'A' (HA) | 1015 ± 15 | −33 | 142 | 6.0 | Warnock et al. 2000 |
| Chequamegon (C) | c. J | −12 | 178 | 5.0 | McCabe & Van der Voo 1983 |
| Jacobsville (J) | <F | −09 | 183 | 4.0 | Roy & Robertson 1978 |
| Freda (F) | <N | 02 | 179 | 4.0 | Henry et al. 1977 |
| Nonesuch (No) | c. 1050? | 08 | 178 | 4.0 | Henry et al. 1977 |
| Lake Shore traps (LS) | 1087 ± 2 | 22 | 181 | 5.0 | Diehl & Haig 1994 |
| Unkar intrusions (Ui) | c. 1090 | 32 | 185 | 8.0 | Weil et al. 2003 |
| Portage Lake (PL) | 1095 ± 3 | 27 | 181 | 2.0 | Halls & Pesonen 1982 |
| Upper Nth Shore (uNS) | 1097 ± 2 | 32 | 184 | 5.0 | Halls & Pesonen 1982 |
| Upper Osler R (uO-r) | 1105 ± 2 | 43 | 195 | 6.0 | Halls 1974 |
| Logan sills (Lo) | 1108 ± 1 | 49 | 220 | 4.0 | Buchan et al. 2000 |
| Abitibi dykes (Ab) | 1141 ± 1 | 43 | 209 | 14.0 | Ernst & Buchan 1993 |
| Upper Bylot (uB) | c. 1200? | 08 | 204 | 3.0 | Fahrig et al. 1981 |
| Sudbury dykes (Sud) | 1235 +7/−3 | −03 | 192 | 3.0 | Palmer et al. 1977 |
| Mackenzie dykes (Mac) | 1267 ± 2 | 04 | 190 | 5.0 | Buchan & Halls 1990 |
| | | | | | |
| *Baltica* | | | | | |
| Hunnedalen dykes (Hun) | c. 850 | 41 | 042 | 10.0 | Walderhaug et al. 1999 |
| Rogaland anorth (Rog) | c. 900? | 42 | 020 | 9.0 | Brown & McEnroe 2004 |
| Gällared Gneiss (GG) | ≪985? | 44 | 044 | 6.0 | Pisarevsky & Bylund 1998 |
| Hakefjorden (Hak) | 916 ± 11 | −05 | 069 | 4.0 | Stearn & Piper 1984 |
| Goteborg–Slussen (Got) | 935 ± 3 | 07 | 062 | 12.0 | Pisarevsky & Bylund 2006 |
| Dalarna dykes (Dal) | 946 ± 1 | −05 | 059 | 15.0 | Pisarevsky & Bylund 2006 |
| Karlshamn–Fajo (Karls) | c. 950 | −13 | 067 | 16.0 | Pisarevsky & Bylund 2006 |
| Laanila Dolerite (Laa) | c. 1045 | 02 | 032 | 15.0 | Mertanen et al. 1996 |
| Bamble intrus. (Bam) | c. 1070 | −03 | 037 | 15.0 | Brown & McEnroe 2004 |

(Continued)

**Table 2.** *Continued*

| Craton/Rock unit* | Age (Ma)[†] | Unrot. °N | °E | A95° | Pole reference |
|---|---|---|---|---|---|
| *Australia* | | | | | |
| Mundine Well (MW) | 755 ± 3 | 45 | 135 | 4.0 | Wingate & Giddings 2000 |
| Browne Fm (Br) | c. 830? | 45 | 142 | 7.0 | Pisarevsky et al. 2007 |
| Bangemall sills (Bang) | 1070 ± 6 | 34 | 095 | 8.0 | Wingate et al. 2002 |
| Lakeview dolerite | c. 1140 | −10 | 131 | 17.5 | Tanaka & Idnurm 1994 |
| Mernda Morn mean | c. 1200 | −48 | 148 | 15.5 | this study[‡] |
| *Congo* | | | | | |
| Mbozi complex (Mb) | ≥748 ± 6 | 46 | 325 | 9.0 | Meert et al. 1995[†] |
| Gagwe lavas (Gag) | 795 ± 7 | −25 | 273 | 10.0 | Meert et al. 1995 |
| Kabanga-Musongati (KM) | 1235 ± 5 | 17 | 293 | 5.0 | Meert et al. 1994 |
| *São Francisco* | | | | | |
| Bahia dykes W-dn (B-w) | see text | 09 | 281 | 7.0 | D'Agrella-Filho et al. 2004 |
| Bahia dykes E-up (B-e) | see text | −12 | 291 | 6.0 | D'Agrella-Filho et al. 2004 |
| *Kalahari* | | | | | |
| Namaqua mean (Nam) | c. 1000 | −09 | 150 | 18.0 | Gose et al. 2006 |
| Kalkpunt Fm (KP) | c. 1090 | 57 | 003 | 7.0 | Briden et al. 1979[†] |
| Umkondo mean (Um) | 1110 ± 3 | 64 | 039 | 3.5 | Gose et al. 2006 |
| *India* | | | | | |
| Malani Rhyolite (MR) | c. 770 | 75 | 071 | 8.5 | Torsvik et al. 2001 |
| Harohalli dykes (Har) | 815/1190 | 25 | 078 | 8.5 | Malone et al. 2008 |
| Majhgawan kimb (Majh) | 1074 ± 14? | 37 | 213 | 15.5 | Gregory et al. 2006 |
| Wajrakarur kimb (Waj) | c. 1100 | 45 | 059 | 11.0 | Miller & Hargraves 1994[†] |
| *South China* | | | | | |
| Liantuo mean (Li) | 748 ± 12 | −04 | 341 | 13.0 | Evans et al. 2000 |
| Xiaofeng dykes (X) | 802 ± 10 | −14 | 271 | 11.0 | Li et al. 2004 |
| *Tarim* | | | | | |
| Beiyixi volcanics (Bei) | 755 ± 15 | −18 | 014 | 4.5 | Huang et al. 2005[†] |
| Aksu dykes (Ak) | 807 ± 12 | −19 | 308 | 6.5 | Chen et al. 2004 |
| *Svalbard* | | | | | |
| Svanbergfjellet (Svan) | c. 790? | −26 | 047 | 6.0 | Maloof et al. 2006 |
| Grusdievbreen u. (Gru2) | c. 800? | 01 | 073 | 6.5 | Maloof et al. 2006 |
| Grusdievbreen l. (Gru1) | c. 805? | −20 | 025 | 11.5 | Maloof et al. 2006 |
| *North China* | | | | | |
| Nanfen Fm (Nan) | c. 790? | −17 | 121 | 11.0 | Zhang et al. 2006 |
| Cuishuang Fm (Cui) | c. 950? | −41 | 045 | 11.5 | Zhang et al. 2006 |
| *Siberia* | | | | | |
| Kandyk Fm (Kan) | c. 990 | −03 | 177 | 4.0 | Pavlov et al. 2002 |
| Milkon Fm (Mil) | 1025? | −06 | 196 | 4.0 | Pisarevsky & Natapov 2003 |
| Malgina Fm (Mal) | 1040? | −25 | 231 | 3.0 | Gallet et al. 2000 |
| *Amazonia* | | | | | |
| Aguapei sills (Agua) | c. 980 | −64 | 279 | 8.5 | D'Agrella-Filho et al. 2003[†] |
| Fortuna Fm (FF) | 1150? | −60 | 336 | 9.5 | D'Agrella-Filho et al. 2008 |
| Nova Floresta sills (NF) | c. 1200 | −25 | 345 | 5.5 | Tohver et al. 2002 |

*Abbreviations for rock units correspond to the poles depicted in Figures 2–10.
[†]Ages are queried where highly uncertain or estimated in part by position on the APW path. Ages are cited fully in Buchan et al. (2000) or Pisarevsky et al. (2003a), or the pole references, except where otherwise noted: Zig-Zag Dal – Midsommersø from Upton et al. (2005); Western Channel diabase from Hamilton & Buchan (2007); Mbozi complex from Mbede et al. (2004); Kunene anorthosite from Mayer et al. (2004) and Drüppel et al. (2007); Harohalli dykes from Malone et al. (2008); Wajrakarur kimberlites from Kumar et al. (2007); Beiyixi volcanics from Xu et al. (2005); Kalkpunt Formation estimated from Pettersson et al. (2007).
[‡]Combined calculation of Fraser dyke VGP (Pisarevsky et al. 2003b) with Ravensthorpe dykes of Giddings (1976) and one additional dyke at Narrogin (A. V. Smirnov & D. A. D. Evans, unpublished data).

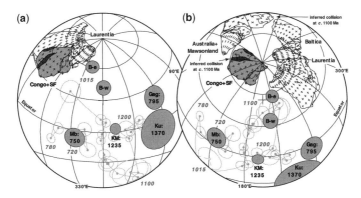

**Fig. 5.** Non-'key' palaeomagnetic poles from Congo + São Francisco during the interval *c*. 1370–750 Ma (Table 2). The 1235 and 755 Ma poles from Congo are used to generate two possible superpositions onto the Laurentian APW path (in the North American reference frame). One polarity option (**a**) generates a complete overlap between the two cratons and is thus not allowed, whereas the other (**b**) produces the 'ANACONDA' juxtaposition described in the text. Note that the Bahia dykes poles fall on the *c*. 1100 Ma segment of the Laurentian path, and a testable prediction of ANACONDA is that the existing Ar/Ar ages of 1080–1010 Ma from these dikes are too young (see text). Euler parameters for the rotation of Congo + São Francisco and its poles from the present African reference frame to Laurentia are (20.0°N, 334.9°E, +162.5°CCW) in panel (A), and (−24.0°N, 249.0°E, +128.5°CCW) in panel (B).

for the asymmetries (Pesonen & Nevanlinna 1981). A joint palaeomagnetic and U–Pb restudy of these dykes is currently underway (Catelani *et al.* 2007). Other poles shown in Figure 5b are from the Kunene anorthosite (Piper 1974), which with new precise U–Pb ages on consanguineous felsic rocks of 1380–1370 Ma (Mayer *et al.* 2004; Drüppel *et al.* 2007) deserves refinement with modern palaeomagnetic techniques; and the Gagwe lavas (Meert *et al.* 1995) with an updated Ar/Ar age of 795 ± 7 Ma (Deblond *et al.* 2001). If the Gagwe pole is primary, then the ANACONDA fit requires a large APW loop shared among all elements of Rodinia at *c*. 800 Ma. Such a loop has not traditionally been accepted for Laurentia, as some anomalous results of that age have been interpreted instead as suffering from local vertical-axis rotation (e.g. Little Dal lavas; Park & Jefferson 1991). However, a large APW loop in the Laurentian path at *c*. 800 Ma has now been demonstrated by high-quality palaeomagnetic results from Svalbard (Maloof *et al.* 2006; see above) and is generally supported by data from several other cratons that imply large APW shifts between *c*. 800 and *c*. 750 Ma (Li *et al.* 2004).

The ANACONDA reconstruction juxtaposes several intriguingly similar geological features among the São Francisco, Congo and northern Laurentian cratons. Extensive Palaeoproterozoic–Mesoproterozoic orogeny in the southern Angola–Congo craton (Seth *et al.* 2003) adjoins crust with a similarly aged interval of deformation in arctic Canada (Thorkelson *et al.* 2001; MacLean & Cook 2004) and suggests initial amalgamation of ANACONDA at *c*. 1.6–1.5 Ga. Post-collisional

igneous activity at *c*. 1.38 Ga is recorded by the Kunene complex and coeval Hart River magmatism in Canada (Thorkelson *et al.* 2001) and Midsommersø dolerite and cogenetic Zig-Zag Dal basalt (Upton *et al.* 2005). Along the southeastern margin of Congo, potassic magmatism at 1360–1330 Ma (Vrana *et al.* 2004) correlates broadly in age with the *c*. 1350 Ma Mashak volcanics in the southern Urals (Maslov *et al.* 1997).

The ANACONDA reconstruction also identifies potential long-sought counterparts to the giant 1.27 Ga Mackenzie/Muskox/Coppermine large igneous province in Canada, with correlatives in Greenland and Baltica (Ernst & Buchan 2002). In eastern Africa, the late-Kibaran intrusive complexes described above were once thought to be of the same age (Tack *et al.* 1994), but are now known to be either older (Tohver *et al.* 2006) or younger (De Waele *et al.* 2008). Nonetheless, a dyke of similar tectonic setting in Burundi is dated by Ar/Ar at *c*. 1280–1250 Ma (Deblond *et al.* 2001), and Tohver *et al.* (2006) raise the possibility that the *c*. 1380 Ma zircons in the layered intrusions are xenocrysts. Thus, more comprehensive geochronology of this region is warranted. In Brazil, the Niquelândia and related mafic-ultramafic complexes have numerous age constraints, the most recent study suggests emplacement ages at 1250 ± 20 Ma (Pimentel *et al.* 2004). The latter intrusions lie within the late Neoproterozoic Brasilia belt, adjacent to the São Francisco craton, and in the present model are considered not grossly allochthonous relative to that craton (Pimentel *et al.* 2006). Concordance of the Laurentian and Congo + São Francisco APW paths younger than this age indicates that extension at 1270–1250 Ma failed to separate the cratons, rather than opening a postulated 'Poseidon' ocean (Jackson & Iannelli 1981).

These magmatic loci could represent early stages of the rifting that is required by palaeomagnetic data to have rotated Baltica away from Congo and toward southern Greenland in the late Mesoproterozoic, as discussed above. Baltica's rotation, coupled with arrival of Australia at c. 1100 Ma as discussed above, isolated a craton-sized tract of remnant ocean at the end of the Mesoproterozoic. This 'hole', in the present revised Rodinia model, is intriguing for several reasons. First, it predicts a Mediterranean-style slab rollback to account for arc magmatism and arc-continent collision in the Irumide belt at c. 1050–1020 Ma (De Waele et al. 2008) as well as c. 1000–800 Ma tectonic events in Greenland (Watt & Thrane 2001) and northern Norway (Kirkland et al. 2006). Second, the large c. 800 Ma evaporite basin hosting the Shaba–Katanga copperbelt in southern Congo (Jackson et al. 2003) may have continued onto Laurentia as the evaporitic upper part of the Amundsen basin and its correlative units in the Mackenzie Mountains (Rainbird et al. 1996). This composite evaporitic basin could represent a lithospheric sag precursor to rifting and separation of ANACONDA – accompanied by the Chuos, Grand Conglomerat and Rapitan glaciogenic deposits (Evans 2000) – between c. 750 and 700 Ma. Finally, it demonstrates how the palaeomagnetic APW-matching method can generate a more refined palaeogeographic framework for supercontinent reconstructions; all previous models of Rodinia, using tectonostratigraphic comparisons or 'closest-approach' palaeomagnetic reconstructions, have placed the cratons together as tightly as possible – essentially ruling out even the possibility of Mediterranean-style remnant-ocean tectonism in the pre-Pangaean world.

As a final note, recall that the palaeomagnetic data from Australia were discordant in the present revised Rodinia model at c. 1140 Ma, requiring collision of Australia + Mawsonland to become a part of Rodinia at c. 1100 Ma. Proximity of Mawsonland to Tanzania in the Rodinia fit (Fig. 5b) implies convergence and inferred collision there as well. One difficulty with this inference is the lack of any direct evidence for c. 1100 Ma tectonism in the central Transantarctic Mountains (Goodge et al. 2001), despite some tenuous Nd-isotopic support for Mesoproterozoic activity there (Borg & DePaolo 1994). However, if Kokonyangi et al. (2006) are correct in proposing a c. 1080 Ma suture between Tanzania + Bangweulu and Congo cratons at the Kibaran orogen, then there is the intriguing possibility presented by this reconstruction, that Tanzania + Bangweulu was originally a fragment of Australia + Mawsonland,

becoming 'orphaned' during mid-Neoproterozoic Rodinia breakup.

The juvenile Hf and Nd signatures of 1.4 Ga A-type granites preserved as clasts and detrital zircons in Transantarctic Mountains sediments (Goodge et al. 2008) have been used to support a connection with western Laurentia in the SWEAT juxtaposition. However, Goodge et al. (2008; their fig. 3a) illustrate other regions of the world with comparable magmatism of the same age: Cathaysia, eastern Congo, southern Amazonia and southwestern Baltica. If the revised Rodinia position for Australia + Mawsonland (Fig. 4b) is correct, then the general proximity of 1.4 Ga A–type granite terrains in Congo and Baltica make them the most attractive candidates as the originally contiguous extensions of the Antarctic magmatic province in pre-Rodinian times.

## Kalahari

As noted in a recent review of late Mesoproterozoic palaeomagnetic data from southern Africa (Gose et al. 2006), the two anchors of the Kalahari APW path are the c. 1100 Ma Umkondo grand-mean pole of highest reliability, followed by various moderately well grouped poles from c. 1000 Ma terranes within the Namaqua–Natal orogen. These two anchor poles permit Kalahari to have been a member of Rodinia throughout the intervening time; they are separated by about 80° or 100° of great-circle arc, depending on relative geomagnetic polarity, the larger value being similar to that separating Laurentian poles on the 'master' Rodinian APW path of the same pair of ages (Logan and Grenville loops, respectively). The precise dating of both Umkondo and early Keweenawan poles at 1108 Ma, with a pronounced geomagnetic polarity bias in both units, constrains the hemisphere option of relative reconstructions between the two cratons, such that the Grenville and Namaqua orogens cannot face each other (Hanson et al. 2004; Gose et al. 2006) as earlier proposed. In addition to this polarity constraint, if the c. 1000 Ma poles are aligned then Kalahari can occupy only one of two positions in relative palaeolatitude and palaeolongitude to Laurentia plus its surrounding cratons in Rodinia. The standard depiction of Kalahari's geon-10 APW path (e.g. Powell et al. 2001; Evans et al. 2002; Meert & Torsvik 2003; Tohver et al. 2006; Gose et al. 2006; Li et al. 2008) is shown in Figure 6a in simplified form, using the two anchor poles plus that of the Kalkpunt redbeds of eastern Namaqualand (Briden et al. 1979). Although Powell et al. (2001) suggested an age of 1065 Ma for the Kalkpunt red beds, recent U–Pb dating of a

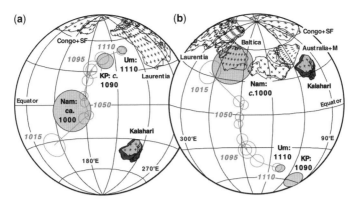

**Fig. 6.** Superposition of Kalahari's 1110 Ma Umkondo palaeomagnetic pole with polarity constraint tied to the early Keweenawan Logan sills pole from Laurentia (Hanson *et al.* 2004; Gose *et al.* 2006), plus younger poles from the Namaqua and Natal Provinces in various polarity options (poles listed in Table 2). Panel (**a**) shows a large distance between Kalahari and the rest of Rodinia, whereas panel (**b**) illustrates the preferred reconstruction herein. Euler parameters for the rotation of Kalahari and its poles from the present African reference frame to Laurentia, in present North American coordinates, are (81.5°N, 187.4°E, −163.4°CCW) in panel (A) and (32.5°N, 307.5°E, +76.5°CCW) in panel (B).

conformably underlying rhyolite indicates an age of only slightly younger than *c.* 1095 Ma (Pettersson *et al.* 2007). The APW arc length transcribed by this polarity option of the three poles is somewhat shorter than that of the Keweenawan APW track of Laurentia, but dating of the Namaqua–Natal poles is forgiving enough to allow for the precise mismatch. However, the reconstruction derived by this APW superposition is one in which Kalahari is widely separated from all other cratons in Rodinia – even if one were to choose more standard models involving Australia and other cratons to the SW of Laurentia. The Kalahari reconstruction shown in Figure 6a is, however, similar to that of Piper (2000, 2007), but as noted above this reconstruction produces numerous palaeomagnetic mismatches when high-quality individual results are considered rather than broad means.

The alternative polarity option for the Namaqua-Natal poles, while preserving the sanctity of geomagnetic polarity matching of Umkondo and early Keweenawan poles, produces the reconstruction shown in Figure 6b. The reconstruction juxtaposes Kalahari's northern margin adjacent to the Vostok (western) margin of Mawsonland (Fig. 6). This polarity option for Namaqua–Natal poles provides a more acceptable APW arc length relative to the Keweenawan APW track and Grenville APW loop from Laurentia, but it introduces a different problem: the Kalkpunt pole now falls on the other side of Umkondo, and reconstructs to a position slightly beyond the apex of the Logan APW loop in the Laurentian reference frame (Fig. 6b). Rather than invoke an additional APW loop, this discrepancy is perhaps best explained by local

vertical-axis rotation of the Koras Group, which is only locally exposed within a region of large strike-slip shear zones (Pettersson *et al.* 2007). About 30 degrees of local rotation are required to bring the Kalkpunt pole into alignment with poles from the younger Keweenawan lavas and intrusions at *c.* 1095 Ma.

The preferred reconstruction of Kalahari shown in Figure 6b is thus the only possible way to include this craton in the Rodinia assembly as early as 1100 Ma according to existing palaeomagnetic constraints. Other published solutions involve late collision of Kalahari into Rodinia at *c.* 1000 Ma (Pisarevsky *et al.* 2003a; Li *et al.* 2008). The well-known Namaqua-Natal belt of southern Kalahari shows the main phases of deformation at *c.* 1090–1060 Ma (Jacobs *et al.* 2008), which is typically correlated in an opposing collisional sense to the Ottawan orogeny of the Grenville Province in Laurentia. These reconstructions, however, must either violate the geomagnetic polarity match between early Keweenawan and Umkondo poles, or invoke an implausible 180-degree rotation of Kalahari relative to Laurentia as they approached each other in geon 10.

In the preferred reconstruction here (Fig. 6b), the more proximal Mesoproterozoic margin to the Laurentian side of Rodinia is the present northwestern side of Kalahari. Along that margin, a single *c.* 1300–1200 Ma orogen has been hypothesized (Singletary *et al.* 2003), and this orogen was stabilized prior to widespread large igneous province mafic magmatism at about 1110 Ma. In the revised Rodinia reconstruction presented herein, the NW Kalahari orogen is proposed to record collision between Kalahari and the Vostok margin of Australia + Mawsonland (Fig. 6). A complex collisional triple junction, suturing this orogen, the Namaqua belt and the Albany belt in Western Australia, would be partly reworked by subsequent Pan-African (Damaride) and Pinjarran tectonics, and partly buried by Antarctic ice; testing this model by correlating the details of the three collisions will be a challenging enterprise.

## India, South China, Tarim

Relative to Australia, the reconstructed positions of India, South China and Tarim are similar in this Rodinia model to previously published versions (Fig. 7). The reconstruction of India to Australia essentially follows Torsvik *et al.* (2001), that of South China relative to India follows Jiang *et al.* (2003) and that of Tarim to Australia follows Li *et al.* (1996), Powell & Pisarevsky (2002) and Li *et al.* (2008). Palaeomagnetic poles from these cratons are sparse, but they provide important constraints on Rodinia configurations of these blocks and also support the existence of a large loop in the Rodinian APW path at *c.* 800 Ma. Discussing the results in detail, numerous Indian poles (reviewed by Malone *et al.* 2008) are summarized by the three depicted in Figure 7. The most reliable is from the Malani rhyolitic large igneous province of Rajasthan, with various U–Pb ages centred around 770 Ma (Torsvik *et al.* 2001; Malone *et al.* 2008). Aside from this result, which has been reproduced numerous times in the past few decades, the Indian palaeomagnetic poles from Rodinia times are questionable either in quality or in age. The oft-cited pole from Harohalli alkaline dykes was previously assigned an age of *c.* 815 Ma, based on Ar/Ar ages (Radhakrishna & Mathew 1996). However, new U–Pb zircon data from these dykes suggest a much older age of *c.* 1190 Ma (cited in Malone *et al.* 2008). If the latter age is correct, then the Harohalli dykes pole is irrelevant for the present discussion, because this portion of Rodinia is proposed herein to have assembled around 1100 Ma or younger. There is also the pole from the well-dated Majhgawan kimberlite (1074 ± 14 Ma by Ar/Ar; Gregory *et al.* 2006), which is nearly identical to poles from the nearby Rewa and Bhander sedimentary succession (Malone *et al.* 2008). The latter units have age uncertainties on the order of 500 Ma, as summarized by Malone *et al.* (2008). No field stability tests have been performed on either the Majhgawan kimberlite or the Rewa/Bhander units, so there remains the possibility that these poles represent a two-polarity magnetic overprint across north-central India. Such an

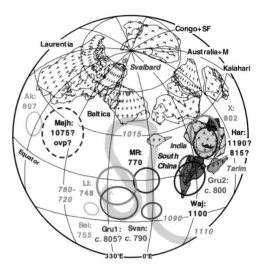

**Fig. 7.** Reconstruction of India (blue), South China (green), and Tarim (peach) near Australia as commonly depicted in Rodinia models. As with Figures 3–6, all cratons and poles have been rotated into the present North American (Laurentian) reference frame (Table 3). The Laurentian APW path is simplified into a curve (light grey) for clarity. Also shown are three poles (red colour) with poor absolute age control but correct stratigraphic order from Svalbard (Maloof *et al.* 2006; Table 2), rotated to a modified position north of Greenland (Table 3) for better APW matching with both the Laurentian poles from geon 7 and the proposed APW loop at *c.* 800 Ma. Palaeomagnetic poles (abbreviated as in Table 2) are shown in the colours of their host cratons. The location of Tarim poles is consistent with the craton reconstruction in darker colour; the lighter-coloured alternative Tarim reconstruction aligns the Aksu dykes pole (Ak) with *c.* 800 Ma poles from other cratons, but results in a mismatch of the Beiyixi volcanics pole (Bei); poles from this alternative reconstruction are not illustrated, for sake of clarity. Queried ages of some of the Indian poles are discussed in text.

interpretation could readily explain the large discordance between the Majhgawan virtual geomagnetic pole (VGP; not averaging geomagnetic secular variation due to brief emplacement of the kimberlite) and those from the nearly coeval Wajrakarur kimberlite field in south-central India (Miller & Hargraves 1994).

If either of these kimberlite poles is primary, then their significant distances from the Laurentian APW path would negate the proposed reconstruction (Fig. 7) at that time. It is possible that collision between India (plus attached Cathaysia block of South China) and NW Australia occurred after 1100–1075 Ma, accounting for the reconstructed pole discrepancy.

Palaeomagnetic poles from China are more straightforward to interpret. In South China (Yangtze craton), the Xiaofeng dykes yield a high palaeolatitude at $802 \pm 10$ Ma (Li *et al.* 2004), and the Liantuo Formation red beds yield a moderate palaeolatitude at $748 \pm 12$ Ma (Evans *et al.* 2000). Similarly, the Aksu dykes in Tarim were emplaced at high palaeolatitude at $807 \pm 12$ Ma (Chen *et al.* 2004), and the Beiyixi volcanics were erupted at lower palaeolatitudes (Huang *et al.* 2005) at $755 \pm 15$ Ma (Xu *et al.* 2005). Matching these two pairs of poles from South China and Tarim, however, results in a large distance between the cratons (not shown in Fig. 7), inconsistent with their strongly compatible Sinian geological histories (Lu *et al.* 2008*a*). Figure 7 shows two alternative positions of Tarim relative to the cratons heretofore discussed. The preferred position is shown in a darker colour, along with the properly rotated pair of Tarim poles. In this position, where Tarim is directly adjacent to both South China and (present NW) India, the 755 Ma Beiyixi pole is aligned with middle-geon-7 poles from Laurentia and other cratons; however, the 807 Ma Aksu dykes pole is discordant (Fig. 7). This could suggest post-800 Ma convergence between Tarim and Rodinia, or it could also be due to unrecognized local vertical-axis rotations of the Aksu area, as suggested by Li *et al.* (2008) to be a general problem for the minimally studied Tarim block.

Alternatively, the Aksu dykes pole could be aligned with the *c.* 800 Ma pole from the coeval Xiaofeng dykes in South China; in which case Tarim reconstructs next to northern Australia (lighter shade of peach colour in Fig. 7) in the same sense as Li *et al.* (1996, 2008). The 755 Ma Beiyixi pole, however, is removed from the Rodinian APW path in this reconstruction. This would suggest either early (pre-755 Ma) rifting of Tarim from Rodinia, or local vertical-axis rotations of the Quruqtagh region where the Beiyixi volcanics are exposed. A third alternative reconstruction of Tarim – adjacent to eastern Australia, based on a proposed radiating dyke swarm at *c.* 820 Ma (Lu *et al.* 2008*a*) – is broadly compatible with the palaeomagnetic data from 755 Ma but, ironically, not *c.* 800 Ma.

The present analysis leaves the position of Tarim somewhat uncertain, but the preferred position is that described first, above, and illustrated with darker peach colour in Figure 7. The main reason for this preference is that new palaeomagnetic results from Cambrian–Ordovician sedimentary rocks in the Quruqtagh area (Zhao *et al.* 2008) are most compatible with the Gondwanaland APW path if Tarim is reconstructed near Arabia, that is separated from Australia by India and South China. If either the northern or eastern Australian

juxtapositions is correct for Tarim in Rodinia, then Tarim would need to rift from that position and re-collide with East Gondwanaland in its peri-Arabian position prior to mid-Cambrian time. Neither Tarim nor northern India records Ediacaran-age orogenic activity that would document such convergence.

Although the *c.* 800 Ma poles just described are far removed from the established Laurentian APW path in the proposed reconstruction, they constitute important independent support from several Rodinian cratons that they – if not the entire supercontinent – experienced an oscillatory pair of rotations at that time. The kinematic evidence for this proposed rotation does not specify a dynamic cause, but inertial-interchange true polar wander (IITPW) events are the most straightforward explanation (Li *et al.* 2004; Maloof *et al.* 2006). When the Svalbard magnetostratigraphic data of Maloof *et al.* (2006) are considered (red colour in Fig. 7), they provide the hitherto unrecognized evidence from Laurentia for the APW loop indicated by India (if Harohalli dykes are *c.* 800 Ma), South China, Tarim and Congo (Fig. 5b). The precise reconstruction of Svalbard relative to Greenland is uncertain, but direct connection between the two areas of Laurentia are strongly supported by lithostratigraphy (Maloof *et al.* 2006). Also, because the Svalbard APW shift is recorded in several widely separated, continuously sampled magnetostratigraphic sections, local vertical-axis rotations cannot account for the directional shifts: the APW loop at *c.* 800 Ma is a genuine feature of the Laurentian palaeomagnetic database that must be included in all Rodinia models.

The reconstruction of India, South China and Tarim, adjacent to Australia as shown in Figure 7, produces some intriguing tectonic juxtapositions, in which compatible histories can be considered as predictions of the model. First, the Sibao orogen in South China (Li *et al.* 1996, 2002) appears to strike directly into northwestern India, where earliest Neoproterozoic tectonomagmatic activity is postulated to be continuous with the Delhi foldbelt in India (Deb *et al.* 2001), under Neoproterozoic sedimentary cover of Rajasthan and north–central Pakistan. If this represents a collisional orogen, then most of cratonic India should have more affinities with the Cathaysia block in South China, colliding with the Yangtze + Tarim craton during final Rodinia assembly. The Tarimian orogeny of similar age (Lu *et al.* 2008*a*) could express a poorly exposed continuation of this collisional belt.

On the other side of India, the *c.* 1000–950 Ma Eastern Ghats orogen (Mezger & Cosca 1999) and its continuation as the Rayner terrane in Antarctica (Kelly *et al.* 2002), extends east of Prydz Bay (Kinny *et al.* 1993; Wang *et al.* 2008), and according to this reconstruction splays into the Edmund foldbelt of Western Australia, which deformed 1070 Ma sills and their host Bangemall basin sedimentary rocks about tight NW–SE axes and led to moderate isotopic disturbance (Occhipinti & Reddy 2009). The full extent of this orogen is probably hidden under the East Antarctic icecap (including the Gamburtsev Subglacial Mountains; Veevers & Saeed 2008), and likely involves smaller Archaean–Palaeoproterozoic cratonic fragments such as the Ruker terrane (Phillips et al. 2006). The orogen is proposed here to involve collision with Kalahari along the latter craton's Namaqua margin at c. 1090–1060 Ma (Jacobs et al. 2008). Tectonothermal events of similar age in the Central Indian Tectonic Zone (Chatterjee et al. 2008; Maji et al. 2008) connect the Delhi and Eastern Ghats/Rayner orogens in poorly understood ways.

The reconstruction also suggests that the precisely coeval igneous events recorded on several cratons at 755 Ma are genetically related: Mundine Well dyke swarm in Australia (Wingate & Giddings 2000), Malani large igneous province in India (Torsvik et al. 2001), Nanhua rift and related provinces in South China (Li et al. 2003) and Beiyixi volcanics in Tarim (Xu et al. 2005). As will be shown below, western Siberia also reconstructs immediately adjacent to Tarim, and the Sharyzhalgai massif contains mafic dykes of precisely the same age (Sklyarov et al. 2003). The Malani region in India is proposed here as the central focus of a hotspot or mantle plume with radiating arms extending across these cratons.

## North China, Siberia

Both North China and Siberia lack coherent late Mesoproterozoic ('Grenvillian') orogens, although possible vestiges can be found along the northern and southern margins of North China (Zhai et al. 2003; but see also discussion in Lu et al. 2008b). Similar carbonate-dominated mid-Proterozoic platform or passive-margin sedimentary successions on both cratons have inspired, along with palaeomagnetic support, hypotheses of a close palaeogeographic connection between the two blocks in Rodinia and in earlier times (Zhai et al. 2003; Wu et al. 2005; Zhang et al. 2006; Li et al. 2008). A recent U–Pb age determination of c. 1380 Ma on ash beds in the upper part of the North China cover succession (Su et al. 2008), however, shows that this succession is almost entirely older than the bulk of 'Riphean' sediments across Siberia (Khudoley et al. 2007). The older age of the North China succession also demonstrates that there are only two moderately reliable palaeomagnetic poles

(Zhang et al. 2006) from the Rodinia interval: the Cuishang Formation (c. 950 Ma?) and the Nanfen Formation (c. 790 Ma?). Ages from both of these formations are very tenuous.

Pisarevsky and Natapov (2003) summarized the Meso-Neoproterozoic stratigraphic record across the Siberian craton, as well as its palaeomagnetic database. The most reliable palaeomagnetic poles define an APW trend that is supported by less reliable results; only the three most reliable are included in this synthesis, but the conclusion is not affected by incorporating the others. The present analysis does not include the high-quality Linok Formation pole from the Turukhansk region (Gallet et al. 2000), because it restores precisely atop that of the likely correlative Malgina Formation in the Uchur-Maya region marginal to the Aldan shield, after restoration of the Devonian Vilyuy rift in central Siberia (Table 1).

Matching of the Siberian APW path from the Uchur-Maya region with the Keweenawan APW track to Grenville loop from Laurentia results in two possibilities, because of geomagnetic polarity options. The first option (not shown) produces the typical reconstruction of Siberia with its southern margin in the vicinity of northern Laurentia (option 'A' of Pisarevsky & Natapov 2003; Pisarevsky et al. 2003a; Li et al. 2008; Pisarevsky et al. 2008), The hypothesized reconstruction of Siberia (Fig. 8) is essentially the same as 'option B' of Pisarevsky & Natapov (2003) and the first option discussed by Meert & Torsvik (2003). Both of those papers concluded that such a reconstruction would probably exclude Siberia because of the great distance from Laurentia, but the present revised Rodinia model covers this gap with Baltica, Australia, India, and North China.

Because there is scant to no evidence for a 'Grenvillian' orogen between India and North China in the proposed reconstruction (Fig. 8), a corollary of the model is that those two cratons were joined in similar fashion since their Palaeoproterozoic consolidations. Zhao et al. (2003) described a series of correlations between southern India and eastern North China, and proposed four possible reconstructions in which those two regions could have been directly juxtaposed. In this paper I present a fifth alternative connection (Fig. 8), which, unlike the previous four, is in agreement with the Rodinia-era palaeomagnetic poles described above. There are no reliable and precisely coeval palaeomagnetic results from the two cratons yet available (Evans & Pisarevsky 2008) to test their earlier hypothesised assembly.

Siberia is also proposed to have been connected to India and North China prior to Rodinia's amalgamation in the late Mesoproterozoic. Pisarevsky & Natapov (2003) summarized the Riphean

**Fig. 8.** Reconstruction of North China (tan) and Siberia (pink) directly adjacent to India on the outer edge of Rodinia. All cratons and their similarly coloured poles have been rotated into the present North American (Laurentian) reference frame (Table 3). The Laurentian APW path is simplified into a curve (light grey), with the additional APW loop at *c.* 800 Ma (see text and Fig. 7) indicated by the dashed segment. Among the North China and Siberia poles shown here, all but the Kandyk Formation (Kan) are poorly constrained in age.

stratigraphic architecture of the present-day margins of Siberia, demonstrating in many areas a clear thickening of strata away from the craton into deeper-water sedimentary facies. The long-lived Mesoproterozoic connections to North China and India (Fig. 8) would be inconsistent with the Siberian stratigraphic record if it could be demonstrated that the Turukhansk, Igarka, or northern Siberian margins faced the open ocean through the Meso-Neoproterozoic transition. However, in the best-documented areas of Turukhansk, there is no preserved record of substantial westward thickening of the Riphean stratigraphy as would be expected for a continent–ocean crustal transition, nor is there any preserved evidence of deep-water facies in the middle Riphean succession (Bartley *et al.* 2001; Pisarevsky & Natapov 2003; Khudoley *et al.* 2007). According to the available information, the present northwestern margin of Siberia is more likely a mid–late Neoproterozoic truncation of a more extensive Rodinian plate with widespread middle Riphean epicratonic cover.

## Amazonia, West Africa, Plata

For the past 25 years, Amazonia has been the craton of choice for proposed colliders with the Grenville

margin of Laurentia at the end of the Mesoproterozoic Era (e.g. Bond *et al.* 1984; Dalziel 1991, 1997; Hoffman 1991; Weil *et al.* 1998; Pisarevsky *et al.* 2003*a*; Tohver *et al.* 2006; Li *et al.* 2008). This is perhaps surprising, given how many alternative possibilities exist due to the global preponderance of late Mesoproterozoic ('Grenvillian') orogens among the world's cratons (Fig. 1). The Laurentia + Amazonia connection is broadly supported by Pb-isotopic signatures (Loewy *et al.* 2003), but without a globally comprehensive dataset such comparisons are merely indicative rather than diagnostic. If any craton-scale tectonic comparisons are to be made, then the progressively younging, accretionary character of Amazonia would fit much better *along strike* of southwestern Laurentia (Santos *et al.* 2008) rather than in the mirrored configuration of the two cratons in typical Rodinia models. The palaeomagnetic evidence in support of the Laurentia + Amazonia connection in Rodinia is not particularly strong, either. Based on new palaeomagnetic data, Tohver *et al.* (2002, 2004) and D'Agrella-Filho *et al.* (2003, 2008) have successively modified the kinematics of the putative Laurentia + Amazonia collision. If confined to such a collisional model, the data now demand two unusual kinematic features: (1) *c.* 5000 km of sinistral motion with Amazonia occupying positions adjacent to Texas and Labrador at 1.2 and 1.0 Ga, respectively; and (2) 90° of anticlockwise rotation of Amazonia relative to Laurentia, so that Amazonia appears to roll like a wheel along the Grenvillian margin during its syncollisional sinistral odyssey. Such odd kinematics could be avoided if the observations were not confined by the initial assumption of a Laurentia + Amazonia collision.

Figure 9 shows the available palaeomagnetic poles from Amazonia during the Rodinia interval. The Nova Floresta (NF) and Fortuna Formation (FF) poles are fully published (Tohver *et al.* 2002; D'Agrella-Filho *et al.* 2008), whereas the Aguapei sills (Agua) result is presented in abstract only (D'Agrella-Filho *et al.* 2003). This latter result is important for constraining the possible position of Amazonia in Rodinia, however, because it, like the Nova Floresta data, is from mafic igneous rocks constrained in age by the Ar/Ar method. The Fortuna Formation red beds are interpreted as gaining their diagenetic hematite remanence at *c.* 1150 Ma, according to SHRIMP U–Pb dating of xenotime (D'Agrella-Filho *et al.* 2008), but that age assignment is questioned here because the likely early-diagenetic xenotime U–Pb age may have little bearing on the timing of hematite pigmentation in the studied sandstone. The reconstruction of Amazonia relative to Laurentia shown in Figure 9 predicts a younger age of *c.* 1020 Ma for

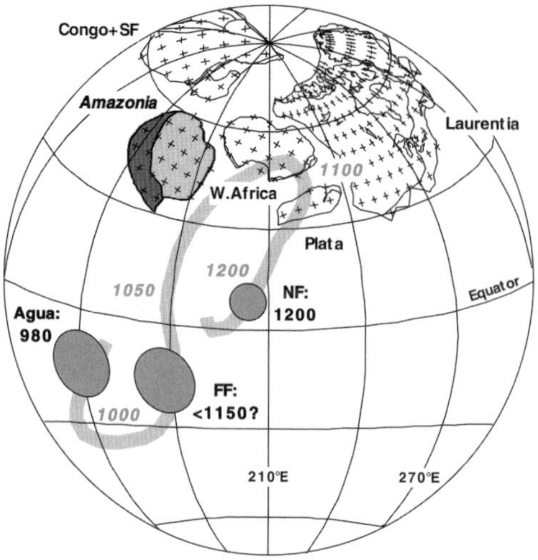

**Fig. 9.** 'COBRA' reconstruction of Amazonia, West Africa, and Plata near western Laurentia. Among these three cratons, only Amazonia has reliable palaeomagnetic constraints from the Rodinia time interval. Cratons are rotated into the present North American reference frame according to Table 3.

the growth of remanence-bearing hematite pigments in the Fortuna Formation.

Because the three Amazonia poles fall roughly along the same great circle, it is possible to consider the alternative polarity assignment relative to the Laurentian APW path; in that case, however, Amazonia reconstructs directly atop Australia and Kalahari (not shown in Fig. 9).

As recently reviewed by Tohver et al. (2006), palaeomagnetic results from the Meso-Neoproterozoic of West Africa are wholly unreliable. For the Plata Craton, Rapalini & Sánchez-Bettucci (2008) similarly show that there are no reliable Rodinian palaeomagnetic constraints. The separation between western Laurentia and Amazonia (Fig. 9), must be filled with cratonic fragments that would form the conjugate rift margin of the Cordilleran miogeocline in either mid-Neoproterozoic or terminal Neoproterozoic times (Bond 1997; Colpron et al. 2002; Harlan et al. 2003). Given that all of the other large cratons of the world have been accounted for in the present Rodinia model, the simplest placements of West Africa, Plata and smaller cratonic fragments in South America (Fuck et al. 2008) are within the gap between Laurentia and Amazonia (Fig. 9). These juxtapositions are collectively referred to as COBRA, named after the general link between the proto-Cordilleran rifted margin of Laurentia with the proto-Brasiliano/ Pharuside rifted margins of the West Gondwanaland cratons.

COBRA unites truncated Archean and Palaeoproterozoic basement provinces among these cratons, suggesting that the amalgamation persisted from the assembly of supercontinent Nuna at 1.8 Ga (Hoffman 1996) until Rodinia fragmentation in mid-Neoproterozoic times. In this reconstruction, 2.1–2.3 Ga terranes in subsurface Yukon-Alberta (Ross 2002) continue into the Birimian (Gasquet et al. 2004) and Maroni–Itacaiunas (Tassinari et al. 2000) provinces of West Africa and Amazonia, respectively. The Archean Wyoming/Medicine Hat craton (Chamberlain et al. 2003) would have been contiguous with the Nico Perez terrane in Uruguay (Hartmann et al. 2001) and Luis Alvez craton in southern Brazil (Sato et al. 2003), constituting parts of an elongate collage of Archean regions extending to the Leo massif in West Africa (Thiéblemont et al. 2004) and the Carajas block in Brazil (Tassinari et al. 2000). Palaeoproterozoic accretion to the south of these provinces includes the Mojave province (Bennett & De Paolo 1987) as the orphaned edge of an extensive region of juvenile 2.2–1.7 Ga terranes in South America (Tassinari et al. 2000; Santos et al. 2000, 2003), characterized by highly radiogenic ($^{207}$Pb-enriched) common-lead isotopic signatures (Wooden & Miller 1990; Tosdal 1996). Detrital zircons of 1.5–1.9 Ga age in the Mesoproterozoic Belt-Purcell basin (Ross et al. 1992; Ross & Villeneuve 2003) find numerous potential sources in extensive granites of that age interval in South America (Tassinari et al. 2000). The 1.3–1.1 Ga Grenville orogen traces southwestward through Sonora (Iriondo et al. 2004) and, according to the COBRA hypothesis, into Brazil and Bolivia, where it bifurcates into the Aguapei and Sunsas belts (Sadowski & Bettencourt 1996). Direct juxtaposition of these provinces in Amazonia with SW North America (Santos et al. 2008) is not allowable palaeomagnetically, by any of the three poles discussed above, regardless of their precise ages within the Meso-Neoproterozoic interval.

COBRA is proposed to have begun rifting at 780 Ma, manifested by the Gunbarrel large igneous province in North America (Harlan et al. 2003), and preceding highly oblique dextral separation (Brookfield 1993) that prolonged rift magmatism to at least 685 Ma (Lund et al. 2003) and delayed passive-margin thermal subsidence to latest Neoproterozoic time (Bond 1997). Precise geochronology of the Gourma–Volta rift basins in West Africa, presently lacking, could provide a direct test of the proposed COBRA fit. Indications of c. 780 Ma mafic magmatism within a possible West African craton fragment in the westernmost Hoggar shield (Caby 2003) and along the distal western São Francisco margin in Brazil (Pimentel et al. 2004) may extend the Gunbarrel province into those regions. Proposing a sequence of rifts in southern South America is difficult due to Phanerozoic cover (compare Ramos 1988 and Cordani *et al.* 2003), but kinematic constraints on a COBRA–West Gondwanaland transition require some events at *c.* 780 Ma and others younger, represented by glaciogenic successions on southernmost Amazonia (Trindade *et al.* 2003) and eastern Rio Plata (Gaucher *et al.* 2003) that are correlated to the Marinoan ice age ending at 635 Ma (Condon *et al.* 2005).

## Nuna to Rodinia to Gondwanaland

Any palaeomagnetic reconstruction of a supercontinent requires concordance of data from constituent cratons into a coherent aggregate APW path. Such a path for the proposed long-lived Rodinia model is shown in Figure 10. The numerous loops and turns could raise objection due to the implied complexity of the supercontinent's motion through its nearly 400 Ma of existence. Nonetheless, all of the loops are generated simply by joining the Laurentian and Baltic APW paths in the proposed reconstruction (Fig. 3). The Laurentia + Baltica juxtapositions, before and after 1100 Ma, are the least controversial aspect of the present Rodinia model, so the APW complexity of Rodinia will be implied by any alternative model incorporating

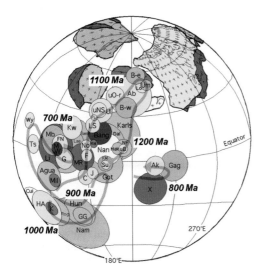

**Fig. 10.** Rodinia master apparent polar wander path in the present North American (Laurentian) reference frame. Palaeomagnetic poles are coloured according to the host cratons as depicted here and in Figure 1, and are tabulated with abbreviations in Table 2. Late Mesoproterozoic ('Grenvillian') orogenic belts are shaded grey.

these relationships. Adding all other cratons' reliable palaeomagnetic data from 1100–750 Ma, in the 'radically' revised Rodinia proposed herein, has resulted in no additional APW loops.

The complexity of Rodinia's aggregate motion is largely due to oscillatory swings in the APW path between 1200 and 900 Ma, and the newly recognized 800 Ma loop. The 1100–1000 Ma segment, in particular, covers >10 000 km, thus averaging rates of latitudinal motion exceeding 10 cm/yr. This would be fast enough for oceanic plates of the modern world, but it is exceptionally fast for a plate containing a supercontinent, presumably with numerous lithospheric keels, to slide over the asthenosphere. An alternative explanation for the majority of Rodinia's motion, and one which more easily accommodates its oscillatory nature, is that of TPW. Evans (2003) incorporated Rodinia's latitude shifts as due to oscillatory TPW about a prolate axis of the geoid inherited from the previous supercontinent, Nuna (a.k.a. Columbia).

Because the Siberian craton is surrounded on many sides by *c.* 1700–1500 Ma rifted passive margins (Pisarevsky & Natapov 2003), it is likely to have lain near the centre of Nuna. In contrast, Figure 11 shows Siberia at the edge of Rodinia. This would suggest that the kinematic evolution between Nuna and Rodinia was partly 'extroverted' (Murphy & Nance 2003, 2005). However, proximity of the Amazonia, West Africa, Congo + São Francisco and Plata cratons in the proposed Rodinia (Fig. 11) suggests long-lived connections from the Palaeoproterozoic (similarities noted by Rogers 1996, and inspiration for his conjectured 'Atlantica' assemblage of that age), rearranging only moderately to form portions of Rodinia and Gondwanaland. The relationships among this group of cratons, as well as the longstanding proximity between Laurentia and Baltica (through nearly the entire latter half of Earth history) suggest a more 'introverted' kinematic style of supercontinental evolution.

According to the Rodinia model proposed herein, assembly took place rapidly at *c.* 1100 Ma, although there are earlier collisions of cratons that persisted into Rodinia time. Table 3 lists the postulated ages of assembly for each craton for the rotation parameters given relative to Laurentia. Some of the proposed connections date from the time of cratonization, typically the Palaeoproterozoic amalgamations of Archean craton, set within and among juvenile terranes. The relevant cratons in this category are those (West Africa, Plata, Amazonia) proposed to reconstruct near Laurentia's proto-Cordilleran margin, where mid–late Neoproterozoic rifting cut across truncated basement fabrics. The hypothesized one-billion-year shared history of these blocks constitutes a powerful

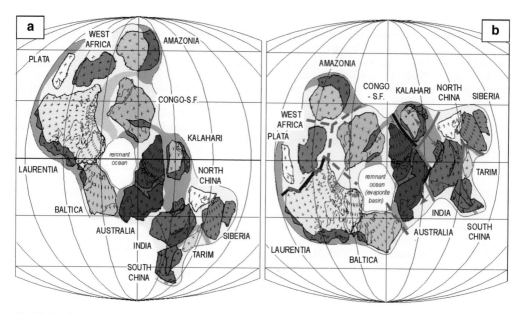

**Fig. 11.** Rodinia radically revised, reconstructed to palaeolatitudes soon after assembly and shortly prior to breakup. (**a**) 1070 Ma reconstruction showing extents of late Mesoproterozoic ('Grenvillian') orogens, both exposed and inferred (dark grey), and the earlier collision between Laurentia and Congo (light grey). The figure is slightly anachronistic, as several blocks are proposed to have collided at 1050 Ma, and South China and Tarim could have joined as late as *c.* 850 Ma (Table 3); however, palaeogeographic reconstructions for those younger ages would involve substantial components of Rodinia over the poles, rendering the Mollweide projection uninformative. (**b**) 780 Ma reconstruction showing incipient breakup rift margins (red) and transform offsets (black). Ridge segments are dashed where not precisely constrained in location.

prediction for future palaeomagnetic tests among these data-scarce blocks.

Because Table 3 lists all rotations relative to Laurentia, it does not provide information on relationships between non-Laurentian cratons or portions thereof, yet some of these may have similarly ancient connections. For example, as discussed above, one hypothesis resulting from the proposed configuration of Congo and Mawsonland, is that the Tanzania–Bangweulu block was originally part of Mawsonland and transferred to Congo via collision at *c.* 1100 Ma and rifting in the mid-Neoproterozoic. Palaeomagnetism of the three Palaeoproterozoic blocks (Congo, Tanzania, Mawsonland) can ultimately test this proposition. Another example is the proposed Palaeo-Mesoproterozoic connection among India, North China, and perhaps also Siberia.

Where Table 3 presents a range of ages, these are set by the limits of palaeomagnetic concordance versus discordance when rotated by the given Euler parameters. Parenthetical values indicate a best estimate based on geological histories of either collision or rifting, or by 'piggy-back' of an intervening collision or rift with Laurentia. The identical-aged *c.* 1050 Ma onset of proposed Euler

reconstructions listed for India, North China and Siberia are an example of this latter case; all these cratons are proposed to have sutured, as a unified plate, to the Rodinia assemblage along the Eastern Ghats–Rayner orogen.

Although this model of Rodinia includes widespread assembly of the supercontinent at 1100–1050 Ma, probably the most contentious of its implications is that all the large cratons are accounted for, and there are no sizable blocks left to play colliding roles in any of the Sveconorwegian, Grenville and Sunsas orogens. Instead, these three orogens are placed along strike of each other, facing the Mirovian Ocean. All three orogens are characterized by an extensive prehistory of accretionary tectonism along the same margins, with successively younger age provinces progressing outward from Archaean cratonic nuclei. The great width and longevity of these three accretionary systems is reminiscent of Panthalassan or circum-Pangaean orogens of the Phanerozoic. The model proposed here requires originally farther oceanward extents of the three orogens as younger juvenile material would have accreted during the early Neoproterozoic. Then, mid-Mirovian spreading ridges would have propagated into the orogens and

**Table 3.** *Rotation parameters to Laurentia: Rodinia radically revised*

| Craton | Maximum age | Minimum age | Euler rotn. | | |
|---|---|---|---|---|---|
| | | | °N | °E | °CCW |
| Baltica[†] | 1120–1070 | 615–555 (610) | 81.5 | 250.0 | −50.0 |
| Australia | 1140–1070 | ≤755 (720) | −12.5 | 064.0 | 134.5 |
| Congo (ex.Tanz) | ≥1375 (1500) | 755–550 (720) | −24.0 | 249.0 | 128.5 |
| Kalahari | 1235–1110 | ≤1000 (790?) | 32.5 | 307.5 | 76.5 |
| India | 1100–770 (1050) | ≤770 (750) | 04.0 | 066.0 | 93.0 |
| South China | ≥800 (850) | ≤750 (750) | 17.5 | 067.0 | 178.0 |
| Tarim | ≥755 (850?) | ≤755 (750) | −41.5 | 161.0 | 57.0 |
| Svalbard | cratonization | Devonian? | 86.0 | 120.0 | −48.0 |
| North China | ≥950 (1050) | ≤790 (750?) | 42.0 | 072.0 | −113.0 |
| Siberia (Aldan)[‡] | ≥1040 (1050) | ≤990 (750?) | 32.5 | 002.5 | −91.5 |
| Amazonia | ≥1200 (crat.) | ≤980 (630?) | 63.0 | 313.0 | −139.0 |
| West Africa | cratonization | (≤780) | −54.0 | 246.5 | 115.0 |
| Plata | cratonization | (≤780) | −18.5 | 239.0 | 125.5 |

[†]Alignments are similar to those discussed in Buchan *et al.* (2000) and Pisarevsky *et al.* (2003*a*).
[‡]Reconstruction is similar to option 'B' of Pisarevsky & Natapov (2003) and that presented by Meert & Torsvik (2003).

removed the outboard, youngest terrains as 'ribbon' continents. The protracted record of tectonism in the Scottish Highlands, inlcuding the Knoydartian orogeny at c. 850–800 Ma with further phases possibly as young as c. 750–700 Ma (reviewed by Cawood *et al.* 2007) could represent the only intact remnants of a once-extensive accretionary orogenic belt that lay outboard of the present Grenville orogen.

The present location of these postulated ribbon terrains is unknown, but the kinematic histories of more recent examples suggest that they would either be transported strike-slip along the circum-Mirovian subduction girdle around Rodinia (such as present-day Baja California or the more extreme possibility of thousands of kilometres in a 'Baja–British Columbia' evolution), or separated far into Mirovia toward an unprescribed fate (such as present Zealandia). Using these analogies, we might expect to find them today as dismembered basement units within the Avalonian–Cadomian orogen (Evans 2005; Murphy *et al.* 2000; Keppie *et al.* 2003) or Borborema–Pharuside strike-slip-dominated orogenic system (Caby 2003), or perhaps partly to completely recycled into the mantle by subduction-erosion (Scholl & von Huene 2007).

Figure 11b shows the incipient breakup of Rodinia at 780 Ma, according to the revised Rodinia model. A first stage of disaggregation at c. 780–720 Ma around the western and northern margins of Laurentia, liberated the Congo, West African, Amazonian and Plata cratons that would eventually recombine to form West Gondwanaland between c. 640 and c. 530 Ma (Trompette 1997; Brito Neves *et al.* 1999; Piuzana *et al.*

2003; Valeriano *et al.* 2004; John *et al.* 2004; Tohver *et al.* 2006). Although the interval between rifting and collision in the Brasiliano foldbelts was brief ('young, short-lived' orogenic cycle of Trompette 1997), the predominant strike-slip component of motion during assembly allowed those belts to contain oceanic (Mirovian) terranes as old as c. 900–750 Ma (Pimentel & Fuck 1992; Babinski *et al.* 1996). The prominent dextral shear zones of the Borborema Province in northeastern Brazil, continue into west–central Africa (Vauchez *et al.* 1995; Cordani *et al.* 2003). These bound enigmatic terranes recording unusual 'Grenvillian' tectonothermal events that are otherwise largely absent in cratonic South America (Fuck *et al.* 2008), bearing witness to the large amount of strike-slip offset accommodating the assembly of West Gondwanaland.

On the other side of the proposed Rodinia, the kinematic evolution toward East Gondwanaland follows more conventional reconstructions, which comes as little surprise because the relative positions of Australia, India, South China, and Tarim are similar to those earlier models. India migrated sinistrally along the Pinjarra orogen to arrive at its Gondwanaland position relative to Australia by c. 550 Ma (Powell & Pisarevsky 2002). South China and Tarim would have lain along the same tectonic plate during that time, arriving at acceptable positions for their palaeomagnetic reconstruction into Gondwanaland (Zhang 2004; Zhao *et al.* 2008). In the palaeogeographic co-ordinate system of 780 Ma (Fig. 11b), Siberia would have rifted to the east, separating from East Gondwanaland fragments. It is debatable whether North China was part of Palaeozoic Gondwanaland; if not, it too may have rifted far away with Siberia. Kalahari would have migrated to the north in the reconstructed co-ordinate system of Figure 11b, joining the West Gondwanaland cratons as they drifted away from Laurentia + Baltica. Final disaggregation of Rodinia occurred *c.* 610–550 Ma, the age of extensive mafic magmatism in eastern Laurentia (reviewed by Cawood *et al.* 2001; Puffer 2002) and Norway (Svenningsen 2001).

Global palaeogeography at the end of the Neoproterozoic Era remains one of the most challenging problems in palaeomagnetic reconstruction, more difficult even than the quest for Rodinia. This is due to four factors: (1) lack of high-precision biostratigraphy in the Precambrian to correlate successions and to date palaeomagnetic poles from sedimentary rocks; (2) scarcity of datable volcanic successions on the large cratons, relative to geon 7; (3) likelihood that most cratons were travelling independently during the transition between Rodinia and Gondwanaland, thus disallowing the APW superposition method used in this paper; and (4) abnormally high dispersion of palaeomagnetic poles from each craton indicating either rapid plate tectonics, rapid TPW, or a non-uniformitarian geomagnetic field during that time. The most complete model incorporating the global tectonic record and palaeomagnetic data is by Collins & Pisarevsky (2005), but this model still needed to resort to separate options of a low- versus high-latitude subset of the Laurentian palaeomagnetic data. If TPW is responsible for the large dispersions in palaeomagnetic poles, which if read literally would typically imply oscillatory motions conforming to the IITPW model of Evans (2003), then there is some hope to produce reconstructions using the long-lived prolate nonhydrostatic geoid as the reference axis, rather than the geomagnetic-rotational reference frame (Raub *et al.* 2007). This alternative method, however, produces reconstructions that are highly sensitive to small errors in magnetization ages, depending on the rapidity of the putative TPW oscillations. Regardless of which class of interpretations will ultimately prove valid, questions such as the widths of Iapetan separation following Rodinian juxtaposition of Amazonia with eastern Laurentia (e.g. Cawood *et al.* 2001), must be considered premature until more precisely dated palaeomagnetic poles are obtained.

## Concluding Remarks

Nearly two decades have elapsed since McMenamin & McMenamin (1990, p. 95) coined the name 'Rodinia' for the late Proterozoic supercontinent and 'Mirovia' for its encircling palaeo-ocean. Within a year of these monikers' establishment,

a general model for Rodinia was conceived (Hoffman 1991), which, despite numerous challenges from both tectonostratigraphy and palaeomagnetism, has remained largely intact in the latest consensus model (Li *et al.* 2008). However, in order to accommodate the palaeomagnetic data in particular, this latest model has shortened the duration of Rodinia's existence to merely 75 Ma (900–825 Ma). Such brevity is acceptable and actualistic in terms of the short-lived Pangaea landmass, but it appears at odds with Hoffman's (1991) original implication of globally widespread 1300–1000 Ma orogens and *c.* 750 Ma rifted margins, as representing Rodinia's assembly and breakup, respectively.

Herein, I have proposed a Rodinia model that is both long-lived according to the original concept, and compatible with the most reliable palaeomagnetic data from the Meso-Neoproterozoic interval, with minimal number of APW loops. My model is a radical departure from all previous models (e.g. Li *et al.* 2008). Which existing Rodinia model, if any, will approximate the 'true' form of the Neoproterozoic supercontinent? As Wegener [1929 (1966, p. 17)] wrote: 'the earth at any one time can only have had one configuration.' How will we test the current Rodinia models and achieve a long-lasting consensus that converges toward the true palaeogeography?

Dalziel (1999) identified six criteria for validity of a 'credible' supercontinent: (1) account for all rifted passive margins at the time of breakup; (2) accurately map continental promontories and embayments, that is in spherical geometry; (3) display sutures related to assembly; (4) match older tectonic fabrics where appropriate; (5) show compatibility with palaeomagnetic data; and (6) be compatible with realistic kinematic evolution forward in time toward Pangaea. The revised Rodinia model proposed herein satisfies all six of these conditions, if one allows for a special consideration involving conjugate rifted and collisional margins, as follows: the past few years of palaeogeographic reconstruction of the Mesozoic–Cenozoic world have led to increasing recogniztion of ribbon-shaped continental fragments with lengths on the order of thousands of kilometres (e.g. Lomonosov Ridge; Lawver *et al.* 2002; Lord Howe Rise/Zealandia; Gaina *et al.* 2003). Farther back in time, the Cimmeride continental ribbon formed the Permian rift conjugate to the >5000 km northern passive margin of Gondwanaland (Stampfli & Borel 2002). Tectonic shuffling and reworking of Cimmeride blocks within the Alpine-Himalayan orogenic collage has largely obscured their original geometric continuity. The Lomonosov Ridge and Lord Howe Rise/Zealandia ribbons are yet to migrate to their final dispositions within accretionary orogens, but they are unlikely to arrive in pristine form. I propose that similar effects may hamper our ability to quantify the passive margin lengths of any Precambrian continental ribbons. In the case of Zealandia, separation from Australia + Antarctica roughly followed the geometry of the Terra Australis orogen (Cawood 2005), thus ending a Wilson cycle without a continent-continent collision. In the Transantarctic Mountains, the rift propagated far enough inboard to bring some of the oldest, most internal segments of the belt (Cambrian–Ordovician Ross orogen) directly in contact with the oceanic passive margin. Would future palaeogeographers interpret this record as one of Cambrian–Ordovician continent–continent collision, followed by tectonic stability inside a supercontinent, and subsequent Mesozoic breakup of that supercontinent? This example highlights the difficulty in robustly characterizing tectonic histories of Precambrian orogens without a palaeogeographic framework. With focused effort on obtaining key geochronologic and palaeomagnetic data from the Rodinian time interval, we may be able to provide that framework.

These ideas have evolved through the last five years, benefiting from discussions with S. Bryan, P. Cawood, A. Collins, B. Collins, D. Corrigan, I. Dalziel, B. de Waele, R. Ernst, I. Fitzsimons, R. Fuck, C. Gower, P. Hoffman, K. Karlstrom, A. Kröner, Z.-X. Li, A. Maloof, B. Murphy, V. Pease, M. Pimentel, S. Pisarevsky, the late C. Powell, T. Rivers, D. Thorkelson, E. Tohver and M. Wingate. P. Hoffman, I. Dalziel, C. MacNiocaill, and R. Van der Voo provided helpful criticisms on earlier versions of the manuscript; P. Cawood and S. Pisarevsky gave expedient and constructive reviews of this iteration. This work was supported by a Fellowship in Science and Engineering from the David and Lucile Packard Foundation.

# References

ALKMIM, F. F., MARSHAK, S. & FONSECA, M. A. 2001. Assembling West Gondwana in the Neoproterozoic: clues from the São Francisco craton region, Brazil. *Geology*, **29**, 319–322.

BABINSKI, M., CHEMALE, F., JR., HARTMANN, L. A., VAN SCHMUS, W. R. & DA SILVA, L. C. 1996. Juvenile accretion at 750–700 Ma in southern Brazil. *Geology*, **24**, 439–442.

BARTLEY, J. K., SEMIKHATOV, M. A., KAUFMAN, A. J., KNOLL, A. H., POPE, M. C. & JACOBSEN, S. B. 2001. Global events across the Mesoproterozoic-Neoproterozoic boundary: C and Sr isotopic evidence from Siberia. *Precambrian Research*, **111**, 165–202.

BENNETT, V. C. & DEPAOLO, D. J. 1987. Proterozoic crustal history of the western United States as determined by neodymium isotopic mapping. *Geological Society of America Bulletin*, **99**, 674–685.

BERRY, R. F., HOLM, O. H. & STEELE, D. A. 2005. Chemical U–Th–Pb monazite dating and the Proterozoic history of King Island, southeast Australia. *Australian Journal of Earth Sciences*, **52**, 461–471.

BINGEN, B., DEMAIFFE, D. & VAN BREEMEN, O. 1998. The 616 Ma old Egersund basaltic dike swarm, SW Norway, and late Neoproterozoic opening of the Iapetus Ocean. *Journal of Geology*, **106**, 565–574.

BLEWETT, R. S. & BLACK, L. P. 1998. Structural and temporal framework of the Coen Region, north Queensland: implications for major tectonothermal events in east and north Australia. *Australian Journal of Earth Sciences*, **45**, 597–609.

BOGDANOVA, S. V., BINGEN, B., GORBATSCHEV, R., KHERASKOVA, T. N., KOZLOV, V. I., PUCHKOV, V. N. & VOLOZH, YU. A. 2008. The East European Craton (Baltica) before and during the assembly of Rodinia. *Precambrian Research*, **160**, 23–45.

BOND, G. C. 1997. New constraints on Rodinia breakup ages from revised tectonic subsidence curves. *Geological Society of America Abstracts with Programs*, **29**(6), 280.

BOND, G. C., NICKESON, P. A. & KOMINZ, M. A. 1984. Breakup of a supercontinent between 625 Ma and 555 Ma: new evidence and implications for continental histories. *Earth and Planetary Science Letters*, **70**, 325–345.

BORG, S. G. & DEPAOLO, D. J. 1994. Laurentia, Australia, and Antarctica as a Late Proterozoic supercontinent: constraints from isotopic mapping. *Geology*, **22**, 307–310.

BRIDEN, J. C., DUFF, B. A. & KRÖNER, A. 1979. Palaeomagnetism of the Koras Group, Northern Cape Province, South Africa. *Precambrian Research*, **10**, 43–57.

BRITO NEVES, B. B., CAMPOS NETO, M. C. & FUCK, R. A. 1999. From Rodinia to Western Gondwana: an approach to the Brasiliano-Pan African Cycle and orogenic collage. *Episodes*, **22**, 155–166.

BROOKFIELD, M. E. 1993. Neoproterozoic Laurentia-Australia fit. *Geology*, **21**, 683–686.

BROWN, L. L. & MCENROE, S. A. 2004. Palaeomagnetism of the Egersund-Ogna anorthosite, Rogaland, Norway, and the position of Fennoscandia in the Late Proterozoic. *Geophysical Journal International*, **158**, 479–488.

BUCHAN, K. L. & HALLS, H. C. 1990. Paleomagnetism of Proterozoic mafic dyke swarms of the Canadian Shield. *In*: PARKER, A. J., RICKWOOD, P. C. & TUCKER, D. H. (eds) *Mafic dykes and emplacement mechanisms*. Balkema, Rotterdam, 209–230.

BUCHAN, K. L., MERTANEN, S., PARK, R. G., PESONEN, L. J., ELMING, S.-Å., ABRAHAMSEN, N. & BYLUND, G. 2000. Comparing the drift of Laurentia and Baltica in the Proterozoic: the importance of key palaeomagnetic poles. *Tectonophysics*, **319**, 167–198.

BUCHAN, K. L., ERNST, R. E., HAMILTON, M. A., MERTANEN, S., PESONEN, L. J. & ELMING, S.-Å. 2001. Rodinia: the evidence from integrated palaeomagnetism and U–Pb geochronology. *Precambrian Research*, **110**, 9–32.

BURRETT, C. & BERRY, R. 2000. Proterozoic Australia-Western United States (AUSWUS) fit between Laurentia and Australia. *Geology*, **28**, 103–106.

CABY, R. 2003. Terrane assembly and geodynamic evolution of central-western Hoggar: a synthesis. *Journal of African Earth Sciences*, **37**, 133–159.

CATELANI, E. L., EVANS, D. A. D., SMIRNOV, A. V., TRINDADE, R. I. F., D'AGRELLA-FILHO, M. S. & HEAMAN, L. M. 2007. *Revisiting the Meso-Neoproterozoic dykes of Bahia, Brazil*. EOS, Transactions of the American Geophysical Union, Joint Assembly, Acapulco, Mexico.

CAWOOD, P. A. 2005. Terra Australis Orogen: rodinia breakup and development of the Pacific and Iapetus margins of Gondwana during the Neoproterozoic and Paleozoic. *Earth-Science Reviews*, **69**, 249–279.

CAWOOD, P. A. & KORSCH, R. J. 2008. Assembling Australia: Proterozoic building of a continent. *Precambrian Research*, **166**, 1–38.

CAWOOD, P. A. & PISAREVSKY, S. A. 2006. Was Baltica right-way-up or upside-down in the Neoproterozoic? *Journal of the Geological Society, London*, **163**, 1–7.

CAWOOD, P. A., MCCAUSLAND, P. J. A. & DUNNING, G. R. 2001. Opening Iapetus: Constraints from the Laurentian margin of Newfoundland. *Geological Society of America Bulletin*, **113**, 443–453.

CAWOOD, P. A., NEMCHIN, A. A., STRACHAN, R., PRAVE, T. & KRABBENDAM, M. 2007. Sedimentary basin and detrital zircon record along East Laurentia and Baltica during assembly and breakup of Rodinia. *Journal of the Geological Society, London*, **164**, 257–275.

CHAMBERLAIN, K. R., FROST, C. D. & FROST, B. R. 2003. Early Archean to Mesoproterozoic evolution of the Wyoming Province: archaean origins to modern lithospheric architecture. *Canadian Journal of Earth Sciences*, **40**, 1357–1374.

CHATTERJEE, N., CROWLEY, J. L. & GHOSE, N. C. 2008. Geochronology of the 1.55 Ga Bengal anorthosite and Grenvillian metamorphism in the Chhotanagpur gneissic complex, eastern India. *Precambrian Research*, **161**, 303–316.

CHEN, Y., XU, B., ZHAN, S. & LI, Y. 2004. First mid-Neoproterozoic paleomagnetic results from the Tarim Basin (NW China) and their geodynamic implications. *Precambrian Research*, **133**, 271–281.

CLARK, D. J., HENSEN, B. J. & KINNY, P. D. 2000. Geochronological constraints for a two-stage history of the Albany-Fraser Orogen, Western Australia. *Precambrian Research*, **102**, 155–183.

COGNÉ, J. P. 2003. PaleoMac: a Macintosh™ application for treating paleomagnetic data and making plate reconstructions. *Geochemistry Geophysics Geosystems*, **4**(1), doi: 10.1029/2001GC000227.

COLLINS, A. S. & PISAREVSKY, S. A. 2005. Amalgamating eastern Gondwana: the evolution of the Circum-Indian Orogens. *Earth-Science Reviews*, **71**, 229–270.

COLPRON, M., LOGAN, J. M. & MORTENSEN, J. K. 2002. U–Pb zircon age constraint for late Neoproterozoic rifting and initiation of the lower Paleozoic passive margin of western Laurentia. *Canadian Journal of Earth Sciences*, **39**, 133–143.

CONDIE, K. C. 2002. The supercontinent cycle: are there two patterns of cyclicity? *Journal of African Earth Sciences*, **35**, 179–183.

CONDON, D., ZHU, M., BOWRING, S., WANG, W., YANG, A. & JIN, Y. 2005. U–Pb ages from the Neoproterozoic Doushantuo Formation, China. *Science*, **308**, 95–98.

CORDANI, U. G., D'AGRELLA-FILHO, M. S., BRITO-NEVES, B. B. & TRINDADE, R. I. F. 2003. Tearing up Rodinia: the Neoproterozoic palaeogeography of South American cratonic fragments. *Terra Nova*, **15**, 350–359.

D'AGRELLA-FILHO, M. S., PACCA, I. G., RENNE, P. R., ONSTOTT, T. C. & TEIXEIRA, W. 1990. Paleomagnetism of Middle Proterozoic (1.01 to 1.08 Ga) mafic dykes in southeastern Bahia State – São Francisco Craton, Brazil. *Earth and Planetary Science Letters*, **101**, 332–348.

D'AGRELLA-FILHO, M. S., TRINDADE, R. I. F., SIQUEIRA, R., PONTE-NETO, C. F. & PACCA, I. I. G. 1998. Paleomagnetic constraints on the Rodinia supercontinent: implications for its Neoproterozoic break-up and the formation of Gondwana. *International Geology Review*, **40**, 171–188.

D'AGRELLA-FILHO, M. S., PACCA, I. G., ELMING, S.-Å., TRINDADE, R. I., GERALDES, M. C. & TEIXEIRA, W. 2003. Paleomagnetic and rock magnetic studies of Proterozoic mafic rocks from western Mato Grosso State, Amazonian craton. *Geophysical Research Abstracts*, **5**, no. 05619.

D'AGRELLA-FILHO, M. S., PACCA, I. I. G., TRINDADE, R. I. F., TEIXEIRA, W., RAPOSO, M. I. B. & ONSTOTT, T. C. 2004. Paleomagnetism and $^{40}Ar/^{39}Ar$ ages of mafic dikes from Salvador (Brazil): new constraints on the São Francisco craton APW path between 1080 and 1010 Ma. *Precambrian Research*, **132**, 55–77.

D'AGRELLA-FILHO, M. S., TOHVER, E., SANTOS, J. O. S., ELMING, S.-Å., TRINDADE, R. I. F., PACCA, I. I. G. & GERALDES, M. C. 2008. Direct dating of paleomagnetic results from Precambrian sediments in the Amazon craton: evidence for Grenvillian emplacement of exotic crust in SE Appalachians of North America. *Earth and Planetary Science Letters*, **267**, 188–199.

DALZIEL, I. W. D. 1991. Pacific margins of Laurentia and East Antarctica-Australia as a conjugate rift pair: evidence and implications for an Eocambrian supercontinent. *Geology*, **19**, 598–601.

DALZIEL, I. W. D. 1997. Neoproterozoic-Paleozoic geography and tectonics: review, hypothesis, environmental speculation. *Geological Society of America Bulletin*, **109**, 16–42.

DALZIEL, I. W. D. 1999. Global paleotectonics; reconstructing a credible supercontinent. *Geological Society of America Abstracts with Programs*, **31**(7), 316.

DALY, M. C., LAWRENCE, S. R., DIEMU-TSHIBAND, K. & MATOUANA, B. 1992. Tectonic evolution of the Cuvette Centrale, Zaire. *Journal of the Geological Society, London*, **149**, 539–546.

DEB, M., THORPE, R. I., KRSTIC, D., CORFU, F. & DAVIS, D. W. 2001. Zircon U–Pb and galena Pb isotope evidence for an approximate 1.0 Ga terrane constituting the western margin of the Aravalli-Delhi orogenic belt, northwestern India. *Precambrian Research*, **108**, 195–213.

DEBLOND, A., PUNZALAN, L. E., BOVEN, A. & TACK, L. 2001. The Malagarazi Supergroup of southeast

Burundi and its correlative Bukoba Supergroup of northwest Tanzania: neo- and Mesoproterozoic chronostratigraphic constraints from Ar–Ar ages on mafic intrusive rocks. *Journal of African Earth Sciences*, **32**, 435–449.

DE WAELE, B., JOHNSON, S. P. & PISAREVSKY, S. A. 2008. Palaeoproterozoic to Neoproterozoic growth and evolution of the eastern Congo Craton: its role in the Rodinia puzzle. *Precambrian Research*, **160**, 127–141.

DIEHL, J. F. & HAIG, T. D. 1994. A paleomagnetic study of the lava flows within the Copper Harbor Conglomerate, Michigan: new results and implications. *Canadian Journal of Earth Sciences*, **31**, 369–380.

DRÜPPEL, K., LITTMANN, S., ROMER, R. L. & OKRUSCH, M. 2007. Petrology and isotope geochemistry of the Mesoproterozoic anorthosite and related rocks of the Kunene Intrusive Complex, NW Namibia. *Precambrian Research*, **156**, 1–31.

ERNST, R. E. & BUCHAN, K. L. 1993. A paleomagnetic study of the lava flows within the Copper Harbor Conglomerate, Michigan: new results and implications. *Canadian Journal of Earth Sciences*, **30**, 1886–1897.

ERNST, R. E. & BUCHAN, K. L. 2002. Maximum size and distribution in time and space of mantle plumes: evidence from large igneous provinces. *Journal of Geodynamics*, **34**, 309–342.

EVANS, D. A. D. 2000. Stratigraphic, geochronological, and paleomagnetic constraints upon the Neoproterozoic climatic paradox. *American Journal of Science*, **300**, 347–433.

EVANS, D. A. D. 2003. True polar wander and supercontinents. *Tectonophysics*, **362**, 303–320.

EVANS, D. A. D. 2005. Do Avalonian terranes contain the lost Grenvillian hinterland? *Geological Association of Canada Abstracts*, **30**, 54–55.

EVANS, D. A. D. 2006. Proterozoic low orbital obliquity and axial-dipolar geomagnetic field from evaporite palaeolatitudes. *Nature*, **444**, 51–55.

EVANS, D. A. D. & PISAREVSKY, S. A. 2008. Plate tectonics on early Earth? Weighing the paleomagnetic evidence. *In*: CONDIE, K. C. & PEASE, V. (eds) *When Did Plate Tectonics Begin on Earth?* Geological Society of America, Special Paper, **440**, 249–263.

EVANS, D. A. D., LI, Z. X., KIRSCHVINK, J. L. & WINGATE, M. T. D. 2000. A high-quality mid-Neoproterozoic paleomagnetic pole from South China, with implications for ice ages and the breakup configuration of Rodinia. *Precambrian Research*, **100**, 313–334.

EVANS, D. A. D., BEUKES, N. J. & KIRSCHVINK, J. L. 2002. Paleomagnetism of a lateritic paleo-weathering horizon and overlying Paleoproterozoic redbeds from South Africa: implications for the Kaapvaal apparent polar wander path and a confirmation of atmospheric oxygen enrichment. *Journal of Geophysical Research*, **107**(B12), doi: 10.1029/2001JB000432.

FAHRIG, W. F., CHRISTIE, K. W. & JONES, D. L. 1981. Paleomagnetism of the Bylot basins: evidence for Mackenzie continental tensional tectonics. *In*: CAMPBELL, F. H. A. (ed.) *Proterozoic Basins of Canada*. Geological Survey of Canada, Paper, **81-10**, 303–312.

FIORETTI, A. M., BLACK, L. P., FODEN, J. & VISONA, D. 2005. Grenville-age magmatism at the South Tasman Rise (Australia): a new piercing point for the reconstruction of Rodinia. *Geology*, **33**, 769–772.

FITZSIMONS, I. C. W. 2003. Proterozoic basement provinces of southern and southwestern Australia, and their correlation with Antarctica. *In*: YOSHIDA, M., WINDLEY, B. F. & DASGUPTA, S. (eds) *Proterozoic East Gondwana: Supercontinent Assembly and Breakup*. Geological Society, London, Special Publications, **206**, 35–55.

FUCK, R. A., BRITO NEVES, B. B. & SCHOBBENHAUS, C. 2008. Rodinia descendants in South America. *Precambrian Research*, **160**, 108–126.

GAINA, C., MÜLLER, R. D., BROWN, B. J. & ISHIHARA, T. 2003. Microcontinent formation around Australia. *In*: HILLS, R. R. & MÜLLER, R. D. (eds) *Evolution and Dynamics of the Australian Plate*. Geological Society of America, Special Paper, **372**, 405–416.

GALLET, Y., PAVLOV, V. E., SEMIKHATOV, M. A. & PETROV, P. YU. 2000. Late Mesoproterozoic magnetostratigraphic results from Siberia: paleogeographic implications and magnetic field behavior. *Journal of Geophysical Research*, **105**(B7), 16481–16499.

GASQUET, D., BARBEY, P., ADOU, M. & PAQUETTE, J. L. 2004. Structure, Sr-Nd isotope geochemistry and zircon U–Pb geochronology of the granitoids of the Dabakala area (Côte d'Ivoire): evidence for a 2.3 Ga crustal growth event in the Palaeoproterozoic of West Africa? *Precambrian Research*, **127**, 329–354.

GAUCHER, C., BOGGIANI, P. C., SPRECHMANN, P., SIAL, A. N. & FAIRCHILD, T. 2003. Integrated correlation of the Vendian to Cambrian Arroyo del Soldado and Corumbá Groups (Uruguay and Brazil): palaeogeographic, palaeoclimatic and palaeobiologic implications. *Precambrian Research*, **120**, 241–278.

GIDDINGS, J. W. 1976. Precambrian palaeomagnetism in Australia I: basic dykes and volcanics from the Yilgarn Block. *Tectonophysics*, **30**, 91–108.

GLASMACHER, U. A., BAUER, W. *ET AL*. 2001. The metamorphic complex of Beloretzk, SW Urals, Russia—a terrane with a polyphase Meso- to Neoproterozoic thermo-dynamic evolution. *Precambrian Research*, **110**, 185–213.

GOODGE, J. W., FANNING, C. M. & BENNETT, V. C. 2001. U–Pb evidence of *c.* 1.7 Ga crustal tectonism during the Nimrod Orogeny in the Transantarctic Mountains, Antarctica: implications for Proterozoic plate reconstructions. *Precambrian Research*, **112**, 261–288.

GOODGE, J. W., VERVOORT, J. D. *ET AL*. 2008. A positive test of East Antarctica – Laurentia juxtaposition within the Rodinia supercontinent. *Science*, **321**, 235–240.

GOSE, W. A., HANSON, R. E., DALZIEL, I. W. D., PANCAKE, J. A. & SEIDEL, E. K. 2006. Paleomagnetism of the 1.1 Ga Umkondo large igneous province in southern Africa. *Journal of Geophysical Research*, **111**, B09101, doi: 10.1029/2005JB03897.

GOWER, C. F., KAMO, S. & KROGH, T. E. 2008. Indentor tectonism in the eastern Grenville Province. *Precambrian Research*, **167**, 201–212.

GOWER, C. F., RYAN, A. B. & RIVERS, T. 1990. Mid-Proterozoic Laurentia-Baltica: an overview of its geological evolution and a summary of the contributions made by this volume. *In*: GOWER, C. F., RIVERS, T. & RYAN, A. B. (eds) *Mid-Proterozoic Laurentia-Baltica*. Geological Association of Canada, Special Paper, **38**, 1–20.

GREGORY, L. C., MEERT, J. G., PRADHAN, V., PANDIT, M. K., TAMRAT, E. & MALONE, S. J. 2006. A paleomagnetic and geochronologic study of the Majhgawan kimberlite, India: implications for the age of the Upper Vindhyan Supergroup. *Precambrian Research*, **149**, 65–75.

GRUNOW, A. M., KENT, D. V. & DALZIEL, I. W. D. 1991. New paleomagnetic data from Thurston Island: implications for the tectonics of West Antarctica and Weddell Sea opening. *Journal of Geophysical Research*, **96**(B11), 17935–17954.

HALLS, H. C. 1974. A paleomagnetic reversal in the Osler Volcanic Group, northern Lake Superior. *Canadian Journal of Earth Sciences*, **11**, 1200–1207.

HALLS, H. C. & PESONEN, L. J. 1982. Paleomagnetism of Keweenawan rocks. *In*: WOLD, R. J. & HINZE, W. J. (eds) *Geology and Tectonics of the Lake Superior Basin*. Geological Society of America, Memoir, **156**, 173–201.

HAMILTON, M. A. & BUCHAN, K. L. 2007. U–Pb geochronology of the Western Channel Diabase, Wopmay Orogen: implications for the APWP for Laurentia in the earliest Mesoproterozoic. *Geological Association of Canada Abstracts*, **32**, 35–36.

HANSON, R. E., CROWLEY, J. L. *ET AL*. 2004. Coeval large-scale magmatism in the Kalahari and Laurentian cratons during Rodinia assembly. *Science*, **304**, 1126–1129.

HARLAN, S. S., GEISSMAN, J. W. & SNEE, L. W. 1997. Paleomagnetic and $^{40}$Ar/$^{39}$Ar geochronologic data from late Proterozoic mafic dykes and sills, Montana and Wyoming. *USGS Professional Paper*, **1580**, 1–16.

HARLAN, S. S., HEAMAN, L., LeCHEMINANT, A. N. & PREMO, W. R. 2003. Gunbarrel mafic magmatic event: a key 780 Ma time marker for Rodinia plate reconstructions. *Geology*, **31**, 1053–1056.

HARTMANN, L. A., CAMPAL, N., SANTOS, J. O. S., MCNAUGHTON, N. J., BOSSI, J., SCHIPILOV, A. & LAFON, J.-M. 2001. Archean crust in the Rio de la Plata Craton, Uruguay—SHRIMP U–Pb zircon reconnaissance geochronology. *Journal of South American Earth Sciences*, **14**, 557–570.

HARTZ, E. H. & TORSVIK, T. H. 2002. Baltica upside down: a new plate tectonic model for Rodinia and the Iapetus Ocean. *Geology*, **30**, 255–258.

HEAMAN, L. M., LeCHEMINANT, A. N. & RAINBIRD, R. H. 1992. Nature and timing of Franklin igneous events, Canada: implications for a Late Proterozoic mantle plume and the breakup of Laurentia. *Earth and Planetary Science Letters*, **109**, 117–131.

HENRY, S. G., MAUK, F. J. & VAN DER VOO, R. 1977. Paleomagnetism of the upper Keweenawan sediments: the Nonesuch Shale and Freda Sandstone. *Canadian Journal of Earth Sciences*, **14**, 1128–1138.

HOFFMAN, P. F. 1991. Did the breakout of Laurentia turn Gondwanaland inside-out? *Science*, **252**, 1409–1412.

HOFFMAN, P. F. 1996. Tectonic genealogy of North America. *In*: VAN DER PLUIJM, B. A. & MARSHAK, S. (eds) *Earth Structure: an Introduction to Structural Geology and Tectonics*. McGraw-Hill, New York, 459–464.

HUANG, B., XU, B., ZHANG, C., LI, Y. & ZHU, R. 2005. Paleomagnetism of the Baiyisi volcanic rocks (*c.* 740 Ma) of Tarim, Northwest China: a continental fragment of Neoproterozoic Western Australia? *Precambrian Research*, **142**, 83–92.

HYODO, H. & DUNLOP, D. J. 1993. Effect of anisotropy on the paleomagnetic contact test for a Grenville dike. *Journal of Geophysical Research*, **98**, 7997–8017.

IDNURM, M. & GIDDINGS, J. W. 1995. Paleoproterozoic-Neoproterozoic North America-Australia link: new evidence from paleomagnetism. *Geology*, **23**, 149–152.

IRIONDO, A., PREMO, W. R. *ET AL*. 2004. Isotopic, geochemical, and temporal characterization of Proterozoic basement rocks in the Quitovac region, northwestern Sonora, Mexico: implications for the reconstruction of the southwestern margin of Laurentia. *Geological Society of America Bulletin*, **116**, 154–170.

IRVING, E., BAKER, J., HAMILTON, M. & WYNNE, P. J. 2004. Early Proterozoic geomagnetic field in western Laurentia: implications for paleolatitudes, local rotations and stratigraphy. *Precambrian Research*, **129**, 251–270.

JACKSON, G. D. & IANNELLI, T. R. 1981. Rift-related cyclic sedimentation in the Neohelikian Borden Basin, northern Baffin Island. *In*: CAMPBELL, F. H. A. (ed.) *Proterozoic basins of Canada*. Geological Survey of Canada, Paper, **81-10**, 269–302.

JACKSON, M. P. A., WARIN, O. N., WOAD, G. M. & HUDEC, M. R. 2003. Neoproterozoic allochthonous salt tectonics during the Lufilian orogeny in the Katangan Copperbelt, central Africa. *Geological Society of America Bulletin*, **115**, 314–330.

JACOBS, J., PISAREVSKY, S., THOMAS, R. J. & BECKER, T. 2008. The Kalahari Craton during the assembly and dispersal of Rodinia. *Precambrian Research*, **160**, 142–158.

JACOBS, J. & THOMAS, R. J. 2004. Himalayan-type indenter-escape tectonics model for the southern part of the late Neoproterozoic–early Paleozoic East African-Antarctic orogen. *Geology*, **32**, 721–724.

JIANG, G., SOHL, L. E. & CHRISTIE-BLICK, N. 2003. Neoproterozoic stratigraphic comparison of the lesser Himalaya (India) and Yangtze block (south China): paleogeographic implications. *Geology*, **31**, 917–920.

JOHANSSON, Å, GEE, D. G., LARIONOV, A. N., OHTA, Y. & TEBENKOV, A. M. 2005. Grenvillian and Caledonian evolution of eastern Svalbard–a tale of two orogenies. *Terra Nova*, **17**, 317–325.

JOHN, T., SCHENK, V., MEZGER, K. & TEMBO, F. 2004. Timing and PT evolution of whiteschist metamorphism in the Lufilian Arc-Zambezi Belt orogen (Zambia): implications for the assembly of Gondwana. *Journal of Geology*, **112**, 71–90.

JOHNSON, S. P., RIVERS, T. & DE WAELE, B. 2005. A review of the Mesoproterozoic to early Palaeozoic magmatic and tectonothermal history of south-central

Africa: implications for Rodinia and Gondwana. *Journal of the Geological Society, London*, **162**, 433–450.

KAMO, S. L. & GOWER, C. F. 1994. Note: U–Pb baddeleyite dating clarifies age of characteristic paleomagnetic remanence of Long Range dykes, southeastern Labrador. *Atlantic Geology*, **30**, 259–262.

KAMO, S. L., GOWER, C. F. & KROGH, T. E. 1989. Birthdate for the Iapetus Ocean? A precise U–Pb zircon and baddeleyite age for the Long Range dikes, southeast Labrador. *Geology*, **17**, 602–605.

KARLSTROM, K. E., HARLAN, S. S., WILLIAMS, M. L., MCLELLAND, J., GEISSMAN, J. W. & AHALL, K. I. 1999. Refining Rodinia: geologic evidence for the Australia-Western U.S. connection in the Proterozoic. *GSA Today*, **9**, 1–7.

KARLSTROM, K. E., ÅHALL, K.-I., HARLAN, S. S., WILLIAMS, M. L., MCLELLAND, J. & GEISSMAN, J. W. 2001. Long-lived (1.8–1.0 Ga) convergent orogen in southern Laurentia, its extensions to Australia and Baltica, and implications for refining Rodinia. *Precambrian Research*, **111**, 5–30.

KELLY, N. M., CLARKE, G. L. & FANNING, C. M. 2002. A two-stage evolution of the Neoproterozoic Rayner Structural Episode: new U–Pb sensitive high resolution ion microprobe constraints from the Oygarden Group, Kemp Land, East Antarctica. *Precambrian Research*, **116**, 307–330.

KEPPIE, J. D., NANCE, R. D., MURPHY, J. B. & DOSTAL, J. 2003. Tethyan, Mediterranean, and Pacific analogues for the Neoproterozoic–Paleozoic birth and development of peri-Gondwanan terranes and their transfer to Laurentia and Laurussia. *Tectonophysics*, **365**, 195–219.

KHUDOLEY, A. K., KROPACHEV, A. P., TKACHENKO, V. I., RUBLEV, A. G., SERGEEV, S. A., MATUKOV, D. I. & LYAHNITSKAYA, O. YU. 2007. Mesoproterozoic to Neoproterozoic evolution of the Siberian craton and adjacent microcontinents: an overview with constraints for a Laurentian connection. *In*: LINK, P. K. & LEWIS, R. S. (eds) *Proterozoic Geology of Western North America and Siberia*. SEPM (Society for Sedimentary Geology) Special Publication, **86**, 209–226.

KINNY, P. D., BLACK, L. P. & SHERATON, J. W. 1993. Zircon ages and the distribution of Archaean and Proterozoic rocks in the Rauer Islands. *Antarctic Science*, **5**, 193–206.

KIRKLAND, C. L., DALY, J. S. & WHITEHOUSE, M. J. 2006. Granitic magmatism of Grenvillian and late Neoproterozoic age in Finnmark, Arctic Norway – Constraining pre-Scandian deformation in the Kalak Nappe Complex. *Precambrian Research*, **145**, 24–52.

KLERKX, J., LIÉGEOIS, J. P., LAVREAU, J. & CLAESSEN, W. 1987. Crustal evolution of northern Kibaran belt in eastern and central Africa. *In*: KRÖNER, A. (ed.) *Proterozoic Lithospheric Evolution*. American Geophysical Union, Geodynamics Series, **17**, 217–233.

KOKONYANGI, J. W., KAMPUNZU, A. B., ARMSTRONG, R., YOSHIDA, M., OKUDAIRA, T., ARIMA, M. & NGULUBE, D. A. 2006. The Mesoproterozoic Kibaride belt (Katanga, SE D.R. Congo). *Journal of African Earth Sciences*, **46**, 1–35.

KRÖNER, A. & CORDANI, U. G. 2003. African, southern Indian and South American cratons were not part of the Rodinia supercontinent: evidence from field relationships and geochronology. *Tectonophysics*, **375**, 325–352.

KUMAR, A., HEAMAN, L. M. & MANIKYAMBA, C. 2007. Mesoproterozoic kimberlites in south India: a possible link to *c* 1.1 Ga global magmatism. *Precambrian Research*, **154**, 192–204.

LAWVER, L. A., GRANTZ, A. & GAHAGAN, L. M. 2002. Plate kinematic evolution of the present Arctic region since the Ordovician. *In*: MILLER, E. L., GRANTZ, A. & KLEMPERER, S. L. (eds) *Tectonic Evolution of the Bering Shelf–Chukchi Sea–Arctic Margin and Adjacent Landmasses*. Geological Society of America, Special Paper, **360**, 333–358.

LECHEMINANT, A. N. & HEAMAN, L. M. 1989. Mackenzie igneous events, Canada: middle Proterozoic hotspot magmatism associated with ocean opening. *Earth and Planetary Science Letters*, **96**, 38–48.

LI, Z.-X. 2000. Palaeomagnetic evidence for unification of the North and West Australian cratons by *c*. 1.7 Ga: new results from the Kimberley Basin of northwestern Australia. *Geophysical Journal International*, **142**, 173–180.

LI, Z.-X. & POWELL, C. MCA. 2001. An outline of the palaeogeographic evolution of the Australasian region since the beginning of the Neoproterozoic. *Earth-Science Reviews*, **53**, 237–277.

LI, Z.-X., ZHANG, L. & POWELL, C. MCA. 1995. South China in Rodinia: part of the missing link between Australia-East Antarctica and Laurentia? *Geology*, **23**, 407–410.

LI, Z.-X., ZHANG, L. & POWELL, C. MCA. 1996. Positions of the East Asian cratons in the Neoproterozoic supercontinent Rodinia. *Australian Journal of Earth Sciences*, **43**, 593–604.

LI, Z.-X., LI, X.-H., ZHOU, H. & KINNY, P. D. 2002. Grenvillian continental collision in south China: new SHRIMP U–Pb zircon results and implications for the configuration of Rodinia. *Geology*, **30**, 163–166.

LI, Z.-X., LI, X. H., KINNY, P. D., WANG, J., ZHANG, S. & ZHOU, H. 2003. Geochronology of Neoproterozoic syn-rift magmatism in the Yangtze Craton, South China and correlations with other continents: evidence for a mantle superplume that broke up Rodinia. *Precambrian Research*, **122**, 85–109.

LI, Z.-X., EVANS, D. A. D. & ZHANG, S. 2004. A 90° spin on Rodinia: possible causal links between the Neoproterozoic supercontinent, superplume, true polar wander and low-latitude glaciation. *Earth and Planetary Science Letters*, **220**, 409–421.

LI, Z.-X., BOGDANOVA, S. V. ET AL. 2008. Assembly, configuration, and break-up history of Rodinia: a synthesis. *Precambrian Research*, **160**, 179–210.

LINK, P. K., CHRISTIE-BLICK, N. ET AL. 1993. Middle and Late Proterozoic stratified rocks of the western US Cordillera, Colorado Plateau, and Basin and Range province. *In*: REED, J. C., JR., BICKFORD, M. E., HOUSTON, R. S., LINK, P. K., RANKIN, D. W., SIMS, P. K. & VAN SCHMUS, W. R. (eds) *Precambrian: Conterminous U.S.* Boulder, Colorado, Geological Society of America, The Geology of North America, **C-2**, 463–595.

LOEWY, S. L., CONNELLY, J. N., DALZIEL, I. W. D. & GOWER, C. F. 2003. Eastern Laurentia in Rodinia: constraints from whole-rock Pb and U/Pb geochronology. *Tectonophysics*, **375**, 169–197.

LU, S., LI, H., ZHANG, C. & NIU, G. 2008a. Geological and geochronological evidence for the Precambrian evolution of the Tarim Craton and surrounding continental fragments. *Precambrian Research*, **160**, 94–107.

LU, S., ZHAO, G., WANG, H. & HAO, G. 2008b. Precambrian metamorphic basement and sedimentary cover of the North China Craton: a review. *Precambrian Research*, **160**, 77–93.

LUND, K., ALEINIKOFF, J. N., EVANS, K. V. & FANNING, C. M. 2003. SHRIMP U–Pb geochronology of Neoproterozoic Windermere Supergroup, central Idaho: implications for rifting of western Laurentia and synchroneity of Sturtian glacial deposits. *Geological Society of America Bulletin*, **115**, 349–372.

MACLEAN, B. C. & COOK, D. G. 2004. Revisions to the Paleoproterozoic Sequence A, based on reflection seismic data across the western plains of the Northwest Territories, Canada. *Precambrian Research*, **129**, 271–289.

MAIER, W. D., PELTONEN, P. & LIVESEY, T. 2007. The ages of the Kabanga North and Kapalagulu intrusions, western Tanzania: a reconnaissance study. *Economic Geology*, **102**, 147–154.

MAJI, A. K., GOON, S., BHATTACHARYA, A., MISHRA, B., MAHATO, S. & BERNHARDT, H.-J. 2008. Proterozoic polyphase metamorphism in the Chhotanagpur Gneissic Complex (India), and implication for transcontinental Gondwanaland correlation. *Precambrian Research*, **162**, 385–402.

MALONE, S. J., MEERT, J. G. ET AL. 2008. Paleomagnetism and detrital zircon geochronology of the upper Vindhyan sequence, Son Valley and Rajasthan, India: a c. 1000 Ma closure age for the Purana basins? *Precambrian Research*, **164**, 137–159.

MALOOF, A. C., HALVERSON, G. P., KIRSCHVINK, J. L., SCHRAG, D. P., WEISS, B. P. & HOFFMAN, P. F. 2006. Combined paleomagnetic, isotopic, and stratigraphic evidence for true polar wander from the Neoproterozoic Akademikerbreen Group, Svalbard. *Geological Society of America Bulletin*, **118**, 1099–1124.

MASLOV, A. V., ERDTMANN, B.-D., IVANOV, K. S., IVANOV, S. N. & KRUPENIN, M. T. 1997. The main tectonic events, depositional history and the palaeogeography of the southern Urals during the Riphean–early Palaeozoic. *Tectonophysics*, **276**, 313–335.

MAYER, A., HOFMANN, A. W., SINIGOI, S. & MORAIS, E. 2004. Mesoproterozoic Sm–Nd and U–Pb ages for the Kunene Anorthosite Complex of SW Angola. *Precambrian Research*, **133**, 187–206.

MBEDE, E. I., KAMPUNZU, A. B. & ARMSTRONG, R. A. 2004. Neoproterozoic inheritance during Cainozoic rifting in the western and southwestern branches of the East African rift system: evidence from carbonatite and alkaline intrusions. *Conference abstract, The East African Rift System: Development, Evolution and Resources*, Addis Ababa, Ethiopia, June 20–24, 2004.

MCCABE, C. & VAN DER VOO, R. 1983. Paleomagnetic results from the upper Keweenawan Chequamegon

Sandstone: implications for red bed diagenesis and Late Precambrian apparent polar wander of North America. *Canadian Journal of Earth Sciences*, **20**, 105–112.

MCELHINNY, M. W., POWELL, C. MCA. & PISAREVSKY, S. A. 2003. Paleozoic terranes of eastern Australia and the drift history of Gondwana. *Tectonophysics*, **362**, 41–65.

MCMENAMIN, M. A. S. & MCMENAMIN, D. L. S. 1990. *The Emergence of Animals: The Cambrian Breakthrough*. New York, Columbia University Press, 1–217.

MEERT, J. G. & STUCKEY, W. 2002. Revisiting the paleomagnetism of the 1.476 Ga St. Francois Mountains igneous province, Missouri, *Tectonics*, **21**(2), doi: 10.1029/2000TC001265.

MEERT, J. G. & TORSVIK, T. H. 2003. The making and unmaking of a supercontinent: rodinia revisited. *Tectonophysics*, **375**, 261–288.

MEERT, J. G., HARGRAVES, R. B., VAN DER VOO, R., HALL, C. H. & HALLIDAY, A. N. 1994. Paleomagnetism and $^{40}Ar/^{39}Ar$ studies of Late Kibaran intrusives in Burundi, East Africa: implications for Late Proterozoic supercontinents. *Journal of Geology*, **102**, 621–637.

MEERT, J. G., VAN DER VOO, R. & AYUB, S. 1995. Paleomagnetic investigation of the Late Proterozoic Gagwe lavas and Mbozi complex, Tanzania and the assembly of Gondwana. *Precambrian Research*, **74**, 225–244.

MERTANEN, S., PESONEN, L. J. & HUHMA, H. 1996. Palaeomagnetism and Sm–Nd ages of the Neoproterozoic diabase dykes in Laanila and Kautokeino, northern Fennoscandia. *In*: BREWER, T. S. (ed.) *Precambrian Crustal Evolution in the North Atlantic Region*. Geological Society, London, Special Publications, **112**, 331–358.

MEZGER, K. & COSCA, M. A. 1999. The thermal history of the Eastern Ghats Belt (India) as revealed by U–Pb and $^{40}Ar/^{39}Ar$ dating of metamorphic and magmatic minerals: implications for the SWEAT correlation. *Precambrian Research*, **94**, 251–271.

MILLER, K. C. & HARGRAVES, R. B. 1994. Paleomagnetism of some Indian kimberlites and lamproites. *Precambrian Research*, **69**, 259–267.

MOORES, E. M. 1991. Southwest US–East Antarctic (SWEAT) connection: a hypothesis. *Geology*, **19**, 425–428.

MURPHY, J. B. & NANCE, R. D. 2003. Do supercontinents introvert or extrovert?: Sm–Nd isotope evidence. *Geology*, **31**, 873–876.

MURPHY, J. B. & NANCE, R. D. 2005. Do supercontinents turn inside-in or inside-out? *International Geology Review*, **47**, 591–619.

MURPHY, J. B., STRACHAN, R. A., NANCE, R. D., PARKER, K. D. & FOWLER, M. B. 2000. Proto-Avalonia: a 1.2–1.0 Ga tectonothermal event and constraints for the evolution of Rodinia. *Geology*, **28**, 1071–1074.

MURTHY, G., GOWER, C., TUBRETT, M. & PÄTZOLD, R. 1992. Paleomagnetism of Eocambrian Long Range dykes and Double Mer Formation from Labrador, Canada. *Canadian Journal of Earth Sciences*, **29**, 1224–1234.

MYERS, J. S., SHAW, R. D. & TYLER, I. M. 1996. Tectonic evolution of Proterozoic Australia. *Tectonics*, **15**, 1431–1446.

OCCHIPINTI, S. A. & REDDY, S. M. 2009. Neoproterozoic reworking of the Palaeoproterozoic Capricorn Orogen of Western Australia and implications for the amalgamation of Rodinia. *In*: MURPHY, J. B. (ed.) *Ancient Orogens and Modern Analogues*. Geological Society, London, Special Publications, **327**, 445–456.

PALMER, H. C., MERZ, B. A. & HAYATSU, A. 1977. The Sudbury dikes of the Grenville Front region: paleomagnetism, petrochemistry, and K–Ar age studies. *Canadian Journal of Earth Sciences*, **14**, 1867–1887.

PARK, J. K. & JEFFERSON, C. W. 1991. Magnetic and tectonic history of the Late Proterozoic Upper Little Dal and Coates Lake Groups of northwestern Canada. *Precambrian Research*, **52**, 1–35.

PARK, J. K., NORRIS, D. K. & LAROCHELLE, A. 1989. Paleomagnetism and the origin of the Mackenzie Arc of northwestern Canada. *Canadian Journal of Earth Sciences*, **26**, 2194–2203.

PATCHETT, P. J., BYLUND, G. & UPTON, B. G. J. 1978. Palaeomagnetism and the Grenville orogeny: new Rb–Sr ages from dolerites in Canada and Greenland. *Earth and Planetary Science Letters*, **40**, 349–364.

PAVLOV, V. E., GALLET, Y., PETROV, P. YU., ZHURAVLEV, D. Z. & SHATSILLO, A. V. 2002. Uy series and late Riphean sills of the Uchur-Maya area: isotopic and palaeomagnetic data and the problem of the Rodinia supercontinent. *Geotectonics*, **36**, 278–292.

PAYNE, J. L., HAND, M., BAROVICH, K. M., REID, A. & EVANS, D. A. D. 2009. Correlations and reconstruction models for the 2500–1500 Ma evolution of the Mawson Continent. *In*: REDDY, S. M., MAZUMDER, R., EVANS, D. A. D. & COLLINS, A. S. (eds) *Palaeoproterozoic Supercontinents and Global Evolution*. Geological Society, London, Special Publications, **323**, 319–355.

PESONEN, L. J. & NEVANLINNA, H. 1981. Late Precambrian Keweenawan asymmetric reversals. *Nature*, **294**, 436–439.

PESONEN, L. J., ELMING, S.-Å. *ET AL*. 2003. Palaeomagnetic configuration of continents during the Proterozoic. *Tectonophysics*, **375**, 289–324.

PETTERSSON, Å., CORNELL, D. H., MOEN, H. F. G., REDDY, S. & EVANS, D. 2007. Ion-probe dating of 1.2 Ga collision and crustal architecture in the Namaqua-Natal Province of southern Africa. *Precambrian Research*, **158**, 79–92.

PHILLIPS, G., WILSON, C. J. L., CAMPBELL, I. H. & ALLEN, C. M. 2006. U–Th–Pb detrital zircon geochronology from the southern Prince Charles Mountains, East Antarctica—Defining the Archaean to Neoproterozoic Ruker Province. *Precambrian Research*, **148**, 292–306.

PIMENTEL, M. M. & FUCK, R. A. 1992. Neoproterozoic crustal accretion in central Brazil. *Geology*, **20**, 375–379.

PIMENTEL, M. M., FERREIRA FILHO, C. F. & ARMSTRONG, R. A. 2004. SHRIMP U–Pb and Sm–Nd ages of the Niquelândia layered complex: meso-(1.25 Ga) and Neoproterozoic (0.79 Ga) extensional events in central Brazil. *Precambrian Research*, **132**, 133–153.

PIMENTEL, M. M., FERREIRA FILHO, C. F. & ARMELE, A. 2006. Neoproterozoic age of the Niquelândia Complex, central Brazil: further ID-TIMS U–Pb and Sm–Nd isotopic evidence. *Journal of South American Earth Sciences*, **21**, 228–238.

PIPER, J. D. A. 1974. Magnetic properties of the Cunene anorthosite Complex, Angola. *Physics of the Earth and Planetary Interiors*, **9**, 353–363.

PIPER, J. D. A. 1980. Analogous Upper Proterozoic apparent polar wander loops. *Nature*, **283**, 845–847.

PIPER, J. D. A. 2000. The Neoproterozoic supercontinent: rodinia or Palaeopangaea? *Earth and Planetary Science Letters*, **176**, 131–146.

PIPER, J. D. A. 2007. The Neoproterozoic supercontinent Palaeopangaea. *Gondwana Research*, **12**, 202–227.

PISAREVSKY, S. A. & BYLUND, G. 1998. Palaeomagnetism of a key section of the Protogine Zone, southern Sweden. *Geophysical Journal International*, **133**, 185–200.

PISAREVSKY, S. A. & BYLUND, G. 2006. Palaeomagnetism of 935 Ma mafic dykes in southern Sweden and implications for the Sveconorwegian Loop. *Geophysical Journal International*, **166**, 1095–1104.

PISAREVSKY, S. A. & NATAPOV, L. M. 2003. Siberia and Rodinia. *Tectonophysics*, **375**, 221–245.

PISAREVSKY, S. A., WINGATE, M. T. D. & HARRIS, L. B. 2003a. Late Mesoproterozoic (*c*. 1.2 Ga) palaeomagnetism of the Albany–Fraser orogen: no pre-Rodinia Australia–Laurentia connection. *Geophysical Journal International*, **155**, F6–F11.

PISAREVSKY, S. A., WINGATE, M. T. D., POWELL, C. MCA., JOHNSON, S. & EVANS, D. A. D. 2003b. Models of Rodinia assembly and fragmentation. *In*: YOSHIDA, M., WINDLEY, B. F. & DASGUPTA, S. (eds) *Proterozoic East Gondwana: Supercontinent Assembly and Breakup*. Geological Society, London, Special Publications, **206**, 35–55.

PISAREVSKY, S. A., WINGATE, M. T. D., STEVENS, M. K. & HAINES, P. W. 2007. Palaeomagnetic results from the Lancer 1 stratigraphic drillhole, Officer Basin, Western Australia, and implications for Rodinia reconstructions. *Australian Journal of Earth Sciences*, **54**, 561–572.

PISAREVSKY, S. A., NATAPOV, L. M., DONSKAYA, T. V., GLADKOCHUB, D. P. & VERNIKOVSKY, V. A. 2008. Proterozoic Siberia: a promontory of Rodinia. *Precambrian Research*, **160**, 66–76.

PIUZANA, D., PIMENTEL, M. M., FUCK, R. A. & ARMSTRONG, R. 2003. SHRIMP U–Pb and Sm–Nd data for the Araxa Group and associated magmatic rocks: constraints for the age of sedimentation and geodynamic context of the southern Brasilia Belt, central Brazil. *Precambrian Research*, **125**, 139–160.

POWELL, C. MCA. & PISAREVSKY, S. A. 2002. Late Neoproterozoic assembly of East Gondwana. *Geology*, **30**, 3–6.

POWELL, C. MCA., LI, Z. X., MCELHINNY, M. W., MEERT, J. G. & PARK, J. K. 1993. Paleomagnetic constraints on timing of the Neoproterozoic breakup of Rodinia and the Cambrian formation of Gondwana, *Geology*, **21**, 889–892.

POWELL, C. MCA., JONES, D. L., PISAREVSKY, S. & WINGATE, M. T. D. 2001. Palaeomagnetic constraints on the position of the Kalahari craton in Rodinia. *Precambrian Research*, **110**, 33–46.

PUFFER, J. H. 2002. A late Neoproterozoic eastern Laurentian superplume: location, size, chemical composition, and environmental impact. *American Journal of Science*, **302**, 1–27.

RADHAKRISHNA, T. & MATHEW, J. 1996. Late Precambrian (850–800 Ma) palaeomagnetic pole for the south Indian shield from the Harohalli alkaline dykes: geotectonic implications for Gondwana reconstructions. *Precambrian Research*, **80**, 77–87.

RAINBIRD, R. H., JEFFERSON, C. W. & YOUNG, G. M. 1996. The early Neoproterozoic sedimentary Succession B of northwestern Laurentia: correlations and paleogeographic significance. *Geological Society of America Bulletin*, **108**, 454–470.

RAINBIRD, R. H., STERN, R. A., KHUDOLEY, A. K., KROPACHEV, A. P., HEAMAN, L. M. & SUKHORUKOV, V. I. 1998. U–Pb geochronology of Riphean sandstone and gabbro from southeast Siberia and its bearing on the Laurentia-Siberia connection. *Earth and Planetary Science Letters*, **164**, 409–420.

RAMOS, V. A. 1988. Late Proterozoic-Early Paleozoic of South America—a collisional history. *Episodes*, **11**, 168–174.

RAPALINI, A. E. & SÁNCHEZ-BETTUCCI, L. 2008. Widespread remagnetization of late Proterozoic sedimentary units of Uruguay and the apparent polar wander path for the Rio de La Plata craton. *Geophysical Journal International*, **174**, 55–74.

RAUB, T. D., KIRSCHVINK, J. L. & EVANS, D. A. D. 2007. True polar wander: linking deep and shallow geodynamics to hydro- and bio-spheric hypotheses. *In*: KONO, M. (ed.) *Treatise on Geophysics, Volume 5: Geomagnetism*. Amsterdam, Elsevier, 565–589.

RENNE, P. R., ONSTOTT, T. C., D'AGRELLA-FILHO, M. S., PACCA, I. G. & TEIXEIRA, W. 1990. $^{40}Ar/^{39}Ar$ dating of 1.0–1.1 Ga magnetizations from the São Francisco and Kalahari cratons: tectonic implications for Pan-African and Brasiliano mobile belts. *Earth and Planetary Science Letters*, **101**, 349–366.

ROEST, W. R. & SRIVASTAVA, S. P. 1989. Seafloor spreading in the Labrador Sea: a new reconstruction. *Geology*, **17**, 1000–1004.

ROGERS, J. J. W. 1996. A history of continents in the past three billion years. *Journal of Geology*, **104**, 91–107.

ROSS, G. M. 2002. Evolution of Precambrian continental lithosphere in Western Canada: results from Lithoprobe studies in Alberta and beyond. *Canadian Journal of Earth Sciences*, **39**, 413–437.

ROSS, G. M. & VILLENEUVE, M. 2003. Provenance of the Mesoproterozoic (1.45 Ga) Belt basin (western North America): another piece in the pre-Rodinia paleogeographic puzzle. *Geological Society of America Bulletin*, **115**, 1191–1217.

ROSS, G. M., PARRISH, R. R. & WINSTON, D. 1992. Provenance and U–Pb geochronology of the Mesoproterozoic Belt Supergroup (northwestern United States): implications for age of deposition and pre-Panthalassa plate reconstructions. *Earth and Planetary Science Letters*, **113**, 57–76.

ROY, J. L. & ROBERTSON, W. A. 1978. Paleomagnetism of the Jacobsville Formation and the apparent polar wander path for the interval ~1100 to ~670 m.y. for North America. *Journal of Geophysical Research*, **83**, 1289–1304.

ROYER, J.-Y., MÜLLER, R. D., GAHAGAN, L. M., LAWVER, L. A., MAYES, C. L., NÜRNBERG, D. & SCLATER, J. G. 1992. *A global isochron chart*. University of Texas Institute for Geophysics Technical Report, no. 117.

SADOWSKI, G. R. & BETTENCOURT, J. S. 1996. Mesoproterozoic tectonic correlations between eastern Laurentia and the western border of the Amazon Craton. *Precambrian Research*, **76**, 213–227.

SALMINEN, J., PESONEN, L. J., MERTANEN, S., VUOLLO, J. & AIRO, M.-L. 2009. Palaeomagnetism of the Salla Diabase Dyke, northeastern Finland, and its implication for the Baltica–Laurentia entity during the Mesoproterozoic. *In*: REDDY, S. M., MAZUMDER, R., EVANS, D. A. D. & COLLINS, A. S. (eds) *Palaeoproterozoic Supercontinents and Global Evolution*. Geological Society, London, Special Publications, **323**, 199–218.

SANTOS, J. O. S., HARTMANN, L. A., GAUDETTE, H. E., GROVES, D. I., McNAUGHTON, N. J. & FLETCHER, I. R. 2000. A new understanding of the provinces of the Amazon craton based on integration of field mapping and U–Pb and Sm–Nd geochronology. *Gondwana Research*, **3**, 453–488.

SANTOS, J. O. S., HARTMANN, L. A., BOSSI, J., CAMPAL, N., SCHIPILOV, A., PIÑEYRO, D. & McNAUGHTON, N. J. 2003. Duration of the Trans-Amazonian cycle and its correlation within South America based on U–Pb SHRIMP geochronology of the La Plata craton, Uruguay. *International Geology Review*, **45**, 27–48.

SANTOS, J. O. S., RIZZOTTO, G. J. *ET AL.* 2008. Age and autochthonous evolution of the Sunsás Orogen in West Amazon Craton based on mapping and U–Pb geochronology. *Precambrian Research*, **165**, 120–152.

SATO, K., SIGA, O., JR., NUTMAN, A. P., BASEI, M. A. S., MCREATH, I. & KAULFUSS, G. 2003. The Atuba complex, southern South American platform: archean components and Paleoproterozoic to Neoproterozoic tectonothermal events. *Gondwana Research*, **6**, 251–263.

SCHMIDT, P. W., WILLIAMS, G. E., CAMACHO, A. & LEE, J. K. W. 2006. Assembly of Proterozoic Australia: implications of a revised pole for the c. 1070 Ma Alcurra Dyke Swarm, central Australia. *Geophysical Journal International*, **167**, 626–634.

SCHOLL, D. W. & VON HUENE, R. 2007. Crustal recycling at modern subduction zones applied to the past issues of growth and preservation of continental basement crust, mantle geochemistry, and supercontinent reconstruction. *In*: HATCHER, R. D., JR., CARLSON, M. P., MCBRIDE, J. H. & MARTINEZ-CATALAN, J. R. (eds) *4-D Framework of Continental Crust*. Geological Society of America Memoir, **200**, 9–32.

SEARS, J. W. & PRICE, R. A. 2003. Tightening the Siberian connection to western Laurentia. *Geological Society of America Bulletin*, **115**, 943–953.

SETH, B., ARMSTRONG, R. A., BRANDT, S., VILLA, I. M. & KRAMERS, J. D. 2003. Mesoproterozoic U–Pb and

Pb–Pb ages of granulites in NW Namibia: reconstructing a complete orogenic cycle. *Precambrian Research*, **126**, 147–168.

SINGLETARY, S. J., HANSON, R. E. ET AL. 2003. Geochronology of basement rocks in the Kalahari Desert, Botswana, and implications for regional Proterozoic tectonics. *Precambrian Research*, **121**, 47–71.

SKLYAROV, E. V., GLADKOCHUB, D. P., MAZUKABZOV, A. M., MENSHAGIN, YU. V., WATANABE, T. & PISAREVSKY, S. A. 2003. Neoproterozoic mafic dike swarms of the Sharyzhalgai metamorphic massif, southern Siberian craton. *Precambrian Research*, **122**, 359–376.

SMETHURST, M. A., KHRAMOV, A. N. & TORSVIK, T. H. 1998. The Neoproterozoic and Palaeozoic palaeomagnetic data for the Siberian platform: from Rodinia to Pangea. *Earth-Science Reviews*, **43**, 1–24.

SRIVASTAVA, S. P. & ROEST, W. R. 1989. Seafloor spreading history II–IV. *In: Map sheets L17-2–L17-6* (Atlantic Geoscience Centre, Geological Survey of Canada, Ottawa).

STAMPFLI, G. M. & BOREL, G. D. 2002. A plate tectonic model for the Paleozoic and Mesozoic constrained by dynamic plate boundaries and restored synthetic oceanic isochrons. *Earth and Planetary Science Letters*, **196**, 17–33.

STEARN, J. E. F. & PIPER, J. D. A. 1984. Palaeomagnetism of the Sveconorwegian mobile belt of the Fennoscandian Shield. *Precambrian Research*, **23**, 201–246.

SU, W., ZHANG, S. ET AL. 2008. SHRIMP U–Pb ages of K-bentonite beds in the Xiamaling Formation: implications for revised subdivision of the Meso- to Neoproterozoic history of the North China Craton. *Gondwana Research*, **14**, 543–553.

SVENNINGSEN, O. M. 2001. Onset of seafloor spreading in the Iapetus Ocean at 608 Ma: precise age of the Sarek Dyke Swarm, northern Swedish Caledonides. *Precambrian Research*, **110**, 241–254.

TACK, L., LIÉGEOIS, J. P., DEBLOND, A. & DUCHESNE, J. C. 1994. Kibaran A-type granitoids and mafic rocks generated by two mantle sources in a late orogenic setting (Burundi). *Precambrian Research*, **68**, 323–356.

TANAKA, H. & IDNURM, M. 1994. Palaeomagnetism of Proterozoic mafic intrusions and host rocks of the Mount Isa Inlier, Australia: revisited. *Precambrian Research*, **69**, 241–258.

TASSINARI, C. C. G., BETTENCOURT, J. S., GERALDES, M. C., MACAMBIRA, M. J. B. & LAFON, J. M. 2000. The Amazonian Craton. *In*: CORDANI, U. G., MILANI, E. J., THOMAZ FILHO, A. & CAMPOS, D. A. (eds) *Tectonic evolution of South America*. 31st International Geological Congress, Rio de Janeiro, 41–95.

THIÉBLEMONT, D., GOUJOU, J. C., EGAL, E., COCHERIE, A., DELOR, C., LAFON, J. M. & FANNING, C. M. 2004. Archean evolution of the Leo Rise and its Eburnean reworking. *Journal of African Earth Sciences*, **39**, 97–104.

THORKELSON, D. J., MORTENSEN, J. K., CREASER, R. A., DAVIDSON, G. J. & ABBOTT, J. G. 2001. Early Proterozoic magmatism in Yukon, Canada: constraints on the evolution of northwestern Laurentia. *Canadian Journal of Earth Sciences*, **38**, 1479–1494.

THORKELSON, D. J., ABBOTT, J. A., MORTENSEN, J. K., CREASER, R. A., VILLENEUVE, M. E., MCNICOLL, V. J. & LAYER, P. W. 2005. Early and Middle Proterozoic evolution of Yukon, Canada. *Canadian Journal of Earth Sciences*, **42**, 1045–1071.

TOHVER, E., VAN DER PLUIJM, B. A., VAN DER VOO, R., RIZZOTTO, G. & SCANDOLARA, J. E. 2002. Paleogeography of the Amazon craton at 1.2 Ga: early Grenvillian collision with the Llano segment of Laurentia. *Earth and Planetary Science Letters*, **199**, 185–200.

TOHVER, E., BETTENCOURT, J. S., TOSDAL, R., MEZGER, K., LEITE, W. B. & PAYOLLA, B. L. 2004. Terrane transfer during the Grenville orogeny: tracing the Amazonian ancestry of southern Appalachian basement through Pb and Nd isotopes. *Earth and Planetary Science Letters*, **228**, 161–176.

TOHVER, E., D'AGRELLA-FILHO, M. S. & TRINDADE, R. I. F. 2006. Paleomagnetic record of Africa and South America for the 1200–500 Ma interval, and evaluation of Rodinia and Gondwana assemblies. *Precambrian Research*, **147**, 193–222.

TORSVIK, T. H., CARTER, L. M., ASHWAL, L. D., BHUSHAN, S. K., PANDID, M. K. & JAMTVEIT, B. 2001. Rodinia refined or obscured: palaeomagnetism of the Malani igneous suite (NW India). *Precambrian Research*, **108**, 319–333.

TORSVIK, T. H., MÜLLER, R. D., VAN DER VOO, R., STEINBERGER, B. & GAINA, C. 2008. Global plate motion frames: toward a unified model. *Reviews of Geophysics*, **46**, RG3004, doi: 10.1029/2007RG000227.

TOSDAL, R. M. 1996. The Amazon-Laurentian connection as viewed from the Middle Proterozoic rocks in the central Andes, western Bolivia and northern Chile. *Tectonics*, **15**, 827–842.

TRINDADE, R. I. F., FONT, E., D'AGRELLA-FILHO, M. S., NOGUEIRA, A. C. R. & RICCOMINI, C. 2003. Low-latitude and multiple geomagnetic reversals in the Neoproterozoic Puga cap carbonate, Amazon craton. *Terra Nova*, **15**, 441–446.

TROMPETTE, R. 1997. Neoproterozoic (c. 600 Ma) aggregation of Western Gondwana: a tentative scenario. *Precambrian Research*, **82**, 101–112.

UPTON, B. G. J., RÄMÖ, O. T., HEAMAN, L. M., BLICHERT-TOFT, J., KALSBEEK, F., BARRY, T. L. & JEPSEN, H. F. 2005. The Mesoproterozoic Zig-Zag Dal basalts and associated intrusions of eastern North Greenland: mantle plume-lithosphere interaction. *Contributions to Mineralogy and Petrology*, **149**, 40–56.

VALERIANO, C. M., MACHADO, N., SIMONETTI, A., VALLADARES, C. S., SEER, H. J. & SIMÕES, L. S. A. 2004. U–Pb geochronology of the southern Brasilia belt (SE-Brazil): sedimentary provenance, Neoproterozoic orogeny and assembly of West Gondwana. *Precambrian Research*, **130**, 27–55.

VAN DER VOO, R. 1990. The reliability of paleomagnetic data. *Tectonophysics*, **184**, 1–9.

VAUCHEZ, A., NEVES, S., CABY, R., CORSINI, M., EGYDIO-SILVA, M., ARTHAUD, M. & AMARO, V. 1995. The Borborema shear zone system, NE Brazil. *Journal of South American Earth Sciences*, **8**, 247–266.

VEEVERS, J. J. & SAEED, A. 2008. Gamburtsev Subglacial Mountains provenance of Permian-Triassic sandstones in the Prince Charles Mountains and offshore Prydz Bay: Integrated U–Pb and $T_{DM}$ ages and host-rock affinity from detrital zircons. *Gondwana Research*, **14**, 316–342.

VOLOZH, YU. A., ANTIPOV, M. P., BRUNET, M.-F., GARAGASH, I. A., LOBKOVSKII, L. I. & CADET, J.-P. 2003. Pre-Mesozoic geodynamics of the Precaspian Basin (Kazakhstan). *Sedimentary Geology*, **156**, 35–58.

VRANA, S., KACHLIK, V., KRÖNER, A., MARHEINE, D., SEIFERT, A. V., ZACEK, V. & BABUREK, J. 2004. Ubendian basement and its late Mesoproterozoic and early Neoproterozoic structural and metamorphic overprint in northeastern Zambia. *Journal of African Earth Sciences*, **38**, 1–21.

WALDERHAUG, H. J., TORSVIK, T. H., EIDE, E. A., SUNDVOLL, B. & BINGEN, B. 1999. Geochronology and palaeomagnetism of the Hunnedalen dykes, SW Norway: implications for the Sveconorwegian apparent polar wander loop. *Earth and Planetary Science Letters*, **169**, 71–83.

WALDERHAUG, H. J., TORSVIK, T. H. & HALVERSON, E. 2007. The Egersund dykes (SW Norway): a robust Early Ediacaran (Vendian) palaeomagnetic pole from Baltica. *Geophysical Journal International*, **168**, 935–948.

WANG, Y., LIU, D., CHUNG, S.-L., TONG, L. & REN, L. 2008. SHRIMP zircon ageconstraints from the Larsemann Hills region, Prydz Bay, for a late Mesoproterozoic to early Neoproterozoic tectono-thermal event in East Antarctica. *American Journal of Science*, **308**, 573–617.

WARNOCK, A. C., KODAMA, K. P. & ZEITLER, P. K. 2000. Using thermochronometry and low-temperature demagnetization to accurately date Precambrian paleomagnetic poles. *Journal of Geophysical Research*, **105**, 19435–19453.

WATT, G. R. & THRANE, K. 2001. Early Neoproterozoic events in East Greenland. *Precambrian Research*, **110**, 165–184.

WEGENER, A. 1929. *The Origin of Continents and Oceans.* [English translation by J. BIRAM, 1966. New York, Dover].

WEIL, A. B., VAN DER VOO, R., MAC NIOCAILL, C. & MEERT, J. G. 1998. The Proterozoic supercontinent Rodinia: paleomagnetically derived reconstructions for 1100 to 800 Ma. *Earth and Planetary Science Letters*, **154**, 13–24.

WEIL, A. B., GEISSMAN, J. W., HEIZLER, M. & VAN DER VOO, R. 2003. Paleomagnetism of Middle Proterozoic mafic intrusions and Upper Proterozoic (Nankoweap) red beds from the Lower Grand Canyon Supergroup, Arizona. *Tectonophysics*, **375**, 199–220.

WEIL, A. B., GEISSMAN, J. W. & VAN DER VOO, R. 2004. Paleomagnetism of the Neoproterozoic Chuar Group, Grand Canyon Supergroup, Arizona: implications for Laurentia's Neoproterozoic APWP and Rodinia break-up. *Precambrian Research*, **129**, 71–92.

WEIL, A. B., GEISSMAN, J. W. & ASHBY, J. M. 2006. A new paleomagnetic pole for the Neoproterozoic

Uinta Mountain supergroup, Central Rocky Mountain States, USA. *Precambrian Research*, **147**, 234–259.

WHITE, R. W., CLARKE, G. L. & NELSON, D. R. 1999. SHRIMP U–Pb zircon dating of Grenville-age events in the western part of the Musgrave Block, central Australia. *Journal of Metamorphic Geology*, **17**, 465–481.

WINGATE, M. T. D. & EVANS, D. A. D. 2003. Palaeomagnetic constraints on the Proterozoictectonic evolution of Australia. *In*: YOSHIDA, M., WINDLEY, B. & DASGUPTA, S. (eds) *Proterozoic East Gondwana: Super Continent Assembly and Break-up.* Geological Society, London, Special Publications, **206**, 77–91.

WINGATE, M. T. D. & GIDDINGS, J. W. 2000. Age and palaeomagnetism of the Mundine Well dyke swarm, Western Australia: implications for an Australia-Laurentia connection at 755 Ma. *Precambrian Research*, **100**, 335–357.

WINGATE, M. T. D., PISAREVSKY, S. A. & EVANS, D. A. D. 2002. Rodinia connections between Australia and Laurentia: no SWEAT, no AUSWUS? *Terra Nova*, **14**, 121–128.

WOODEN, J. L. & MILLER, D. M. 1990. Chronologic and isotopic framework for Early Proterozoic crustal evolution in the eastern Mojave Desert region. *Journal of Geophysical Research*, **95**, 20133–20146.

WU, H., ZHANG, S., LI, Z.-X., LI, H. & DONG, J. 2005. New paleomagnetic results from the Yangzhuang Formation of the Jixian System, North China, and tectonic implications. *Chinese Science Bulletin*, **50**, 1483–1489.

XU, B., JIAN, P., ZHENG, H., ZOU, H., ZHANG, L. & LIU, D. 2005. U–Pb zircon geochronology and geochemistry of Neoproterozoic volcanic rocks in the Tarim Block of northwest China: implications for the breakup of Rodinia supercontinent and Neoproterozoic glaciations. *Precambrian Research*, **136**, 107–123.

YANG, Z. & BESSE, J. 2001. New Mesozoic apparent polar wander path for south China: tectonic consequences. *Journal of Geophysical Research*, **106**, 8493–8520.

ZHAI, M., SHAO, J., HAO, J. & PENG, P. 2003. Geological signature and possible position of the North China Block in the supercontinent Rodinia. *Gondwana Research*, **6**, 171–183.

ZHANG, S. 2004. South China's Gondwana connection in the Paleozoic: paleomagnetic evidence. *Progress in Natural Science*, **14**, 85–90.

ZHANG, S., LI, Z.-X. & WU, H. 2006. New Precambrian palaeomagnetic constraints on the position of the North China Block in Rodinia. *Precambrian Research*, **144**, 213–238.

ZHAO, G., SUN, M. & WILDE, S. A. 2003. Correlations between the Eastern Block of the North China Craton and the South Indian Block of the Indian Shield: an Achaean to Palaeoproterozoic link. *Precambrian Research*, **122**, 201–233.

ZHAO, X. X., PARK, Y. H. & WANG, C. S. 2008. Tarim affinities of the Western Australia and Gondwana: new constraints from Late Cambrian and Middle Ordovician paleomagnetic poles. *The Gondwana 13 Program and Abstracts*, **267**.

# The Grenville Province as a large hot long-duration collisional orogen – insights from the spatial and thermal evolution of its orogenic fronts

TOBY RIVERS

*Department of Earth Sciences, Memorial University of Newfoundland, PO Box 4200, St. John's NL, A1B 3X5, Canada (e-mail: trivers@mun.ca)*

**Abstract:** The proposition that the Grenville Province is a remnant of a large hot long-duration collisional orogen is examined through a comparative study of its present orogenic front, the Grenville Front, and a former front, the Allochthon Boundary Thrust. Structural, metamorphic and geochronologic data for both boundaries and their hanging walls from the length of the Grenville Province are compared. Cumulative displacement across the Grenville Front was minor (10 s of km) whereas that across the Allochthon Boundary Thrust was major (100 s of km), consistent with the observation that the latter boundary separates rocks with a different age, and *P–T* character, of metamorphism.

On an orogen scale, Grenvillian metamorphism can be subdivided into two spatially and temporally distinct orogenic phases, a relatively high *T* Ottawan (*c.* 1090–1020 Ma) phase in the hanging wall of the Allochthon Boundary Thrust, and a relatively lower *T* Rigolet (*c.* 1000–980 Ma) phase in the hanging wall of the Grenville Front. It is argued that the structural setting and ≥50 My duration of Ottawan metamorphism are compatible with some form of channel flow beneath an orogenic plateau, with the Allochthon Boundary Thrust forming the base of the channel. Channel flow ceased at *c.* 1020 Ma when the Allochthon Boundary Thrust was reworked as part of a system of normal-sense shear zones, and following a hiatus of *c.* 20 My the short-lived Rigolet metamorphism took place in the former foreland and involved the development of a new orogenic front, the Grenville Front. Taken together, this suggests the Grenville Orogen developed as a large hot long-duration orogen during the Ottawan orogenic phase, but following gravitational collapse of the plateau the locus of thickening migrated into the foreland and active tectonism was restricted to a subjacent small cold short-duration orogen. The foreland-ward migration of the orogenic front from the Allochthon Boundary Thrust to the Grenville Front, the contrasting *P–T–t* character of the metamorphic rocks in their hanging walls, and the evidence for orogenic collapse followed by renewed growth, provide insights into the complex evolution of a long-duration collisional orogen.

## Introduction

From a mechanical perspective, the evolution of a collisional orogen can be conceptually divided into constructive (or growth) and destructive (or decay) phases separated by a climax representing the time of maximum volume. The constructive phase is characterized by increasing crustal thickness and potential energy (Hodges 1998), whereas during the destructive phase the stored potential energy gradually decays as the orogenic crust thins by gravitational collapse and erosion (Dewey 1988; Rey *et al.* 2001).

Modern collisional orogens vary greatly in width and duration, for example from <150 km and <30 My for the Pyrenees, to >1000 km and >50 My for the Himalaya–Tibet Orogen. This variation has also been reproduced in thermal-mechanical experiments (e.g. Beaumont *et al.* 2001*a, b*, 2004, 2006; Jamieson *et al.* 2002, 2004, 2007), which show that for similar model parameters, for example, crustal rheology, radioactive heat-producing potential, convergence velocity, erosion rate, mantle heat flow, etc., orogen width, duration of orogenesis and the maximum temperature attained are all positively correlated. On this basis, Beaumont *et al.* (2006) suggested that collisional orogens form a continuum between two end-members: (i) small cold short-duration orogens (SCOs), for which the essential structural characteristics can be simulated with mechanical models; and (ii) large hot long-duration orogens (LHOs), for which coupled thermal-mechanical models are necessary to produce realistic simulations. Moreover, as is implicit in the experiments and is developed here with respect to the Grenville Province, LHOs also become increasingly complex over time. This is a result of the protracted constructive phase, which leads to widening of the orogen, and because the accompanying regional metamorphism is of high temperature due to the long time available for conductive and radioactive

*From*: MURPHY, J. B., KEPPIE, J. D. & HYNES, A. J. (eds) *Ancient Orogens and Modern Analogues.*
Geological Society, London, Special Publications, **327**, 405–444.
DOI: 10.1144/SP327.17 0305-8719/09/$15.00 © The Geological Society of London 2009.

heating, thereby promoting rheological weakening and the potential for mid-crustal flow.

Orogenic fronts define the contemporary limits of orogens, their contractional character providing linkage to the hinterland where the greatest thickening and heating occur. This paper focuses on the structural/metamorphic character and timing of two orogenic fronts in the Grenville Province and the insight they provide on the evolution of large hot long-duration collisional orogens.

## Factors affecting orogenic fronts in large hot long-duration orogens

As inferred for the Himalaya–Tibet Orogen (e.g. Clark & Royden 2000; Grujic et al. 2002; Godin et al. 2006a; Searle et al. 2006) and modelled in numerical experiments, LHOs develop in orthogonal collisional settings and are characterized by a plateau in the orogenic hinterland that forms when the crust is thickened beyond its load-bearing capacity. The resultant forced flow of melt-weakened crust reduces local topography on the plateau and under appropriate erosional conditions leads to exhumation of a mid-crustal 'channel' at the orogenic front. On the basis of their numerical models, Beaumont et al. (2006) distinguished three crustal flow modes: (i) homogeneous channel flow of low-viscosity mid crust; (ii) heterogeneous channel flow of low-viscosity mid-crust and weak segments of lower crust, both modes driven by gravitational forcing; and (iii) extrusion of higher viscosity hot fold nappes by tectonic forcing.

It follows from the modelling of Beaumont et al. (2006) that in a stable LHO, defined as one in which the volume of crust remains constant, the amount of new crust added by thrusting is balanced by that removed by erosion. Since plateaux are regions of low erosion rate, most of the mass is removed at orogenic fronts where rates are higher (Grujic et al. 2006). Thus, emergence of the mid-crustal channel at the orogenic front is the principal method by which stable LHOs reduce their mass. Destabilization of the mass balance would occur if, for instance, the volume of crust added to the orogen exceeded the volume flowing through the channel, or if the viscosity of the flowing mid-crust in the channel was changed beyond critical values by heating or cooling. Such scenarios could lead to overthickening and orogenic collapse, and/or widening and the formation of a new orogenic front. In this contribution, it is argued that a record of these processes is preserved in the Grenville Province where there is evidence for collapse following thickening in the orogenic interior, and for outward growth of the orogen into its former foreland and the formation of a new orogenic front.

## Orogenic fronts of the Grenville Province

The Grenville Orogen, part of which is exposed in the Grenville Province in the SE Canadian Shield, is a late Mesoproterozoic–early Neoproterozoic feature widely interpreted to have developed during assembly of Rodinia as a result of collision between Laurentia and Amazonia (Li et al. 2008). It is a prime candidate for a large hot long-duration orogen, because it is >600 km wide (minimum estimate because the SE margin was removed by Neoproterozoic rifting), largely underlain by Grenvillian granulite- and upper-amphibolite-facies rocks at the present erosion surface implying it was hot, and because the Grenvillian Orogeny lasted for ≥100 My (from c. 1090–980 Ma; Rivers 1997; Carr et al. 2000; Rivers 2008). The orogenic fronts of the Grenville Province discussed in this paper are an older, more internal boundary known as the Allochthon Boundary Thrust and a younger external boundary known as the Grenville Front. Their $P-T$ evolution and crustal-scale architecture are used to develop a tectonic model for the Grenville Orogen that is compared to a thermal-mechanical model for the western Grenville Province developed by Jamieson et al. (2007).

## General features of the Grenville Province

The Grenville Province is composed of rocks of Archaean to Mesoproterozoic age that generally young towards the SE, reflecting the accretionary growth of southeastern Laurentia during the Palaeo- and Mesoproterozoic (Hoffman 1989; Rivers 1997; Karlstrom et al. 2001). Thus Archaean rocks derived from the Superior Province extend into the northern Grenville Province and underlie much of it at depth and some Palaeoproterozoic units can be traced across the Grenville Front, whereas the youngest units, c. 1.3–1.2 Ga accreted terranes, occur in the upper and mid-crust in the SE (Figs 1 & 2; e.g. Gower et al. 1980; Rivers & Chown 1986; Rivers et al. 1989). The collisional suture with Amazonia does not outcrop in the exposed Grenville Province, but on the basis of Pb isotopic data it may lie close to the Grenville inliers in the SE Appalachians (Loewy et al. 2003; Fig. 1 inset). The Interior Magmatic Belt (Fig. 1; Rivers 1997), characterized by syn- to post-Grenvillian plutons, defines the thermal core of the orogen.

## Crustal and metamorphic architecture and timing of metamorphism

The Grenville Orogen is a crustal-scale thrust stack (modified by later extension – see below) composed

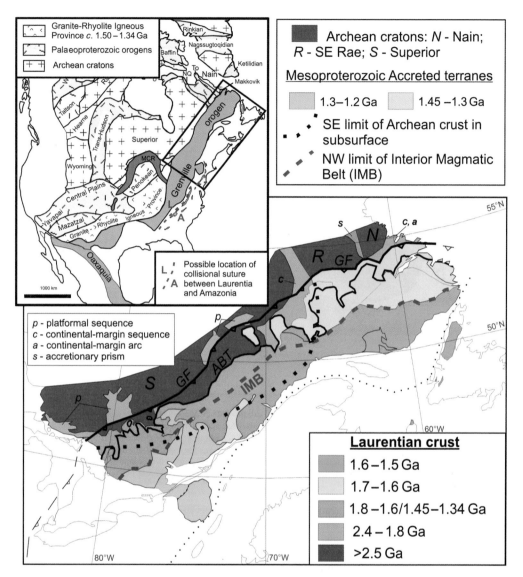

**Fig. 1.** Sketch map of the Grenville Province showing age of pre-Grenvillian rocks (modified from Rivers *et al.* in review). Note continuity of Archaean and Palaeoproterozoic units from the foreland into the Grenville Orogen and presence of Mesoproterozoic accreted terranes in the southwestern Grenville Province. *Inset*: sketch map of North America showing that the Grenville Front truncates older orogenic provinces (after Hoffman 1989) and possible location of the orogenic suture. ABT, Allochthon Boundary Thrust; GF, Grenville Front; IMB, Interior Magmatic Belt.

of structurally-bound terranes and domains that are grouped into belts on a regional scale (the term terrane is used in the sense of a metamorphic terrane with a distinctive Grenvillian metamorphic history; it does not necessarily connote an exotic origin with respect to Laurentia). The lowest belt in the stack, the Parautochthonous Belt (Rivers *et al.* 1989), is bounded by the Grenville Front at its base and the Allochthon Boundary Thrust at its

roof (Fig. 2). As elaborated below, the Grenville Front is a contractional structure that carries the Parautochthonous Belt over the adjacent foreland, but continuity of several units across it indicates it is not the site of major tectonic transport (Rivers *et al.* 1989). The Allochthon Boundary Thrust is a major ductile shear zone that, as its name implies, was initiated as a contractional structure, but was reworked in extension locally (Culshaw *et al.*

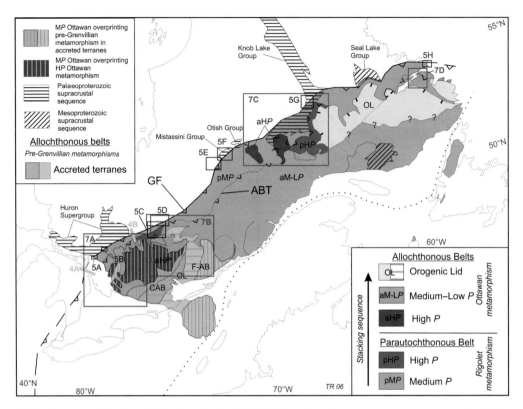

**Fig. 2.** Sketch map of the Grenville Province showing the distribution of late Mesoproterozoic metamorphism (modified from Rivers *et al.* submitted). Grenvillian metamorphism is subdivided into Ottawan (*c.* 1190–1120 Ma) and Rigolet (*c.* 1005–980 Ma) orogenic phases. Metamorphism associated with accretion of the Composite Arc Belt (CAB; *c.* 1240–1225 Ma) and Frontenac–Adirondack Belt (F–AB; *c.* 1190–1140 Ma) was pre-Grenvillian. Blue lines labelled 4A–B are locations of *Lithoprobe* deep-seismic reflection experiments shown in Figure 4; boxes labelled 5A–H and 7A–D show locations of detailed studies of the Grenville Front and Allochthon Boundary Thrust in Figures 5 and 7.

1997; Ketchum *et al.* 1998). As shown below, the Allochthon Boundary Thrust marks a major break in the distribution of lithologic units, structural fabrics, and in the age and character of metamorphism. In order to emphasise these contrasts, Rivers *et al.* (1989) referred to the terranes in its hanging wall as allochthonous, indicating that they are far-travelled (but not exotic to Laurentia).

The Grenville Province is largely underlain by high-grade rocks formed during the Grenvillian Orogeny, implying the present erosion surface provides a view of the mid- and lower orogenic crust. In order to analyse its crustal-scale structure, Grenvillian metamorphism is subdivided on the basis of age and baric character (Fig. 2).

*Age of Grenvillian metamorphism.* Grenvillian metamorphism took place from *c.* 1090–980 Ma, but there are important temporal variations that correspond to the spatial division into the para-autochthonous and allochthonous belts (Rivers

1997, 2008; Rivers *et al.* 2002). Apart from local accretionary metamorphic events that pre-dated collisional orogenesis and are not considered in detail here, the main regional metamorphism and associated crustal thickening in the allochthonous terranes in the interior Grenville Province took place from *c.* 1090–1020 Ma, referred to as the Ottawan event, whereas in the Para-autochthonous Belt at the northwestern margin the main metamorphism took place from *c.* 1005–980 Ma (Rigolet event; Fig. 2). Both these metamorphic events are of regional extent but for the most part their effects are geographically separated, and moreover there was a hiatus between them following an important period of extension at *c.* 1020 Ma at the end of the Ottawan phase. The tectonic significance of these two regional metamorphisms and associated contractional events has spawned a confusing terminology. It is possible they represent independent collisions, in which case each would merit the status of an orogeny, as proposed by McLelland *et al.* (2001).

Alternatively they may be manifestations of a single long-duration collisional orogeny, an interpretation supported by comparison with models of LHOs and preferred by Rivers (1997, 2008), Gower & Krogh (2002), Rivers *et al.* (2002), and further developed herein. Thus in this paper the Ottawan and Rigolet are referred to as orogenic phases of the long-duration Grenvillian Orogeny. In any case, the distribution of metamorphic ages in Figure 2 implies that in Ottawan times the contemporary orogenic front lay close to the Allochthon Boundary Thrust and that in Rigolet times the orogen advanced into its foreland and the Grenville Front was established.

*Baric subdivisions.* In light of their different ages, metamorphism in the Parautochthonous Belt and allochthonous belts is considered separately, symbolized by the prefixes p and a respectively. With respect to the Parautochthonous Belt, the Rigolet metamorphism is principally medium pressure (pM*P*) or Barrovian in character, but there are two known areas of high pressure metamorphism (pH*P*, >1200 MPa) characterized by eclogite- and H*P* granulite-facies assemblages. With respect to the allochthonous belts, Ottawan metamorphism is subdivided into a High-Pressure (aH*P*) Belt, also characterized by eclogite, relict eclogite and H*P* granulite assemblages, a Medium- to Low-Pressure (aM-L*P*) Belt underlain by granulite- to amphibolite-facies and locally greenschist-facies assemblages, and an Orogenic Lid that largely escaped metamorphic reworking (White *et al.* 2000; Rivers *et al.* 2002; Rivers & Indares 2006; Rivers 2008). More detailed mapping is necessary to subdivide the M-L*P* belt into discrete M*P* and L*P* segments.

## Geophysical character

The levelled magnetic and Bouguer gravity anomaly maps of the Grenville Province are shown in Figure 3. In terms of magnetic expression (Fig. 3a), the Parautochthonous Belt has a distinctive subdued signature in the central Grenville Province, contrasting with the E-trending pattern in the adjacent Superior Province despite the continuity of Archaean crust across the Grenville Front. Although not studied in detail, the contrast may be related to the breakdown of magnetite during the Rigolet metamorphic overprint (Toft *et al.* 1993). In contrast to the Parautochthonous Belt, the overlying allochthonous belts exhibit a short-wavelength, variable-intensity magnetic signature that defines the complex regional structure, the high intensity components of the signal suggesting magnetite was stable during the Ottawan metamorphism.

With respect to the Bouguer gravity map (Fig. 3b), apart from several large negative anomalies corresponding to anorthosite complexes, much of the interior Grenville Province is in approximate isostatic equilibrium, confirming seismic studies that it is underlain by crust of average thickness (*c.* 35 km). In contrast, the Bouguer signature of the Parautochthonous Belt varies from positive in the west to negative under the *c.* 1000 km-long Grenville Front Gravity Low (GFGL). Assuming the underlying crust is of average density, this implies there are important variations in crustal thickness under the northern Grenville Province, from *c.* 30 km in the west to ≥50 km under the GFGL (Hynes 1994), estimates independently confirmed by seismic reflection studies (Green *et al.* 1988; Martignole & Calvert 1996; Hynes *et al.* 2000). The significance of this is discussed later.

# The northwestern Grenville Province – Rigolet deformation and metamorphism

In this section, the deep seismic, structural, metamorphic and age characteristics of the Grenville Front and Parautochthonous Belt are assembled from published studies. The results of two *Lithoprobe* deep seismic experiments in Ontario and western Québec are reviewed first, followed by a discussion of the surface geology from case studies along the length of the Grenville Province.

## Seismic images of the Grenville Front

The *Lithoprobe* seismic study over Lake Huron, Ontario (Green *et al.* 1988; White *et al.* 2000) supported field evidence indicating that the Grenville Front in this region marks the northern limit of a 30 km-wide, SE-dipping, crustal-scale shear zone, the Grenville Front Tectonic Zone (GFTZ; Wynne-Edwards 1972). The seismic section (Fig. 4a) shows the front truncates a prominent subhorizontal mid-crustal reflector in the foreland and reaches *c.* 30 km depth at the SE end of the line. The straightened granitoid and mafic gneisses exposed in the GFTZ exhibit upper amphibolite-facies assemblages, reverse-sense kinematic indicators and down-dip elongation lineations, implying exhumation from the SE.

In the seismic experiment near Témiskamingue, western Québec (Fig. 4b; Kellett *et al.* 1994), not only was reflectivity much weaker but a very different image of the crustal structure was obtained. The Grenville Front in this area is located at a steep fault, one of several reworked Archaean structures with sinistral splays, and exhibits a ramp-flat profile with prominent flats at *c.* 10 and 20 km depth. Reflectors were poorly imaged near the surface, probably due to the steep dips, and moreover Kellett *et al.* (1994) noted that Grenvillian fabrics

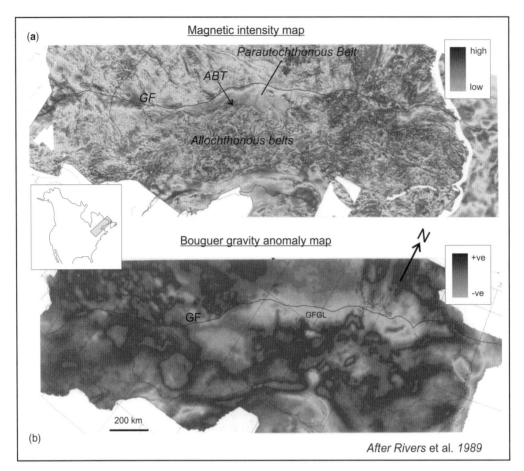

**Fig. 3.** Levelled potential field maps of the Grenville Province (after Rivers *et al.* 1989); grey scales indicate relative intensities. (**a**) aeromagnetic intensity; red lines in Archaean Superior Province highlight E–W structural trends; (**b**) Bouguer gravity anomaly. ABT, Allochthon Boundary Thrust; GF, Grenville Front; GFGL, Grenville Front Gravity Low.

at the surface were limited, except in the vicinity of faults.

The two seismic sections are located *c.* 450 km apart (Fig. 2), implying the tectonic character of the Grenville Front and northern Parautochthonous Belt is heterogeneous at this scale. For instance, there is no structure comparable to the GFTZ at the surface in the western Québec section, the architecture of the Grenville Front changes from moderately-dipping to a ramp-flat geometry, and reaches *c.* 30 km depth at the SE end of the Ontario section, but only *c.* 20 km in western Québec.

## Grenville Front and Parautochthonous Belt – Case studies

Maps and cross-sections at various scales are used with relevant *P–T* and geochronological data to illustrate the structural architecture, metamorphic character and thermal history of the Grenville Front and northern Parautochthonous Belt. Emphasis is placed on units that can be traced across the Grenville Front and only experienced Grenvillian orogenesis, and for one area two maps are presented to illustrate the range of features in different units. Study areas are discussed from west to east, locations are shown in Figure 2, and details of the *P–T* estimates and geochronologic data are given in Table 1.

*Grenville Front Tectonic Zone near Killarney, Ontario.* Part of the Mesoproterozoic (*c.* 1238 Ma) Sudbury dyke swarm crosses the Grenville Front from the foreland into the GFTZ near Killarney, Ontario, making it a suitable marker for Grenvillian deformation and metamorphism (Fig. 5a; Bethune 1997). The foreland in this part of the Southern

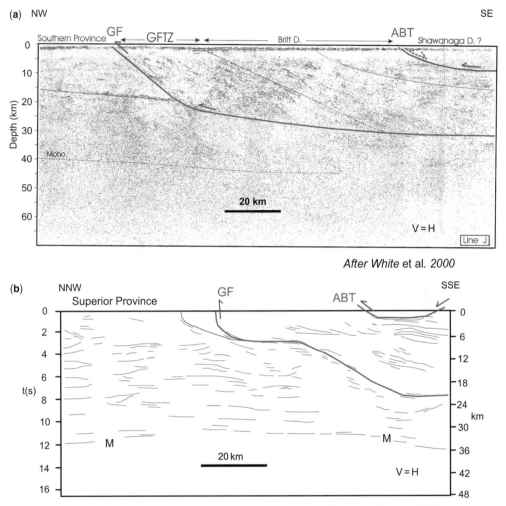

**Fig. 4.** *Lithoprobe* deep seismic reflection transects across the Grenville Front and Allochthon Boundary Thrust. (**a**) part of GLIMPCE-J line in Ontario. (**b**) part of line 15 in western Québec. The Grenville Front (GF) and Grenville Front Tectonic Zone (GFTZ) are crustal-scale contractional structures reaching depths of 25–30 km. The Allochthon Boundary Thrust (ABT) is a sub-horizontal contractional structure that was reworked in extension locally.

(Penokean) Province is underlain by the Palaeoproterozoic Huron Supergroup and Nipissing gabbro sills with E-trending structures and greenschist-facies assemblages cut by a younger NE-trending belt of Palaeoproterozoic granite. Reworked equivalents of all these units occur in the GFTZ, the location of the Grenville Front in the granitoid belt suggesting it followed a pre-existing discontinuity. The SE-trending Sudbury dykes exhibit straight segments and are undeformed in the foreland, some are truncated at the Grenville Front and those that continue into the GFTZ are disrupted, folded and partially transposed into NE trends (Fig. 5a).

Petrographic analysis of Sudbury dykes in the GFTZ has shown that metamorphic mineralogy in coronas separating primary olivine and plagioclase changes from amphibolite- to granulite-facies with distance from the Grenville Front. Peak $P$–$T$ estimates (i.e. $P$ at maximum $T$) for garnet-bearing coronas range from $c$. 630 MPa/710 °C for amphibolite-facies assemblages $c$. 7 km SE of the Grenville Front to $c$. 830 MPa/730 °C for granulite-facies assemblages $c$. 16 km SE of the Front, implying the depth of exhumation increases towards the SE (Bethune 1997; Bethune & Davidson 1997). Circa 1 Ga zircon overgrowths on igneous

**Table 1.** *Compilation of Rigolet mineral assemblages and estimated metamorphic conditions, U–Pb ages of metamorphism, and $^{40}Ar/^{39}Ar$ Hbl cooling ages in the Parautochthonous Belt (PB)*

| Location W → E, pMP/pHP | Assemblages | Estimated P–T–t of metamorphism | References | Comments |
|---|---|---|---|---|
| 20 km transect from GF near Killarney (Ontario) SE into GFTZ [pMP] | Granitic ogn, pgn, amphibolite – Ttn and Hbl-bearing assemblages | 11 U–Pb Ttn analyses lie on chord with UI at 1450 Ma at GF and LI at 978 Ma 20 km SE of GF. Ar Hbl plateau ages from 993–979 Ma | Haggart et al. 1993; Krogh 1994 | c. 0.98 Ga Ttn rims on 1.45 Ga Ttn (or Pb loss?) Hbl plateau ages 15 km SE of GF (partially reset older ages near GF) |
| c. 30 km SE GFTZ, Georgian Bay, (Ontario) [pMP] | Garnet amphibolite: Grt-Pl-Hbl-Bt-Qtz-Ilm-Spl | ≥1100 MPa/750°C c. 990–980 Ma. Ar Hbl ages from 980–960 Ma | Culshaw et al. 1991; Jamieson et al. 1995; Reynolds et al. 1995 | Quasi-isothermal decompression path in southern GFTZ |
| NW GFTZ (between Killarney and Sudbury, Ontario) [pMP] | Metabasite: Grt-Cpx-Pl-Opx-Qtz-Hbl-Spl | 630 MPa/710°C c. 7 km SE of GF, 830 MPa/730°C c. 16 km SE of GF. 1.0 Ga overgrowths of metamorphic Zrn on igneous Bdy | Davidson & van Breemen 1988; Dudás et al. 1994; Bethune 1997; Bethune & Davidson 1997 | Metamorphic field gradient SE of GF |
| GFTZ SE of Val d'Or (W. Québec) [pMP and pHP] | Metabasite: Grt-Cpx-Pl-Opx-Qtz-Hbl | 1200–1500 MPa/800°C to 900 MPa/700°C c. 1 Ga (Mnz) Ar Hbl ages c. 995 Ma | Indares & Martignole 1989; Gariépy et al. 1990; Childe et al. 1993; Martignole & Reynolds 1997; Martignole & Martelat 2005 | Quasi-isothermal decompression path. Grenvillian Mnz in southern PB |
| Gagnon terrane (E. Québec) [pMP and pHP] | Pelite: Qtz-Ms-Bt-Grt-Ky-Pl-L | c. 1400 MPa/>800°C 995–985 Ma | Indares 1995; Jordan et al. 2006 | Bt dehydration melting followed by melt crystallization |
| Gagnon terrane (near Wabush, W. Labrador) [pMP] | Pelite: Mfg from Qtz-Chl-Bt to Qtz-Ms-Bt-Grt-Ky-Pl-L | c. 600 MPa/450°C 10 km SE of GF to c. 1100 MPa/750°C 30 km SE of GF. c. 1 Ga (Zrn, Mnz) Ar Hbl ages c. 960–940 Ma | Rivers 1983a, b; Dallmeyer & Rivers 1983; van Gool et al. 2008; Rivers et al. 1995, 2002; Cox & Rivers in press | Inverted metamorphic field gradient in thrust stack SE of GF |
| Molson Lake terrane [pHP] | Prg-Cpx-Opx-Grt coronas separating Ol and Pl (Ky inclusions in Pl; ≤51% Jd in Cpx) | 1200–1400 MPa, 700–750°C in NW, 1600–1800 MPa, 800–850°C in SE. Zrn LI c. 1005 Ma; Ar Hbl cooling ages 940–905 Ma | Dallmeyer & Rivers 1983; Rivers & Mengel 1988; Connelly & Heaman 1993; Indares & Rivers 1995; Connelly et al. 1995 | Coronitic textures. Metamorphic field gradient from NW to SE. |
| Smokey Archipelago (E. Labrador) [pMP] | Mafic to intermediate orthogneiss: Grt-Hbl-Bt-Pl-Qtz | 560–630°C, P undetermined c. 1040–1030 Ma (Zrn: LI) | Owen et al. 1986, 1988; Krogh et al. 2002 | Negligible Grenvillian Zrn and Mnz near GF |

Abbreviations: pMP, medium pressure; pHP, high pressure; GF, Grenville Front; GFTZ, Grenville Front Tectonic Zone; mfg, metamorphic field gradient; ogn, orthogneiss; pgn, paragneiss; Bdy, Baddeleyite; Bt, biotite; Cpx, clinopyroxene; Grt, garnet; Hbl, hornblende; Ilm, ilmenite; Jd, jadeite; Ky, kyanite; L, leucosome (granitic liquid); Mnz, monazite; Ms, muscovite; Ol, olivine; Opx, orthopyroxene; Pl, plagioclase; Qtz, quartz; Spl, spinel; Ttn, titanite; Zrn, zircon. LI, UI, lower and upper intercepts on concordia.

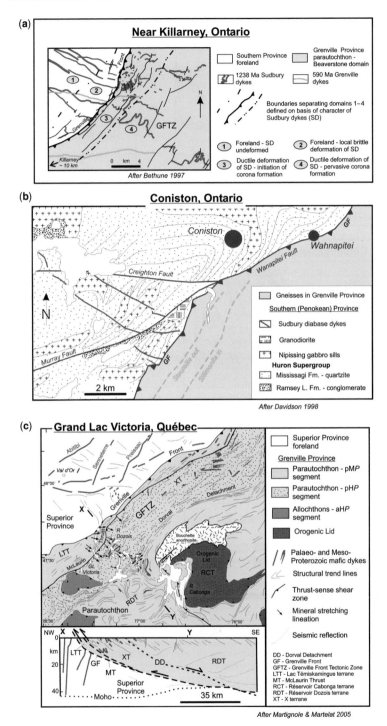

**Fig. 5.** Geological maps and cross-sections illustrating features of the Grenville Front and adjacent Parautochthonous Belt, northwestern Grenville Province; for locations see Figure 2. (**a**) Sudbury dykes (SD) in the Grenville Front Tectonic Zone (GFTZ) of Ontario. (**b**) Truncation of the Huron Supergroup at the Grenville Front near Coniston Ontario. (**c**) Mafic dykes in the Grenville foreland and the GFTZ in western Québec.

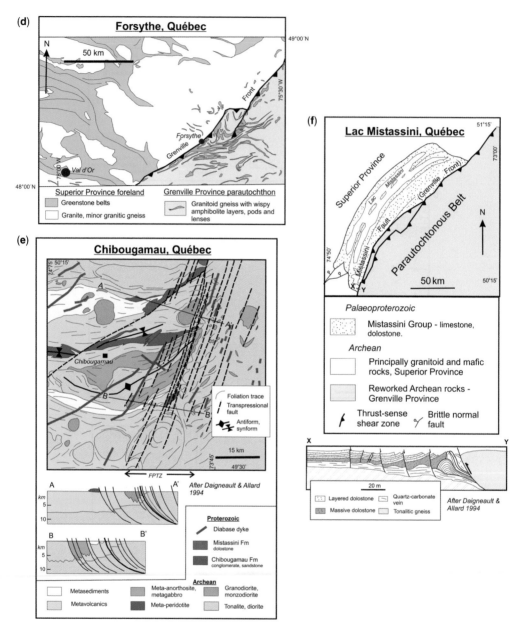

**Fig. 5.** (*Continued*) (**d**) Regional map of the Grenville Front near Forsythe, Québec (redrawn from 1:400 000 scale on-demand map of the Ministère des richesses naturelles, QC). (**e**) Grenville Front near Chibougamau, Québec. (**f**) Grenville Front near Lac Mistassini, Québec (redrawn from 1:400 000 scale map of the Ministère des richesses naturelles, QC).

baddeleyite indicate corona reactions took place during the Rigolet orogenic phase (Davidson & van Breemen 1988).

*Grenville Front Tectonic Zone near Coniston, Ontario.* The greenschist-facies Huron Supergroup

with Nipissing gabbro sills is also present in the Grenville foreland near Coniston, some 80 km farther NE (Fig. 5b). However, at this location both units and the Sudbury dykes are essentially truncated close to the Grenville Front, which is the site of an abrupt rise in metamorphic grade indicated

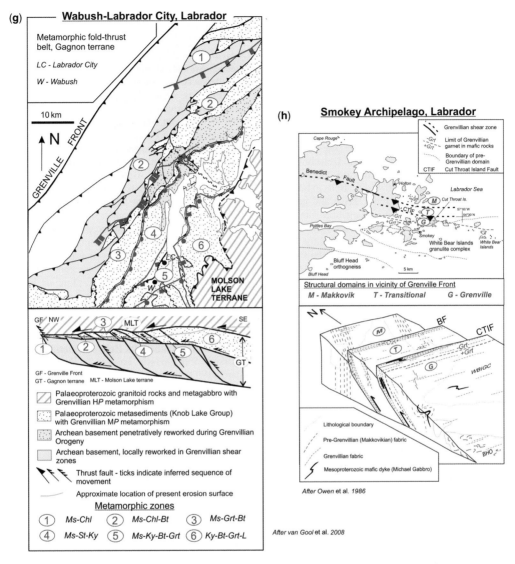

**Fig. 5.** (*Continued*) (**g**) Grenville Front and Para-autochthonous Belt in Gagnon terrane, western Labrador; cross-sectional model shows inferred tectonic architecture before erosion. (**h**) Grenville Front near Smokey, eastern Labrador.

by staurolite and sillimanite isograds mapped in thin slivers of pelitic schist derived from the Huron Supergroup (Davidson 1998). Farther SE, the Huron Supergroup and Nipissing gabbro have not been identified, their absence first documented by Quirke & Collins (1930) in their paper entitled 'The disappearance of the Huronian' and now interpreted to be due to exhumation above the present erosion surface. Davidson (1998) located the Grenville Front at the steep Wanapitei Fault (Fig. 5b), in which slivers of quartzite of the Huron Supergroup exhibit mylonitic and ultramylonitic microstructures. The Wanapitei Fault reworks part of the

Palaeoproterozoic Murray Fault system, indicating the Grenville Front followed a pre-existing structure at this locality too. The 'disappearance' of a thick supracrustal sequence together with the rapid rise in metamorphic grade implies the Grenville Front is the site of profound exhumation over a narrow zone a few kilometres wide at this locality.

*Grenville Front Tectonic Zone near Grand Lac Victoria, western Québec.* The Grenville foreland in western Québec is composed of belts of Neoarchaean supracrustal and granitoid rocks cut by Proterozoic mafic dykes, including the Senneterre and

Preissac (*c.* 2214 Ma) and Abitibi (*c.* 1140 Ma) swarms, the deformed and disaggregated remnants of which continue into the Grenvillian Parautochthonous Belt (Fig. 5c; Martignole & Martelat 2005). In this area, the GFTZ consists of stacked slices of reworked Archaean rocks, including amphibolite-facies metasediments (in Lac Témiskamingue terrane) structurally overlain by granulite-facies migmatite and orthogneiss (in X-terrane). X-terrane is itself structurally overlain by amphibolite-facies gneisses of the Réservoir Dozois terrane along the Dorval Detachment. In terms of their Grenvillian structure, the Archaean rocks of the Parautochthonous Belt exhibit NE trends parallel to the Grenville Front and gently to moderately SE-plunging elongation lineations developed during NW-directed thrusting. Lineations rotate into ENE-trends in the vicinity of the Dorval Detachment, a crustal-scale transtensional structure that offsets the Moho at the SE end of the cross-section (Fig. 5c; Martignole & Martelat 2005).

Grenvillian coronas in mafic dyke remnants in Lac Témiskamingue terrane at the base of the thrust stack exhibit amphibolite-facies assemblages, whereas those in the overlying X-terrane are granulite-facies. Peak $P-T$ estimates of *c.* 900 MPa/700°C and 1500 MPa/750–800°C, respectively have been determined (Martignole & Martelat 2005), implying the latter are high-pressure granulites, with variably retrograded samples defining a quasi-isothermal decompression path. Several geochronological studies of reworked gneisses from this area using the ID-TIMS method have yielded Archaean ages (e.g. Gariépy *et al.* 1990; Childe *et al.* 1993), but *in-situ* chemical dating of monazite has provided evidence for 1 Ga overgrowths on Archaean grains and for Grenvillian monazite as young as *c.* 960 ± 30 Ma (Martignole & Martelat 2005).

*Grenville Front Tectonic Zone near Forsythe, Québec.* The area shown in Figure 5d overlaps with that shown in Figure 5c, but focuses on the architecture of Neoarchaean mafic rocks where the Val d'Or strand of the Abitibi greenstone belt is intersected by the Grenville Front. The sketch-map illustrates the dramatic change in outcrop pattern from wide WNW-trending greenschist-facies belts on the Superior side to thin wispy, NE-trending layers, pods and lenses of polydeformed amphibolite on the Grenville side. Granitoid rocks on the Superior side commonly exhibit igneous textures and original contact relationships, whereas quartzofeldspathic gneisses on the Grenville side preserve little evidence of either. These features suggest the exhumed, reworked roots of the greenstone belt occur at the erosion surface SE of the Grenville Front, implying significant reverse displacement

and penetrative ductile Grenvillian reworking within the Parautochthonous Belt.

*Grenville Front Tectonic Zone near Chibougamau, Québec.* Another strand of the Abitibi greenstone belt intersects the Grenville Front near Chibougamau (Fig. 5e). At this location, in contrast to near Forsythe, the greenstone belt is effectively terminated at the Front, replaced to the SE by km- to dm-scale pods and lenses of chemically similar amphibolite, meta-anorthosite and metaperidotite in tonalitic to dioritic gneiss (Bandyayera *et al.* 2006). Daigneault & Allard (1994) described the Grenville Front at this location as a 20 km-wide amphibolite-facies shear zone with SE-plunging mineral lineations and reverse (SE-side-up) ductile strain indicated by rotated porphyroblasts, $s-c$ fabrics, shear bands and asymmetric pressure shadows. This shear zone is overprinted by a family of narrow lower-grade shear zones (shown as faults on Fig. 5e), also with SE-plunging lineations indicating reverse and minor sinistral strike-slip kinematics. Taken together, the two sets of structures document progressive exhumation of the greenstone belt towards the NW, with the result that the Parautochthonous Belt is underlain by its reworked orthogneissic basement. Daigneault & Allard (1994) defined a 'Foreland–Parautochthon Transition Zone' across which cumulative exhumation of at least 5–10 km was inferred. As at Killarney and Grand Lac Victoria, Proterozoic mafic dykes that cross-cut Archaean structures in the foreland are truncated, segmented and reoriented subparallel to the Grenville Front.

*Grenville Front near Lac Mistassini, Québec.* Fifty kilometres NE of Chibougamau, the Grenville foreland is underlain by the Palaeoproterozoic Mistassini Group, an unmetamorphosed and essentially flat-lying platformal carbonate sequence. As shown in the northern-most part of Figure 5e and in Figure 5f, the Mistassini Group is truncated at the Grenville Front, which is placed locally at the Mistassini Fault, one of the late, narrow, reverse-sense shear zones noted above, that at this location exhumes the reworked gneissic basement of the greenstone belt over the Mistassini Group (Daigneault & Allard 1994). As shown in a cross-section from the southern end of its exposure in a former mine (Fig. 5f), folding in the Mistassini Group is restricted to the immediate footwall of the Mistassini Fault and the lack of metamorphism implies a structurally high level of the foreland is preserved here.

*Grenville Front near Wabush – Labrador City, western Labrador.* In western Labrador, the Grenville foreland is underlain by a Palaeoproterozoic continental-margin sequence (the Knob Lake Group, Fig. 2) and its Archaean basement. Both cover and basement lithologies can be traced across the Grenville Front into the parautochthonous Gagnon terrane (Fig. 3), where they are imbricated in a metamorphic foreland-fold-thrust belt (Fig. 5g; Rivers 1983a, b; Rivers et al. 1993; van Gool et al. 2008). The NE grain of the thrust belt, the SE plunge of the mineral elongation lineations, and the asymmetry of ductile microstructures all indicate NW-vergence. The Gagnon terrane is overlain by the Molson Lake terrane, which is characterized by a high-pressure Grenvillian metamorphic signature (Indares & Rivers 1995). Within the Gagnon terrane, thrusting was thin-skinned in the Palaeoproterozoic cover sequence and thick-skinned in the underlying Archaean basement, leading to a dual-level thrust belt (Fig. 5g). In this part of the Gagnon terrane, grade of metamorphism ranges from greenschist facies near the Grenville Front, to upper amphibolite facies farther SE where partial melting was widespread, the higher grade rocks structurally overlying the lower grade rocks in a classic inverted metamorphic sequence. $P-T$ estimates range from c. 600 MPa/450 °C near the Grenville Front to c. 1100 MPa/750 °C some 30 km from the front near the southeastern limit of the Gagnon terrane (van Gool et al. 2008). Geochronological data, including metamorphic monazite ages, crystallization ages of cross-cutting pegmatite, and several zircon lower intercepts, indicate metamorphism in both Gagnon and Molson Lake terranes took place at c. 1 Ga (Rivers et al. 2002; van Gool et al. 2008; Cox & Rivers in press). About 200 km farther SW in the Gagnon terrane in eastern Québec, peak $P-T$ rises to c. 1400 MPa/800 °C and the metamorphism has been dated at 995–985 Ma (Indares 1995; Jordan et al. 2006).

*Grenville Front in the Smokey Archipelago, eastern Labrador.* The Grenville Front in the Smokey Archipelago at the east of the exposed Grenville Province was studied by Owen et al. (1986, 1988). Figure 5h shows the NNE-trends in the Palaeoproterozoic Makkovik Province in the foreland are truncated at the Grenville Front by E-trending Grenvillian structures. In detail, there are two narrow, sub-parallel, thrust-sense shear zones, the Benedict and Cut Throat Island faults, the latter being considered the local Grenville Front. North of the Benedict Fault (domain M), Makkovik structures are essentially unaffected by Grenvillian overprinting; in domain T (transition) between the Benedict and Cut Throat Island faults both Makkovik and Grenvillian fabrics are present, and south of Cut Throat Island Fault Grenvillian structures become more prominent (domain G), as exhibited by polyphase folding and boudinage of the c. 1.43 Ga Michael

Gabbro. Grenvillian metamorphic grade reached epidote-amphibolite facies in domain T and mid-amphibolite facies in domain G, where Michael Gabbro dykes exhibit garnet amphibolite margins. Owen et al. (1988) estimated Grenvillian temperature reached 560–630 °C in domain G. Krogh et al. (2002) showed that zircon, monazite and titanite in the vicinity of the Grenville Front yield pre-Grenvillian U–Pb ages, and Grenvillian lower intercept ages of c. 1040–1030 Ma were obtained in the vicinity of the Cut Throat Island Fault. This suggests this shear zone was active in Ottawan time, but whether it was also active in Rigolet time, as elsewhere along the front, has not been determined.

### Summary of characteristics

The seismic sections indicate the Grenville Front is the limit of SE-dipping Grenvillian fabrics and the structural/metamorphic studies imply reverse-sense kinematics consistent with the increasing grade of Grenvillian metamorphism in that direction. However, as noted, the total amount of differential exhumation across the front is limited by the continuity of several supracrustal units across it, implying the vertical component of transport could not have exceeded a few tens of kilometres. This is compatible with the observation that it follows and reworks pre-existing structures locally and that the intensity of Grenvillian strain was quite variable (Fig. 6). In terms of its architecture, although the front is everywhere a moderately-dipping, reverse-sense shear zone, its width and soling depth vary significantly and it feeds into a mid-crustal foreland fold-thrust belt locally.

Also variable is the extent of exhumed high-grade Grenvillian metamorphic rocks from the mid and lower crust. Such rocks are absent near Témiskamingue and in the Smokey Archipelago (Fig. 6b, f), suggesting these areas represent relatively high levels of the Parautochthonous Belt, whereas a deeper crustal level is exposed in the GFTZ. In detail, two styles of crustal architecture are apparent in the GFTZ: a crustal-scale shear zone in which the depth of exhumation increases systematically towards the SE (e.g. near Killarney and Chibougamau; Fig. 6a, d), and two lithologically- and metamorphically-distinct stacked terranes derived from discrete crustal levels (near Grand Lac Victoria and Wabush; Fig. 6c, e).

The SE-plunging elongation lineations and reverse-sense kinematics imply tectonic transport was approximately orthogonal to the Grenville Front, except in local NNE-trending transfer zones where strike-slip structures record a component of sinistral transpression. However, the depth from which the overthrust slice was derived, the extent

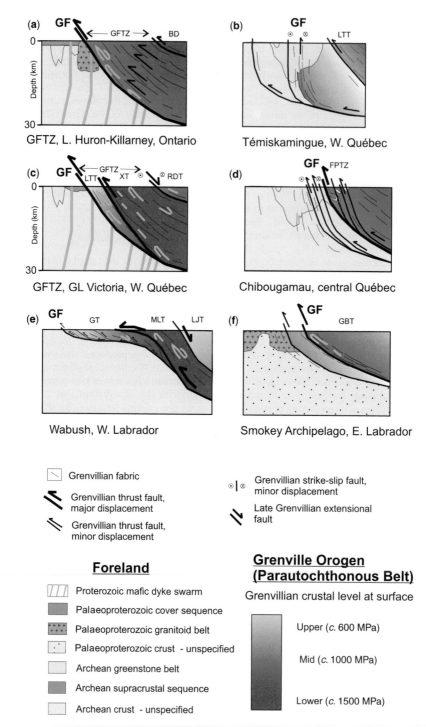

**Fig. 6.** Summary figure showing schematic crustal-scale structural signatures of the Grenville Front.

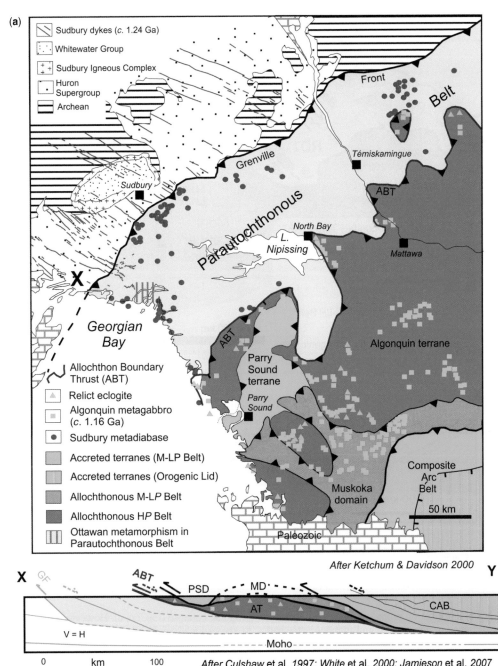

**(a)**

Sudbury dykes (c. 1.24 Ga)

Whitewater Group

Sudbury Igneous Complex

Huron Supergroup

Archean

Front Belt

Grenville

Témiskamingue

Sudbury

ABT

Parautochthonous

North Bay

L. Nipissing

Mattawa

X

Georgian Bay

Allochthon Boundary Thrust (ABT)

Relict eclogite

Algonquin metagabbro (c. 1.16 Ga)

Sudbury metadiabase

Accreted terranes (M-LP Belt)

Accreted terranes (Orogenic Lid)

Allochthonous M-LP Belt

Allochthonous HP Belt

Ottawan metamorphism in Parautochthonous Belt

ABT

Parry Sound terrane

Parry Sound

Algonquin terrane

Composite Arc Belt

Muskoka domain

50 km

Paleozoic

*After Ketchum & Davidson 2000*

**Y**

**X**  GF

ABT

PSD

MD

CAB

AT

V = H

Moho

0    km    100

*After Culshaw et al. 1997; White et al. 2000; Jamieson et al. 2007*

**Fig. 7.** Geological maps and crustal-scale cross-sections illustrating features of the Allochthon Boundary Thrust and adjacent terranes and domains. (**a**) In Ontario showing the distribution of mafic rocks on either side of the Allochthon Boundary Thrust (1.24 Ga Sudbury diabase in footwall and 1.16 Ga Algonquin gabbro and retrogressed eclogite in hanging wall). Note lobate shape of Muskoka domain (MD). (**b**) In western Québec. (**c**) In eastern Québec – western Labrador showing the distribution of *c*. 1.43 Ga Shabogamo Gabbro in the footwall of the Allochthon Boundary Thrust: the hanging wall Lac Joseph and Hart Jaune terranes (LJT, HJT) contain *c*. 1.65 and 1.45 Ma granite and gabbronorite bodies not present in the footwall. (**d**) In eastern Labrador showing the distribution of *c*. 1.43 Ga Michael Gabbro in the footwall of the Allochthon Boundary Thrust: GBT, Groswater Bay terrane; LMT, Lake Melville terrane; MMT, Mealy Mountains terrane.

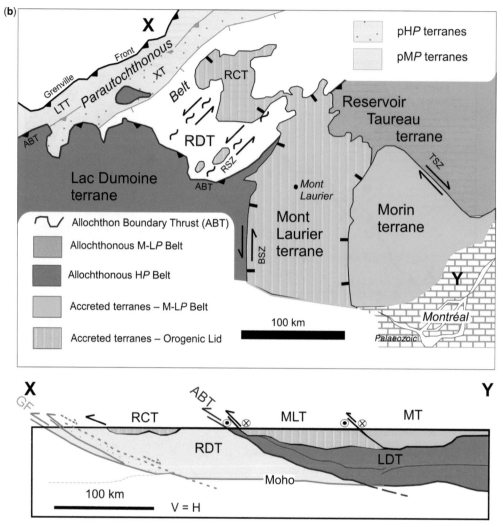

After Martignole et al. 2000; Martignole & Martelat 2005

**Fig. 7.** (*Continued*).

to which deformation propagated into the foreland, the presence or absence of ramp-flat geometry and the intensity of Grenvillian strain are all variable, presumably controlled by local features such as footwall rheology, pre-existing structural weaknesses and the amount of shortening. For instance, where present, high-pressure metamorphic terranes occur above the structural base of the thrust stack suggesting their emplacement drove thrust propagation into the foreland, and they are structurally overlain by normal-sense detachment zones that carry lower grade rocks in their hanging walls, implying the overthickened crust underwent

orogenic collapse (Fig. 6c, e). However, in contrast to the Allochthon Boundary Thrust (see below), the Grenville Front itself was not reworked in extension.

In summary, peak pressure estimates for metamorphic rocks in the hanging wall of the Grenville Front imply exhumation from depths of c. 20–45 km, zircon and monazite ages of c. 1000–980 Ma indicate metamorphism was during the Rigolet orogenic phase except possibly in the easternmost Grenville Province, and the $^{40}Ar/^{39}Ar$ hornblende ages of c. 990–940 Ma (Table 1) imply rapid post-peak cooling.

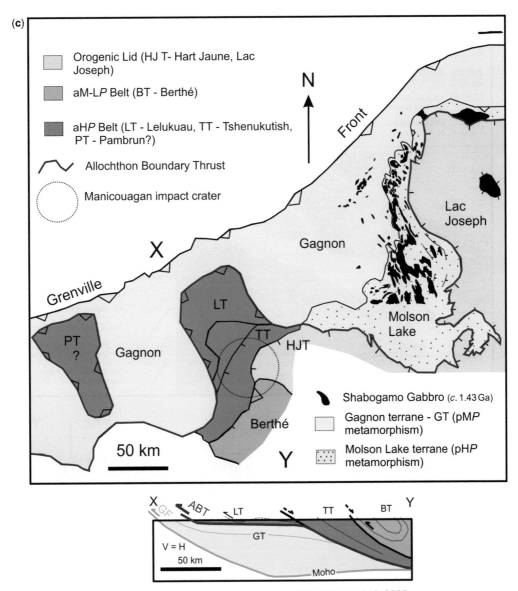

**(c)**

Orogenic Lid (HJ T- Hart Jaune, Lac Joseph)

aM-L*P* Belt (BT - Berthé)

aH*P* Belt (LT - Lelukuau, TT - Tshenukutish, PT - Pambrun?)

Allochthon Boundary Thrust

Manicouagan impact crater

Shabogamo Gabbro (*c.* 1.43 Ga)

Gagnon terrane - GT (p*MP* metamorphism)

Molson Lake terrane (p*HP* metamorphism)

50 km

*After Hynes* et al. *2000*

**Fig. 7.** (*Continued*).

## The interior Grenville Province – Ottawan deformation and metamorphism

In this study, the interior Grenville Province is defined as that part of the orogen in the hanging wall of the Allochthon Boundary Thrust. However, the thrust itself, despite its recognized importance, has received little detailed examination except by Ketchum *et al.* (1998) and Ketchum & Davidson (2000), and is principally known from regional studies (e.g. Rivers *et al.* 1989, 2002; Wardle *et al.* 1990; Culshaw *et al.* 1997; Carr *et al.* 2000; Corrigan *et al.* 2000; Hynes *et al.* 2000; Martignole *et al.* 2000). Following an examination of seismic images of the Allochthon Boundary Thrust, the geological signature of rocks in its hanging wall is reviewed in four case studies.

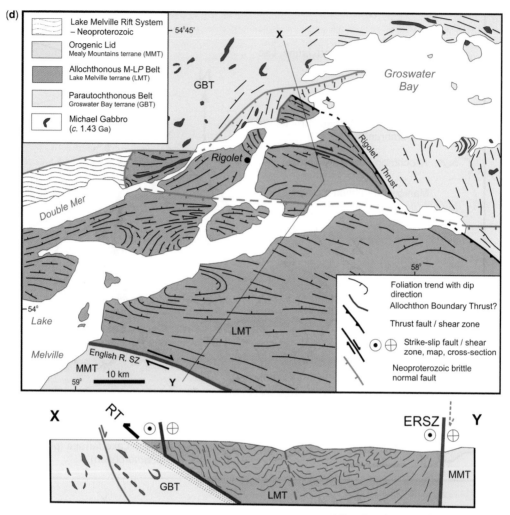

After Corrigan et al. 2000

**Fig. 7.** (*Continued*).

## Seismic images of the Allochthon Boundary Thrust

Seismic images of the northwestern extremity of the Allochthon Boundary Thrust in Figure 2 indicate it is a gently SE-dipping structure located above the Grenville Front. In the seismic section in Ontario (Fig. 4a) it is not the site of a change in reflectivity and so it is not readily picked without surface information, whereas in that of western Québec (Fig. 4b) it forms a strong package of reflectors at the base of a klippe. On the basis of seismic sections SE of those shown, the Allochthon Boundary Thrust is inferred to be a crustal-scale structure that penetrates the mid and lower crust in the SE Grenville Province.

## Allochthon Boundary Thrust and the allochthonous terranes – Case studies

*Allochthon Boundary Thrust near Georgian Bay, Ontario.* In the western Grenville Province, the Allochthon Boundary Thrust exhibits a lobate map pattern and is manifest in the field as a km-wide shear zone with amphibolite-facies high-strain fabrics and reverse-sense kinematics locally overprinted in extension. Lithological contacts are 'buried' in the wide high-strain zone (Culshaw *et al.* 1997), but are apparent at the regional scale. For instance, Figure 7a shows that it separates 1.24 Ga Sudbury dykes metamorphosed at *c.* 1000 Ma in its footwall from 1.16 Ga Algonquin gabbro and relict eclogite metamorphosed at $c. \geq 1090$ Ma in its hanging wall (Ketchum & Davidson 2000). Moreover, it also coincides with a change in Nd model ages (Dickin & Guo 2001), all features consistent with it being the site of major tectonic transport. On the basis of structural constraints, Culshaw *et al.* (1997) estimated it had accommodated a *minimum* of 100 km displacement. The cross-section in Figure 7a shows that it reaches the mid-crust, implying that the relict eclogite-facies (lower-crustal) rocks in its hanging wall must have been derived from farther SE.

The structural succession in the hanging wall of the Allochthon Boundary Thrust at this location comprises: (i) the a*HP* Algonquin terrane characterized by $\geq 1090$ Ma relict eclogite, overlain by (ii) the a*M*-*LP* Muskoka domain characterized by *c*. 1080 Ma upper amphibolite-facies and granulite-facies assemblages, which in turn is overlain by (iii) the Composite Arc Belt, part of which was reworked under *MP* Ottawan metamorphic conditions and part of which comprises the Orogenic Lid.

As indicated in Figure 2, the Allochthon Boundary Thrust approximately coincides with the northwestern limit of Ottawan metamorphism in much of the Grenville Province, but Figure 7a shows that Ottawan metamorphism locally occurs NW of (structurally below) the Allochthon Boundary Thrust. Finally, a key feature of the Allochthon Boundary Thrust in this area was its profound ductile reworking in extension at *c*. 1020 Ma as part of the Allochthon Boundary Detachment system of shear zones (Ketchum *et al.* 1998). It appears to have been inactive since that time.

*Allochthon Boundary Thrust in western Québec.* Although less studied than in Ontario, the Allochthon Boundary Thrust in western Québec (Fig. 7b) also exhibits a lobate outcrop pattern and separates rocks with remnants of Proterozoic dyke swarms present in the foreland in its footwall (Fig. 5c) from mafic rocks with relict H*P* Ottawan assemblages in its immediate hanging wall (Lac Dumoine terrane and a related klippe). Although the a*M*–*LP* terranes do not physically overlie the a*HP* terranes in this map, they may do in three dimensions. Other features contrasting with Figure 7a include the presence of several transpressional terrane boundaries and the seismic interpretation that the Allochthon Boundary Thrust reaches the Moho in this section.

*Allochthon Boundary Thrust, eastern Québec – western Labrador.* In the vicinity of the Manicouagan Reservoir (Fig. 7c), the Allochthon Boundary Thrust separates terranes containing the *c*. 1.43 Ga Shabogamo Gabbro metamorphosed at *c*. 1000 Ma in its footwall from terranes containing *c*. 1.65 Ga gabbronorite and anorthosite metamorphosed at *c*. 1060 Ma in its hanging wall (Hynes *et al.* 2000;

Indares *et al.* 2000; Rivers *et al.* 2002). The original thrust-sense boundary of the Allochthon Boundary Thrust is preserved at the base of the a*HP* Lelukuau terrane, but elsewhere it was extensively reworked in extension associated with the emplacement of the Orogenic Lid (Hart Jaune and Lac Joseph terranes). Well-preserved eclogite-facies rocks in the Lelukuau terrane have yielded metamorphic ages of *c*. 1060–1040 Ma and subsequent extension has been dated at *c*. 1015 Ma (Indares *et al.* 1998, 2000). The seismic section shows that the Allochthon Boundary Thrust reaches the Moho in this area.

*Allochthon Boundary Thrust near Rigolet, eastern Labrador.* The structural interpretation of the Allochthon Boundary Thrust in eastern Labrador (Fig. 7d) is unconstrained by seismic information and complicated by several ages of faulting. In addition to brittle normal faults associated with the Neoproterozoic Lake Melville graben, the Rigolet thrust, a ductile shear zone assumed to be the location of the Allochthon Boundary Thrust during field studies, appears to be of Palaeoproterozoic age (Corrigan *et al.* 2000; F. Korhonen, personal communication 2008), although many of the rocks in its hanging wall have undergone penetrative Grenvillian metamorphism (see below). As a result, the exact location of the Allochthon Boundary Thrust is not well constrained, but it may lie along an unnamed transpressional shear zone a few kilometres south of the Rigolet Thrust. Despite this uncertainty, its location can be approximately mapped from the distributions of the *c*. 1.43 Ga Michael Gabbro with *c*. 1000 Ma metamorphism in its footwall (in Groswater Bay terrane) and Palaeoproterozoic gabbro bodies with *c*. 1080 Ma metamorphism in its hanging wall (in Lake Melville terrane). In the south of the map area, Lake Melville terrane is separated from the Mealy Mountains terrane, part of the Orogenic Lid, by the transtensional English River shear zone.

Two features of this area set it apart from those described previously: (i) high-pressure rocks do not occur, the Lake Melville terrane being part of the a*M*-*LP* Belt; and (ii), regional E-trending folds of ductile gneisses occur in the hanging wall of the Allochthon Boundary Thrust (cross-section, Fig. 7d). Corrigan *et al.* (2000) interpreted the latter structures to have formed when the early Ottawan (*c*. 1080 Ma) gneissic fabric underwent shortening at *c*. 1050 Ma between two relatively rigid blocks, the contemporaneous foreland to the north and the down-dropped Mealy Mountains terrane, part of the Orogenic Lid to the south.

## Summary of structural characteristics

Its surface manifestation as a ductile shear zone whose location is independent of the local footwall

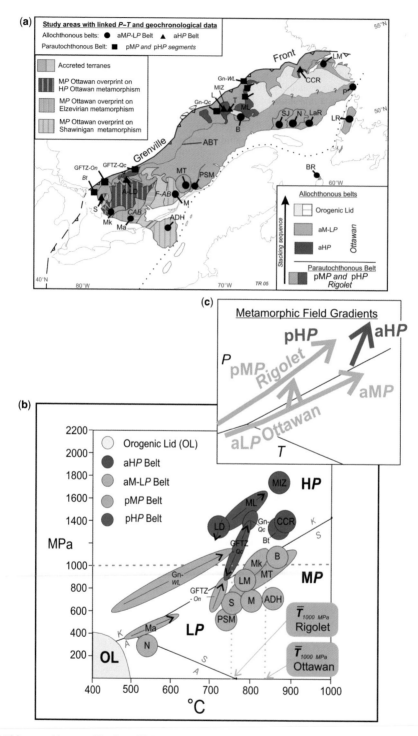

**Fig. 8.** (**a**) Metamorphic map of the Grenville Province showing locations of detailed studies with *P*–*T* estimates linked to geochronology of the metamorphic rocks. (**b**) *P*–*T* diagram showing estimated Ottawan *P*–*T* conditions in the allochthonous belts and Rigolet metamorphic conditions in the Parautochthonous Belt. (**c**) Summary of Ottawan and Rigolet metamorphic field gradients. *Abbreviations in A and B:* B, Berthé terrane; BR, Blair River outlier;

geology, together with the contrasting distributions of Mesoproterozoic dyke swarms on either side, suggest the Allochthon Boundary Thrust is the site of a large, but unquantified amount of subhorizontal displacement. This inference is supported by the presence of high-grade Ottawan metamorphic rocks in its hanging wall, including some with eclogite-facies assemblages implying substantial tectonic transport on such gently-dipping surfaces. Although there is local evidence for Ottawan metamorphism in the footwall of the Allochthon Boundary Thrust, available evidence suggests that it marks the approximate limit of high-grade metamorphism in Ottawan times.

Following its initiation as a thrust-sense shear zone, the Allochthon Boundary Thrust was locally reworked as a normal-sense structure in late Ottawan times at c. 1020 Ma (Ketchum et al. 1998; Indares et al. 2000; Rivers et al. 2002), and elsewhere normal-sense displacement occurred on shear zones a few kilometres to the SE. Considering the Grenville Province as a whole, the normal-sense displacements began locally as early as c. 1060 Ma but became widespread at c. 1020 Ma, signalling gravitational collapse of the Ottawan crust, the regional folding in eastern Labrador suggesting this occurred in an overall compressional setting.

## Metamorphic characteristics of hanging wall rocks

In this section, the metamorphic signature of the Allochthon Boundary Thrust and the overlying allochthonous belts in the four areas are integrated with data from elsewhere in the Grenville Province. Locations mentioned in the text are shown in Figure 8a.

*Allochthon Boundary Thrust.* The wide shear zone defining the Allochthon Boundary Thrust exhibits upper amphibolite-facies assemblages formed by retrogression and annealing of the eclogite- and granulite-facies rocks in the adjacent hanging wall terranes. As a result, there is no obvious metamorphic contrast across it in the field.

*Allochthonous High Pressure Belt.* Three separate areas of outcrop of the aHP Belt have been identified in the western, central and eastern Grenville Province, in each of which the HP rocks lie close to or in the immediate hanging wall of the Allochthon Boundary Thrust (Fig. 2). In addition as noted, Ottawan HP rocks also occur locally in its footwall in the western Grenville Province. Estimates of peak metamorphic conditions, time of metamorphism and references to the original literature are given in Table 2. In the Cape Caribou River Allochthon in the east, peak *P–T* estimates in HP granulite-facies assemblages are c. 1400 MPa/875°C and the metamorphism is dated at c. 1050 Ma, whereas eclogite and HP granulite from Lelukuau terrane in the central Grenville Province have yielded peak *P–T* conditions of c. 1700–1900 MPa/850–920°C at c. 1060–1040 Ma. In Algonquin–Lac Dumoine terrane in the western Grenville Province, the eclogite assemblages were pervasively overprinted during exhumation, their former presence being inferred from textural and mineralogical evidence (Davidson 1990). As a result, no *P–T* data are available, with the best estimate for the age of eclogite metamorphism being ≥1090 Ma. Finally the Ottawan metamorphic rocks that occur locally in the footwall of the Allochthon Boundary Thrust have yielded peak *P–T* conditions of 1200–1400 MPa/850–870°C and a minimum monazite age of c. 1035 Ma. Taken together, these data imply differential exhumation of the HP terranes and that timing of Ottawan HP metamorphism was diachronous.

*Allochthonous Medium to Low Pressure Belt.* Regional Grenvillian metamorphic conditions and geochronological information for the aM–LP Belt are given in Table 3. With respect to the MP segment of this belt, peak Ottawan *P–T* conditions for these upper amphibolite- and granulite-facies assemblages cluster in the range c. 800–1100 MPa and 800–850°C suggesting the present erosion surface exposes metamorphic rocks formed at 25–33 km depth. The *P–T* data plot in the sillimanite field, compatible with its presence in assemblages from this belt, and imply an elevated geothermal

---

**Fig. 8.** (*Continued*) BT, Britt domain; CAB, Composite Arc Belt; CCR, Cape Caribou River allochthon; FAB, Frontenac–Adirondack Belt; GFTZ, Grenville Front Tectonic Zone (*On*, Ontario; *Qc*, western Québec); Gn, Gagnon terrane (*Qc*, eastern Québec; *WL*, western Labrador); L, Lelukuau terrane; LD, Lac Dumoine terrane; LM, Lake Melville terrane; LR, Long Range outlier; LaR, La Romaine domain; Ma, Mazinaw domain; MIZ, Manicouagan Imbricate Zone; Mk, Muskoka domain; ML, Molson Lake terrane; MT, Mékinac–Taureau terrane; N, Natashquan domain; P, Pinware terrane; PSM, Portneuf–St-Maurice domain; S, Shawanaga domain; SJ, St-Jean domain; T, Tshenukutish terrane. *Abbreviations and symbols in B and C:* aHP and aM–LP, allochthonous High-Pressure and Medium- to Low-Pressure belts with Ottawan metamorphism; OL, Orogenic Lid; pMP and pHP, parautochthonous Medium-Pressure and High-pressure belts with Rigolet metamorphism. Circles in B represent approximate uncertainties in *P–T* estimates, ellipses represent ranges of *P–T* conditions in metamorphic field gradients, arrows indicate pro- or retrograde character.

**Table 2.** *Compilation of Ottawan metamorphic conditions, U–Pb metamorphic ages, and $^{40}Ar/^{39}Ar$ hornblende and muscovite cooling ages in the aHP Belt*

| Location W → E | Assemblages | Estimated P–T and time of metamorphism | References | Comments |
|---|---|---|---|---|
| Shawanaga domain [*western HP segment*] | Cpx-Opx-Grt-Pl (retrograde *Prg*), Spl, Spr, Crn replacing Grt-Omp) | U/Pb Zrn: Eclogite facies ≥1090 Ma? Granulite facies at c. 1080 Ma Amphibolite facies at c. 1020 Ma. Ar Hbl cooling age 970 Ma | Davidson 1990, 1998; Bussy et al. 1995; Ketchum & Krogh 1997, 1998; Ketchum et al. 1998; Reynolds et al. 1995 | Granulite-facies overprint associated with exhumation of HP Belt; amphibolite-facies overprint with extension on Shawanaga shear zone |
| Southern GFTZ and northern Britt domain [*western HP segment*] | Grt-Hbl-Pl ± Qtz ± Cpx ± Opx ± Bt ± Ilm ± Mt | HP granulite; 1380 MPa/870°C. U–Pb Mnz: 1037 ± 1, 1035 ± 1 Ma. Ar Hbl c. 975–965 Ma; Ms c. 925–905 Ma | Corrigan et al. 1994; Jamieson et al. 1995 | Structurally beneath the ABT. Mnz dates are minimum ages for the granulite-facies metamorphism |
| Lac Dumoine terrane [*western HP segment*] | Cpx-Opx-Grt-Pl (retrograde *Prg*, Spr, Crn replacing Grt-Omp) | 1350 MPa, 720°C 1070 Ma | Indares & Martignole 1990a; Indares & Dunning 1997 | Retrograde P–T conditions |
| Manicouagan Imbricate Zone (Lelukuau and Tshenukutish terranes) [*central HP segment*] | Cpx-Opx-Grt-Pl (Ky inclusions in Pl); Grt-Omp replacing igneous Ol, Px and Pl | 1700–1900 MPa, 750–920°C U–Pb Zrn: 1060–1040 Ma | Indares 1997, 2003; Cox et al. 1998; Cox & Indares 1999a, b; Indares et al. 1998, 2000; Indares & Dunning 2004; Yang & Indares 2005 | Eclogite and HP granulite with coronitic and granoblastic textures. HP granulite in lithologies with low bulk Na$_2$O. Steep retrograde dP/dT paths |
| Cape Caribou River Allochthon [*eastern HP segment*] | Prg-Cpx-Opx-Grt-Pl | 1400 MPa, 875°C c. 1050 Ma (U–Pb Zrn) | Krauss & Rivers 2004; Cox et al. unpublished | HP granulite assemblages |

ABT, Allochthon Boundary Thrust; Mineral abbreviations: *Cpx*, clinopyroxene; *Crn*, corundum; *Grt*, garnet; *Hbl*, hornblende; *Ilm*, ilmenite; *Jd*, jadeite; *Ky*, kyanite; *Mnz*, monazite; *Ms*, muscovite; *Mt*, magnetite; *Ol*, olivine; *Omp*, omphacite; *Opx*, orthopyroxene; *Pl*, plagioclase; *Prg*, pargasite; *Px*, pyroxene; *Spl*, spinel; *Spr*, sapphirine; *Zrn*, zircon.

**Table 3.** *Compilation of Ottawan metamorphic conditions, U–Pb metamorphic ages, and $^{40}Ar/^{39}Ar$ hornblende and muscovite cooling ages in the aM–LP Belt*

| Location W → E; MP/LP | Lithology and assemblages | Metamorphic facies, Estimated P–T and time of metamorphism | References | Comments |
|---|---|---|---|---|
| Parry Sound terrane MP | Pelite, orthogneiss, marble | Granulite- to Ur amphibolite; c. 1080 Ma; Ar Hbl c. 890–880 Ma | Reynolds et al. 1995; Wodicka et al. 1996 | Ottawan overprint on Shawinigan metamorphism |
| Shawanaga domain MP | Principally felsic orthogneiss, minor paragneiss | Ur amphibolite; c. 1050–1020 Ma. Ar Hbl c. 970 Ma | Culshaw et al. 1994, 1997; Reynolds et al. 1995; Ketchum et al. 1998 | Ottawan metamorphism at c. 1050 Ma, extensional shearing at c. 1020 Ma |
| Muskoka domain MP | Metabasite: Pl-Grt-Cpx-Opx-Hbl-Qtz Felsic granulite: Kfs-Pl-Qtz-Opx-Hbl-Bt | 1000–1150 MPa, 750–850°C; c. 1080–1050 Ma. Ar Hbl c. 1000 and 970 Ma | Cosca et al. 1991; Timmermann et al. 1997, 2002; Slagstad et al. 2004 | Cross-cutting granulite vein yielded 1065 Ma (Zrn) and 1045 Ma (Tm) ages; implies hot crust for 20–30 My. |
| Mazinaw terrane LP | Pelite: Qtz-Bt-Ms-Ky-Sil-Pl-Trm | Greenschist to Ur amphibolite; c. 300–600 MPa, 550–650°C; 1050–1000 Ma. Ar Hbl c. 940 Ma | Moore & Thompson 1980; Cosca et al. 1991; Corfu & Easton 1995 | Metamorphic field gradient from greenschist to Ur amphibolite facies |
| Adirondack Highlands terrane MP | Pelite: Crd-Grt-Sil-Spl-Opx-Pl-Bt-Kfs-Ilm-Trm-Psm-L | Granulite; 700–800 MPa, 675–915°C; Ottawan rims date anatexis at c. 1050–1020 Ma (U–Pb Zrn). Local outermost rims yield 1012–990 Ma ages. Ar Hbl c. 950–900 Ma | Bohlen 1987; Onstott & Peacock 1987; Mezger et al. 1991; McLelland et al. 1996, 2001; Spear & Markussen 1997; Alcock et al. 2004; Darling et al. 2004; Johnson et al. 2004; Bickford et al. 2008 | Ottawan overprint on Shawinigan (c. 1180–1140 Ma) granulite-facies metamorphism. Ottawan anatexis at 1050 Ma, orogenic collapse on Carthage-Colton shear zone pre-1020 Ma. Local Zrn growth syn-Rigolet |
| Morin terrane MP | Pelite: Qtz-Pl-Kfs-Bt-Grt-Sil Metabasite: Pl-Hbl-Grt-Opx-Cpx-Qtz-Bt-L | Granulite; 600–700 MPa, 650–775°C Peak metamorphism not directly dated. Ar Hbl c. 1050–990 Ma | Indares & Martignole 1990b; Martignole & Friedman 1998; Peck et al. 2005; Martignole et al. 2006 | Peck et al. (2005) argued assemblages are Ottawan based on reactions in skarn adjacent to Morin anorthosite |
| Mékinac–Taureau terrane MP | Granulite-facies orthogneiss | Granulite; 700–1150 MPa, 725–900°C; c. 1087 Ma (Zrn; thrusting) – 1050 Ma (Zrn; extension) | Corrigan & van Breemen 1997 | FW of extensional Tawachiche shear zone |
| Portneuf–St.-Maurice LP | Pelite: Opx-Ath-Crd-Bt-Pl, Crd-Spl-Grt-Sil-Qtz | Ur amphibolite to granulite; 500–600 MPa, 750°C; c. 1056 Ma (Zrn) | Herd et al. 1986; Corrigan & van Breemen 1997 | Montauban Group, HW of extensional Tawachiche shear zone |

(Continued)

**Table 3.** *Continued*

| Location W → E; MP/LP | Lithology and assemblages | Metamorphic facies, Estimated P–T and time of metamorphism | References | Comments |
|---|---|---|---|---|
| Berthé terrane (Gabriel complex) MP | Pelite: Qtz-Pl-Grt-Sil-Kfs-Bt-L  Metabasite: Opx-Cpx-Pl-Grt | Granulite; 1000–1100 MPa c.1050–1040 Ma (Zrn) | Indares & Dunning 2004 | Structurally juxtaposed against Grenvillian HP Belt and Orogenic Lid |
| St-Jean domain MP? | Pelite: Qtz-Kfs-Grt-Sil-Bt  Metabasite: Opx-Cpx-Pl-Grt | Ur amphibolite to granulite (P–T unquantified); 1080–1070 and 1060–1045 Ma (Zrn & Mnz), 950 Ma (Rt) | Gobeil et al. 2003; Wodicka et al. 2003 | Structurally overlies Natashquan domain |
| Natashquan domain LP | Wakeham Group: pelite – Qtz-Ms-Grt-Bt-And ± St ± Sil | Greenschist to amphibolite; 350 MPa/550 °C; 1052 Ma (Rt), c. 1030, 1010–990 Ma (Mnz), 972, 938 Ma (Ttn) | Indares & Martignole 1993; Madore et al. 1999; Gobeil et al. 2003; Wodicka et al. 2003 | P–T estimate from amphibolite-facies sample in metamorphic field gradient |
| Natashquan domain – La Romaine segment LP? | Wakeham Group: pelite (altered volcanic rocks) – Grt-Bt-Sil-Pl-Kfs-L, Qtz-Crd-Kfs | Ur amphibolite, (P–T unquantified) c. 1019 Ma (Zrn), 1010–1000 (Mnz), 990 Ma (Ttn) | Bonnet et al. 2005; Corriveau & Bonnet 2005; van Breemen & Corriveau 2005 | Ur amphibolite- to granulite-facies in continuation of Wakeham Group |
| Pinware terrane MP? | Orthogneiss | Amphibolite; (P–T unquantified) 1036–1020 Ma (Zrn) | Wasteneys et al. 1997; Gower & Krogh 2002 | Pinware terrane MP? |
| Lake Melville terrane MP | Pelitic paragneiss  Grt-Bt-Ky-Sil-Kfs-L | Granulite to amphibolite, 800 MPa, 820 °C; 1088, 1057 and 1046 Ma (Mnz) | Corrigan et al. 2000 | Single-grain Mnz ages |
| Blair River inlier, Cape Breton, Nova Scotia MP? | Mafic and felsic orthogneiss | Granulite (P–T unquantified), 1040 Ma (Zrn) | Miller et al. 1996 | |
| Long Range inlier, west Newfoundland MP? | Mafic and felsic orthogneiss | Granulite to amphibolitefacies (P–T unquantified) ≥1032 and 1022 Ma (Zrn) | Heaman et al. 2002 | |

MP, medium pressure; LP, low pressure; ? indicates uncertain affinity; FW/HW, footwall/hanging wall; Lr, lower; Me, middle; Ur, upper. Mineral abbreviations: *And*, andalusite; *Ath*, anthophyllite; *Bt*, biotite; *Cpx*, clinopyroxene; *Crd*, cordierite; *Grt*, garnet; *Hbl*, hornblende; *Ilm*, ilmenite; *Ky*, kyanite; *Kfs*, K feldspar; *L*, leucosome (former granitic liquid); *Mnz*, monazite; *Ms*, muscovite; *Opx*, orthopyroxene; *Pl*, plagioclase; *Qtz*, quartz; *Psm*, prismatine; *Rt*, rutile; *Sil*, sillimanite; *Spl*, spinel; *St*, staurolite; *Trm*, tourmaline; *Ttn*, titanite; *Zrn*, zircon.

gradient. High-grade M*P* Ottawan metamorphism took place from *c*. 1080–1050 Ma at slightly different times in individual terranes (Table 3).

With respect to the L*P* segments, the Portneuf–St.-Maurice domain exhibits L*P* granulite-facies assemblages, whereas the L*P* signature of Natashquan domain is indicated by the andalusite → sillimanite transition and widespread cordierite (Madore *et al.* 1999; Bonnet *et al.* 2005; Corriveau & Bonnet 2005). Although the kyanite → sillimanite transition occurs in Mazinaw terrane, the metamorphic field gradient passes close to the Al-silicate triple point indicating a L*P* evolution (Moore & Thompson 1980). Estimated peak *P–T* conditions range from *c*. 600–700 MPa/750 °C in the Portneuf–St.-Maurice domain, to *c*. 350–400 MPa/550 °C in Natashquan domain and Mazinaw terrane, implying metamorphism at depths of *c*. 18–20 and 10–12 km, respectively.

*Orogenic Lid.* The Orogenic Lid is composed of terranes that lack evidence for penetrative Grenvillian metamorphism and deformation, except in their basal shear zones. Typical evidence includes pre-Grenvillian $^{40}$Ar/$^{39}$Ar hornblende ages and miarolitic cavities and chilled margins in *c*. 1080 Ma intrusions (Rivers 2008), implying the rocks were in the upper crust during Ottawan times. The Orogenic Lid occurs in two segments, in the NE Grenville Province where it is largely composed of Palaeoproterozoic (Labradorian) gneisses, and in the SW within part of the accreted terranes. The presence of pre-Grenvillian $^{40}$Ar/$^{39}$Ar hornblende ages in both segments implies it was not heated above *c*. 500 °C during the Ottawan orogenic phase.

# Comparison of Rigolet and Ottawan metamorphisms

Estimated *P–T* conditions during the Rigolet and Ottawan phases of the Grenvillian Orogeny are shown in Figure 8b.

## Rigolet metamorphism

Most peak *P–T* estimates for Rigolet metamorphism fall in the kyanite field, compatible with mineral assemblages and suggesting a relatively high-pressure Barrovian character. The data are subdivided into pM*P* and pH*P* groups with the most deeply exhumed parts of the exposed Parautochthonous Belt formed at pressures of *c*. 1400–1700 MPa implying exhumation from 40–50 km depth. In western Québec, retrogression from peak conditions followed a steep d*P*/d*T* path suggesting rapid exhumation, an interpretation supported by geochronological data from throughout the Parautochthonous Belt, indicating a small difference

between U–Pb zircon and $^{40}$Ar/$^{39}$Ar hornblende ages (Table 1). The different slopes of the *P–T* gradients in the Parautochthonous Belt may be a function of the architecture of the Grenville Front, with the steeper gradients in the GFTZ related to a crustal-scale ramp and the gentler gradients to the presence of flats in a fold-thrust belt where horizontal tectonics dominated.

## Ottawan metamorphism

Quantitative estimates of peak Ottawan metamorphic conditions form three clusters in Figure 8b, consistent with the aH*P*, aM*P* and aL*P* groups and implying derivation from relatively discrete crustal levels. The Orogenic Lid, which was heated to < 500 °C, is interpreted as Ottawan upper crust. With regard to the aH*P* Belt, the highest peak *P–T* estimates are *c*. 1700–1800 MPa/800–875 °C, the high temperatures at these pressures being characteristic of orogenic eclogite formed near the base of doubly thickened crust (*c*. 50–55 km; O'Brien *et al.* 1990; Indares *et al.* 2000; Rivers *et al.* 2002). With respect to the aM*P* segment, all *P–T* estimates fall in the sillimanite field, compatible with mineral assemblages (Table 3), and the *P* estimates of *c*. 800–1100 MPa imply exhumation from 25–33 km depth, that is near the middle of doubly thickened crust. The peak *T* estimates cluster around 800–875 °C, implying an elevated thermal regime in the Ottawan mid-crust. Moreover, they overlap with those for the aH*P* segment, indicating that the temperature did not increase significantly from *c*. 30 km depth (in the aM*P* segment) to *c*. 55 km depth (in the aH*P* segment). This may impute a role for advection of magmatic heat, compatible with the location of the aM*P* segment within the Interior Magmatic Belt (Fig. 1) and local observations of syntectonic mafic intrusions (Indares & Dunning 2004). Pressure estimates for the Ottawan L*P* terranes range from *c*. 350–600 MPa, which when considered with their associated peak temperatures imply formation in a similar L*P*–H*T* metamorphic regime to the M*P* segment, but at shallower depths.

## Summary and implications for crustal rheology

The *P–T* estimates for Ottawan and Rigolet assemblages are compatible with the dominant Al-silicate species present indicating they are probably robust and the differences between them are real. This has been quantified in Figure 8b by selecting 1000 MPa (*c*. 30 km depth) as a reference pressure, from which it can be seen that the *T* at this depth was *c*. 800–900 °C during the Ottawan metamorphism, but *c*. 650–750 °C during the Rigolet metamorphism. This implies the two metamorphisms developed under contrasting geothermal gradients (Fig. 8c; i.e. *c.* 27–30 °C km$^{-1}$ during the Ottawan orogenic phase versus *c.* 23–27 °C km$^{-1}$ during the Rigolet orogenic phase, assuming an average crustal density of 3000 kg m$^{-3}$).

Another important contrast between the two metamorphisms is the rate of cooling. The large difference ( $\geq$100 My) between U–Pb zircon/monazite ages of peak metamorphism and $^{40}$Ar/$^{39}$Ar hornblende cooling ages for Ottawan metamorphic rocks (Tables 2 & 3) implies cooling in the interval *c.* 850 and 500 °C was slow. In contrast, the small difference ($\leq$15 My) for Rigolet metamorphic rocks (Table 1) implies much more rapid cooling.

In summary, the Ottawan metamorphism: (i) took place in the orogen interior from *c.* 1090–1020 Ma; (ii) was associated with stable magnetite; (iii) involved a relatively high geothermal gradient that may have been enhanced by advected magmatic heat; and (iv) was followed by slow cooling. In contrast, Rigolet metamorphism (v) took place at the orogen margin from *c.* 1000–980 Ma; (vi) resulted in assemblages in which magnetite was not stable (at least in the central Grenville Province); (vii) was characterized by a lower geothermal gradient and the absence of advected magmatic heat (it is situated outside the Interior Magmatic Belt); and (viii) was followed by fast cooling. Ottawan and Rigolet metamorphisms thus took place at different times, in geographically separate parts of the Grenville Orogen, were of different *P–T* character, developed under different oxygen fugacities and geothermal gradients, and followed different cooling paths.

The estimated peak temperatures in the Ottawan mid-crust are pertinent to the discussion of channel flow that follows. The low viscosities necessary for channel flow require extensive dehydration melting of micas in felsic lithologies (Rosenberg & Handy 2005), a process initiated at temperatures of 750–800 °C. Figure 8b shows that temperatures in this range or greater occurred throughout the a*MP* segment, compatible with field evidence for abundant leucosome in felsic lithologies. For instance, Slagstad *et al.* (2004) documented Ottawan *T* estimates of 750–850 °C and widespread field evidence for partial melting in the Muskoka domain, the map pattern of which resembles a ductile nappe (Fig. 7a), leading them to suggest it had undergone viscous flow. Since *P–T* conditions in Muskoka domain were not atypical of the *MP* segment (Fig. 8b) and large-scale nappes also formed elsewhere at this time (e.g. Adirondack Highlands terrane), it is likely that large segments of the Ottawan mid-crust were rheologically weak and capable of flowing when subjected to a small differential stress.

## Cross-sections of the Grenville Orogen

Several deep-seismic images of the Grenville Orogen produced in the course of the *Lithoprobe* program are shown in Figure 9, augmented by crustal-scale cross-sections derived from structural studies (modified from Rivers *et al.* 2002). The overall SE-dipping structural grain of the northern Grenville Province, in part defined by the Grenville Front and Allochthon Boundary Thrust, is evident in the cross-sections. Both structures exhibit ramp-flat profiles at the crustal scale and there is a correlation between the presence of ramps and the occurrence of Archaean rocks in the footwall, suggesting the latter were competent and formed a buttress during crustal shortening.

The architecture of the Grenville Orogen is a result of thrust- and normal-sense tectonics. The distribution of metamorphic rocks in Figure 9 suggests it principally developed during thrusting, but several contractional boundaries, including the Allochthon Boundary Thrust, locally carry rocks with a lower grade of Grenvillian metamorphism in their hanging wall than footwall, consistent with normal-sense reworking. On the basis of available data, ductile normal-sense shearing in the interior Grenville Province began as early as *c.* 1060 Ma, became widespread in the late Ottawan (*c.* 1020 Ma) and continued in narrow shear zones until late- to post-Rigolet times (Cosca *et al.* 1991; Corrigan & van Breemen 1997; Ketchum *et al.* 1998). With regard to the northwestern margin of the province, there is no evidence the Grenville Front was reactivated as a normal-sense structure, but the shear zone that offsets the Moho beneath the Parautochthonous Belt in western Québec (Fig. 5c) implies crustal-scale extension of post-Rigolet age. The significance of this is discussed later.

## Comparison with thermal–mechanical models for large hot orogens

Jamieson *et al.* (2007) compared the crustal-scale structure of the western Grenville Province with that simulated in numerical experiments in which the kinematic and intrinsic properties of the scale models were specified and the thermal–mechanical response of the crust was calculated. Their experiments were continued for over 100 My$_{emt}$ (elapsed model time) to simulate a LHO, and weak lower-crustal blocks were incorporated close to the suture zone as a proxy for accreted terranes at the continental margin. The experiments exhibited plateau development in the hinterland after *c.* 40 My$_{emt}$ due to decreased viscosity of the hot mid-crust and initiation of heterogeneous channel flow shortly thereafter. The first-order crustal

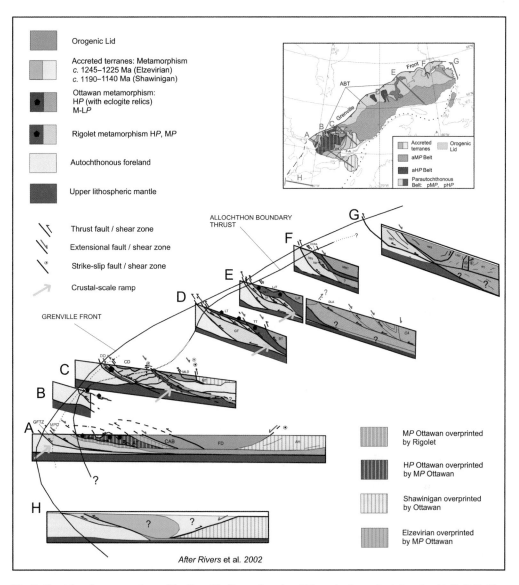

**Fig. 9.** Crustal-scale cross-sections of the Grenville Orogen based on *Lithoprobe* deep seismic imaging (A, B, C, D, G), *COCORP* deep seismic imaging (H), and inferred from surface geological data (E, F). Cross-sections are coloured according to the inferred age of metamorphism (modified after Rivers *et al.* 2002).

structure and $P-T$ distribution of the Georgian Bay and western Québec sections were closely simulated after $97.5\,My_{emt}$ and $82.5\,My_{emt}$ respectively, leading Jamieson *et al.* (2007) to conclude that the interior of the western Grenville Province was the site of channel flow under an orogenic plateau, that is the Grenville Province is part of a LHO.

Although discontinuities cannot be simulated with the finite-element method used in the modelling, zones of very high strain in the experiments

may be analogous to ductile shear zones in nature. Thus, with regard to the two structural boundaries that are the focus of this contribution, Jamieson *et al.* (2007) equated the base of the mid-crustal channel in the experiments to the Allochthon Boundary Thrust and a high-strain zone near the orogen margin to the Grenville Front. With respect to the former boundary, examination of their results shows that lower crustal rocks from the pro-side of the model orogen underwent tectonic transport of several hundred km towards the retro-side before partial exhumation into a mid-crustal channel, return flow and final exhumation at the pro-margin, which is compatible with the inference of major tectonic transport on the Allochthon Boundary Thrust (Fig. 7). With respect to the high-strain zone that may simulate the Grenville Front, it exhumes mid-crustal rocks from 20–30 km depth compatible with the GFTZ in the western Grenville Province.

The effect of the variable lower crustal rheology is well illustrated by these experiments, prompting a search for specific analogues in the Grenville Province. For instance, weak lower crustal blocks that enter the model orogen early detach from the subjacent mantle lithosphere, become exhumed to the mid-crust and are transported in a channel towards the erosion front; such a scenario may be analogous to the evolution of the aHP Algonquin–Lac Dumoine terrane in the western Grenville Province. In contrast, strong lower crustal blocks that enter the model orogen late remain attached to the subjacent mantle lithosphere, driving reverse flow of the mid-crust above it, providing a possible analogue for Archaean crust in the footwall of crustal-scale ramps in the northern Grenville Province. Thus, as argued by Jamieson *et al.* (2007), these experiments offer important insight into the causes and modes of flow and their role on the first-order architecture and thermal structure of LHOs in general, and the Grenville Orogen in particular.

The data assembled in this paper provide support for the interpretation of Jamieson *et al.* (2007) that the first-order architecture of the interior Grenville Orogen is a result of some form of channel flow beneath an orogenic plateau following prolonged crustal thickening. However, several features of the Grenville Front and the Allochthon Boundary Thrust discussed above are not readily incorporated into their model and point to the need for some second-order modifications. These include the following:

(1) The present architecture of the Grenville Province is a result of both thrust- and normal-sense structures, but only thrust-sense structures are simulated in their numerical experiments. This omission is particularly important with respect to Allochthon Boundary Thrust and the formation of the Orogenic Lid;

(2) Although there is a progressive younging of the age of deformation and metamorphism towards the foreland in the model orogen, there is no apparent division into two orogenic phases separated by a hiatus in their experiments;

(3) The contrasting metamorphic field gradients of the Ottawan and Rigolet orogenic phases are not explained, nor the presence of

high-pressure thrust sheets of Rigolet age beneath the Allochthon Boundary Thrust; and

(4) The contrasting rheological behaviour of the western and central segments of the aHP Belt is not explained.

These points are discussed in the next section.

## Ottawan orogenic collapse and formation of the Orogenic Lid

As noted above, normal-sense displacements on Ottawan shear zones began locally at *c.* 1060 Ma, were widespread by 1020 Ma, and sole in the mid-crust. Assuming the high *T* ($\geq$800 °C) and abundant leucosome of the aMP segment signal the Ottawan mid-crust was ductile, and considering numerical simulations of LHOs 'inevitably' result in plateau formation, the normal-sense displacements are interpreted to imply gravitational collapse of the plateau on a mid-crustal detachment system. A direct consequence of collapse was juxtaposition of the upper and mid-crust (Orogenic Lid and aM-LP Belt). In eastern Labrador, regional folding of the ductile *MP* footwall of the detachment at *c.* 1050 Ma between the more rigid Orogenic Lid and contemporaneous foreland implies compression continued after normal faulting, suggesting collapse took place in an overall contractional setting due to failure of melt-weakened mid-crust (fixed boundary collapse; Rey *et al.* 2001). Moreover from an orogen-scale perspective, the widespread reworking of the Allochthon Boundary Thrust as part of a system of detachment zones at *c.* 1020 Ma appears to have halted channel flow, paving the way for a new flow path during the Rigolet phase. Since there is no evidence for reactivation of the Allochthon Boundary Thrust after 1020 Ma, it is inferred that orogenic collapse terminated the channel flow regime that had existed since early Ottawan times, thereby promoting a fundamental change in the architectural and thermal structure of the orogen. Preservation of the Orogenic Lid above the Allochthon Boundary Thrust is a result of this change, its lack of penetrative Ottawan deformation suggesting it formed part of the orogenic suprastructure.

## Significance of hiatus between Ottawan and Rigolet orogenic phases

The paucity of metamorphic ages between *c.* 1020 and 1000 Ma points to a hiatus in the tectonic development of the Grenville Orogen. Moreover, when tectonism resumed it was focused in the Parautochthonous Belt, the former orogenic foreland. As noted, the significance of this hiatus has engendered discussion and the interpretation proposed here is in the context of the LHO paradigm. Given the role of the Allochthon Boundary Thrust as the base of a mid-crustal channel for 70 My (from *c*. 1090–1020 Ma), its termination implies a fundamental change in the orogen. Specifically, it is inferred that juxtaposition of upper- and mid-crustal rocks during orogenic collapse cooled the mid-crust and caused widespread freezing of leucosomes rendering re-activation of the channel mechanically unfavourable. As a result, ongoing shortening led to initiation of a new shear zone system and eventually the formation of the Grenville Front and related structures in the Parautochthonous Belt. In this context, the *c*. 20 My hiatus may represent the time taken for migration of the locus of high-grade metamorphism from the hinterland to the foreland, development of a crustal-scale critical wedge, and conductive heating of the thickened crust.

## Significance of contrasting Ottawan and Rigolet metamorphic field gradients

Evidence for the contrasting Ottawan and Rigolet metamorphic field gradients is summarized in Figure 8. Concerning the relatively high Ottawan gradient, the mid-crust in a LHO is hot principally because the low erosion rate on the plateau results in a protracted period of conductive heating, but factors such as radioactive self-heating, mantle heat flow and advection of magmatic heat also contribute. In this case, conductive heating during the 70 My duration of the Ottawan orogenic phase was clearly critical, but radioactive heating may have also been significant considering the hanging wall of the Allochthon Boundary Thrust is largely composed of mid- to late-Proterozoic (i.e. relatively young) granitoid rocks. Moreover, the presence of mantle- and crustal-derived intrusions of Ottawan age in the Interior Magmatic Belt (Fig. 1) implies the contemporary mantle heat flow and advected magmatic heat would have been high at least locally and augmented the regional gradient.

With respect to the cooler Rigolet gradient, the incorporation of cold crust from the former foreland and the limited time for conductive heating were probably crucial factors. Peak temperatures at mid-crustal depths, although appropriate for partial melting, were too short-lived to initiate channel flow. Ancillary factors include the inferred low self-heating potential of the Archaean crust, which underlies a large part of the Parautochthonous Belt, and the location of the latter outside the Interior Magmatic Belt implying mantle heat flow and advected magmatic heat would have been low. The negligible role of advected magmatic heat on the Rigolet gradient is highlighted by the presence of many small intrusions of late- to post-Rigolet age (*c*. 980–950 Ma) in the Interior Magmatic Belt (e.g. Gower *et al.* 1991), outside the area of penetrative Rigolet metamorphism.

## High-pressure thrust sheets in the Parautochthonous Belt

In the model of Jamieson *et al.* (2007), the Rigolet orogenic phase is associated with the detachment and exhumation of H*P* nappes above a strong lower-crustal indentor after 90–100 My$_{emt}$. Exhumation occurs well inboard from the contemporary orogen margin (i.e. they would be aH*P* nappes in the terminology used in this paper) and they form as the channel propagates outwards into the contemporary foreland. Moreover the Grenville Front is the site of exhumed mid-crustal rocks; no H*P* rocks occur (i.e. pH*P* in the terminology of this paper). Thus, when compared with the Grenville Province, their model raises two issues: (i) the timing does not fit that in the Grenville Province, where aH*P* terranes in the hanging wall of the Allochthon Boundary Thrust were exhumed in mid- to late-Ottawan times, well before the Rigolet stage; (ii) there is no mechanism in the experiments to cause the exhumation of pH*P* terranes with Rigolet metamorphism. With respect to the first issue, earlier exhumation of aH*P* terranes could probably be achieved in the experiments by inserting strong lower crustal blocks closer to the orogenic suture, so this may not require fundamental modification to their model. However, the second issue constitutes a more significant problem and it remains unclear to this author whether a single-stage model, in which the mid-crustal channel remains operational throughout contractional orogenesis, can explain the exhumation of pH*P* terranes near the orogen margin late in the collisional process.

## Contrasting rheological behaviour of the western and central aH*P* belts

The contrasting rheological expression of the aH*P* terranes in the western and central Grenville Province has been noted. In the central Grenville Province, mafic lithologies in the Manicouagan Imbricate Zone form coherent units and their H*P* mineral assemblages are well-preserved despite ductile deformation and locally intense shearing, whereas in the western Grenville Province, mafic units in Algonquin-Lac Dumoine occur as disaggregated pods and lenses, the terrane as a whole has undergone pervasive ductile flow, and the H*P* minerals are overprinted by granulite-facies assemblages.

Since exhumation of H*P* rocks implies passage through the mid-crust, its temperature and the duration of mid-crustal residence may be critical factors controlling preservation. In this case, *T* estimates for the M*P* segments in the two areas are similar (Fig. 8), whereas mid-crustal residence time was inferred to have been short for the Manicouagan Imbricate Zone, but prolonged for the Algonquin–Lac Dumoine terrane (Davidson 1990;

Indares *et al.* 2000). The heterogeneous flow model of Jamieson *et al.* (2007) provides additional context for this contrast. In the central Grenville Province, Archaean crust penetrates over 100 km into the orogen at the surface and over twice that distance at depth (Fig. 1). Assuming it behaved in a similar manner to a strong indentor in the experiments of Jamieson *et al.* (2007), it would have promoted rapid exhumation of weaker, inboard lower crustal blocks by reverse flow, as hot nappes above it. On the other hand, the absence of Archaean rocks, and hence a strong indentor, in the western Grenville Province would have suppressed exhumation, but promoted heterogeneous flow in a mid-crustal channel, compatible with the granulite-facies overprint and ductile disaggregation of the mafic units. Hence it is suggested that the contrasting evolutions of aH*P* terranes do not relate to intrinsic differences between them, but rather to the rheological character of the crust that entered the orogen behind them.

## Refinements to the heterogeneous channel flow model for the Grenville Orogen

In light of the above, several second-order refinements to the heterogeneous channel flow model of Jamieson *et al.* (2007) are proposed. As discussed, the preferred model involves two stages of thrusting and crustal thickening in different parts of the orogen separated by an important episode of extension and crustal thinning, all of which took place in an overall compressional orogen.

The model is illustrated schematically in Figure 10. The Ottawan stage (Fig. 10a) involved growth of an orogenic plateau with channel flow of low-viscosity mid-crust and extrusion of hot lower-crustal nappes, both of which were exhumed at the orogenic front. This stage closely resembles the model of Jamieson *et al.* (2007) with the caveats that criteria to distinguish between heterogeneous channel flow and homogeneous channel flow with coeval extrusion of hot nappes are not presently available, and that there is evidence for local normal-sense shear zones in the orogenic hinterland at this time. The first significant deviation from the model of Jamieson *et al.* (2007) occurred at *c.* 1020 Ma (Fig. 10b), when the Allochthon Boundary Thrust was reworked as part of the normal-sense Allochthon Boundary Detachment system of shear zones. It is inferred that this resulted in cooling the channel, increasing its viscosity above the critical value for flow thereby terminating channel flow despite ongoing subduction of Laurentian sub-continental lithospheric mantle. In the Rigolet stage (Fig. 10c), the orogen adjusted to loss of the material flow-path through the hinterland by advancing into its foreland, leading to burial and

rapid exhumation of parautochthonous lower- and mid-crustal rocks in the hanging wall of the Grenville Front. Metamorphism and deformation at this time were focused between the Grenville Front and Allochthon Boundary Detachment, with the hinterland, although still hot, being essentially inactive except for local small-volume plutonism, minor normal-sense adjustments in the upper crust, and local growth of metamorphic rims on zircon and monazite. There was thus an evolution over an interval of *c.* 20 My from a large hot long-duration orogen in the Ottawan phase to a subjacent, small cold short-duration orogen in the Rigolet phase.

Schematic particle and *P*–*T*–*t* paths are shown in Figure 11. The particle paths (Fig. 11a) illustrate transport of crust from the hinterland towards the foreland in Ottawan times by viscous flow in a mid-crustal channel and as hot lower-crustal nappes. Lithologic linkages between the Orogenic Lid and adjacent mid-crustal terranes suggest independent transport of the upper crust was limited. After orogenic collapse terminated channel flow, a shorter particle path developed during the Rigolet phase in the former foreland, bypassing the hinterland beneath the Allochthon Boundary Detachment. This model, in addition to accommodating the different times, *P*–*T* gradients and particle paths of the Ottawan and Rigolet metamorphisms, is compatible with evidence for greater horizontal displacement on the Allochthon Boundary Thrust than the Grenville Front. Moreover, the change in particle paths not only signalled destabilization of Ottawan channel flow, but is also compatible with migration of the metamorphic core of the orogen into its former foreland. In numerical experiments, such 'tunnelling' of the mid-crust towards the foreland is enhanced when there is no erosion at the orogen margin (thereby suppressing exhumation; Beaumont *et al.* 2006) suggesting this may have been a time of relatively subdued topography in the Grenville Orogen.

There are few published *P*–*T*–*t* paths for the Grenville Province, so the paths in Figure 11b were constrained by: (i) *P*–*T* estimates of peak metamorphic conditions; (ii) U–Pb zircon and monazite determinations of the time of peak metamorphism; (iii) compatibility with the Al-silicate phase diagram; and (iv) the $^{40}Ar/^{39}Ar$ hornblende cooling ages (data in Tables 1–3). The Ottawan and Rigolet paths in Figure 11b are distinctive, supporting the inference they developed under different geothermal gradients. The $^{40}Ar/^{39}Ar$ hornblende cooling ages, representing the time of cooling through *c.* 500 °C, are mostly in the range 980–940 Ma throughout the northern Grenville Province, implying slow cooling in the allochthonous belts after the Ottawan orogenic phase, but much faster cooling in the Parautochthonous Belt after the Rigolet orogenic phase. Slow cooling following

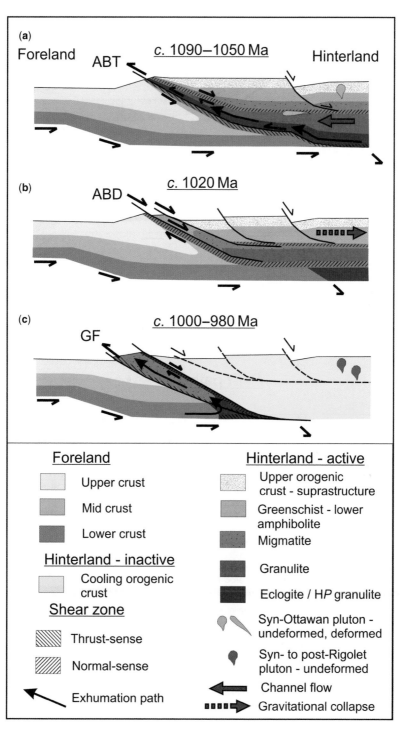

**Fig. 10.** Schematic diagram illustrating inferred distribution of metamorphic facies and tectonic evolution of the Grenville Province. (**a**) early Ottawan, formation of orogenic plateau with some form of channel flow and/or extrusion of hot nappes on Allochthon Boundary Thrust (ABT), local orogenic collapse. (**b**) late Ottawan, widespread orogenic collapse, reworking of ABT as Allochthon Boundary detachment (ABD), cessation of channel flow. (**c**) Rigolet, orogen migrates into foreland, formation of Grenville Front (GF), hinterland becomes tectonically inactive. See text for an additional explanation.

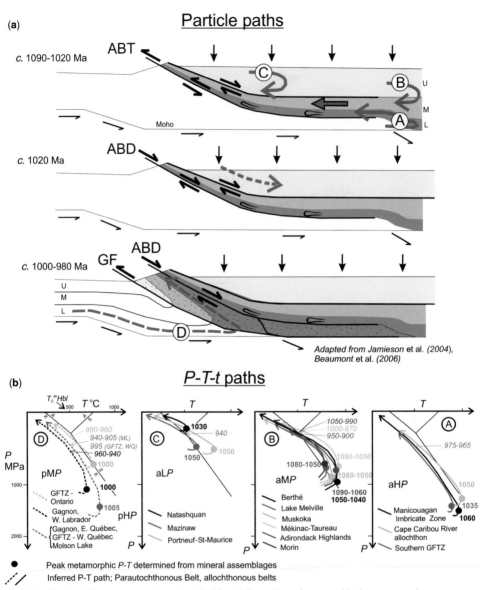

**Fig. 11.** Schematic particle paths and *P–T–t* paths for the Grenville Orogen (adapted from a figure in Jamieson *et al.* 2004). (**a**) Solid red arrows labelled A–C represent particle paths for the aH*P*, aM*P*, aL*P* segments, dashed red arrow labelled D represents the path for the pM*P* and pH*P* segments. Path A probably involved extrusion of a lower-crustal nappe whereas paths B–C may have involved some form of mid-crustal channel flow. Path D took place beneath the former channel and involved rapid burial and thrust exhumation. ABT, ABD and GF are locations of Allochthon Boundary Thrust, Allochthon Boundary Detachment and Grenville Front. (**b**) *P–T–t* diagrams are based on peak *P–T* estimates, U–Pb zircon or monazite ages of peak metamorphism, Al-silicate phase present, and ⁴⁰Ar/³⁹Ar hornblende cooling ages (Tc ≈ 500 °C; Tables 1–3).

the Ottawan phase took place beneath the residual orogenic plateau, whereas fast cooling in the Parautochthonous Belt is compatible with a short particle path that bypassed the warm hinterland (Fig. 11a).

## Discussion

In the introduction it was proposed that evolution of a collisional orogen could be conceptually divided into constructive and destructive stages separated by a climax representing the time of maximum size. This study suggests the Grenville Orogen underwent a more nuanced evolution with interruption of the long-duration constructive stage by orogenic collapse due to mid-crustal weakening, followed by internal reorganization and renewed growth near the orogen margin before the orogen as a whole entered its destructive stage. Imaging this evolution through the Grenville Front and Allochthon Boundary Thrust and their immediate hanging walls has highlighted several critical features. In particular, the evidence for long-distance transport of mid and lower crust on the Allochthon Boundary Thrust for ≥50 My during the Ottawan orogenic phase by some form of channel flow and/or as hot nappes suggests a mass balance between the material added to the orogen as a result of collision and that removed by erosion at the orogen margin was established during this period. Destabilization of this mass balance at the end of the Ottawan phase, probably due to rheological weakening of the mid-crust, led to orogenic collapse and initiated a fundamental change in structural style as crustal thickening migrated into the former foreland. New and shorter particle paths were established in a subjacent small cold orogen, resulting in formation of the Grenville Front and related structures in the Parautochthonous Belt, and the hinterland essentially became tectonically inactive. Thus, despite their superficially similar appearances in the field and on seismic sections, the tectonic signatures of the Grenville Front and the Allochthon Boundary Thrust are fundamentally different in several important respects and together they encapsulate much of the first-order tectonic evolution of the Grenville Orogen.

Many aspects of this model require additional documentation and testing. The difficulty in assembling unambiguous evidence for the former existence of an orogenic plateau and channel flow in ancient orogens was addressed by Godin et al. (2006a). In this case it is argued that compatible features include the long duration of high temperature metamorphism in the orogenic interior and preservation of the rheologically strong orogenic suprastructure in crustal-scale graben. The inference that orogenic collapse triggered a fundamental change in crustal flow paths, a central tenet of the

model proposed here, is based on field evidence and empirical deduction, but was not predicted by published numerical simulations of LHOs. In effect, it is inferred that widespread fixed boundary collapse in a convergent setting in late Ottawan times paved the way for a new tectonic style dominated by thrusting rather than flow in the Rigolet orogenic phase. Although normal-sense structures have been identified in the field, the gravitational collapse aspect of orogenic evolution remains understudied given its significance in the proposed model. Another problematic issue concerns the mechanism by which the Allochthon Boundary Thrust was exhumed at the orogenic front during the Ottawan orogenic phase. Numerical models of channel flow suggest that elevated rates of erosion are required to exhume the channel, such as result from the monsoons in the Himalaya–Tibet Orogen (e.g. Beaumont et al. 2004; Jamieson et al. 2004), but it may be problematic to extend a rainfall-driven erosion model to a mountain chain located in the middle of a supercontinent, as implied by the Rodinia reconstruction of Li et al. (2008).

Turning to the decay phase of orogenesis, the presence of crust of average thickness (c. 35 km) throughout the present-day hinterland of the Grenville Province implies isostatic equilibrium was eventually attained as the mantle lithosphere underwent visco-elastic rebound to a gravitationally stable configuration (Fig. 3). However, the Parautochthonous Belt near the orogen margin is underlain by anomalously thick (c. 50 km) crust in the east and anomalously thin (c. 30 km) crust in the west, indicating equilibrium was not established at these locations. In the case of thick crust under the Grenville Front Gravity Low, the lack of isostatic compensation has been attributed to suppression of flexural rebound due to cool mantle lithosphere (Hynes 1994), although further study is needed to determine whether a correlation exists between crustal thickness and exposed structural level at the Grenville Front. On the other hand, the presence of thin crust under the Parautochthonous Belt in the western Grenville Province suggests the mantle lithosphere was sufficiently hot to undergo flexural rebound there. In this case, a causal relationship with the crustal-scale Dorval Detachment in cross-section C discussed previously and the Boundary Shear in cross-section A (the two comprising the Mid-Parautochthon Detachment; Fig. 9; Rivers et al. 2002) is likely, specifically that rebound led to extension and crustal thinning. In summary, the regionally compensated gravity signature beneath the hinterland where the mantle lithosphere was hot and cooling was slow, and the non-uniform gravitational compensation beneath the Parautochthonous Belt where the temperature of the mantle lithosphere was variable and cooling was rapid provides another contrast between the Ottawan and Rigolet orogenic phases.

Finally, it is appropriate to briefly compare the proposed tectonic model for the Grenville Orogen with that for the Himalaya–Tibet Orogen, the most intensely investigated LHO. In addition to evidence for the long duration of orthogonal collision and the striking similarities between crustal-scale structures, such as the Allochthon Boundary Thrust and High Himal Thrust, and the Grenville Front and Main Boundary Thrust, both orogens grew by advancing into their respective forelands. In the case of the Himalaya–Tibet Orogen the tectonic evolution has been interpreted in terms of critical wedge theory, with plateau formation and channel flow taking place in a stable wedge with a critical angle of taper ($\varphi_c$), extension on the South Tibetan Detachment leading to the wedge becoming sub-critical ($\varphi < \varphi_c$), and renewed thickening of the hinterland by thrusting and crustal-scale folding leading to re-establishment of critical taper ($\varphi = \varphi_c$), which in turn permitted advance of the wedge into its foreland (Hodges 1996, 2006; Godin et al. 2006b). However, in the case of the Grenville Province, the apparent lack of evidence for penetrative deformation in the hinterland after orogenic collapse (Fig. 10c) suggests a comparable model may not be viable. Much depends on the tectonic significance of the sparse zircon and monazite rim ages in the aM–LP Belt. If the hinterland did not undergo renewed thickening after 1020 Ma, but essentially became tectonically inactive as suggested by the available evidence and indicated in Figure 10c, the advance into the foreland at c. 1000 Ma must have had a different driving force. One possible explanation is that the basal friction and internal strength of the lower and mid crust in the orogenic wedge increased significantly due to freezing of the abundant leucosome and subsequent escape of magmatic fluids, rendering the wedge stronger than the adjacent hot foreland. Such an interpretation would be compatible with establishment of the inferred short particle paths during the Rigolet orogenic phase (Fig. 11).

The metamorphic evolution provides another relevant point of comparison. Goscombe et al. (2006) showed that metamorphism in the Himalaya–Tibet Orogen can be subdivided into two temporally and geographically distinct regimes, an older MP–HT regime in the orogenic hinterland and a younger HP–MT regime nearer the foreland. As in the Grenville Province, the older, higher T regime in the hinterland structurally overlies the younger, lower T regime nearer the foreland along a crustal-scale thrust (the High Himal Thrust), and Goscombe et al. (2006) argued that the contrasting P–T character of the two metamorphic regimes resulted from development under different geothermal gradients.

Although the duration of both metamorphisms was shorter than in the Grenville Orogen and there was no inference they were separated by a period of orogenic collapse, the parallels are nonetheless striking and suggest that successive metamorphic regimes with contrasting P–T–t paths in the orogenic hinterland and foreland may be a fundamental feature of LHOs.

## Conclusions

The aim of this paper has been to show that an examination of orogenic fronts can provide unique insight into the evolution of the orogen as a whole. In the case of the Grenville Province, an archetypical candidate for a large hot long-duration orogen, the fronts associated with the Ottawan and Rigolet orogenic phases document two contrasting tectonic styles that took place at different times and in different places, pointing to an evolution from a large hot long-duration orogen with some form of channel flow under a plateau during the Ottawan phase to a subjacent small cold short-duration orogen during the Rigolet phase. The two orogenic phases are components of a prolonged collision that lasted c. 110 My and are separated by a period of midcrustal collapse that terminated channel flow and caused the orogen to migrate into its former foreland. This evolution is qualitatively compatible with observed differences in P–T–t paths and cooling histories between the allochthonous and parautochthonous belts, and with the inference that the Allochthon Boundary Thrust formed the base of a mid-crustal channel in the orogenic hinterland whereas the Grenville Front represents the limit of orogenic crust derived from the immediate foreland and exhumed by thrusting. Although exhibiting first-order compatibility with a published heterogeneous channel flow model for the western Grenville Province (Jamieson et al. 2007), the model proposed in this paper incorporates several important modifications based on diverse second-order geological constraints from the length of the Grenville Province.

The field data on which this paper is based were produced by a generation of students of the Grenville Orogen. In addition, there is an obvious intellectual debt to the finite-element modelling of C. Beaumont, R. Jamieson and colleagues and the unique insight it provides into the physical basis of orogenesis. This is acknowledged with gratitude, but in no way implies they agree with everything in this paper. I thank L. Godin and A. Hynes for constructive reviews, R. Law for pointing out the similar two-stage metamorphic evolution of the Himalaya–Tibet Orogen, and B. Murphy, D. Keppie and A. Hynes for their editorial handiwork. My research has benefitted from the Lithoprobe program and many years of funding by NSERC.

# References

ALCOCK, J., ISACHSEN, C., LIVI, K. & MULLER, P. 2004. Unraveling growth history of zircon in anatectites from the northeast Adirondack Highlands, New York: constraints on pressure-temperature-time paths. *In*: TOLLO, R. P., CORRIVEAU, L., MCLELLAND, J. & BARTHOLOMEW, M. J. (eds) *Proterozoic Tectonic Evolution of the Grenville Orogen in North America*. Geological Society of America Memoir, **197**, 267–284.

BANDYAYERA, D., CADÉRON, S., ROY, P. & RIVERS, T. 2006. Archaean volcano-sedimentary formations in the Chibougamau and Caopatina segments and their extensions in the Grenvillian Parautochtonous Belt, Québec, Canada. *Geological Association of Canada – Mineralogical Association of Canada, Annual Meeting Montréal*, Abstract volume, 9–10.

BEAUMONT, C., JAMIESON, R. A., NGUYEN, M. H. & LEE, B. 2001a. Himalayan tectonics explained by extrusion of a low-viscosity crustal channel coupled to focused surface denudation. *Nature*, **414**, 738–742.

BEAUMONT, C., JAMIESON, R. A., NGUYEN, M. H. & LEE, B. 2001b. Mid-crustal channel flow in large hot orogens: results from coupled thermal-mechanical models. *In: Slave – Northern Cordillera Lithospheric Evolution (SNORCLE) and Cordilleran Tectonics Workshop, Lithoprobe Secretariat, UBC, Vancouver, BC, Lithoprobe Report*, **79**, 112–180.

BEAUMONT, C., JAMIESON, R. A., NGUYEN, M. H. & MEDVEDEV, S. 2004. Crustal channel flows: 1. Numerical models with applications to the tectonics of the Himalayan-Tibetan orogen. *Journal of Geophysical Research*, **109**, B06406; doi: 10.1029/2003JB002809.

BEAUMONT, C., NGUYEN, M. H., JAMIESON, R. A. & ELLIS, S. 2006. Crustal flow modes in large hot orogens. *In*: LAW, R. D., SEARLE, M. P. & GODIN, L. (eds) *Channel Flow, Ductile Extrusion and Exhumation in Continental Collision Zones*. Geological Society, London, Special Publications, **268**, 91–145.

BETHUNE, K. M. 1997. The Sudbury dyke swarm and its bearing on the tectonic development of the Grenville Front, Ontario, Canada. *Precambrian Research*, **85**, 117–146.

BETHUNE, K. M. & DAVIDSON, A. 1997. Grenvillian metamorphism of the Sudbury diabase dyke-swarm: from protolith to two-pyroxene–garnet coronite. *Canadian Mineralogist*, **35**, 1191–1220.

BICKFORD, M. E., MCLELLAND, J. M., SELLECK, B. W., HILL, B. M. & HEUMANN, M. J. 2008. Timing of anatexis in the eastern Adirondack Highlands: implications for tectonic evolution during *c.* 1050 Ma Ottawan orogenesis. *GSA Bulletin*, **120**, 950–961, doi: 10.1130/B26309.1.

BOHLEN, S. R. 1987. Pressure, temperature, time paths, and a tectonic model for the evolution of granulites. *Journal of Geology*, **95**, 617–632.

BONNET, A.-L., CORRIVEAU, L. & LA FLÈCHE, M. R. 2005. Chemical imprint of highly metamorphosed volcanic-hosted hydrothermal alterations in the La Romaine supracrustal belt, eastern Grenville Province, Quebec. *Canadian Journal of Earth Sciences*, **42**, 1783–1814.

BUSSY, F., KROGH, T. E., KLEMENS, W. P. & SCHWERDTNER, W. M. 1995. Tectonic and metamorphic events in the westernmost Grenville Province, central Ontario: new results from high-precision geochronology. *Canadian Journal of Earth Sciences*, **32**, 660–671.

CARR, S. D., EASTON, R. M., JAMIESON, R. A. & CULSHAW, N. G. 2000. Geologic transect across the Grenville orogen of Ontario and New York. *Canadian Journal of Earth Sciences*, **37**, 193–216.

CHILDE, F., DOIG, R. & GARIÉPY, C. 1993. Monazite as a metamorphic chronometer, south of the Grenville Front, western Quebec. *Canadian Journal of Earth Sciences*, **30**, 1056–1065.

CLARK, M. K. & ROYDEN, L. H. 2000. Topographic ooze: building the eastern margin of Tibet by lower crustal flow. *Geology*, **28**, 703–706.

CONNELLY, J. N. & HEAMAN, L. M. 1993. U–Pb geochronological constraints on the tectonic evolution of the Grenville Province, western Labrador. *Precambrian Research*, **63**, 123–142.

CONNELLY, J. N., RIVERS, T. & JAMES, D. T. 1995. Thermotectonic evolution of the Grenville Province of western Labrador. *Tectonics*, **14**, 202–217.

CORFU, F. & EASTON, R. M. 1995. U–Pb geochronology of the Mazinaw terrane, an imbricate segment of the Central Metasedimentary Belt, Grenville Province, Ontario. *Canadian Journal of Earth Sciences*, **32**, 959–976.

CORRIGAN, D. & VAN BREEMEN, O. 1997. U–Pb age constraints for the lithotectonic evolution of the Grenville Province along the Mauricie transect, Québec. *Canadian Journal of Earth Sciences*, **34**, 299–316.

CORRIGAN, D., CULSHAW, N. G. & MORTENSEN, J. K. 1994. Pre-Grenvillian evolution and Grenvillian overprinting of the Parautochthonous Belt in the Key Harbour area, Ontario: U–Pb and field constraints. *Canadian Journal of Earth Sciences*, **31**, 583–596.

CORRIGAN, D., RIVERS, T. & DUNNING, G. 2000. U–Pb constraints for the plutonic and tectonometamorphic evolution of Lake Melville terrane, Labrador and implications for basement reworking in the northeastern Grenville Province. *Precambrian Research*, **99**, 65–90.

CORRIVEAU, L. & BONNET, A.-L. 2005. Pinwarian (1.50 Ga) volcanism and hydrothermal activity at the eastern margin of the Wakeham Group, Grenville Province, Quebec. *Canadian Journal of Earth Sciences*, **42**, 1749–1782.

COSCA, M. A., SUTTER, J. F. & ESSENE, E. J. 1991. Cooling and inferred uplift/erosion history of the Grenville orogen, Ontario: constraints from $^{40}$Ar/$^{39}$Ar thermochronology. *Tectonics*, **10**, 959–977.

COX, R. & INDARES, A. 1999a. High pressure and temperature metamorphism of the mafic and ultramafic Lac Espadon suite, Manicouagan Imbricate Zone, eastern Grenville Province. *Canadian Mineralogist*, **37**, 335–357.

COX, R. & INDARES, A. 1999b. Transformation of Fe-Ti gabbro to coronite, eclogite and amphibolite in the Baie du Nord segment, eastern Grenville Province. *Journal of Metamorphic Geology*, **17**, 537–555.

COX, R. & RIVERS, T. in press. Reconnaissance laser ablation (LA)-ICP-MS U–Pb zircon and monazite geochronology in reworked Archaean basement and Palaeoproterozoic supracrustal rocks, Gagnon terrane, Grenville Province of Québec and Labrador. *Precambrian Research*.

COX, R., INDARES, A. & DUNNING, G. 1998. Petrology and U–Pb geochronology of mafic, high-P metamorphic coronites from the Tshenukutish domain, eastern Grenville Province. *Precambrian Research*, **90**, 59–83.

CULSHAW, N. G., REYNOLDS, P. H. & CHECK, G. 1991. A $^{40}$Ar/$^{39}$Ar study of post-tectonic cooling in the Britt domain of the Grenville Province, Ontario. *Earth & Planetary Science Letters*, **105**, 405–415.

CULSHAW, N. G., KETCHUM, J. W. F., WODICKA, N. & WALLACE, P. 1994. Ductile extension following thrusting in the southwestern Grenville Province, Ontario. *Canadian Journal of Earth Sciences*, **31**, 160–175.

CULSHAW, N. G., JAMIESON, R. A., KETCHUM, J. W. F., WODICKA, N., CORRIGAN, D. & REYNOLDS, P. H. 1997. Transect across the northwestern Grenville orogen, Georgian Bay, Ontario: polystage convergence and extension in the lower orogenic crust. *Tectonics*, **16**, 966–982.

DAIGNEAULT, R. & ALLARD, G. O. 1994. Transformation of Archaean structural inheritance at the Grenvillian Foreland – Parautochthon transition zone, Chibougamau Quebec. *Canadian Journal of Earth Sciences*, **31**, 470–488.

DALLMEYER, R. D. & RIVERS, T. 1983. Recognition of extraneous argon components through incremental-release $^{40}$Ar/$^{39}$Ar analysis of biotite and hornblende across the Grenvillian metamorphic gradient in southwestern Labrador. *Geochimica et Cosmochimica Acta*, **47**, 413–428.

DARLING, R. S., FLORENCE, F. P., LESTER, G. W. & WHITNEY, P. R. 2004. Petrogenesis of prismatine-bearing metapelitic gneiss along the Moose River, west-central Adirondacks, New York. *In*: TOLLO, R. P., CORRIVEAU, L., MCLELLAND, J. & BARTHOLOMEW, M. J. (eds) *Proterozoic tectonic evolution of the Grenville orogen in North America*. Geological Society of America Memoir, **197**, 325–336.

DAVIDSON, A. 1990. Evidence for eclogite metamorphism in the southwestern Grenville Province. *Geological Survey of Canada*, **90–1C**, 113–118.

DAVIDSON, A. 1998. An overview of the Grenville Province geology, Canadian Shield. *In*: LUCAS, S. B. & ST-ONGE, M. R. (eds) *Geology of the Precambrian Superior and Grenville Provinces and Precambrian Fossils in North America*. Geological Survey of Canada, Geology of Canada.

DAVIDSON, A. & VAN BREEMEN, O. 1988. Baddeleyite-zircon relationships in coronitic metagabbro, Grenville Province, Ontario: implications for geochronology. *Contributions to Mineralogy and Petrology*, **100**, 291–299.

DEWEY, J. F. 1988. Extensional collapse of orogens. *Tectonics*, **7**, 1123–1139.

DICKIN, A. P. & GUO, A. 2001. The location of the Allochthon Boundary Thrust and the Archean–Proterozoic suture in the Mattawa area of the Grenville Province: Nd isotopic evidence. *Precambrian Research*, **107**, 31–43.

DUDÁS, F. Ö., DAVIDSON, A. & BETHUNE, K. M. 1994. Age of the Sudbury diabase dykes and their metamorphism in the Grenville Province. *In*: *Radiogenic age and isotope studies: Report 8*. Geological Survey of Canada, Current Research, **1994–F**, 97–106.

GARIÉPY, C., VERNER, D. & DOIG, R. 1990. Dating Archaean metamorphic minerals southeast of the Grenville Front, western Quebec, using Pb isotopes. *Geology*, **18**, 1078–1081.

GOBEIL, A., BRISEBOIS, D., CLARK, T., VERPAELST, P., MADORE, L., WODICKA, N. & CHEVÉ, S. 2003. Géologie de la moyenne Côte-Nord. *In*: BRISEBOIS, D. & CLARK, T. (eds) *Géologie et resources minerales des la partie est de la Province de Grenville*. Ministère de Richesses naturelles, Québec, DV2002–2003, 59–117.

GODIN, L., GRUJIC, D., LAW, R. D. & SEARLE, M. P. 2006*a*. Channel flow, ductile extrusion and exhumation in continental collision zones: an introduction. *In*: LAW, R. D., SEARLE, M. P. & GODIN, L. (eds) *Channel Flow, Ductile Extrusion and Exhumation in Continental Collision Zones*. Geological Society, London, Special Publications, **268**, 1–23.

GODIN, L., GLEESON, T. P., SEARLE, M. P., ULLRICH, T. D. & PARRISH, R. R. 2006*b*. Locking of southward extrusion in favour of rapid crustal-scale buckling of the Greater Himalayan sequence, Nar Valley, central Nepal. *In*: LAW, R. D., SEARLE, M. P. & GODIN, L. (eds) *Channel Flow, Ductile Extrusion and Exhumation in Continental Collision Zones*. Geological Society, London, Special Publications, **268**, 269–292.

GOSCOMBE, B., GRAY, D. & HAND, M. 2006. Crustal architecture of the Himalayan metamorphic front in eastern Nepal. *Gondwana Research*, **10**, 232–255, doi: 10.1016/j.gr2006.05.003.

GOWER, C. F. & KROGH, T. E. 2002. A U–Pb geochronological review of the Proterozoic history of the eastern Grenville Province. *Canadian Journal of Earth Sciences*, **39**, 795–829.

GOWER, C. F., RYAN, A. B., BAILEY, D. G. & THOMAS, A. 1980. The position of the Grenville Front in eastern and central Labrador. *Canadian Journal of Earth Sciences*, **17**, 784–788.

GOWER, C. F., HEAMAN, L. M., LOVERIDGE, W. D., SCHÄRER, U. & TUCKER, R. D. 1991. Grenvillian granitoid plutonism in the eastern Grenville Province. *Precambrian Research*, **51**, 315–336.

GREEN, A. G., MILKREIT, B. *ET AL.* 1988. Crustal structure of the Grenville Front and adjacent terranes. *Geology*, **16**, 788–792.

GRUJIC, D., HOLLISTER, L. & PARRISH, R. R. 2002. Himalayan metamorphic sequence as an orogenic channel: insight from Bhutan. *Earth and Planetary Science Letters*, **198**, 177–191.

GRUJIC, D., COUTLAND, I., BOOKHAGEN, B., BLYTHE, A. & DUNCAN, C. 2006. Climatic forcing of erosion, landscape, and tectonics in the Bhutan Himalayas. *Geology*, **34**, 801–804.

HAGGART, M. J., JAMIESON, R. A., REYNOLDS, P. H., KROGH, T. E., BEAUMONT, C. & CULSHAW, N. G. 1993. Last gasp of the Grenville Orogeny: thermochronology of the Grenville Front Tectonic Zone near Killarney, Ontario. *Journal of Geology*, **101**, 575–589.

HEAMAN, L. M., ERDMER, P. & OWEN, J. V. 2002. U–Pb geochronologic constraints on the crustal evolution of the Long Range Inlier, Newfoundland. *Canadian Journal of Earth Sciences*, **39**, 845–865.

HERD, R. K., ACKERMAN, D., WINDLEY, B. F. & RONDOT, J. 1986. Sapphirine–garnet rocks, St. Maurice area, Québec: petrology and implications for tectonics and metamorphism. *In*: MOORE, J. M., DAVIDSON, A. & BAER, A. J. (eds) *The Grenville Province*. Geological Association of Canada, Special Paper, **31**, 241–253.

HODGES, K. V. 1998. The thermodynamics of Himalayan orogenesis. *In*: TRELOAR, P. J. & O'BRIEN, P. J. (eds) *What Drives Metamorphism And Metamorphic Reactions?* Geological Society, London, Special Publications, **138**, 7–22.

HODGES, K. V. 2006. A synthesis of the channel flow – extrusion hypothesis as developed for the Himalayan – Tibetan orogenic system. *In*: LAW, R. D., SEARLE, M. P. & GODIN, L. (eds) *Channel Flow, Ductile Extrusion and Exhumation in Continental Collision Zones*, Geological Society, London, Special Publications, **268**, 71–90.

HODGES, K. V., PARRISH, R. R. & SEARLE, M. P. 1996. Tectonic evolution of the central Annapurna Range, Nepalese Himalaya. *Tectonics*, **15**, 1264–1291.

HOFFMAN, P. F. 1989. Precambrian geology and tectonic history of North America. *In*: BALLY, A. W. & PALMER, A. R. (eds) *The Geology of North America – an overview*. Geological Society of America, Decade of North American Geology, A, 447–512.

HYNES, A. 1994. Gravity, flexure, and the deep structure of the Grenville front, eastern Quebec and Labrador. *Canadian Journal of Earth Sciences*, **31**, 1002–1011.

HYNES, A., INDARES, A., RIVERS, T. & GOBEIL, A. 2000. Lithoprobe line 55: integration of out-of-plane seismic results with surface structure, metamorphism, and geochronology, and the tectonic evolution of the eastern Grenville Province. *Canadian Journal of Earth Sciences*, **37**, 341–358.

INDARES, A. 1995. Metamorphic interpretation of high-pressure–temperature metapelites with preserved growth zoning in garnet, eastern Grenville Province, Canadian Shield. *Journal of Metamorphic Geology*, **13**, 475–486.

INDARES, A. 1997. Garnet-clinopyroxenites and garnet-kyanite restites from the Manicouagan Imbricate Zone: a case of high-*P*–high-*T* metamorphism in the Grenville Province. *Canadian Mineralogist*, **35**, 1161–1171.

INDARES, A. 2003. Metamorphic textures and *P–T* evolution of high-*P* granulites from the Lelukuau terrane, NE Grenville Province. *Journal of Metamorphic Geology*, **21**, 35–48.

INDARES, A. & DUNNING, G. 1997. Coronitic metagabbro and eclogite from the Grenville Province of western Quebec: interpretation of U–Pb geochronology and metamorphism. *Canadian Journal of Earth Sciences*, **34**, 891–901.

INDARES, A. & DUNNING, G. 2004. Crustal architecture above the high-pressure belt of the Grenville Province

in the Manicouagan area: new structural, petrologic and U–Pb age constraints. *Precambrian Research*, **130**, 199–228.

INDARES, A. & MARTIGNOLE, J. 1989. The Grenville Front south of Val d'Or. *Tectonophysics*, **157**, 221–239.

INDARES, A. & MARTIGNOLE, J. 1990a. Metamorphic constraints on the evolution of the gneisses from the parautochthonous and allochthonous polycyclic belts, Grenville Province, western Quebec. *Canadian Journal of Earth Sciences*, **27**, 357–370.

INDARES, A. & MARTIGNOLE, J. 1990b. Metamorphic constraints on the tectonic evolution of the allochthonous monocyclic belt, Grenville Province, western Quebec. *Canadian Journal of Earth Sciences*, **27**, 371–386.

INDARES, A. & MARTIGNOLE, J. 1993. *Étude régionale du Supergroupe de Wakeham, Province de Grenville*. Ministère de l'Énergie et des Resources, Québec, MB; 91–**21**, 1–73.

INDARES, A. & RIVERS, T. 1995. Textures, metamorphic reactions and thermobarometry of eclogitized meta-gabbros: a Proterozoic example. *European Journal of Mineralogy*, **7**, 43–56.

INDARES, A., DUNNING, G., COX, R., GALE, D. & CONNELLY, J. 1998. High-pressure, high-temperature rocks from the base of thick continental crust: geology and age constraints from the Manicouagan Imbricate Zone, eastern Grenville Province. *Tectonics*, **17**, 426–440.

INDARES, A., DUNNING, G. & COX, R. 2000. Tectono-thermal evolution of deep crust in a Mesoproterozoic continental collision setting: the Manicouagan example. *Canadian Journal of Earth Sciences*, **37**, 325–340.

JAMIESON, R. A., CULSHAW, N. G. & CORRIGAN, D. 1995. North-west propagation of the Grenville orogen: Grenville structure and metamorphism near Key Harbour, Georgian Bay, Ontario, Canada. *Journal of Metamorphic Geology*, **13**, 185–207.

JAMIESON, R. A., BEAUMONT, C., NGUYEN, M. H. & LEE, B. 2002. Interaction of metamorphism, deformation, and exhumation in large convergent orogens. *Journal of Metamorphic Geology*, **20**, 1–16.

JAMIESON, R. A., BEAUMONT, C. & MEDVEDEV, S. 2004. Crustal channel flows: 2. Numerical models with implications for metamorphism in the Himalayan-Tibetan orogen. *Journal of Geophysical Research*, **109**, B06407, doi: 10.1029/2003JB002811.

JAMIESON, R. A., BEAUMONT, C., NGUYEN, M. H. & CULSHAW, N. G. 2007. Synconvergent ductile flow in variable-strength continental crust: numerical models with application to the western Grenville Province. *Tectonics*, **26**, TC5005, doi: 10.1029/2006TC002036.

JOHNSON, E. L., GOERGEN, E. T. & FRUCHEY, B. L. 2004. Right lateral oblique slip movements followed by post-Ottawan (1050–1020) Ma orogenic collapse along the Carthage-Colton shear zone: data from the Dana Hill metagabbro body, Adirondack Mountains, New York. *In*: TOLLO, R. P., CORRIVEAU, L., MCLELLAND, J. & BARTHOLOMEW, M. J. (eds) *Proterozoic tectonic evolution of the Grenville orogen in North America*. Geological Society of America Memoir, **197**, 357–378.

JORDAN, S. L., INDARES, A. & DUNNING, G. 2006. Partial melting of metapelites in the Gagnon terrane below the high-pressure belt in the Manicouagan area (Grenville Province): pressure–temperature ($P–T$) and U–Pb age constraints and implications. *Canadian Journal of Earth Sciences*, **43**, 1309–1329.

KARLSTROM, K. E., ÅHÄLL, K.-I., HARLAN, S. S., WILLIAMS, M. L., MCLELLAND, J. & GEISSMAN, J. W. 2001. Long-lived (1.8–1.0 Ga) convergent orogen in southern Laurentia, its extensions to Australia and Baltica, and implications for refining Rodinia. *Precambrian Research*, **111**, 5–30.

KELLETT, R. L., BARNES, A. E. & RIVE, M. 1994. The deep structure of the Grenville Front: a new perspective from western Quebec. *Canadian Journal of Earth Sciences*, **31**, 282–292.

KETCHUM, J. W. F. & DAVIDSON, A. 2000. Crustal architecture and tectonic assembly of the Central Gneiss Belt, southwestern Grenville Province, Canada: a new interpretation. *Canadian Journal of Earth Sciences*, **37**, 217–234.

KETCHUM, J. W. F. & KROGH, T. E. 1997. U–Pb constraints on high-pressure metamorphism in the Central Gneiss Belt, southwestern Grenville orogen. *Geological Association of Canada – Mineralogical Association of Canada, Program with Abstracts*, **22**, A78.

KETCHUM, J. W. F. & KROGH, T. E. 1998. U–Pb constraints on high-pressure metamorphism in the southwestern Grenville orogen. *Goldschmidt Conference 1998, Abstracts volume, Mineralogical Magazine*, **62A**, 775–776.

KETCHUM, J. W. F., HEAMAN, L. M., KROGH, T. E., CULSHAW, N. G. & JAMIESON, R. A. 1998. Timing and thermal influence of late orogenic extension in the lower crust: a U–Pb geochronological study from the southwest Grenville orogen, Canada. *Precambrian Research*, **89**, 25–45.

KRAUSS, J. B. & RIVERS, T. 2004. High-pressure granulites in the Grenvillian Grand Lake thrust system, Labrador: pressure-temperature conditions and tectonic evolution. *In*: TOLLO, R. P., CORRIVEAU, L., MCLELLAND, J. & BARTHOLOMEW, M. J. (eds) *Proterozoic tectonic evolution of the Grenville orogen in North America*. Geological Society of America Memoir, **197**, 105–134.

KROGH, T. E. 1994. Precise U–Pb ages for Grenvillian and pre-Grenvillian thrusting of Proterozoic and Archaean metamorphic assemblages in the Grenville Front tectonic zone, Canada. *Tectonics*, **13**, 963–982.

KROGH, T. E., KAMO, S., GOWER, C. F. & OWEN, J. V. 2002. Augmented and reassessed U–Pb geochronological data from the Labradorian–Grenvillian front in the Smokey archipelago, eastern Labrador. *Canadian Journal of Earth Sciences*, **39**, 831–843.

LI, Z. X., BOGDANOVA, S. V. ET AL. 2008. Assembly, configuration, and break-up of Rodinia: a synthesis. *Precambrian Research*, **160**, 179–210.

LOEWY, S. L., CONNELLY, J. N., DALZIEL, I. W. D. & GOWER, C. F. 2003. Eastern Laurentia in Rodinia: constraints from whole-rock Pb and U/Pb geochronology. *Tectonophysics*, **375**, 169–197.

MADORE, L., VERPAELST, P., BRISEBOIS, D., HOCQ, M. & DION, D. J. 1999. *Géologie de la région du lac Allard (SNRC 12L/11)*. Ministère des Richesses naturelles, Québec, RG 96–05, 30.

MARTIGNOLE, J. & CALVERT, A. J. 1996. Crustal-scale shortening and extension across the Grenville Province of western Quebec. *Tectonics*, **15**, 376–386.

MARTIGNOLE, J. & FRIEDMAN, R. 1998. Geochronological constraints on the last stages of terrane assembly in the central part of the Grenville Province (Quebec). *Precambrian Research*, **92**, 145–164.

MARTIGNOLE, J. & MARTELAT, J.-E. 2005. Proterozoic mafic dykes as monitors of HP granulite facies metamorphism in the Grenville Front Tectonic Zone (western Quebec). *Precambrian Research*, **138**, 183–207.

MARTIGNOLE, J. & REYNOLDS, P. 1997. $^{40}$Ar/$^{39}$Ar thermochronology along a western Québec transect of the Grenville Province, Canada. *Journal of Metamorphic Geology*, **15**, 283–296.

MARTIGNOLE, J., CALVERT, A. J., FRIEDMAN, R. & REYNOLDS, P. 2000. Crustal evolution along a seismic section across the Grenville Province (western Quebec). *Canadian Journal of Earth Sciences*, **37**, 291–306.

MARTIGNOLE, J., JI, S. & NANTEL, S. 2006. *A section through the Morin Terrane, Grenville Province (Québec)*. Geological Association of Canada – Mineralogical Association of Canada, Joint Annual Meeting, Montréal 2006, Field Trip A2 Guidebook, 21p.

MCLELLAND, J., DALY, J. S. & MCLELLAND, J. M. 1996. The Grenville orogenic cycle (c. 1350–1000 Ma): an Adirondack perspective. *Tectonophysics*, **265**, 1–28.

MCLELLAND, J., HAMILTON, M., SELLECK, B., MCLELLAND, J., WALKER, D. & ORRELL, S. 2001. Zircon U–Pb geochronology of the Ottawan Orogeny, Adirondack Highlands, New York: regional and tectonic implications. *Precambrian Research*, **109**, 39–72.

MEZGER, K., RAWNSLEY, C. M., BOHLEN, S. R. & HANSON, G. N. 1991. U–Pb garnet, sphene, monazite, and rutile ages: implications for the duration of high-grade metamorphism and cooling histories, Adirondack Mtns, New York. *Journal of Geology*, **99**, 415–428.

MILLER, B. V., DUNNING, G. R., BARR, S. M., RAESIDE, R. P., JAMIESON, R. A. & REYNOLDS, P. H. 1996. Magmatism and metamorphism in a Grenvillian fragment: U–Pb and $^{40}$Ar/$^{39}$Ar ages from the Blair River Complex, northern Cape Breton Island, Nova Scotia. *Geological Society of America Bulletin*, **108**, 127–140.

MOORE, J. M. & THOMPSON, P. H. 1980. The Flinton Group: a late Precambrian metasedimentary succession in the Grenville Province. *Canadian Journal of Earth Sciences*, **17**, 221–229.

O'BRIEN, P. J., CARSWELL, D. A. & GEBAUER, D. 1990. Eclogite formation and distribution in the European Variscides. *In*: CARSWELL, D. A. (ed.) *Eclogite Facies Rocks*. Blackie, Glasgow & London, 204–224.

ONSTOTT, T. C. & PEACOCK, M. W. 1987. Argon retentivity of hornblendes: a field experiment in a slowly cooled metamorphic terrane. *Geochimica et Cosmochimic Acta*, **51**, 2891–2903.

OWEN, J. V., RIVERS, T. & GOWER, C. F. 1986. The Grenville Front on the Labrador coast. *In*: MOORE, J. M., DAVIDSON, A. & BAER, A. J. (eds) *The*

*Grenville Province*. Geological Association of Canada, Special Paper, **31**, 95–106.

OWEN, J. V., DALLMEYER, R. D., GOWER, C. F. & RIVERS, T. 1988. Metamorphic conditions and $^{40}Ar/^{39}Ar$ contrasts across the Grenville Front zone, coastal Labrador, Canada. *Lithos*, **21**, 13–35.

PECK, W. H., DEANGELIS, M. T., MEREDITH, M. T. & MORIN, E. 2005. Polymetamorphism of marbles in the Morin terrane, Grenville Province, Quebec. *Canadian Journal of Earth Sciences*, **42**, 1949–1965.

QUIRKE, T. T. & COLLINS, W. H. 1930. *The disappearance of the Huronian*. Geological Survey of Canada, Memoir, **160**.

REY, P., VANDERHAEGUE, O. & TEYSSIER, C. 2001. Gravitational collapse of the continental crust: definition, regimes and modes. *Tectonophysics*, **342**, 435–449.

REYNOLDS, P. H., CULSHAW, N. G., JAMIESON, R. A., GRANT, S. L. & MCKENZIE, K. J. 1995. $^{40}Ar/^{39}Ar$ traverse – Grenville Front Tectonic Zone to Britt domain, Grenville Province, Ontario, Canada. *Journal of Metamorphic Geology*, **13**, 209–221.

RIVERS, T. 1983a. The northern margin of the Grenville Province in western Labrador – anatomy of an ancient orogenic front. *Precambrian Research*, **22**, 41–73.

RIVERS, T. 1983b. Progressive metamorphism of pelitic and quartzofeldspathic assemblages in the Grenville Province of western Labrador – tectonic implications of bathozone 6 assemblages. *Canadian Journal of Earth Sciences*, **20**, 1791–1804.

RIVERS, T. 1997. Lithotectonic elements of the Grenville Province: review and tectonic implications. *Precambrian Research*, **86**, 117–154.

RIVERS, T. 2008. Assembly and preservation of lower, mid and upper orogenic crust in the Grenville Province – Implications for the evolution of large, hot long-duration orogens. *Precambrian Research*, **167**, 237–259.

RIVERS, T. & CHOWN, E. H. 1986. The Grenville orogen in eastern Quebec and western Labrador – definition, identification and tectonometamorphic relationships of autochthonous, parautochthonous and allochthonous terranes. *In*: MOORE, J. M., DAVIDSON, A. & BAER, A. J. (eds) *The Grenville Province*. Geological Association of Canada, Special Paper, **31**, 31–50.

RIVERS, T. & INDARES, A. 2006. Assembly of a 60 km crustal section through the Grenville orogen from eclogites through granulites to the non-metamorphic lid – implications for Grenvillian tectonic evolution. *Geological Association of Canada – Mineralogical Association of Canada, Abstracts*, **31**, 128.

RIVERS, T. & MENGEL, F. C. 1988. Contrasting assemblages and petrogenetic evolution of corona and non-coronaa gabbros in the Grenville province of western Labrador. *Canadian Journal of Earth Sciences*, **25**, 1629–1648.

RIVERS, T., MARTIGNOLE, J., GOWER, C. F. & DAVIDSON, A. 1989. New tectonic divisions of the Grenville Province, southeast Canadian shield. *Tectonics*, **8**, 63–84.

RIVERS, T., VAN GOOL, J. A. M. & CONNELLY, J. N. 1993. Contrasting tectonic styles in the northern Grenville Province: implications for the dynamics of orogenic fronts. *Geology*, **21**, 1127–1130.

RIVERS, T., KETCHUM, J., INDARES, A. & HYNES, A. 2002. The High Pressure Belt in the Grenville Province: architecture, timing and exhumation. *Canadian Journal of Earth Sciences*, **39**, 867–893.

RIVERS, T., CULSHAW, N., HYNES, A., INDARES, A., JAMIESON, R. & MARTIGNOLE, J. in review. The Grenville Orogen. *In*: PERCIVAL, J. & COOK, F. (eds) *Variations in Tectonic Styles Revisited: a Lithoprobe Perspective*. Geological Association of Canada, Special Paper.

ROSENBERG, C. L. & HANDY, M. R. 2005. Experimental deformation of partially melted granite revisited: implications for the continental crust. *Journal of Metamorphic Geology*, **23**, 19–28.

SEARLE, M. P., LAW, R. D. & JESSUP, M. J. 2006. Crustal structure, restoration and evolution of the Greater Himalaya in Nepal-South Tibet: implications for channel flow and ductile extrusion of the middle crust. *In*: LAW, R. D., SEARLE, M. P. & GODIN, L. (eds) *Channel Flow, Ductile Extrusion and Exhumation in Continental Collision Zones*. Geological Society, London, Special Publications, **268**, 355–378.

SLAGSTAD, T., HAMILTON, M. A., JAMIESON, R. A. & CULSHAW, N. G. 2004. Timing and duration of melting in the mid orogenic crust: constraints from U–Pb (SHRIMP) data, Muskoka and Shawanaga domains, Grenville Province, Ontario. *Canadian Journal of Earth Sciences*, **41**, 1339–1365.

SPEAR, F. S. & MARKUSSEN, J. C. 1997. Mineral zoning, P-T-X-M phase relations, and metamorphic evolution of some Adirondack granulites. *Journal of Petrology*, **38**, 757–783.

TIMMERMANN, H., PARRISH, R. R., JAMIESON, R. A. & CULSHAW, N. G. 1997. Time of metamorphism beneath the Central Metasedimentary Belt Boundary Thrust Zone, Grenville Orogen, Ontario: accretion at 1080 Ma? *Canadian Journal of Earth Sciences*, **34**, 1023–1029.

TIMMERMANN, H., JAMIESON, R. A., PARRISH, R. R. & CULSHAW, N. G. 2002. Coeval migmatites and granulites, Muskoka domain, southwestern Grenville Province, Ontario. *Canadian Journal of Earth Sciences*, **39**, 239–258.

TOFT, P. B., SCOWEN, P. A. H., ARKANI-HAMED, J. & FRANCIS, D. 1993. Demagnetization by hydration in deep-crustal rocks in the Grenville province of Quebec, Canada: implications for magnetic anomalies of continental collision zones. *Geology*, **21**, 999–1002.

VAN BREEMEN, O. & CORRIVEAU, L. 2005. U–Pb age constraints on arenaceous and volcanic rocks of the Wakeham Group, eastern Grenville Province. *Canadian Journal of Earth Sciences*, **42**, 1677–1697.

VAN GOOL, J., RIVERS, T. & CALON, T. 2008. The Grenville Front zone, Gagnon terrane, southwestern Labrador: configuration of a mid-crustal foreland fold-thrust belt. *Tectonics*, **27**, TC1004, doi: 10.1029/2006TC002095.

WARDLE, R. J., RYAN, A. B., PHILIPPE, S. & SCHÄRER, U. 1990. Proterozoic crustal development in the Goose Bay region, Grenville Province, Labrador, Canada. *In*: GOWER, C. F., RIVERS, T. & RYAN, A. B. (eds) *Mid-Proterozoic Laurentia–Baltica*.

Geological Association of Canada Special Paper, **38**, 197–214.

WASTENEYS, H. A., KAMO, S. L., MOSER, D., KROGH, T. E., GOWER, C. F. & OWEN, J. V. 1997. U–Pb geochronological constraints on the geological evolution of Pinware terrane and adjacent areas, Grenville Province, southeastern Labrador, Canada. *Precambrian Research*, **81**, 101–128.

WODICKA, N., PARRISH, R. R. & JAMIESON, R. A. 1996. The Parry Sound domain: a far-travelled allochthon? New evidence from U–Pb zircon geochronology. *Canadian Journal of Earth Sciences*, **33**, 1087–1104.

WODICKA, N., DAVID, J., PARENT, M., GOBEIL, A. & VERPAELST, P. 2003. Géochronologie U–Pb et Pb–Pb de la region de Sept-Îles–Natashquan, Province de Grenville, moyenne Côte-Nord. *In*: BRISEBOIS, D. & CLARK, T. (eds) *Géologie et resources minerales de la partie est de la Province de Grenville*. Ministère des Richesses naturelles, Québec, DV2002–2003, 59–117.

WYNNE-EDWARDS, H. R. 1972. The Grenville Province. *In*: PRICE, R. A. & DOUGLAS, R. J. W. (eds) *Variations in Tectonic Styles in Canada*. Geological Association of Canada Special Paper, **11**, 263–334.

YANG, P. & INDARES, A. D. 2005. Mineral zoning, phase relations, and *P–T* evolution of high-pressure granulites from the Lelukuau terrane, northeastern Grenville Province, Quebec. *Canadian Mineralogist*, **43**, 443–462.

# Neoproterozoic reworking of the Palaeoproterozoic Capricorn Orogen of Western Australia and implications for the amalgamation of Rodinia

SANDRA A. OCCHIPINTI[1,2] & STEVEN M. REDDY[1]*

[1]*The Institute for Geoscience Research, Dept of Applied Geology, Curtin University of Technology, GPO Box U1987, Perth, WA 6845, Australia*

[2]*Now at: AngloGoldAshanti, Level 13, St. Martin's Tower, 44 St. George's Terrace, WA 6000, Australia*

*\*Corresponding author (e-mail: S.Reddy@curtin.edu.au)*

**Abstract:** Argon isotopic data from mica from the southern Capricorn region of Western Australia record complex intra- and inter-grain systematics that reflect modification due to a range of processes. However, $^{40}Ar/^{39}Ar$ age distributions, though complex, generally show early Neoproterozoic ages in the west, increasing to Mesoproterozoic ages in the east. Palaeoproterozoic ages associated with cooling after the *c.* 1.8 Ga Capricorn Orogen or *c.* 1.6 Ga Mangaroon Orogen are not preserved. These data reflect cooling from a *c.* 300 °C thermal overprint that took place prior to 960 Ma that is related to the enigmatic Edmundian Orogeny. These data, combined with sediment provenance data from the Early Neoproterozoic Officer Basin and U–Pb age data from the nearby Pinjarra Orogen, indicate that the late Mesoproterozoic–Neoproterozoic Pinjarra and Edmundian events are dynamically linked and reflect tectonic activity on the western margin of the amalgamated West Australian Craton. The temporal framework for this event suggest a link to the evolving Rodinian supercontinent and reflect the oblique collision of either Greater India or Kalahari cratons with the West Australian Craton. These results illustrate that the temporal evolution of poorly preserved orogens can be constrained by low-temperature thermochronology in the adjacent cratons.

**Supplementary material:** Summary of $^{40}Ar/^{39}Ar$ results reported in detail is available at http://www.geolsoc.org.uk/SUP18357.

The cratonic cores of continental interiors are commonly typified by ancient high-grade metamorphic rocks that have seen little tectonic activity since the Archaean. In contrast, the margins of these cratons often record a complex geological evolution involving cycles of rifting, accretion and collision due to the global reorganization of continental fragments during the repeated dispersal and amalgamation of supercontinents. Geological analysis of craton margins therefore provides a valuable means of constraining ancient supercontinent cycles. However, processes such as continental rifting, subduction erosion and crustal reworking that may take place at cratonic margins can mask or destroy the evidence of earlier tectonic activity, thereby limiting the ability of such areas to successfully assist in the reconstruction of tectonic histories and palaeogeography. An alternative approach is to attempt to identify and characterize the far-field effects of craton margin tectonism within the craton, and use these data to constrain the temporal evolution of processes taking place at the margin. This approach is highlighted by presenting mica $^{40}Ar/^{39}Ar$ data from the Palaeoproterozoic Capricorn Orogen of West Australia to provide temporal constraints on tectonic activity along the western margin of the West Australian Craton during the formation of Rodinia.

## Geological background

The West Australian Craton comprises the Archaean Pilbara and Yilgarn cratonic blocks and a series of tectonically complex basement rocks and basins of the Palaeoproterozoic Capricorn Orogen (Myers 1993) (Fig. 1). To the south, the West Australian Craton passes into the Albany–Fraser Belt, a complex series of high-grade metamorphic rocks that were strongly deformed during the Mesoproterozoic collision of the West Australian Craton with the South Australian–East Antarctic continent (Clark *et al.* 2000). The eastern margin of the West Australian Craton is overlain and completely hidden by sediments of the Proterozoic Officer Basin, the Phanerozoic Canning Basin and the Tertiary Eucla Basin (Trendall & Cockbain 1990).

*From*: MURPHY, J. B., KEPPIE, J. D. & HYNES, A. J. (eds) *Ancient Orogens and Modern Analogues.*
Geological Society, London, Special Publications, **327**, 445–456.
DOI: 10.1144/SP327.18    0305-8719/09/$15.00 © The Geological Society of London 2009.

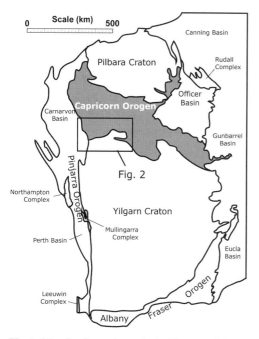

**Fig. 1.** Map showing major geological features of the West Australian Craton.

The geological evolution of the western margin of the West Australian Craton is also enigmatic because basin formation associated with the rifting and dispersal of Australia and Greater India during the Cretaceous breakup of Gondwana (Song & Cawood 2000) masks the earlier history. Despite this, the presence of Mesoproterozoic and Neoproterozoic rocks of the Pinjarra Orogen (Fig. 1) attests to an eventful geological evolution of the western margin of the craton (Myers 1990; Fitzsimons 2003) following its amalgamation in the Palaeoproterozoic era.

The Capricorn Orogen lies between the Archaean Pilbara and Yilgarn cratons and contains a series of terranes that comprise early to late Archaean granite and granitic gneiss, Palaeoproterozoic metasedimentary and mafic meta-igneous rocks, granite and granitic gneiss (Fig. 2). These units are locally overlain by various sedimentary units deposited in a range of settings between the Palaeoproterozoic era to the Permian period (Cawood & Tyler 2004, and references therein).

The Capricorn Orogen comprises rocks deformed and metamorphosed during the 2000–1950 Ma Glenburgh Orogeny and the 1830–1780 Ma Capricorn Orogeny (Occhipinti et al. 1998; Occhipinti et al. 2004; Sheppard et al. 2004). Mineral assemblages throughout the range of basement rocks in the Capricorn Orogen indicate a regional greenschist facies metamorphic and deformation overprint associated with tectonic activity associated with the latest stages of Capricorn

orogenesis (Occhipinti & Reddy 2004; Reddy & Occhipinti 2004; Sheppard et al. 2004). Magmatism, metamorphism and deformation during the 1680–1620 Ma Mangaroon Orogeny is also heterogeneously distributed through the region, becoming more significant to the north (Sheppard et al. 2005).

In the central and eastern parts of the Capricorn Orogen a regionally extensive series of sediments and volcanics (the Bangemall Supergroup) unconformably overlie basement rocks. Dolerite sills that intruded the base of this stratigraphic sequence yield dates of $1465 \pm 3$ Ma (Wingate et al. 2002) and $1070 \pm 6$ Ma (Wingate et al. 2004). Some of these dykes (of unknown age) were deformed and metamorphosed with the sediments and the underlying basement at low metamorphic grades during the enigmatic Edmundian Orogeny, before being cut by northerly trending dolerite dykes of the 750 Ma Mundine Well dyke swarm (Wingate & Giddings 2000; Martin & Thorne 2004). These overprinting relationships loosely constrain the Edmundian event at 1070–750 Ma.

Despite the relatively well-constrained Palaeoproterozoic evolution of the Capricorn Orogen, there are no published $^{40}\text{Ar}/^{39}\text{Ar}$ data from the orogen. In this paper, the results of a regional $^{40}\text{Ar}/^{39}\text{Ar}$ study document a widespread Late Mesoproterozoic–Neoproterozoic reworking of the Capricorn Orogen. The data, combined with previously published sedimentological data from the Officer Basin, and U–Pb zircon data from basement rocks of the Pinjarra Orogen constrain tectonic activity on the western margin of the West Australian Craton.

## Analytical procedure

A regional suite of samples from different terranes of the Capricorn Orogen and immediately adjacent Yilgarn Craton have been analysed by the $^{40}\text{Ar}/^{39}\text{Ar}$ dating technique. Details of the analysed samples, including sample localities and rock unit descriptions, are given as Supplementary Data. In many cases the analysed samples have an igneous origin and are granitic in composition, though a few samples are amphibolite–granulite facies metasedimentary rocks. In all samples the primary mineralogical assemblage has been retrogressed to greenschist facies metamorphic assemblage (see Supplementary Data). In many cases this reflects the Capricorn Orogeny phase of the tectonic evolution (c. 1800 Ma) (Occhipinti et al. 2004), although recent studies have illustrated potentially younger metamorphic overprints within the north of the Capricorn region (Sheppard et al. 2005).

Sample preparation and analytical procedure have been described in detail elsewhere (Occhipinti

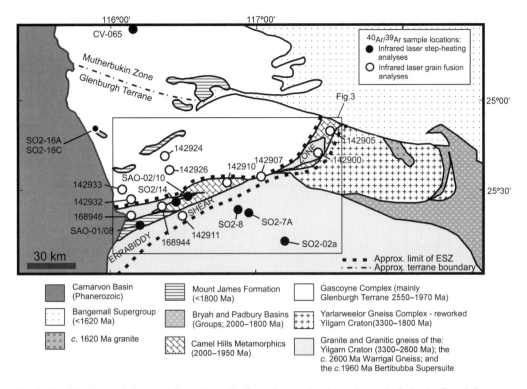

**Fig. 2.** Simplified geological map of the southern Capricorn Orogen showing major geological units. Sample locations for $^{40}Ar/^{39}Ar$ analyses from the Yilgarn craton (Narryer terrane), the Errabiddy Shear Zone and the Gascoyne Complex are shown.

2004; Reddy *et al.* 2004) so only a brief summary is presented here. $^{40}Ar/^{39}Ar$ analyses were conducted on muscovite and biotite using two different analytical approaches: Infrared laser total grain fusion and infrared laser step-heating. In both cases, samples were crushed and inclusion-free mica grains were selected after examination with a binocular microscope. Depending on grain size, single or multiple grain aliquots were used to ensure sufficient Ar for measurement. All samples were cleaned in methanol, then de-ionized water in an ultrasonic bath. Once dry, the samples were packed in aluminium foil and loaded into an aluminium package with other samples. The package was then Cd-shielded (0.4 mm) and irradiated in the H5 position of the McMaster University Reactor, Hamilton, Canada for 90 hours. Biotite age standard Tinto B, with a K–Ar age of 409 Ma (Rex & Guise 1995) was placed at 5 mm intervals throughout the aluminium package to monitor the neutron flux gradient. Tinto B is a standard that has seen widespread use in the literature (Kelley *et al.* 1994; Sherlock & Kelley 2002; Reddy *et al.* 2004; Downes *et al.* 2006). Correction factors are as follows: $(^{36}Ar/^{37}Ar)Ca = 0.000255$, $(^{39}Ar/^{37}Ar)Ca = 0.00065$ and $(^{40}Ar/^{39}Ar)K = 0.0015$. Corrections for $(^{38}Ar/^{39}Ar)K$ and $(^{38}Cl/^{39}Ar)K$ were not undertaken because of the Proterozoic age and low Cl characteristics of the samples and the short amount of time between irradiation and the time of analyses.

Following irradiation, Ar was extracted using a CW–Nd–YAG laser, fired through a Merchantek computer-controlled X–Y–Z sample chamber stage and microscope system. A defocused 200 μm beam (9.7–11 Amps for 120 s) was used for infrared laser analyses. Data were corrected for mass spectrometer discrimination and nuclear interference reactions. Errors quoted on the $^{40}Ar/^{39}Ar$ ages are $1\sigma$, and ages were calculated using usual decay constants (Steiger & Jager 1977). J values are noted on the supplementary data tables. Background Ar levels were monitored prior to and after each analysis and the mean of the two blanks was used to correct each sample analysis. Ar data were corrected for mass spectrometer discrimination, $^{37}Ar$ decay, and $^{38}Ar$ decay.

## Results

Total-fusion data and step-heating age spectra are presented as accompanying supplementary data and are shown in Figures 4, 5 and 6. The $^{40}Ar/^{39}Ar$ data are also summarized in Table 1 and on the

**Table 1.** Summary of $^{40}Ar/^{39}Ar$ results reported in detail in the supplementary data. Bolded italicized ages are those interpreted as the best estimates of isotopic closure based on detailed analysis of the composition, grains size, Ar relationships by Occhipinti (2004). WM, white mica (muscovite); B. biotite; IRF, Infrared fusion; IRSH, Infrared step heated; SGA, single grain analyses; MGA, multiple grain analyses (number of grains indicated in parentheses). All ages are quoted at 1σ. For 'sample 142900' mean ages were not calculated because the age range was considerable

| Area | Sample name | Mineral | Analytical method | % Atmospheric Ar (range) | Weighted mean age (Ma) | Unweighted mean age (Ma) | Plateaux ages, unweighted mean ages using select data (Ma) | Max age (Ma) | Min age (Ma) | Grain diameter (µm) |
|---|---|---|---|---|---|---|---|---|---|---|
| Narryer Terrane | SO2-2a | B GR1 | IRSH | 0–1.60 | 1434 ± 2 | 1440 ± 38 | | 1591 ± 20 | ***1389 ± 10*** | 313 |
| | | B GR2 | IRSH | 0–1.0 | 1249 ± 2 | 1255 ± 44 | | 1341 ± 6 | ***1196 ± 5*** | 262 |
| | SO2-7a | WM | | | ***834 ± 2*** | ***834 ± 21*** | | 874 ± 5 | 773 ± 15 | 332 |
| | SO2-08 | B | IRSH | 0–1.8 | 1532 ± 2 | 1536 ± 49 | ***1552 ± 30*** (93.3% $^{39}$Ar) | 1604 ± 25 | 1452 ± 6 | 578, 260 |
| Errabiddy Shear Zone | | | | | | | | | | |
| East | 142900 | WM | IRF | 0.03–6.53 | N/A | N/A | | ***1694 ± 16*** | ***881 ± 4*** | 050–200 |
| West | 142905 | B | IRF | 0–5.01 | 1623 ± 1 | ***1626 ± 28*** | | 1690 ± 13 | 1582 ± 5 | 050–150 |
| | 142907 | B | IRF | 1.47–14.01 | 927 ± 1 | ***927 ± 23*** | | 994 ± 6 | 893 ± 10 | 050–125 |
| | 142907 | WM | IRF | 0–1.51 | 826 ± 1 | 837 ± 70 | | ***961 ± 4*** | ***724 ± 3*** | 100–200 |
| | 142910 | B | IRF | 1.78–9.11 | 961 ± 1 | ***962 ± 10*** | | 984 ± 6 | 973 ± 5 | 075–145 |
| | 142910 | WM | IRF | 0–5.81 | 912 ± 1 | 913 ± 20 | | 959 ± 4 | ***888 ± 5*** | 050–200 |
| | 142911 | B | IRF | 0–12.22 | 921 ± 1 | 921 ± 20 | | 964 ± 8 | ***888 ± 7*** | 050–115 |
| | 168944 | B | IRF | 1.72–7.09 | 1220 ± 2 | ***1224 ± 26*** | | 1263 ± 30 | 1178 ± 5 | 050–110 |
| | 168944 | WM | IRF | 0–2.43 | 941 ± 2 | 931 ± 44 | | 1021 ± 4 | ***896 ± 9*** | 050–200 |
| | 168946 | B | IRF | 2.56–10.87 | 902 ± 2 | ***905 ± 22*** | | 935 ± 4 | 877 ± 4 | 030–150 |
| | SO2/10 | B GR1 | IRSH | 0–40.08 | 828 ± 2 | 843 ± 61 | ***874 ± 11*** (65.6% $^{39}$Ar) | 882 ± 7 | 722 ± 4 | 410 |
| | | B GR2 | | 2.90–23.9 | 789 ± 1 | 794 ± 46 | ***820 ± 14*** (52.5% $^{39}$Ar) | 844 ± 4 | 701 ± 3 | 650 |
| | SO2/14 | WM GR1 | IRSH | 0–13.56 | 884 ± 1 | 866 ± 43 | ***901 ± 24*** (91.7% $^{39}$Ar) | 942 ± 4 | 801 ± 7 | 260 |
| | | WM GR2 | | 0–5.92 | 902 ± 2 | ***896 ± 15*** | ***903 ± 10*** (91.59% $^{39}$Ar) | 915 ± 4 | 875 ± 13 | 260 |
| | | B | IRSH | 4.67–10.924 | 662 ± 1 | ***662 ± 26*** | | 696 ± 3 | 627 ± 3 | 650 |
| | SAO-01/08 | WM | IRSH | 0–2.16 | 873 ± 1 | ***872 ± 6*** | | 881 ± 5 | 861 ± 5 | 740 |
| | 142924 | WM | IRF | 0.09–1.70 | 911 ± 1 | 914 ± 43 | | 999 ± 4 | ***855 ± 3*** | 050–350 |
| Gascoyne Complex | 142926 | B | IRF | 3.56–12.04 | 1001 ± 1 | ***1001 ± 16*** | | 1032 ± 4 | 964 ± 4 | 075–175 |
| | 142932 | B | IRF | 0–8.68 | 936 ± 2 | ***933 ± 33*** | | 999 ± 6 | 889 ± 21 | 100–200 |
| | 142932 | WM | IRF | 0–2.61 | 898 ± 2 | 906 ± 30 | | 963 ± 7 | ***880 ± 4*** | 150–200 |
| | 142933 | B | IRF | 0.10–4.24 | 1102 ± 1 | 1106 ± 105 | | 1176 ± 5 | ***967 ± 6*** | 60–290 |
| | SO2-16A | WM GR1 | IRSH | 0.13–13.29 | 886 ± 2 | 940 ± 137 | ***895 ± 7*** (98.4% $^{39}$Ar) | 1269 ± 16 | 932 ± 3 | 480 |
| | | GR2 | | 0–1.04 | 950 ± 2 | 966 ± 93 | ***925 ± 10*** (95.5% $^{39}$Ar) | 1174 ± 5 | 906 ± 4 | 400 |
| | SO2-16C | WM | IRSH | 0–9.05 | 1045 ± 1 | 1061 ± 76 | | 1182 ± 17 | 914 ± 4 | 890 |
| | CV-065 | WM | IRSH | 0–15.97 | 834 ± 2 | 845 ± 85 | ***832 ± 1*** (plateau age) | 1071 ± 60 | 750 ± 8 | Not measured |
| | | B GR1 | IRSH | 0–26.97 | 836 ± 2 | 781 ± 140 | ***865 ± 12*** (61.9% $^{39}$Ar) | 881 ± 4 | 506 ± 12 | Not measured |
| | | B GR2 | IRSH | 0–19.32 | 852 ± 2 | 850 ± 87 | ***895 ± 6*** (61.1% $^{39}$Ar) | 903 ± 4 | 8 ± 4 | Not measured |

regional geological map (Fig. 3). Overall the data show that apparent $^{40}Ar/^{39}Ar$ ages measured in biotite are often older than those measured in muscovite from the same samples and the same range of grain sizes (Figs 4 & 5). In addition, there is commonly a wide range of $^{40}Ar/^{39}Ar$ ages within individual samples. In some cases this is directly correlated to grain size and indicates the potential presence of excess argon, particularly in some of the biotite samples, and heterogeneous Ar loss in others. In all samples, measured $^{36}Ar$ differs little from background $^{36}Ar$ levels. As a result, the use of isotope correlation diagrams ($^{36}Ar/^{40}Ar$ vs $^{39}Ar/^{40}Ar$) is precluded as a means of recognizing excess $^{40}Ar$. A detailed analysis of the complexity recorded at the individual sample level in the $^{40}Ar/^{39}Ar$ data has previously been described and analysed in considerable detail with respect to composition (associated with mineral and fluid inclusion contamination), grain size and excess $^{40}Ar$ and Ar loss (Occhipinti 2004). A summary of these data and age interpretations derived by Occhipinti (2004), taking these variables into account, are given in Table 1. Readdressing the complexity is beyond the scope of this paper and we focus on the broad patterns that emerge from the data and their tectonic significance.

Age data for the eastern Errabiddy Shear Zone (Fig. 3), the western Errabiddy Shear Zone, and the Gascoyne Complex for single and multiple grain total-fusion show a wide range of apparent $^{40}Ar/^{39}Ar$ ages that show complex relationships

with grain size but define several distinct age peaks (Figs 4 & 5). By far the biggest peaks, and therefore age distributions, for both muscovite and biotite, are of early Neoproterozoic age (Fig. 5). In detail age variations correspond to different regions. Biotite analysed by infrared total fusion (Fig. 6a) show well-defined peaks between 1650 and 1580 Ma (Eastern Errabiddy Shear Zone, $n = 10$ of 11 analyses), 960 and 880 Ma (Western Errabiddy Shear Zone, $n = 40$ of 51 analyses) and 1020 and 930 Ma (Gascoyne complex, $n = 24$ of 38 analyses). Smaller peaks are present between 1270 and 1160 Ma (Western Errabiddy Shear Zone, $n = 11$ of 51 analyses), and older ages up to c. 1350 Ma are recorded from the Gascoyne Complex ($n = 9$). Muscovite total fusion ages between c. 1690–880 Ma are recorded for the Eastern Errabiddy Shear Zone, but do not define statistically valid peaks. For the Western Errabiddy Shear Zone, 15 out of 34 ages are between 920–880 Ma. This is broadly consistent with the greatest number of ages measured in the Gascoyne Complex between 900–870 Ma ($n = 7$ of 11 analyses) (Fig. 5).

Step-heating experiments on both biotite and muscovite yield complex age spectra (Fig. 6) that often correlate with compositional (Ca and Cl) variations (Occhipinti 2004). Generally the spectra do not record statistically valid plateaus. However, the distribution of Mesoproterozoic and Neoproterozoic ages recorded by total fusion analyses are mimicked in the step-heating data and, despite the

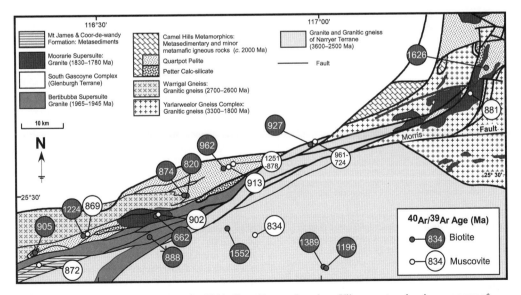

**Fig. 3.** Simplified geological map of the Errabiddy Shear Zone and northern Yilgarn craton showing summary of $^{40}Ar/^{39}Ar$ ages of biotite and muscovite. Note that age data from CV-065 and SO2_16A/C are not shown. Ages represent interpretation after analysis of compositional, excess Ar and Ar loss in detailed by Occhipinti (2004).

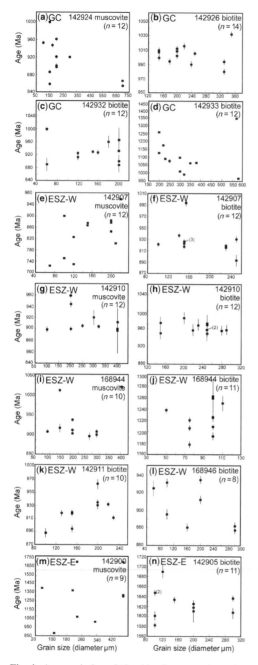

**Fig. 4.** Age–grainsize relationships for muscovite and biotite analysed by infrared laser total grain fusion. Samples are from the Gascoyne Complex (**a–d**), the western Errabiddy Shear Zone (**e–l**) and eastern Errabiddy Shear Zone (**m–n**). *n* = number of analyses for each sample. Where analyses plot over each other the number of analyses is noted in parentheses besides the analysis symbol. Note that error bars are often small and are hidden behind the symbol used to represent the analyses.

complexity, attest to Neoproterozoic isotopic resetting. It is noticeable that biotite from the Narryer Terrane yields Mesoproterozoic ages that are considerably different to those measured in both the Gascoyne Complex and Errabiddy Shear Zone (Fig. 6). However a single muscovite spectrum from the Narryer Terrane also yields Neoproterozoic ages and is similar to ages further north.

## Discussion

Despite complexities in grain size–age relationships (Fig. 4) and Ar age–composition spectra (Occhipinti 2004), general patterns in $^{40}Ar/^{39}Ar$ age distributions indicate differences in the thermal history of different parts of the orogen at a time substantially postdating the last major orogenic (Palaeoproterozoic) event in the region. The apparent overlap of c. 960–820 Ma ages in the single-grain fusion and step-heating data from the western Errabiddy Shear Zone and Gascoyne Complex indicate a previously unrecognized isotopic resetting associated with regional heating of the western Capricorn Orogen during the early Neoproterozoic era. The extent of this resetting event (17 000 km$^2$) is significant and indicates a regional, not local, thermal perturbation. Closure temperature models for micas suggest temperatures of 350–270 °C (calculated using a cooling history assumption of 10 °C/Ma) are required to cause this resetting. This temperature range is consistent with the low metamorphic grades seen in Mesoproterozoic sedimentary rocks in the region. Older ages from the Eastern Errabiddy Shear Zone indicate a lesser degree of isotopic resetting but still indicate some Ar isotopic disturbance significantly after the c. 1800 Ma greenschist facies metamorphism associated with Palaeoproterozoic Capricorn orogenesis. The pattern of isotopic resetting generally decreases towards the east, indicating that the cause of resetting was more proximal to the west. Mesoproterozoic aged biotites in the Narryer Terrane may also reflect incomplete resetting after the Capricorn event. However, a single muscovite age of 834 Ma may indicate a component of excess $^{40}Ar$ in the biotite data. New Ar data from the Archaean rocks of the Narryer Terrane to the south of this study area indicates both a Capricorn (c. 1750 Ma) overprint and a younger, weaker and heterogeneously distributed overprint (Spaggiari et al. 2008).

The Edmundian Orogeny is associated with localized north and east trending folds and faults recorded throughout the Capricorn region, particularly in the Bangemall Supergroup. The age of this deformation event is poorly constrained but has been bracketed by age data on deformed and younger

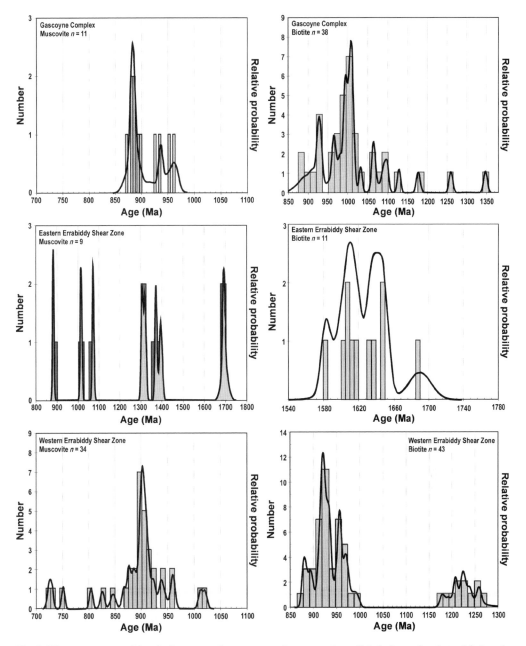

**Fig. 5.** Histograms summarizing the frequency of ages measured on muscovite and biotite by total grain total fusion of single and occasional multiple grain fusion of small grains (usually <200 μm) for different regions of the south Capricorn region. Distributions were calculated using a bin size of 20 Ma.

undeformed dolerite dykes that yield ages of 1070 and 750 Ma, respectively (Wingate & Giddings 2000; Martin & Thorne 2004; Wingate *et al.* 2004). The regional causes of Edmundian orogenesis has remained unclear. However, the 960–820 Ma age range for regional mica resetting is consistent with

an Edmundian thermal perturbation associated with early Neoproterozoic tectonism.

The Neoproterozoic Officer Basin forms part of the Centralian Superbasin (Walter *et al.* 1995). Palaeocurrent data from the base of the northwestern part of the Officer Basin, the Sunbeam Group,

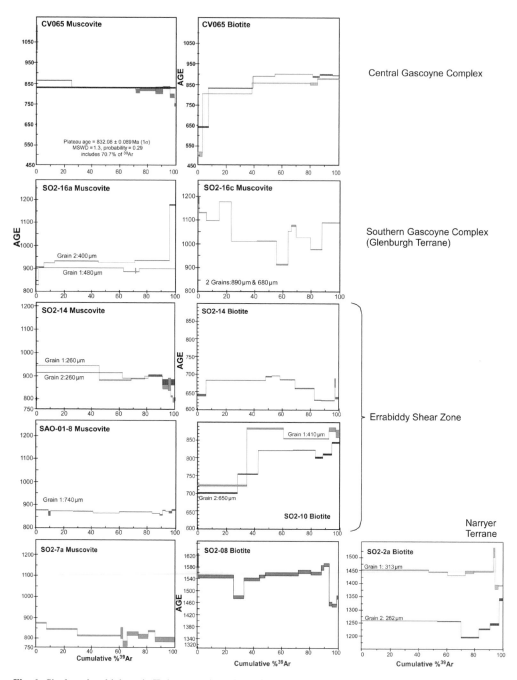

**Fig. 6.** Single and multiple grain IR-laser step-heated data from samples from the Central and Southern Gascoyne Complex, the Errabiddy Shear Zone and the Yilgarn Craton (Narryer Terrane).

indicates that during the early Neoproterozoic era, sediments were derived from the west (Fig. 7). Sediment characteristics are consistent with sourcing from the eroding Bangemall Supergroup and a rapid increase in sediment supply (Williams 1992)

and deposition in non-marine to shallow-marine conditions (Grey *et al.* 2005). Detrital zircon populations from the Tarcunyah Group of the Officer Basin, immediately north of the Sunbeam Group, also record U–Pb ages consistent with derivation

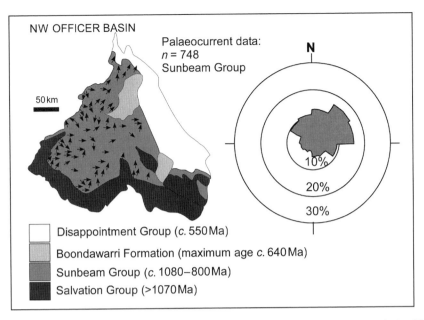

**Fig. 7.** Summary of palaeocurrent data from the Sunbeam Group of the Neoproterozoic Officer Basin (modified after Williams 1992). Arrows show the direction of flow inferred from palaeocurrent indicators and, along with petrological evidence, indicate provenance of sediments from the eroding Edmundian Orogeny.

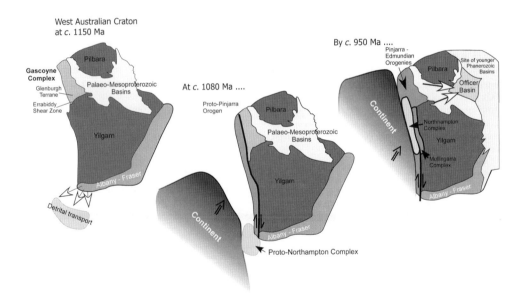

**Fig. 8.** Simplified model for development of the western margin of the West Australian Craton, during the Mesoproterozoic and Neoproterozoic based on data presented here and by Fitzsimons (2003). 1150 Ma: deposition of sediments now preserved in the proto-Mullingarra and Northampton Complexes. 1080 Ma: metamorphism of these sediments and continuing dextral transpressional deformation along the western part of Western Australia lead to northward migration of the Mullingarra and Northampton complexes to their current positions. 950 Ma: Continuing deformation on the western margin of the West Australian Craton caused the initiation of the Edmundian Orogeny in the Capricorn Orogen. Thermal effects associated with collisional activity give rise to resetting of Ar isotopes in mica. Increasing uplift leads to increased erosion of the evolving orogen and led to deposition of sedimentary detritus into the Neoproterozoic Officer Basin.

from the Gascoyne Complex (Bagas 2003). Asymmetric sediment dispersion within the Sunbeam Group, coarsening upward successions, and the lack of any active volcanism associated with its development, together support the possibility that the sediments were deposited in an intracontinental foreland basin-like setting (Jordan 1995; Miall 1995) with an evolving orogen situated to the west (Williams 1992). Although, alternative interpretations of deposition environment are possible from sedimentological observations (Bagas 2003; Grey *et al.* 2005), the available data from the age-equivalent parts of the Officer Basin are consistent with Neoproterozoic tectonism localized in the western Capricorn Orogen.

To the west of the amalgamated West Australian Craton, the Pinjarra Orogen (Fig. 1) comprises high grade metamorphic rocks of the Northampton and Mullingarra complexes in which granulite and amphibolite facies metamorphism has been dated at 1079 and 1058 Ma, respectively (Bruguier *et al.* 1999; Fitzsimons 2003). Zircon provenance data from paragneiss units within these two complexes indicate the Mesoproterozoic Albany–Fraser Orogen as a likely sediment source and, combined with the limited extent of high-grade metamorphism, has been used to suggest that the Northampton and Mullingarra Complexes are allochtonous and are derived from much further south (Fitzsimons 2003). Mafic dykes that are emplaced both within the Northampton Complex and the Yilgarn Craton, indicate that tectonic juxtaposition of the two complexes with the West Australian Craton must have taken place prior to *c.* 750 Ma (Wingate & Giddings 2000). Consequently, there is some evidence for tectonism within the Pinjarra Orogen between 1080 and 750 Ma (Fitzsimons 2003). From the geometry of the 750 Ma mafic dykes and brittle-ductile northerly trending dextral shear zones that post-date their emplacement (Byrne & Harris 1993), Fitzsimons (2003) argues that dextral transpressional within the Pinjarra Orogen took place around 750 Ma. Although broadly consistent with this model, the Ar data presented here indicate much earlier tectonic activity along this western margin of the West Australian Craton.

The coincidence of thermal resetting of Capricorn micas, the provenance of sediments from the Officer Basin and zircon data from the high-grade metamorphic complexes of the Pinjarra Orogen are consistent with tectonic activity along the western margin of the West Australian Craton during the early Neoproterozoic. In this scenario the Pinjarra and Edmundian orogenies are linked in that they represent the local and far-field effects of this tectonism respectively. On a larger scale, this tectonism is likely to represent convergence of continental material with the West Australian Craton during the early Neoproterozoic assembly of the supercontinent Rodinia (Fig. 8). Depending on which Rodinian reconstruction is preferred, either Kalahari or Greater India are potential candidates for collision with the West Australian Craton during the early Neoproterozoic (Pisarevsky *et al.* 2003; Li *et al.* 2008). Irrespective of which of these may or may not be correct, the isotopic resetting associated with Pinjarra/Edmundian tectonism within the Capricorn Orogen suggests that deformation and metamorphism of the region took place considerably earlier than the 750 Ma minimum age of orogenesis deduced from the earlier studies (Wingate & Giddings 2000; Fitzsimons 2003; Martin & Thorne 2004; Wingate *et al.* 2004). Consequently, future testing of Rodinian reconstructions requires that early Neoproterozoic tectonism be used for temporal correlation with potential collisional candidates.

## Conclusions

$^{40}$Ar/$^{39}$Ar dating of micas from the Palaeoproterozoic Capricorn Orogen yield cooling ages indicating early Neoproterozoic lower greenschist facies metamorphic conditions of regional extent. Variations in the degree of resetting suggest higher temperatures in the west of the region and point towards the thermal event being associated with tectonic activity on the western margin of the amalgamated West Australian Craton. This interpretation is supported by evidence from the Neoproterozoic Officer Basin and the Pinjarra Orogen. The results indicate the potential of low temperature thermochronology to provide within-craton evidence of far-field tectonic activity that may be poorly preserved at the craton margins. In this case the results suggest a dynamic link between the Pinjarra and Edmundian orogenies associated with collisional orogenesis during the early Neoproterozoic amalgamation of Rodinia.

The authors thank the Australian Research Council (A00106036) for funding this research. The Tectonics Special Research Centre and the Geological Survey of Western Australia, particularly I. Tyler and S. Sheppard, are thanked for their support. P. Betts, D. Corrigan, D. Foster and an anonymous reviewer are thanked for their comments on the manuscript. This paper is TIGeR publication 96.

## References

BAGAS, L. 2003. *Zircon Provenance in the Basal Part of the Northwestern Officer Basin, Western Australia.* Western Australia Geological Survey, Annual Review, 2002–2003.

BRUGUIER, O., BOSCH, D., PIDGEON, R. T., BYRNE, D. I. & HARRIS, L. B. 1999. U–Pb chronology of the

Northampton Complex, Western Australia – evidence for Grenvillian sedimentation, metamorphism and deformation and geodynamic implications. *Contributions to Mineralogy and Petrology*, **136**, 258–272.

BYRNE, D. & HARRIS, L. B. 1993. Structural controls on the base-metal vein deposits of the Northampton Complex, Western Australia. *Ore Geology Reviews*, **8**, 89–115.

CAWOOD, P. A. & TYLER, I. M. 2004. Assembling and reactivating the Proterozoic Capricorn Orogen: lithotectonic elements, orogenies, and significance. *Precambrian Research*, **128**, 201–218.

CLARK, D. J., HENSEN, B. J. & KINNY, P. D. 2000. Geochronological constraints for a two-stage history of the Albany-Fraser Orogen, Western Australia. *Precambrian Research*, **102**, 155–183.

DOWNES, P. J., WARTHO, J. A. & GRIFFIN, B. J. 2006. Magmatic evolution and ascent history of the Aries Micaceous Kimberlite, central Kimberley Basin, Western Australia: evidence from zoned phlogopite phenocrysts, and UV laser $^{40}Ar/^{39}Ar$ analysis of phlogopite-biotite. *Journal of Petrology*, **47**, 1751.

FITZSIMONS, I. C. W. 2003. Proterozoic basement provinces of southern and southwestern Australia, and their correlation with Antarctica. *In*: YOSHIDA, M., WINDLEY, B. F. & DASGUPTA, S. (eds) *Proterozoic East Gondwana: Supercontinent Assembly and Breakup*. Geological Society, London, Special Publications, **206**, 93–130.

GREY, K., HOCKING, R. M. *ET AL*. 2005. *Lithostratigraphic nomenclature of the Officer Basin and correlative parts of the Paterson Orogen, Western Australia*. Western Australia Geological Survey Report, **93**.

JORDAN, T. E. 1995. Retroarc foreland and Related Basins. *In*: BUSBY, C. J. & INGERSOLL, R. V. (eds) *Tectonics of Sedimentary Basins*. Blackwell Science, Cambridge, MA, 331–362.

KELLEY, S. P., ARNAUD, N. O. & TURNER, S. P. 1994. High-spatial-resolution $^{40}Ar/^{39}Ar$ investigations using an ultra-violet laser probe extraction technique. *Geochimica et Cosmochimica Acta*, **58**, 3519–3525.

LI, Z. X., BOGDANOVA, S. V. *ET AL*. 2008. Assembly, configuration, and break-up history of Rodinia: a synthesis. *Precambrian Research*, **160**, 179–210.

MARTIN, D. & THORNE, A. M. 2008. Tectonic setting and basin evolution of the Bangemall Supergroup in the northwestern Capricorn Orogen. *Precambrian Research*, **128**, 385–409.

MIALL, A. D. 1995. Collision-Related Foreland Basins. *In*: BUSBY, C. J. & INGERSOLL, R. V. (eds) *Tectonics of Sedimentary Basins*. Blackwell Science, Cambridge, MA, 393–424.

MYERS, J. S. 1990. Pinjarra Orogen. *Geology and Mineral Resources of Western Australia*. Geological Survey of Western Australia, Perth, 265–274.

MYERS, J. S. 1993. Precambrian history of the West Australian craton and adjacent orogens. *Annual Review of Earth and Planetary Sciences*, **21**, 453–485.

OCCHIPINTI, S. A. 2004. *Tectonic Evolution of the Southern Capricorn Orogen, Western Australia*. Curtin University of Technology.

OCCHIPINTI, S. A. & REDDY, S. M. 2004. Deformation in a complex crustal-scale shear zone: Errabiddy Shear Zone, Western Australia. *In*: ALSOP, G. I., HOLDSWORTH, R. E., McCAFFREY, K. J. W. & HAND, M. (eds) *Flow Processes in Faults and Shear Zones*. Geological Society, London, Special Publications, **224**, 229–248.

OCCHIPINTI, S. A., SHEPPARD, S., NELSON, D. R., MYERS, J. S. & TYLER, I. M. 1998. Syntectonic granite in the southern margin of the Palaeoproterozoic Capricorn Orogen, Western Australia. *Australian Journal of Earth Sciences*, **45**, 822–822.

OCCHIPINTI, S. A., SHEPPARD, S., PASSCHIER, C., TYLER, I. M. & NELSON, D. R. 2004. Palaeoproterozoic crustal accretion and collision in the southern Capricorn Orogen: the Glenburgh Orogeny. *Precambrian Research*, **128**, 237–255.

PISAREVSKY, S. A., WINGATE, M. T. D., POWELL, C. M., JOHNSON, S. & EVANS, D. A. D. 2003. Models of Rodinia assembly and fragmentation. *In*: YOSHIDA, M., WINDLEY, B. E. & DASGUPTA, S. (eds) *Proterozoic East Gondwana: Supercontinent Assembly and Breakup*. Geological Society, London, Special Publications, **206**, 35–55.

REDDY, S. M. & OCCHIPINTI, S. A. 2004. High-strain zone deformation in the southern Capricorn Orogen, Western Australia: kinematics and age constraints. *Precambrian Research*, **128**, 295–314.

REDDY, S. M., COLLINS, A. S., BUCHAN, C. & MRUMA, A. H. 2004. Heterogeneous excess argon and Neoproterozoic heating in the Usagaran Orogen, Tanzania, revealed by single grain $^{40}Ar/^{39}Ar$ thermochronology. *Journal of African Earth Sciences*, **39**, 165–176.

REX, D. C. & GUISE, P. G. 1995. Evaluation of argon standards with special emphasis on time scale measurements. *In*: ODIN, G. S. (ed.) *Phanerozoic Time Scale*. IUGS Subcommision on Geochronology, 21–23.

SHEPPARD, S., OCCHIPINTI, S. A. & TYLER, I. M. 2004. A 2005–1970 Ma Andean-type batholith in the southern Gascoyne Complex, Western Australia. *Precambrian Research*, **128**, 257–277.

SHEPPARD, S., OCCHIPINTI, S. A. & NELSON, D. R. 2005. Intracontinental reworking in the Capricorn Orogen, Western Australia: the 1680–1620 Ma Mangaroon Orogeny. *Australian Journal of Earth Sciences*, **52**, 443–460.

SHERLOCK, S. & KELLEY, S. P. 2002. Excess argon evolution in HP-LT rocks: a UVLAMP study of phengite and K-free minerals, NW Turkey. *Chemical Geology*, **183**, 619–636.

SONG, T. & CAWOOD, P. A. 2000. Structural styles in the Perth Basin associated with the Mesozoic break-up of Greater India and Australia. *Tectonophysics*, **317**, 55–72.

SPAGGIARI, C. V., WARTHO, J.-A. & WILDE, S. A. 2008. Proterozoic Deformation in the Northwest of the Archean Yilgarn Craton, Western Australia. *Precambrian Research*, **165**, 354–384.

STEIGER, R. H. & JAGER, E. 1977. Subcommision on Geochronology: convention on the use of decay constants in geo- and cosmochronology. *Earth and Planetary Science Letters*, **36**, 359–362.

TRENDALL, A. F. & COCKBAIN, A. E. 1990. Introduction, Chapter Four: basins. *Memoir of the Geological Survey of Western Australia*, **3**, 291–293.

WALTER, M. R., VEEVERS, J. J., CALVER, C. R. & GREY, K. 1995. Neoproterozoic stratigraphy of the Centralian Superbasin, Australia. *Precambrian Research*, **73**, 173–195.

WILLIAMS, I. R. 1992. Geology of the Savory Basin, Western Australia, *Western Australia Geological Survey, Bulletin*, **141**, 1–115.

WINGATE, M. T. D. & GIDDINGS, J. W. 2000. Age and palaeomagnetism of the Mundine Well dyke swarm, Western Australia: implications for an Australia-Laurentia connection at 755 Ma. *Precambrian Research*, **100**, 335–357.

WINGATE, M. T. D., PISAREVSKY, S. A. & EVANS, D. A. D. 2002. Rodinia connections between Australia and Laurentia: no SWEAT, no AUSWUS? *Terra Nova*, **14**, 121–128.

WINGATE, M. T. D., PIRAJNO, F. & MORRIS, P. A. 2004. Warakurna large igneous province: a new Mesoproterozoic large igneous province in west-central Australia. *Geology*, **32**, 105–108.

# The Palaeoproterozoic Trans-Hudson Orogen: a prototype of modern accretionary processes

D. CORRIGAN*, S. PEHRSSON, N. WODICKA & E. DE KEMP

*Geological Survey of Canada, 615 Booth Street, Ottawa, Ontario, Canada, K1A 0E8*

*Corresponding author (e-mail: dcorriga@NRCan.gc.ca)*

**Abstract:** The Trans-Hudson Orogen (THO) of North America is one of the earliest orogens in Earth's history that evolved through a complete Wilson Cycle. It represents *c.* 150 Ma of opening of the Manikewan Ocean, from *c.* 2.07–1.92 Ga, followed by its demise in the interval 1.92–1.80 Ga, during the final phase of growth of the Supercontinent Columbia (Nuna). It is characterized by three lithotectonic divisions: (i) Churchill margin (or peri-Churchill); (ii) Reindeer Zone; and (iii) Superior margin (or peri-Superior). The peri-Churchill realm records progressive outward continental growth by accretion of Archaean to Palaeoproterozoic micro-continents (Hearne, Meta Incognita/Core Zone, Sugluk) and eventually arc terranes (La Ronge–Lynn Lake) to the Slave-Rae nuclei, with attendant development of orogenies and basin inversions related to the specific accretion events (1.92–1.89 Ga Snowbird; 1.88–1.865 Ga Foxe; 1.87–1.865 Ga Reindeer orogenies). The Reindeer Zone is characterized by primitive to evolved oceanic arcs, back-arc basins, oceanic crust and ocean plateaus that formed during closure of the Manikewan Ocean, and accretion of a micro-continent (Sask Craton) and smaller Archaean crustal fragments. The terminal phase of the Trans–Hudson orogeny represents collision between the Superior craton, the Reindeer Zone and the composite western Churchill Province during the interval 1.83–1.80 Ga.

## Introduction

The Palaeoproterozoic Era (2.5–1.6 Ga) forms a unique period of Earth's evolution, highlighted by profound changes in continental plate configuration and tectonic processes (e.g. Griffin *et al.* 2008; Hou *et al.* 2008), ocean and atmospheric compositions (e.g. Anbar & Knoll 2002) and the biosphere (e.g. Konhauser *et al.* 2002). By 2.5 Ga, Archaean cratons had grown by tectonic and magmatic accretion into large, stable continental masses buoyed by thick, depleted lithospheric roots that could accommodate the deposition of laterally extensive passive margin sequences (e.g. Artemieva & Mooney 2001; Bédard 2006; Percival 2007; Francis 2003). Wilson-Cycle tectonics also appeared in the Palaeoproterozoic, resulting in the oldest documented obducted ophiolites and development of tectonostratigraphic sequences akin to those observed in the Phanerozoic (e.g. Hoffman 1988). Within the Canadian Shield, the Trans-Hudson Orogen (THO, Fig. 1; Hoffman 1988; Lewry & Collerson 1990) represents the largest and most completely preserved segment amongst a number of Palaeoproterozoic collisional belts that formed during the amalgamation of supercontinent Nuna (also known as Columbia) during the interval 2.0–1.8 Ga (Rogers & Santosh 2002; Pesonen *et al.* 2003; Hou *et al.* 2008). Before the opening of the North Atlantic Ocean, the THO formed part of a nearly continuous orogenic system that extended from the present-day location of South Dakota and perhaps even further to the Grand Canyon area (Bickford & Hill 2007), over a distance of *c.* 3000 km across the Canadian Prairies, Hudson Bay, Baffin Island, Greenland and Scandinavia (Hoffman 1988). It may originally have been even more extensive, as both ends are now truncated by younger orogenic belts. In Canada, the THO is generally well exposed and remained relatively intact after its formation, providing the opportunity to study its evolution both parallel to and across orogenic strike.

The THO is the site of closure of the Manikewan Ocean (Stauffer 1984), the oceanic plate that once existed between the Superior Craton and the Rae Craton (Fig. 2a), stretching from South Dakota to the Ungava Peninsula in northern Québec (present-day geometry). It is likely that this ocean once continued eastwards past the Ungava Peninsula, but its extent from there is unclear. Closure of the Manikewan Ocean was traditionally associated with formation of juvenile crust and its eventual accretion to the Superior craton and composite western Churchill Province (Gibb & Walcott 1971; Gibb 1975). Recent studies have highlighted that it also involved accretion of a number of smaller intervening Archaean to earliest-Palaeoproterozoic continental fragments, including the Sask Craton (Lewry *et al.* 1994; Hajnal *et al.* 1996). From a tectonostratigraphic perspective, the THO preserves a relatively complete Wilson-Cycle, from early *c.* 2.45–1.92 Ga rift to drift sedimentary assemblages

*From*: MURPHY, J. B., KEPPIE, J. D. & HYNES, A. J. (eds) *Ancient Orogens and Modern Analogues.*
Geological Society, London, Special Publications, **327**, 457–479.
DOI: 10.1144/SP327.19   0305-8719/09/$15.00 © The Geological Society of London 2009.

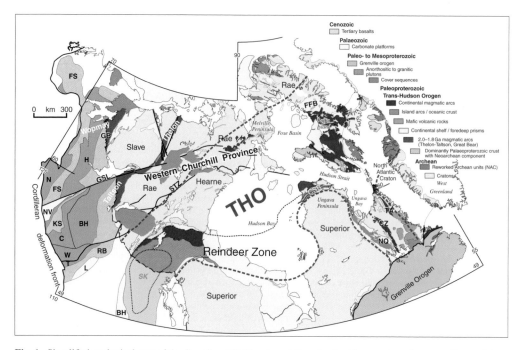

**Fig. 1.** Simplified geological map of the Canadian Shield and west Greenland, with extension of western Precambrian terranes shown beneath Phanerozoic cover (pale colours SW of thick hashed line). The area affected by Trans–Hudson age tectonothermal reactivation (1.83–1.80 Ga) is delimited by the thick red hashed line. Limit of the Reindeer Zone is shown (thick stippled blue line). Minimum limit of Trans-Hudson age tectonothermal overprint (thick hashed red line). Terrane names and major faults: BH, Buffalo Head terrane; BK, Black Hills; C, Chinchaga belt; CZ, Core Zone; FS, Fort Simpson terrane; FFB, Foxe Fold Belt; GB, Great Bear; GSL, Great Slave Lake shear zone; H, Hottah terrane; KS, Ksituan belt; L, Lacombe belt; N, Nahanni belt; NO, New Québec; NV, Nova terrane; RB, Rimbey Belt; STZ, Snowbird Tectonic Zone; T, Thorsby basin; TA, Tasiuyak domain; W, Wabamun belt.

deposited along Archaean craton margins, to formation of *c.* 2.00–1.88 Ga oceanic and pericratonic arcs and back-arc basins, as well as younger (1.88–1.83 Ga) continental arcs, foredeep and collisional basins, and eventual terminal collision (Ansdell 2005). This paper provides a synopsis of the accretionary and collisional evolution of the THO in Canada. The objective is to provide the reader with an appreciation of the complexity of this composite orogenic system, the scale and age of colliding crustal fragments, and the type of geodynamic processes that were in place during the Palaeoproterozoic. This paper considers the whole of the Canadian THO, in comparison to Ansdell (2005), Corrigan *et al.* (2007) and St-Onge *et al.* (2006) that provide overviews of selected parts of the orogen. The evolution of the eastern Churchill Province is not discussed in detail in this paper, but utilizes the overview published by Wardle *et al.* (2002).

## Tectonic context

Historically, the usage of the term 'Trans-Hudson Orogen' has varied widely. It was introduced by

Hoffman (1981) to identify the Palaeoproterozoic collision zone that formed between the Superior Province and the Archaean domains of the western Churchill Province (Green *et al.* 1985). This usage differs from the one proposed by Lewry *et al.* (1985) and Lewry & Collerson (1990), who preferably based their definition on timing and contiguous extent of peak 'Hudsonian-age' tectonothermal overprint, defined at *c.* 1.83–1.80 Ga. Within that framework, the latter authors refer to Hoffman's THO as the Reindeer Zone, an area represented by the Palaeoproterozoic juvenile internides (Fig. 1). They include within the definition of the THO a vast region represented by reactivated Archaean rocks of the western Churchill Province encompassing nearly all of the Hearne Craton, as well as a large portion of the Rae Craton (Fig. 1).

The THO as defined by Lewry & Collerson (1990) is more inclusive and helps understand the orogen in its entirety in terms of ocean closure, terrane accretion, continental margin reactivation and syn- to post-orogenic tectonothermal overprint. The THO is herein more broadly defined as an area of contiguous or nearly-contiguous structural and metamorphic overprint caused by *c.* 1.83–1.80 Ga collision between the Superior Craton, the Reindeer Zone, and a previously amalgamated continental collage north of the Manikewan Ocean consisting of the Slave, Rae and Hearne cratons, with the latter two collectively referred to as the 'western Churchill Province' (Fig. 1). This zone of tectonometamorphic overprint reaches a maximum width of *c.* 1800 km across strike of the orogen, from the reactivated edge of the Superior Craton to the northwestern limit of the orogenic front in the western Churchill Province (Lewry & Collerson 1990; Corrigan 2002; Pehrsson *et al.* 2004). The southeastern Churchill Province, which separates the Superior Craton from the North Atlantic Craton, is commonly interpreted as a branch of the THO, although the actual correlation between distinct tectonostratigraphic units from the western and southeastern 'arms' is still a matter of debate (e.g. Bourlon *et al.* 2002; St-Onge *et al.* 2002; Wardle *et al.* 2002).

The lifetime of 'Manikewan Ocean' can be estimated on the basis of the history of the bounding cratons involved in the THO. Rift-related mafic sequences dated at *c.* 2.07 Ga along the margins of the Hearne and Superior cratons, including the 2075 ± Ma Courtenay Lake porphyry (Ansdell *et al.* 2000) and 2072 ± 3 Ma Cauchon dykes (Heaman & Corkery 1996), respectively, provide a maximum age for formation of this oceanic crust. Mafic and ultramafic rocks interpreted as remnants of obducted ophiolite from the Watts Group in the Cape Smith Belt are dated at *c.* 2.0 Ga (Parrish 1989), in agreement with the postulated presence of oceanic crust during that period (Scott *et al.* 1992).

On the basis of geophysical, geochronological and tracer isotope data, at least three large continental fragments or micro-continents are identified in the Manikewan Ocean 'realm' and identified as the Sask, Sugluk and Meta Incognita–Core Zone blocks in Figure 2a. The predominantly buried Sask Craton (Lewry *et al.* 1994) has been identified in three relatively small basement windows in the Flin Flon–Glennie complex and, on the basis of seismic images, is interpreted to underlie a large part of the Reindeer Zone at middle to lower crustal levels (Hajnal *et al.* 2005). Where exposed, the Sask Craton footwall is separated from overlying juvenile Proterozoic rocks by high metamorphic grade mylonite zones with top-to-the SW sense of shear (Ashton *et al.* 2005). Tracer isotope and U–Pb geochronological data obtained mainly from the limited surface exposures and drill cores suggest that the Sask Craton is predominantly formed of 2.4–2.5 Ga felsic to mafic igneous rocks intruded into 3.1–2.8 Ga calc-alkaline orthogneiss (Chiarenzelli *et al.* 1998; Bickford *et al.* 2005; Rayner *et al.* 2005*b*).

The Sugluk block [introduced by Hoffman (1985) as 'Sugluk Terrane'] contains crystalline basement rocks that fall in a narrow Mesoarchaean age range of *c.* 2.8–3.2 Ga (whole rock Rb–Sr on sedimentary rocks, Doig 1987; whole rock Sm–Nd, Dunphy & Ludden 1998; U–Pb on zircon, Wodicka & Scott 1997; Rayner *et al.* 2008) and in contrast to the Meta Incognita micro-continent, has no known 2.4–2.5 Ga magmatism. Its main exposure lies in the northwestern edge of Ungava Peninsula where it has been variably named 'Sugluk Terrane' (Hoffman 1985, 1990), 'Narsajuaq Terrane' (St-Onge & Lucas 1990) and 'Narsajuaq arc' (St-Onge *et al.* 2006). On Ungava Peninsula, the Sugluk block is intruded by juvenile to variably enriched granitoids ranging in age from 1.86–1.80 Ga, collectively referred to as the Narsajuaq suite (Dunphy & Ludden 1998). Aeromagnetic lineaments (Fig. 3) and Bouger gravity anomalies characteristic to this crustal block suggest that it potentially extends northwards to the southern edge of Baffin Island, where it is presently in structural contact with the Meta Incognita micro-continent. It appears to extend westwards under Hudson Bay, corresponding in large part to the 'Hudson protocontinent' defined by Roksandic (1987). The Sugluk block described herein forms the continental crust onto which the Narsajuaq arc was built, as described by Dunphy & Ludden (1998), and is of different age and nature than the Meta Incognita continental crust which hosted Narsajuaq-age intrusions on Baffin Island (e.g. Rayner *et al.* 2007). We therefore consider it a separate crustal entity.

The Meta Incognita micro-continent (e.g. St-Onge *et al.* 2000, 2006) underlies much of southern Baffin Island and comprises basement rocks consisting of Neoarchaean crust (*c.* 2.7–2.6 Ga) and early Palaeoproterozoic (*c.* 2.40–2.15 Ga) granitoid intrusions (N. Wodicka, unpublished data) that are structurally overlain by a clastic–carbonate shelf sequence (Lake Harbour Group), which was deposited during the 2.01–1.90 Ga interval (Scott *et al.* 2002; Wodicka *et al.* 2008). In contrast with the Sugluk block, evidence for older Mesoarchaean crust is provided indirectly from U–Pb inheritance in U–Pb zircon ages (e.g. Rayner *et al.* 2007) and $T_{DM}$ model ages (Whalen *et al.* 2008). On the basis of U–Pb age, magnetic and gravity anomalies, as well as lithological associations, the Meta Incognita micro-continent may extend southeastwards (present-day co-ordinates) and correlates, at least in part, with the Core Zone of the New Québec orogen (Scott & St-Onge 1998; Bourlon *et al.* 2002; Wardle & Hall 2002).

Closure of the Manikewan Ocean had initiated by at least *c.* 1905 Ma, which is the oldest age obtained for oceanic arc rhyolite in the Reindeer

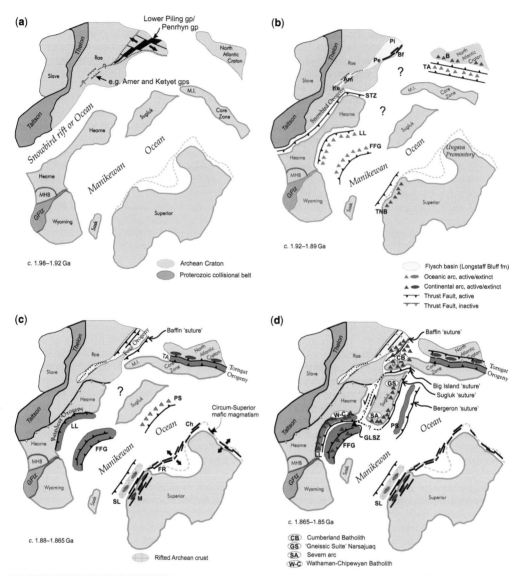

**Fig. 2.** Series of cartoons illustrating a map view of the geological evolution of the THO, with the main tectonic elements shown: (**a**) main lithotectonic elements during the interval 1.98–1.92 Ga, prior to the onset of convergence; (**b**) the interval 1.92–1.89 Ga; (**c**) the interval 1.88–1.865 Ga; (**d**) the interval 1.865–1.85 Ga; (**e**) the interval 1.85–1.83 Ga; and (**f**) the interval 1.83–1.80 Ga. The blue coloured hashed lines represent possible extensions of the Superior Craton at Moho. Although not discussed in text, evolution of North Atlantic Craton (shown here to include Archaean basement rocks and their cover on Hall and Cumberland peninsulas on Baffin Island (e.g. Scott 1999; Jackson & Berman 2000) and tectonostratigraphic elements of the New Quebec and Torngat orogens are after Wardle *et al.* (2002). Abbreviations: Am, Amer group; B, Burwell arc; BeS, Bergeron suture; BS, Baffin suture; BiS, Big Island suture; Bf, Bravo formation; Ch, Chukotat; FFG, Flin Flon–Glennie Complex; FR, Fox River belt; GFtz, Great Falls tectonic zone; GLsz, Granville Lake structural zone; Hoare Bay group; KB, Kisseynew Basin; Ke, Ketyet group; LL, La Ronge–Lynn Lake belts; LTSz, Lac Tudor shear zone; M, Molson dykes; MHB, Medicine Hat block; M.I., Meta Incognita micro-continent; Pe, Penrhyn group; Pi, Piling group; PS, Parent arc and Spartan forearc; PT, Pelican thrust; SL, Snow Lake belt; SS, Sugluk suture; STZ, Snowbird Tectonic Zone; TA, Tasiuyak domain; TNB, Thompson Nickel Belt.

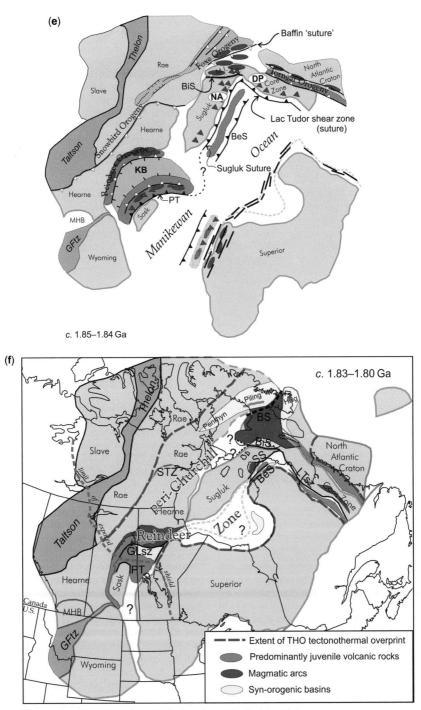

**Fig. 2.** (*Continued*).

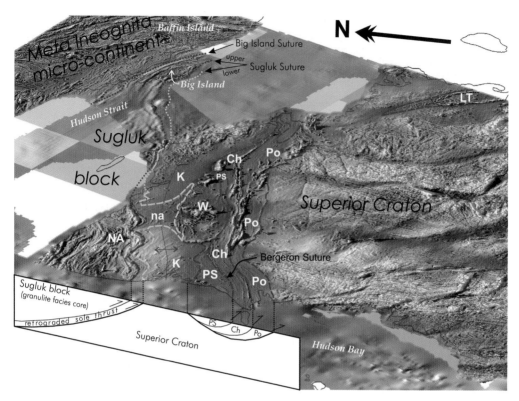

**Fig. 3.** Oblique view of greyscale, shaded relief total field aeromagnetic anomaly map of the Ungava Peninsula and southern Baffin Island area (illumination from NW). Red line is contour of shoreline. Thick blue lines are main tectonostratigraphic boundaries. Thin blue line is trace of the Kovik antiform hinge. The cross section on bottom (drawn to scale) is after Hoffman (1985) and St-Onge *et al.* (2002). Abbreviations are: Ch, Chukotat group; K, Kovik antiform; LT, Labrador Trough; NA, Narsajuaq continental arc (emplaced in Sugluk block); na, metamorphically retrograded equivalent of Narsajuaq Arc rocks in Sugluk suture zone; Po, Povungnituk group; PS, Parent arc and Spartan fore-arc; W, Watts group oceanic crust. See text for explanations.

Zone (Baldwin *et al.* 1987; Stern *et al.* 1995; Corrigan *et al.* 2001*a*). However, the more primitive arcs likely formed earlier, perhaps *c.* 1.92 Ga, which is the oldest age of detrital zircon found in arc-derived clastic sediments within the La Ronge and Lynn lake belts (Ansdell 2005; Corrigan *et al.* 2005). Note that contrary to Tran *et al.* (2008), we do not attribute the inclusion of *c.* 1.91–1.92 Ga detrital zircons in the Hearne margin cover sequence (lower Wollaston Group) to a putative 'Rottenstone arc', but rather to local derivation from *c.* 1.91–1.92 Ga intrusions in the Peter Lake Domain (e.g. Porter Bay complex; Rayner *et al.* 2005*a*). Taken at face value, these geochronological constraints suggest that the Manikewan Ocean had formed over a time span of at least 150 Ma before it started to close, which is comparable in duration to most modern oceans. The *c.* 1.83–1.80 Ga terminal collision with the Superior Province, which historically defines the THO *sensu stricto*, was thus preceded by episodes of accretion and convergence

dating back to *c.* 1.92 Ga. This 100 Ma time frame of THO evolution is herein subdivided into a series of orogenies, in much the same manner as the Appalachian orogen, for example, providing a framework in which to better characterize distinct periods of intraoceanic or pericratonic crustal accretion, magmatic accretion in continental arc batholiths and terminal continent-continent collision.

The early period, termed the Snowbird orogeny (*c.* 1.92–1.89 Ga, Berman *et al.* 2007), is recognized as a phase of amalgamation of the Hearne and Rae cratons, concurrent with oceanic arc formation within the Manikewan Ocean (Ansdell 2005; Corrigan *et al.* 2005, 2007) and subduction beneath the western Superior Craton margin (Percival *et al.* 2005). The next major phase, *c.* 1.88–1.865 Ga, involves formation of the Glennie–Flin Flon–Snow Lake intraoceanic collage (Lucas *et al.* 1994) and accretion of the La Ronge–Lynn Lake arcs to the southeastern Hearne Craton margin (Bickford *et al.* 1990, 1994; Corrigan *et al.* 2005), with the latter event referred to herein as the Reindeer Orogeny. Penecontemporaneous with the latter, but structurally unrelated, was the accretion of the Meta Incognita micro-continent to the eastern Rae Craton (St-Onge *et al.* 2006) with attendant tectonothermal reworking (Foxe Orogeny).

The third phase, informally named herein the *c.* 1.865–1.845 Ga Wathaman Orogeny, is marked by voluminous magmatic accretion along the southeastern margin of the western Churchill Province (Thériault *et al.* 2001; Corrigan *et al.* 2005) and accretion of the Sugluk block to the latter (this paper, and Berman *et al.* 2007). A fourth phase of micro-continent accretion and continental arc magmatism followed during the interval 1.84–1.82 Ga, as the Sask Craton collided with the Flin Flon–Glennie Complex and arc magmatism (Narsajuaq *sensu stricto*) continued in the eastern Manikewan Ocean. Terminal collision with the Superior Craton eventually closed the Manikewan Ocean and caused the widespread tectonothermal reworking characterizing the 1.83–1.80 Ga 'Hudsonian orogeny' (cf. Stockwell 1961), or Trans-Hudson Orogen *sensu stricto*. The following sections consider each of these orogenies in turn, providing a brief description of accretionary or collisional effects both within the Manikewan Ocean and along the cratonic margins.

## Initiation of Manikewan Ocean closure and Snowbird orogeny: (1.92–1.89 Ga)

The time period 1.92–1.89 Ga marks the beginning of Manikewan Ocean closure, with attendant effects both within the internal Reindeer Zone, Rae and Hearne Provinces, and Superior Craton margin (Fig. 2a). We consider these effects in turn, focusing on distinct intraoceanic, intracratonic or pericratonic effects, respectively. The original extent of the Manikewan Ocean is equivocal, although some estimates based on palaeomagnetic data suggest an original width of at least 5000 km (Symons & Harris 2005).

### Juvenile crust formation: the western Reindeer Zone

The Reindeer Zone of the THO consists mainly of juvenile crust of Palaeoproterozoic age that was formed during the closure of the Manikewan Ocean (Lewry & Collerson 1990). An overwhelming proportion of this crust is preserved in the western part of the Reindeer Zone in the Canadian Prairies, west and NW of the Superior Craton. By contrast, a relatively smaller fraction is identified in the eastern part of the Reindeer Zone, now preserved mainly as plutonic rocks in the Sugluk

block and in a klippe (Cape Smith Belt) thrust onto the Ungava Promontory of the Superior Craton in northern Québec (Hoffman 1985). Not much is known about the nature of the Reindeer Zone beneath the Hudson Bay basin, although a potentially substantive magmatic arc (Severn arc) has been interpreted to exist on the basis of continuity of a large region of aeromagnetic high appearing to correspond on land to the Wathaman–Chipewyan batholith (Green *et al.* 1985; Hoffman 1990). In the western Reindeer Zone, juvenile crust occurs in two separate belts, which are: (i) an intra-oceanic assemblage (Flin Flon–Glennie Complex; Ashton 1999) composed of *c.* 1.91–1.88 Ga primitive to evolved island arc, ocean floor, ocean plateau and associated sedimentary and plutonic rocks that developed during closure of the Manikewan Ocean (e.g. Stauffer 1984; Syme & Bailes 1993; Lucas *et al.* 1996; Ansdell 2005); (ii) laterally discontinuous belts of ocean arc, back-arc, ocean crust and associated sediments, and sub-arc plutonic rocks that evolved as pericratonic arcs during the interval 1.91–1.88 Ga before accretion to the southeastern Hearne Craton margin (northwestern Reindeer Zone; Maxeiner *et al.* 2004; Corrigan *et al.* 2005; Zwanzig 2000); and (iii) magmatic arcs that were emplaced in continental arc and oceanic successor arc settings (Fumerton *et al.* 1984; Whalen *et al.* 1999).

### Flin Flon–Glennie Complex

From east to west the Flin Flon–Glennie Complex (Fig. 4) comprises the Snow Lake arc assemblage, the Amisk collage, Hanson Lake block and the Glennie Domain, interpreted as part of a single crustal entity formed at *c.* 1.87 Ga as a result of intraoceanic accretion (Lewry & Collerson 1990; Lucas *et al.* 1996). The various tectonostratigraphic assemblages forming the complex are fold-repeated and thrust-stacked, and structurally overlie the Archaean to earliest-Palaeoproterozoic Sask Craton above a basal décollement (Pelican Thrust) that was active as early as *c.* 1.84 Ga (Ashton *et al.* 2005). The lithotectonic evolution of Flin Flon–Glennie complex is generally regarded as having involved five main development stages (Lucas *et al.* 1996; Stern *et al.* 1999), consisting of: (i) 1.92–1.88 Ga formation of juvenile arcs, back-arc basins, ocean plateaus (Fig. 2b); (ii) 1.88–1.87 Ga intraoceanic accretion (Fig. 2c); (iii) 1.87–1.84 Ga post-accretion development of successor arc intrusions and inter-arc basins (Fig. 2d); and (iv) 1.84–1.83 Ga terminal collision stage (Fig. 2d, e), first with the Sask Craton at *c.* 1.84–1.83 Ga and later, at 1.83–1.80 Ga, with the Superior Craton (Bleeker 1990; Ellis & Beaumont 1999; Ashton *et al.* 2005). In general, the pre-accretionary period records an evolution from primitive arc tholeiites to evolved

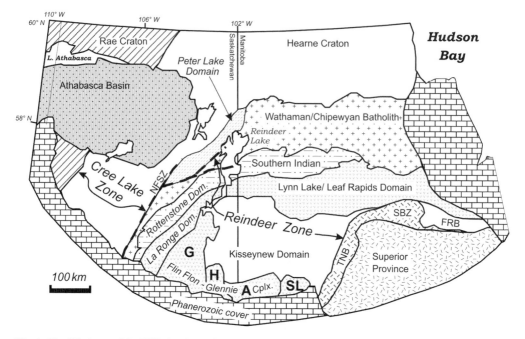

**Fig. 4.** Simplified map of the THO showing main lithotectonic domains of the Reindeer Zone and bounding domains. Cree Lake zone is the reactivated portion of the Hearne Craton. Abbreviations: A, Amisk Lake block; FRB, Fox River belt; G, Glennie domain; H, Hanson Lake block; NFSZ, Needle Falls shear zone; SBZ, Superior Boundary Zone; SL, Snow Lake belt; TNB, Thompson Nickel Belt.

calc-alkaline arc rocks (Maxeiner *et al.* 1999; Bailes & Galley 1999). Although the arcs are mostly juvenile, Nd isotopic data suggest variable input from Archaean crust, perhaps recycled in sediments or as rifted crustal fragments locally incorporated in the early arc assemblages (Stern *et al.* 1995). In the Flin Flon Domain, Archaean crust is actually preserved as thrust-imbricated slices interleaved with juvenile arc and back-arc crust (David & Syme 1994; Lucas *et al.* 1996; Syme *et al.* 1999).

The Snow Lake Belt, which is located between the Flin Flon–Glennie Complex and the Superior Craton, also comprises remnants of an oceanic arc volcano-plutonic complex and has historically been linked to the latter (e.g. Galley *et al.* 1986; Lucas *et al.* 1994; Bailes & Galley 1999). However, recent models suggest that the Snow Lake Belt may have evolved as a pericratonic arc outboard of the Superior Craton (Fig. 2b), implying a possibly independent evolution from the remainder of the Flin Flon–Glennie Complex prior to terminal collision (Percival *et al.* 2004; Corrigan *et al.* 2007).

### The La Ronge–Lynn Lake domains

The La Ronge Domain (Saskatchewan) and the Lynn Lake/Leaf Rapids domains (Manitoba) form the core of a more or less continuous curvilinear,

accreted terrane *c.* 50 km wide by 750 km long that is exposed from Lac La Ronge in north-central Saskatchewan to north-central Manitoba (Fig. 4). The La Ronge Domain hosts the 1905–1876 Ma (Van Schmus *et al.* 1987; Corrigan *et al.* 2001*a*) Central Metavolcanic Belt, interpreted as a composite volcanic, plutonic and sedimentary assemblage formed in an ensimatic island arc setting transitional into an ensialic setting (Lewry 1981; Watters & Pearce 1987; Thomas 1993). Arc assemblages contain a lower succession of ultramafic tectonite, mafic to minor intermediate volcanic rocks, pelite, and exhalative horizons, as well as oxide- and silicate-facies iron formation (collectively named the Lawrence Point Assemblage), overlain by an upper succession of felsic volcanic and pyroclastic rocks, interbedded with and laterally grading into volcaniclastic rocks (Reed Lake Assemblage) (Maxeiner *et al.* 2004, 2005). The stratigraphic top of the Reed Lake Assemblage conformably grades into a thick, laterally continuous and homogeneous succession of graphite-bearing, mostly semi-pelitic to pelitic gneiss interpreted as a remnant of either a forearc basin or accretionary prism formed on the northern flank of the La Ronge and Lynn Lake arcs, prior to *c.* 1.88 Ga collision with the Hearne Craton margin (Corrigan *et al.* 2005).

The Lynn Lake Domain (Lewry *et al.* 1990) broadly correlates with the La Ronge Domain in the sense that both contain tectonically imbricated and complexly folded volcanic, plutonic and sedimentary rocks. However, there are some major differences as well. Zwanzig *et al.* (1999) subdivided supracrustal rocks of the Lynn Lake Domain (Wasekwan Group) into a northern belt, predominantly consisting of mafic volcanic and volcaniclastic rocks of tholeiitic affinity and a southern belt consisting of tholeiitic to calc-alkaline volcanic rocks with minor amounts of mid-oceanic ridge basalt (MORB). Rare felsic volcanic rocks from the northern belt have yielded U–Pb zircon ages ranging from $1915 + 7/-6$ Ma to $1886 \pm 2$ Ma (Baldwin *et al.* 1987; Beaumont-Smith & Böhm 2002). These rocks are significantly older than felsic to intermediate volcanic rocks of the southern belt, which have so far yielded U–Pb zircon ages of $1881 + 3/-2$ Ma, $1864 \pm 2$ Ma, $1856 \pm 2$ Ma and *c.* 1842 Ma (Beaumont-Smith & Böhm 2002, 2003). Recent Nd isotope data suggest that igneous rocks of the northern belt, which yield $\varepsilon_{Nd}$ values of $-2.3$ to $+1.3$ and $T_{DM}$ model ages of *c.* 2.5 Ga, have been contaminated by Archaean crust or sediments with a high proportion of Archaean detritus (Beaumont-Smith & Böhm 2003). In contrast, igneous rocks of the southern belt yield $\varepsilon_{Nd}$ values of $+4$ to $+5$ and $T_{DM}$ model ages <2.2 Ga, suggesting a juvenile origin (Beaumont-Smith & Böhm 2003). Both the northern belt and parts of the southern volcanic belt are intruded by tonalitic plutons of the 1876–1871 Ma Pool Lake intrusive suite (Baldwin *et al.* 1987; Turek *et al.* 2000).

The Rusty Lake Belt (Baldwin 1987) forms part of the Leaf Rapids Domain (Fig. 4). It comprises remnants of an evolved arc with mafic flows and sills, minor dacitic and rhyolitic flows, volcaniclastic rocks, polymictic conglomerate and metaturbidite. Precise U–Pb geochronology in the Ruttan Block has yielded an $1883 \pm 2$ Ma zircon age for a quartz-feldspar porphyry rhyolite, and a unimodal, 1880–1920 Ma age-range detrital source for meta-turbidites directly overlying the volcanic edifice, consistent with the age range of arc volcanic rocks from the Reindeer Zone internides (Rayner & Corrigan 2004).

Mafic to felsic plutonic rocks intruding the metavolcanic rocks of the La Ronge, Lynn Lake and Rusty Lake belts range in age from *c.* 1894–1871 Ma and are interpreted as the arc plutonic root (Baldwin *et al.* 1987; Turek *et al.* 2000; Corrigan *et al.* 2005). These essentially form the core of the Rottenstone and Southern Indian domains (Fig. 4). The supracrustal and plutonic rocks of the northwestern Reindeer Zone are unconformably overlain by fluvial to littoral facies siliciclastic metasedimentary rocks (Park Island

Assemblage and lateral equivalents in the Rottenstone and Southern Indian domains; Corrigan *et al.* 1998; 2001*b*), interpreted as a foreland or molasse basin deposited on the previously accreted arc assemblages (Corrigan *et al.* 2005). All the above assemblages are intruded by the *c.* 1863–1850 Ma Wathaman–Chipewyan Batholith, implying that they had been tectonically accreted to the Hearne Craton margin prior to continental arc magmatism.

## Pericratonic Accretion: 1.88–1.865 Ga Reindeer and Foxe orogenies

The period 1.88–1.865 Ga was characterized by increased accretionary activity in pericratonic and intra-oceanic settings (Fig. 2c), including accretion of the La Ronge–Lynn Lake–Rusty Lake arcs to the southeastern Hearne Craton margin (Reindeer Orogeny) and accretion of the Meta Incognita micro-continent to the eastern Rae Craton (Foxe Orogeny). It was also characterized by the development of a juvenile oceanic arc in the eastern Manikewan Ocean (Parent and Spartan Groups of the Cape Smith Belt: Picard *et al.* 1990; St-Onge *et al.* 1992; Barette 1994) and by renewed extension along the northern and eastern Superior Craton margin, resulting in voluminous mafic to ultramafic magmatism (Baragar & Scoates 1981; Hulbert *et al.* 2005). This time period also coincided with collision between the Core Zone, Tasiuyak arc and North Atlantic Craton, forming the Burwell arc and Torngat Orogen (Wardle *et al.* 2002).

## *Pericratonic arc accretion on southeastern Hearne Craton margin: the Reindeer Orogeny*

The northwestern Reindeer Zone (Fig. 4) consists of an autochthonous cover sequence of volcanic and sedimentary rocks (Lower Wollaston Group), an allochthonous sequence of oceanic arc fragments (La Ronge, Lynn Lake and Rusty Lake belts), their plutonic root (Rottenstone and South Indian domains), an accretionary prism (Crew Lake, Milton Island and Zed Lake belts) and flysch/molasse deposits (Upper Wollaston Group, Park Island and Partridge Breast belts) that either formed on or accreted to the southeastern margin of the Hearne Craton during the interval 1.88–1.865 Ga (Corrigan *et al.* 2005). The upper age limit of this time bracket corresponds to the waning stages of ocean arc-type magmatic activity in the arc assemblages (Corrigan *et al.* 2005), and the lower age bracket as the earliest intrusion of continental arc type magmas (e.g. Van Schmus *et al.* 1987).

A direct consequence of tectonic accretion of the arc terranes on the Hearne margin along north-vergent thrust faults was the deposition of a foreland or molasse basin whose relics now form the Upper Wollaston Belt (Yeo 1998), as well as inliers within the northwestern Reindeer Zone (e.g. the Park Island and Partridge Breast assemblages Corrigan *et al.* 1998, 1999, 2001*b*). Both these units contain abundant detrital zircons originating from the juvenile La Ronge–Lynn Lake–Rusty Lake supracrustal belt, with a peak at *c.* 1.92–1.88 Ga (Tran *et al.* 2008; D. Corrigan, unpublished data) and are intruded by the *c.* 1.865–1.85 Ga Wathaman–Chipewyan Batholith, suggesting deposition sometime during the interval 1.88–1.865 Ga (Corrigan *et al.* 2005). Thermal effects of this accretion event have not been directly dated, and may have been reset by ubiquitous Hudsonian-age (*c.* 1.83–1.80 Ga) peak metamorphism. However, a penetrative deformation event in the form of early tight folds and layer-parallel foliation ($D_1$) has been recorded in protoliths that are younger than *c.* 1.88 Ga and that are cross-cut by Wathaman–Chipewyan age intrusions (Corrigan *et al.* 1998). The actual extent of the Reindeer Orogeny north and west of the Wollaston Belt is as yet unconstrained.

## The Foxe Orogeny: accretion of Meta Incognita micro-continent

Farther NE along the western Churchill Province margin, the Foxe Orogeny (Hoffman 1990) formed as a result of collision between the eastern Rae Craton and the Meta Incognita micro-continent (Fig. 2c) during the time interval 1.88–1.865 Ga (St-Onge *et al.* 2006). Collision resulted in closure of the Penrhyn and Piling basins, which had formed earlier during the interval $\leq$2.16 $\geq$ 1.88 Ga (Wodicka *et al.* 2007*a*, 2008). The Penrhyn and Piling sequences comprise: (i) shelf-margin assemblages (e.g. Lower Piling Group) deposited after 2.16 Ga but prior to 1.98–1.94 Ga; (ii) a continental to incipient oceanic rift sequence (e.g. Bravo Lake formation) formed between *c.* 1.98 and 1.88 Ga, consisting of mafic and ultramafic flows and sills, iron formation, and volcaniclastic and siliciclastic rocks (Johns *et al.* 2006; Modeland & Francis 2010); and (iii) deep-water basinal strata including sulphidic black shale and metapelite (e.g. Astarte River Formation) and a thick sequence of greywacke (e.g. Longstaff Bluff Formation) that was rapidly deposited, buried and folded before being intruded by hypersolvus granites at *c.* 1.90 Ga (Wodicka *et al.* 2002*a*, 2007*a*). We interpret the Penrhyn and Piling groups as a sequence of supracrustal rocks

that were deposited in a continental margin prism setting (e.g. Morgan *et al.* 1975, 1976) that formed at the edge of the Rae Craton during the approximate interval 2.16–1.88 Ga, possibly evolving in an incipient oceanic basin analogous to the Red Sea (e.g. Bohannon & Eittreim 1991). This basin was rapidly filled by a flysch-type, thick greywacke sequence during the interval 1.92–1.90 Ga and was folded and metamorphosed at greenschist to low-pressure granulite facies as a result of northerly-vergent thrusting between *c.* 1.88 and 1.865 Ga (St-Onge *et al.* 2006; Gagné *et al.* 2009). It has been postulated that this orogenic event was caused by docking and subsequent collision of the Meta Incognita micro-continent (see below) with the southeastern margin of the Rae Craton (Corrigan 2002; St-Onge *et al.* 2006). There are no constraints at present on the precise nature of the suture between the Rae Craton and Meta Incognita micro-continent. It has been named the Baffin Suture (St-Onge *et al.* 2002) and its present position is approximate, based primarily on the distribution of Piling Group versus Lake Harbour Group and Hoare Bay Group tecto-nostratigraphic marker units. The northward vergence of structures in the Piling Group and evidence for moderate tectonic burial of the latter suggests that the Meta Incognita micro-continent may have formed the upper plate in this collisional zone. This thrust vergence is concordant to that inferred for the Snowbird Tectonic Zone (Ross 2002; Berman *et al.* 2007), although there is presently insufficient data to determine if the latter and the Baffin Suture actually link. During this period, renewed deformation and magmatism was also recorded further in the hinterland region in the Committee Bay region (Berman *et al.* 2005), potentially as a far-field effect of the Foxe and/or Reindeer orogenies.

## Intraoceanic Accretion: the Flin Flon–Glennie Complex

Intra-oceanic tectonic accretion is a well-documented evolutionary process in the development of the Flin Flon–Glennie Complex. It was first recognized in the Flin Flon belt (Lewry *et al.* 1990), which contains an imbricate stack of oceanic arc rocks, ocean island basalt, ocean plateau and associated plutonic and sedimentary rocks that were juxtaposed along a series of thrust faults soon after their formation on the Manikewan ocean floor. U–Pb ages on plutonic rocks that cut the earliest faults suggest that these volcanic assemblages were structurally juxtaposed at *c.* 1.87 Ga (Lucas *et al.* 1996). Although most of the arc crustal fragments are juvenile, some have negative $\varepsilon_{Nd}$ values, which suggest that they may have been built on fragments of Archaean crust (David *et al.* 1996; Stern *et al.* 1999). Some of this Archaean crust is actually preserved as thin, fault-bounded blocks imbricated between the arc slices (David & Syme 1994). Formation of this early intra-oceanic collage played a role in the further development of the Reindeer Zone, as the Flin Flon–Glennie complex behaved as a coherent crustal block when it eventually collided with the northern Reindeer Zone and later rifted from the latter during the opening of the Kisseynew back-arc basin (see below).

## Magmatic accretion along Superior Craton margin

During the interval 1.88–1.865 Ga, as the Reindeer Zone internides and western Churchill Province margin were in an overall contractional geodynamic setting, most of the Superior Craton margin facing the Manikewan Ocean was undergoing renewed extension, perhaps as a result of pull from the subducting Manikewan oceanic plate. Extension, dated between *c.* 1.88 and 1.87 Ga, resulted in the generation of voluminous mafic to ultramafic bodies emplaced as dykes, sills and flows (Fig. 2c), forming the extensive Circum-Superior Belt (Baragar & Scoates 1981). Most of the mafic and ultramafic magma generated from this event was emplaced through attenuated Superior Craton crust and parautochthonous Palaeoproterozoic cover, as suggested by the slightly contaminated $\varepsilon_{Nd}$ values (Chauvel *et al.* 1987; Hegner & Bevier 1991). Dated members of this magmatic 'event' include the Molson dykes (Heaman *et al.* 1986; Halls & Heaman 1997), the Fox River Sill (Hulbert *et al.* 2005), the Chukotat Sills in the Cape Smith Belt (Wodicka *et al.* 2002b; Mungall 2007) and mafic-ultramafic sills in the Labrador trough (Wardle *et al.* 1990; Wodicka *et al.* 2002b). Historically, these mafic/ultramafic bodies have been interpreted as the result of plume interaction with the margin of the Superior craton (e.g. Ernst & Buchan 2001). However, although a plume source may be possible, recent models point to the penecontemporaneous relationship between emplacement of continental arc type calc-alkaline plutons in the Thompson Nickel and Snow Lake belts, and intrusion of the *c.* 1.88 Ga Molson Lake dykes to infer an ensialic back-arc setting for the generation of the mafic/ultramafic magmas (Percival *et al.* 2004, 2005; Corrigan *et al.* 2007). In the Cape Smith Belt, the Chukotat sills, dated at *c.* 1.89–1.87 Ga, overlap in age with the oldest magmatic rocks of the Parent arc (1874 + 4/ − 3 Ma; Machado *et al.* 1993), suggesting an overall active geodynamic setting during their emplacement.

## Development of a large continental magmatic arc on southeastern margin of western Churchill Province: the 1.865–1.85 Ga Wathaman Orogeny

The period 1.865–1.85 Ga records yet another fundamental step in the evolution of the THO, with the emplacement of voluminous, mainly felsic plutonic rocks in an Andaean-type continental margin setting (Fig. 2d). The larger batholiths (Wathaman–Chipewyan and Cumberland batholiths) were emplaced in the previously accreted southeastern margin of the western Churchill Province (Bickford *et al.* 1994; Thériault *et al.* 2001; Corrigan *et al.* 2005). We speculate that the subduction zone related to this event may have been continuous for at least 2000 km, along the northern edge of the Manikewan oceanic plate (Fig. 2d). Contemporaneous subduction-related magmatism was also occurring, albeit in smaller volumes, within the Flin Flon–Glennie Complex and in the Sugluk Block.

### Wathaman–Chipewyan Batholith

The Wathaman–Chipewyan Batholith is a major intrusive complex that extends *c.* 700 km from north-central Saskatchewan to northeastern Manitoba (Lewry *et al.* 1981; Fumerton *et al.* 1984; Meyer *et al.* 1992; MacHattie 2001). It was emplaced during a relatively narrow time interval between *c.* 1862 and 1850 Ma, in a continental arc setting (Fumerton *et al.* 1984). The main body of the batholith is composed of K–feldspar megacrystic, biotite ± hornblende ± titanite-bearing granite, granodiorite and quartz monzonite, with lesser amounts of diorite and layered gabbro (Corrigan *et al.* 1999). Along its northern flank, the batholith intrudes basement and cover sequences of the Hearne Craton margin. On its southern flank it intrudes previously accreted juvenile terranes including the La Ronge, Lynn Lake and Rusty Lake arcs (Corrigan *et al.* 2005). These intrusive contacts were zones of focused strike-slip deformation during terminal Trans-Hudson collision at *c.* 1.82 Ga (Lafrance & Varga 1996; Corrigan *et al.* 2005). Nd tracer isotope data across the batholith show progressive contamination towards the Hearne Craton, with $\varepsilon_{Nd}$ values ranging from +2 near the southern margin of the batholith to −7 along the northern margin (MacHattie 2001). This suggests that the batholith is a stitching intrusion that masks the suture between accreted terranes and Hearne Craton margin. Intrusive relationships with surrounding rocks also imply that the juvenile arcs must have been already accreted to the Hearne Margin prior to 1.865 Ga, and thus implies that the batholith was produced as a result of sub-duction flip towards the N–NW, beneath the accreted margin (Corrigan *et al.* 2007).

The waning stage of Wathaman–Chipewyan Batholith magmatism at *c.* 1.85 Ga is interpreted to have occurred when the Flin Flon–Glennie complex collided with the previously accreted La Ronge–Lynn Lake–Rusty Lake arcs (northwestern Reindeer Zone), essentially consuming the interven-ing oceanic lithosphere (Corrigan *et al.* 2005). The suture produced by this collision is interpreted as the Granville Lake Structural Zone (Zwanzig 2000; Corrigan *et al.* 2005, 2007), which currently correlates with a strong, crustal-scale seismic aniso-tropy that links at depth with a 5 km vertical offset in the Moho and corresponding gravity anomaly (White *et al.* 2000).

## Cumberland Batholith and 'gneissic suite' of the Narsajuaq Arc

The *c.* 1.865–1.85 Ga Cumberland Batholith (Figs 1 & 2d) is contemporaneous with the Wathaman–Chipewyan Batholith and has also been interpreted as the result of continental arc magma-tism (Thériault *et al.* 2001; St-Onge *et al.* 2006). It forms a volumetrically large complex that was emplaced mainly within the Meta Incognita micro-continent, but extends northwards into the south-eastern margin of the Rae Craton in central Baffin Island (St-Onge *et al.* 2006). The Cumberland Bath-olith is dominated by rocks of felsic composition and in contrast to the Wathaman–Chipewyan Bath-olith, includes only minor mafic to intermediate compositions. Moreover, the former yields only negative $\varepsilon_{Nd}$ values ranging from $-2$ to $-12$ (Thériault *et al.* 2001; Whalen *et al.* 2008), suggesting either assimilation of a greater pro-portion of Archaean crust, or a post-collisional setting (e.g. Whalen *et al.* 2008). The age and distri-bution of neodymium $T_{DM}$ model ages across the Cumberland Batholith led Whalen *et al.* 2008 to suggest that the latter images the Meta Incognita micro-continent and its bounding Archaean cratons (i.e. southern Rae margin in the north and the Sugluk block in the south). If correct, this would suggest impingement of the Sugluk block beneath Meta Incognita micro-continent as early as *c.* 1.865 Ga. Whalen *et al.* (2008) postulate that the unusually large volume of magma forming the Cumberland Batholith could have been the product of slab breakoff. A similar origin could potentially be inferred for the Wathaman-Chipewyan Batholith, as suggested by the unusually large volume of magma produced and the above-mentioned geophysical anomalies associated with the Granville Lake Structural Zone.

Plutonic rocks of similar age range (but as young as 1844 Ma) and of more intermediate composition, occur within the Sugluk block on Ungava Peninsula, and have been named the 'gneissic suite of the Nar-sajuaq arc' (Dunphy & Ludden 1998). They consist of highly transposed, granulite to amphibolite facies orthogneiss of tonalitic to dioritic composition, which have $\varepsilon_{Nd}$ values ranging from $+4$ to $-10.7$ and have been interpreted as the magmatic product of subduction beneath a tectonically attenuated con-tinental margin containing an Archaean crustal component (Dunphy & Ludden 1998). We agree with this model and postulate that this Archaean crust forms the core of the Sugluk block (Fig. 2d). Genetically linking the Cumberland Batholith with the gneissic suite would require that the Sugluk block and Meta Incognita micro-continent were welded by 1.865 Ga, which contradicts current models that bring these two crustal blocks together only at *c.* 1.82 Ga (e.g. St-Onge *et al.* 2000). There-fore, we postulate that the penecontemporaneous gneissic suite was produced by a different subduc-tion zone presently defined by the Sugluk suture (Hoffman 1985) (Figs 2d & 3). Cumberland Batho-lith magmatism ended when the Sugluk block and Meta Incognita micro-continent collided, shortly after *c.* 1.85 Ga. This event may also have produced rapid uplift and cooling of granulite facies rocks dated at *c.* 1.84 Ga in the Meta Incognita micro-continent (M1 thermal event in Lake Harbour group; see St-Onge *et al.* 2007), as the Sugluk block began to underthrust. One possible expla-nation that would reconcile the apparent dichotomy in timing of collision is that the Sugluk and Meta Incognita blocks did begin to collide at *c.* 1.865 Ga, and that the bounding fault (Big Island suture) was reactivated at *c.* 1.82 Ga.

On Ungava Peninsula, magmatism related to the 1863–1844 Ma gneissic suite is restricted to the Sugluk block (Dunphy & Ludden 1998). We postu-late that this subduction-related magmatism stopped when the juvenile Parent–Spartan–Watts collage collided with the Sugluk block at *c.* 1.84 Ga, reacti-vating the Bergeron suture on the trailing edge of the accreted collage and initiating the Narsajuaq arc *sensu stricto* (Fig. 2e).

## Far-field c. 1.85 Ga reactivation in the Snowbird Orogen

Berman *et al.* (2007) identified a large region within the northeastern part of the Snowbird Orogen that was subjected to NW-vergent thrusting and low pressure, moderate temperature metamorphism at *c.* 1.85 Ga, and speculated that it was possibly pro-duced by far-field reactivation during accretion of the Hudson protocontinent (Sugluk block in this paper). Farther SW along the Snowbird Orogen, exhumation of high-pressure rocks by east-verging transpression along the Legs Lake shear zone at *c*. 1.85 Ga is documented by Mahan *et al.* (2006). These events are contemporaneous with the waning stages of Wathaman–Chipewyan magmatism and beginning of infringement of the subduction zone that produced the Wathaman–Chipewyan arc (Granville Lake Structural Zone) by the Flin Flon–Glennie Complex (Fig. 2d). This specific docking event may have been the trigger that produced a far-field reactivation within the thermally softened Snowbird Orogeny. We also agree with the postulation of Berman *et al.* (2005, 2007) that some of that reactivation, particularly in the northeastern part of the Snowbird Orogeny, may have been specifically produced by accretion of the Sugluk block. Moreover, we propose that this collision was predominantly of sinistral transpressional nature and that the resulting fault zone may have been part of a continuous structure, linking the Granville Lake Structural Zone with the Big Island suture to the NE (Fig. 2d). The oblique nature of convergence may explain the absence of continental arc magmatism in the area located between the Wathaman–Chipewyan and Cumberland batholiths. Another possibility is that the Granville Lake Structural Zone connects with the Sugluk Suture, as postulated in Hoffman (1990), and that the postulated Severn arc is part of that system.

## Final magmatic accretion: the Narsajuaq arc (*sensu stricto*) and De Pas Batholith (*c*. 1.84–1.82 Ga)

Published definitions of the Narsajuaq arc are overly inclusive, comprising all plutons ranging in age from 1898–1800 Ma that occur within the Ungava Peninsula in northern Québec (Dunphy & Ludden 1998), as well as *c*. 1842–1821 Ma meta-plutonic rocks, mostly anatectic melts, that intrude the Meta Incognita micro-continent on Baffin Island (e.g. Scott 1997; Wodicka & Scott 1997). The definition has been expanded to include a *c*. 1.87–1.86 Ga island arc and associated sediments (Parent and Spartan Group) built on *c*. 2.0 Ga oceanic crust (Watts Group) (St-Onge *et al.* 1992, 2006). We propose herein to restrict the definition of the Narsajuaq arc to specifically comprise *c*. 1.84–1.82 Ga continental arc-derived plutons that occur along the exposed southeastern margin of the Sugluk block (i.e. the younger suite of Dunphy & Ludden 1998), as well as the *c*. 1842–1821 Ma plutonic suite identified along the southern margin of the Meta Incognita micro-continent and historically correlated with the Narsajuaq arc plutons on Ungava Peninsula (e.g. Scott

1997; Wodicka & Scott 1997). This new definition restricts the age of arc magmatism to a duration of *c*. 20 Ma and allows the interpretation of this suite as the product of continental arc magmatism in an upper plate formed of the previously amalgamated Sugluk and Meta Incognita micro-continents, after the extinction of the 1.865–1.85 Ga Cumberland Batholith, during the time interval 1.84–1.82 Ga (Fig. 2e).

This new definition also permits a potential link between the Narsajuaq arc *sensu stricto* and the subduction system (Sugluk suture and Lac Tudor shear zone) that produced the 1.84–1.82 Ga De Pas Batholith (Thomas & Kearey 1980; van der Leeden *et al.* 1990; St-Onge *et al.* 2002). The De Pas batholith intrudes along the western margin of the Core Zone, a potential extension of Meta Incognita micro-continent south of Ungava Bay (e.g. figure 10 in Wardle *et al.* 2002). Narsajuaq arc magmatism waned at *c*. 1.82 Ga when the Manikewan Ocean eventually closed and the Superior Craton began its terminal collision with the accreted orogen internides.

## Back-arc opening and inversion in western Manikewan Ocean: 1.85–1.84 Ga Kisseynew Domain

The Kisseynew Domain (Fig. 4) forms the youngest tectonostratigraphic entity in the Reindeer Zone and consists in large part of migmatitic meta-turbidite of the Burntwood Group, interpreted to have been deposited in a *c*. 1.85–1.84 Ga back-arc basin formed between the La Ronge–Lynn Lake and Flin Flon–Glennie arc complexes (Zwanzig 1990; Ansdell *et al.* 1995) (Fig. 2e). This model implies that *c*. 1.85–1.84 Ga magmatism in the Flin Flon–Glennie Complex was the result of rifting of the latter from the NW Reindeer Zone, perhaps as the result of hinge roll-back and opening of the Burntwood back-arc basin (Ansdell *et al.* 1995; Corrigan *et al.* 2007). The Flin Flon–Glennie complex was indeed quite magmatically active during that period, relative to the northwestern Reindeer Zone. This phase of plutonism, however, is generally more mafic and includes large, Cu-Ni ±PGE-hosting mafic to ultramafic plutons, such as the 1847 ± 6 Ma Namew Lake complex (Cumming & Krstic 1991) and 1840 ± 4 Ma gabbros that intrude the Bear Lake Block in the Flin Flon belt (Bailes & Theyer 2006).

The Burntwood Group is unconformably and/or disconformably overlain along most of its perimeter by fluvial-alluvial sediments deposited between *c*. 1845 and 1835 Ma, collectively interpreted as syn-collisional molasse basins (Stauffer 1990). Closure of the Burntwood back-arc basin may have occurred at *c.* 1.84 Ga, shortly after its deposition, perhaps as a result of collision of the Sask craton with the Flin Flon–Glennie Complex (Fig. 2e, f). This age coincides with the development of the Pelican Thrust (Pelican suture in Fig. 2e) at the interface between the Sask Craton and overlying Glennie Domain (Ashton *et al.* 2005). Inversion of the Burntwood basin is also reflected in the upward coarsening of sediments from turbidite to intraformational conglomerate (Corrigan & Rayner 2002).

## Terminal Collision: docking of Superior Craton at 1.83–1.80 Ga

Terminal collision in the THO (Fig. 2f) began with early collision of the Superior Craton with the Reindeer Zone, resulting in final consumption of the Manikewan Ocean and widespread tectonothermal overprint across the Reindeer Zone and a large part of the western Churchill Province (Lewry *et al.* 1985; Lewry & Collerson 1990). Docking of the Superior Craton appears to have been slightly diachronous, with earliest impingement occurring in the Thompson Nickel Belt at *c.* 1.83 Ga (Bleeker 1990) and subsequent collision with the Cape Smith Belt at *c.* 1.82 Ga (St-Onge *et al.* 2002, 2006). Collision with the Core Zone was even later, at *c.* 1.81 Ga (Wardle *et al.* 2002), and likely the result of indentation of the Ungava Peninsula and clockwise rotation of the Meta Incognita–Core Zone crustal block (e.g. Hoffman 1990). Along most of its periphery, the Superior Craton remained the lower crustal block during collision, forming the footwall to accreted juvenile terranes, such as the Cape Smith klippe (Hoffman 1985; Lucas 1989). One exception to this general rule is the Thompson Nickel Belt, which began as a normal, SE-vergent Proterozoic-on-Archaean imbricate stack, but eventually developed retroshears that thrust the Superior Craton basement onto the orogen internides during 1.82–1.77 Ga sinistral transpression (Ellis & Beaumont 1999; White *et al.* 2002). The initial crustal architecture, with the Reindeer Zone overthrust onto the Superior Craton, may explain why most of the eastern portion of the Kisseynew Basin is currently underlain by Archaean crust (Percival *et al.* 2007).

As opposed to the Superior Craton, where Hudsonian tectonothermal overprint is limited to a relatively narrow zone along its periphery, the western Churchill Province is by comparison reactivated at a much greater scale. Two nearly complete transects across the orogen, in the western and eastern segments, show widely distributed U–Pb monazite and titanite ages representing peak- or post-peak metamorphic temperatures, that range

between *c.* 1.83 and 1.76 Ga. Schneider *et al.* (2007, and references therein) presented a comprehensive set of new and previously published thermochronometric data in the western Reindeer Zone in Saskatchewan and Manitoba, that show an essentially homogeneous distribution of *c.* 1.83–1.76 Ga monazite and titanite U–Pb ages that range geographically from the eastern margin of the Thompson Nickel Belt to the western edge of the Hearne Craton, near the Snowbird Tectonic Zone, a distance of *c.* 600 km (Fig. 2f). Similarly, Wodicka *et al.* (2007*b*, *c*) published a compilation of U–Pb monazite and titanite ages from the Ungava–Baffin segment of the THO, showing a broad distribution of *c.* 1.83–1.78 Ga peak- to post-peak metamorphic cooling ages spanning from the Cape Smith Belt, across Baffin Island, to the northernmost extents of reactivated Rae Craton north of the Piling Group along Isortoq Fault (Jackson & Berman 2000; Bethune & Scammell 2003), a total distance of *c.* 800 km (Fig. 2f). Although the thermochronological dataset from the core of the western Churchill Province is sparser, there is sufficient information to infer a more or less uniform Hudsonian tectonothermal overprint, although locally at low metamorphic grade, across most of its southwestern half (Berman *et al.* 2000) (Fig. 2f).

## Summary and conclusions

From its incipient intra-oceanic contractional phase to terminal collision with the Superior Craton, the composite THO represents *c.* 120 Ma of gradual ocean basin closure, tectonic and magmatic accretion, intracratonic basin opening and closure, deformation and metamorphism. Much in the same way as the Grenville or Appalachian orogens, for example, the THO can be defined as a series of specific, short duration 'orogenies' that comprise specific accretionary phases within a much larger orogenic system (e.g. Rivers 1997; Rivers & Corrigan 2000; Hanmer *et al.* 2000; Murphy *et al.* 1999; van Staal 2005; Hibbard *et al.* 2007).

The THO can be broadly subdivided into three lithotectonic divisions, which are the Churchill margin (or peri-Churchill), the Reindeer Zone and the Superior margin (or peri-Superior) 'realms'. The initial demise of the Manikewan Ocean and formation of juvenile arcs, beginning at *c.* 1.92 Ga, was penecontemporaneous with amalgamation of microcontinents (Hearne, Sugluk and Meta Incognita/Core Zone) along the Rae margin, and accretion of juvenile arc complexes (La Ronge–Lynn Lake–Rusty Lake arcs) outboard of that continental collage, by *c.* 1.88 Ga. The collisions between the Rae Craton and the micro-continents and arc complexes resulted in basin inversions and attendant tectonothermal 'events' that are divisible into the *c*. 1.92–1.89 Ga Snowbird Orogeny and the *c*. 1.88–1.865 Ga Foxe and Reindeer orogenies. Along the peri-Superior realm facing the Manikewan Ocean, early convergent activity appears to be limited to the Thompson Nickel Belt where *c*. 1.90–1.88 Ga calc-alkaline plutons intruding the crystalline basement and Palaeoproterozoic cover sequence suggest continent margin magmatism, which in turn implies subduction beneath the Superior Craton margin during that period. The emplacement of voluminous, margin-parallel mafic and ultramafic dykes and sills (Molson suite) at *c*. 1.88 Ga may signify development of an ensialic back-arc basin and establishment of a pericratonic active arc (Snow Lake) outboard of that margin (e.g. Percival *et al.* 2005). Within the Reindeer Zone, oceanic arcs formed first in the western part of Manikewan Ocean, with the *c*. 1.91–1.87 Ga amalgamation of the Flin Flon–Glennie Complex, followed at *c*. 1.87–1.86 Ga by development of arcs (Parent and Spartan arc complex) in the eastern part (or 1.90–1.86 Ga if the Cape Smith suite 'plutonic root' is included).

An important, intermediate phase in the development of the THO was the establishment of a subduction complex along the peri-Churchill margin, beginning at *c*. 1.865 Ga and leading to the generation of voluminous continental arc batholiths. This occurred in two distinct phases with the first one at 1.865–1.85 Ga producing the Wathaman–Chipewyan batholith in the Hearne Craton margin, as well as the penecontemporaneous Cumberland batholith in the Meta Incognita microcontinent and 1.865–1.844 Ga 'older gneiss' suite in the Sugluk block. A younger phase, at *c*. 1.84–1.82 Ga, was subsequent to accretion of the Sugluk block to the peri-Churchill margin, leading to back-stepping of the subduction zone and production of the Narsajuaq and De Pas continental magmatic arcs and related anatectic melts. All subduction-related calc-alkaline magmatism ended at *c*. 1.82 Ga, with the final consumption of the Manikewan Ocean.

Within this context, the THO *sensu stricto* represents the latest and most extensive of these orogenic events, resulting from terminal collision between the peri-Churchill collage, the Reindeer Zone internides and the Superior Craton. The broad asymmetry in extent of tectonothermal reactivation observed between the Superior Craton and the western Churchill Province (Fig. 1), may be explained by the fact that the former represents a unique, previously cratonized, cold and rigid lithospheric block, whereas the latter comprises a crustal mosaic of smaller cratons that were assembled shortly before terminal collision and hence represents hotter, weaker lithosphere with

numerous structural breaks. The actual width of tectonothermal reactivation in the western Churchill Province, in the order of *c*. 900 km at its widest extent, is similar in scale to the Parautochthonous Belt of the Grenville Orogen, which extends for a maximum distance of up to 800 km across strike of the orogen (Figs 1 and 2f). It is noteworthy that the Grenville Parautochthonous Belt, like the western Churchill Province, also evolved as a long-lived accretionary margin prior to terminal Grenvillian collision between the Laurentian and Amazonian cratons (Rivers & Corrigan 2000), and hence was also thermally softened prior to terminal collision.

An interesting observation from the THO is the apparent absence of large extensional detachments, core complexes, or other structural features indicative of orogenic collapse (although, see Schneider *et al.* 2007, for alternative view). This is consistent with the rarity of ultrahigh- to high-pressure metamorphic rocks of *c*. 1.83 Ga and younger age, which suggests that the orogen, at least in its terminal phase, may have been wide but of relatively modest thickness. Most late Trans-Hudson age (i.e. *c*. 1.85–1.80 Ga) palaeo-pressures at the present-day surface are in the range of 3–7 kb (e.g. Gordon 1989; Kraus & Menard 1997; Orrell *et al.* 1999; Berman *et al.* 2007; Gagne *et al.* in press), with higher pressures of *c*. 8–9 kb found uniquely in the Meta Incognita micro-continent and Cape Smith Belt (Bégin 1992; St-Onge & Ijewliw 1996; St-Onge *et al.* 2007), and perhaps parts of the southwestern Hearne Craton (Annesley *et al.* 2005).

Assuming a present-day crustal thickness of *c*. 40 km for most of the THO (White *et al.* 2005), this suggests predominant crustal thicknesses of *c*. 52–58 km during peak orogenesis, which is substantially less than thicknesses of *c*. 70–80 km estimated for most of the Himalayan or Grenville orogens (Hirn *et al.* 1984; Indares & Dunning 2001; Jamieson *et al.* 2002; Li & Mashele 2008). The relatively broad and modestly thick nature of the THO may have been caused by early thermal softening in the Reindeer Zone and western Churchill Province, which would in turn have favored lateral growth rather than vertical thickening (e.g. Molnar & Lyon Caen 1988).

An exception to the observed homogeneity of thermochronometric ages and palaeo-pressure and temperature conditions across the western THO is found in the Flin Flon–Glennie Complex. At that location, greenschist facies rocks of the Amisk block are juxtaposed against mid- to upper-amphibolite rocks of the Hanson Lake block and Glennie Domain, along the north–south trending Tabbernor and Sturgeon–Weir fault system, with the Amisk block (eastern side) down-dropped with respect to the western side (Ashton *et al.* 2005). We speculate that this faulting may have been caused by southern extrusion of the underlying Sask Craton, caused by impingement of the Superior Craton, which collided at *c.* 1.83 Ga, *c.* 10 Ma after the Sask Craton had collided with the Reindeer Zone internides (e.g. Hajnal *et al.* 1996).

One of the many features that have become apparent from the series of time-slice cartoons illustrated in Figure 2 is the gradual southward growth of the western Churchill Province, with consecutive accretion of: (i) the Hearne Craton; (ii) La Ronge–Lynn Lake Domains and Meta Incognita micro-continent; (iii) Flin Flon–Glennie Complex and Sugluk block; (iv) Sask Craton; and (v) Superior Craton; beginning at 1.92 Ga and ending at *c.* 1.80 Ga. Moreover, once the Superior Craton stabilized at *c.* 1.78 Ga (Bleeker 1990), overall accretion and convergence at the North American scale did not stop but simply migrated to the southern margin of the Wyoming–Trans-Hudson–Superior collage, which then evolved in an open, Andean-type setting for *c.* 600 Ma until terminal collision during the *c.* 1.2–1.0 Ga Grenville Orogeny (Hoffman 1988; Rivers & Corrigan 2000; Whitmeyer & Karlstrom 2007). Another feature which stands out from this orogen-scale compilation is the apparent continuity of first-order structures, such as subduction zones, between the western and eastern regions of the Manikewan Ocean, not only in timing of displacement but also inferred subduction polarities. One of the palaeo-subduction zones or sutures that might have been thus linked is the Granville Lake Structural Zone–Big Island Suture (Fig. 2d). This subduction system is of similar scale to most operating today in the SW Pacific, for example.

The model presented herein for the Palaeoproterozoic evolution of the Ungava–Baffin segment of the THO varies somewhat from previously published models (cf. Dunphy & Ludden 1998; St-Onge *et al.* 2006). For one part, we interpret the Sugluk Suture (Hoffman 1990) as the northernmost extent of exposed Superior Craton (Fig. 3). This suture connects in Ungava Bay with the Lac Tudor Shear Zone, which is interpreted as the western margin of the Core Zone (Wardle *et al.* 2002). The slightly older Big Island Suture (new name introduced herein) separates the Sugluk block from the Meta Incognita micro-continent, and merges with the Sugluk–Lac Tudor 'suture' (Fig. 2e). Within this framework, the Bergeron Suture forms the sole thrust that carried the juvenile Watts–Parent–Spartan oceanic crust and arc assemblage onto the edge of the Superior Craton, where it is currently preserved as a structural klippe (cf. Hoffman 1985). The Bergeron suture likely merged with the Sugluk suture prior to folding of the basement-cover sequence in orogen-parallel folds (Fig. 3), as both terranes were carried in a piggy-back and out-of-sequence style onto the continental margin (Hoffman 1985; Lucas 1989).

Another important point that is worth addressing is the pre-collisional evolution of the western margin of the Superior Craton, and in particular the nature of the Snow Lake arc. Historically, the latter has been associated with the intra-oceanic Flin Flon–Glennie Complex (e.g. Lewry *et al.* 1990). However, given its relatively higher degree of contamination by Archaean age crust, and the recent recognition of early arc magmatism on the Superior Craton margin (Percival *et al.* 2004, 2005), the question may be asked whether the Snow Lake arc did actually evolve outboard, as part of the Flin Flon–Glennie Complex, or as a peri-cratonic arc built on a rifted slice of the Superior Craton. If the latter, then one has to consider that the Snow Lake–Amisk block boundary may be a fundamental, late-orogenic suture within the Flin Flon–Glennie Complex, and not part of the earlier *c.* 1.87 Ga thrust stack of genetically related ocean arc slices.

The authors owe a debt of gratitude to the influence of numerous geologists who have through the years unravelled the story of different parts of this large and complex orogen, and in particular to P. Hoffman who began to painstakingly put it all together during the Decade in North-American Geology (DNAG) years. We thank B. Murphy for the invitation to contribute to this volume, as well as R. Berman, P. Bickford and K. Ansdell for providing insightful reviews. This paper is Geological Survey of Canada Contribution no. 20090130.

# References

ANBAR, A. D. & KNOLL, A. H. 2002. Proterozoic ocean chemistry and evolution: a bioinorganic bridge? *Science*, **297**, 1137–1142.

ANNESLEY, I. R., MADORE, C. & PORTELLA, P. 2005. Geology and thermotectonic evolution of the western margin of the Trans-Hudson Orogen: evidence from the eastern sub-Athabasca basement, Saskatchewan. *Canadian Journal of Earth Sciences*, **42**, 573–597.

ANSDELL, K. M. 2005. Tectonic evolution of the Manitoba-Saskatchewan segment of the Palaeoproterozoic Trans-Hudson Orogen, Canada. *Canadian Journal of Earth Sciences*, **42**, 741–759.

ANSDELL, K. M., LUCAS, S. B., CONNORS, K. & STERN, R. A. 1995. Kisseynew metasedimentary gneiss belt, Trans-Hudson Orogen (Canada): back-arc origin and collisional inversion. *Geology*, **23**, 1039–1043.

ANSDELL, K. M., MACNEIL, A., DELANEY, G. D. & HAMILTON, M. A. 2000. Rifting and development of the Hearne passive margin: age constraint from the Cook Lake area, Saskatchewan. *Geological Association of Canada – Mineralogical Association of Canada, GeoCanada 2000, Extended abstract #777 (CD-ROM)*.

ARTEMIEVA, L. M. & MOONEY, W. D. 2001. Thermal thickness and evolution of Precambrian lithosphere: a global study. *Journal of Geophysical Research*, **106**, 16,387–16,414.

ASHTON, K. E. 1999. *A proposed lithotectonic domain reclassification of the southeastern Reindeer Zone in Saskatchewan*. Summary of Investigations 1999, Volume 1, Saskatchewan Geological Survey, Saskatchewan Energy and Mines, Miscellaneous Report **99-4**, 92–100.

ASHTON, K. E., LEWRY, J. F., HEAMAN, L. M., HARTLAUB, R. P., STAUFFER, M. R. & TRAN, H. T. 2005. The Pelican Thrust Zone: basal detachment between the Archaean Sask Craton and Palaeoproterozoic Flin Flon–Glennie Complex, western Trans-Hudson Orogen. *Canadian Journal of Earth Sciences*, **42**, 685–706.

BAILES, A. H. & GALLEY, A. G. 1999. Evolution of the Palaeoproterozoic Snow Lake arc assemblage and geodynamic setting for associated volcanic-hosted massive sulphide deposits, Flin Flon Belt, Manitoba, Canada. *Canadian Journal of Earth Sciences*, **36**, 1789–1805.

BAILES, A. H. & THEYER, P. 2006. *Wonderland gabbro: U–Pb age and geological interpretations, Manitoba (NTS 63K 12NE)*. Report of Activities, Manitoba Science, Technology, Energy and Mines, Manitoba Geological Survey, 35–41.

BALDWIN, D. A. 1987. *Proterozoic subaerial felsic volcanism in the Rusty Lake greenstone belt*. PhD thesis, University of Manitoba, Winnipeg, Manitoba.

BALDWIN, D. A., SYME, E. C., ZWANZIG, H. V., GORDON, T. M., HUNT, P. A. & STEVENS, R. D. 1987. U–Pb zircon ages from the Lynn Lake and Rusty Lake metavolcanic belts, Manitoba: two ages of Proterozoic magmatism. *Canadian Journal of Earth Sciences*, **24**, 1053–1063.

BARAGAR, W. R. A. & SCOATES, R. F. J. 1981. The Circum-Superior belt; a Proterozoic plate margin? *In*: KRONER, A. (ed.) *Precambrian Plate Tectonics*. Elsevier, Amsterdam, 297–330.

BARETTE, P. D. 1994. Lithostratigraphy and map-scale structure in the western Cape Smith Belt, northern Quebec: a tentative correlation between two tectonic domains. *Canadian Journal of Earth Sciences*, **31**, 986–994.

BEAUMONT-SMITH, C. J. & BÖHM, C. O. 2002. *Structural analysis and geochronological studies in the Lynn Lake greenstone belt and its gold-bearing shear zones (NTS 64C10, 11, 12, 14, 15 and 16)*. Report of Activities 2002, Manitoba Industry, Trade and Mines, Manitoba Geological Survey, 159–170.

BEAUMONT-SMITH, C. J. & BÖHM, C. O. 2003. *Tectonic evolution and gold metallogeny of the Lynn lake greenstone belt, Manitoba (NTS 64C10, 11, 12, 14, 15 and 16)*. Report of Activities 2003, Manitoba Industry, Trade and Mines, Manitoba Geological Survey, 39–49.

BÉDARD, J. H. 2006. A catalytic delamination-driven model for coupled genesis of Archaean crust and subcontinental lithospheric mantle. *Geochimica and Cosmochimica Acta*, **70**, 1188–1214.

BÉGIN, N. J. 1992. Contrasting mineral isograd sequences in metabasites of the Cape Smith Belt, northern Quebec, Canada: three new bathograds for mafic rocks. *Journal of Metamorphic Geology*, **10**, 685–704.

BERMAN, R. G., RYAN, J. J. ET AL. 2000. The case of multiple metamorphic events in the western Churchill Province: evidence from linked thermobarometric and *in-situ* SHRIMP data, and jury deliberations. GAC–MAC 2000 Program with Abstracts, Volume 25; GeoCanada 2000 Conference CD, Abstract 836.

BERMAN, R. G., SANBORN-BARRIE, M., STERN, R. & CARSON, C. 2005. Tectonometamorphism *c.* 2.35 and 1.85 Ga in the Rae domain, western Churchill Province Nunavut, Canada: insights from structural, metamorphic and *in-situ* geochronological analysis of the southwestern Committee Bay belt. *Canadian Mineralogist*, **43**, 409–442.

BERMAN, R. G., DAVIS, W. J. & PEHRSSON, S. 2007. Collisional Snowbird Tectonic Zone resurrected: growth of Laurentia during the 1.9 Ga accretionary phase of the Hudsonian orogeny. *Geology*, **35**, 911–914.

BETHUNE, K. M. & SCAMMELL, R. J. 2003. Distinguishing between Archaean and Palaeoproterozoic tectonism, and evolution of the Isortoq fault zone, Eqe Bay area, north-central Baffin Island. *Canadian Journal of Earth Sciences*, **40**, 1111–1135.

BICKFORD, M. E. & HILL, B. M. 2007. Does the arc accretion model adequately explain the Paleoproterozoic evolution of southern Laurentia?: An expanded interpretation. *Geology*, **35**, 167–170.

BICKFORD, M. E., COLLERSON, K. D., LEWRY, J. F., VAN SCHMUS, W. R. & CHIARENZELLI, J. 1990. Proterozoic collisional tectonism in the Trans-Hudson orogen, Saskatchewan. *Geology*, **18**, 14–18.

BICKFORD, M. E., COLLERSON, K. D. & LEWRY, J. F. 1994. Crustal histories of the Hearne and Rae provinces, southwestern Canadian Shield, Saskatchewan: constraints from geochronologic and isotopic data. *Precambrian Research*, **68**, 1–21.

BICKFORD, M. E., MOCK, T. D., STEINHART III, W. E., COLLERSON, K. D. & LEWRY, J. F. 2005. Origin of the Archaean Sask craton and its extent within the Trans-Hudson orogen: evidence from Pb and Nd isotopic compositions of basement rocks and post-orogenic intrusions. *Canadian Journal of Earth Sciences*, **42**, 659–684.

BLEEKER, W. 1990. New structural-metamorphic constraints on Early Proterozoic oblique collision along the Thompson Nickel Belt, Manitoba, Canada. *In*: LEWRY, J. F. & STAUFFER, M. R. (eds) *The Early Proterozoic Trans-Hudson Orogen of North America*. Geological Association of Canada, Special Paper, **37**, 57–73.

BOHANNON, R. G. & EITTREIM, S. L. 1991. Tectonic development of passive continental margins of the southern and central Red Sea with a comparison to Wilkes Land, Antarctica. *Tectonophysics*, **198**, 129–154.

BOURLON, E., MARESCHAL, J. C., ROEST, W. & TELMAT, H. 2002. Geophysical correlations in the Ungava Bay area. *Canadian Journal of Earth Sciences*, **39**, 625–637.

CHAUVEL, C., ARNDT, N. T., KIELINZCUK, S. & THOM, A. 1987. Formation of Canadian 1.9 Ga old continental crust. I: Nd isotopic data. *Canadian Journal of Earth Sciences*, **24**, 396–406.

CHIARENZELLI, J. R., ASPLER, L. B., VILLENEUVE, M. & LEWRY, J. F. 1998. Early Proterozoic evolution of the Saskatchewan craton and its allochthonous cover, Trans-Hudson orogen. *Journal of Geology*, **106**, 247–267.

CORRIGAN, D. 2002. The Trans-Hudson Orogen: old paradigms and new concepts. *Annual Meeting of the Geological Association of Canada – Mineralogical Association of Canada, Program with Abstracts, Saskatoon 2002.*

CORRIGAN, D. & RAYNER, N. 2002. *Churchill River–Southern Indian Lake Targeted Geoscience Initiative (NTS 64B, 64C, 64G, 64H), Manitoba: Update and New Findings.* Report of Activities 2002, Manitoba Industry, Trade and Mines, Manitoba Geological Survey, 144–158.

CORRIGAN, D., MAXEINER, R., BASHFORTH, A. & LUCAS, S. 1998. *Preliminary report on the geology and tectonic history of the Trans-Hudson Orogen in the northwestern Reindeer Zone, Saskatchewan.* Current Research, Part C; Geological Survey of Canada, Paper, **98-1C**, 95–106.

CORRIGAN, D., MACHATTIE, T. G. & CHAKUNGAL, J. 1999. The Wathaman Batholith and its relation to the Peter Lake Domain: insights from recent mapping along the Reindeer Lake transect, Trans-Hudson Orogen. Saskatchewan Geological Survey, Miscellaneous Report, *Summary of Investigations, 1999, Volume 2,* **99-4.2**, 132–142.

CORRIGAN, D., MAXEINER, R. O. & HARPER, C. 2001*a*. Preliminary U–Pb results from the La Ronge-Lynn Lake Bridge project. *In: Saskatchewan Geological Survey, Miscellaneous Report.* Summary of Investigations 2001, Volume 2, **2001-4.2**, 111–115.

CORRIGAN, D., THERRIAULT, A. & RAYNER, N. 2001*b*. Preliminary results from the Churchill River – Southern Indian Lake Targeted Geoscience Initiative. *Report of Activities, Manitoba Industry, Trade and Mines.* Manitoba Geological Survey, 94–107.

CORRIGAN, D., HAJNAL, Z., NÉMETH, B. & LUCAS, S. B. 2005. Tectonic framework of a Palaeoproterozoic arc-continent to continent-continent collisional zone, Trans-Hudson Orogen, from geological and seismic reflection studies. *Canadian Journal of Earth Sciences*, **42**, 421–434.

CORRIGAN, D., GALLEY, A. G. & PEHRSSON, S. 2007. Tectonic evolution and metallogeny of the southwestern Trans-Hudson Orogen. *In:* GOODFELLOW, W. D. (ed.) *Mineral Deposits of Canada: A Synthesis of Major Deposit-types, District Metallogeny, the Evolution of Geological Provinces, and Exploration Methods.* Geological Association of Canada, Mineral Deposits Division, Special Publication, **5**, 881–902.

CUMMING, G. L. & KRSTIC, D. 1991. Geochronology at the Namew Lake Ni–Cu deposit, Flin Flon area, Manitoba, Canada: a Pb/Pb study of whole rocks and ore minerals. *Canadian Journal of Earth Sciences*, **28**, 1328–1339.

DAVID, J. & SYME, E. C. 1994. U–Pb geochronology of late Neoarchean tonalites in the Flin Flon Belt, Trans-Hudson Orogen: surprise at surface. *Canadian Journal of Earth Sciences*, **31**, 1785–1790.

DAVID, J., BAILES, A. H. & MACHADO, N. 1996. Evolution of the Snow Lake portion of the Palaeoproterozoic Flin Flon and Kisseynew belts, Trans-Hudson Orogen, Manitoba, Canada. *Precambrian Research*, **80**, 107–124.

DOIG, R. 1987. Rb–Sr geochronology and metamorphic history of Proterozoic to early Archaean rocks north of the Cape Smith Fold Belt, Quebec. *Canadian Journal of Earth Sciences*, **24**, 813–825.

DUNPHY, J. M. & LUDDEN, J. N. 1998. Petrological and geochemical characteristics of a Palaeoproterozoic magmatic arc (Narsajuaq terrane, Ungava Orogen, Canada) and comparisons to Superior Province granitoids. *Precambrian Research*, **91**, 109–142.

ELLIS, S. & BEAUMONT, C. 1999. Models of convergent boundary tectonics: implications for the interpretation of Lithoprobe data. *Canadian Journal of Earth Sciences*, **36**, 1711–1741.

ERNST, R. & BUCHAN, K. 2001. Large mafic magmatic events through time and links to mantle-plume heads. *Geological Society of America, Special Paper*, **352**, 483–575.

FRANCIS, D. 2003. Cratonic mantle roots, remnants of a more chondritic Archaean mantle? *Lithos*, **71**, 135–152.

FUMERTON, S. L., STAUFFER, M. R. & LEWRY, J. F. 1984. The Wathaman Batholith: largest known Precambrian pluton. *Canadian Journal of Earth Sciences*, **21**, 1082–1097.

GAGNÉ, S., JAMIESON, R. A., MACKAY, R., WODICKA, N. & CORRIGAN, D. 2009. Texture, composition, and age variations in monazite from lower amphibolite to granulite facies, Longstaff Bluff Formation, Baffin Island, Canada. *Canadian Mineralogist*, **47**, 847–869.

GALLEY, A. G., ZIEHLKE, D. V., FRANKLIN, J. M., AMES, D. E. & GORDON, T. M. 1986. Gold mineralization in the Snow Lake–Wekusko Lake region, Manitoba. *In:* CLARK, A. L. (ed.) *Gold in the Western Shield.* Canadian Institute of Mining and Metallurgy, Special Volume, **38**, 379–398.

GIBB, R. A. 1975. Collisional tectonics in the Canadian Shield. *Earth and Planetary Sciences Letters*, **27**, 378–382.

GIBB, R. A. & WALCOTT, R. I. 1971. A Precambrian suture in the Canadian Shield. *Earth and Planetary Science Letters*, **10**, 417–422.

GORDON, T. M. 1989. Thermal evolution of the Kisseynew sedimentary gneiss belt, Manitoba: metamorphism at an early Proterozoic accretionary margin. *In:* DALY, J. S., CLIFF, R. A. & YARDLEY, B. W. D. (eds) *Evolution of Metamorphic Belts.* Geological Society, London, Special Publications, **43**, 233–244.

GREEN, A. G., HAJNAL, Z. & WEBER, W. 1985. An evolutionary model for the western Churchill Province and western margin of the Superior Province in Canada and the north-central United States. *Tectonophysics*, **116**, 281–322.

GRIFFIN, W. L., O'REILLY, S. Y., ALFONSO, J. C. & BEGG, G. C. 2008. The composition and evolution of lithospheric mantle: A re-evaluation and its tectonic implications. *Journal of Petrology*, doi: 10.1093/petrology/egn033.

HAJNAL, Z., LUCAS, S., WHITE, D., LEWRY, J., BEZDAN, S., STAUFFER, M. R. & THOMAS, M. D. 1996. Seismic reflection images of high-angle faults and linked detachments in the Trans-Hudson Orogen. *Tectonics*, **15**, 427–439.

HAJNAL, Z., LEWRY, J. F. *ET AL.* 2005. The Sask Craton and Hearne Province margin: seismic reflection studies in the western Trans-Hudson Orogen. *Canadian Journal of Earth Sciences*, **42**, 403–419.

HALLS, H. C. & HEAMAN, L. M. 1997. New constraints on the Palaeoproterozoic segment of the Superior Province apparent polar wander path from U–Pb dating of Molson Dykes, Manitoba. Geological Association of Canada – Mineralogical Association of Canada, *Joint Annual Meeting, Program with Abstracts*, **22**, A61.

HANMER, S., CORRIGAN, D., PEHRSSON, S. & NADEAU, L. 2000. SW Grenville Province, Canada: the case against post-1.4 Ga accretionary tectonics. *Tectonophysics*, **319**, 33–51.

HEAMAN, L. M. & CORKERY, T. 1996. U–Pb geochronology of the Split Lake Block, Manitoba: preliminary results. *In*: HAJNAL, Z. & LEWRY, J. (eds) *Trans-Hudson Lithoprobe Transect*. Lithoprobe Report, **55**, 60–67.

HEAMAN, L. M., MACHADO, N., KROGH, T. E. & WEBER, W. 1986. Presise U–Pb zircon ages for the Molson dyke swarm and the Fox River sill: constraints for Early Proterozoic crustal evolution in northern Manitoba, Canada. *Contributions to Mineralogy and Petrology*, **94**, 82–89.

HEGNER, E. & BEVIER, M. L. 1991. Nd and Pb isotopic constraints on the origin of the Purtiniq ophiolites and Early Proterozoic Cape Smith Belt, northern Quebec, Canada. *Chemical Geology*, **91**, 357–371.

HIBBARD, J. P., VAN STAAL, C. R. & RANKIN, D. W. 2007. A comparative analysis of pre-Silurian crustal building blocks of the northern and southern Appalachians Orogen. *American Journal of Science*, **307**, 23–45.

HIRN, A., LÉPINE, J.-C. *ET AL.* 1984. Crustal structure and variability of the Himalayan border of Tibet. *Nature*, **307**, 23–25.

HOFFMAN, P. F. 1981. Autopsy of Athapuscow Aulacogen: a failed arm affected by three collisions. *In*: CAMPBELL, F. H. A. (ed.) *Proterozoic Basins of Canada*. Geological Survey of Canada, Paper, **81-10**, 97–101.

HOFFMAN, P. F. 1985. Is the Cape Smith Belt (northern Quebec) a klippe? *Canadian Journal of Earth Sciences*, **22**, 1361–1369.

HOFFMAN, P. F. 1988. United plates of America, the birth of a craton: early Proterozoic assembly and growth of Laurentia. *Annual Review of Earth and Planetary Science Letters*, **16**, 543–603.

HOFFMAN, P. F. 1990. Subdivision of the Churchill province and extent of the Trans-Hudson Orogen. *In*: LEWRY, J. F. & STAUFFER, M. R. (eds) *The Early Proterozoic Trans-Hudson Orogen of North America*. Geological Association of Canada, Special Paper, **37**, 15–39.

HOU, G., SANTOSH, M., QIAN, X., LISTER, G. S. & LI, J. 2008. Configuration of the late Palaeoproterozoic supercontinent Columbia: insights from radiating mafic dyke swarms. *Gondwana Research*, **14**, 395–409.

HULBERT, L. J., HAMILTON, M. A., HORAN, M. F. & SCOATES, R. F. J. 2005. U–Pb zircon and Re–Os isotope geochronology of mineralized ultramafic intrusions and associated nickel ores from the Thompson Nickel Belt, Manitoba, Canada. *Economic Geology*, **100**, 29–41.

INDARES, A. & DUNNING, G. 2001. Partial melting of high P-T metapelites from the Tshenukutish Terrane (Grenville Province): petrography and U–Pb geochronology. *Journal of Petrology*, **42**, 1547–1565.

JACKSON, G. D. & BERMAN, R. G. 2000. Precambrian metamorphic and tectonic evolution of northern Baffin Island, Nunavut, Canada. *Canadian Mineralogist*, **38**, 399–421.

JAMIESON, R. A., BEAUMONT, C., NGUYEN, M. H. & LEE, B. 2002. Interaction of metamorphism, deformation and exhumation in large convergent orogens. *Journal of Metamorphic Geology*, **20**, 9–24.

JOHNS, S. M., HELMSTAEDT, H. H. & KYSER, T. K. 2006. Palaeoproterozoic submarine intrabasinal rifting, Baffin Island, Nunavut, Canada: volcanic structure and geochemistry of the Bravo Lake Formation. *Canadian Journal of Earth Sciences*, **43**, 593–616.

KONHAUSER, K. O., HAMADE, T., RAISWELL, R., MORRIS, R. C., FERRIS, F. G., SOUTHAM, G. & CANFIELD, D. E. 2002. Could bacteria have formed the Precambrian banded iron formations? *Geology*, **30**, 1079–1082.

KRAUS, J. & MENARD, T. 1997. A thermal gradient at constant pressure: implications for low- to medium-pressure metamorphism in a compressional tectonic setting, Flin Flon and Kisseynew domains, Trans-Hudson Orogen, Central Canada. *Canadian Mineralogist*, **35**, 1117–1136.

LAFRANCE, B. & VARGA, M. 1996. *Structural studies of the Parker Lake Shear Zone and the Reilly Lake Shear Zone, Reindeer Lake*. Summary of Investigations 1996, Saskatchewan Geological Survey, Miscellaneous Report **96-4**, 119–124.

LEWRY, J. F. 1981. Lower Proterozoic arc-microcontinent collisional tectonics in the western Churchill province. *Nature*, **294**, 69–72.

LEWRY, J. F. & COLLERSON, K. D. 1990. The Trans-Hudson Orogen: extent, subdivisions and problems. *In*: LEWRY, J. F. & STAUFFER, M. R. (eds) *The Early Proterozoic Trans-Hudson Orogen of North America*. Geological Association of Canada, Special Paper, **37**, 1–14.

LEWRY, J. F., STAUFFER, M. R. & FUMERTON, S. 1981. A Cordilleran-type batholithic belt in the Churchill Province in northern Saskatchewan. *Precambrian Research*, **14**, 277–313.

LEWRY, J. F., SIBBALD, T. I. I. & SCHLEDEWITZ, D. C. P. 1985. Variation in character of Archaean rocks in the western Churchill province and its significance. *In*: AYRES, L. D., THURSTON, P. C., CARD, K. D. & WEBER, W. (eds) *Evolution of Archaean Supracrustal Sequences*. Geological Association of Canada, Special Paper, **28**, 239–261.

LEWRY, J. F., THOMAS, D. J., MACDONALD, R. & CHIARENZELLI, J. 1990. Structural relations in accreted terranes of the Trans-Hudson Orogen, Saskatchewan: telescoping in a collisional regime? *In*: LEWRY, J. F. & STAUFFER, M. R. (eds) *The Early Proterozoic Trans-Hudson Orogen of North America*. Geological Association of Canada, Special Paper, **37**, 75–94.

LEWRY, J. F., HAJNAL, Z. *ET AL.* 1994. Structure of a Palaeoproterozoic continent-continent collision zone; a Lithoprobe seismic reflection profile across the Trans-Hudson Orogen, Canada. *Tectonophysics*, **232**, 143–160.

LI, A. & MASHELE, B. 2008. Crustal structure in the Pakistan Himalayas. *Joint Meeting of The Geological Society of America, Soil Science Society of America, American Society of Agronomy, Crop Science Society of America, Gulf Coast Association of Geological Societies with the Gulf Coast Section of SEPM. Houston, Texas*, Paper **204-4**.

LUCAS, S. B. 1989. Structural evolution of the Cape Smith thrust belt, and the role of out-of-sequence faulting in the thickening of mountain belts. *Tectonics*, **8**, 655–676.

LUCAS, S. B., WHITE, D. *ET AL.* 1994. Three-dimensional collisional structure of the Trans-Hudson Orogen, Canada. *Tectonophysics*, **232**, 161–177.

LUCAS, S. B., STERN, R. A., SYME, E. C., REILLY, B. A. & THOMAS, D. J. 1996. Intraoceanic tectonics and the development of continental crust: 1.92–1.84 Ga evolution of the Flin Flon belt, Canada. *Geological Society of America Bulletin*, **108**, 602–629.

MACHADO, N., DAVID, J., SCOTT, D. J., LAMOTHE, D., PHILIPPE, S. & GARIÉPY, C. 1993. U–Pb geochronology of the western Cape Smith Belt, Canada: new insights on the age of initial rifting and arc magmatism. *Precambrian Research*, **63**, 211–223.

MACHATTIE, T. G. 2001. *Petrogenesis of the Wathaman Batholith and La Ronge Domain plutons in the Reindeer Lake area, Trans-Hudson Orogen, Saskatchewan.* MSc thesis, Memorial University, St. John's, Newfoundland.

MAHAN, K., GONCALVES, P., WILLIAMS, M. & JERCINOVIC, M. 2006. Dating metamorphic reactions and fluid flow: application to exhumation of high-P granulites in a crustal-scale shear zone, western Canadian Shield. *Journal of Metamorphic Geology*, **24**, 193–217.

MAXEINER, R. O., SIBBALD, T. I. I., SLIMMON, W. L., HEAMAN, L. M. & WATTERS, B. R. 1999. Lithogeochemistry of volcano-plutonic assemblages of the southern Hanson Lake Block and Southeastern Glennie Domain, Trans-Hudson Orogen: evidence for a single island arc complex. *Canadian Journal of Earth Sciences*, **36**, 209–225.

MAXEINER, R. O., HARPER, C., CORRIGAN, D. & MACDOUGALL, D. G. 2004. *La Ronge-Lynn Lake Bridge Project: geology of the Southern Reindeer Lake area. Saskatchewan Industry and Resources.* Open File Report **2003-1**, 2 CD-ROM set and 3 maps.

MAXEINER, R. O., CORRIGAN, D., HARPER, C. T., MACDOUGALL, D. G. & ANSDELL, K. 2005. Palaeoproterozoic arc and ophiolitic rocks on the northwest-margin of the Trans-Hudson Orogen, Saskatchewan, Canada; their contribution to a revised tectonic framework for the orogen. *Precambrian Research*, **136**, 67–106.

MEYER, M. T., BICKFORD, M. E. & LEWRY, J. F. 1992. The Wathaman Batholith: an Early Proterozoic continental arc in the Trans-Hudson orogenic belt, Canada. *Geological Society of America Bulletin*, **104**, 1073–1085.

MODELAND, S. & FRANCIS, D. 2010. Enriched mantle components in Palaeoproterozoic alkaline magmas of the Bravo Lakes formation, central Baffin Island, Nunavut. *Journal of Petrology, in press.*

MOLNAR, P. & LYON-CAEN, H. 1988. Some simple physical aspects of the support, structure, and evolution of mountain belts. *In*: CLARK, S. P. *ET AL.* (eds) *Processes in continental lithospheric deformation.* Geological Society of America Special Paper, **218**, 179–207.

MORGAN, W. C., BOURNE, J., HERD, R. K., PICKETT, J. W. & TIPPETT, C. R. 1975. Geology of the Foxe Fold Belt, Baffin Island, District of Franklin. *Report of Activities, Part A, April–October 1974*, Geological Survey of Canada, Paper 75-01A, 343–347.

MORGAN, W. C., OKULITCH, A. V. & THOMPSON, P. H. 1976. *Stratigraphy, Structure and Metamorphism of the West Half of the Foxe Fold Belt, Baffin Island.* Report of Activities Part B/Report of Activities Part A, Geological Survey of Canada, Paper 76-01B, 387–391.

MUNGALL, J. E. 2007. Crustal contamination of picritic magmas during transport through dikes: the Expo Intrusive Suite, Cape Smith Fold Belt, New Quebec. *Journal of Petrology*, **48**, 1021–1039.

MURPHY, J. B., KEPPIE, D., DOSTAL, J. & NANCE, D. 1999. Neoproterozoic–early Paleozoic evolution of Avalonia. *In*: RAMOS, V. & KEPPIE, D. (eds) *Laurentia–Gondwana Connections before Pangea.* Geological Society of America Special Paper, **336**, 253–266.

ORRELL, S. E., BICKFORD, M. E. & LEWRY, J. F. 1999. Crustal evolution and age of thermotectonic reworking in the western hinterland of the Trans-Hudson Orogen, northern Saskatchewan. *Precambrian Research*, **95**, 187–223.

PARRISH, R. R. 1989. U–Pb geochronology of the Cape Smith Belt and Sugluk Block, northern Quebec. *In*: LUCAS, S. B., PICARD, C. & ST-ONGE, M. R. (eds) *Tectonic, Magmatic and Metallogenic Evolution of the Early Proterozoic Cape Smith Thrust Belt.* Geoscience Canada, **16**, 126–130.

PEHRSSON, S. J. AND THE WESTERN CHURCHILL METALLOGENY PROJECT WORKING GROUP, 2004. *Evolution of thought on the evolution of a craton: new perspectives on the origin and reworking of the western Churchill Province.* PanLithoprobe IV, Lithoprobe celebratory conference, Toronto, Program with abstracts, poster, CD-ROM.

PERCIVAL, J. A. 2007. Geology and metallogeny of the Superior Province. *In*: GOODFELLOW, W. D. (ed.) *Mineral Deposits of Canada: A Synthesis of Major Deposit-types, District Metallogeny, the Evolution of Geological Provinces, and Exploration Methods.* Geological Association of Canada, Mineral Deposits Division, Special Publication, **5**, 309–328.

PERCIVAL, J. A., WHALEN, J. B. & RAYNER, N. 2004. *Pikwitonei–Snow Lake, Manitoba transect, Trans-Hudson Orogen – Superior Margin Metallotect Project: initial geological, isotopic, and U–Pb SHRIMP results.* Report of Activities 2004, Manitoba Industry, Economic Development and Mines, Manitoba Geological Survey, 120–134.

PERCIVAL, J. A., WHALEN, J. B. & RAYNER, N. 2005. *Pikwitonei – Snow Lake, Manitoba transect (parts of*

*NTS 63J, 63O and 63P), Trans-Hudson Orogen – Superior Margin Metallotect Project: new results and tectonic interpretation.* Report of Activities 2005, Manitoba Industry, Economic Development and Mines, Manitoba Geological Survey, 69–91.

PERCIVAL, J. A., RAYNER, N., GROWDON, M. L., WHALEN, J. B. & ZWANZIG, H. V. 2007. *New field and geochronological results for the Osik–Atik–Footprint lakes area, Manitoba (NTS 63O13, 14, 15, 64B2, 3).* Report of Activities 2007, Manitoba Science, Technology, Energy and Mines, Manitoba Geological Survey, 71–81.

PESONEN, L. J., ELMING, S.-Å. ET AL. 2003. Paleomagnetic configuration of continents during the Proterozoic. *Tectonophysics*, **375**, 289–324.

PICARD, C., LAMOTHE, D., PIBOULE, M. & OLIVIER, R. 1990. Magmatic and geotectonic evolution of a Proterozoic oceanic basin system: the Cape Smith Thrust-Fold Belt (New Quebec). *Precambrian Research*, **47**, 223–249.

RAYNER, N. & CORRIGAN, D. 2004. *Uranium-lead geochronological results from the Churchill River – Southern Indian Lake transect, northern Manitoba.* Geological Survey of Canada, Current Research, **2004-F1**, 1–14.

RAYNER, N. M., MAXEINER, R. O. & CORRIGAN, D. 2005a. Progress report on U–Pb geochronology results for the Peter Lake Domain Project. *In*: *Summary of Investigations 2005, Saskatchewan Geological Survey.* Miscellaneous Report, 2005-4.2, CD-ROM, paper A-4, 12.

RAYNER, N., STERN, R. A. & BICKFORD, M. E. 2005b. Tectonic implications of new SHRIMP and TIMS U–Pb geochronology of rocks from the Sask Craton, Peter Lake Domain, and Hearne margin, Trans-Hudson Orogen, Saskatchewan. *Canadian Journal of Earth Sciences*, **42**, 635–657.

RAYNER, N., SANBORN-BARRIE, M., WODICKA, N. & ST-ONGE, M. 2007. New U–Pb geochronological constraints on the timing of deformation and the nature of basement of SW Baffin Island, Nunavut. *In*: CAIRNS, S. & FALCK, H. (compilers), *35th Annual Yellowknife Geoscience Forum Abstracts.* Northwest Territories Geoscience Office, Yellowknife, NT. YKGSF Abstracts **49**, 2007.

RAYNER, N. M., SCOTT, D. J., WODICKA, N. & KASSAM, A. 2008. *New Geochronological Constraints from Mill, Salisbury, and Nottingham Islands, Nunavut.* Geological Survey of Canada, Current Research **2008-22**, 1–14.

RIVERS, T. 1997. Lithotectonic elements of the Grenville Province: review and tectonic implications. *Precambrian Research*, **86**, 117–154.

RIVERS, T. & CORRIGAN, D. 2000. Convergent margin on southeastern Laurentia during the Mesoproterozoic: tectonic implications. *Canadian Journal of Earth Sciences*, **37**, 359–383.

ROGERS, J. J. W. & SANTOSH, M. 2002. Configuration of Columbia, a Mesoproterozoic Supercontinent. *Gondwana Research*, **5**, 5–22.

ROKSANDIC, M. M. 1987. The tectonics and evolution of the Hudson Bay region. *In*: BEAUMONT, C. & TANKARD, A. J. (eds) *Sedimentary Basins and Basin-forming Mechanisms.* Canadian Society of Petroleum Geologists, Memoir, **12**, 507–518.

ROSS, G. M. 2002. Evolution of Precambrian continental lithosphere in Western Canada: results from Lithoprobe studies in Alberta and beyond. *Canadian Journal of Earth Sciences*, **39**, 413–437.

SCHNEIDER, D. A., HEIZLER, M. T., BICKFORD, M. E., WORTMAN, G. L., CONDIE, K. C. & PERILLI, S. 2007. Timing constraints of orogeny to cratonization: thermochronology of the Palaeoproterozoic Trans-Hudson orogen, Manitoba and Saskatchewan, Canada. *Precambrian Research*, **153**, 65–95.

SCOTT, D. J. 1997. Geology, U–Pb and Pb–Pb geochronology of the Lake Harbour area, southern Baffin Island: implications for the Palaeoproterozoic tectonic evolution of north-eastern Laurentia. *Canadian Journal of Earth Sciences*, **34**, 140–155.

SCOTT, D. J. 1999. U–Pb geochronology of the eastern Hall Peninsula, southern Baffin Island, Canada: a northern link between the Archaean of West Greenland and the Palaeoproterozoic Torngat Orogen of northern Labrador. *Precambrian Research*, **93**, 5–26.

SCOTT, D. J. & ST-ONGE, M. R. 1998. Proterozoic assembly of northeast Laurentia revisited: a model based on southward extrapolation of Ungava – Baffin crustal architecture. *In*: WARDLE, R. J. & HALL, J. (eds) *Eastern Canadian Shield Onshore-Offshore Transect (ECSOOT).* Transect Meeting (May 4–5, 1998), The University of British Columbia, Lithoprobe Secreteriat, Report No. **68**, 134–147.

SCOTT, D. J., HELMSTAEDT, H. & BICKLE, M. J. 1992. Purtuniq ophiolite, Cape Smith belt, northern Québec, Canada: a reconstructed section of Early Proterozoic oceanic crust. *Geology*, **20**, 173–176.

SCOTT, D. J., STERN, R. A., ST-ONGE, M. R. & MCMULLEN, S. M. 2002. U–Pb geochronology of detrital zircons in metasedimentary rocks from southern Baffin Island: implications for the Palaeoproterozoic tectonic evolution of Northeastern Laurentia. *Canadian Journal of Earth Sciences*, **39**, 611–623.

STAUFFER, M. R. 1984. Manikewan and early Proterozoic ocean in central Canada, its igneous history and orogenic closure. *Precambrian Research*, **25**, 257–281.

STAUFFER, M. R. 1990. The Missi Formation: an Aphebian molasse deposit in the Reindeer Lake Zone of the Trans-Hudson Orogen, Canada. *In*: LEWRY, J. F. & STAUFFER, M. R. (eds) *The Early Proterozoic Trans-Hudson Orogen of North America.* Geological Association of Canada, Special Paper, **37**, 75–94.

STERN, R. A., SYME, E. C. & LUCAS, S. B. 1995. Geochemistry of 1.9 Ga MORB- and OIB-like basalts from the Amisk collage, Flin Flon belt, Canada: evidence for an intra-oceanic origin. *Geochimica et Cosmochimica Acta*, **59**, 3131–3154.

STERN, R. A., MACHADO, N., SYME, E. C., LUCAS, S. B. & DAVID, J. 1999. Chronology of crustal growth and recycling in the Palaeoproterozoic Amisk collage (Flin Flon Belt), Trans-Hudson Orogen, Canada. *Canadian Journal of Earth Sciences*, **36**, 1807–1827.

ST-ONGE, M. R. & IJLEWLIW, O. J. 1996. Mineral corona formation during high-P retrogression of granulitic rocks, Ungava Orogen, Canada. *Journal of Petrology*, **37**, 553–582.

ST-ONGE, M. R. & LUCAS, S. B. 1990. Evolution of the Cape Smith Belt: early Proterozoic continent underthrusting, ophiolites obduction and thick-skinned folding. *In*: LEWRY, J. F. & STAUFFER, M. R. (eds) *The Early Proterozoic Trans-Hudson Orogen of North America*. Geological Association of Canada, Special Paper, **37**, 313–351.

ST-ONGE, M. R., LUCAS, S. B. & PARRISH, R. R. 1992. Terrane accretion in the internal zone of the Ungava orogen, northern Quebec. Part 1: tectonostratigraphic assemblages and their tectonic implications. *Canadian Journal of Earth Sciences*, **29**, 746–764.

ST-ONGE, M. R., SCOTT, D. J. & LUCAS, S. B. 2000. Early partitioning of Quebec: microcontinent formation in the Palaeoproterozoic. *Geology*, **28**, 323–326.

ST-ONGE, M. R., SCOTT, D. J. & WODICKA, N. 2002. Review of crustal architecture and evolution in the Ungava Paninsula – Baffin Island area: connection to the Lithoprobe ECSOOT transect. *Canadian Journal of Earth Sciences*, **39**, 589–610.

ST-ONGE, M. R., SEARLE, M. P. & WODICKA, N. 2006. Trans-Hudson Orogen of North America and Himalaya-Karakoram-Tibetan Orogen of Asia: Structural and thermal characteristics of the lower and upper plates. *Tectonics*, **25**, TC4006, doi: 10.1029/2005TC001907.

ST-ONGE, M. R., WODICKA, N. & IJEWLIW, O. 2007. Polymetamorphic evolution of the Trans-Hudson Orogen, Baffin Island, Canada: integration of petrological, structural, and geochronological data. *Journal of Petrology*, **48**, 271–302, doi: 10.1093/petrology/egl060.

STOCKWELL, C. H. 1961. Structural provinces, orogenies and time classification of rocks of the Canadian Precambrian Shield. *In*: LOWDON, J. A. (ed.) *Age Determinations by the Geological Survey of Canada, Report 2: Isotopic Ages*. Geological Survey of Canada, Paper **61–17**, 108–118.

SYME, E. C. & BAILES, A. H. 1993. Stratigraphic and tectonic setting of volcanogenic massive sulphide deposits, Flin Flon, Manitoba. *Economic Geology*, **88**, 566–589.

SYME, E. C., LUCAS, S. B., BAILES, A. H. & STERN, R. A. 1999. Contrasting arc and MORB-like assemblages in the Palaeoproterozoic Flin Flon Belt, Manitoba, and the role of intra-arc extension in localizing volcanic-hosted massive sulphide deposits. *Canadian Journal of Earth Sciences*, **36**, 1767–1788.

SYMONS, D. T. A. & HARRIS, M. J. 2005. Accretion history of the Trans-Hudson Orogen in Manitoba and Saskatchewan from paleomagnetism. *Canadian Journal of Earth Sciences*, **42**, 723–740.

THÉRIAULT, R. J., ST-ONGE, M. R. & SCOTT, D. J. 2001. Nd isotopic and geochemical signature of the Palaeoproterozoic Trans-Hudson Orogen, southern Baffin Island, Canada: implications for the evolution of eastern Laurentia. *Precambrian Research*, **108**, 113–138.

THOMAS, D. J. 1993. *Geology of the Star Lake-Otter Lake Portion of the Central Metavolcanic Belt, La Ronge Domain*. Saskatchewan Geological Survey, Miscellaneous Report, **236**, 1–133.

THOMAS, M. D. & KEAREY, P. 1980. Gravity anomalies, block-faulting and Andean-type tectonism in the Eastern Churchill province. *Nature*, **283**, 61–63.

TRAN, H. T., ANSDELL, K., BETHUNE, K. M., ASHTON, K. & HAMILTON, M. A. 2008. Provenance and tectonic setting of Palaeoproterozoic metasedimentary rocks along the eastern margin of the Hearne craton: constraints from SHRIMP geochronology, Wollaston Group, Saskatchewan, Canada. *Precambrian Research*, **167**, 171–185.

TUREK, A., WOODHEAD, J. & ZWANZIG, H. V. 2000. *U–Pb age of the gabbro and other plutons at Lynn Lake (parts of NTS 64C)*. Report of Activities 2000, Manitoba Industry, Trade and Mines, Manitoba Geological Survey, 97–104.

VAN SCHMUS, W. R., BICKFORD, M. E., LEWRY, J. F. & MACDONALD, R. 1987. U–Pb geochronology in the Trans-Hudson Orogen, northern Saskatchewan, Canada. *Canadian Journal of Earth Sciences*, **24**, 407–424.

VAN DER LEEDEN, J., BÉLANGER, M., DANIS, D., GITRARD, R. & MARTELAIN, J. 1990. Lithotectonic domains in the high-grade terrain east of the Labrador Trough (Québec). *In*: LEWRY, J. F. & STAUFFER, M. R. (eds) *The Early Proterozoic Trans-Hudson Orogen of North America*. Geological Association of Canada, Special Paper, **37**, 371–386.

VAN STAAL, C. R. 2005. Northern Appalachians. *In*: SELLEY, R. C., COCKS, R. M. & PILMER, I. R. (eds) *Encyclopedia of Geology*. Oxford, Elsevier, **4**, 81–91.

WARDLE, R. J. & HALL, J. 2002. Proterozoic evolution of the northeastern Canadian Shield: Lithoprobe Eastern Canadian Shield Onshore-Offshore Transect (ECSOOT), introduction and summary. *Canadian Journal of Earth Sciences*, **39**, 563–567.

WARDLE, R. J., RYAN, B., NUNN, G. A. C. & MENGEL, F. C. 1990. Labrador segment of the Trans-Hudson orogen: crustal development through oblique convergence and collision. *Geological Association of Canada Special Paper*, **37**, 353–369.

WARDLE, R. J., JAMES, D. T., SCOTT, D. J. & HALL, J. 2002. The southeastern Churchill Province: synthesis of a Palaeoproterozoic transpressional orogen. *Canadian Journal of Earth Sciences*, **39**, 639–663.

WATTERS, B. R. & PEARCE, J. A. 1987. Metavolcanic rocks of the La Ronge Domain in the Churchill Province, Saskatchewan geochemical evidence for a volcanic arc origin. *In*: PHARAOH, T. C., BECKINSALE, R. D. & RICKARD, D. (eds) *Geochemistry and Mineralization of Proterozoic Volcanic Suites*. Geological Society, London, Special Publications, **33**, 167–182.

WHALEN, J. B., SYME, E. C. & STERN, R. A. 1999. Geochemical and Nd isotopic evolution of Palaeoproterozoic arc-type granitoids magmatism in the Flin Flon Belt, Trans-Hudson Orogen, Canada. *Canadian Journal of Earth Sciences*, **36**, 227–250.

WHALEN, J. B., WODICKA, N. & TAYLOR, B. E. 2008. Cumberland Batholith petrogenesis: implications for Palaeoproterozoic crustal and orogenic processes. *Geochimica Cosmochimica et Acta, Goldschmidt Conference Abstracts 2008*, **72**, A1016.

WHITE, D. J., ZWANZIG, H. V. & HAJNAL, Z. 2000. Crustal suture preserved in the Palaeoproterozoic Trans-Hudson Orogen, Canada. *Geology*, **29**, 527–530.

WHITE, D. J., LUCAS, S. B., BLEEKER, W., HAJNAL, Z., LEWRY, J. F. & ZWANZIG, H. V. 2002. Suture-zone geometry along an irregular Palaeoproterozoic margin: the Superior Boundary Zone, Manitoba, Canada. *Geology*, **30**, 735–738.

WHITE, D. J., THOMAS, M. D., JONES, A. G., HOPE, J., NÉMETH, B. & HAJNAL, Z. 2005. Geophysical transect across a Palaeoproterozoic continent-continent collision zone: the Trans-Hudson Orogen. *Canadian Journal of Earth Sciences*, **42**, 385–402.

WHITMEYER, S. J. & KARLSTROM, K. E. 2007. Tectonic model for the Proterozoic growth of North America. *Geosphere*, **3**, 220–259.

WODICKA, N. & SCOTT, D. J. 1997. *A preliminary report on the U–Pb geochronology of the Meta Incognita Peninsula, southern Baffin Island, Northwest Territories.* Geological Survey of Canada, Paper, **1997-C**, 167–178.

WODICKA, N., ST-ONGE, M. R., SCOTT, D. J. & CORRIGAN, D. 2002a. *Preliminary report on the U–Pb geochronology of the northern margin of the Trans-Hudson Orogen, central Baffin Island, Nunavut.* Radiogenic Age and Isotopic Studies, Report 15, Geological Survey of Canada, Current Research, **2002-F7**, 12.

WODICKA, N., MADORE, L., LARBI, Y. & VICKER, P. 2002b. Géochronologie U–Pb de filons-couches mafiques de la Ceinture de Cape Smith et de la Fosse du Labrador: Dans: L'exploration minérale au Québec: notre savoir, vos découvertes. *Séminaire d'information sur la recherche géologique, Programme et résumés 2002. Ministère des Ressources naturelles, Québec*, **DV 2002-10**, 48.

WODICKA, N., ST-ONGE, M. R., CORRIGAN, D. & SCOTT, D. J. 2007a. Depositional age and provenance of the Piling Group, central Baffin Island, Nunavut: implications for the Palaeoproterozoic tectonic development of the southern Rae margin. *Geological Association of Canada – Mineralogical Association of Canada Annual Meeting, Abstracts*, **32**, 88.

WODICKA, N., BREITSPRECHER, K. & WHALEN, J. B. 2007b. Geochronological compilation of Trans-Hudson Orogen in Ungava Peninsula. *In*: ST-ONGE, M. R., LAMOTHE, D., HENDERSON, I. & FORD, A. (eds) *Atlas géoscientifique numérique, ceinture de Cape Smith et environs, péninsule d'Ungava, Québec-Nunavut/Digital geoscience atlas of the Cape Smith Belt and adjacent domains, Ungava Peninsula, Quebec-Nunavut.* Geological Survey of Canada, Open File, **5117**.

WODICKA, N., WHALEN, J. B., JACKSON, G. D. & HEGNER, E. 2007c. Geochronological compilation of Trans-Hudson Orogen and environs, Baffin Island. *In*: ST-ONGE, M. R., FORD, A. & HENDERSON, I. (eds) *Digital geoscience atlas of Baffin Island (south of 70°N and east of 80°W), Nunavut.* Geological Survey of Canada, Open File, **5116**.

WODICKA, N., ST-ONGE, M. R. & WHALEN, J. B. 2008. Characteristics of two opposing continental margin successions in northeast Laurentia. *Geochimica Cosmochimica et Acta, Goldschmidt Conference Abstracts 2008*, **72**, A1030.

YEO, G. 1998. *A systems tract approach to the stratigraphy of paragneisses in the southeastern Wollaston Domain. Summary of Investigations 1998.* Saskatchewan Geological Survey, Saskatchewan Energy and Mines, Miscellaneous Report, **98-4**, 36–47.

ZWANZIG, H. 1990. Kisseynew gneiss belt in Manitoba: stratigraphy, structure, and tectonic evolution. *In*: LEWRY, J. F. & STAUFFER, M. R. (eds) *The Early Proterozoic Trans-Hudson Orogen of North America.* Geological Association of Canada, Special Paper, **37**, 95–120.

ZWANZIG, H. V. 2000. *Geochemistry and tectonic framework of the Kisseynew Domain – Lynn Lake belt boundary (part of NTS 64P/13).* Report of Activities 2000, Manitoba Industry, Trades and Mines, Manitoba Geological Survey, 102–114.

ZWANZIG, H. V., SYME, E. C. & GILBERT, H. P. 1999. Updated trace element geochemistry of the *c.* 1.9 Ga metavolcanic rocks in the Palaeoproterozoic Lynn Lake Belt. *Manitoba Industry, Trade and Mines, Geological Services, Open File Report*, **99-13**, 1–46, plus accompanying map and disk.

# Index

Page numbers in *italic* denote figures. Page numbers in **bold** denote tables.